Engineering Physics

SECOND EDITION

SANJAY D JAIN
Head, Knowledge Center
Priyadharshini Institute of Engineering and Technology
Nagpur

GIRISH G SAHASRABUDHE
Professor, Department of Physics
Shri Ramdeobaba Kamla Nehru Engineering College
Nagpur

Universities Press

Engineering Physics (Second Edition)

UNIVERSITIES PRESS (INDIA) PRIVATE LIMITED

Registered Office
3-6-747/1/A & 3-6-754/1, Himayatnagar,
Hyderabad 500 029, Telangana, India
e-mail: info@universitiespress.com

Distributed by
Orient Blackswan Private Limited

Registered Office
3-6-752 Himayatnagar, Hyderabad 500 029, Telangana, India

Other Offices
Bengaluru / Bhopal / Chennai / Ernakulam / Guwahati
Hyderabad / Jaipur / Kolkata / Lucknow / Mumbai
New Delhi / Noida / Patna / Vijayawada

First published 2016

Cover and book design

ISBN 978 81 7371 991 2

Typeset in Times New Roman 11/13 by
OSDATA, Hyderabad 500 029

501160

08032016

Printed at
Graphica Printers

Published by

Universities Press (India) Private Limited
3-6-747/1/A & 3-6-754/1, Himayatnagar,
Hyderabad 500 029, Telangana, India

Foreword

Writing a textbook on physics for engineering students has always been a challenging task. It is well known, especially in India, that students who opt for engineering education tend to look at physics as one of the subjects in which they must score well in the engineering entrance examination. After the students join an engineering college or institution, they are lost in the study of engineering sciences and treat physics as a secondary subject. This malady has grown because generally there is no attempt to explain to the students how physics originated, grew and developed as a basis for engineering sciences. The present book *Engineering Physics* by Dr Sanjay Jain and Dr Girish Sahasrabudhe has been written in a style that overcomes this drawback, presenting a clear, lucid and educative account of physics needed by engineering students. The first chapter, Physics and Engineering, describes the relation between physics and engineering in a novel manner. The role of basic experiments, methodology and errors of measurement have been brought out in a practical way. Every chapter in the book has been written with the same freshness and novel approach. Particularly, the chapters on areas of engineering applications such as fibre optics and lasers have been written in a very interesting manner. Although the authors have maintained their novel approach throughout the book, they have kept the text at a practical and logical level so that students will find it very useful for their courses as well as for the examinations.

I am sure that the present book will be a valuable teaching aid not only to students of engineering but to students and teachers of physics also. It will prove to be an ideal companion to those teachers who wish to teach physics beyond simple formulae.

I heartily congratulate Dr Sanjay Jain and Dr Girish Sahasrabudhe for writing such an excellent book. I am confident that their book will augment the present national efforts in India to promote science education in the country.

Ramesh M Singru
Former Professor of Physics
Indian Institute of Technology Kanpur
Kanpur

Foreword

Writing a textbook on physics for engineering students has always been a challenging task. It is well known, especially in India, that students who opt for engineering education tend to look at physics as one of the subjects in which they must score well in the engineering entrance examination. After the students join an engineering college or institution, they are lost in the study of engineering sciences and treat physics as a secondary subject. This malady has grown because generally there is no attempt to explain to the students how physics originated, grew and developed as a basis for engineering sciences. The present book *Engineering Physics* by Dr Sanjay Jain and Dr Girish Sahasrabudhe has been written in a style that overcomes this drawback, presenting a clear, lucid and educative account of physics needed by engineering students. The first chapter, Physics and Engineering, describes the relation between physics and engineering in a novel manner. The role of basic experiments, methodology and errors of measurement have been brought out in a practical way. Every chapter in the book has been written with the same freshness and novel approach. Particularly, the chapters on areas of engineering applications such as fibre optics and lasers have been written in a very interesting manner. Although the authors have maintained their novel approach throughout the book, they have kept the text at a practical and logical level so that students will find it very useful for their courses as well as for the examinations.

I am sure that the present book will be a valuable teaching aid not only for students of engineering but to students and teachers of physics also. It will prove to be an ideal companion to those teachers who wish to teach physics beyond simple formulae.

I heartily congratulate Dr Sanjay Jain and Dr Girish Sahasrabudhe for writing such an excellent book. I am confident that their book will augment the present national efforts in India to promote science education in the country.

Ramesh M Singru
Former Professor of Physics
Indian Institute of Technology Kanpur
Kanpur

Preface to the Second Edition

Physics seeks to know nature. It began and evolved as 'natural philosophy' and grew into a major scientific discipline over the last four centuries. It has occupied a position of unchallenged prestige derived from its fundamental nature, which defines the paradigm of modern scientific thought. Only in recent times, since the emergence of computer science and communication technology as the hub of knowledge pursuit, has this position come to be challenged. However, even in the contemporary phase of the so-called Knowledge Era, physics still enjoys a balanced and cardinal position. This is probably because it engages philosophical thought and offers novel and rich technological applications, most notably in evidence, for example, in an area like quantum computing.

Classical engineering is founded on scientific knowledge in general, and knowledge of physics in particular. As the euphemism goes, physics seeks knowledge and engineering seeks its application. Physicists accumulate tested, verified, systematised knowledge of the physical world and define possibilities strictly, and engineers bring this knowledge to bear on practical problems. No engineer has ever made possible anything that the physicist declared to be not possible. Yet it is true that the engineer has defined and redefined the practice of the possible. It is for the physicist to discover and expand newer dimensions of the possible: it is for the engineer to master these.

The significant role of physics in the development of the engineering profession has been duly recognised by the All India Council of Technical Education (AICTE), which has recommended a basic science component of 15–25% in all engineering curricula. Almost all the premier engineering institutions in India and abroad have full-fledged departments of physics that are engaged in providing engineering students with a strong foundation in physics in the early part of their career, and in research relevant to industry and technology. In fact, basic research at the forefront of physics is considered one of the highest functions in electrical and electronic engineering. However, it is the experience of the authors that the close connection between physics and engineering has been rather obscured these days in our country. In the face of extremely rapid technological changes, students in universities have been led to think of physics as a hurdle in engineering rather than as its foundation. Neglect of basic sciences has been an inevitable consequence, as is now very widely recognised in the academic community. Students with poor assimilation of concepts in basic sciences are naturally found to be capable of only maintenance engineering and not of creatively facing the new technologies.

The most important objective of this book is to rekindle the interest of engineering students in physics and to provide a sound foundation in the concepts. It is a comprehensive attempt to bring out the role of basic discoveries in physics in evolving, guiding and shaping new technologies. The book makes use of the following special features in order to achieve the desired objective.

The story at the beginning of each chapter brings out interesting links in the development of the subject under study in that chapter. Various topics in the chapter are presented as a judicious blend of theory and experiment in consonance with the so-called scientific method. Thus, topics are presented either as experimental verification of a theory or a theoretical explanation of an experiment. Topics based on common principles are presented in charts to bring out the unifying character of physics. These charts

- facilitate the study of different topics together at a glance
- help students appreciate the underlying unification
- enhance understanding of different topics in less time to achieve quick preparation for examinations.

Appropriate problems are integrated into theory to improve clarity of concepts and to gain the required quantitative appreciation. The fact that engineering grows through the utilitarian translation of discoveries and concepts in physics is brought to the fore in the last section of each chapter.

The book is useful as a core text in the first year of a degree course in engineering or technology. It may be recommended as a useful source of basic knowledge of physics for senior students of engineering and for those taking various competitive examinations. The book comprises of sixteen chapters that can be broadly divided as (i) optics, (ii) quantum, atomic and nuclear physics, (iii) physics of materials such as crystals, dielectrics and semiconductors, (iv) electronics, electron ballistics, lasers, optical fibres and nanotechnology and (v) acoustics.

In optics (Chapters 2–5), concepts in wave theory, which dominated the second half of the nineteenth century, such as interference, diffraction and polarisation, are discussed, and the utilisation of these concepts in the development of technologies is outlined. Failure of the wave theory and emergence of the quantum theory at the beginning of the twentieth century, their role in the evolution of atomic and nuclear physics and the concomitant effect of this new knowledge on technology are discussed in Chapters 6–8. In Chapters 9–11, the physics of crystalline solids, dielectric and magnetic materials, conductors, semiconductors and superconductors, and their applications in engineering are discussed. Chapters 12–15 and 17 show how discoveries in semiconductor physics, quantum optics and electron ballistics stimulated the growth of new frontiers such as electron devices, microelectronics, lasers, fibre optics and nanotechnology in the second half of the twentieth century. The physics of sound and the applications of this knowledge in engineering are discussed in Chapter 16.

The main addition to this edition is that of the chapter on Nanotechnology. Engineering Physics syllabi of many universities, including ours, now include an introduction to this new area. Nanotechnology has much to do with new technology rather than any new science. However, with its advent, it has become necessary to re-emphasize, or present from new angles, some of the material already forming a part of these syllabi. Unfortunately, in this edition we have not been able to do this. Also, the new chapter departs from the structure common to other chapters. For example, there is no Story of Nanotechnology appearing distinctively. However, the Introduction serves essentially the same purpose – providing a 'historical' background. This may be justifiable too because of the very young age of the technology compared to the areas covered in other chapters.

Dr Sanjay Jain is grateful to Dr Satish Chaturvedi, Chairman, Lokmanya Tilak Jankalyan Shikshan Sanstha (LTJSS); Smt. Abha Chaturvedi, Secretary, LTJSS; Mr Abhijeet Deshmukh,

Director, LTJSS; and Dr V M Nanoti, Principal, Priyadarshini Institute of Engineering and Technology (PIET), Nagpur; for their kind encouragement and valuable support. He also thanks his colleagues at the department of Applied Physics at PIET for useful discussions. He also wishes to express his deep sense of gratitude to his parents, Shri DN Lambade and Shrimati Sudha Lambade for their kind blessings and support. He is also deeply grateful to his wife, Saroj Jain, for her emotional support and for relieving him of major domestic responsibilities, thus allowing him to concentrate on this work. He also wishes to express his profuse thanks to his son Piyush and daughter Poornima, whose deep love and affection have been a constant source of enthusiasm.

Dr Girish Sahsrabudhe wishes to express his deep sense of gratitude to all his colleagues in the Department of Physics, Shri Ramdeobaba Kamla Nehru Engineering College (SRKNEC), Nagpur, for their sustained support and camaraderie. This support is evident in the chapters, in places too numerous to list. He wishes to especially thank Dr DK Bopardikar and Dr SM Pande, the latter also for a few beautiful drawings used in the text. He is deeply indebted to his parents, wife Neelima and sons Harshad and Mihir for their silent support and appreciation of the effort, which has now taken the form of this book. He would like to dedicate the book to the memory of his father. He is grateful to the management of SRKNEC for the constant support and encouragement without which this book would not have been written.

Both of us wish to thank our large student fraternity who have always inspired us through their proactive responses and curious enquiries to give the best we are capable of. We are thankful to Mr Madhu Reddy, Universities Press, for the publication of our manuscript. We are also grateful to Prof. Ramesh Singru, retired Professor, IIT Kanpur, who has been kind enough to pen the foreword for this book.

We also wish to place on record our sincere thanks to Prof. G Umesh, Physics Department, National Institute of Technology, Karnataka, Surathkal, Mangalore, for his appreciative review of this book and useful suggestions in *Current Science* (Vol. 100, No. 10, 25 May 2011).

Sanjay Jain
Girish Sahasrabudhe

Director, LTISS, and Dr V M Nanoti, Principal, Priyadarshini Institute of Engineering and Technology (PIET), Nagpur, for their kind encouragement and valuable support. He also thanks his colleagues at the department of Applied Physics at PIET for useful discussions. He also wishes to express his deep sense of gratitude to his parents, Shri DN Lambade and Shrimati Sudha Lambade for their kind blessings and support. He is also deeply grateful to his wife, Saroj Jain, for her emotional support and for relieving him of major domestic responsibilities, thus allowing him to concentrate on this work. He also wishes to express his profuse thanks to his son Piyush and daughter Poornima, whose deep love and affection have been a constant source of enthusiasm.

Dr Girish Sahasrabudhe wishes to express his deep sense of gratitude to all his colleagues in the Department of Physics, Shri Ramdeobaba Kamla Nehru Engineering College (SRKNEC), Nagpur, for their sustained support and camaraderie. This support is evident in the chapters, in places too numerous to list. He wishes to especially thank Dr DK Bopardikar and Dr SM Pande, the latter also for a few beautiful drawings used in the text. He is deeply indebted to his parents, wife Neelima and sons Harshad and Mihir for their silent support and appreciation of the effort, which has now taken the form of this book. He would like to dedicate the book to the memory of his father. He is grateful to the management of SRKNEC for the constant support and encouragement without which this book would not have been written.

Both of us wish to thank our large student fraternity who have always inspired us through their proactive responses and curious enquiries to give the best we are capable of. We are thankful to Mr Madhu Reddy, Universities Press, for the publication of our manuscript. We are also grateful to Prof. Ramesh Singh, retired Professor, IIT Kanpur, who has been kind enough to pen the foreword for this book.

We also wish to place on record our sincere thanks to Prof. G Umesh, Physics Department, National Institute of Technology Karnataka, Surathkal, Mangalore, for his appreciative review of this book and useful suggestions in *Current Science* (Vol. 100, No. 10, 25 May 2011).

Sanjay Jain
Girish Sahasrabudhe

Contents

1

Physics and Engineering

1.1 THE STORY OF PHYSICS AND ENGINEERING

The mysteries of nature have always fascinated us. Physics, known earlier as 'natural philosophy', tries to understand and unfold the mysteries of nature. It is thus an eternal science. The word physics is derived from the Greek word 'physicos' meaning nature. Being concerned with matter, energy, and their interactions, physics encompasses the entire material universe in its ambit. For this reason physics is known as a fundamental science or the leader of all sciences. To quote from the McGraw-Hill Encyclopedia of Science and Technology: *Where other sciences are content to describe and relate phenomena in terms of restricted concepts peculiar to their own disciplines, physics always seeks to understand the same phenomena as a special manifestation of the underlying uniform structure of nature as a whole.*

Physics is also known as the 'zeroth order science' since it does not draw upon any other discipline, except mathematics, in its role of elucidating a phenomenon. Its dependence on mathematics is in the sense that mathematics is the language of all sciences. Mathematicians develop abstract theories, which physicists apply to concrete physical situations in nature. Physicists infuse life into mathematical equations!

In the Greek era (third to fifth century) several natural philosophers including Leucippus, Democritus and Epicurus, advocated the atomic point of view. They believed that nature consists of immutable atoms moving in empty space, their different configurations and motion giving rise to different phenomena in nature. Aristotle and Archimedes contributed to the understanding of mechanics in nature. The medieval period was mainly dominated by the influence of Aristotle's physics.

The period ranging from the fifteenth to the seventeenth century witnessed a revolution in scientific thinking, and science emerged as an autonomous discipline—distinct from both technology and philosophy. Nicolaus Copernicus, the Polish astronomer, challenged the established ideas in astronomy and proposed that the earth moved round the sun. Tycho Brahe, the Danish astronomer and Galileo Galilei, the Italian astronomer, made pioneering contributions to this revolution. This revolution culminated in the seminal work of Sir Issac Newton at the end of the seventeenth century. He solved major problems of mechanics and astronomy encountered in this revolution and published them in 1687 in his monumental volume, *Mathematical Principles of Natural Philosophy*. Another revolution was witnessed at the beginning of the twentieth century when quantum mechanics and relativity emerged as more general theories of which Newtonian physics turned out to be a special case.

The history of physics is replete with examples where seemingly different aspects are synthesised and shown to be but different aspects of the same thing. In 1687, Newton unified celestial and terrestrial mechanics by showing that the same laws hold for astronomical objects and objects on the Earth. In the period 1820–30 Oersted and Faraday integrated the then separate sciences of electricity and magnetism. In 1873, Maxwell further achieved a merger of these sciences with optics. In fact, physics as we know it today, is a synthesis of a large number of branches such as acoustics, optics, electricity, magnetism, heat and thermodynamics. This synthesis was mainly due to the underlying interrelations among various forces and energies in nature.

Physics grows through attempts to improve agreement, or to explain the lack of it, between theory and experiment. The results of theoretical models are tested against experiment, whereas the results of experimental investigations are interpreted through theoretical models. Specific theoretical models are at times derived from and at times woven into larger theories. These theories are in turn evaluated against criteria of logic, consistency and completeness. The interplay between theory and experiment is the motivation for research.

Sometimes the experimental data is available but there is no suitable theory to explain it. For example, in AD 140, Claudius Ptolemy had accumulated the experimental data of angles of incidence and angles of refraction for light rays in water. But the phenomenon of refraction and the relation between the two angles were not understood for the next fifteen centuries. It was only in 1621 that Willebrord Snell, a Dutch mathematician, deduced a law connecting the two angles that came to be known as Snell's law of refraction.

Sometimes a theory is available but there is no experimental evidence to support it. For example, the theory of stimulated emission was proposed by Einstein as early as 1917, but it could be translated into experimental reality only in 1960 when Maiman constructed the first laser using stimulated emission as a key to its operation.

Sometimes the results of both theory and experiment are available, but the two do not agree. Such a situation necessitates a closer look at the assumptions of the theory and demands newer insights for its refinement. For example, the observed features of the photoelectric effect, discovered by Hertz in 1887, could not be explained assuming the wave theory of light. Einstein succeeded in explaining these features in 1905 on the basis of the quantum theory of light.

In this chapter we will see how we can learn physics in the same spirit as it has grown over all these years. The utilitarian value of its study in engineering will be discussed in the last section.

1.2 LEARNING PHYSICS

Any course in physics can be made interesting by discussing it in consonance with the objectives of physics that were spelt out in the previous section. Learning physics can be an enjoyable experience for an inquisitive student as it arouses the natural curiosity and quenches the thirst for knowledge of nature. It can be accomplished as a three-step process:

Observation means experiment and *explanation* means theory. *Application* means use of the knowledge accrued through the confluence of theory and experiment to achieve the desired objectives. As this three-step process is also followed at the topmost level of research, it is essential here to distinguish between research in physics and learning physics as a part of the curriculum. In research this process is an open ended one; the researcher takes up an unsolved problem in literature and struggles to find its solution. Thus the output of research is always new knowledge for everyone. In learning, on the other hand, the focus is on new knowledge for oneself.

A student can be looked upon as a *young researcher*. Just as challenges in research mean acquiring newer knowledge, learning means gaining a better understanding of the available knowledge. A student is not so much expected to build new models or discover new experiments, as to develop a better grasp of theories and deftness in experiments. Better grasp of theories brings students closer to understanding what the physicists who developed these theories thought at that time, and deftness in experiments helps them acquire new skills. A student who is good in applications will be able to develop innovative projects.

The knowledge gathered from research gets well established over the years and percolates to lower levels of learning. For example, Einstein had reported the explanation for the observed experimental facts of the Photoelectric effect in 1905 in his research paper, "On a Heuristic Viewpoint Concerning the Production and Transformation of Light", published in the German *Annalen der Physik*. Thus the Photoelectric effect was considered an area of research in the first quarter of the last century. In later years it was studied at the postgraduate level. Today it is studied at pre-college and undergraduate levels.

Most of what we study in a physics course today first appeared in the form of a research paper in a scientific journal. A few of the leading international research journals are *Physical Review*, *Journal of Applied Physics*, *Journal of Physics*, *Nature* and *Japanese Journal of Applied Physics*. Some of the research reported in journals passes the test of time and gains the status of scientific knowledge. This knowledge is disseminated in popular form and language by journals such as *Scientific American* (USA), *Current Science* (India) and *Resonance* (India).

In this book an attempt is made to discuss the subject using the three-step process mentioned above. The following features are used to facilitate learning through this process.

Stories

Each chapter opens with the 'story' that narrates the historical development of the subject to be covered in that chapter. These stories recount the endeavours of physicists (together with other professionals like engineers and doctors) to understand and explain various phenomena. It should be remembered that failures in research have also contributed to the growth of knowledge. These stories also bring to the fore some of the links between different phases in this growth. They should help the student view the various events narrated in a wider perspective.

Theory and Experiment

Necessary theory and related experiments are discussed in relevant chapters. Experiments are integrated into theory to stress that both are equally important, interdependent components of physics. Theory and experiment are often distanced from each other, for example, there are separate books for each, conveying an impression that the two are independent of each other. We will be discussing this feature in greater detail in the coming sections.

Problems

In each chapter appropriate numerical problems have been included at relevant places. These problems serve to offer quantitative appreciation of the parameters involved in a particular model. Many physical concepts can be understood with the help of problems. A problem, for example, can demonstrate the success or failure of a particular model.

Charts

Topics in physics share some common, unifying and founding aspects. Thus physics can be learnt in a comprehensive manner through the use of charts displaying the comparative study of various similar topics. Charts enable one to understand particular topics at a glance, which is an advantage over usual learning from a book in which topics are presented sequentially, one after another.

Applications

Applications of the subject matter covered in a chapter are discussed at the end of each chapter, with more exphasis being laid on the engineering applications.

These features form the essential components of the physics course presented in this book. The mutual interdependence of these components is emphatically brought out.

1.3 THEORY

We appreciate by now that *theory* and *experiment* are the two pivots supporting physics, be it research or learning physics at the undergraduate level. A theory attempts to explain why nature behaves as it does. For example, Newton's theory of gravitation explains why objects thrown up always come down after a definite time with a definite speed. A theory is based on certain postulates/laws/hypotheses and certain assumptions. Basic postulates in physics are the starting point of all theories. Postulates cannot be derived; they are the first principles. For example, Newton's law of gravitation is a starting point in Newton's theory of gravitation. Newton's laws of motion, Coulomb's law, Gauss' law, Planck's laws and Schrodinger's equation are all basic laws and in physics. However, assumptions of a theory are open to examination and refinement. For example, while in Newton's theory of gravitation the range of the force of gravitation is assumed to be infinite, Einstein considers gravitation to be a manifestation of local geometry of space and time.

The second step in the development of a theory is the mathematical formulation or the derivation using some fundamental abstractions or concepts such as time, space, charge and energy, and some mathematical tools such as matrices, integration and differentiation. This formulation leads to a final outcome in the form of a certain equation. Thus the outcome of a theory is a mathematical statement that equates two physically distinguishable concepts. It says that numerical values of both sides are equal, no matter who the experimenter is. It should be understood here that this statement is the outcome of a theory based on certain assumptions. It is not a universal statement. This statement can by checked by performing an experiment. If the equation is satisfied by a wealth of experimental data, then the theory is successful. If the data do not satisfy the equation, then the theory fails and necessitates closer scrutiny of the assumptions on which it is based. If there is more than one theory related to the same experimental data, the one which is more comprehensive and simpler is considered to be more successful. This is because we believe that nature is inherently simple and is understandable in terms of a few basic principles.

1.4 EXPERIMENT

A student, as noted earlier, is not expected to develop any new theory but is expected to study the available theories. However, with respect to the experimental component, students should perform experiments with total involvement and a curious mind. If it is done in this manner then experiments can turn out to be very strong techniques of learning physics. As one starts explaining what one observes, and starts observing what one expects, the process of learning becomes most natural! The following discussion helps us equip ourselves with some common prerequisites for all the experiments discussed in this book.

All experiments basically comprise of the following three components:

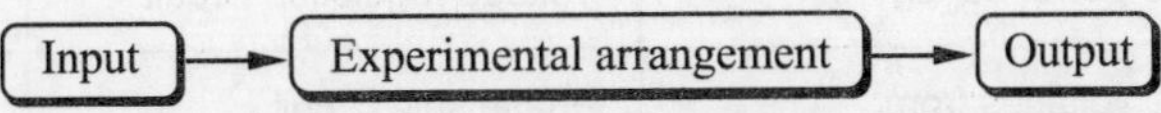

We provide input in some form and study the output. Output is basically the modified input due to the presence of experimental arrangement. For example, in the rectification experiment, the input is AC voltage and the experimental arrangement converts this input to DC voltage; in the biprism experiment the input is monochromatic light and the experimental arrangement modifies this light into a series of dark and bright fringes of a stable interference pattern. In Chart 1.1 we illustrate this aspect for all the experiments studied in this book. The details of the experimental arrangement and how that arrangement produces a particular output (the theory behind that experiment) are discussed in the respective experiments.

The output is generally analysed by instruments that can be categorised as either a *meter* or a *scope* or a *graph*. We can measure a quantity using a meter, see the pattern using a scope and

CHART 1.1 INPUT, EXPERIMENTAL ARRANGEMENT AND OUTPUT IN VARIOUS EXPERIMENTS

S.No	Experiment	Input	Experimental arrangement	Output
1	Biprism	Source of monochromatic radiation, for example, sodium lamp	Biprism	Stationary interference pattern of straight line fringes
2	Newton's rings	Source of monochromatic radiation, for example, sodium lamp	Newton's rings apparatus	Stationary interference pattern of concentric circular rings
3	Wedge-shaped film	Source of monochromatic radiation, for example, sodium lamp	Wedge-shaped film apparatus	Stationary interference pattern of straight line fringes
4	Grating	Source of monochromatic radiation, for example, sodium lamp	Diffraction grating	D_1 and D_2 lines of sodium doublet
5	Birefringence	Source of monochromatic radiation, for example, sodium lamp	Birefringent crystal	Ordinary ray and extraordinary ray
6	Photoelectric effect	Light of varying intensity and frequency, power supply to photocell	Photocell	Photoelectrons that constiute photoelectric current
7	Four probe method	Current through outer pair of probes,temperature of the oven	Four probe apparatus kept in an oven	Voltage between inner pair of probes
8	*p-n* junction	Variable DC power supply	Diodes connected in forward and reverse bias	Current through diode
9	Transistor	Variable DC power supply	Transistors connected in CB and CE mode	Emitter current, base current and collector current
10	Applications of diode	AC power supply, variable DC power supply	Voltage rectifier circuit Voltage regulation circuit	Rectified voltage Regulated voltage
11	CRO	AC voltages from transformer, signal generator	Phase shift circuit	Visual display of input and output
12	Hall effect	Magnetic field; current through Hall probe	Hall probe	Hall voltage

record the pattern using a graph. For example, a spectrometer, spectroscope and spectrograph are the instruments that enable measurement, display and recording of spectra.

The measurement of a physical quantity involves some physical process and is based on some physical principle. For example, the measurement of temperature involves the thermal expansion of mercury and is based on the principle that the change in volume of mercury is linearly proportional to the change in temperature. Chart 1.2 displays the input and output devices used in the set of experiments discussed in this book. The chart also includes the method of reading the instrument and the determination of its least count.

1.4.1 Least Count and Range

The least count of the instruments used is of foremost importance in any experiment. The least count of an instrument is the smallest value it can measure. Obviously, smaller the least count, larger is the accuracy in the measured value. An instrument with a given least count cannot measure a value with an error smaller than half the least count. We will be discussing this in more detail in the next section.

There is also a limit on the maximum value an instrument can measure. Thus each instrument has a certain range, and measurements beyond this range are not possible. Instruments used in any experiment are thus appropriately decided taking into account the quantities involved. If the measured value exceeds the maximum limit of an instrument it means an instrument of higher range is to be used. If it is zero, all it means is that the value is too small to be measured by that particular instrument. An instrument of lower least count can possibly read that value. A point to be noted here is that zero in mathematics means 0.000... up to an infinite number of decimal places. But zero encountered in a physics laboratory just means a value smaller than the measuring ability of an instrument. In a milliammeter with a least count of 0.5 mA, current equal to 0.1 mA will be taken to register as 0 mA; in a milliammeter with a least count of 0.1 mA, current equal to 0.01 mA will be taken to register as 0 mA and so on. Thus the handling of instruments of various ranges gives the students a quantitative 'feel' of the accuracy of measurement and of the fact that no measuring instrument can guarantee infinite accuracy.

1.4.2 Error Analysis

Errors are always present in the measurement of any physical quantity. Broadly speaking, there are two distinct types of errors: subjective or individual and objective or instrumental. Subjective errors are those committed by the experimenter. These occur due to unsystematic procedure, lack of expertise in using instruments, or negligence. Instrumental errors are inherent in the construction of the measuring instrument(s) used. As noted in the last section, an instrument can measure a quantity only with a finite accuracy fixed by its least count. An ordinary 6-inch scale has a least count of 1 mm. One cannot record a measurement of 2.42 cm or 5.78 cm with this scale.

Subjective errors can be reduced by performing the experiment in a careful, systematic and skilful manner. They can also be reduced by designing instruments which require less human

CHART 1.2 INPUT AND OUTPUT DEVICES

	Input devices
OPTICAL	Two commonly used sources of light are the sodium- and mercury-vapour lamps. These are basically evacuated electric discharge tubes containing a mixture of either sodium and neon or mercury and argon. In the sodium vapour lamp the discharge is initiated in neon by electrons emitted from the tungsten filament, followed by evaporation of sodium. Hence the lamp glows red (characteristic of neon) immediately after it is turned on. As sodium gets excited, yellow sodium light is emitted. Though the sodium light contains two wavelengths (doublet) with an interval of 5.97 Å, it is considered a practical monochromatic source with average wavelength λ = 5892 Å. In a mercury vapour lamp, argon ionises first and as the lamp warms up, mercury vapourises gradually. Though mercury emits all wavelengths present in white light, greenish blue colour is dominant.
ELECTRICAL	The commonly used sources provide voltage, current or power in either AC or DC form. The AC supplies are also called oscillator or signal/ function generator. These devices are also categorised as either analogue or digital devices depending on whether the input is available in analogue or digital form.
MAGNETIC	The commonly used source is an electromagnet made of a solenoid wound over an iron core. The current passing through the solenoid produces a magnetic field that is directly proportional to this current.

Output devices

Two commonly used instruments are the spectrometer and the travelling microscope. These are discussed in detail in Expts. 4.1 and 3.2, respectively.

However, the method of reading the instruments and finding the least count has been explained here as it is common to almost all the optical instruments. These instruments consist of a main scale and a vernier scale. The main scale reading (MSR) is the reading on the main scale before the zero on the vernier scale. The vernier scale division (VSD) is the division on the vernier scale that coincides exactly with some division on the main scale.

Vernier scale reading (VSR) = VSD × LC (Note that VSD has no units but VSR has units of LC.)

Total reading (TR) = MSR + VSR.

The least count is the difference between one MSD and one VSD. If N vernier scale divisions coincide with (N–1) main scale divisions, then 1 VSD = [(N–1)/N] MSD.

So, LC = 1 MSD –1 VSD = 1 MSD – [(N–1)/N] MSD = (1/N) MSD.

Thus, LC = smallest division on the main scale/total number of divisions on the vernier scale.

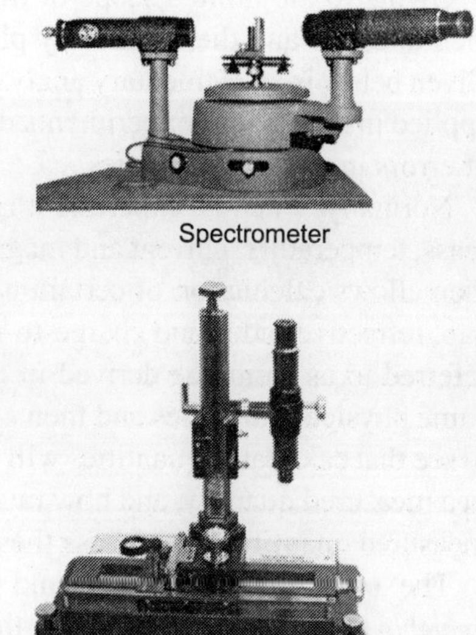

Spectrometer

Travelling microscope

The commonly used instrument is the cathode ray oscilloscope (CRO) which is capable of measuring various electrical parameters and also enables the visual display of electrical waveforms. It is discussed in detail in Expt. 13.1. Other familiar instruments based on the principle of the moving coil galvanometer, such as the voltmeter and the ammeter, are available either as analogue or digital devices.

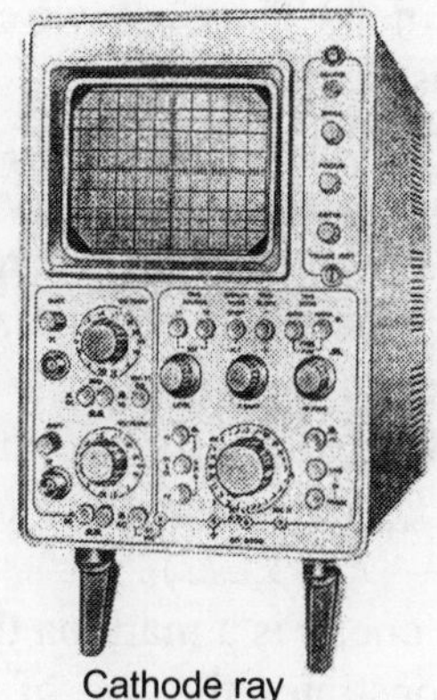

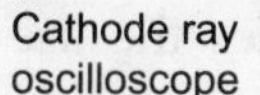

Cathode ray oscilloscope

Analog meter

Digital panel meter

A gaussmeter is commonly used to measure the magnetic field. Some electromagnets are provided with a gaussmeter that is calibrated to read the magnetic field directly. A Hall probe can also be used for measurement of a magnetic field (see Expt. 13.2).

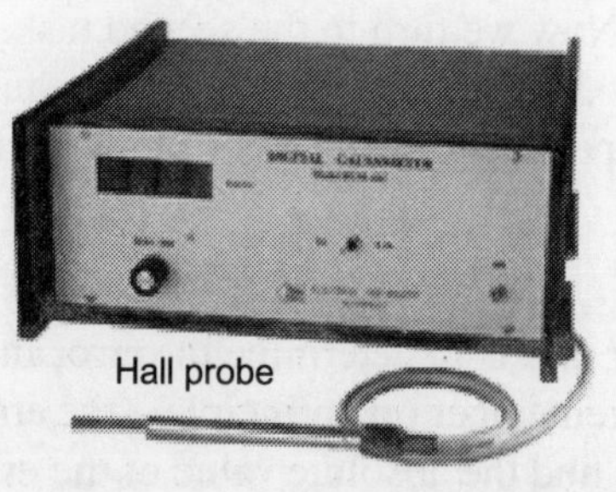

Hall probe

operation. Objective instrumental errors can be reduced only by building better, more accurate, and innovatively designed instruments.

Owing to the limited scope of this book it is sufficient to learn about the estimation of instrumental errors and the limits they place on the values of quantities determined in an experiment. Given below is an elementary analysis of the procedure by which this can be done. This analysis is applied in the various experiments discussed in this book to derive the formulae for the estimation of errors in various quantities.

Normally, when an experiment is performed, certain physical quantities such as length, angle, mass, temperature, current and magnetic field are directly measured. The theory of the experiment then allows calculation of certain other quantities like area, density, resistivity, forbidden energy gap, refractive index and charge-to-mass ratio for electron. This is done using certain expressions referred to as formulae derived in the theory of the experiment. Therefore, we directly *measure* some physical quantities and then *calculate* others using known mathematical formulae. It is easy to see that calculated quantities will also be in error. Our task is to determine how much is the error in a measured quantity, and how much is the error in the quantity calculated using this erroneously measured quantity. We address these tasks one by one.

The main point to understand about making a measurement is that it means specifying *unambiguously* the range in which the measured quantity has its value. The factor that determines this range is the least count of the measuring instrument. So, how can a measuring instrument of a given least count be used to record an observation such that the range in which the measured quantity has its value is unambiguously specified? There are two conventions used while recording an observation. Either one of them can be used consistently.

Consider an experiment where the length of an object is being measured. The first convention is to look at the mark just to the left of the actual reading. If this mark corresponds to the reading x_1 then the observation is recorded as $(x_1 + x_2)/2 \pm$ (LC/2), where x_2 is the reading corresponding to the next mark on the scale and LC is the least count of the scale. This observation establishes unambiguously that the measured value lies in the range $x_1 - x_2$. The other convention is to look at the mark nearest to the actual reading. If this mark corresponds to the reading x then the observation is recorded as $x \pm$ (LC/2). This observation is understood unambiguously as the measured value is in the range $[x -$ (LC/2)$] - [x +$ (LC/2)$]$.

In the second convention, unlike the first one, x is a mark on the scale. The second convention is more generally used. Figure 1.1 demonstrates the use of both these conventions for the measurement of the length of a chalk. The error in measurement is therefore half the least count. The scale cannot give an error smaller than this, as the least count of the scale cannot be transcended. The scale is only so good as to measure lengths with this much error.

Now we turn to the second task. Suppose we have to measure three quantities experimentally: x_1, x_2 and x_3. They are measured using instruments with least counts Δx_1, Δx_2 and Δx_3 respectively. If the theory tells us that we can calculate the quantity y using the formula

$$y = f(x_1, x_2, x_3) \tag{1.1}$$

our task is to determine the error in y. We can use simple calculus to find this error. We have only to remember one principle—the error in y because of errors in x_1, x_2, and x_3 could be *additive*. So we find the absolute value of the error in y because of error Δx_1 in x_1. We do the same for x_2 and

x_3. Then, we add all the errors. Now, calculus tells us that the error in y because of error Δx_1 in x_1 is

$$\Delta y = \left|\frac{\partial f}{\partial x_1}\right| \Delta x_1 \tag{1.2}$$

This expression follows from the meaning of the derivative. The right side is simply the product of the partial derivative of f, that is, of y, with respect to x_1, the rate of change of y with respect to x_1, and the possible change, the maximum error, Δx_1 in x_1. Similar expressions give us the contributions to the error in y by errors Δx_2 in x_2 and Δx_3 in x_3. The net error in y is, therefore,

$$\Delta y = \left|\frac{\partial f}{\partial x_1}\right| \Delta x_1 + \left|\frac{\partial f}{\partial x_2}\right| \Delta x_2 + \left|\frac{\partial f}{\partial x_3}\right| \Delta x_3 \tag{1.3}$$

Equation 1.3 gives the absolute error in y. The fractional or relative error in y is given by $|(\Delta y/y)|$, and $|(\Delta y/y)| \times 100$ gives the percentage error in y. It should be noted that Eq. 1.3 gives the maximum possible error in y. Physically it means that the calculation of y is wrong by an amount which is not more than Δy. Examples 1.1 and 1.2 illustrate, respectively, how the foregoing analysis can be applied to find out the error in the determination of volume and density. In these examples we have assumed that all length measurements are differences between readings of the instrument and error in measurement is then equal to the least count, rather than half the least count.

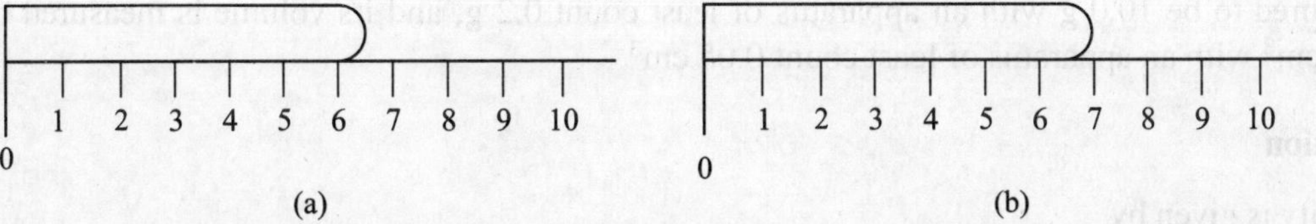

Figure 1.1 Measurement of the length of a chalk using a scale of least count 1 cm. Using the first convention the length is (6.5 ± 0.5) cm in both (a) and (b). This way of recording the observation establishes unambiguously that the measured value is in the range (6–7) cm. Using the second convention the length is (6 ± 0.5) cm in (a) and (7 ± 0.5) cm in (b). These observations establish unambiguously that the measured value is in the range 5.5–6.5 cm in Fig. 1.1a and in the range 6.5–7.5 cm in Fig. 1.1b.

EXAMPLE 1.1

Estimation of error in the determination of volume

Estimate the error in the determination of the volume of a parallelopiped with length 9.3 cm, breadth 8.5 cm and height 5.4 cm, all measured with a scale having a least count equal to 0.1 cm.

Solution

Volume is given by $V = l \times b \times h$

$$= (9.3 \text{ cm}) \times (8.5 \text{ cm}) \times (5.4 \text{ cm}) = 426.9 \text{ cm}^3$$

The absolute error in V is,

$$\Delta V = \left|\frac{\partial}{\partial l}(lbh)\right| \Delta l + \left|\frac{\partial}{\partial b}(lbh)\right| \Delta b + \left|\frac{\partial}{\partial h}(lbh)\right| \Delta h$$

$$= bh\Delta l + lh\Delta b + lb\Delta h$$

$$= [(8.5 \times 5.4)\ 0.1 + (9.3 \times 5.4)\ 0.1 + (9.3 \times 8.5)\ 0.1]\ \text{cm}^3$$

$$= 17.5\ \text{cm}^3$$

Thus the relative error is,

$$\frac{\Delta V}{V} = \frac{17.5}{426.9} = 0.04$$

The percentage error is about 4%.

EXAMPLE 1.2
Estimation of error in the determination of density

Estimate the error in the determination of the density of a cube of magnesium if its weight is measured to be 10.0 g with an apparatus of least count 0.2 g, and its volume is measured to be 5.80 cm^3 with an apparatus of least count 0.05 cm^3.

Solution

Density is given by

$$\rho = \frac{M}{V} = \frac{10.0\ \text{g}}{5.80\ \text{cm}^3} = 1.72\ \text{g/cm}^3$$

The absolute error in ρ is,

$$\Delta\rho = \left|\frac{\partial}{\partial M}(M/V)\right| \Delta M + \left|\frac{\partial}{\partial V}(M/V)\right| \Delta V$$

$$= \frac{\Delta M}{V} + \frac{M\Delta V}{V^2}$$

$$= \frac{0.1\ \text{g}}{5.80\ \text{cm}^3} + \frac{10.0\ \text{g} \times 0.05\ \text{cm}^3}{(5.80\ \text{cm}^3)^2}$$

$$= 0.017\ \text{g/cm}^3 + 0.015\ \text{g/cm}^3 = 0.032\ \text{g/cm}^3$$

Thus the relative error is, $\dfrac{\Delta\rho}{\rho} = \dfrac{0.032}{1.72} = 0.019$ and the percentage error is 1.9%.

1.4.3 Significant Figures

In the last subsection, the general idea of errors of measurement and calculation has been discussed. Approximate numbers are used to represent quantities in error. The error may be a result either of measurement, or, of computation, which uses other quantities in error. Such quantities are thus always written, along with the maximum possible error (that is, the absolute error) in them. There is another way of handling approximate numbers. It uses the concept of significant figures. Though this way of treating numbers is not rigorous like the treatment of errors discussed earlier, it has the advantage of simplicity. This treatment is discussed briefly in the following.

An approximate number is written using a certain number of digits, all, except the last of which are certain. This one digit is in error. The number of digits used is called the number of *significant digits*. For example, if the measured room temperature is reported along with its absolute error as (27.5 ± 0.5)°C, then it can be alternatively expressed that the temperature is 27.5°C with *three* significant digits. It signifies that we are sure about the first two digits but not about the third. The third digit is uncertain by $\pm$ 0.5. Usually, if the absolute error is not known, the last significant digit is regarded as uncertain by $\pm$ 1.

It is thus easy to appreciate that exact numbers have infinite number of significant figures. Numbers such as mathematical constants like π are to be regarded as exact, since in principle, we can find their values to any number of significant digits. For example, π to 50 decimal digits, is 3.1415926535897932384626433832795028841971693993751.

Table 1.1 gives the set of rules to determine the number of significant digits in a given approximate number and its application for three approximate numbers.

Table 1.1 Rules to determine number of significant digits

Rule	Application		
	3.527	2.0031	0.002340
Take the number as it is given. Do not add any zeros after the number.	3.527	2.0031	0.002340
Count all non-zero digits.	4	3	3
Count all zeros between non-zero digits.	0	2	0
Count all zeros after all non-zero digits.	0	0	1
Do not count any zeros before all non-zero digits.	–	–	0
The total count is the number of significant digits.	4	5	4

In order to take care of numbers like in the last column scientific notation is used. In this notation the number is 2.340×10^{-3}. The first part 2.340 (called the *mantissa*) is approximate and contains all the significant digits. The exponent of ten is an exact number. Thus $345.78 = 3.4578 \times 10^2$ has 5 significant digits whereas $0.0005 = 5 \times 10^{-4}$ has 1 significant digit.

There are certain rules of arithmetic of approximate numbers. Although the precise rules are studied in theory of errors, two important rules of thumb are given here.

Rule 1 *(Addition and subtraction):* Given two numbers x and y, the absolute error in both $x+y$ and $x-y$ is the sum of absolute errors in x and in y. Therefore, the number of significant digits in $x \pm y$ is determined indirectly by the number of significant digits in the number with the larger error.

For example, if $x = 3.25$ and $y = 2.1$, then the absolute error in x, that is, Δx, is 0.01 and that in y is 0.1. The absolute errors in both $x+y$ and $x-y$ are $\Delta(x \pm y) = \Delta x + \Delta y = 0.11$. So, the approximate numbers $x + y = 5.3$ and $x - y = 1.1$ have 2 significant digits. It is evident, that is, adding/subtracting two numbers, the larger error dominates, or, the number with the smallest number of certain decimal digits determines the error and indirectly the number of significant digits.

Rule 2 *(Multiplication and Division):* Given two numbers x and y with n and m significant digits respectively, the number of significant digits in their product, xy, and quotient, x/y, is Min (n, m), that is, the smaller of n and m.

For example, with x = 22.00 (4 significant digits) and y = 7.00 (3 significant digits), xy = 154 and x/y = 3.14 have 3 significant digits. In Ex. 1.1, the range of volume obtained by the significant digit calculation is 4.2×10^2 cm^3–4.3×10^2 cm^3, whereas that obtained using the error estimation is 409.4 cm^3–444.4 cm^3.

1.4.4 Method of Least Squares

The results of an experiment or observation are sometimes represented as a graph. When points are plotted there is a confusion regarding how to join the points. For example, for a set of points (x, y), with y linearly dependent on x (see Fig. 1.2), the question arises which of the lines—1, 2, 3 or 4—represents the best plot.

The statistical principle that should be used to plot the line, which represents the data in the best possible manner, is called the *Principle of least squares*. According to this principle, *the line of best fit is that line for which the sum of the squares of deviations between the observed values and the corresponding values on it is minimum*, that is,

$$E = \sum_{i=1}^{N} (y_i - \hat{y}_i)^2 \tag{1.4}$$

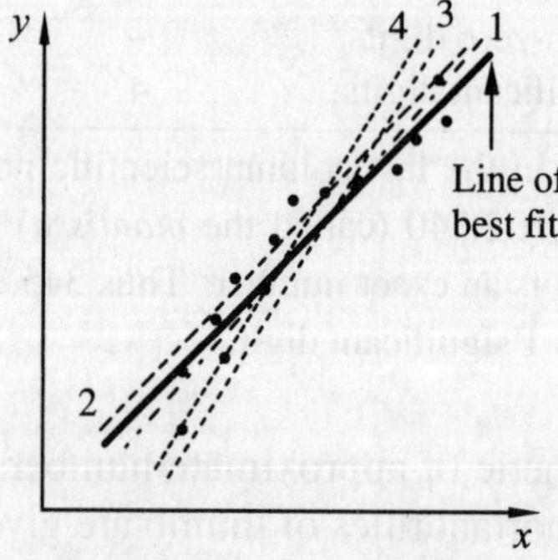

Figure 1.2 Representation of the observed data using the principle of least squares

should be minimum, where N is the number of data points, y_i is the observed value of y at $x = x_i$ and $\hat{y}_i$ is the corresponding value on the line of best fit, that is,

$$\hat{y}_i = mx_i + c \tag{1.5}$$

where m and c are called the parameters of the best fit. These parameters are to be evaluated such that E is minimum.

Thus, $\dfrac{\partial E}{\partial m} = 0$ and $\dfrac{\partial E}{\partial c} = 0$

or $\dfrac{\partial}{\partial m}\left[\sum_i (y_i - mx_i - c)^2\right] = 0$ and $\dfrac{\partial}{\partial c}\left[\sum_i (y_i - mx_i - c)^2\right] = 0$

Carrying out the differentiation with respect to m and c regarding x_i, y_i as constants gives

$$\sum_i x_i y_i - m\sum_i x_i^2 - c\sum_i x_i = 0 \text{ and } \sum_i y_i - m\sum_i x_i - cN = 0$$

These equations are called normal equations. They can be simultaneously solved for the unknowns c and m. Solving them, we get,

$$m = \left[N\sum_i x_i y_i - \sum_i x_i \sum_i y_i\right] \Big/ \left[N\sum_i x_i^2 - \left(\sum_i x_i\right)^2\right] \tag{1.6}$$

and $$c = \left[\sum_i x_i^2 \sum_i y_i - \sum_i x_i \sum_i x_i y_i\right] \Big/ \left[N\sum_i x_i^2 - \left(\sum_i x_i\right)^2\right] \tag{1.7}$$

The summations in these expressions are calculated as shown in Table 1.2. The values of best-fit parameters m and c are then substituted in Eq. 1.5 and this equation is plotted as the line of best fit. It is interesting to note that the line of best fit may not pass through any of the observed points but represents the observed data in the best manner as the sum of the squares of errors is the least for it (see Fig. 1.2).

Table 1.2 Summations required in the calculation of parameters of best fit

S. No.	x_i	y_i	x_i^2	$x_i y_i$
1	x_1	y_1	x_1^2	$x_1 y_1$
2	x_2	y_2	x_2^2	$x_2 y_2$
–	–	–	–	–
–	–	–	–	–
N	x_N	y_N	x_N^2	$x_N y_N$
Sum	$\sum_{i=1}^{N} x_i$	$\sum_{i=1}^{N} y_i$	$\sum_{i=1}^{N} x_i^2$	$\sum_{i=1}^{N} x_i y_i$

1.5 SEEKING AGREEMENT BETWEEN THEORY AND EXPERIMENT

In the preceding sections we saw the meaning of theory and experiment. The goal of physics, as pointed out earlier, is to establish agreement between theory and experiment. Generally we first seek a *qualitative* agreement, that is, whether the observed *trends* agree with those expected by the theory. Then we seek a *quantitative* agreement, that is, whether the observed *values* agree with those expected by the theory. For example, the observation of Compton that wavelength of x-rays increases upon scattering was in qualitative agreement with the quantum theory. Compton further established quantitative agreement by working out a model based on the quantum theory and showing that the theoretical value of increase in wavelength was comparable to the experimental value.

We pursue the same goal in this book by presenting a discussion of this aspect at the end of each experiment. However, it should be noted that the error in an experiment limits the extent to which agreement can be sought, and with the instruments available in college laboratories, only qualitative agreement can be reached in most cases. Students should keep in mind that *it is not the actual agreement but the training of seeking the agreement that is important*. Students can develop insight into the subject by trying to understand the reasons for the departure of the theory from the experiment.

In Section 1.2 we compared the role of an undergraduate student with that of a researcher in the study of physics. Now it can be clearly understood that the journey of an undergraduate student to a young researcher and from there to a mature researcher is the journey from broader and coarser levels of agreement to deeper, finer and subtler levels of agreement between theory and experiment. An undergraduate student may have to discontinue the pursuit of this goal due to limitations of laboratories or other ambitions, but this continues to be the main goal of a professional researcher. We also saw that the agreement can be improved by either refining the experiment or refining the theory. Examples 1.3 and 1.4 illustrate this point. Chart 6.2 illustrates confirmation of de Broglie's theory by the Davisson and Germer experiment.

A very important aspect of the success of physics is the phenomenal extent of agreement that has been sought and achieved. Penrose called theories such as Einstein's theory of relativity ' super theories'. Einstein's theory of relativity enabled a calculation of the precession of perihelion of mercury, which could be verified experimentally with an accuracy of one part in 10^{14}!

EXAMPLE 1.3

Refinement in experiment to improve agreement with theory

The theoretical formula for determining the circumference c of a circle of radius r is $c = 2\pi r$. Discuss the agreement of this formula with the following experimental results. (a) A vernier scale of least count 0.1 cm is used and the measured values are r = 1.1 cm and c = 6.9 cm; (b) An optical technique that provides a least count of 0.001 cm is used and the measured values are r = 1.005 cm and c = 6.316 cm.

Solution

The formula for calculating the circumference is $c = 2\pi r$ or $\frac{c}{r} = 2\pi$ (1.8)

Thus the error in c/r, is,

$$\Delta(c/r) = (c/r^2)\Delta r + \Delta c/r = [(c/r^2) + (1/r)](\text{LC}/2),$$

where LC is the least count.

(a) Using the instrument with least count of 0.1 cm,

$\Delta(c/r) = 0.33$ and observed value of $c/r = 6.27$

Actual value of $c/r = 6.27 \pm 0.33$ and $\Delta(c/r)/(c/r) = 5.3\%$

Thus c/r lies in the range 5.9–6.6 and the measured value is in error by 5.3%. Now the right side of Eq. (1.8) is 2π, which is known to any accuracy. Considering its value only up to the first decimal, we find that the value 6.3 lies in the above range. Thus the experiment agrees with the theory to an accuracy of 5.3%.

(b) Using the instrument with least count of 0.001 cm,

$\Delta(c/r) = 0.004$ and observed value of $c/r = 6.285$

Actual value of $c/r = 6.285 \pm 0.004$ and $\Delta(c/r)/(c/r) = 0.06\%$

Thus c/r lies in the range 6.281–6.289 and the measured value is in error by 0.06%. Considering the value of 2π up to the third decimal, we find that the value 6.283 lies in the above range. Thus the experiment agrees with the theory to an accuracy of 0.06%.

This example thus illustrates how a refinement in the experiment can lead to better agreement with the theory.

EXAMPLE 1.4

Refinement in theory to improve agreement with experiment

The experimental value of velocity of sound in air is known to be 330 m/s. Discuss the agreement of this result with the results of the following theories: (a) Newton assumed that the propagation of sound waves in a gas is an isothermal process and derived the formula for the velocity of sound as $v = \sqrt{P/\rho}$. (b) Laplace argued that compressions and rarefactions in sound waves take place so rapidly that the heat developed during compression, or the cooling produced during rarefaction, has no time to dissipate itself into the surrounding medium and hence the propagation should be assumed to be adiabatic and not isothermal. The formula for the velocity of sound derived by Laplace is $v = \sqrt{\gamma P/\rho}$. Given: $P = 76$ cm of Hg, $\rho = 0.001293$ g/cm^3 for dry air at 0°C and γ for air = 1.41.

Solution

(a) The velocity of sound using Newton's theory is $v = \sqrt{P/\rho}$. Using the given data, we have

$$v = \sqrt{\frac{76 \text{ cm} \times 13.6 \text{ g/cm}^3 \times 980 \text{ cm/s}^2}{0.001293 \text{ g/cm}^3}} = 280 \text{ m/s}$$

Thus the value obtained using Newton's theory is $[|330 - 280|/330] \times 100 = 15\%$ lower than the experimental value.

(b) Using Laplace's theory the velocity is given by $v = \sqrt{\gamma P/\rho}$. Thus,

$$v = \sqrt{\frac{1.41 \times 76 \text{ cm} \times 13.6 \text{ g/cm}^3 \times 980 \text{ cm/s}^2}{0.001293 \text{ g/cm}^3}} = 332 \text{ m/s}$$

Thus the value obtained using Laplace's theory is $[|330 - 332|/330] \times 100 = 0.6\%$ higher than the experimental value.

This example thus illustrates how a refinement in the theory can lead to a better agreement with the experiment.

1.6 APPLICATIONS

The Engineer's Council for professional development defines engineering as the profession in which knowledge of mathematical and physical sciences gained by study, experience and practice is applied with judgement to develop ways to utilise economically the *materials and forces of nature* for the progressive well being of humankind. The relevance of physics to the engineering curriculum is emphatically implied in this definition. Physics leads science in the study of matter, energy and their interactions. Technology deals with innovative applications of scientific knowledge and engineering is concerned with the practical use of appropriate technologies. Relations among science, technology and engineering can be expressed as under.

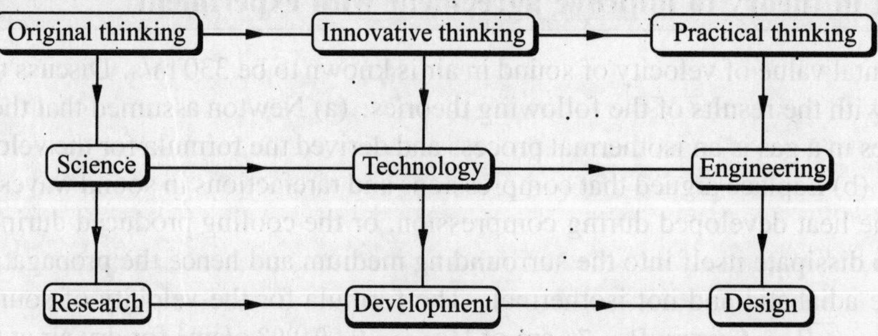

Thus a thorough assimilation of the fundamentals of physics is essential for founding a career in engineering. However, this close connection seems to have been rather obscured these days.

One of the main reasons for this is that many of the fundamental principles and techniques that originated in physics have grown into autonomous fields of investigation that have opened up new disciplines in engineering and technology. The situation can be compared to the growth of a tree in which new roots lead to the development of new branches. Sometimes the tallest branch of a tree seems to be unrelated to its deepest root, but the oneness of the tree exudes the unification that nature manifests. An engineering student aspiring to mature as an engineering professional should not lose sight of the ubiquitous role that physics plays in various engineering curricula.

Chart 1.3 displays the various branches of science and engineering and their roots in physics. This chart convinces us that research in physics holds immense potential in giving birth to and shaping new technologies. *The physics research of today is the engineering of tomorrow!* The following examples help us understand this better. The two greatest discoveries of quantum physics in the twentieth century were the transistor (1948) and laser (1960). The transistor stimulated the growth of microelectronics in the second half of the twentieth century, while the laser is guiding the growth of a fledgling technology, known as *photonics*, which is expected to relegate electronics to second place in the coming years. To further emphasise the point, ideas which arose in the decades following 1960 in the field of quantum computation and quantum information are today guiding technological search for quantum computers and communication channels.

It is also worth noting that countries leading in research in basic sciences like physics have marched way ahead of others in the field of engineering. The more talented students in these countries prefer basic sciences to engineering. Engineering is derived from science; so it is but natural that sound science will ensure progressive engineering.

Today the time lag between research breakthroughs in physics and their applications to engineering has shrunk enormously. Technology is evolving at a rate faster than ever before. Hence it is necessary for engineering students to acquaint themselves with the frontier areas of research in order to get conditioned for the atmosphere of change that they will encounter in their career.

In the last section of each chapter we will try to appreciate how the subject covered in that chapter formed the knowledge base on which technology and engineering flourished.

CHART 1.3 ROOTS OF ENGINEERING IN PHYSICS

This chart has been compiled from the *Encyclopedia Britannica* (*Macropedia*, Vol. 18). The encyclopedia defines engineering as the professional art of applying science to the optimum conversion of the resources of nature to the uses of mankind. It is based principally on physics, chemistry and mathematics and their extensions into materials science, solid and fluid mechanics, thermodynamics, transfer and rate processes and system analysis. The major functions of all branches of engineering in order of decreasing emphasis on science are research, development, design, construction, production, operation and management. Thus, engineering is a very vast field of activity having roots in physics and other basic sciences. Engineering courses are also designed on a similar pattern: students gain knowledge of physics and other sciences in the early part of their career and then apply this knowledge in the chosen branch of engineering. Though early engineers were trained by apprenticeship to become skilled engineers, following World War II, new trends became apparent, due to the rapidly growing body of theoretical knowledge that placed greater emphasis on mathematics, physical sciences and the engineering sciences. Here we attempt to trace the roots of various branches of engineering in physics.

Branch	What it deals with	Roots in physics
Electrical and electronics engineering	Electrical engineering deals with the practical applications of electricity in all its forms, including in the field of electronics. Electronics engineering is that branch of electrical engineering concerned with the uses of the electromagnetic spectrum and with the application of such electronic devices as integrated circuits, transistors and vacuum tubes. The functions performed by electrical and electronics engineers include, among others, basic research at the forefront of physics, other sciences, and applied mathematics, in order to extend knowledge applicable in their own field.	Electrical and electronics engineering have a common origin in physics. This is amply evident from the fact that the first courses in electrical engineering were introduced in the physics department of the Massachusetts Institute of Technology (MIT) in 1882 and were followed shortly by similar courses at Cornell University. By 1890 there were 10 such courses in the physics departments of major US universities. In the late twentieth century, electrical and electronics engineers outnumbered all others in the world. In the United Kingdom, formal electrical engineering education began within the discipline of physics. In the US, however, electrical engineering students directed little attention towards mathematics and the physics of electricity; they instead devoted their time to shops work, detailed study of manufacturers' catalogues and drawings, and visits to electrical power plants. However, by 1932 the growing need for mathematical analysis in the design of radio communications and electronic systems led nearly all engineering schools to include in their undergraduate curricula rigorous courses in higher mathematics and physics. This requirement is now so well established that electrical and electronics courses have the reputation of being difficult in comparison with other branches of engineering. The roots of electrical engineering can be traced to the pioneering works of physicists Andre Ampere (early investigator of electricity, 1775–1836), George Ohm (early investigator of electricity, 1789–1854), Charles Coulomb (Coulomb's law of electrostatics, 1785), Alessandro Volta (electric cell, 1800), Hans Oersted (electromagnetism, 1820), Michael Faraday and Joseph Henry (electromagnetic induction, 1831) and Heinrich Lenz (Lenz' law of electromagnetic induction, 1834). Electrical engineering may be said to have emerged as a discipline in 1864 when James Clerk Maxwell summarised the basic laws of electricity in mathematical form and predicted that radiation of electromagnetic energy would occur in a form that later became known as radio waves. In 1887 Heinrich Hertz experimentally demonstrated the existence of radio waves. Electronics engineering came into prominence through the pioneering researches of Edison (incandescent lamp, 1878), Hendrick Lorentz (electron theory of electric charge, 1895), Sir JJ Thomson (e/m for electron, 1897), Lee de Forest (triode, 1907), VK Zworykin (demonstration of CRO, 1920) and John Bardeen, William H Brattain, and William B Shockley (transistor, 1948). Electrical and electronics engineering later grew into new branches such as telecommunication engineering, space engineering and computer engineering that drew vital inputs from disciplines of physics such as fibre optics, spectroscopy, space physics and semiconductor physics.

CHART 1.3 (Continued)

Branch	What it deals with	Roots in physics
Mechanical engineering	Mechanical engineering deals with machines and the production of power. It is particularly concerned with forces and motion. The foremost function of a mechanical engineer is the understanding of and dealing with the bases of mechanical science. These include the dynamics concerning the relation between forces and motion, such as in vibration; automatic control; thermodynamics, dealing with the relations among the various forms of heat, energy and power; fluid flow; heat transfer; lubrication; and properties of materials.	The invention of the steam engine in the latter part of the eighteenth century, provided a key source of power for the Industrial Revolution, and gave an enormous impetus to the development of machinery of all types. As a result a new major classification of engineering— separate from civil engineering and dealing with tools and machines— developed. It received formal recognition in 1847 in the founding of the Institution of Mechanical Engineers in Birmingham, England. The roots of mechanical engineering can be traced to the pioneering works of physicists Robert Boyle (Boyle's law, 1662), Daniel Bernoulli (1700–1732), Robert Sterling (Sterling engine, entropy and Second law of thermodynamics, 1816), NL Sadi Carnot (concept of engines, 1824), Julius Mayor (equivalence of heat and work, 1842), James Joule (experimental relation between heat and work, 1850), Rudolf Clausius (kinetic theory, 1854; entropy, 1865), Maxwell (distribution of velocities, 1864), Ludwig Boltzman (statistical interpretation of Second law of thermodynamics, 1877), Joseph Stephan (heat transfer mechanisms, 1879), and Lord Kelvin (temperature scale, Second law of thermodynamics, 1824–1907).
Civil engineering	Civil engineering deals with the planning, design, construction and management of the built environment. This environment includes all structures built according to scientific principles from irrigation and drainage systems to rocket launching facilities. The design of civil engineering works may require the application of design theory from many fields, for example, hydraulics, thermo-dynamics, or nuclear physics.	The word 'civil' derives from the Latin for citizen. In 1782, Englishman John Smeaton used the term to differentiate his non-military engineering work from that of the military engineers who predominated at that time. Civil engineering draws inputs from the engineering sciences of mechanics, geology, petrology, strength of materials and fluid mechanics. The roots of civil engineering can be traced in the pioneering works of physicists Daniel Bernoulli (1700–1782), Robert Hooke (Hooke's law, 1679), Issac Newton (*Principia*, 1687), Amontons (laws of friction, 1699), John Bernoulli (*principle* of virtual work, 1717), Coulomb (laws of friction, 1779), Lagrange (principle of virtual work, 1788) and Sir David Brewster (photoelasticity, 1816).
Chemical engineering	Chemical engineering is the development of processes and the design and operation of plants in which materials undergo changes in physical or chemical state on a technical scale.	Applied throughout the process industries, chemical engineering is founded on the principles of chemistry, physics and mathematics. The laws of physical chemistry and physics govern the practicability and efficiency of chemical engineering operations. Energy changes, deriving from thermodynamic considerations, are particularly important. The roots of chemical engineering can thus be found in the overlapping research areas of physics and chemistry.

CHART 1.3 (Continued)

Branch	What it deals with	Roots in physics
Nuclear engineering	Nuclear engineering is concerned with the control and use of energy and radiation released from nuclear reactions. Research and development in nuclear engineering is concerned with the conception and development of new materials, processes, components and systems for nuclear facilities and the development of analytical methods and experimental procedures for use in the development, analysis, design and control of fission and fusion systems.	The roots of nuclear engineering can be traced to the pioneering works of physicists Henry Becquerel (radioactivity, 1896), Marie and Pierre Curie (radium and polonium, 1898) Ernest Rutherford (α-particle scattering, 1911; proton, 1914; splitting of atom, 1919), James Chadwick (neutron, 1932), Otto Hahn (fission, 1938), Enrico Fermi (first fission reaction, 1934; first nuclear reactor, 1942) and CL Cowan and F Reines (detection of neutrino, 1953). Fusion research since about 1950 has concentrated on the issues of plasma physics, specifically the production of high temperature (> 10^8C) plasmas that can be confined at sufficiently high densities for sufficiently long times to produce net energy. Fusion physics research has made steady and impressive progress, and research efforts have begun to address the important engineering issues of fusion.

Apart from those mentioned above there are several other branches of engineering. Some of these branches and their roots in physics are: aerospace engineering — aerodynamics (a branch of theoretical physics); petroleum engineering — petrophysics (determination of fluid and rock characteristics) — quantum engineering — quantum mechanics; environmental engineering — environmental physics; biomedical engineering — medical physics and biophysics; metallurgical engineering — solid state physics; engineering geology—geophysics and petrophysics, and nanoengineering— nanophysics.

There are also many specialised branches. A few examples are: lens design engineering (based on geometrical optics), crystal engineering (based on crystal physics) and band gap engineering (based on semiconductor physics).

2

What is Light?

2.1 THE STORY OF LIGHT

A major aim of physics is to study energy in its different manifestations. Light, which is a form of energy that stimulates vision, has attracted the attention of physicists since ancient times. The Greeks believed that light is made up of tiny particles emitted by a light source. Upon entering the eye, the stream of particles stimulates the sense of vision.

During the time of Newton (1642–1727), almost all the scientists enquired into the nature of light. Newton himself devoted considerable time to the study of light and proposed the corpuscular (particle) theory of light in his famous book, *Optiks*, published in 1704. (This book has been reprinted by Dover Publications, New York, in 1952.) However, Newton's theory could not explain anything beyond reflection and some aspects of refraction of light. Despite this weakness, Newton's model was not much criticised, and dominated for well over a century, mainly because of the compelling authority of Newton as a scientist. A seemingly final blow to Newton's corpuscular theory was struck in 1850 when Foucault measured the speed of light in water and found it to be smaller than its speed in air—an observation contrary to the prediction of Newton's theory that light should travel faster in denser media.

In 1678, a Dutch physicist, Christian Huygens, suggested the wave theory of light, which assumes light to be a wave—similar to a water wave or a sound wave. Maxwell (1873) substantiated the wave theory by propounding that light waves are electromagnetic transverse waves, which can travel through vacuum. Using the equations of electromagnetism, (known today as Maxwell's equations), he obtained the theoretical value for the speed of light as 3.1074×10^8

m/s which was in close agreement with the experimental value measured by Fizeau: 3.14858×10^8 m/s. The coincidence of these two numbers is considered to be a major milestone in the history of physics. Maxwell exclaimed after his discovery: *Let there be electricity and magnetism and there is light!*

Wave theory offered a successful explanation for the properties of light, such as reflection, refraction, interference, diffraction and polarisation (scc Chapters 3, 4 and 5). Thomas Young (1773–1829), Augustine Fresnel (1788–1829), and several other physicists offered experimental proof in support of the wave theory. Thus the spectacular success of the wave theory made scientists believe in the wave nature of light till the end of the nineteenth century.

However, the citadel of wave theory started developing cracks in the beginning of the twentieth century when newly gathered experimental evidence challenged the wave theory. Experimental observations of black body radiation, the Compton effect, line spectra and photoelectric effect could not be accounted for by the wave theory and necessitated a fresh approach to the understanding of the physical nature of light.

Max Planck (1900) and Albert Einstein (1905) made pioneering efforts in this direction and postulated the quantum theory of light, which assumes that the energy of a light beam is present in quantised bundles of energy called *photons*. Quantum theory could offer convincing explanation for these new experiments, and today it is regarded as one of the most successful theories in physics.

The success of the wave theory in explaining some of the properties of light, and the success of the quantum theory in explaining some others led to an intriguing question. What is light? Is it a wave? Or is it a particle? These seemingly contradictory viewpoints were reconciled in the statement of wave–particle dualism of radiation, according to which light is both. Sometimes it behaves like a wave (for example, in interference, diffraction and polarisation), and sometimes as a particle (for example, in the photoelectric effect and black body radiation).

Research into the nature of light provided valuable clues towards the behaviour of particles in the microworld. It was the brilliance of a young French physicist, Louis de Broglie, which captured these clues and led to the extension of the idea of wave–particle dualism of radiation to matter. This and the other developments of quantum physics will be studied in Chapter 6.

Thus, the subject of light has had a long and varied scientific history, and the understanding of the nature of light will continue to grow in future. We will end this short narration with a quotation from the McGraw Hill Encyclopaedia of Science and Technology, Volume 10: *The study of light deals with some of the most fundamental properties of the physical world and is intimately linked with the study of properties of sub-microscopic particles on the one hand and with the properties of the entire universe on the other. The creation of electromagnetic radiation from matter and creation of matter from radiation, both of which have been achieved, provide a fascinating insight into the unity of physics.*

2.2 GEOMETRICAL AND PHYSICAL OPTICS

The branch of science dealing with light, its origin and propagation, its effects and other phenomena associated with it is known as *optics*. If the dimensions of the obstacles which light encounters are much larger than its wavelength, then spreading effects like diffraction (see Chapter 4) can be ignored and we are in the realm of geometrical optics. Here we consider light to consist of rays that obey the laws of reflection and refraction. The effect of optical instruments such as prisms, lenses and mirrors—with dimensions much larger than the wavelength of light—on the paths traced by these rays is dealt with in geometrical optics.

If the condition for geometrical optics is not satisfied, then we cannot assume that light is present simply as rays and its physical nature has to be taken into account. This branch, known as physical optics, is further divided into wave optics and quantum optics, depending on the model of light assumed. The different properties of light can be grouped together and classified under these fields as shown in Table 2.1. We will be dealing mainly with physical optics in this and the subsequent chapters.

Table 2.1 Light under three-fold division of optics

Geometrical optics	**Wave optics**	**Quantum optics**
Rectilinear propagation	Interference	Atomic orbits
Finite speed	Diffraction	Probability densities
Reflection	Electromagnetic character	Energy levels
Refraction	Polarisation	Quanta
Dispersion	Double refraction	Lasers

2.3 WAVE EQUATION AND WAVE PARAMETERS

What is common to daily chores like watching television, cooking food and listening to music; to setting bone fractures and strained muscles; or to disasters like earthquakes and floods? The answer is 'waves'. These examples are concerned respectively with radio waves, microwaves (or thermal waves), sound waves, x-rays and infrared waves, seismic waves and water waves. These are some of the many types of waves that are common. Apart from these perceptible ones, there are numerous imperceptible waves associated with particles in the sub-microscopic world (for example, de Broglie waves, Chapter 6). Thus the world around us is filled with waves. Waves are everywhere and everything waves!

A wave is the propagation of a 'disturbance'. Generally there is transfer of energy associated with a wave. However, there are waves in which, over a time period, no net energy is carried in any direction. These are called *stationary waves*. If we go by the equation which governs the propagation of a wave, we can classify waves into two types: *classical waves* and *non-classical waves*. The former are essentially of two types: *mechanical* and *electromagnetic*. The latter are matter waves of de Broglie, which we will discuss in Chapter 6. Here we will try to understand the basic physical features of a classical wave and its mathematical treatment. Many of the ideas are applicable to matter waves too, with appropriate changes in their physical interpretation.

The form of the disturbance, or the 'shape' of the wave, can vary widely depending on the situation (see Fig. 2.1). For example, if you give a short and sharp jerk to the free end of a string anchored to a wall at the other end, a hump or a wave travels on it. Such a wave is called a *pulse*. If two sound waves of slightly different pitch mix with each other, the resulting pressure profile has ups and downs called *beats*. The wave of an earthquake may have a very complicated profile of pressure.

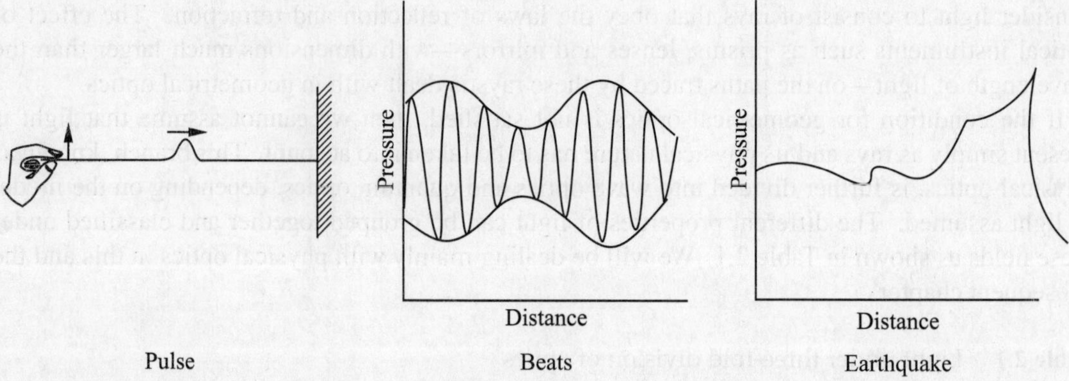

Figure 2.1 Different waveshapes

The question that arises is: How can we study waves of many different possible shapes in a unified way? The problem is simplified if we note two related points. The first is that a *single mathematical equation governs propagation of disturbance in classical waves.* The second is that *it is possible to understand complicated waves as mixtures of simple waves.*

Mechanical waves differ from electromagnetic waves in one important respect. Mechanical waves need a material medium for propagation, whereas electromagnetic waves can travel through vacuum. In case of mechanical waves the disturbance is propagated due to elasticity and inertia of the medium, without actual displacement of the medium as a whole. Elasticity provides the force which tends to restore the medium to its original position, and inertia enables the transfer of momentum from one part of the medium to its adjoining part.

The equation governing the propagation of disturbance in a classical wave is

$$\frac{\partial^2 f}{\partial x^2} = \left(\frac{1}{v^2}\right)\frac{\partial^2 f}{\partial t^2} \tag{2.1}$$

where $f \equiv f(x, t)$ is the function representing the disturbance, where f could be the physical displacement of a particle of the medium carrying the wave; it could also be the electric or the magnetic field. More generally it could even be a derived macroscopic quantity like pressure. v is the velocity with which the disturbance travels. Equation 2.1 is known as the one-dimensional wave equation. It can be derived from equations of mechanics for a mechanical wave and from equations of electrodynamics for a light wave. If a classical wave travels in two- or three-dimensional space, second order partial derivatives of f with respect to y or with respect to y and z get added to the left hand side of Eq. 2.1.

The solution of Eq. 2.1 has a very general form, which follows from the way partial derivatives appear in it. Suppose we take

$$f(x, t) = F[k(x - vt) + \phi] = F(\xi),\ \xi = k(x - vt) + \phi \tag{2.2}$$

that is, suppose the disturbance function is a function of the combination $k(x - vt) + \phi$ of x and t. Using the chain rule of differentiation,

$$\frac{\partial F}{\partial t} = \left(\frac{dF}{d\xi}\right)\left(\frac{\partial \xi}{\partial t}\right) = -kv\left(\frac{dF}{d\xi}\right)$$

and $$\frac{\partial^2 F}{\partial t^2} = k^2 v^2 \left(\frac{d^2 F}{d\xi^2}\right)$$

Also $$\frac{\partial^2 F}{\partial x^2} = k^2 \left(\frac{d^2 F}{d\xi^2}\right)$$

This means $F(\xi)$ satisfies the wave equation 2.1 for any function F! Now, it is easy to see that the shape of the wave—the graph of F versus ξ—can be any shape since F can be any function. This is the general solution of Eq. 2.1. It should be noted that although $F[k(x - vt) + \phi]$ is a solution of Eq. 2.1 for any function F, there are solutions which cannot be represented in this form. Example 2.1 makes this point clear.

EXAMPLE 2.1

All solutions of the wave equation are not of the form $f(x \pm vt)$

Show that $y = \sin(kx)\cos(\omega t)$ is a solution of a wave equation if $\omega/k = v$.

Solution

Differentiating y twice, we get

$$\frac{\partial^2 y}{\partial x^2} = -k^2 \sin(kx)\cos(\omega t)$$

and $$\frac{\partial^2 y}{\partial t^2} = -\omega^2 \sin(kx)\cos(\omega t)$$

Since $v = \frac{\omega}{k}$

$$\frac{\partial^2 y}{\partial x^2} = \frac{1}{v^2} \times \frac{\partial^2 y}{\partial t^2}$$

Hence, $y = sin(kx)cos(\omega t)$ is the solution of a wave equation.

However, it is not of the form $f(x \pm vt)$ since it is a product of a function of x and one of t. Every function of the form $f(x \pm vt)$ is a wave, but every wave is not of this form. In fact $\sin(kx)\cos(\omega t)$ is a mixture of two waves: $f(x - vt)$ and $f(x + vt)$ (see Ex. 3.2).

The simplest function from many points of view, as explained below, is the harmonic sine function,

$$f = A\sin[k(x - vt) + \phi] \tag{2.3}$$

It can be verified easily that this solution does satisfy Eq. 2.1 by differentiating f twice with respect to x and t. Equation 2.3 is plotted in Fig. 2.2 for $\phi = 0, t = 0$ and $\phi = 0, x = 0$.

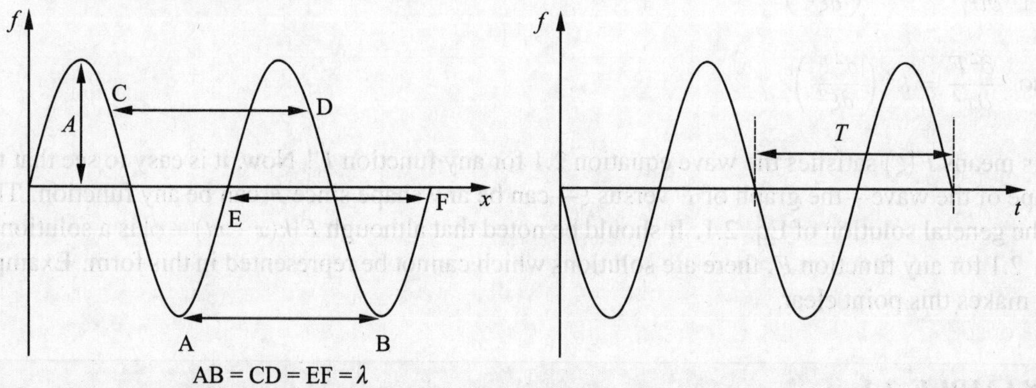

Figure 2.2 Picturisation of a wave: (a) Space profile: shape of disturbance at fixed time and (b) Time profile: SHM at a fixed point in space

The shape of the disturbance at fixed time (Fig. 2.2a) gives the space profile of the wave. The time profile (Fig. 2.2b) picturises the simple harmonic motion (SHM) at a fixed point in space. The simplicity of this wave allows us to define several characteristics discussed below.

Amplitude

The maximum size of the disturbance is known as the amplitude A of the wave. It can be seen from Eq. 2.3 that the maximum value of f is A, as the maximum value of the sine function is one.

Phase angle

The argument of the function on the right-hand side of Eq. 2.3 is known as the phase angle or simply the phase of the disturbance. It is expressed either in degrees or in radians. The constant ϕ is called the initial phase. Without loss of generality, we can assume $\phi = 0$. Thus we can write Eq. 2.3 as

$$f = A\sin[k(x - vt)] \tag{2.4}$$

A wavefront is defined as the locus of the points, which are in the same phase. According to Huygen's theory, each point on a wavefront acts as a new source of disturbance and the wavelets emanating from these points spread out in all directions with the speed of the wave. Thus a point source will produce a spherical wavefront, a line source, a cylindrical wavefront and a plane source will produce a plane wavefront at small distances from the source. The corresponding waves are termed, respectively, as spherical waves, cylindrical waves and plane waves. However, at large distances from the source, a small portion of any of these wavefronts can be considered to be a plane and the wave can be assumed to be a plane wave. The name of a wave is often derived from the shape of its wavefront. For example, plane waves have flat wavefronts, whereas spherical waves have expanding spheres as wavefronts. Equation 2.4 represents the equation of a plane wave for which the amplitude remains constant throughout, that is, it represents an unattenuated wave. For a spherical wave, however, the equation can be shown to be of the form,

$$f = \frac{A}{r} \sin k(r - vt)$$

where r is the radial distance, showing that the amplitude falls off as $1/r$.

Wavelength

It is the distance the wave has to travel along the direction of propagation to change the phase by 2π. It is usually denoted by λ. In Fig. 2.2a this is the distance AB, CD or EF. The wavelength may be understood as the space period. Clearly the wavelength is the normal distance between two consecutive wavefronts of the same phase, that is, whose phases differ by 2π.

Time period

It is the time required for the wave to travel a distance of one wavelength λ so that

$$\lambda = vT \tag{2.5}$$

It is thus the time required to complete one cycle (see Fig. 2.2b). The time period may be understood as time wavelength.

Frequency

The quantity $\nu = \dfrac{1}{T}$ (2.6)

is known as the frequency of the wave and represents the number of cycles completed in one second.

Angular frequency

It represents the radian angle change in the phase of the wave per second and is thus given by

$$\omega = 2\pi\nu = \frac{2\pi}{T} \tag{2.7}$$

Wave number and wave vector

The constant k is known as the wave number or phase constant. It is the magnitude of wave vector, **k**, and represents the number of wavelengths in 2π metres and is given by

$$k = \frac{2\pi}{\lambda} \tag{2.8}$$

According to Eq. 2.8 k is the ratio of phase change 2π to λ—the distance over which the phase changes by 2π. Thus k is the rate of change of phase with path travelled. If a wave moves a distance Δx, its phase changes by

$$\Delta\phi = k\Delta x = \frac{2\pi\Delta x}{\lambda} \tag{2.9}$$

Similarly, if the phase of a wave changes by $\Delta\phi$, it has travelled a path

$$\Delta x = \frac{\Delta\phi}{k} = \frac{\Delta\phi\lambda}{2\pi} \tag{2.10}$$

We can write Eq. 2.4 using Eqs. 2.5, 2.7 and 2.8, as

$$f = A\sin(kx - \omega t) \tag{2.11}$$

Equation 2.11 can be expressed in the vector form as

$$f = A\sin(\mathbf{k.r} - \omega t) \tag{2.12}$$

where **r** is the position vector and **k** is a wave vector along any arbitrary direction. Figure 2.3 shows the plane wavefronts of a plane wave satisfying the equation **k.r** $-\omega t$ = constant.

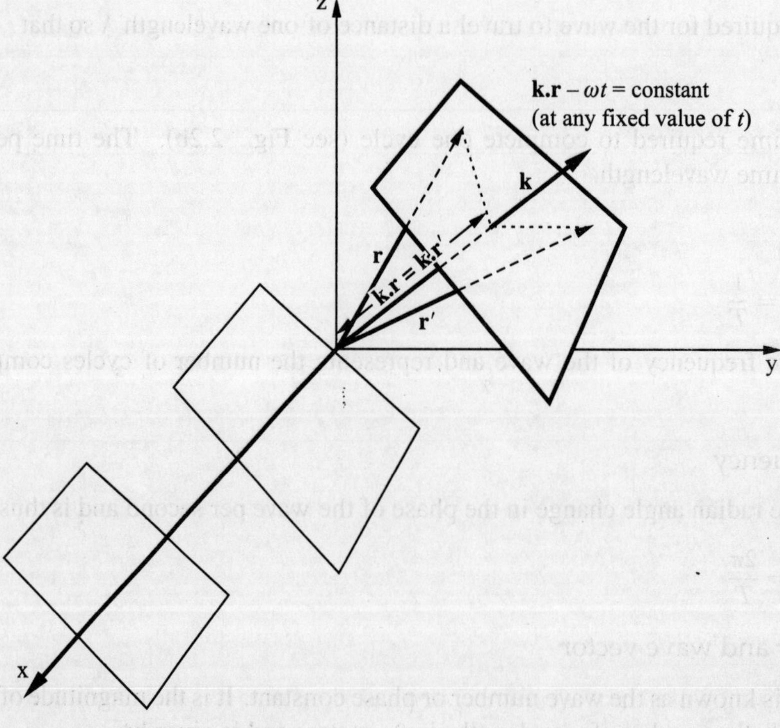

Figure 2.3 Wavefronts of a plane wave

Velocity

The velocity v (also known as the phase velocity) of a wave is defined as the velocity of its wavefronts. To obtain this velocity we assume the wavefront to move a distance Δx in time Δt and write,

$$\varphi(x,t) = \varphi(x+\Delta x, t+\Delta t) = \text{phase of the wavefront}$$

where $\varphi(x,t) = kx - \omega t + \phi$ is the phase function.

From this it follows that

$$v = \frac{\Delta x}{\Delta t} = \frac{\omega}{k} = \nu\lambda \tag{2.13}$$

Polarisation

Polarisation of a wave specifies the direction of the disturbance that constitutes the wave. For waves like pressure waves, the disturbance is a scalar quantity. Polarisation is not defined for scalar waves of this type. In other cases, the meaning of direction of disturbance depends on the type of disturbance. Thus, for a mechanical wave, it means the direction in which particles move locally; for a light wave, it means the direction of the electric field. Waves are classified as either transverse or longitudinal, depending on whether the polarisation is perpendicular to or in the same direction as the direction of propagation of the wave. Thus the following equations for the displacement of a particle represent longitudinal and transverse waves respectively.

$$\mathbf{y}_1 = \mathbf{i}A\sin(kx - \omega t) \tag{2.14}$$

$$\mathbf{y}_2 = \mathbf{j}A\sin(kx - \omega t) \tag{2.15}$$

i and **j** being the unit vectors along the x-and y-axes.

Intensity

It is defined as the energy transported across a unit area by the wave per unit time, that is, the power delivered by the wave across a unit area perpendicular to the direction of its propagation. It is the product of the velocity of the wave and the energy density. The energy density is the total energy per unit volume. The intensity I of a wave is proportional to the square of its amplitude, that is,

$$I \propto A^2 \tag{2.16}$$

Thus the intensity for spherical waves falls off as $1/r^2$.

Coherence

The wave assumed in Eq. 2.11 maintains its sinusoidal character at all times and for all points along the $+x$ direction. This is known as coherence. For a real wave, the time for which the displacement is truly sinusoidal is short, and is called the coherence time of the wave and the product of coherence time and velocity is known as the coherence length. Coherence of a wave depends on the mechanism of its emission and will be discussed in more detail in the next chapter.

Examples 2.2 and 2.3 help us to appreciate the various aspects of waves studied in this section. Example 2.4 illustrates why $f(x - vt)$ and $f(x + vt)$ represent, respectively, the waves travelling along the positive and negative x-axis. Example 2.5 illustrates that the commonly used sinc, cosine and exponential functions are solutions of the wave equation.

EXAMPLE 2.2
Space profile and time profile of a wave

Consider the wave $y = 2\sin(10\pi t - \pi x/40 + \pi/4)$. Plot (a) the space profile at $t = T/4$ and (b) the time profile at $x = \lambda/8$.

Solution

Comparing the given equation with $y = A\sin(\omega t - kx + \phi)$, and assuming the units of x and t to be metres and seconds, we get

$$\omega = 10\pi \text{ rad/s and } k = \frac{\pi}{40} \text{ rad/m}$$

Thus, $\dfrac{2\pi}{T} = 10\pi$ or $T = \dfrac{1}{5}$ s and $\dfrac{2\pi}{\lambda} = \dfrac{\pi}{40}$ or $\lambda = 80$ m.

The functions to be plotted are

$$y_1(x) = y\left(x, t = \frac{T}{4} = \frac{1}{20}\right)$$

$$= 2\sin\left[10\pi \times \frac{1}{20} - \frac{\pi x}{40} + \frac{\pi}{4}\right]$$

$$= 2\sin\left(\frac{3\pi}{4} - \frac{\pi x}{40}\right)$$

and $$y_2(t) = y\left(x = \frac{\lambda}{8} = 10, t\right)$$

$$= 2\sin\left[10\pi t - \left(\frac{\pi}{40}\right)10 + \frac{\pi}{4}\right] = 2\sin(10\pi t).$$

Thus we get the points shown in the table. These points are plotted to obtain the space profile and the time profile as shown in Figs. 2.4 and 2.5.

x (m)	$y_1(x)$ (m)	t (s)	$y_2(t)$ (m)
0	$\sqrt{2}$	0	0
10	2	0.025	$\sqrt{2}$
20	$\sqrt{2}$	0.05	2
30	0	0.075	$\sqrt{2}$
40	$-\sqrt{2}$	0.1	0
50	-2	0.125	$-\sqrt{2}$
60	$-\sqrt{2}$	0.15	-2
70	0	0.175	$-\sqrt{2}$
80	$\sqrt{2}$	0.2	0

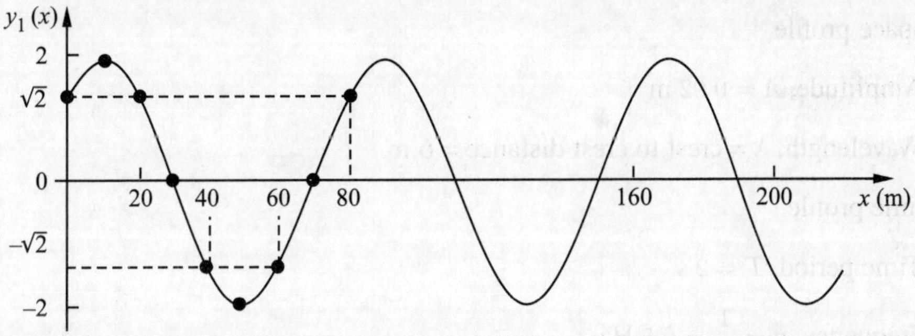

Figure 2.4 Space profile at $t = T/4$

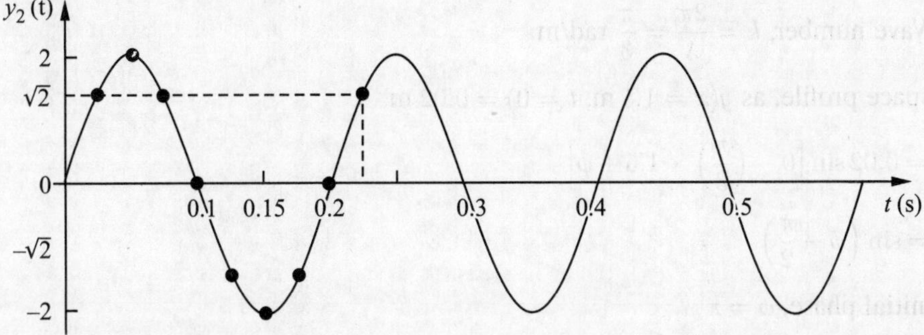

Figure 2.5 Time profile at $x = \lambda/8$

EXAMPLE 2.3
Wave parameters

From the space profile and the time profile of a wave shown in Fig. 2.6 determine the various parameters discussed in Section 2.3.

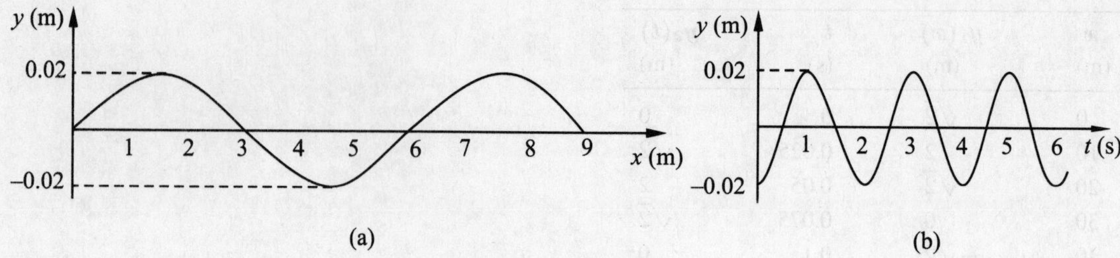

Figure 2.6 (a) Space profile at $t = 0$; (b) Time profile at $x = ?$

Solution

Let the wave function be $y = A\sin(\omega t - kx + \phi)$.

From the space profile

(i) Amplitude, $A = 0.02$ m

(ii) Wavelength, λ = crest to crest distance = 6 m

From the time profile,

(iii) Time period, $T = 2$ s

So, (iv) Frequency, $\nu = \dfrac{1}{T} = 0.5$ Hz

(v) Angular frequency, $\omega = 2\pi\nu = \pi$ rad/s

and (vi) Wave number, $k = \dfrac{2\pi}{\lambda} = \dfrac{\pi}{3}$ rad/m

From the space profile, as $y(x = 1.5\text{ m}, t = 0) = 0.02$ m

$$0.02 = 0.02\sin\left[0 - \left(\frac{\pi}{3}\right) \times 1.5 + \phi\right]$$

$$1 = \sin\left(\phi - \frac{\pi}{2}\right)$$

So, (vii) Initial phase, $\phi = \pi$

The wave function is thus $y = 0.02\sin\left(\pi t - \dfrac{\pi x}{3} + \pi\right)$.

From the time profile

$$y(x, t = 0) = -0.02 = 0.02\sin\left(0 - \frac{\pi x}{3} + \pi\right)$$

Thus $-\pi x/3 + \pi = -\pi/2, -5\pi/2$, giving $x = 9/2$ m, $21/2$ m.

(viii) The velocity of the wave $v = \dfrac{\omega}{k} = \dfrac{\pi\text{ rad/s}}{(\pi/3\text{ rad/m})} = 3$ m/s [also $v = \frac{\lambda}{T} = \frac{6 \text{ m}}{2\text{s}} = 3$ m/s]

(ix) Polarisation: Not defined.

(x) Intensity, $I \alpha (0.02)^2 \alpha 4 \times 10^{-4}$ m^2

(xi) Wavefronts, $x =$ constant-planes parallel to yz-plane.

(xii) Coherence: The wave is completely coherent since it is monochromatic. It maintains its sinusoidal character at all times and for all points along the x-direction.

EXAMPLE 2.4

Waves travelling along the positive and negative direction of the x-axis

Show that $f(x - vt)$ and $f(x + vt)$ represent waves travelling along the positive and negative directions of the x-axis, respectively.

Solution

We already know that $f(x \pm vt)$ is a function for any wave. Now, the form of the wave $f(x - vt)$ implies that the phase of the wave is a function, say $\varphi(x - vt)$, of $x - vt$. Suppose that the phase at x_0 at time t_0 is φ_0, that is, $\varphi_0 = \varphi(x_0 - vt_0)$. Hence the phase at a nearby point $x_0 + \delta x_0$ just ahead in future at $t_0 + \delta t_0$ is $\varphi[x_0 + \delta x_0 - v(t_0 + \delta t_0)]$. This will obviously be φ_0 if $\delta x_0 = v\delta t_0$. This means that the wavefront which passes through the point P (Fig. 2.7) with co-ordinate x_0 at time t_0 has moved in time δt_0 and passes through point Q with x coordinate $x_0 + \delta x_0$. Thus in future, that is, for positive δt_0, we need to go along positive x-axis, that is, positive $\delta x_0 = v\delta t_0$, to keep up with the wavefront. The wavefront thus travels along the x-axis with velocity v. Similar argument for $f(x + vt)$ shows $\delta x_0 = -v\delta t_0$ so that the wavefronts move along the negative x-direction (Fig. 2.8).

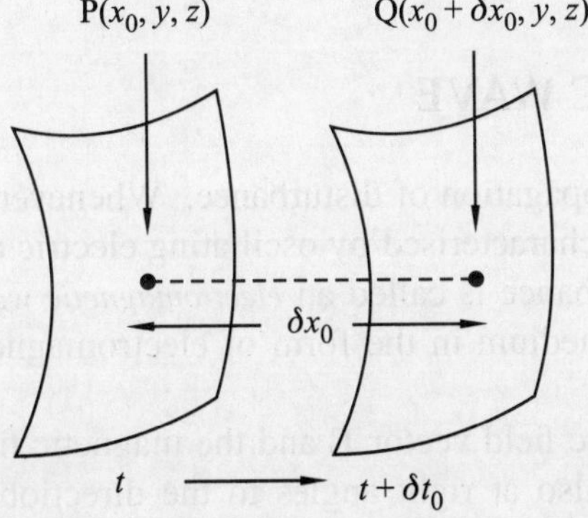

Figure 2.7 $f(x - vt)$: wave along + x-axis

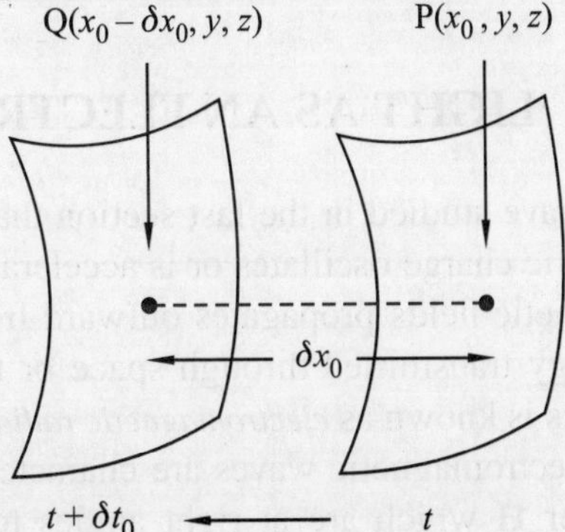

Figure 2.8 $f(x + vt)$: wave along – x-axis

EXAMPLE 2.5

Common solutions of the wave equation

Show that $y = A\sin(kx - \omega t), y = A\cos(kx - \omega t)$ and $y = A\exp[i(kx - \omega t)]$ are solutions of the wave equation.

Solution

The wave equation is

$$\frac{\partial^2 y}{\partial x^2} = \left(\frac{1}{v^2}\right)\frac{\partial^2 y}{\partial t^2}$$

where v is the velocity given by $v = \nu\lambda = \omega/k$

For $y = A\sin(kx - \omega t)$,

$$\frac{\partial^2 y}{\partial x^2} = -k^2 y \text{ and } \frac{\partial^2 y}{\partial t^2} = -\omega^2 y$$

For $y = A\cos(kx - \omega t)$,

$$\frac{\partial^2 y}{\partial x^2} = -k^2 y \text{ and } \frac{\partial^2 y}{\partial t^2} = -\omega^2 y$$

For $y = A\exp[i(kx - \omega t)]$,

$$\frac{\partial^2 y}{\partial x^2} = -k^2 y \text{ and } \frac{\partial^2 y}{\partial t^2} = -\omega^2 y$$

Substitution in the wave equation shows that $A\sin(kx - \omega t), A\cos(kx - \omega t)$ and $A\exp[i(kx - \omega t)]$ are its solutions.

2.4 LIGHT AS AN ELECTROMAGNETIC WAVE

We have studied in the last section that a wave is the propagation of disturbance. Whenever an electric charge oscillates or is accelerated, a disturbance characterised by oscillating electric and magnetic fields propagates outward from it. This disturbance is called an *electromagnetic wave.* Energy transmitted through space or through material medium in the form of electromagnetic waves is known as *electromagnetic radiation*.

Electromagnetic waves are characterised by the electric field vector **E** and the magnetic field vector **H** which are at right angles to each other and also at right angles to the direction of propagation of the wave (Fig. 2.9). These two vectors are always in phase and are interdependent. An electric field varying in time gives rise to a magnetic field varying in space and time, which in turn produces an electric field varying in space and time and so on.

Thus an electromagnetic wave is transverse and is also said to be linearly polarised as the electric field vector is always along a particular direction and the magnetic field vector is perpendicular to this direction. The three components of **E** and **H** vectors for a wave depicted in Fig. 2.9 can be written as

$$E_x = 0, E_y = E_0 \sin(\omega t - kx), E_z = 0 \tag{2.17}$$

$$H_x = 0, H_y = 0, H_z = H_0 \sin(\omega t - kx) \tag{2.18}$$

After Maxwell's pioneering work in electromagnetism it was realised that there exist a large variety of waves originating from oscillating charges and thus could be identified as electromagnetic waves. These waves, as we know today, constitute different parts of the electromagnetic spectrum and differ widely in their frequency and wavelength (see Chart 2.1). However, they are unified in that they share the same fundamental physics and travel with the same speed in vacuum, given by

$$v = \frac{1}{\sqrt{\mu_0 \epsilon_0}} \tag{2.19}$$

where μ_0 and ϵ_0 are the permeability and permittivity of free space respectively. For any other medium the speed is $v = 1/\sqrt{\mu\epsilon}$, where μ and ϵ are the corresponding constants for that medium.

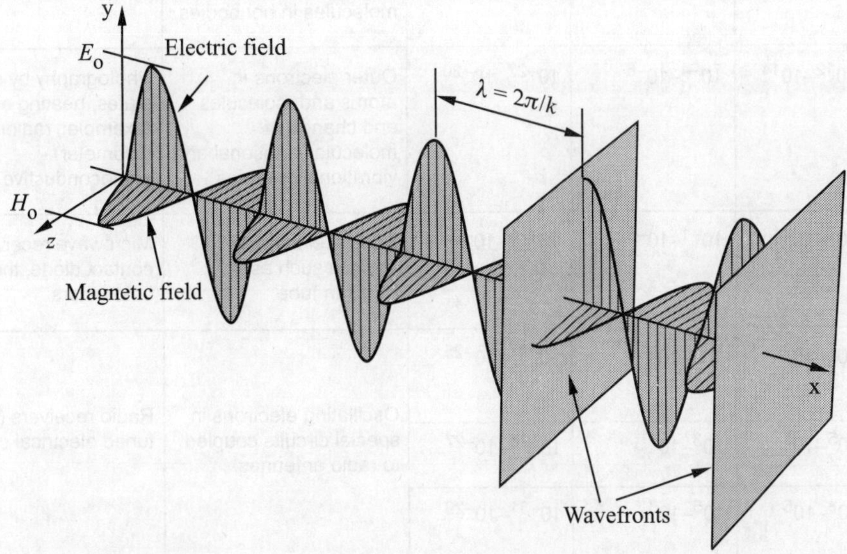

Figure 2.9 An electromagnetic wave propagating along the $+z$-direction. The lines represent instantaneous electric and magnetic field vectors, **E** and **H**.

Since the speed of light measured by Fizeau agreed with Eq. 2.19, and the transverse character of electromagnetic waves coincided with the experimental observation that light has transverse wave nature (Young's experiments on polarisation proved it), Maxwell concluded that light is an electromagnetic wave. Visible light covers a very small portion of the electromagnetic spectrum, extending in wavelength from roughly 4000 Å to 7000 Å. We will assume the electromagnetic wave character of light in the next three chapters.

CHART 2.1 ELECTROMAGNETIC SPECTRUM

Name	Frequency (Hz)	Wavelength (m)	Photon energy (J)	Generation	Detection
γ-rays	10^{19}–10^{22}	10^{-11}–10^{-14}	10^{-15}–10^{-12}	Changes of energy levels in radioactive nuclei	Photography, ionisation chamber, phosphorescence (example: ZnS)
X-rays	10^{18}–10^{20}	10^{-10}–10^{-12}	10^{-16}–10^{-14}	Rapid deceleration of fast-moving electrons by a target (example: tungsten)	Photography, ionisation chamber, phosphorescence (example: ZnS)
Ultra-violet	10^{16}–10^{18}	10^{-8}–10^{-10}	10^{-18}–10^{-16}	Sun; electrons in excited atoms as those in sparks, arcs and gas discharge tubes	Photography, photoelectric cell, fluorescence
Visible	10^{15}	10^{-7}	10^{-19}	Rearrangement of outer electrons in atoms and molecules in hot bodies	Eye, photography, photocell
Infrared	10^{12}–10^{14}	10^{-4}–10^{-6}	10^{-22}–10^{-20}	Outer electrons in atoms and molecules and change of molecular-rotational and vibrational energies	Photography by special plates, heating effects (example: radiometer or bolometer), photoconductive cells
Micro-waves	10^{9}–10^{11}	10^{-1}–10^{-3}	10^{-25}–10^{-23}	Special electronic devices such as Klystron tube	Microwave receivers, point contact diode, thermistor bolometers
Short radio waves	10^{7}–10^{9}	10^{1}–10^{-1}	10^{-27}–10^{-25}	Oscillating electrons in special circuits coupled to radio antennas	Radio receivers (example: tuned electrical circuit)
Standard broadcast	10^{5}–10^{7}	10^{3}–10	10^{-25}–10^{-27}		
Long waves	10^{3}–10^{5}	10^{5}–10^{3}	10^{-31}–10^{-29}		

2.5 APPLICATIONS

We see light everyday and everywhere. But physicists have seen *into* light and their inquisitive efforts in this direction over the last three centuries have led to the discovery of many new and interesting facets of the enigma that is light. These developments had a concomitant and profound impact on technology. Though it is not possible to discuss all these developments, we will pick up relevant aspects of this story in their proper context. Here we discuss the aspects related to the points covered in this chapter.

In Section 2.2 we studied about geometrical and physical optics. The developments in optics since the time of the ancient Greeks up to the end of the eighteenth century were based mainly on the study of geometrical optics. The study was aimed at improving the fidelity of images by reducing the aberrations of light, using optical systems that included mainly mirrors and lenses. This led to the invention of useful devices such as eye-piece, camera, binoculars, microscope, telescope and projector.

Though the lens as a magnifier was used as early as the middle of the fifteenth century, efforts of several workers, including those of Leeuwenhoek (Dutch microscopist who developed the theory of aberration-free lenses by about 1830), led to improvement in magnification. By the end of the nineteenth century, microscopes with large magnifying power were available. However, their resolution was limited by the numerical aperture and the wavelength of light used. This limitation was overcome by the brilliant discovery of matter waves by de Broglie in 1924 (see Chapter 6).

Joseph Nicephore Niepce and Louis-Jaques-Mande Daguerre of France (1820–30) were among the pioneers who developed the camera. The development of the zoom and telephoto lenses helped to refine the design and performance of the camera. By the twentieth century, a variety of cameras for specific purposes such as aerial photography, document copying, movie, television and video were developed. Galileo (beginning of seventeenth century) and Kepler (1610) made pioneering contributions to the development of a telescope. In addition to these devices many other optical instruments and devices such as tracking instruments, gunsights, periscopes and rangefinders were developed for use in war and navigation.

Though geometrical optics had been relegated to the background, after the emergence of physical optics, its importance as a tool in solving optical problems has been widely recognised and is still applied with remarkable success in problems in interference, diffraction (Chapters 3 and 4), laser resonators and waveguides (Chapters 14 and 15).

Ideas of wave motion introduced in Section 2.3 are of fundamental significance to both science and technology as waves are ubiquitous in nature. The importance of waves can be judged from the definitions of two of the most important branches of physics, namely, acoustics and optics, as the study of the generation, propagation and reception of mechanical waves and electromagnetic waves (in the visible range), respectively. Acoustic waves are discussed in Chapter 16. Sound waves and light waves behave in much the same way. Example 2.6 illustrates this point. Thus there had been a happy interflow of ideas between acoustics and optics leading to their mutual development. This development is discussed in Section 16.8.

EXAMPLE 2.6

Sound waves and light waves

Compare the magnitudes of the wavelength and wave number for (a) sound wave generated by a tunning fork oscillating with a frequency of 440 Hz, assuming the velocity of sound in air as 340 m/s, and (b) red light of frequency 5×10^{14} Hz.

Solution

Using the expressions $\lambda = v/\nu$ and $k = 2\pi/\lambda$ we get the following comparative results.

	Sound wave	Light wave
Frequency, ν	440 Hz	5×10^{14} Hz
Velocity, v	340 m/s	3×10^8 m/s
Wavelength, λ	0.772 m	6×10^{-7} m = 6000 Å
Wave number, k	8.14/m	1.05×10^7/m
Wave equation	$y = A\sin(8.14\,x - 2.76 \times 10^3 t)$	$y = A\sin(1.05 \times 10^7 x - 3.14 \times 10^{15} t)$

In Section 2.4 we studied the electromagnetic character of light as theorised by Maxwell in 1873. Maxwell's discovery proved to be a momentous breakthrough in physics, which facilitated the refinement of technologies existing then, and the emergence of many new technologies for many years to come. It signalled two main routes for this progress. First, it inspired investigation of the interaction of various components of the electromagnetic spectrum with matter. Second, it inspired attempts to find, produce and control electromagnetic waves of frequency other than those of light. The first route led to the furtherance of our understanding of the structure of matter and growth of knowledge in solid state, atomic, nuclear and quantum physics. The second route resulted in the discovery of x-rays by Roentgen (1895) and the development of radio and television.

Maxwell's discovery provided a definite direction of research to the physicists. This direction was to look for the effects which light exhibited in the other parts of the electromagnetic spectrum. Thus, taking a cue from optical astronomy, the fields of gamma ray astronomy, x-ray astronomy, IR astronomy and radio astronomy were developed. Similarly, attempts were made to duplicate the available knowledge of microscopy, holography, luminescence, diffractometry and spectroscopy in the optical range in the other ranges of the electromagnetic spectrum. Apart from these, applications characteristic of a particular radiation, due to its particular energy range, were developed for appropriate use in technology.

Gamma rays which are emitted from radioactive nucleii could be used as nuclear labels and found wide application in nuclear engineering. They also proved to be a useful tool in ecological studies for detecting small quantities of tin or arsenic in plant or animal tissues, in industry for identifying trace elements which are present as impurities, and in medical diagnostics for carrying out tracer studies on tumours.

X-rays have wide ranging applications in almost all the fields of research and modern life. The property of x-rays that they could transmit through flesh but cannot pass through bones was exploited by physicians for imaging the internal organs of the human body. The property that x-rays are diffracted by obstacles of the dimensions of the order of 1 to 100 Å (see Chapter 4) was exploited by engineers to study the composition of materials. However, the field which was enriched most by the discovery of x-rays was scientific research. X-ray diffraction (XRD) studies of crystals have given birth to a new branch of physics, called x-ray crystallography. XRD studies facilitate the determination of crystal structure, electron distribution, range of order and disorder in a solid, limits of solid solubility, strain size and many other characteristics which are extremely useful in solid state chemistry and materials science.

Ultraviolet (UV) radiation has been mainly used in the development of fluorescent and phosphorescent lamps, dyes and pigments, in UV photography, and in the treatment of rickets and other skin diseases.

In the infrared (IR) range, the imaging devices that converted an invisible image into a visible image found wide application in meteorology (both for imagery and imaging of cloud patterns), medical diagnostics (breast cancer screening), forestry (detection of location of forest fires) and analytical chemistry (as a spectroscopic tool). IR technology has also been used in military and space engineering for night vision applications and satellite stabilisation near the earth's surface. In addition to these, IR lamps have been developed and used for comfort healing, drying (tobacco leaves, tea leaves and printing inks), and heating (food, rooms and farm animals).

In the radio waves range an application device known as RADAR (radio detection and ranging) is the most significant development. RADAR uses the same principle as SONAR (sending waves and deciphering the knowledge of objects on the basis of reflected waves or echoes). In nature bats use ultrasonic waves to 'see' objects as they do not possess anatomical eyes. RADAR is very useful in meteorology (forecasting, cloud physics and precipitation physics), military engineering and space applications. The defeat of the Germans (in the battle of Britain at Luftwaffe) by the British army is attributed to the robust RADAR network established by the British during World War II.

Radio waves have a historical significance. It was in this range that Hertz demonstrated the practical application of Maxwell's theory by producing waves (also known as Hertzian waves) by passing rapidly alternating current through a wire. Marconi developed the radio wave antenna and broadcast the first paid radiogram from the Isle of Wight in 1898. The importance of radio wave communication was realised dramatically in the disasters caused by sinking of passenger ships, Republic (1909) and Titanic (1912), when thousands of lives were saved on the basis of radio calls made by passengers seeking help.

Wireless communication using radio waves has benefited immensely by these early developments, namely, the constant research outputs of the study of radio wave propagation through various layers of atmosphere, particularly the ionosphere. This resulted in the proliferation of radio broadcasting stations in USA, so much so that, very few stations have been granted licence by the FCC (Federal Communications Commission) since 1970, mainly due to the scarcity of unoccupied frequencies and the large number of stations already competing in all larger communities. Efforts are being made to overcome these and other problems in radio wave communication by replacing it by optical wave communication. Photonics, which deals with these problems, is a frontier area of research today and has potential for developing into a major technology in the near future.

EXERCISES

1. The speed of an electromagnetic wave travelling in a transparent non-magnetic material is given by $v = 1/\sqrt{\epsilon_o \epsilon_r \mu_o}$, where ϵ_r is the dielectric constant of the substance. Determine the speed of light in water, which has a dielectric constant of 1.7 at optical frequencies.
2. When light travels, what is it that oscillates? What is it that is transported across a given region?
3. What is the fundamental source of electromagnetic radiation?
4. As light travels from one medium to another, does the wavelength of the light change? Does the frequency change? Does the speed change? Explain.
5. Light of wavelength λ is incident on a slit of width d. What are the conditions under which the ray approximation is valid?
6. A travelling wave is given by a displacement function $y = A\sin(Bx - Ct)$, where $B = 0.30\ \text{m}^{-1}$ and $C = 0.020\ \text{s}^{-1}$. Plot the $y - t$ graph at both $x = 1.0$ m and $x = 12$ m.
7. Equation $y = A\sin(kx + \omega t)$ describes a wave, where $k = 2.0\ \text{m}^{-1}, \omega = 1.0\ \text{s}^{-1}$, and $A = 0.30$ m. Determine the wavelength, the frequency, the maximum amplitude, the period and the wave number.
8. A harmonic wave of wavelength 0.17 m moves with speed 16 m/s. Determine the frequency f and the wave number k of this wave.
9. Show that the general form of a travelling wave, $f(x - vt)$, satisfies the wave equation.
10. Give some examples of transverse and longitudinal waves. In each of them discuss the nature of motion of particles of the medium.

3

Interference

3.1 THE STORY OF INTERFERENCE OF LIGHT WAVES

In the last chapter we overviewed some early developments in the field of optics. The beginning of the nineteenth century provided major evidence in favour of the wave theory of light, which had been developed by Huygens way back in 1678. In 1801 Thomas Young, an English physicist, provided the first strong experimental proof by demonstrating the interference of light waves coming out of two pinholes. He also reported the first experimental determination of the wavelength of light. The value obtained by him, $\lambda = 5500$ Å, agrees well with that obtained by using today's sophisticated instruments (5700 Å). Young, a child prodigy (he had read the Bible twice by the age of four), was a physician whose early research on the human eye inspired him to enquire into the nature of light. He is called the founder of the wave theory for his valuable contribution to this field. In 1802 Young proposed the *superposition principle*, which forms the conceptual basis for explanation of the properties of light such as interference, diffraction and polarisation.

However, even scientists display inertia of old established ideas. Physicists are no exception to Newton's second law! Young had to face his share of criticism. One of the staunch votaries of Newton (for his corpuscular theory of light) was critical of Young's paper: *We wish to raise our feeble voice against innovations that can have no other effect than to check the progress of science, and renew all those wild phantoms of the imagination which Bacon and Newton put to flight from her temple. This paper contains nothing that deserves the name of either experiment or discovery.*

Several researchers proved Young's critics wrong by providing fresh experimental evidence in support of the wave theory. Notable among them was Augustin Fresnel, a French engineer. He demonstrated the interference effect using a biprism in 1814–15. Fresnel also made pioneering contribution to the study of diffraction of light. After this, many investigators contributed to the study of interference by developing several useful and elegant designs for optical instruments—called *interferometers*—required for this study..

AA Michelson, a brilliant American physicist, built one of the first interferometers in 1881. Michelson graduated in 1873 from the Naval Academy, USA. He later worked there as an instructor in physics and chemistry, and was often chided by the superintendent of the academy for wasting his time in useless experiments. Years later Michelson became the first Nobel prize winner in science from America, receiving the award in 1907 for his researches on light. Twyman, Green, Jamin, Haidinger, Mach Zehnder, Stokes and Rayleigh were other physicists who made important contributions to the science of interferometry. Interestingly, this study had notable contributions from professionals in engineering and medicine. This is true for many other areas of research in applied physics.

Research in interference of light also motivated physicists to probe into one of the most important aspects of waves—coherence. Scientists discovered that only coherent waves can produce stable interference patterns and that light waves have poorer coherence than microwaves and radio waves. The question uppermost in their minds was how the coherence of light can be enhanced. Einstein, the all time genius, had an ingenious answer. In 1917 Einstein suggested theoretically that a highly coherent light beam could be obtained if the natural mechanism of emission of light is manipulated in such a way that the emission is stimulated rather than spontaneous (see Chapter 14). Research along these lines culminated in the development of the first practical laser by Maiman in 1960. Before the advent of laser, optics was often called incoherent optics; the age of lasers is the age of coherent optics.

3.2 SUPERPOSITION OF WAVES

In Section 2.3 we studied the basic parameters of a wave. One or more of these parameters (except the frequency and coherence time which are the characteristics of the emission process) will undergo a change when a wave encounters either matter—leading to phenomena such as scattering or absorption—or a field—leading to effects such as electrooptic effect or acoustooptic effect. Of interest to us here is what happens to a (light) wave when it encounters another (light) wave. What is the result of superposition of two waves?

The answer is implicit in the statement that Eq. 2.4 is the solution of Eq. 2.1, that is, if y_1 and y_2 are solutions of Eq. 2.1, then $y = y_1 + y_2$ is also a solution. The superposition principle thus states that the *resultant disturbance is the algebraic sum of the disturbances of the separate constituent waves*. It should be noted here that the superposition principle is a consequence of the linearity of the wave equation (see Ex. 3.1). In case of high amplitude waves, there is a departure from linearity and the superposition principle is not applicable. Example 3.2 illustrates how the superposition principle can be applied in two special cases.

EXAMPLE 3.1

Superposition principle as a consequence of linearity of wave equation

Suppose the wave equation (see Eq. 2.1) is changed to $\partial^2 y/\partial x^2 = (1/v^2)(\partial y/\partial t)^2$. Will the superposition principle still hold?

Solution

The wave equation is

$$\frac{\partial^2 y}{\partial x^2} = \left(\frac{1}{v^2}\right)\frac{\partial^2 y}{\partial t^2}$$

Thus if y_1 and y_2 are the solutions, that is,

$$\frac{\partial^2 y_1}{\partial x^2} = \left(\frac{1}{v^2}\right)\frac{\partial^2 y_1}{\partial t^2} \text{ and } \frac{\partial^2 y_2}{\partial x^2} = \left(\frac{1}{v^2}\right)\frac{\partial^2 y_2}{\partial t^2}$$

then

$$\frac{\partial^2 y}{\partial x^2} = \frac{\partial^2 (y_1 + y_2)}{\partial x^2} = \frac{\partial^2 y_1}{\partial x^2} + \frac{\partial^2 y_2}{\partial x^2} = \left(\frac{1}{v^2}\right)\left[\frac{\partial^2 y_1}{\partial t^2} + \frac{\partial^2 y_2}{\partial t^2}\right]$$

$$= \left(\frac{1}{v^2}\right)\left[\frac{\partial^2 (y_1 + y_2)}{\partial t^2}\right] = \left(\frac{1}{v^2}\right)\left[\frac{\partial^2 y}{\partial t^2}\right]$$

that is, $y = y_1 + y_2$ is also the solution.

But if the wave equation is changed as stated in the question, then $y = y_1 + y_2$ is not the solution, because of the presence of the quadratic term, that is, because,

$$\left[\frac{\partial y}{\partial t}\right]^2 = \left[\frac{\partial (y_1 + y_2)}{\partial t}\right]^2 \neq \left(\frac{\partial y_1}{\partial t}\right)^2 + \left(\frac{\partial y_2}{\partial t}\right)^2$$

EXAMPLE 3.2

Application of the superposition principle

Find the result of superposition in the following cases: (a) two waves of the same frequency and same amplitude but travelling in opposite directions, and (b) two waves propagating in the same direction with the same amplitude but slightly different frequencies.

Solution

(a) Two waves with the same frequency and amplitude but travelling in opposite directions can be represented by the equations

$$y_1 = A\sin(kx - \omega t) \text{ and } y_2 = A\sin(kx + \omega t)$$

The resultant displacement is

$$y = y_1 + y_2 = 2A\sin kx \cos \omega t$$

which represents a standing wave (Fig. 3.1). The points such as A, B, C, D, E and F are called *nodes* for which the displacement is zero at all times and the points P, Q, R, S and T are called *antinodes* at which the displacement is maximum.

(b) The two waves can be represented by the equations

$$y_1 = A\sin(kx - \omega t) \text{ and } y_2 = A\sin[(k + dk)x - (\omega + d\omega)t]$$

The resultant displacement is

$$y = y_1 + y_2 = 2A\sin[(k + \tfrac{1}{2}dk)x - (\omega + \tfrac{1}{2}d\omega)t]\cos(\tfrac{1}{2}d\omega t - \tfrac{1}{2}dkx)$$

which represents the equation of beats. It represents a wave with an amplitude $2A\cos[(1/2)d\omega t - (1/2)dkx]$ (Fig. 3.2) which varies with time with a frequency $d\omega/2$ that is called the frequency

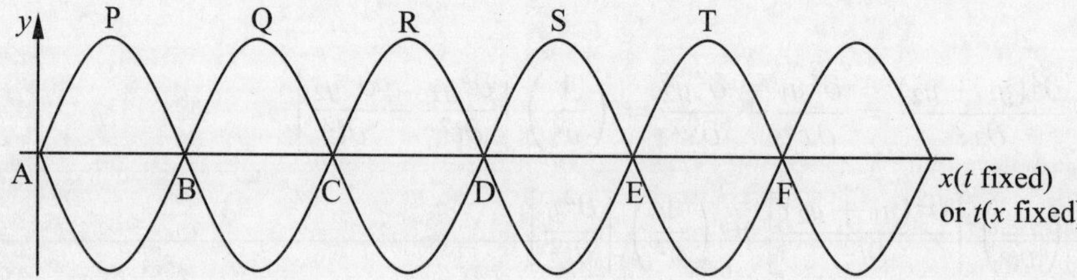

Figure 3.1 Standing wave

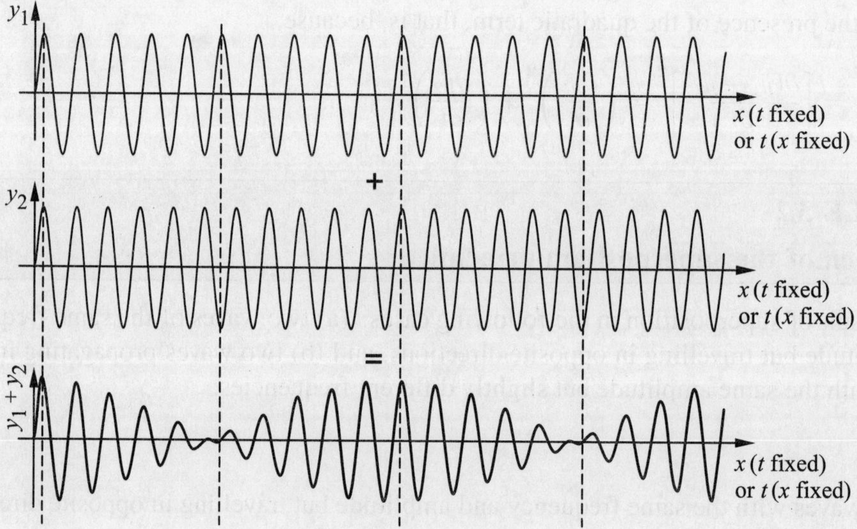

Figure 3.2 Beats ($y_1 + y_2$ is not plotted on the same scale as y_1 and y_2)

of beats. That the amplitude of the resultant of two waves of slightly different frequencies peaks at some points and drops at some other points is useful in the construction of a wave packet —a concept that will be introduced in Section 6.6.

The phenomenon of beats and standing waves are observable in case of sound waves due to their large wavelengths. However it is very difficult to carry out these experiments with light waves due to their short wavelengths. Interference pattern produced by standing light waves was demonstrated for the first time by Ives and Wiener in 1890.

Consider two waves which have the same frequency ω but different initial phases ϕ_1 and ϕ_2. They travel different distances x_1 and x_2 before interfering in the region of observation at time t. The two waves can be represented as

$$y_1 = A_1 \sin(kx_1 - \omega t + \phi_1) \tag{3.1}$$

$$y_2 = A_2 \sin(kx_2 - \omega t + \phi_2) \tag{3.2}$$

If $\alpha_1 = kx_1 + \phi_1$ and $\alpha_2 = kx_2 + \phi_2$ (3.3)

we can write,

$$y_1 = A_1 \sin(\alpha_1 - \omega t) \tag{3.4}$$

and $y_2 = A_2 \sin(\alpha_2 - \omega t)$ (3.5)

Applying the principle of superposition, the resultant displacement, y, is given by

$$y = y_1 + y_2 = A_1 \sin(\alpha_1 - \omega t) + A_2 \sin(\alpha_2 - \omega t)$$

or $y = (A_1 \sin\alpha_1 + A_2 \sin\alpha_2)\cos\omega t - (A_1 \cos\alpha_1 + A_2 \cos\alpha_2)\sin\omega t$ (3.6)

Let $A_1 \sin\alpha_1 + A_2 \sin\alpha_2 = A \sin\alpha$ (3.7)

and $A_1 \cos\alpha_1 + A_2 \cos\alpha_2 = A \cos\alpha$ (3.8)

On squaring and adding Eqs. 3.7 and 3.8

$$A^2 = A_1^2 + A_2^2 + 2A_1A_2\cos(\alpha_1 - \alpha_2) \tag{3.9}$$

and $\tan\alpha = \dfrac{(A_1 \sin\alpha_1 + A_2 \sin\alpha_2)}{(A_1 \cos\alpha_1 + A_2 \cos\alpha_2)}$ (3.10)

Thus $y = A\sin(\alpha - \omega t)$ (3.11)

Thus the resultant wave has the same frequency as that of the constituent waves but has a different amplitude and phase. The amplitude of the resultant depends on the phase difference between the two waves $\alpha_1 - \alpha_2$.

Using Eq. 3.9, for $\alpha_1 - \alpha_2 = 2n\pi$, where n is an integer,

$$A = A_1 + A_2 \tag{3.12}$$

and for $\alpha_1 - \alpha_2 = (2n+1)\pi$

$$A = A_1 - A_2 \tag{3.13}$$

The corresponding conditions for path difference are

$$\Delta = n\lambda \tag{3.12a}$$

and $\Delta = \left(n + \frac{1}{2}\right)\lambda$ (3.13a)

Thus for waves arriving in phase at any point, the resultant amplitude is the sum of two amplitudes; and for those arriving out of phase, the resultant is the difference of the two amplitudes (Fig. 3.3). *This modulation of the distribution of light energy due to superposition of waves is known as the phenomenon of interference.* Equations 3.12 and 3.13 are the conditions for constructive and destructive interference, respectively, leading to relative brightness and darkness in the interference pattern. In this chapter we will deal mainly with two-beam interference rather than multiple-beam interference.

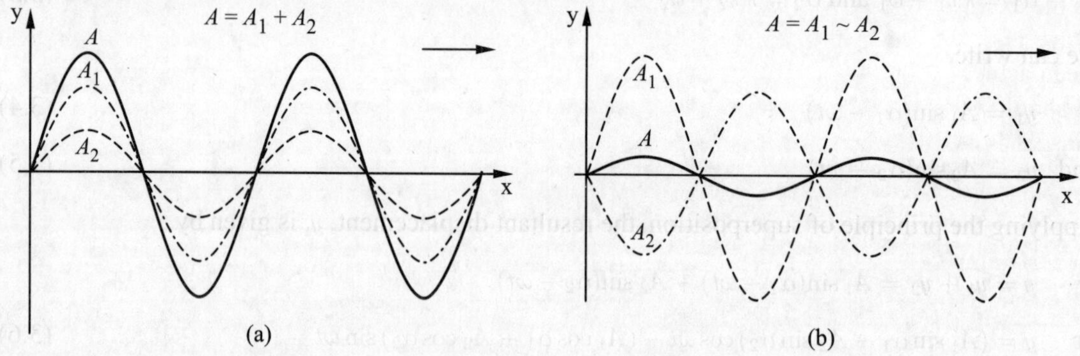

Figure 3.3 Superposition of two harmonic waves: (a) in phase and (b) out of phase

The practical observability of a stationary interference pattern is limited by several factors. In the mathematical model considered here we have assumed the superposition of harmonic plane waves represented by Eqs. 3.1 and 3.2, leading to the results of Eqs. 3.12–3.13a, indicating the formation of an interference pattern consisting of dark and bright regions. However, it should be understood that actual waves in light beams differ much from the waves of Eqs. 3.1 and 3.2. The main reason for this is that the waves in these equations are totally coherent, whereas those in light beams are not. Let us try to understand this.

3.3 COHERENCE

Coherence was defined in Section 2.3. The concept captures the uniformity of harmonic character over extended spatial regions and time intervals. To understand the concept of coherence, we have to understand the mechanism of emission of light. There is a fundamental difference between the process of wave generation in case of light and that in higher-wavelength radiation such as radio waves and microwaves. The latter are generated by charge oscillations over macroscopic distances such as, for example, those in an alternating current carried by an antenna of macroscopic size. Light waves (and other low-wavelength radiations such as ultraviolet rays, γ-rays and x-rays), on the other hand, are emitted due to quantum transitions, described as photons (see Chapter 6), in individual atoms or molecules.

These photon waves are emitted independently of each other. Atoms emitting these waves have no coordination with one another. The energies, momenta, polarisations, phases and directions of these waves are different. Each wave lasts for about 10^{-9} to 10^{-10}s—the time required for an excited atom to come to the ground state. This is *incoherence*. Light waves, which are mixtures of such individual photon waves, are very different from the ideal waves considered in the previous section. They are modelled as wave trains or wave packets. Figure 3.4 shows the comparison of an ideal and a real light wave. The electric field associated with a real light wave remains nicely sinusoidal and its phase can be predicted within 10^{-9} s, the coherence time, τ_c, of the wave. The coherence length, given by the relation

$$l_c = c\tau_c \tag{3.14}$$

is the extent in space over which the wave is coherent. Thus the coherence length for a real light wave is of the order of a few centimetres. Coherence times of the order of microseconds have been achieved with lasers, the corresponding coherence length being of the order of kilometres! It should be noted that an ideally coherent source is an abstraction. A pure wave like the one in Eq. 3.1 is only an approximate model for real light beams.

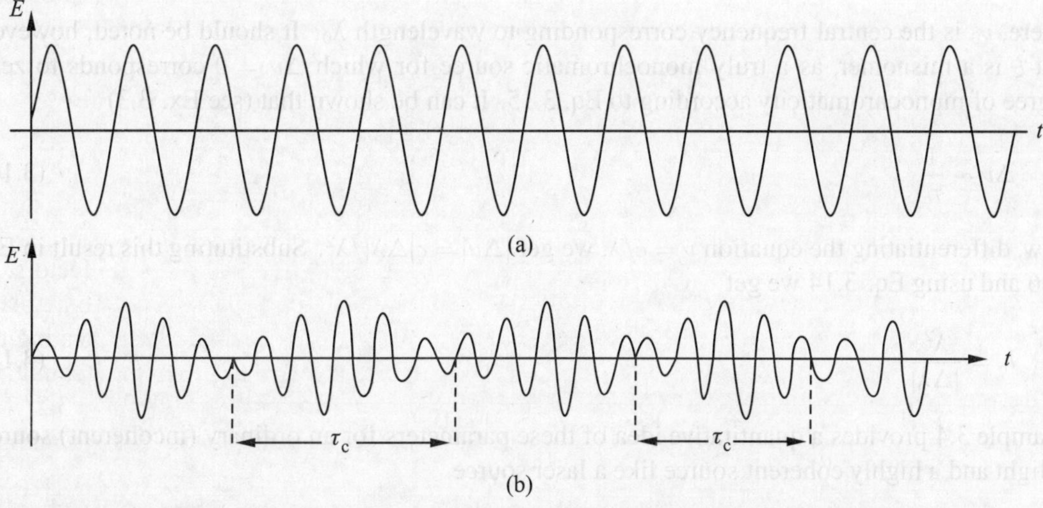

Figure 3.4 Comparison of (a) an ideal light wave ($\tau_c \approx \infty$) and (b) a real light wave ($\tau_c \approx 10^{-9}$s)

However, the above considerations bring out an interesting fact: Even with such a small value of τ_c, a single wave train from an ordinary light source with frequency 10^{14} Hz will contain

$$\frac{\tau_c}{T} = \frac{10^{-9}\ \text{s}}{10^{-14}\ \text{s}} = 100,000 \text{ oscillations.}$$

Alternatively, the wave train extends about 30 cm in space and contains

$$\frac{l_c}{\lambda} = \frac{30 \times 10^{-2}\ \text{m}}{[(3 \times 10^8 \text{m/s})/10^{14}\ \text{s}^{-1}]} = 100,000 \text{ oscillations}$$

(On this scale even an ordinary light beam would seem to be coherent!)

A wave train or a wave packet representing a real light wave is anharmonic and hence non-monochromatic. However, it can be obtained mathematically by synthesising a large number of harmonic, monochromatic waves using a technique given by Fourier. Experimentally this corresponds to the fact that spectral lines obtained because of emission are never infinitely sharp but have a spread of frequencies, $\Delta\nu$, called the natural linewidth of the spectral line; $\Delta\lambda$ is the corresponding spread in wavelength (Fig. 3.5).

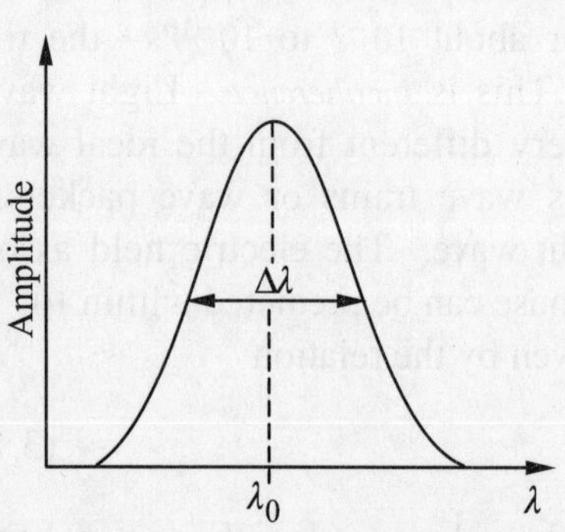

Figure 3.5 A wavepacket showing the natural linewidth of a spectral line

The degree of monochromaticity, which is the measure of spectral purity is given by

$$\xi = \frac{\Delta\nu}{\nu_0} \tag{3.15}$$

where, ν_0 is the central frequency corresponding to wavelength λ_0. It should be noted, however, that ξ is a misnomer, as a truly monochromatic source for which $\Delta\nu = 0$ corresponds to zero degree of monochromaticity according to Eq. 3.15. It can be shown that (see Ex. 3.3)

$$\Delta\nu \sim \frac{1}{\tau_c} \tag{3.16}$$

Now, differentiating the equation $\nu = c/\lambda$, we get $|\Delta\nu| = c|\Delta\lambda|/\lambda^2$. Substituting this result in Eq. 3.16 and using Eq. 3.14 we get

$$l_c = \frac{\lambda^2}{|\Delta\lambda|} \tag{3.17}$$

Example 3.4 provides a quantitative idea of these parameters for an ordinary (incoherent) source of light and a highly coherent source like a laser source.

EXAMPLE 3.3

To show $\Delta\nu = 1/\tau_c$ and to use it for explaining the incoherence of sodium doublet

Show that $\Delta\nu = 1/\tau_c$. Use it to explain the incoherence of the sodium doublet ($\lambda_1 = 5890$ Å and $\lambda_2 = 5896$ Å).

Solution

The phase of a pure harmonic wave changes by 2π in one time period. Now, as the two sources—of two frequencies ν and $\nu + \Delta\nu$—oscillate, the one with frequency $\nu + \Delta\nu$ has a shorter period, so its phase changes by 2π, slightly before the phase of the other one does. This mismatch of phase will increase as time passes. The phases will become completely mixed in a time which contains one period of frequency $\nu + \Delta\nu$ more than those of frequency ν.

Thus the coherence time, $\tau_c \approx \dfrac{N+1}{\nu + \Delta\nu} = \dfrac{N}{\nu}$

or $\quad \nu = N\Delta\nu$

that is., $\Delta\nu = \dfrac{\nu}{N} = \dfrac{1}{\tau_c}$

For a sodium doublet $\nu_1 = 5.0934 \times 10^{14}$ Hz and $\nu_2 = 5.0882 \times 10^{14}$ Hz.

Thus, $\Delta\nu = 0.0052 \times 10^{14}$ Hz

or $\quad \tau_c = \dfrac{1}{\Delta\nu} = 1.92 \times 10^{-12}$ s

In this coherence time we have

$$1.92 \times 10^{-12} \text{ s} \times 5.0934 \times 10^{14} \text{ Hz} = 977.9 \approx 978 \text{ cycles of wavelength } 5890 \text{ Å}$$

and 1.92×10^{-12} s $\times\ 5.0882 \times 10^{14}$ Hz $= 976.9 \approx 977$ cycles of wavelength 5896 Å

EXAMPLE 3.4

Comparison of $l_c, \Delta\nu, \tau_c$ and ξ for ordinary source and laser source

The line width of the orange line of krypton, Kr^{86} at $\lambda = 6058$ Å is 0.01 Å, and that of a laser source of the same wavelength is 0.00015 Å. Calculate (a) coherence length, (b) bandwidth, (c) coherence time and (d) degree of monochromaticity for both these sources of light.

Solution

Data for $\longrightarrow$	**Orange line of krypton**	**Laser source**
λ	6058 Å	6058 Å
$\Delta\lambda$	0.01 Å	0.00015 Å
Formula	**Calculated results**	
$l_c = \lambda^2/\Delta\lambda$	36.70 cm	24.47 m
$\Delta\nu = c(\Delta\lambda/\lambda^2) = c/l_c$	8.17×10^8 Hz	1.23×10^7 Hz
$\tau_c = l_c/c$	0.12×10^{-8} s	8.13×10^{-8} s
$\xi = \Delta\nu/\nu_0, \nu_0 = c/\lambda$	1.65×10^{-6}	2.5×10^{-8}

Thus the coherence time and coherence length are much larger for the laser source compared to that for an ordinary source of light.

Armed with a proper idea of coherence and the nature of real light beams, we now turn back to our discussion of interference. If we use two independent sources of light S_1 and S_2 illuminating two pinholes as shown in Fig. 3.6, then the waves coming out will maintain a definite phase relationship for time less than $\tau_c \sim 10^{-9}$ s. This means the difference $\phi_1 - \phi_2$, as well as the total phase difference $\alpha_1 - \alpha_2 = \phi_1 - \phi_2 + k(x_1 - x_2)$, will keep changing every billionth of a second. Such rapid and random changes in phase imply, according to Eqs. 3.12 and 3.13, that a bright point becomes dark and then bright again and so on in time far less than the persistence of vision of a human eye (0.1 s). Thus we fail to notice these changes and only uniform illumination is seen on the screen. Theoretically, one can 'snap' the interference pattern if a camera with exposure time (that is, the time of shutter opening) less than 10^{-9} s is devised, provided the polarisations of the two waves are made equal by placing polaroids in front of S_1 and S_2.

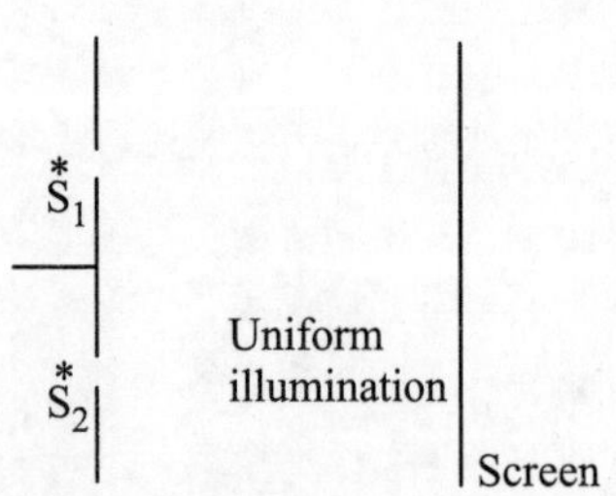

Figure 3.6 Interference pattern cannot be observed with two independent sources of light

For two coherent sources of light (Fig. 3.7), the difference $\phi_1 - \phi_2$ is constant, that is, independent of time. So the total phase difference $\alpha_1 - \alpha_2$ at a particular point on the screen will depend on $x_1 - x_2$, the difference in physical paths travelled by the two waves before reuniting at this point. Thus the point will appear either relatively brighter or darker depending on whether Eq. 3.12 or Eq. 3.13 is satisfied. As the phase difference is constant in time but different at different points on the screen, a stationary interference pattern will be observed. In a particular arrangement used by Young (Fig. 3.7), straight-line fringes are observed.

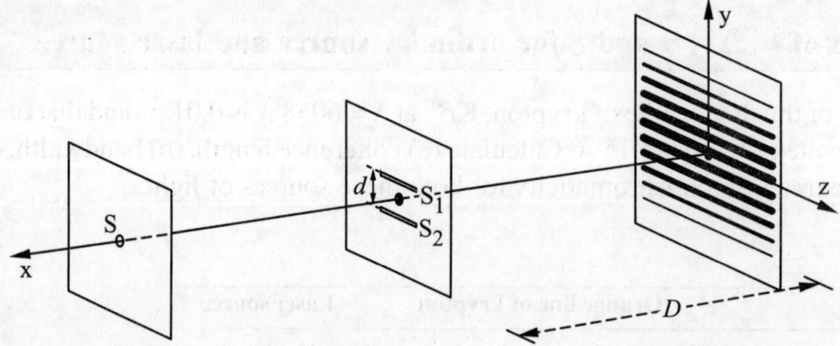

Figure 3.7 Stationary interference pattern due to two coherent sources of light (Young's experimental arrangement)

It could be gathered here that like the coherence of the interfering waves the resolution of the instrument used for observation, that is, a sensor (a detector or a recorder), will also play a role in 'sensing' the pattern. The sensors (instruments like photocell, bolometer, photographic emulsion and the eye) sense the intensity of the wave. Intensity is measured by these instruments as a time averaged quantity, defined as

$$< I(t) >= \left(\frac{1}{\tau}\right)\int_{-\tau/2}^{\tau/2} I(t)dt \tag{3.18}$$

where, τ is the averaging time or the resolution time of an instrument. For example, τ is 0.1 s for a normal eye and 0.001 s for a camera with an exposure time of one millisecond.

The practical significance of Eq. 3.9 can now be brought out by taking the averages of both sides, remembering that the intensity is proportional to the square of the amplitude. The constant of proportionality can be ignored as we will be concerned with only the relative intensities. Thus we get

$$I = I_1 + I_2 + 2\sqrt{I_1 I_2} < \cos(\alpha_1 - \alpha_2) > \tag{3.19}$$

where I is the resultant intensity and I_1 and I_2 are the intensities of the interfering waves of Eqs. 3.1 and 3.2. For the incoherent sources S_1 and S_2 considered in Fig. 3.6, $< \cos(\alpha_1 - \alpha_2) >$, obtained using definition of Eq. 3.18, will be zero, because $\alpha_1 - \alpha_2$ changes within $\tau_c \sim 10^{-9}$ s which is much smaller than the averaging time τ for any practical sensor. Thus for two incoherent sources

$$I = I_1 + I_2 \tag{3.20}$$

and uniform illumination is observed on the screen as explained earlier.

For coherent waves, however, there will be redistribution of intensities, maxima and minima in intensity corresponding to +1 and –1 values of $< \cos(\alpha_1 - \alpha_2) >$, respectively. Thus we have

$$I_{\max} = (\sqrt{I_1} + \sqrt{I_2})^2 \tag{3.21a}$$

and $$I_{\min} = (\sqrt{I_1} - \sqrt{I_2})^2 \tag{3.21b}$$

For N waves Eq. 3.9 can be written as

$$A^2 = \sum_{i=1}^{N} A_i^2 + 2\sum_{\substack{i=1\\ j>i}}^{N} A_i A_j \cos(\alpha_i - \alpha_j) \tag{3.22}$$

Thus for N incoherent sources the resultant intensity will be

$$I = NI_0 \tag{3.23}$$

(as $A^2 = NA_0^2$) where I_0 is the intensity of each source.

For N coherent sources,

$$I_{\max} = N^2 I_0 \tag{3.24a}$$

[as $A^2 = NA_0^2 + 2A_0^2 N(N-1)/2 = N^2 A_0^2$]

and $$I_{\min} = 0 \tag{3.24b}$$

Thus we see that the *amplitudes are squared first and then added to get the resultant intensity for incoherent sources, but they are added first and then squared to get the resultant intensity at bright points for coherent sources.* It is also interesting to note that the spatial average of intensities for coherent waves is still the same as that for incoherent waves as $(N^2I_0 + 0)/N = NI_0$. *The total energy is conserved; it is merely redistributed for coherent sources.* Thus for two coherent sources, absence of light energy at dark points ($I_{\min} = 0$) is compensated by doubling of the sum of two intensities at the bright points ($I_{\max} = 4I_0$).

Having understood the phenomenon of stable interference, as being the result of superposition of coherent waves, we now proceed to examine some specific cases in the next section. We will also explore the close connection between coherence and interference in this section.

3.4 INTERFERENCE

We learnt about the necessity of coherent sources for obtaining a stationary interference pattern in the previous section. However, until the advent of lasers in 1960, coherent light sources were not available. Moreover, even with a laser source it is difficult to obtain a steady pattern if the sources are independent. Young got around this difficulty by suggesting an ingenious idea that we derive two sources from a single parent source.

These two sources will be coherent even if the parent source is incoherent. Young's idea served as a key principle for obtaining coherent sources for many future experimenters. This is mainly accomplished either through the division of amplitude or the division of wavefront of the parent wave. The latter is achieved by allowing the beam to fall on either a pair of slits (Young's experiment), a mirror (Lloyd's arrangement), or a biprism (Fresnel's experiment). In the former method, a beam is divided by partial reflection at two (or more) surfaces (thin films, Newton's rings, Michelson's interferometer).

3.4.1 Procedure for Studying Interference

Interference phenomenon in any experimental arrangement can be analysed by following a simple two-step procedure:

1. Identify the two rays that interfere;
2. Calculate the total path difference between them and use conditions in Eqs. 3.12a and 3.13a.

However, in calculating the total path difference we have to be careful in including various contributions as explained below.

Geometrical and optical paths

Geometrical path is the actual distance covered by a light ray. If l is the geometrical path travelled by a ray of light in the medium of refractive index μ, then the optical path is given by

$$\Delta = \mu l \tag{3.25}$$

This is because the velocity v and the wavelength λ' of light decrease in the medium, while the frequency remains the same. In vacuum, $c = \nu\lambda$, whereas in a medium of refractive index μ, $v = \nu\lambda'$, $v = c/\mu$ and $\lambda' = \lambda/\mu$). Thus we have,

$$l = vt = \frac{ct}{\mu} \quad \text{or} \quad \mu l = ct \tag{3.26}$$

Also $l = N\lambda'$ (where N is the number of waves in the medium) $= \dfrac{N\lambda}{\mu}$ or $\mu l = N\lambda$ (3.27)

Thus the optical path μl represents the distance travelled by light in vacuum in the time in which it travels the distance l in the medium of refractive index μ. Alternatively it represents the distance in vacuum which will accommodate the same number of waves as the actual physical path in the medium. It follows that change in phase of the wave as it travels a distance l in a medium of refractive index μ is the same as when the wave travels a distance μl in vacuum. (The optical path is more because light travels more slowly in the medium and consequently has a shorter wavelength.) Example 3.5 makes the concept of optical path clear. The total path of a wave traversing several media will be $\sum \mu l$ (see Ex. 3.6).

EXAMPLE 3.5
Optical path in the medium

A monochromatic ray of wavelength 5890 Å in vacuum traverses a 5.89 μm thick film of a material of refractive index 1.35. (a) Find the optical path in the medium; (b) Verify the comments made in Section 3.4.1 by calculating (i) the number of waves and (ii) the phase difference in the medium.

Solution

(a) Optical path in the medium,

$$\Delta = \mu l = 1.35 \times 5.89 \ \mu\text{m} = 7.9515 \ \mu\text{m}$$

(b) (i) Number of waves in the medium,

$$N = \frac{l}{\lambda/\mu} = \frac{5.89 \ \mu\text{m}}{5890 \ \text{Å}/1.35} = 13.5$$

The distance in vacuum which will accommodate the same number of waves

$$= 13.5 \times \lambda = 13.5 \times 5890 \ \text{Å} = 79515 \ \text{Å} = 7.9515 \ \mu\text{m} = \text{optical path}$$

(ii) The phase difference in the medium,

$$\phi = \left(\frac{2\pi}{\lambda/\mu}\right) l = 2\pi \times N = 2\pi \times 13.5 = 27\pi \equiv \pi$$

This is the same as when the wave travels a distance μl in vacuum because in this case

$$\phi = \left(\frac{2\pi}{\lambda}\right)\mu l$$

EXAMPLE 3.6
Total optical path

A beam of sodium light (λ = 5890 Å) passes through air ($\mu \sim 1$), then through water of refractive index 1.33, then through oil of index 1.39 and finally through glass of index 1.64. In each medium it traverses a thickness of 100 cm. Find (a) wavelength of light, (b) velocity of light, (c) time of travel, (d) number of waves and (e) optical path in these media. Calculate the total optical path traversed by the beam of light.

Solution

Data for ⟶	**Air**	**Water**	**Oil**	**Glass**
Thickness	100 cm			
Refractive index	1.00	1.33	1.39	1.64
Formulae	**Calculated results**			
Wavelength $\lambda' = \lambda/\mu$ (Å)	5890	4428.6	4237.4	3591.5
Velocity $v = c/\mu$ (m/s)	3×10^8	2.26×10^8	2.16×10^8	1.83×10^8
Time of travel $t = l/v$ (s)	33.3×10^{-10}	44.2×10^{-10}	46.3×10^{-10}	54.6×10^{-10}
Number of waves $N = l/\lambda'$	1.7×10^6	2.3×10^6	2.4×10^6	2.8×10^6
Optical path $\Delta = \mu l$ (cm)	100	133	139	164

The total optical path traversed by the beam of light = 100 + 133 + 139 + 164 = 536 cm.

Path change due to reflection

There is a phase change when a light ray is reflected at a rarer–denser interface (Fig. 3.8). If a ray travelling in a medium of refractive index μ_1 encounters the boundary with another medium with refractive index μ_2 then the reflected ray looses phase by π if $\mu_1 < \mu_2$. This phase change is equivalent to an extra path of $\lambda/2$, as long as the appropriate λ—that in medium with refractive index μ_1—is taken. There is no phase change associated with reflection at denser–rarer interface or with transmission.

The phenomenon of phase change on reflection explained above occurs for any type of wave. Figure 3.9 shows this phenomenon for a pulse on the string. A pulse travelling on a string reflects partially when it encounters a 'joint' to another string of a different density. It reflects with reversal of displacement—change of phase by π—if the other string is denser and with no phase change if the other string is rarer. A voltage wave on a transmission line undergoes a similar phase change by π—the voltage in the reflected wave is of opposite sign—if it encounters an open circuit. The open circuit is infinite impedance, which is like a boundary with an infinitely denser medium. The short circuit is zero impedance and is like a boundary with infinitely rarer medium that reflects a voltage wave with no phase change.

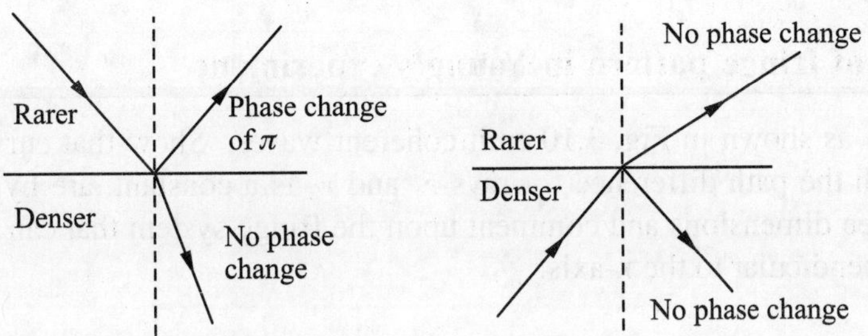

Figure 3.8 Phase changes due to reflection

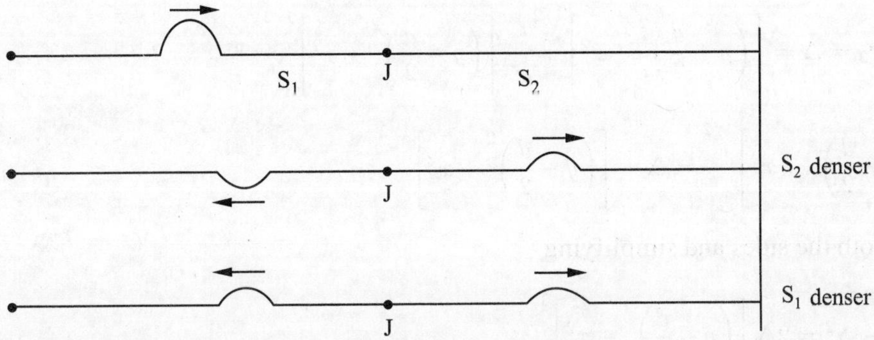

Figure 3.9 Phase changes for a pulse on the string

3.4.2 Interference in Different Cases

Although the procedure mentioned above is simple, the complex geometry of the interferometer may make the analysis involved. This procedure is applied in a 'production line' spirit here to study interference in different cases as shown in Chart 3.1. These cases are analysed in a unified way through the following approach.

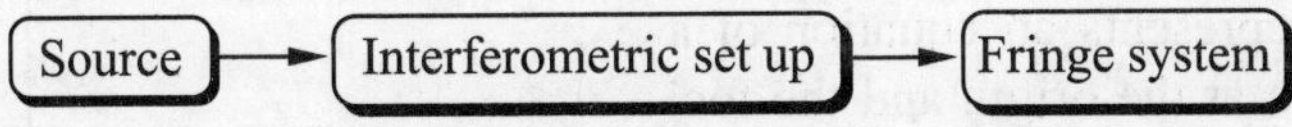

The fringe system resulting for a particular interferometric set up is studied for different options of a source such as monochromatic/white light. The effect of coherence on the fringe system is also discussed. The fringe system may be studied by using a sensor such as a recorder, a detector or simply the eye. The rays between which interference is sought are labelled as ray 1 and ray 2.

The cases discussed in Chart 3.1 can also be studied in a laboratory. Experiments 3.1–3.3 describe the study of interference using a biprism, Newton's rings apparatus and wedge-shaped arrangement respectively.

EXAMPLE 3.7
Determination of fringe pattern in Young's experiment

Two point sources as shown in Fig. 3.10 emit coherent waves. Show that curves—such as that given—over which the path difference for rays r_1 and r_2 is a constant, are hyperbolas. Extend the analysis to three dimensions and comment upon the fringe system that can be observed on a distant screen perpendicular to the x-axis.

Solution

The path difference between rays r_1 and r_2 is

$$r_1 - r_2 = \Delta = \left[\left(y + \frac{d}{2}\right)^2 + x^2\right]^{1/2} - \left[\left(y - \frac{d}{2}\right)^2 + x^2\right]^{1/2}$$

or $$\left[\left(y + \frac{d}{2}\right)^2 + x^2\right]^{1/2} = \Delta + \left[\left(y - \frac{d}{2}\right)^2 + x^2\right]^{1/2}$$

Squaring both the sides and simplifying

$$2yd - \Delta^2 = 2\Delta\left[\left(y - \frac{d}{2}\right)^2 + x^2\right]^{1/2}$$

Squaring both the sides and simplifying

$$y^2\left(\frac{d^2}{\Delta^2} - 1\right) - x^2 = \frac{d^2 - \Delta^2}{4}$$

or $$\frac{y^2}{\Delta^2/4} - \frac{x^2}{(d^2 - \Delta^2)/4} = 1$$

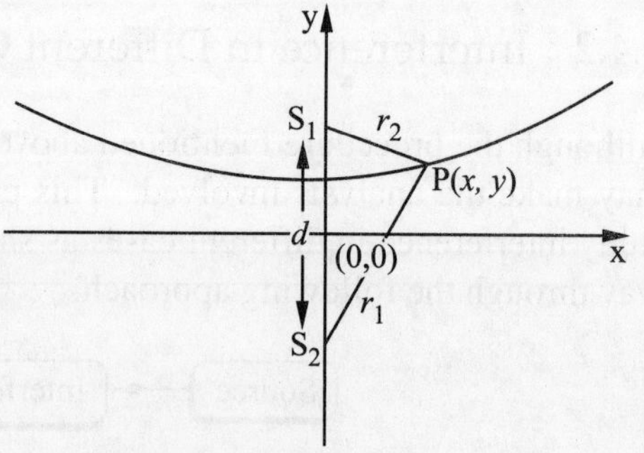

Figure 3.10

As $d >> \Delta$, $d^2 - \Delta^2$ will be a positive quantity. Thus this equation represents an equation of a hyperbola with centre at the origin and the foci on the y-axis. It is clear from Fig. 3.10 that the whole picture is symmetric with respect to rotation about the line joining the point sources. Thus we could have taken the x-axis in any direction in the plane perpendicular to the line joining the sources. Therefore, in three dimensions, the locus of the point P will be a hyperboloid, the figure of revolution of the hyperbola.

CHART 3.1 INTERFERENCE IN DIFFERENT CASES

Case		Young's experiment	Constant thickness film
Diagram		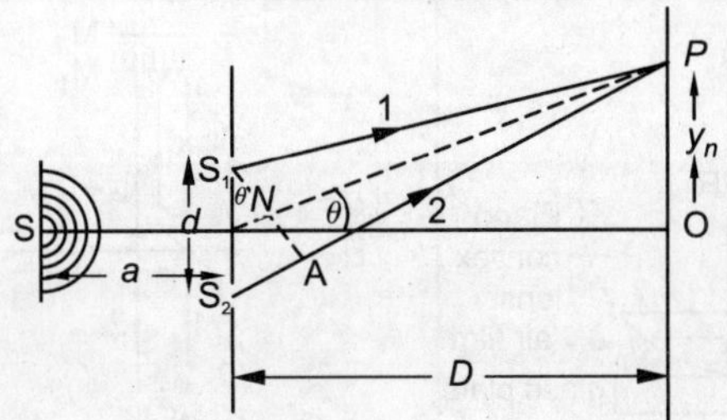 S, S_1, S_2, N, A, θ, d, a, 1, 2, P, y_n, O, D	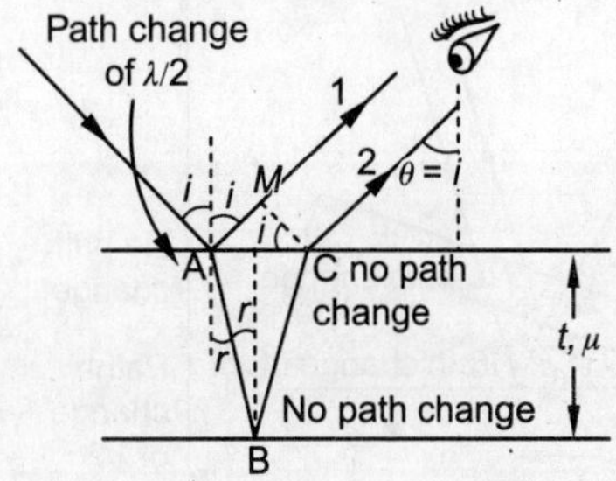 Path change of $\lambda/2$, 1, 2, $\theta = i$, i, i, M, i, A, C no path change, r, r, t, μ, No path change, B
Path difference Δ		$\Delta = S_2P - S_1P = d\sin\theta$ Thus, $\Delta = dy_n/D$ (1) (as $D >> d$, $\angle S_2S_1A$ and $\angle PNO$ are very small and practically equal, that is, 1 and 2 are almost parallel and S_1A is perpendicular to both).	$\Delta = \mu(AB + BC) - AM \pm \lambda/2$ Now, $AB = BC = t/\cos r$ and $AM = AC \sin i = 2t \tan r \sin i$ $= 2\mu t \sin^2 r/\cos r$ (because $\mu = \sin i/\sin r$). Thus, $\Delta = 2\mu t \cos r \pm \lambda/2$ (6)
Fringe system using	(A) Monochromatic light	$\Delta = dy_n/D = n\lambda$ for the nth bright fringe. Thus, $\beta = (y_{n+p} - y_n)/p = \lambda D/d$ (2) is the fringe width. The fringes are evenly spaced, as β is independent of n, under the condition that θ is small. It can be shown, by solving Eq. $S_2P - S_1P = \Delta$, that the locus of points corresponding to minima (called nodal lines) as well as maxima is a hyperbola. For large values of D, it reduces to a straight line. Thus the fringe pattern is nonlocalised and consists of equidistant straight lines (see Ex. 3.7).	$2\mu t \cos r \pm \lambda/2 = n\lambda$ or $(n + ½)\lambda$ for the nth bright or dark fringe respectively. Thus the conditions are $2\mu t \cos r = n\lambda$ for minima and $= (n + ½)\lambda$ for maxima. As rays 1 and 2 do not intersect at finite distances, interference fringes are not formed at finite distances, and fringes are localised at infinity. These rays make an angle $\theta(= i)$ with the perpendicular drawn from the eye to the face of the film. All rays entering the eye at the same angle of view correspond to fixed Δ and so the fringes observed are called fringes of equal inclination (also known as Haidinger's fringes).
	(B) White light	The central fringe produced at point O will be bright as all wavelengths constructively interfere here.	Coloured fringes will be observed. The colours in the fringe pattern will vary with the angle of inclination.
Role of coherence		The path difference should be much smaller than the coherence length, that is, $\Delta << l_c$ (temporal coherence). As the coherence length is equal to $\lambda^2/\Delta\lambda$, the maximum number of fringes is decided by $n_{max}\lambda = \lambda^2/\Delta\lambda$ or $n_{max} = \lambda/\Delta\lambda$ (3) (see Ex. 3.8). The spatial extent of the source also decides the clarity of interference fringes. It can be shown that the maximum spatial extent of the source which can produce a distinct interference pattern is given by $l = a\lambda/2d$ (4) or $\phi = l/a = \lambda/2d$ (5) is the angle subtended by the source at point N for a distinct interference pattern.	$\Delta = 2\mu t \cos r + \lambda/2$ should be small compared to the coherence length, that is, $\Delta = 2\mu t \cos r + \lambda/2 << l_c$. Using Eq. 3.17, we get $2\mu t \cos r << \lambda^2/\Delta\lambda - \lambda/2$. Thus, for normal incidence, $2\mu t << \lambda^2/\Delta\lambda - \lambda/2$ or $t << (1/2\mu)(\lambda^2/\Delta\lambda - \lambda/2)$ or $t << (\lambda/2\mu)(\lambda/\Delta\lambda - 1/2)$. As $\lambda/\Delta\lambda >> ½$, we have $t << \lambda^2/2\mu\Delta\lambda$ (7) A film satisfying this criterion will show interference fringes. This explains the need for a 'thin' film, for example, for sodium light, $t << 1$ mm (see Ex. 3.9).

Wedge-shaped film	Newton's rings apparatus	Michelson's interferometer
1 2 No path change α t_n Path change of $\lambda/2$ x_n	1 2 2R No path change Plano convex lens A t_n t_n air film Path change of $\lambda/2$ B glass plate r_n r_n	M'_2 M_1 x_2 x_1 S P x_2 M_2 1 2
It is the same as in the case of a film of constant thickness, that is, $2\mu t_n \cos r \pm \lambda/2$. For near normal incidence, $r = 0$ and so $\Delta = 2\mu t_n \pm \lambda/2$ (8) (This expression can be easily deduced from the figure also.) For small α, $t_n = \alpha x_n$ or $\Delta = 2\mu\alpha x_n \pm \lambda/2$.	It is the same as in the case of a wedge-shaped film, that is, $\Delta = 2\mu t_n \pm \lambda/2$ (10) for near normal incidence. (This expression can be easily deduced from the figure also). Now, $(2R - t_n)t_n = r_n^2$. As $t_n << R$, $r_n^2 = 2Rt_n$. Thus, $\Delta = \mu r_n^2/R \pm \lambda/2$	$\Delta = x_1 \sim x_2$ (12) as the waves emanating from P will appear to get reflected from two parallel mirrors separated by a distance $(x_1 \sim x_2)$.
$2\mu\alpha x_n = n\lambda$ for the n th dark fringe and $2\mu\alpha x_{n+p} = (n+p)\lambda$ for $(n+p)$th dark fringe. Hence the fringe width, $\beta = (x_{n+p} - x_n)/p = \lambda/2\mu\alpha$ (9) for small values of α. Thus the fringes are equidistant. As the locus of points of constant thickness is a straight line, the fringe pattern consists of equidistant straight lines. These are called the fringes of equal thickness or isopachic or contour fringes. (see Ex. 3.10).	$\mu r_n^2/R = n\lambda$ for the *n*th dark fringe. The fringe pattern is decided by the relation t_n = constant. As the locus of points of constant thickness is a circle, the fringe pattern consists of concentric dark and bright rings. The central spot is dark as $\Delta = \lambda/2$ here. The radius of the *n*th dark ring, $r_n = \sqrt{Rn\lambda/\mu}$ (11) As $r_n \propto \sqrt{n}$, the rings get closer as one moves away from the centre [true for bright rings also for which $r_n \propto \sqrt{(n + ½)}$].	The fringe pattern depends on the orientation and position of mirror M_1 relative to that of the image M'_2 of M_2 in M_1. When these two are parallel circular fringes of equal inclination are observed. If M_1 and M'_2 are inclined, the air film enclosed is wedge-shaped and straight-line fringes are observed.
Coloured fringes will be observed as constant thickness fringes for near normal incidence, for example, colours in soap bubbles and oil slicks.	Coloured fringes will be produced around the central black spot. However only few fringes will be observed as for points far away from the central spot many wavelengths in the visible region will interfere constructively to produce a uniform white illumination.	Coloured fringes are observed. A central dark fringe is bordered on both sides by few coloured fringes which are followed by uniform white illumination.
Explanation is similar as in the case of a uniform thickness film.	Explanation is similar as in the case of a uniform thickness film.	Well defined fringes are observed for $\Delta = x_1 \sim x_2 << l_c$. There is no definite distance at which the interference pattern disappears. As the distance increases the contrast of the fringes becomes poorer and poorer and finally the fringe system disappears.

The fringe shape will then be given by the curve in which the screen intersects the hyperboloid of constant path difference. If the screen is placed perpendicular to the x-axis, that is, parallel to yz-plane, this curve is a hyperbola. It is important to realise that the interference fringes are hyperbolic curves for point sources at S_1 and S_2. If we take parts of these hyperbolae near the x-axis, these are almost straight. Now if the point sources at S_1 and S_2 are made slits parallel to the third perpendicular direction—the z-axis—then the straight central portions of the hyperbolic curves will enhance each other to give straight-line fringes parallel to the slits (Fig. 3.11).

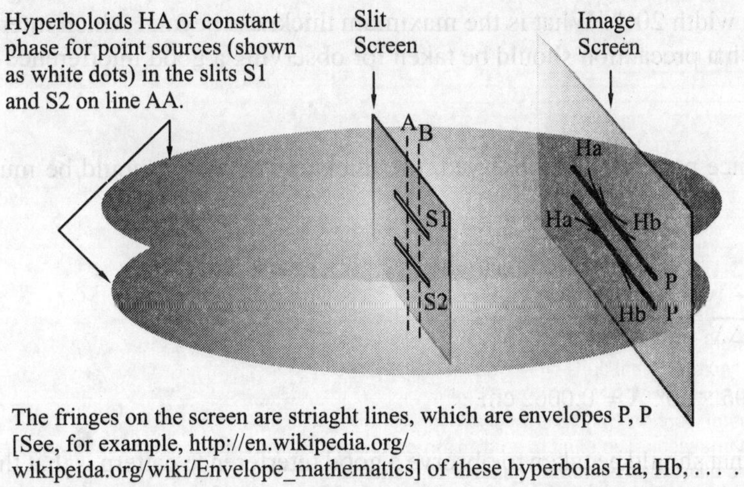

Figure 3.11 The spherical Huygens' waves propagating to the right of the slit screen from point sources in the slits on line **AA** have equal phase difference Δ on the **hyperbolas Ha** [See Example 3.7]. The hyperbolas Ha are intersections of the image screen **S** with the **hyperboloids HA**, which are loci of constant phase difference Δ for the spherical waves from point sources on **AA**. The same holds for point sources on line **BB** and hyperbolas Hb, Hb. For the sake of clarity, the hyperboloids of constant phase for point sources on **BB** are not shown.

EXAMPLE 3.8
Maximum observable fringes

Find the maximum number of fringes observable with light of wavelength 6058 Å and line width equal to (a) 0.01 Å for a krypton source and (b) 0.00015 Å for a laser source.

Solution

The maximum number of fringes is given by $n_{\max} = \dfrac{\lambda}{\Delta\lambda}$

Thus (a) $n_{\max} = \dfrac{6058}{0.01} = 605,800$ for a krypton source,

and (b) $n_{\max} = \dfrac{6058}{0.00015} = 40,386,667$ for a laser source.

Thus the maximum number of observable fringes is much larger for a laser source. This is due to its high degree of coherence.

EXAMPLE 3.9
Observability of interference pattern

A film of uniform thickness and refractive index 1.4 is illuminated with sodium light of wavelength 5890 Å and line width 20 Å. What is the maximum thickness beyond which interference will not be observed? What precaution should be taken for observing a good interference pattern?

Solution

For an interference pattern to be observed, the thickness of film t should be much smaller than $\frac{\lambda^2}{2\mu\Delta\lambda}$, that is,

$$t_{\max} = \frac{\lambda^2}{2\mu\Delta\lambda} = \frac{(5890)^2\ \text{Å}^2}{(2 \times 1.4 \times 20)\text{Å}}$$

$$= 6.195 \times 10^5\ \text{Å} = 0.0062\ \text{cm}$$

The precaution that should be taken to observe a good interference pattern is that the film thickness should be smaller compared to the value of $t_{\max}$. However, for extremely thin films ($t < \lambda/4$), complete darkness would be observed.

EXAMPLE 3.10
Wedge angle

Two plane glass plates are placed on top of one another and on one side a paper is introduced to form a thin wedge of air. Calculate the wedge angle if a beam of wavelength 6000 Å is incident normally to produce a fringe pattern having 100 fringes per centimetre.

Solution

We know that the fringe width β for a wedge is given by $\beta = \dfrac{\lambda}{2\mu\alpha}$.

Here $\mu = 1$, $\lambda = 6000\text{Å}$ and $\beta = \dfrac{1}{100}\ \text{cm} = 0.01\ \text{cm}$

Hence, the wedge angle, $\alpha = \dfrac{6000 \times 10^{-8}\ \text{cm}}{2 \times 0.01\text{cm}} = 0.003\ \text{rad}$

EXPERIMENT 3.1 BIPRISM

Aim

To determine the wavelength of sodium light using Fresnel's biprism

Apparatus

An optical bench, a biprism, a short-focus convex lens, sodium vapour lamp, a slit, micrometer eyepiece and four uprights

Formula

$$\lambda = \frac{\beta d}{D} \tag{3.28}$$

where, λ is the wavelength of light, d is the distance between two virtual sources formed by the biprism, β is the mean fringe width and D is the distance between the slit and the eyepiece.

Theory

This experiment is based on the phenomenon of interference in case of Young's double slit arrangement described in Chart 3.1. A biprism used for producing two coherent sources (Fig. 3.12) is usually made from a small piece of glass by grounding the base angles to about 0.5° so that the angle of the biprism is about 179°. Light waves coming from the two virtual sources S_1 and S_2 interfere and the fringes can be viewed through the eyepiece (Fig. 3.13). Biprism fringes are much brighter than those obtained in Young's experiment as the amount of light passing through the biprism is much greater here than in Young's double slit apparatus.

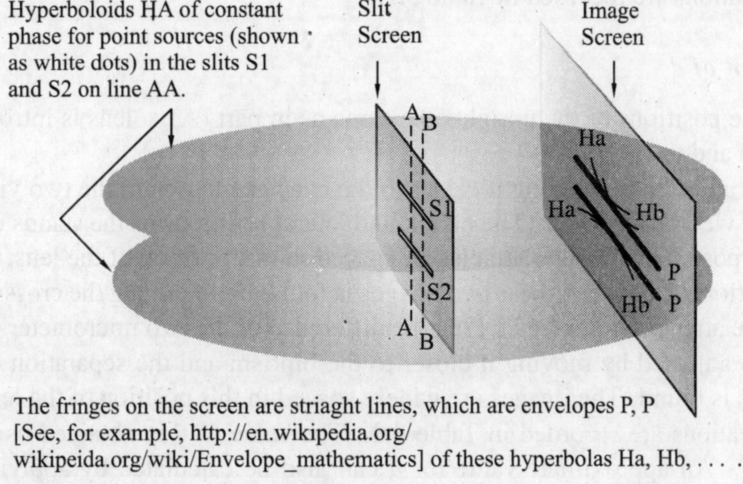

Figure 3.12 Ray diagram

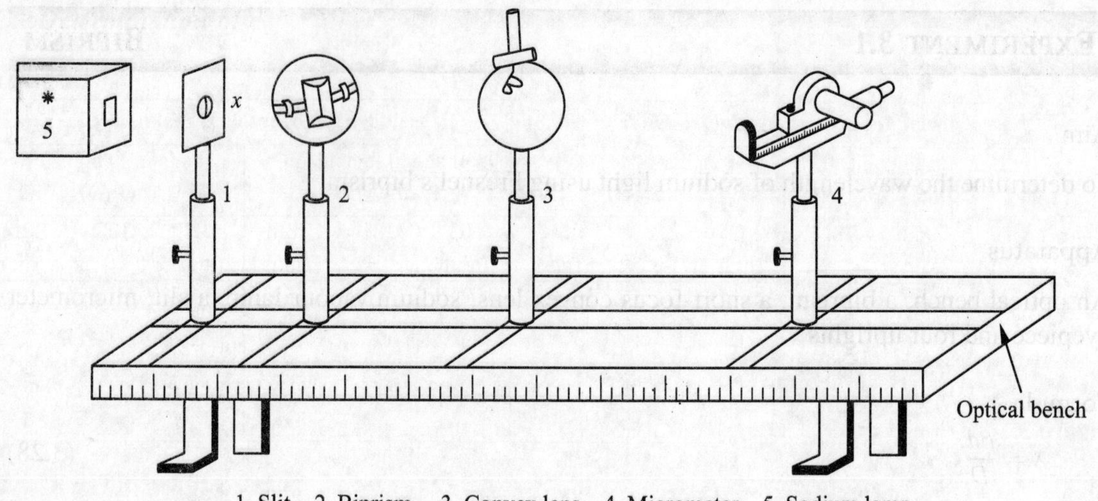

Figure 3.13 Experimental arrangement

Procedure

A. *Measurement of* β

1. Using the experimental arrangement shown in Fig. 3.13 (without the lens), a fringe pattern (Fig. 3.14) is obtained by proper alignment of the uprights and the biprism.
2. The vertical crosswire is focussed on one of the extreme left bright fringes (say the zeroth fringe) and the readings of main scale and circular scale of the micrometer are noted.
3. The readings are taken for every third bright fringe by shifting the crosswire.
4. The observations are recorded in Table 3.1.

B. *Measurement of* d

1. Keeping the positions of the uprights the same as in part (A), a lens is introduced between the biprism and the eyepiece.
2. The lens is adjusted by moving it closer to the eyepiece to obtain the two virtual images in the field of view (Fig. 3.15). (The height and lateral positions of the stands can be adjusted for this purpose.) The images are closely spaced in this position of the lens.
3. The separation d_1, between these two images is found by focussing the crosswire on the two images, one after the other, and taking the difference of the two micrometer readings.
4. The lens is adjusted by moving it closer to the biprism and the separation d_2 between the two images is found. The images are widely spaced in this position of the lens.
5. The observations are recorded in Table 3.2. The value of d is obtained using the relation $d = \sqrt{d_1 d_2}$. An approximate value for d can also be calculated by applying the relation $d = 2a(\mu - 1)\alpha$ (see Exs. 3.11 and 3.12).

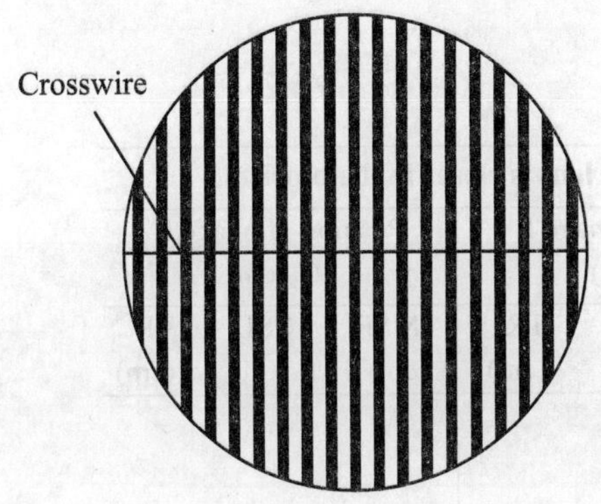

Figure 3.14 Fringe pattern

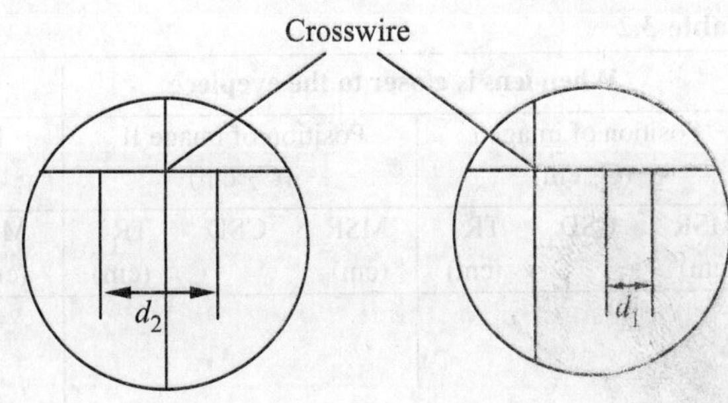

Figure 3.15 Measurement of d

C. *Measurement of* D

The positions of the uprights holding the slit and the eyepiece are noted on the optical bench and the difference between the two readings gives the value of D.

Observations

A. *Least count of micrometer*

Pitch = smallest division on the main scale = X cm

Number of divisions on circular scale = Y

Least count = (X/Y) cm

B. *Determination of* β

Table 3.1

Fringe number	Reading of micrometer (a cm)			Fringe number	Reading of micrometer (b cm)		
	MSR (cm)	CSD	TR (cm)		MSR (cm)	CSD	TR (cm)
0				15			
3				18			
6				21			
9				24			
12				27			

C. Determination of d

Table 3.2

When lens is closer to the eyepiece						When lens is closer to the biprism					
Position of image I (C_1 cm)			Position of image II (C_2 cm)			Position of image I (C_1' cm)			Position of image II (C_2' cm)		
MSR (cm)	CSD	TR (cm)	MSR (cm)	CSD	TR (cm)	MSR (cm)	CSD	TR (cm)	MSR (cm)	CSD	TR (cm)

D. Determination of D

Position of slit = D_1 = — cm

Position of eyepiece = D_2 = — cm

Calculations

β_i = separation of 15 fringes = $(a \sim b)$ cm, that is, $\beta_1 = (a_0 \sim b_{15})$ cm; $\beta_2 = (a_3 \sim b_{18})$ cm; ... $\beta_5 = (a_{12} \sim b_{27})$ cm

β' = mean separation of 15 fringes = $\dfrac{\beta_1 + \beta_2 + \beta_3 + \beta_4 + \beta_5}{5}$ = — cm

β = mean fringe width = $\dfrac{\beta'}{15}$ = — cm

$d_1 = (C_1 \sim C_2)$ = — cm; $d_2 = (C_1' \sim C_2')$ = — cm; $d = \sqrt{d_1 d_2}$ = — cm; $D = (D_2 \sim D_1)$ = — cm

Thus, $\lambda = \dfrac{\beta d}{D}$ = — cm = — Å

Estimation of error

The error $\Delta\lambda$ in the estimation of wavelength can be determined by adopting the method outlined in Section 1.4.2. Thus,

$$\Delta\lambda = \frac{d}{D}\Delta\beta + \frac{\beta}{D}\Delta d + \frac{\beta d}{D^2}\Delta D \tag{3.29}$$

Using $\beta = (x_{n+p} - x_n)/p$, where p is the number of fringes, $d = \sqrt{d_1 d_2}$ and $D = D_2 \sim D_1$ and simplifying, we get,

$$\Delta\lambda = \frac{2d}{Dp}\Delta x + \frac{\beta}{D}\left[\sqrt{\frac{d_2}{d_1}} + \sqrt{\frac{d_1}{d_2}}\right]\Delta x + \frac{2\beta d}{D^2}\Delta X \tag{3.30}$$

where Δx is the error in the reading of the micrometer and ΔX is the error in the reading of the optical bench.

For example, for the following measured data, $p = 15, \beta = 0.033$ cm, $D = 81.8$ cm, $d_1 = 0.213$ cm, $d_2 = 0.111$ cm

Least count of the micrometer = 0.001 cm and the least count of the optical bench = 0.1 cm.

$$\Delta x = \frac{0.001}{2} = 0.0005 \text{ cm}; \Delta X = \frac{0.1}{2} = 0.05 \text{ cm}; d = \sqrt{0.213 \times 0.111} = 0.154 \text{ cm}$$

Thus

$$\Delta\lambda = \frac{2 \times 0.154 \times 0.0005}{81.8 \times 15} + \frac{0.033}{81.8}\left[\sqrt{\frac{0.111}{0.213}} + \sqrt{\frac{0.213}{0.111}}\right] \times 0.0005 + \frac{2 \times 0.033 \times 0.154 \times 0.05}{81.8^2}$$

$$= (1.255 \times 10^{-7}) \text{ cm} + (4.250 \times 10^{-7}) \text{ cm} + (7.595 \times 10^{-8}) \text{ cm}$$

$$= 12.55 \text{ Å} + 42.50 \text{ Å} + 7.6 \text{ Å} = 62.65 \text{ Å}$$

Now, $\lambda = \dfrac{\beta d}{D} = \dfrac{0.033 \times 0.154}{81.8} = 6212.7$ Å

Thus the relative error is $\dfrac{\Delta\lambda}{\lambda} = \dfrac{62.65}{6212.7} = 0.01$ and the percentage error is 1%.

Result

The wavelength of sodium light is found to be — ± — Å. The error in the determination of $\lambda = (\Delta\lambda/\lambda) \times 100 =$ — %.

Discussion

We know that the most stringent requirement for obtaining a stable interference pattern is two coherent sources of light. This experiment demonstrates how this requirement can be fulfilled using a biprism. We also appreciate from this experiment that the regular fringe pattern formed (that is, for which the fringe width, $\beta = \lambda d/D$, is constant) allows us to determine the wavelength of a light wave.

In the following table the theory and the experiment are compared.

Theory	**Experiment**
Straight-line fringes should be obtained if extended sources are used in Young's experimental arrangement (see Chart 3.1 and Ex. 3.7).	Straight-line fringes are observed (see Fig. 3.14).
Fringes should be evenly spaced as β is independent of n.	The separation of fringes is found to be the same within experimental limits.

It can be seen from this table that there is a qualitative agreement between theory and experiment.

It should be borne in mind that the above expression for β is derived under the assumption $D >> d$. This condition is easily satisfied as D is typically of the order of a metre whereas the order of d is a few millimetres. However, the inherent coherence of the sodium light (see Ex. 3.3) also plays a role in the fringe pattern, as discussed in Chart 3.1. The temporal coherence decides the maximum number, $n_{\max}$, of observable fringes. For a sodium doublet, $\Delta\lambda = 6$ Å. Thus, $n_{\max} = \lambda/\Delta\lambda = 5890$ Å/6 Å ≈ 982. In reality the number of fringes observed may be one or even two orders of magnitude less than this. The reason for this is to be found in the quality of instrumentation, such as, for example, the optical flatness of faces of biprism and actual degree of incoherence in the light source.

The contrast of the fringe pattern is decided by the spatial coherence of the light incident on the biprism. For a broad (more spatially incoherent) source, no good fringe pattern is seen. As the slit width is decreased, spatial coherence increases and fringes start becoming more and more distinct. If sodium light is replaced by a laser source having much larger temporal and spatial coherence, many more clear and distinct fringes can be obtained.

The example discussed in the estimation of error shows that the error contribution due to error in measurement of d is one order higher than that due to error in measurement of D. This is in spite of the fact that the least count used in the measurement of d is two orders of magnitude better—0.001 cm to 0.1 cm. Thus the accuracy in the measurement of d_1 and d_2 is of crucial importance.

EXAMPLE 3.11

Expression for *d* in biprism experiment

In the Fresnel's biprism experiment, show that $d = 2(\mu - 1)a\alpha$ where a represents the distance from the source to the base of the prism, α is the base angle of the biprism, shown in Fig. 3.12, and μ is the refractive index of the material of the biprism.

Solution

The refractive index of the prism is given by

$$\mu = \frac{\sin(\alpha + \delta_m)/2}{\sin(\alpha/2)}$$

Here δ_m is the angle of minimum deviation. The subscript m can be dropped as prisms are always used in minimum or near minimum deviation position. For a biprism, α is too small, so we get

$$\mu = \frac{(\alpha + \delta)}{\alpha} \quad \text{or} \quad \delta = (\mu - 1)\alpha$$

From Fig. 3.12, $\frac{d}{2} = a \tan\delta \sim a\delta$ or $d = 2(\mu - 1)a\alpha$

EXAMPLE 3.12
Experimental determination of d

In order to determine the distance between the virtual images in the Fresnel biprism experiment, a convex lens is placed between the biprism and the eyepiece. Show that if $D > 4f$, two positions of the lens will be obtained where the image of the slits will be formed at the eyepiece; here f is the focal length of the convex lens and D is the distance between the slit and the eyepiece. If d_1 and d_2 are the distances between the images, as measured by the micrometer eyepiece, then show that $d = \sqrt{d_1 d_2}$. What would happen if $D < 4f$?

Solution

According to the principle of conjugate foci (or the principle of optical interchangability), a lens offers two different magnifications if the object and image distances are interchanged, that is, if

$$\frac{d_1}{d} = \frac{u}{v} \text{ then } \frac{d_2}{d} = \frac{v}{u} \text{ (Fig. 3.16)}$$

Thus, $d_1 d_2 = d^2$ or $d = \sqrt{d_1 d_2}$.

If $D < 4f$ only one of these cases will be observed.

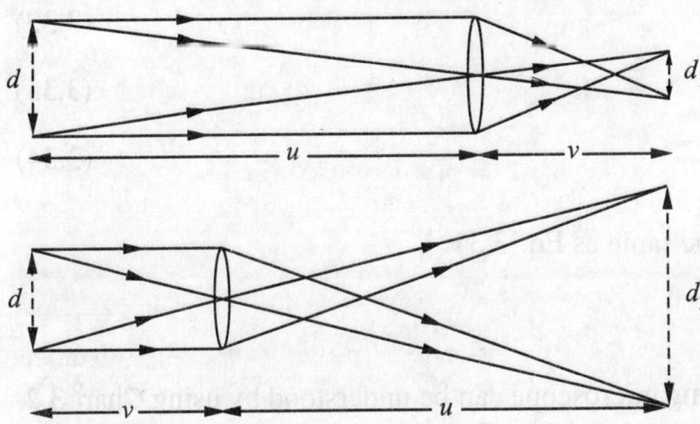

Figure 3.16 Principle of conjugate foci

EXPERIMENT 3.2 NEWTON'S RINGS

Aim

To determine the radius of curvature of a plano convex lens using Newton's rings apparatus

Apparatus

Newton's rings apparatus consisting of a plano convex lens affixed on a plane glass plate, travelling microscope and a sodium vapour lamp

Formula

$$R = \frac{(D_{n+p}^2 - D_n^2)\mu}{4p\lambda} = \frac{S}{4\lambda} \tag{3.31}$$

where R is the radius of curvature of the spherical surface of the given lens, D_n is the diameter of the nth dark ring, D_{n+p} is the diameter of the $(n+p)$th dark ring, λ is the wavelength of sodium light = 5890 Å, μ = refractive index of air = 1 and $S = (D_{n+p}^2 - D_n^2)/p$ is the slope of the graph of D_n^2 versus n.

Theory

This experiment is based on the phenomenon of interference in the case of Newton's rings arrangement as described in Chart 3.1. An interference fringe pattern consisting of concentric circular rings is formed due to interference between the rays reflected from the upper and lower surfaces of the thin air film present between the plano convex lens and the plane glass plate (see Fig. 3.17b and Chart 3.1). Circular rings are formed because the loci of points of constant thickness are circles. The central spot is dark because the thickness at the point of contact of the lens and plate is zero, and the path difference of $\lambda/2$ is introduced due to reflection at rarer–denser interface. The radius of the nth dark ring is given by (see Eq. 11 of Chart 3.1).

$$r_n = \sqrt{Rn\lambda} \quad (\mu = 1 \text{ for air}) \tag{3.32}$$

Thus, $$D_n = 2\sqrt{Rn\lambda} \tag{3.33}$$

and $$D_{n+p} = 2\sqrt{R(n+p)\lambda} \tag{3.34}$$

Thus we get, $R = \dfrac{(D_{n+p}^2 - D_n^2)}{4p\lambda}$, which is the same as Eq. 3.31.

Procedure

1. The various adjustments of the travelling microscope can be understood by using Chart 3.2. By proper alignment of the microscope, glass plate G and the Newton's rings apparatus (Fig. 3.17a), an interference pattern consisting of concentric alternate dark and bright rings (Fig. 3.17c) is obtained.
2. The vertical cross wire of the microscope eyepiece is then positioned tangentially to the 22nd left dark ring and the readings on the main scale and vernier scale are recorded.
3. The cross wire is then shifted to the 18th ring and readings are taken.
4. In this way readings are taken for every fourth ring starting from the 22nd left ring and reaching the 22nd ring on the right-hand side.
5. The readings are tabulated in Table 3.3. Care should be taken not to reverse the direction of the scale at any time to avoid the backlash error.

Observations and calculations

A. Least count of the microscope

Pitch = Smallest division on the main scale = P (cm)

Number of divisions on the vernier scale = Q

Least count $= \dfrac{P}{Q}$ (cm)

CHART 3.2 THE TRAVELLING MICROSCOPE

A travelling microscope is an optical instrument which is used to measure small distances and thus enables us to study various patterns formed due to interference of light. It mainly consists of a movable compound microscope attached with a scale that facilitates readings at different positions of the microscope.

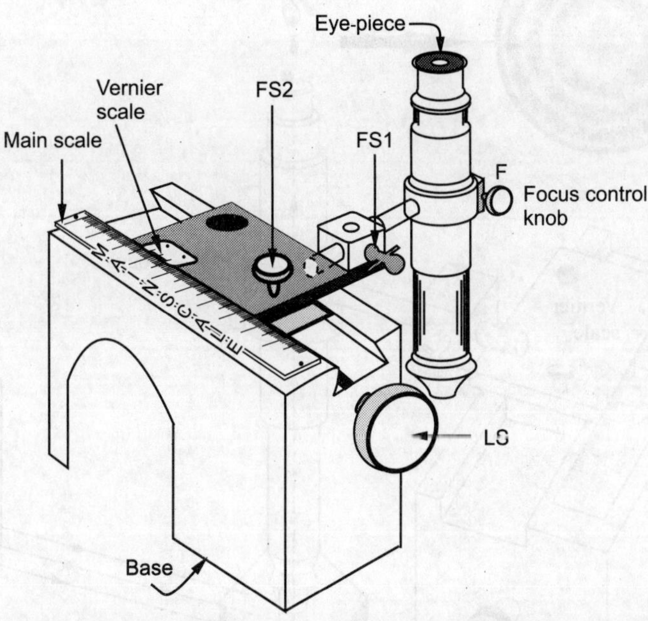

Scale reading	Various adjustments

Vernier scale reading: 35
35th vernier division coincides with division mark 10.20 on main scale

Main scale reading: 8.45
Last main scale division mark before vernier zero

0 10 20 30 40 50

20 30 40

9.5 10.5

8.5 9.5 10.5

6 7 8 9 10 11 12

10

The method of taking readings and determination of least count is the same as explained in Chart 1.2. Thus for the case shown in the above figure,

Least count = Pitch/No. of divisions on the vernier scale
= 0.05 cm/50 = 0.001 cm

and TR = 8.45 cm + 35 × 0.001 cm = 8.485 cm

Note: The screw LS should always be moved in the same direction while taking readings. This is because reversing the direction of the screw introduces a *back lash error* which arises due to mechanical play in the threads of the screw.

1. The rack and pinion arrangement attached to the focus control knob F enables the adjustment of the distance between the object and the objective lens of the microscope, that is, it helps to bring the object into the focal plane of this lens.

2. FS1 and FS2 are the fixing screws. The screw FS2 is normally tightly closed so that the rotation of the long screw LS produces sliding of the vernier scale over the main scale. If larger distances are to be measured FS2 can be loosened and the scale can be mechanically slid by hand.

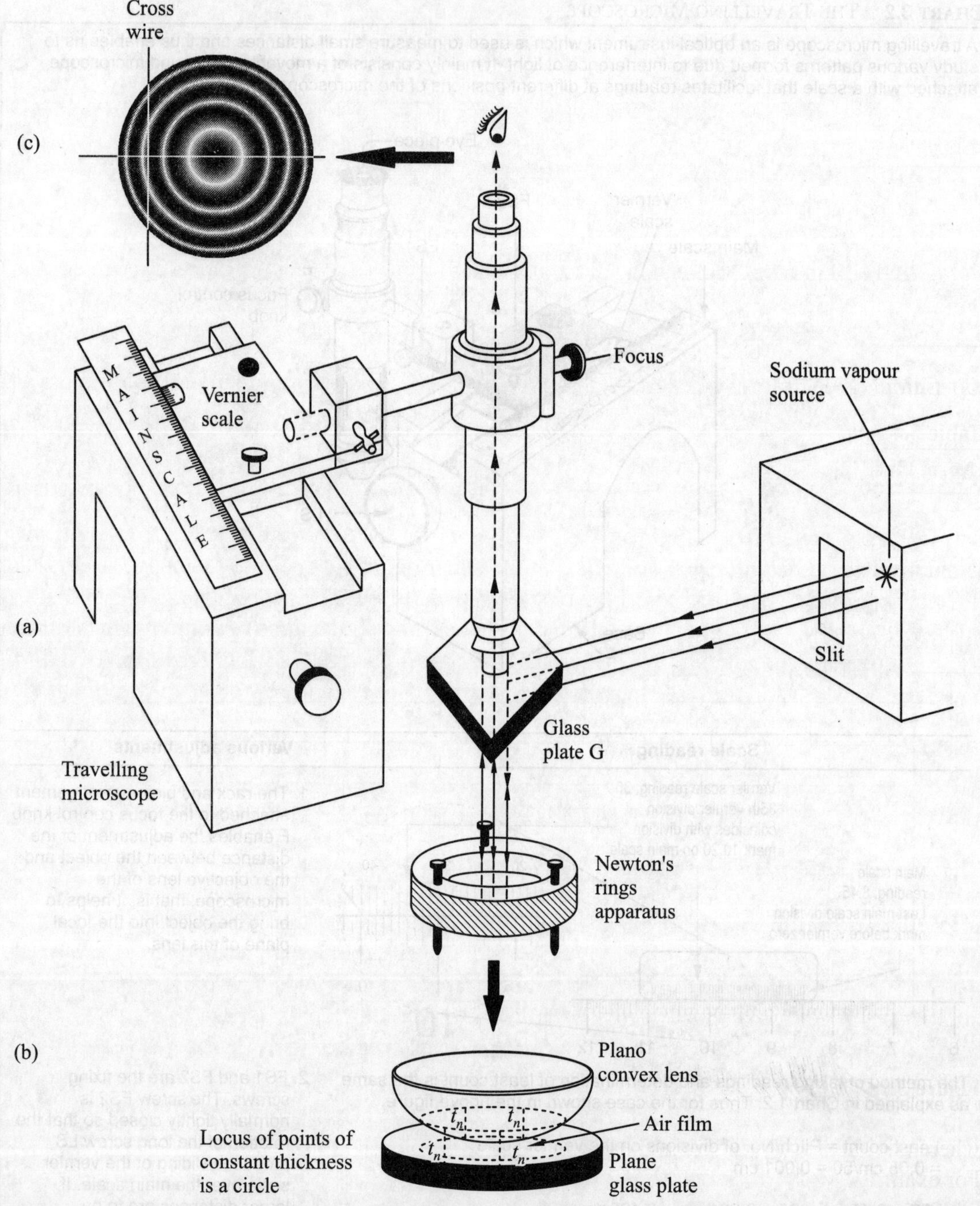

Figure 3.17 Newton's rings: (a) Experimental arrangement, (b) Magnified view of air film and (c) Fringe pattern

B. Readings for ring diameters

Table 3.3

Left-hand side				**Right-hand side**			
No. of the ring	MSR (cm)	VSD	TR (cm) A	No. of the ring	MSR (cm)	VSD	TR (cm) B
22				22			
18				18			
14				14			
10				10			
6				6			
2				2			

C. Calculation of D_n^2

Table 3.4

No. of the ring (n)	$D_n = A \sim B$ **(cm)**	D_n^2 **(cm²)**
22		
18		
14		
10		
6		
2		

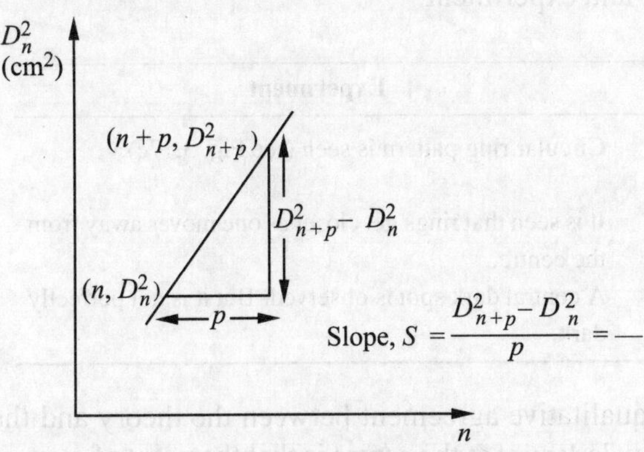

Figure 3.18 Variation of D_n^2 with n

A graph of D_n^2 versus n is plotted (Fig. 3.18) and the value of the slope, $S = (D_{n+p}^2 - D_n^2)/p$ is calculated. Then R is determined using Eq. 3.31.

Estimation of error

The error ΔR in the estimation of R can be determined by adopting the method outlined in Section 1.4.2. Thus,

$$\Delta R = \frac{\Delta S}{4\lambda} = (D_{n+p} + D_n)\frac{\Delta x}{p\lambda} \qquad (3.35)$$

where Δx is the error in the reading of the microscope.

For example, for the following measured data,

$D_{22} = 1.003$ cm, $D_{10} = 0.648$ cm, and the least count of the microscope = 0.001 cm,

we have, $\Delta x = \dfrac{0.001}{2} = 0.0005$ cm, $D_{n+p} + D_n = 1.651$ cm and $p = 12$.

Thus, $\Delta R = \dfrac{1.651 \times 0.0005}{12 \times 5890 \times 10^{-8}} = 1.168$ cm

Now, $R = \dfrac{D_{n+p}^2 - D_n^2}{4p\lambda}$

$$R = \frac{(1.003^2 - 0.648^2)}{(4 \times 12 \times 5890 \times 10^{-8})} = 207.3 \text{ cm}$$

Thus the relative error in R is $\dfrac{\Delta R}{R} = \dfrac{1.168 \text{ cm}}{207.3 \text{ cm}} = 0.006$ or the percentage error in R is 0.6%.

Result

The radius of curvature of the given lens is found to be — ± — cm. The error in the determination of $R = \dfrac{\Delta R}{R} \times 100 =$ — %.

Discussion

In the biprism experiment the parent wavefront is divided by the biprism to produce two wavefronts that serve as coherent sources. This experiment demonstrates how coherent waves can be obtained by the division of amplitude. The amplitude of the incident parent wave (see figure in Chart 3.1) gets divided at points A (on the upper surface of the air film) and B (on the lower surface of the air film). Thus rays 1 and 2 are derived from a single incident ray and are thus coherent.

In the following table we compare the theory and experiment.

Theory	Experiment
Concentric circular rings should be obtained using Newton's rings apparatus (see Chart 3.1).	Circular ring pattern is seen (see Fig. 3.17c).
The ring diameters are proportional to $\sqrt{n}$ for dark rings and to $\sqrt{(n+1/2)}$ for bright rings.	It is seen that rings get closer as one moves away from the centre.
The central spot should be dark as phase change of π is introduced due to reflection at denser medium.	A central dark spot is observed. But it is not perfectly dark.

It can be seen from this table that there is a qualitative agreement between the theory and the experiment. The reason for not getting a perfectly dark spot at the centre is that the point of contact is not perfect due to the presence of minute dust particles.

Just as in the biprism experiment, the fringe pattern here is also governed by the inherent coherence of the sodium light and the quality of instrumentation used. Factors such as the departure of lens surface from strictly spherical shape and the angular resolution of the microscope may affect the fringe pattern. For a good fringe pattern the thickness t of the air film should be much smaller compared to l_{coh}— the coherence length for sodium light. As one moves away from the centre, t starts approaching l_{coh} and the fringe pattern starts disappearing. The limiting thickness is $t << \lambda^2/2\mu\Delta\lambda << (5890)^2\text{Å}^2/(2 \times 6)\text{Å} << 0.029$ cm.

We know that (see Chart 3.1) the thickness t_n of air film corresponding to the nth ring is given by the expression $t_n = r_n^2/2R$. If we were to plot values of t_n corresponding to this relation we would be able to map the shape of the air film. This would help students to gain an insight into the science of interferometry.

It should be noted that different straight lines may be plotted by different students using the same data for n and D_n^2, thus giving different values of R. A calculation based on the method of least squares (see Section 1.4.3) gives the best possible result one can arrive at given a set of observations.

EXPERIMENT 3.3 WEDGE-SHAPED FILM

Aim

To determine the thickness of a thin foil using a wedge-shaped arrangement

Apparарus

Sodium vapour lamp, travelling microscope, two plane glass plates and a thin foil

Formula

$$t = \frac{L\lambda}{2\beta} \tag{3.36}$$

where t is the thickness of the foil, L is the length of the air wedge, λ is the wavelength of sodium light and β is the fringe width.

Theory

This experiment is based on the phenomenon of interference in case of a wedge-shaped film described in Chart 3.1. A wedge is formed by placing a thin foil between two plane glass plates as shown in Fig. 3.19a. The thickness of the air film enclosed between these plates increases linearly from 0 (at the point of contact) to t (at the other end where the foil is placed) (Fig. 3.19b). The rays reflected from the top and the bottom surfaces of the air film interfere giving rise to straight-line fringes as explained in Chart 3.1. The fringe width β is given by $\lambda/2\mu\alpha$, where α is the wedge angle and μ is the refractive index of air, which is equal to 1. As $\alpha = t/L$, we get $t = L\lambda/2\beta$.

Procedure

A. Measurement of β

1. A thin wedge of air is created by inserting a thin foil between two well-cleaned plane glass plates. These plates are held in contact at the other end by securing them together using a sticking tape.

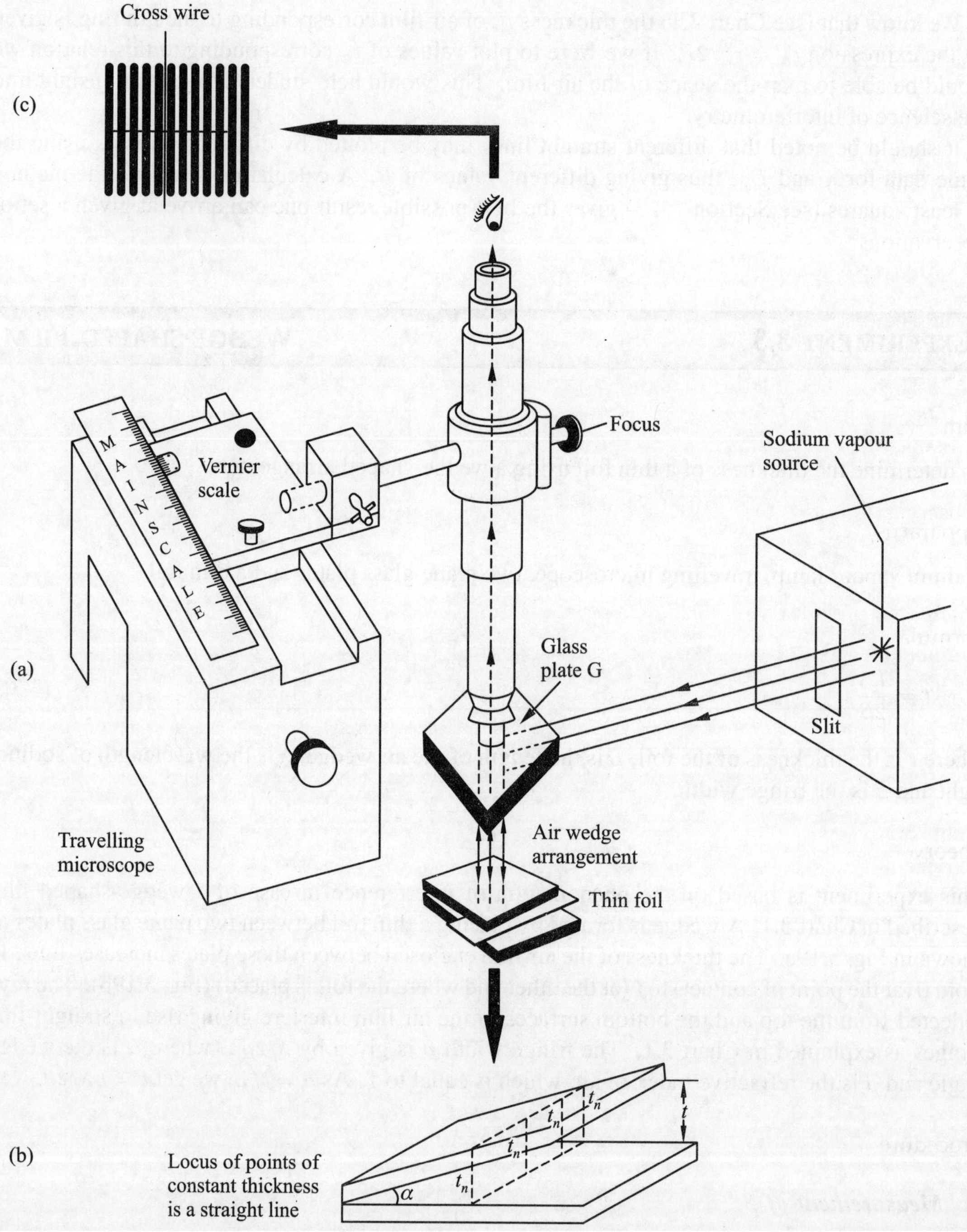

Figure 3.19 Wedge-shaped film: (a) Experimental arrangement, (b) Magnified view of air film and (c) Fringe pattern

2. In the experimental arrangement shown in Fig. 3.19a, the glass plate G is adjusted at an angle of 45° with the direction of incident light. The microscope is then correctly adjusted to obtain a clear fringe pattern (Fig. 3.19c).
3. The vertical cross wire is focussed on one of the dark fringes (say the zeroth fringe) on the left side and the readings of main scale and vernier scale are noted.
4. The readings are taken for every third fringe by shifting the cross wire to the right.
5. The observations are recorded in Table 3.5.

B. *Measurement of* L

1. The cross wire is focused on the edge of the wedge (that is, the end at which the plates are in contact) and the reading is noted.
2. The cross wire is focused on the other end, that is, at the inner edge of the foil and the reading is noted.
3. The observations are recorded in Table 3.6.

Observations

A. *Least count of microscope*

Pitch = smallest division on the main scale = X cm

Number of divisions on vernier scale = Y

Least count = $\frac{X}{Y}$ cm

B. *Determination of* β

Table 3.5

Fringe number	Reading of microscope (a cm)			Fringe number	Reading of micrometer (b cm)		
	MSR (cm)	VSD	TR (cm)		MSR (cm)	VSD	TR (cm)
0				15			
3				18			
6				21			
9				24			
12				27			

C. *Determination of* L

Table 3.6

Reading at the edge of the wedge (L_1 cm)			Reading at the inner edge of the foil (L_2 cm)		
MSR (cm)	VSD	TR (cm)	MSR (cm)	VSD	TR (cm)

Calculations

β_i = separation of 15 fringes = $(a \sim b)$ cm, that is, $\beta_1 = (a_0 \sim b_{15})$ cm, $\beta_2 = (a_3 \sim b_{18})$ cm, —, $\beta_5 = (a_{12} \sim b_{27})$ cm

$$\beta' = \text{mean separation of 15 fringes} = \frac{\beta_1 + \beta_2 + \beta_3 + \beta_4 + \beta_5}{5} = \text{— cm}$$

$$\beta = \text{mean fringe width} = \frac{\beta'}{15} = \text{— cm}$$

$$L = (L_1 \sim L_2) = \text{— cm}$$

$$\text{Thus, } t = \frac{L\lambda}{2\beta} = \frac{(\quad)\text{ cm} \times (5890 \times 10^{-8})\text{ cm}}{2(\quad)\text{ cm}} = \text{— cm} = \text{— Å}$$

Estimation of error

The error Δt in the estimation of t can be determined by adopting the method outlined in section 1.4.2. Thus,

$$\Delta t = \frac{\lambda}{2}\left[\frac{L}{\beta^2}\Delta\beta + \frac{1}{\beta}\Delta L\right]$$

$$\text{or} \quad \Delta t = \frac{\lambda}{\beta}\left[\frac{L}{\beta p} + 1\right]\Delta x \tag{3.37}$$

where, Δx is the error in the reading of microscope.

For example, for the following measured data,

$\beta = 0.024$ cm, $L = 2.53$ cm, $p = 8$, least count of microscope = 0.001 cm, we have,

$$\Delta x = \frac{0.001}{2} = 0.0005 \text{ cm and thus,}$$

$$\Delta t = \frac{5890 \times 10^{-8}\text{ cm}}{0.024\text{ cm}}\left[\frac{2.53\text{ cm}}{8 \times 0.024\text{ cm}} + 1\right] \times 0.0005\text{ cm}$$

$$= 0.174 \times 10^{-4}\text{ cm}$$

$$\text{Now, } t = \frac{L\lambda}{2\beta} = \frac{2.53\text{ cm} \times 5890 \times 10^{-8}\text{ cm}}{2 \times 0.024\text{ cm}} = 31.04 \times 10^{-4}\text{ cm}$$

$$\text{Thus the relative error in } t = \frac{\Delta t}{t} = \frac{0.174 \times 10^{-4}\text{ cm}}{31.04 \times 10^{-4}\text{ cm}} = 0.006 \text{ or the percentage error in } t = 0.6\%.$$

Result

The thickness of the given thin foil is found to be — ± — cm. The error in the determination of $t = \frac{\Delta t}{t} \times 100 = \text{— }\%$.

Discussion

This experiment brings out the application potential of the phenomenon of interference in measurement of very small dimensions. As the phenomenon involves few wavelengths of light, interferometric methods offer the advantage of measuring minute dimensions that are beyond the capabilities of mechanical methods.

The following table compares the theory and experiment.

Theory	Experiment
Straight-line fringes should be obtained using a wedge-shaped film apparatus (see Chart 3.1).	Straight-line fringes are observed (see Fig. 3.19c).
Fringes should be evenly spaced as β is independent on n.	The separation of fringes is found to be same within experimental limits.

It can be seen from this table that there is a qualitative agreement between the theory and the experiment.

The regularity of fringe pattern and constancy of $\beta(= \lambda L/2t)$ is again a factor (just as in the biprism experiment) that enables us to find the small thickness, t. The condition $t << l_{\text{coh}}$ (that is, $t << 0.029$ cm; see discussion in Expt. 3.2) is very well satisfied in this experiment. Hence a distinct interference pattern consisting of a large number of fringes is observed. The fringe pattern is better compared to that observed in Expts. 3.1 and 3.2. For example, for the measured data considered for the estimation of error, the number of fringes is $L/\beta = 2.53$ cm/0.024 cm ≈ 105.

The example discussed in the estimation of error shows that the error is dominated by the first term in the bracket in Eq. 3.37. That is, it is dominated by the measurement of fringe width. This is to be expected because the width of the fringe, which is much smaller than L, is being measured with the same accuracy as L.

3.5 APPLICATIONS

In this chapter we studied the phenomenon of interference as a superposition of light waves coming from coherent sources. The practical applications of the fundamental principles involved in this study led to the development of many useful devices and techniques. As the superposition principle (Section 3.2) holds for any periodic disturbance, it was also applied with keen interest to study interference in cases other than light waves (see Ex. 3.13).

EXAMPLE 3.13

Interference for different types of waves

All waves exhibit the phenomenon of interference. The double slit experiment is set up to demonstrate this. Calculate how far apart the two slits must be so that zeroth order and first order maxima subtend an angle of 4×10^{-2} rad at the double slits for the following waves (a) sound waves of $\lambda = 0.75$ m, (b) ultrasonic waves of $\lambda = 0.1$ m, (c) microwaves of $\lambda = 2.9$ cm,

(d) IR of $\lambda = 10$ μm and (e) light waves of $\lambda = 5890$ Å.

Solution

$\theta = 4 \times 10^{-2}$ rad $= 2.2918312°$

Now, for first order maximum, $d \sin\theta = \lambda$

Thus, $\quad d = \lambda/\sin\theta = \lambda/0.03999$

Hence $\quad d = 18.75$ m for sound waves

$= 2.50$ m for ultrasonic waves

$= 0.72$ m for microwaves

$= 250.1$ μm for IR waves

$= 14.73$ μm for light waves

In acoustics, the superposition of an incident wave and the reflected wave leading to a standing wave pattern (using a device called as the acoustic interferometer), provides a useful technique for the determination of the wavelength of ultrasound in the medium. This in turn provides information about the velocity of ultrasound in the medium and the elastic constants of the medium (see Section 16.7). Interference also produces dramatic effects in sound waves. It is possible, for example, to generate sound waves to arrive at a person's ear so as to cancel out disturbing noise. The study of interference effects involving waves in other parts of the electromagnetic spectrum, such as, radio waves (see Ex. 3.14), x-rays and gamma rays, has resulted in the development of wide-ranging applications in engineering and technology.

EXAMPLE 3.14
Intensity distribution due to dipole antenna

A certain radio station, operating at a frequency of 1200 kHz, has two identical vertical dipole antennae spaced 500 m apart. What is the intensity distribution in the resulting radiation pattern?

Solution

The interferometric geometry is similar to Young's double slit arrangement. Hence the path difference between the radiation emitted by the dipole antennae, is given by $\Delta = d\sin\theta$. Now the intensity distribution is given by (see Eq. 3.19),

$$I = I_1 + I_2 + 2\sqrt{I_1 I_2}\cos(\alpha_1 - \alpha_2)$$

Using $\alpha = \alpha_1 - \alpha_2$ and $I_1 = I_2 = I_0$, for the present case we have

$$I = 2I_0(1 + \cos\alpha)$$

As the path difference Δ corresponds to a phase difference $(2\pi/\lambda)\Delta$, we have

$$\alpha = \frac{2\pi}{\lambda} d \sin\theta$$

Now, $\lambda = \dfrac{3 \times 10^8 \text{ m/s}}{1.2 \times 10^6 \text{ Hz}} = 250$ m and $d = 500$ m

Thus $\alpha = \dfrac{2\pi}{250} \times 500 \sin\theta = 4\pi \sin\theta$

Hence $I = 2I_0[1 + \cos(4\pi \sin\theta)]$

Directions along which the intensity is maximum are given by

$$d \sin\theta = n\lambda$$

Thus, $\theta = \sin^{-1}\left[\dfrac{n\lambda}{d}\right] = \sin^{-1}\left(\dfrac{n}{2}\right) = 0, 30°, 90°.$

Directions for which intensity is minimum are given by

$$\theta = \sin^{-1}\left[\frac{(n + 1/2)\lambda}{d}\right] = 14.5$$

In Section 3.3 we studied the concept of coherence and its close correlation with interference. Research on these lines culminated in the development of a laser (see Section 3.1). The technological applications of lasers will be discussed in Section 14.6. Coherence has been well researched using both simple calculus (classical coherence theory) and advanced Hilbert space formulations (quantum coherence theory). The knowledge gained through this research has found application in areas such as statistical radiometry, spectral representation, radio astronomy and others. Though it is not possible to discuss all the applications of interference here, those with more relevance are dealt with in the following.

3.5.1 Measurement of Length

Interferometric techniques for measurement of length are more precise and accurate compared to mechanical ones, as the wavelength of light ($\sim 5 \times 10^{-7}$ cm) is used as the scale. These techniques provide an accuracy of 1 part in 10^9 in the intercomparison of lengths.

The thickness of a thin transparent plate can be measured using the double-slit arrangement (Fig. 3.20a). The introduction of a plate of thickness t causes an additional path difference $(\mu - 1)t$ (μ being the refractive index of the material of the plate), which results in the shift of the central fringe from O to O′. Thus the thickness can be determined either by counting the number p of fringes shifted and using the relation

$$(\mu - 1)t = p\lambda \tag{3.38}$$

or by measuring the distance OO′ and using the relation

$$(\mu - 1)t = \frac{d}{D}\mathrm{OO}' \tag{3.39}$$

Using white light the shift of the central fringe is easy to measure. The thickness of paper or a thin wire can also be measured using an air wedge arrangement. Experiment 3.3 describes how the thickness of a thin foil can be measured using such an arrangement.

The thickness of a thin film deposited on a glass plate (for example, coatings on optical components) can be measured using thin film interference (Fig. 3.20b). It can be deduced directly by noting the reflected colour (that is, λ), and using the relation $2\mu t = \lambda$. (Here the refractive index of the film is less than that of the substrate and the first order colour is observed.)

Measurements using the multiple-beam interference can be employed for finding the thickness t of the deposite metal film. Theory of this phenomenon relates the shift δ in the fringe pattern to the film thickness t. Small changes in length can also be measured using Newton's ring apparatus which can be further used for determining properties of materials such as elastic constants and thermal expansivity (see Ex. 3.15).

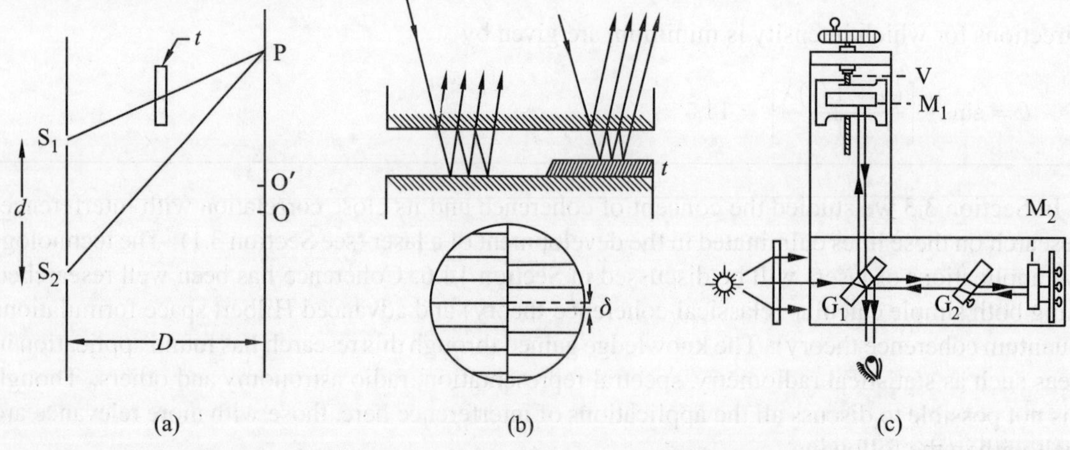

Figure 3.20 Measurement of (a) thickness of a thin plate using Young's arrangement, (b) thickness of a thin film using thin film interference and (c) small distances using Michelson's interferometer

EXAMPLE 3.15

Linear expansivity using Newton's rings apparatus

In the arrangement for Newton's rings experiment a plano-convex lens rests on a metal cylinder 5.0 mm long which stands on a flat glass plate. The lens is illuminated from above with light of wavelength 5900 Å and Newton's rings are formed in the space between the lens and the plate. The temperature of the metal cylinder is now raised by 150 K, and the rings move such that the 25th ring now occupies the position where the 5th ring originally was. Calculate the linear expansivity of the metal of the cylinder.

Solution

Newton's rings are formed due to the interference of rays 1 and 2 (Fig. 3.21). Let the nth dark ring be observed at thickness t_n of the air film, that is,

$$2\mu t_n = n\lambda \tag{3.40}$$

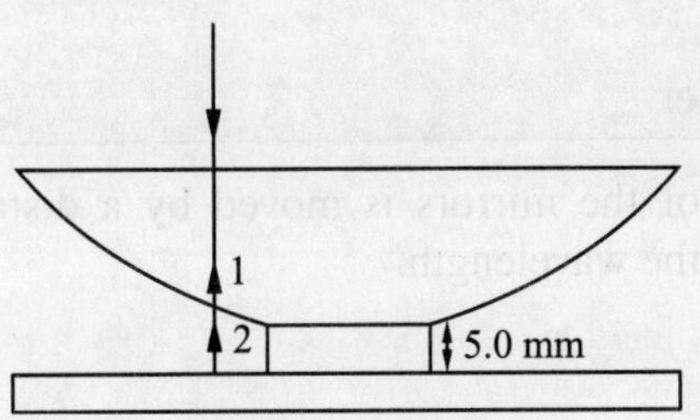

Fig. 3.21 Linear expansivity using Newton's rings apparatus

As the temperature is raised, t_n increases due to the increase Δl in the length l of the cylinder, so that

$$2\mu(t_n + \Delta l) = (n + p)\lambda \tag{3.41}$$

where p is the number of rings shifted due to increase in t_n.

From Eqs. 3.40 and 3.41, we get,

$$2\mu\Delta l = p\lambda$$

Thus $\Delta l = \dfrac{20 \times 5900\ \text{Å}}{2 \times 1} = 59000\ \text{Å} = 0.0059\ \text{mm}$

Linear expansivity of the metal of the cylinder is given by

$$\alpha = \frac{1}{l} \times \frac{\Delta l}{\Delta T} = \frac{1}{5\ \text{mm}} \times \frac{0.0059\ \text{mm}}{150\ \text{K}} = 7.9 \times 10^{-6}/\text{K}$$

Accurate measurement of distance is obtained using the Michelson's interferometer (Fig. 3.20c). The compensating plate G_2 is introduced to make the path in glass the same for two interfering rays. (The beam between G_1 and M_1 travels thrice through G_1 whereas that between G_1 and M_2 travels once through G_1.) The path difference (see Chart 3.1) $x_1 \sim x_2$ between the two interfering rays can be changed by displacing the mirror M_1 slowly using the screw V. Taking $x_1 \sim x_2 = d$, we can write,

$$2d_1 = n\lambda \tag{3.42}$$

and $2d_2 = (n + p)\lambda$ (3.43)

for two different positions of M_1. Therefore,

$$2(d_2 - d_1) = p\lambda \text{ or } d_2 - d_1 = p\left(\frac{\lambda}{2}\right) \tag{3.44}$$

(see Ex. 3.16).

Thus the distance moved equals the number of fringes moved multiplied by half the wavelength. Michelson exploited this accuracy to redefine the metre. Using the orange line of krypton with wavelength $\lambda = 6057.80211$ Å, *the metre was redefined on 14 October 1960 by the General Conference on Weights and Measures as, 1 metre = 1,650,763.73 wavelengths, giving an accuracy of 1 part in* 10^9.

EXAMPLE 3.16

Measurement of λ by Michelson's interferometer

In the Michelson's interferometer arrangement if one of the mirrors is moved by a distance 0.065 mm, 200 fringes cross the field of view. Calculate the wavelength.

Solution

In the Michelson's interferometer arrangement

$$2d = p\lambda$$

$$\text{Thus, } \lambda = \frac{2d}{p} = \frac{2 \times 0.065 \text{ mm}}{200}$$

$$= 6.5 \times 10^{-4} \text{ mm} = 6500 \text{ Å}$$

3.5.2 Measurement of Refractive Index

Refractive index can be measured using Eq. 3.38 for known thickness with the help of Young's arrangement (called the Rayleigh interferometer) shown in Fig. 3.22. First the chambers C_1 and C_2 are evacuated. The gas under study is then admitted slowly to one of the chambers and the shift p of the central fringe is noted. Knowing t, the length of the chamber, μ, can be determined.

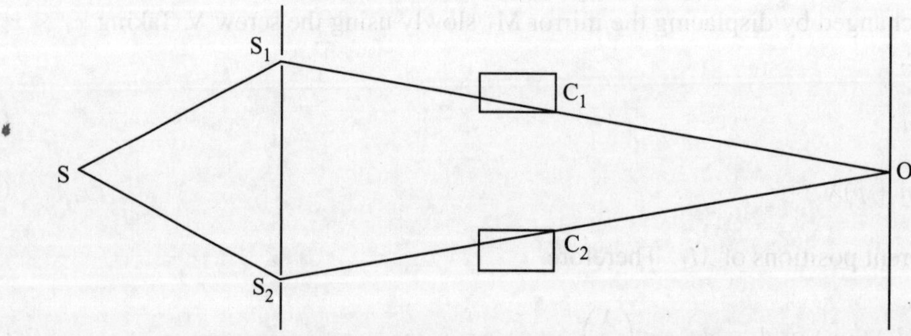

Figure 3.22 Rayleigh interferometer for measuring the refractive index of a gas

Several types of interferometers have been developed for the determination of refractive index, and are called refractometers. Refractometers developed by Jamin, Mach–Zehnder and Rayleigh are used for this purpose. Mach–Zehnder interferometer is more suitable for studying the variation of refractive index. It is used in researching wind tunnel aerodynamics to study airflow around an obstacle such as the model of an airplane wing. (Pressure around the obstacle is determined by measuring the distribution of refractive index.) Refractive index of a medium can also be measured using Newton's rings apparatus for known R and λ, applying Eq. 3.31 (see Ex. 3.17).

EXAMPLE 3.17

Measurement of μ by Newton's rings apparatus

The diameter of the 10th bright ring in a Newton's rings apparatus changes from 1.75 cm to 1.59 cm as a liquid is introduced between the lens and the plate. Find the refractive index of the liquid.

Solution

The diameter of the nth bright ring is given by

$$D_n = 2\sqrt{R\left(n+\frac{1}{2}\right)\frac{\lambda}{\mu}}$$

where μ is the refractive index of the medium between the lens and the plate.

Thus, $(D_{10})_{\text{air}} = 2\sqrt{R(10.5)(\lambda/1)} = 1.75$ cm

and $(D_{10})_{\text{liquid}} = 2\sqrt{R(10.5)(\lambda/\mu)} = 1.59$ cm

Hence $\mu = \dfrac{(D_{10}^2)_{\text{air}}}{(D_{10}^2)_{\text{liquid}}} = \left(\dfrac{1.75}{1.59}\right)^2 = 1.211$

3.5.3 Non-reflecting/High-reflecting Films

One of the simple but important applications of thin film interference is to either suppress or enhance the light reflected from an optical component. This is achieved by coating the surface of the component with a thin film of suitable thickness and refractive index (Fig. 3.23). A non-reflecting film (or an anti-reflecting film) enhances the transmitted intensity by bringing about destructive interference of the reflected rays. In this case the surface of the component is coated with a film of refractive index less than that of glass and thickness is given by the relation

$$2\mu_f t = \frac{\lambda}{2}$$

or $$t = \frac{\lambda}{4\mu_f} \tag{3.45}$$

This expression also gives the thickness of a high reflection film, provided μ_f, the refractive index of the film material, is greater than that of glass.

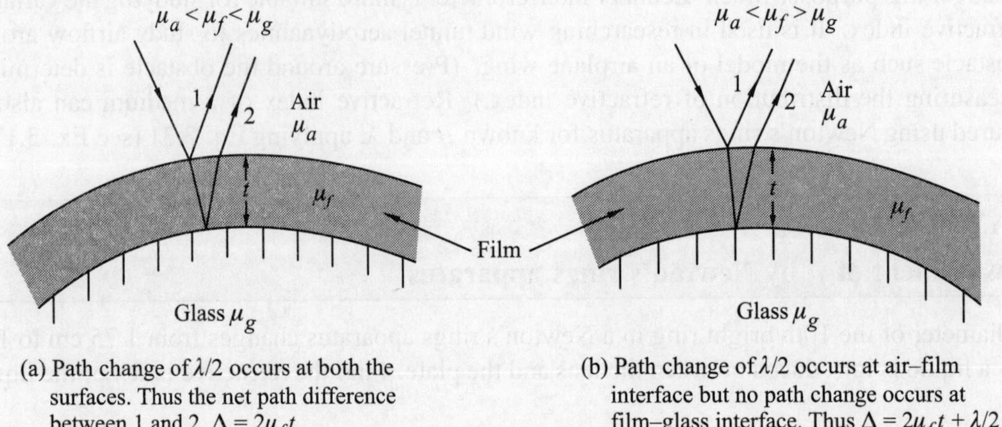

Figure 3.23 Thin film coating (a) For anti-reflection film, $\Delta = 2\mu_f t$; thus for first order minimum, $2\mu_f t = \lambda/2$, or $t = \lambda/4\mu_f$ and (b) For high reflection film, $\Delta = 2\mu_f t + \lambda/2$; thus for first order maximum, $2\mu_f t + \lambda/2 = \lambda$, or $t = \lambda/4\mu_f$

It can be shown that the refractive index of the anti-reflection film should be given by

$$\mu_f = \sqrt{\mu_a \mu_g} = \sqrt{\mu_g} \text{ (as } \mu_a = \text{refractive index of air} = 1) \tag{3.46}$$

where μ_g is the refractive index of the glass. This condition makes the amplitudes of the two reflected rays equal, thus minimising the reflections. The materials generally used as coating non-reflecting films are magnesium fluoride ($\mu_f = 1.38$) and cryolite ($\mu_f = 1.36$). Zinc sulphide ($\mu_f = 2.32$) is used as a high reflection coating. High reflection films (also called as dielectric mirrors as the coating is non-conducting) are used as beam splitting mirrors to divide a beam of light into two parts of a given intensity ratio. Anti-reflection films are often used to coat lenses in telescopes, cameras and other optical instruments (see Ex. 3.18).

EXAMPLE 3.18
Thickness of anti-reflection coating

Expensive cameras often use lenses coated with anti-reflection films. Typically, the lens surface (refractive index of glass = 1.5) is covered with a thin layer of MgF_2 ($\mu_f = 1.38$). Find the minimum thickness of this layer for destructive interference to occur between the light reflected from the two surfaces of the layer when 5500 Å light is used. Why would a layer of lucite ($\mu_f = 1.48$) be less suitable for this purpose?

Solution

The minimum thickness of the film that will produce destructive interference between the reflected rays (rays 1 and 2 in Fig. 3.23a) is

$$t = \frac{\lambda}{4\mu_f} = \frac{5500}{(4 \times 1.38)} \sim 996.4 \text{ Å}$$

The resultant reflected intensity is given by (see Ex. 3.14)

$$I = 2I_0(1 + \cos\alpha) \tag{3.47}$$

where, I_0 is the incident intensity and

$$\alpha = \left(\frac{2\pi}{\lambda}\right) \times \text{path difference} = \left(\frac{2\pi}{\lambda}\right) \times 2\mu_f t$$

$$= 4\pi\left(\frac{t}{\lambda}\right) \times \mu_f = 4\pi\left(\frac{996.4 \text{ Å}}{5500 \text{ Å}}\right) \times \mu_f$$

Thus $\alpha = \pi$ for MgF_2 and 1.072 π for lucite.

Substituting these values in Eq. 3.47 we get, $I = 0$ for MgF_2 and $I = 0.051I_0$ for lucite. Thus, 5.1% of the incident light is reflected using lucite, making it less suitable for coating.

3.5.4 Test of Flatness of a Surface

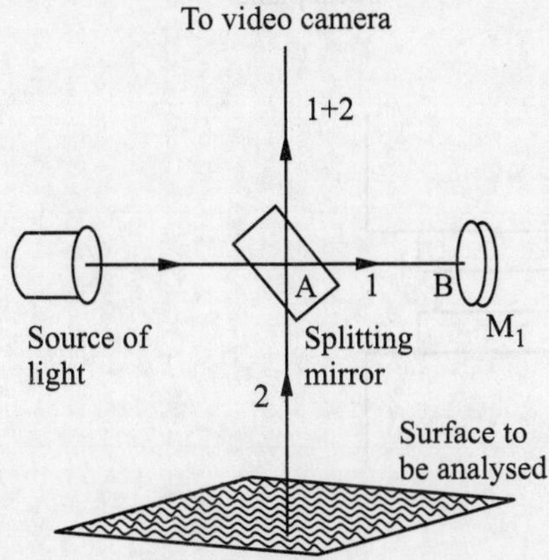

Figure 3.24 Surface analysis using Michelson interferometer

The surface to be tested is placed on an optical flat at an angle and the wedge is illuminated with a monochromatic light. If the surface under test is perfect, the fringes will be straight and equidistant; otherwise, they will be curved and the distance between them will vary. However, minute polishing defects cannot be detected by this method and multiple-beam interference methods have to be employed for this purpose. Fringe-by-fringe examination of the pattern helps to reveal these defects.

Figure 3.24 shows the use of Michelson interferometer for surface analysis using a technique known as 'computer analysed optical interferometry'. Rays 1 and 2 are reflected from the mirror M_1 and the surface to be analysed. These two rays interfere and a video camera records the intensity of the resultant beam. Three such records, called the *interferograms*, are obtained by scanning the surface—varying the distance AB slightly for each scanning. These records are then fed to a computer where, using an appropriate program, a three-dimensional representation of the surface is obtained. This technique can be used to detect variations of the order of a few angstroms in the surface texture. Thus, it is widely used to check the surface finish of devices such as videotapes, photographic films, computer discs and ball bearings.

3.5.5 Interference Filters

An interference filter uses an interferometer constructed by Fabry and Perot, which is based on multiple-beam interference (Fig. 3.25a). It consists of a set of two plane parallel glass plates, thinly silvered on the inner surfaces and enclosing an air film of thickness d. The thickness can be varied by displacing one plate with reference to the other fixed plate. For normal incidence the path difference between the interfering rays is $2d$. For a large difference d, several wavelengths, given by $\lambda = 2d/n$, are observed for the incident visible light. (Such a spectrum is called a channelled spectrum.) However, the number of wavelengths can be reduced by decreasing d. Thus the filter serves as a monochromator.

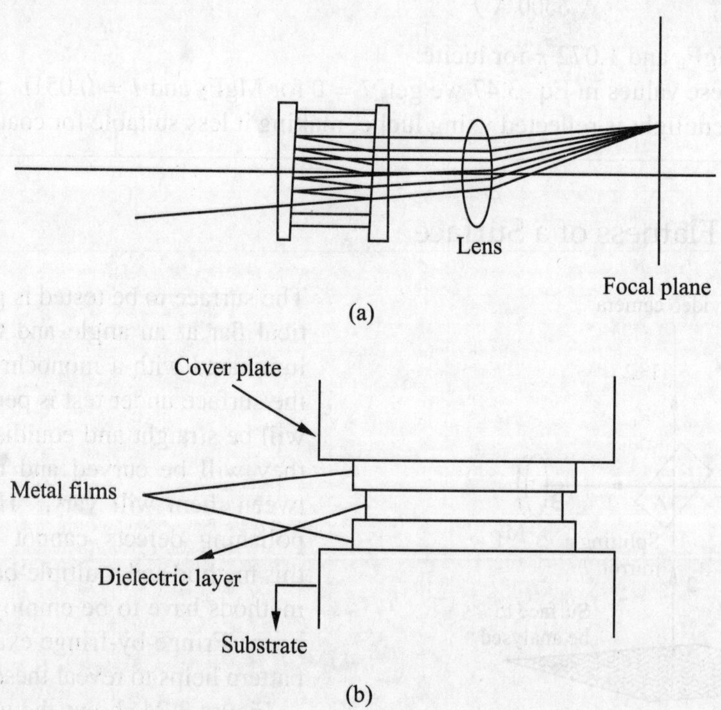

Figure 3.25 (a) The Fabry–Parot interferometer and (b) The interference filter

Interference filters are constructed using the vacuum deposition technique. A thin silver film is deposited on the substrate (Fig. 3.25b). A thin layer of dielectric material, such as cryolite, is then deposited, over which another thin metallic film is coated, and the structure is then covered by another glass plate to protect it from any damage. By varying the thickness of the dielectric layer a particular wavelength can be filtered.

Apart from the applications mentioned above, interferometry also finds application in various branches of engineering and technology.

Astronomy: It is used to measure binary star separations and the diameters of stars.

Spectroscopy: It is used to study hyperfine structure, and the width and shape of spectral lines. Investigation of the hyperfine structure throws light on the properties of atomic nuclei, whereas the width and shape provide information about the conditions of temperature and pressure in the light source.

Microscopy: It is used to visualise transparent or reflecting specimens. Interference microscopes are especially useful to biologists, who otherwise resort to colouring the living objects in order to perceive them.

Metallurgy: It is used to study reflecting specimens such as metallic surfaces or metallised replica of surfaces.

EXERCISES

1. What is a thin film? Obtain an expression for the path difference in the case of interference in thin films due to reflected light.
2. A wedge-shaped film is illuminated by monochromatic light. Obtain an expression for the fringe width of the interference pattern formed. Explain why the fringe at the apex of the wedge is always dark.
3. Obtain an expression for the fringe width in case of interference pattern formed by a wedge-shaped thin film. How is it used for testing if a surface is optically flat?
4. Describe the Newton's rings experiment to determine the wavelength of incident monochromatic light.
5. Why are circular fringes obtained in the Newton's rings arrangement? Why is the central spot dark as seen by reflection? Why do rings away from the centre appear crowded?
6. In the Newton's rings experiment why should the plano convex lens have a larger radius of curvature?
7. State any two applications of the Newton's ring experiment. In this experiment why are the rings unevenly spaced?
8. What will happen if the convex lens in the Newton's ring apparatus is lifted by $\lambda/4$, where λ is the wavelength of the light used?
9. What is an anti-reflection coating? Deduce an expression for minimum thickness of an anti-reflection coating.
10. Why does a coated lens look purple in reflected light?

PROBLEMS

1. Calculate the minimum thickness of a soap bubble film ($\mu = 1.33$) that results in constructive interference in the reflected light if the film is illuminated with light of wavelength 600 nm in free space.
2. A drop of liquid of volume 0.2 ml is dropped on the surface of water in a tank of area 1 m^2. The film spreads uniformly over the whole surface. White light incident normally on the surface is observed through a spectroscope. The spectrum is seen to contain one dark band, which at the centre has a wavelength of 5500 Å in air. Find the refractive index of the given liquid.
3. A wedge-shaped film is illuminated by light of wavelength 465 nm. The angle of the wedge is 40°. Calculate the fringe width.
4. A cellophone film with index of refraction 1.5 has a wedge shape with the thickness on two opposite sides being t_2 and t_1. If with normally incident light of $\lambda = 6000$ Å, the number of fringes appearing on the film is 10, calculate the difference $t_2 - t_1$.
5. In a Newton's rings arrangement the radius of the plano-convex lens used is 120 cm. The diameter of the 16th ring was found to be 0.64 cm and that of the 6th ring was 0.36 cm. Determine the wavelength of light used.
6. A film of uniform thickness and refractive index 1.4 is illuminated with sodium light of wavelength 5890 Å and line width 20 Å. What is the maximum thickness beyond which interference pattern will not be observed? What precaution should be taken for observing a good interference pattern?

QUESTIONS

1. What is interference of light?
2. What are the conditions under which interference is observable?
3. What is superposition of waves?
4. We represent partial differentiation with respect to a variable by a ∂ sign with variable in subsript. So, for a function $f(x,t)$ of x and t, $\partial_t f(x,t) \equiv \frac{\partial f(x,t)}{\partial t}$ and $\partial_{x,s} f(x,t) \equiv \frac{\partial^2 f(x,t)}{\partial x^2}$, and so on ... . Now, consider the following equations in which v and γ are known constants and $g(x)$ is some known function of x – $\mathcal{A} : \partial_{x,x} f(x,t) = (1/v^2)\partial_{t,t} f(x,t)$; $\mathcal{B} : \partial_{x,x} f(x,t) = (1/v^2)\partial_{t,t} f(x,t) + g(x)$; $\mathcal{C} : \partial_{x,x} f(x,t) = \gamma(\partial_t f(x,t))^2$. Solutions of which of these equations will satisfy the superposition principle? Why, or why not?
5. When sunlight reflects from a soap bubble floating in air you see coloured interference fringes. Is this an example of interference by division of wavefront, or by division of amplitude? Explain.
6. Add the waves $y_1 = 3 : 0 \cos(20\pi t - 2x)$ and $y_2 = \cos(20\pi t - 2x + \pi/3)$ at $x = 0$ using the geometrical (phasor) method.
7. A region is illuminated by two sources of light. The intensity at each point is found to be equal to $I_1 + I_2$, where I_1 is intensity of light at a point when source 2 is absent. I_2 is similarly defined. Are the sources coherent or incoherent? Explain.
8. If a source of light gives out mixed light waves of wavelengths between $\lambda - \delta\lambda/2$ and $\lambda + \delta\lambda/2$ where $\delta\lambda \ll \lambda$, then the coherence length of emitted light is $\lambda^2/\delta\lambda$. Coherence length is the maximum distance up to which phase of the wave is predictable. Suppose that a Young double-slit experiment is arranged with this light. What would be the condition on the distance between the two slits in order to obtain a stable observable interference pattern on the screen?
9. We know that electron is a wave. Can we think of a Young double-slit interference experiment with electron beams? If we are dealing with 100 eV electrons, what is the order of slit separation you need?
10. Examine the intensities given by Eq(3) when $\alpha_2 - \alpha_1$ takes on values $0, 2\pi, 4\pi, \ldots$. Are these different? Now, examine the intensity distribution pattern shown on the right in the doubleslit interference figure. Does it show the same intensity at all maxima? If we want to claim that the superposition principle explains interference phenomena, how do you explain the answers to these two questions?
11. What is thin film interference?
12. What is (are) the criterion (criteria) for deciding whether the film is thin, or thick?
13. Suppose you take a glass of water and add a drop of an oil on the water surface. The drop spreads. Now you observe the water surface in re ected sunlight. What do you expect to see? Add more drops of oil onto the same surface and repeat your observation. Continue adding a few drops each time till you can see clearly a layer of oil. Describe what is likely to be observed at each stage.
14. Consider a wedge-shaped film formed by placing a glass plate on top of another identical plate with their edges in contact on one side and edges separated by a rigid cylindrical wire on the opposite side. You can observe straight fringes of a fixed fringe width. Now, suppose you push the cylindrical wire slowly toward the edges in contact, always maintaining it parallel to those edges. What do you expect to happen to fringe pattern?

15. In Newton's rings set-up we see rings merging into each other toward the periphery of the ring system. What could be the reasons?
16. If the plano-convex lens in Newton's rings set-up is replaced by a piece of a cylinder cut parallel to its axis, what type of fringe pattern will you see?
17. Parallel rays of the Sun fall on a soap bubble floating in air and the reflected light shows coloured fringes. Discuss their shapes and colours. Are these fringes of equal inclination, or thickness, or none of these? Discuss.
18. If a transparent liquid is introduced between the plate and the plano-convex lens in a Newton's rings set-up, what do you expect to happen?
19. If a Newton's rings experiment is performed with white light what will happen to the ring pattern?
20. In the Newton's ring set-up, or in a wedge-shaped film formed between two glass plates, why do the waves reflected from the top and the bottom surfaces of the lens, or the upper glass plate not disturb the interference fringes observed?
21. Consider a wedge shaped film illuminated by white light containing wavelengths between 400 *nm* and 700 *nm*. Estimate the thickness of the film for which you can still see fringes with sufficient clarity. What is the basic idea involved in this consideration? What effct will this have on the fringe pattern you see?
22. Suppose you observe fringes in sufficiently monochromatic light reflected from a wedge-shaped film. Since interfering waves are generally not of the same amplitude, how will this show in the observed fringes?

4

Diffraction

4.1 THE STORY OF DIFFRACTION

In the last chapter we studied the experimental and theoretical aspects of the phenomenon of interference of light. For the last three and a half centuries physicists have keenly studied another phenomenon of light—diffraction. Interference and diffraction are both 'superposition effects'. The terms are often used interchangeably.

Diffraction is concerned with the bending of light and blurring of images. It was discovered as early as 1565 by Francesco Grimaldi, an Italian mathematician. Grimaldi and Newton tried to explain diffraction as a corpuscular effect caused by the deflection of light corpuscles in passing close to the edge of an obstacle. However, their explanation was far from convincing and could not account for the systematic diffraction patterns observed for various obstacles.

In 1818, Augustin Fresnel, a young military engineer from France, offered a brilliant explanation for diffraction. He used Huygen's wave theory of light. Newton's supporters, however, were not happy with Fresnel's ideas. Poisson, a French mathematician and one of the staunch votaries of Newton, even challenged Fresnel that if wave theory and Fresnel's model were true, then one should obtain a bright spot at the centre of the image of a circular object. An experiment was performed by Dominique Arago to settle this controversy. To the utter dismay of Newton's supporters, and to the delight of Fresnel, a bright spot was indeed observed! The prediction and subsequent discovery of the Fresnel bright spot is considered as the first conclusive proof of the wave nature of light.

Research in diffraction pertained mainly to the interpretation of the diffraction patterns of known obstacles such as slits and straight edges. Apart from Fresnel, Fraunhofer (1787–1826), a Bavarian optician, and MA Cornu (1841–1902), a French experimental physicist, made pioneering contribution to this research. These studies inspired researchers to try out newer diffracting obstacles. Discoveries of a zone plate by Lord Rayleigh in 1871 and of a diffraction grating by Rowland (1848–1901) in 1882 are two prominent examples of the several diffracting obstacles discovered by researchers in this sequel.

Thus diffraction studies played a historical role in providing incontrovertible evidence of the wave nature of light. They also evoked interest in similar work in case of other waves like mechanical waves and other parts of the electromagnetic spectrum. The same diffraction studies later proved to be of immense importance in recognising the wave nature of matter (see Chapter 6). The knowledge gathered out of this research had many practical applications (Section 4.4).

4.2 THE PHENOMENON OF DIFFRACTION

In the last chapter we studied the formation of a stable interference pattern due to superposition of waves originating from a single parent source. In Young's experiment (see Fig. 3.7) these waves are obtained by deriving two sources S_1 and S_2 from a single source S. What would happen if the light from source S is made to fall directly on the screen? An obvious answer would be that no interference pattern would be seen. Moreover, if one assumes rectilinear propagation of light, the geometrical replica (shadow, if an opaque object is placed in the path of light) of the obstacle is expected to be seen. However, the observations differ from the expectations: the light is found to deviate around the boundaries of the obstacle. This produces some brightness (blurring) in the region of the geometrical shadow (or some darkness where complete brightness is expected) (Fig. 4.1). *This bending of light round the corners and spreading of light into the geometrical shadow of an object is known as diffraction.* The word diffraction is derived from the Latin word *diffractus*, meaning 'breaking to pieces'.

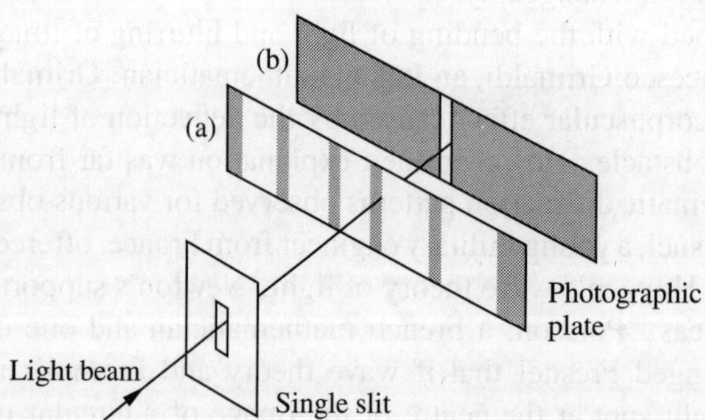

Figure 4.1 (a) Observed diffraction pattern for a single slit illuminated by light; (b) The expected pattern if rectilinear propagation of light is assumed

Diffraction effects are more pronounced and a diffraction pattern can be seen if the size of the diffracting obstacle is of the same order as the wavelength of the light used (Fig. 4.2). Long edges, mirror stops in optical instruments, source slit of the spectrometer are some typical diffracting obstacles. Some examples of diffraction patterns in nature are coloured rings around a street light in frosty weather, coloured bands viewed by reflection from a compact disc, spreading of light round your eyelashes, pattern seen when a distant source of light is seen through a crack between two fingers, and when a streetlight is seen through a cloth umbrella. Figure 4.3 shows diffraction patterns formed by various objects.

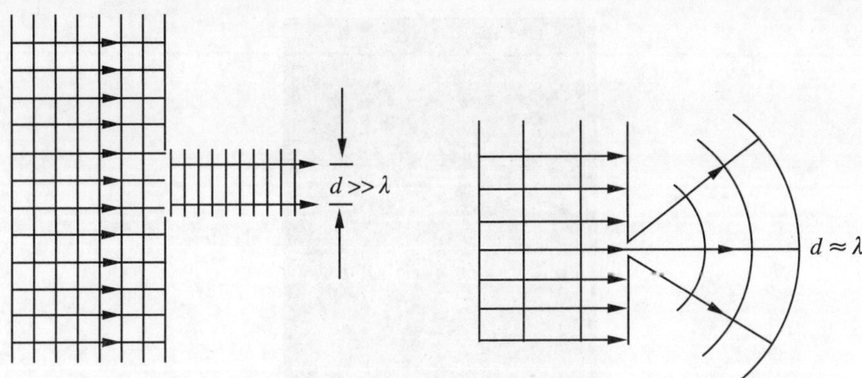

Figure 4.2 Diffraction pattern is pronounced if the dimension d of the diffracting obstacle is of the same order as the wavelength, λ, of the light used

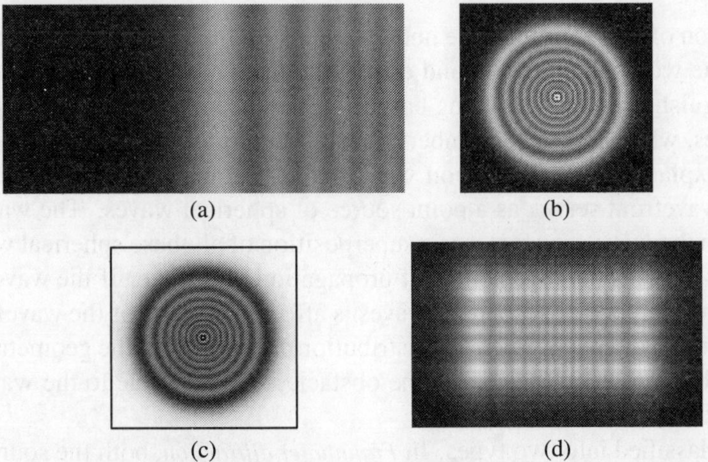

Figure 4.3 Diffraction patterns produced by several obstacles: (a) straight edge, (b) opaque disc, (c) circular aperture, (d) rectangular aperture (Schematic patterns based on http://www.ph.unimelb.edu.au/~ssk/fresnel/edge.html, http://www2.andrews.edu/~murdick/write/fresnel/fresnel.html, http://www.physics.udel.edu/~mfdecamp/lectures/Physics313_08S/Labs/313Lab07FresnelDiffraction.pdf and http://www.geocities.com/piratord/Difra/difra.html)

Diffraction is shown not just by light waves but by all kinds of waves. The wavelength of the waves and the dimensions of the diffracting obstacle decide how observable the diffraction effects are. Thus the diffraction of sound waves, due to their long wavelength (of the order of metres), makes sound audible across closed rooms. Diffraction of water waves can be seen when water gushes out through a small opening (see Fig. 4.4). Diffraction of light waves is not so apparent due to their small wavelength (4000–7000Å). X-rays with too small a wavelength (1–100Å) are even more difficult to diffract, but are able to reveal minute internal structure of solids as atoms themselves serve as diffracting obstacles (see Chapter 9).

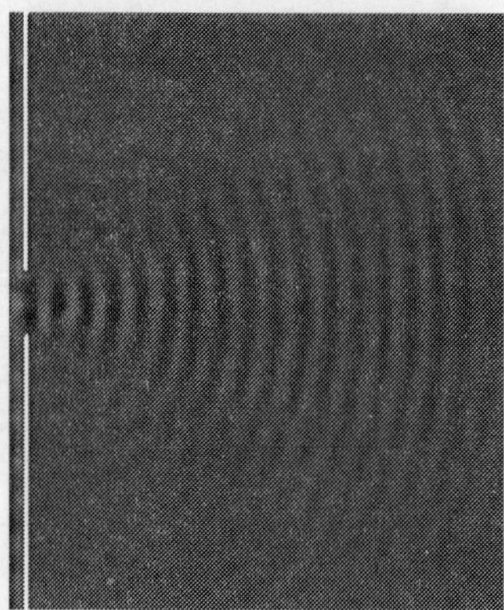

Figure 4.4 Diffraction of water waves (adapted from PIRA Paul Falstad applets from http://www.falstad.com/ripple/)

The phenomenon of diffraction can be understood along the same lines as interference. In fact, as noted above, the words interference and diffraction have been used interchangeably. They are sometimes distinguished on the following basis: interference refers to the superposition of waves from a few sources, whereas a large number of waves superpose to produce diffraction.

A theoretical explanation of diffraction was offered by Fresnel using Huygen's wave model. Each point on a wavefront serves as a point source of spherical waves. The wave disturbance at any point to which the wave propagates is a superposition of all these spherical waves as they pass through this point. This is the mechanism of propagation of a wave. If the wave now encounters an obstacle, this interference of secondary waves is affected as parts of the wavefront are blocked. This leads to different patterns of intensity distribution depending on the geometry of the situation. The pattern is stable, provided the size of the obstacle is comparable to the wavelength of light. This pattern is known as a *diffraction pattern*.

Diffraction is classified into two types. In *Fraunhofer diffraction*, both the source and the screen are infinitely far from the obstacle. In *Fresnel diffraction*, either the source or the screen, or both, are at a finite distance from the obstacle. Figure 4.5 shows these two types of diffraction. The Fresnel–Huygen model can be used to analyse both types of diffraction. However, treatment of Fresnel diffraction is mathematically far more complex than the Fraunhofer case. We will limit ourselves to the discussions of Fraunhofer diffraction in this book. We study here the diffraction patterns produced by a single slit, a double slit and an array of N slits. Chart 4.1 illustrates a comparison of these cases.

As diffraction is due to the superposition of a large number of waves, we use the mathematical treatment of Section 3.2. For a special case where the amplitudes (A_i, A_j) of the interfering waves are equal (A_0) and the phase angles are $\alpha_1 = 0, \alpha_2 = \alpha, \alpha_3 = 2\alpha - - - - -, \alpha_n = (n-1)\alpha$, we get (see Ex. 4.1),

$$A = A_0 \frac{\sin(n\alpha/2)}{\sin(\alpha/2)} \tag{4.1}$$

and the corresponding resultant intensity is given by

$$I = \frac{i_0 \sin^2(n\alpha/2)}{\sin^2(\alpha/2)} \tag{4.2}$$

i_0 being the intensity of each secondary source.

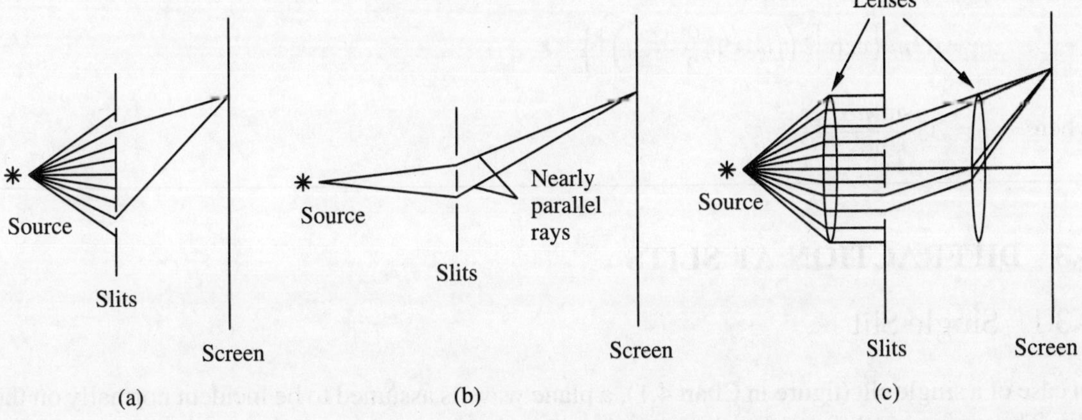

Figure 4.5 (a) Fresnel diffraction, (b) and (c) Fraunhofer diffraction. In (c) the condition that source and screen are at large distances from the obstacle is achieved by means of lenses.

EXAMPLE 4.1

Superposition of n waves

Find out the resultant of the waves $y_1 = A_0 \sin(\alpha_1 - \omega t), y_2 = A_0 \sin(\alpha_2 - \omega t), \ldots, y_n = A_0 \sin(\alpha_n - \omega t)$, if $\alpha_1 = 0, \alpha_2 = \alpha, \alpha_3 = 2\alpha, \ldots, \alpha_n = (n-1)\alpha$.

Solution

Using the superposition principle,

$$\begin{aligned} y &= y_1 + y_2 + \ldots + y_n \\ &= A_0[\sin(\alpha_1 - \omega t) + \sin(\alpha_2 - \omega t) + \ldots + \sin(\alpha_n - \omega t)] \end{aligned}$$

$$= A_0[\sin(0 - \omega t) + \sin(\alpha - \omega t) + \ldots + \sin((n-1)\alpha - \omega t)]$$

$$= A_0 Im\{\exp(-i\omega t)[1 + \exp(i\alpha) + \ldots + \exp i(n-1)\alpha]\}$$

where Im denotes the imaginary part of the exponential function [$\exp(ix) = \cos x + i \sin x$; $\cos x$ is the real part and $\sin x$ is the imaginary part of the exponential function].

Thus, $$y = A_0 Im\left[\exp(-i\omega t)\frac{1 - \exp(in\alpha)}{1 - \exp(i\alpha)}\right]$$

$$= A_0 Im\left\{\exp(-i\omega t)\frac{\exp(in\alpha/2)[\exp(-in\alpha/2) - \exp(in\alpha/2)]}{(\exp i\alpha/2)[\exp(-i\alpha/2) - \exp(i\alpha/2)]}\right\}$$

$$= A_0 Im\left\{\exp\left[i\left((n-1)\frac{\alpha}{2} - \omega t\right)\right] \times \frac{\sin(n\alpha/2)}{\sin(\alpha/2)}\right\}$$

$$= A Im\left\{\exp\left[i\left((n-1)\frac{\alpha}{2} - \omega t\right)\right]\right\}$$

where $A = A_0 \dfrac{\sin(n\alpha/2)}{\sin(\alpha/2)}$

4.3 DIFFRACTION AT SLITS

4.3.1 Single Slit

In case of a single slit (figure in Chart 4.1), a plane wave is assumed to be incident normally on the slit. The wave gets diffracted and in order to calculate the resulting diffraction pattern we divide the slit into a large number of equally spaced point sources $A_1, A_2, \ldots, A_n$. Let Δ be the distance between two adjacent points. The slit width can thus be given as

$$d = (n-1)\Delta \tag{4.3}$$

The waves emitted by these sources will have approximately the same amplitudes when they reunite at point P on the screen as the distance of P from the slit is very large compared to d. However, their phases differ as they travel different distances before reaching P. The phase difference between the two waves coming from adjacent sources, say A_1 and A_2, is given by

$$\alpha = \frac{2\pi}{\lambda}\Delta \sin\theta \tag{4.4}$$

where $\Delta \sin\theta$ is the path difference $A_2 A_2'$ between the two waves, θ being the angular position of P relative to the direction of the incident wave at the centre of the slit. The resultant intensity at P due to the sources $A_1, A_2, \ldots A_n$ is obtained using Eqs. 4.2 and 4.4 as

$$I = i_0 \frac{\sin^2[(n\pi/\lambda)\Delta \sin\theta]}{\sin^2[(\pi/\lambda)\Delta \sin\theta]} \tag{4.5}$$

CHART 4.1 DIFFRACTION IN DIFFERENT CASES

	Figure	Diffraction pattern	Expression for intensity, I and conditions for extrema
Single slit	**Figure 1** 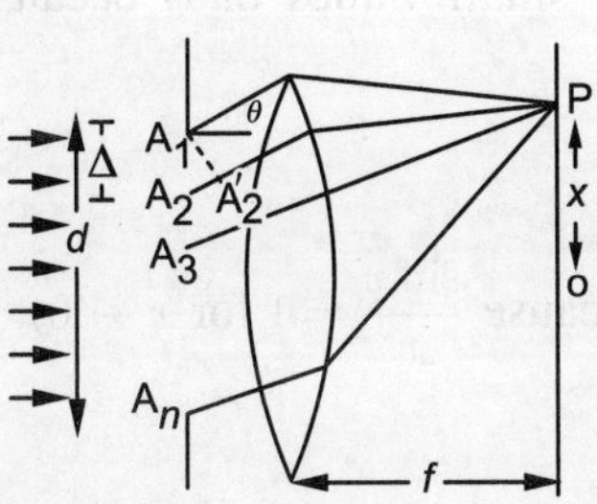	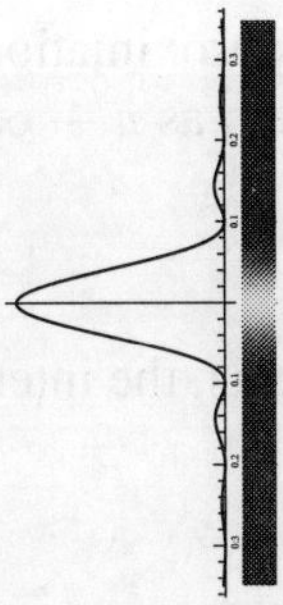	$I = \frac{I_0 \sin^2\beta}{\beta^2}$ $\beta = \left(\frac{\pi}{\lambda}\right) d\sin\theta$ Maxima: $\tan\beta = \beta$ Minima: $d\sin\theta = m\lambda$
Double slit	**Figure 2** 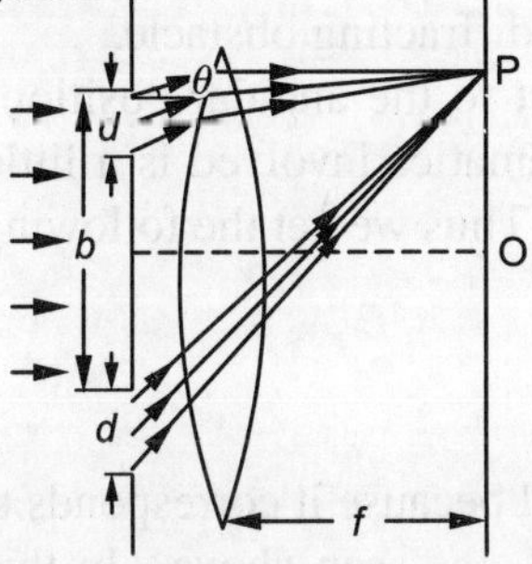	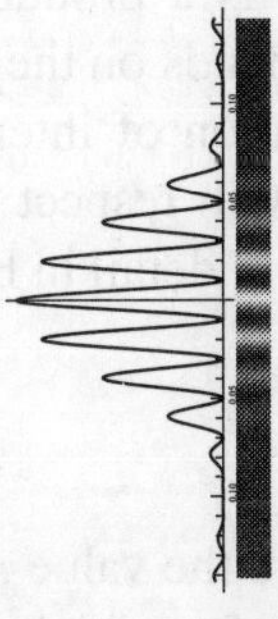	$I = \left[4I_0 \frac{\sin^2\beta}{\beta^2}\right][\cos^2\gamma]$ $\gamma = \left(\frac{\pi}{\lambda}\right) b\sin\theta$ Maxima: $d\sin\theta$ $= \left(m + \frac{1}{2}\right)\lambda$ and $b\sin\theta = p\lambda$ Minima: $d\sin\theta = m\lambda$ or $b\sin\theta = \left(p + \frac{1}{2}\right)\lambda$
Multiple slits	**Figure 3** 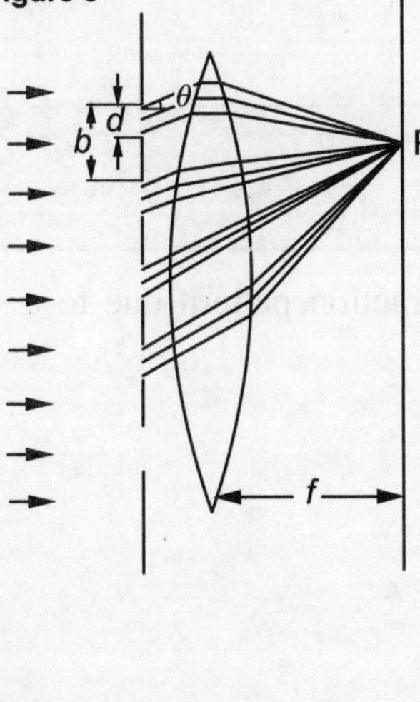	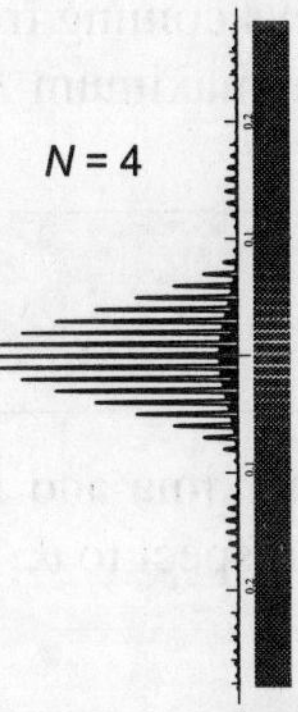 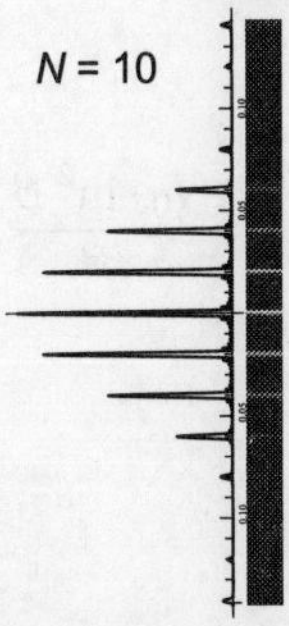 $N = 4$ $N = 10$	$I = \frac{I_0 \sin^2\beta}{\beta^2} \times \frac{\sin^2(N\gamma)}{\sin^2\gamma}$ Maxima: $b\sin\theta = m\lambda$ Minima: $d\sin\theta = m\lambda$ or $Nb\sin\theta = p\lambda$, $p \neq N, 2N, \ldots$

Now, in the limit $n \to \infty$ and $\Delta \to 0$, such that we always have $n\Delta = d$ (4.6)

Eq. 4.5 becomes $I = i_0 \dfrac{\sin^2[(\pi/\lambda)d\sin\theta]}{[(\pi/n\lambda)d\sin\theta]^2}$ (4.7)

where we have used the approximation $\sin\phi \sim \phi$, for small values of ϕ because the angle $[(\pi/n\lambda)d\sin\theta]$ approaches zero as $n \to \infty$. Substituting,

$$I_0 = (nA_0)^2 = n^2 i_0 \tag{4.8}$$

for the forward intensity, that is, the intensity at $\theta = 0$ (because $\dfrac{\sin x}{x} = 1$ for $x \to 0$), we get,

$$I = I_0 \frac{\sin^2[(\pi/\lambda)d\sin\theta]}{[(\pi/\lambda)d\sin\theta]^2} \tag{4.9}$$

Thus the resultant intensity is a product of two terms: the forward intensity and the angular modulation factor which depends on the geometry of the diffracting obstacle.

We can analyse the variation of intensity with respect to the angular position on the screen by differentiating Eq. 4.9 with respect to θ. The mathematics involved is a little complicated. However, this is worked out in detail in Exs. 4.2 and 4.3. Thus we get the following condition for minimum

$$d\sin\theta = m\lambda \tag{4.10}$$

where, $m = \pm 1, \pm 2, \ldots$. Here the value $m = 0$ is excluded because it corresponds to the direction of incidence, that is, $\theta = 0$, for which the intensity is I_0, as seen above. In this case there is no path difference for the rays coming from different points, and the interference is constructive, producing a central principal maximum.

EXAMPLE 4.2

Conditions for extrema in single slit diffraction

Obtain the conditions for maxima and minima in a diffraction pattern due to a single slit by differentiating Eq. 4.9 with respect to θ.

Solution

We have

$$I = I_0 \frac{\sin^2[(\pi/\lambda)d\sin\theta]}{[(\pi/\lambda)d\sin\theta]^2} = \frac{I_0 \sin^2\beta}{\beta^2}, \text{ where } \beta = \frac{\pi d\sin\theta}{\lambda}$$

Thus $\dfrac{dI}{d\theta} = \left(\dfrac{dI}{d\beta}\right)\left(\dfrac{d\beta}{d\theta}\right)$

$$= I_0 \left[\frac{2\sin\beta\cos\beta}{\beta^2} - \frac{2\sin^2\beta}{\beta^3}\right]\left[\frac{\pi d\cos\theta}{\lambda}\right]$$

$$= I_0 \left[\frac{\pi d\cos\theta}{\lambda} \times \frac{\sin 2\beta}{\beta^3}\right](\beta - \tan\beta)$$

$$= I_0 \left(\frac{\lambda^2}{\pi^2 d^2} \times \frac{\cot\theta}{\sin^2\theta}\right)\sin 2\beta(\beta - \tan\beta)$$

The condition for extrema is that $dI/d\theta$ must vanish. Thus three possibilities arise:

(i) $\cot\theta = 0$, (ii) $\sin(2\beta) = 0$ and (iii) $\beta - \tan\beta = 0$

The first of these implies $\theta = \pi/2$, which is of no interest as it means the direction perpendicular to the direction of incidence. The second, $\sin 2\beta = 0$, is satisfied when

$$\frac{2\pi d\sin\theta}{\lambda} = m\pi$$

or $\quad \sin\theta = \dfrac{m\lambda}{2d}$ for any integer m.

When this condition is satisfied the last factor in $dI/d\theta$ becomes $[(m\pi/2) - \tan(m\pi/2)]$.

Now, this factor is $m\pi/2$ when m is even. However, it becomes infinite when m is odd. This means that when m is odd, $dI/d\theta$ becomes $0 \times \infty$, an indeterminate form. The limit of this form is infinite because $\tan\beta \to \infty$ faster than $\sin 2\beta \to 0$ as the angle approaches $\pi/2$. Consequently, $dI/d\theta \neq 0$ when $\sin\theta = m\lambda/2d$ for odd values of m, and m should be restricted to even values for $dI/d\theta$ to vanish under the condition $\sin\theta = m\lambda/2d$.

Writing $m = 2p$, $dI/d\theta$ vanishes whenever $\sin\theta = p\lambda/d$ for any integer p.

The third possibility for $dI/d\theta$ to vanish is

$$\beta = \tan\beta \quad \text{or} \quad \frac{\pi d\sin\theta}{\lambda} = \tan\left(\frac{\pi d\sin\theta}{\lambda}\right)$$

The solution of this equation, that is, angles for which the equation is satisfied, are generally obtained graphically.

Therefore the intensity extrema of interest appear whenever

$$\sin\theta = \frac{p\lambda}{d}, \text{ where } p \text{ is any integer,} \quad \text{or} \quad \frac{\pi d\sin\theta}{\lambda} = \tan\left(\frac{\pi d\sin\theta}{\lambda}\right)$$

The first of these conditions gives intensity minima except when $p = 0$, which corresponds to the principal maximum in the forward direction $\theta = 0$. The second condition gives intensity maxima. This is shown by evaluating the second derivative in Ex. 4.3.

EXAMPLE 4.3
To find conditions for maxima and minima by evaluating $d^2I/d\theta^2$

Show that the condition $\beta - \tan\beta = 0$ corresponds to maxima other than the principal maximum. Also show that the condition $\sin 2\beta = 0$ corresponds to minima.

Solution

From Ex. 4.2 we have,

$$\frac{dI}{d\theta} = I_0\left(\frac{\pi d\cos\theta}{\lambda} \times \frac{\sin 2\beta}{\beta^3}\right)(\beta - \tan\beta), \text{ where } \beta = \frac{\pi d \sin\theta}{\lambda}$$

Differentiating this equation with respect to θ, we get,

$$\frac{d^2I}{d\theta^2} = I_0\left[\left(-\frac{\pi d\sin\theta}{\lambda} \times \frac{\sin 2\beta}{\beta^3}\right)(\beta - \tan\beta) + \left(\frac{\pi^2 d^2\cos^2\theta}{\lambda^2}\right)\left(\frac{2\cos 2\beta}{\beta^3} - \frac{3\sin 2\beta}{\beta^4}\right)(\beta - \tan\beta) + \left(\frac{\pi^2 d^2\cos^2\theta}{\lambda^2} \times \frac{\sin 2\beta}{\beta^3}\right)(1 - \sec^2\beta)\right]$$

where we have used $\dfrac{d\beta}{d\theta} = \dfrac{\pi d\cos\theta}{\lambda}$

Now,

$$\left.\frac{d^2I}{d\theta^2}\right|_{\beta=\tan\beta} = \left(\frac{\pi^2 d^2\cos^2\theta}{\lambda^2} \times \frac{\sin 2\beta}{\tan^3\beta}\right)(-\tan^2\beta)$$

$$= -\left(\frac{\pi^2 d^2\cos^2\theta}{\lambda^2} \times \frac{\sin 2\beta}{\tan\beta}\right)$$

$$= -2\left(\frac{\pi^2 d^2\cos^2\theta}{\lambda^2}\right)\cos^2\beta < 0$$

Hence the condition $\beta - \tan\beta = 0$ is the condition for maxima.

Now,

$$\left.\frac{d^2I}{d\theta^2}\right|_{\sin 2\beta = 0, \beta\neq 0} = \left.\left(\frac{\pi^2 d^2\cos^2\theta}{\lambda^2} \times \frac{2\cos 2\beta}{\beta^3}\right)(\beta - \tan\beta)\right|_{\beta = m\pi,\ \text{m is even, m} \neq 0}$$

$$= \left(\frac{\pi^2 d^2 \cos^2\theta}{\lambda^2} \times \frac{2}{m^3\pi^3}\right)(m\pi)$$

$$= \left[\frac{2d^2 \cos^2\theta}{m^2\lambda^2}\right] > 0$$

Hence, $\sin 2\beta = 0$, with $\beta = m\pi$, m is even, and $m \neq 0$ gives the condition for minima.

Equation 4.10 can also be derived in an alternative way using simple qualitative arguments. If we divide the slit into two halves as shown in Fig. 4.6 and consider point sources A_1, $A_2, \ldots$ and B_1, $B_2, \ldots$ in the upper and lower half respectively, then the path difference between waves from A_1 and B_1; A_2 and B_2, and so on, will be $(d/2)\sin\theta$. Hence, point P will be dark if $(d/2)(\sin\theta) = (2n+1)\lambda/2, n = 0, 1, 2\ldots$ or $d\sin\theta = (2n+1)\lambda$. If we divide the slit into $4, 8\ldots$ parts, we get the conditon $(d/2^p)(\sin\theta) = (2n+1)\lambda/2, p = 1, 2\cdots$ and $n = 0, 1, 2\ldots$ or $d\sin\theta = m\lambda$, where m is any non-zero integer, for dark fringes.

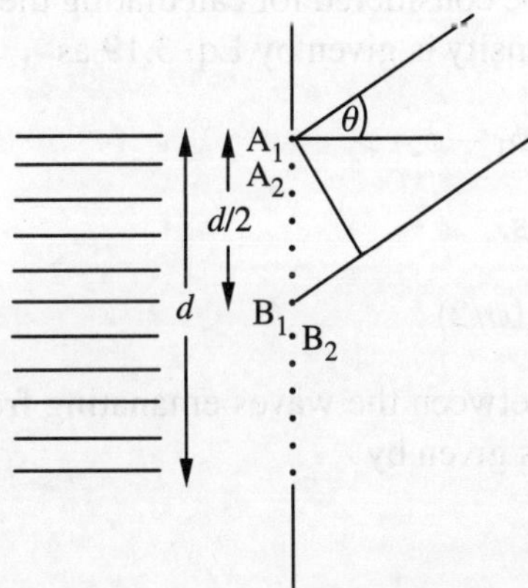

Figure 4.6 Division of slit into two halves for deriving the condition for the first minima

Maximum intensity is obtained from Eq 4.9, using the condition (see Exs. 4.2 and 4.3),

$$\tan\left(\frac{\pi d}{\lambda}\right)\sin\theta = \left(\frac{\pi d}{\lambda}\right)\sin\theta \tag{4.11}$$

Equation 4.11 is a transcendental equation, the roots of which are obtained by determining the points of intersection of the curves $y = (\pi d/\lambda)\sin\theta$ and $y = \tan[(\pi d/\lambda)\sin\theta]$. The roots are $(\pi d/\lambda)\sin\theta = 1.430\pi$, $(\pi d/\lambda)\sin\theta = 2.459\pi$, and so on.

Thus the diffraction pattern due to a single slit consists of a series of minima and maxima on both sides of the central maximum (corresponding to $d\sin\theta = 0$). The fringe pattern due to diffraction differs from that due to interference (see Chart 4.1) in that the diffraction fringes are not equally spaced, whereas those in interference are equally spaced. In diffraction the intensity of the central maximum is maximum and decreases on either side as the order of diffraction increases, whereas in case of interference the maxima are of the same intensity.

If x is the distance of the first maximum from the centre of the screen, and f is the focal length of the converging lens (see Figure 1 of Chart 4.1), then the width of the central maximum is given by

$$2x = \frac{2f\lambda}{d} \tag{4.12}$$

because $\tan\theta = x/f$ and $\tan\theta = \sin\theta$ for small angles, and hence $\sin\theta = x/f = \lambda/d$. Thus the central maximum is wider for thinner slits and larger wavelengths. The divergence angle, $\Delta\theta$, which contains most of the light energy is given by

$$\Delta\theta \approx \frac{\lambda}{d} \tag{4.13}$$

4.3.2 Double Slit

If a two slit arrangement is used (Figure 2 of Chart 4.1) then both effects—single slit diffraction and two slit interference—are to be considered for calculating the resultant on the screen. For two slit interference, the resultant intensity is given by Eq. 3.19 as

$$I = I_1 + I_2 + 2\sqrt{I_1 I_2} < \cos(\phi_1 - \phi_2) >$$

which in the present case becomes,

$$I = 2I_0'[1 + \cos\phi] = 4I_0'\cos^2(\phi/2) \tag{4.14}$$

where ϕ is the phase difference between the waves emanating from the two slits. If the distance between the two slits is b then ϕ is given by

$$\phi = \frac{2\pi}{\lambda} b\sin\theta \tag{4.15}$$

Here I_0' is the intensity due to the individual slit decided by the diffraction effect and given by Eq. 4.9. Thus the resultant intensity in case of a two-slit arrangement is given by

$$I = \frac{4I_0\sin^2[(\pi/\lambda)d\sin\theta]\cos^2[(\pi/\lambda)b\sin\theta]}{[(\pi/\lambda)d\sin\theta]^2} \tag{4.16}$$

Thus I is the product of two terms. The first term represents the diffraction pattern due to a single slit and the second term represents the interference pattern due to two point sources. Equation 4.16 tells us that the minimum in intensity can be observed if $(\pi/\lambda)d\sin\theta = m\pi, m = 1, 2, \ldots$ or if $(\pi/\lambda)b\sin\theta = (p + 1/2)\pi$, p being an integer.

Thus, the conditions for minima are

$$d\sin\theta = m\lambda \tag{4.17}$$

or $$b\sin\theta = \left(p + \frac{1}{2}\right)\lambda \tag{4.18}$$

The conditions for maxima are

$$d \sin\theta = \left(m + \frac{1}{2}\right)\lambda \tag{4.19}$$

and $b \sin\theta = p\lambda$ (4.20)

As b is generally many times the value of d, these equations predict that many interference maxima and minima will be enveloped by two diffraction minima on either side of the central maximum (Ex. 4.4) and also by the secondary diffraction maxima on both sides. This prediction matches with the observed diffraction–interference pattern for a double slit arrangement (see Chart 4.1).

EXAMPLE 4.4

Number of interference minima in a double slit diffraction pattern

In a double slit arrangement where the slit width is equal to 8.8×10^{-2} mm and the separation between two slits is equal to 0.7 mm, how many interference minima will occur between the two diffraction minima on either side of the central maximum? (Assume $\lambda = 6328$ Å)

Solution

Here $d = 8.8 \times 10^{-3}$ cm and $b = 0.07$ cm.

Thus $\dfrac{\lambda}{d} = \dfrac{(6328 \times 10^{-8})}{(8.8 \times 10^{-3})} = 7.19 \times 10^{-3}$

and $\dfrac{\lambda}{b} = \dfrac{(6328 \times 10^{-8})}{0.07} = 0.904 \times 10^{-3}$

Now, the condition for first diffraction minima is $d \sin\theta = \lambda$

Thus $\theta = \sin^{-1}(\lambda/d) = \sin^{-1}(7.19 \times 10^{-3}) = 0.412°$

The condition for interference minima is given by,

$$\sin\theta = \frac{(p + 1/2)\lambda}{b}$$

or $\theta = \sin^{-1}\left[\left(p + \dfrac{1}{2}\right) \times 0.904 \times 10^{-3}\right]$

Substituting $p = 0, 1, 2, 3, 4, 5, ,6$ and 7, we get $\theta_1, \theta_2, \theta_3, \theta_4, \theta_5, \theta_6, \theta_7, \theta_8$ as 0.026°, 0.078°, 0.129°, 0.181°, 0.233°, 0.285°, 0.337° and 0.388°, respectively. For $p \geq 8$, the value of θ exceeds 0.412°.

Hence there will be 16 minima between the first two diffraction minima.

4.3.3 Multiple Slits

If a multiple-slit arrangement consisting of N parallel slits, each of width d, with b as the distance between consecutive slits (Figure 3 of Chart 4.1), is used, then the above analysis can be extended using Eq. 4.2 to show that the resultant intensity is given by

$$I = \frac{I_0 \sin^2[(\pi/\lambda)d\sin\theta]}{[(\pi/\lambda)d\sin\theta]^2} \times \frac{\sin^2[(N\pi/\lambda)b\sin\theta]}{\sin^2[(\pi/\lambda)b\sin\theta]} \tag{4.21}$$

Thus I is the product of two terms. The first term represents the diffraction pattern due to a single slit and the second term represents the interference pattern due to N equally spaced point sources. This equation reduces to Eq. 4.9 for $N = 1$, representing a single slit diffraction pattern and to Eq. 4.16 for $N = 2$, representing a pattern for a two-slit arrangement (see Ex. 4.5).

EXAMPLE 4.5

Special cases of Equation 4.21

Show that Eqs. 4.9 and 4.16 are special cases of Eq. 4.21 for $N = 1$ (that is, single slit) and $N = 2$ (that is, double slit) respectively.

Solution

Equation 4.21 is

$$I = \frac{I_0 \sin^2[(\pi/\lambda)d\sin\theta]}{[(\pi/\lambda)d\sin\theta]^2} \times \frac{\sin^2[(N\pi/\lambda)b\sin\theta]}{\sin^2[(\pi/\lambda)b\sin\theta]}$$

when $N = 1$, the second term equals 1 and so

$$I = I_0 \frac{\sin^2[(\pi/\lambda)d\sin\theta]}{[(\pi/\lambda)d\sin\theta]^2}, \text{ which is Eq. 4.9.}$$

When $N = 2$, the second term

$$= \frac{\sin^2[(2\pi/\lambda)b\sin\theta]}{\sin^2[(\pi/\lambda)b\sin\theta]}$$

$$= \frac{\{2\sin[(\pi/\lambda)b\sin\theta]\cos[(\pi/\lambda)b\sin\theta]\}^2}{\sin^2[(\pi/\lambda)b\sin\theta]} = 4\cos^2[(\pi/\lambda)b\sin\theta]$$

Thus we get

$$I = 4I_0\cos^2[(\pi/\lambda)b\sin\theta]\frac{\sin^2[(\pi/\lambda)d\sin\theta]}{[(\pi/\lambda)d\sin\theta]^2}, \text{ which is the same as Eq. 4.16.}$$

Equation 4.21 gives the maximum value for

$$\frac{\pi}{\lambda}\, b\sin\theta = m\pi \tag{4.22}$$

This is because

$$\lim_{\gamma\to m\pi}\frac{\sin(N\gamma)}{\sin\gamma} = \lim_{\gamma\to m\pi}\frac{N\cos(N\gamma)}{\cos\gamma} = \pm N$$

where $\gamma = \left(\frac{\pi}{\lambda} b\sin\theta\right)$

and so the interference term has maximum value equal to N^2. These maxima are called the principal maxima because the resultant intensity here is N^2 times the intensity due to each slit, that is, the resultant amplitude is the sum of the amplitudes of the waves from all the N slits. Equation 4.21 gives the minimum value for

$$\left(\frac{\pi}{\lambda}\right) d\sin\theta = m\pi \quad\text{or}\quad \left(\frac{N\pi}{\lambda}\right) b\sin\theta = p\pi$$

($p \neq N, 2N, \ldots$ as it leads to the condition for principal maxima given in Eq. 4.22), that is,

$$d\sin\theta = m\lambda, m = 1, 2, \ldots \tag{4.23}$$

or $$Nb\sin\theta = p\lambda, p \neq N, 2N, \ldots \tag{4.24}$$

Thus minimum intensity will be observed for angles given by

$$b\sin\theta = \lambda/N, 2\lambda/N, \ldots (N-1)\lambda/N, (N+1)\lambda/N, (N+2)\lambda/N, \ldots$$

$$\ldots, (2N-1)\lambda/N, (2N+1)\lambda/N, (2N+2)\lambda/N, \ldots \tag{4.25}$$

and there will be $(N - 1)$ minima between two principal maxima. Between two consecutive minima there will be maxima that are known as secondary maxima. Secondary maxima will be much weaker in intensity compared to the principal maxima. If the condition for interference maxima (Eq. 4.22) and the diffraction minima (Eq. 4.23) are satisfied simultaneously, then that maximum will not appear (or will be very weak in intensity). These are referred to as missing orders in the diffraction pattern.

A multiple slit diffraction pattern can be studied in the laboratory using an arrangement known as an optical diffraction grating. Experiment 4.1 demonstrates the use of a grating in the determination of the wavelengths of spectral lines of sodium.

EXPERIMENT 4.1 GRATING

Aim

To determine the wavelengths of the sodium doublet using a plane transmission grating

Apparatus

Sodium vapour lamp, spectrometer, optical diffraction grating, reading lens, spirit level

Formula

$$(a + b) \sin \theta_n = n\lambda \tag{4.26}$$

where $(a + b)$ is the grating element given by

$$a + b = (2.54/N) \text{ cm} \tag{4.27}$$

where N is the number of lines ruled per inch on the grating, θ_n is the angle of diffraction of order n and λ is the wavelength of sodium light.

Theory

Diffraction grating: An arrangement of a large number of parallel equidistant slits of the same width is called a *diffraction grating*. If the slits are made on a glass plate or on a sheet of any transparent material we get a *transmission grating*. If a mirror is used in place of a transparent plate a *reflection grating* is obtained. A grating is made by ruling lines using a diamond scriber. The lines so ruled are opaque to light, whereas the spaces between lines serve as slits which transmit and diffract light. Laboratory gratings are the replicas of the original one. They are prepared by pouring a solution of cellulose acetate on the original grating and allowing it to harden, thus forming a strong thin film, detachable from the parent grating. This thin film is then mounted on a glass plate. The distance between centres of two adjacent transparent regions is called the *grating period* or *grating constant*. The grating period, $a + b$, is thus equal to the sum of the slit width a and the width of the ruled opaque region b next to that slit. It should be noted that the symbol b was used in Section 4.3 for grating constant.

Diffraction pattern due to a grating: The theory of diffraction due to a grating has already been covered under multiple slit diffraction in Section 4.3. The positions of the principal maxima are thus given by (see Eq. 4.22)

$$(a + b) \sin \theta_n = n\lambda; n = 0, \pm 1, \pm 2, \cdots$$

This equation is known as grating equation and signifies that the waves emanating from each of the slits add in phase along the direction θ_n and thus produce a large intensity. The zeroth order principal maxima occurs at $\theta_n = 0$ and is the brightest. It is flanked on both sides by maxima of higher orders. The maximum number of orders that are observable with a particular grating is given by $(a + b)/\lambda$, as the maximum value of $\sin \theta_n$ in Eq. 4.26 is 1 (see Ex. 4.6). Along the directions other than those at which the principal maxima are seen, either the condition for minima is satisfied or that for the secondary maxima, which is too weak to be observed, is satisfied.

Characteristics of a grating

(i) Dispersion: If we differentiate Eq. 4.26 we get

$$(a + b) \cos \theta_n \Delta\theta_n = n\Delta\lambda.$$

Hence $$\frac{\Delta\theta_n}{\Delta\lambda} = \frac{n}{(a+b)\cos\theta_n} \tag{4.28}$$

$\Delta\theta_n/\Delta\lambda$ is called the *dispersion* and is a measure of the angular separation produced by a grating between two incident monochromatic waves whose wavelengths differ by a small interval $\Delta\lambda$.

(ii) Resolving power: It is defined as

$$R = \lambda/\Delta\lambda \tag{4.29}$$

where λ is the mean wavelength of the two spectral lines that can be just resolved (that is, just separated) and $\Delta\lambda$ is the difference between these wavelengths. Smaller the value of $\Delta\lambda$, closer the wavelengths that can be resolved, and higher the resolving power. Using the Rayleigh's criterion of resolution (see Section 4.4 and Ex. 4.7) it can be shown that

$$\Delta\theta_n = \frac{\lambda}{[N(a+b)cos\theta_n]} \tag{4.30}$$

Using Eqs. 4.28, 4.29 and 4.30, we get,

$$R = nN \tag{4.31}$$

Equations 4.28 and 4.31 tell us that both dispersion and resolving power are more for higher orders. Whereas a smaller grating period increases the dispersion of a grating, a larger number of rulings increases its resolving power. Example 4.8 illustrates the comparison of dispersion and resolving power of gratings.

Description of the apparatus: Chart 4.2 describes the construction and various adjustments of a spectrometer. Figure 4.7 shows the experimental set up. Monochromatic light emitted by the sodium vapour lamp is incident normally on the grating and the resulting diffraction pattern is seen through the spectrometer.

Procedure

A. Initial arrangement of spectrometer

1. The sodium lamp is switched on and the light is allowed to fall on the slit of the collimator.
2. The telescope and collimator are brought in one line and the light coming through the collimator is seen through the telescope.
3. The width of the collimator slit is adjusted so that the slit is narrow but intense.
4. The plane of the spectrometer is levelled using a spirit level.
5. The focussing arrangement of telescope and collimator is used to obtain a sharp and intense image of the slit on the vertical cross wire of the eyepiece.

B. Adjustment of spectrometer and grating for normal incidence

1. The telescope is rotated through 90°. This is achieved by taking the initial reading of the telescope, say x, when the cross wire coincides with the image of the slit, and then rotating the telescope either clockwise or anti-clockwise to obtain the reading $(x + 90)°$ or $(x - 90)°$. The telescope is clamped in this position (Fig. 4.8a).

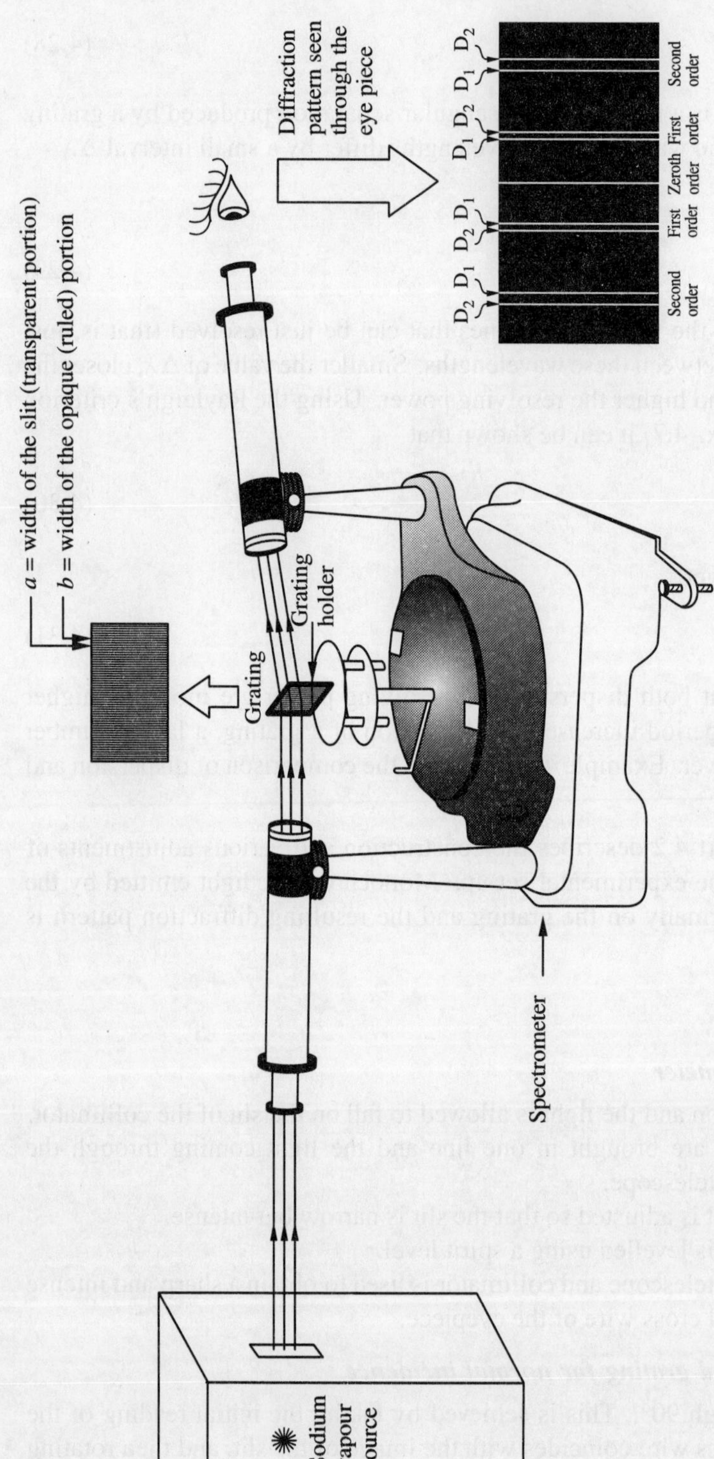

Figure 4.7 Experimental arrangement to study diffraction of light

2. Now the grating is placed on the prism table and adjusted in such a way that the reflected image of the slit formed by the grating is seen through the telescope. The grating is now making an angle of 45° with the direction of incident beam (Fig. 4.8b).
3. The grating is then rotated through 45° towards the telescope so that it is now perpendicular to the incident beam. This is achieved, as explained above, by noting the initial reading of the position of the prism table, say y, and rotating it to get the reading $(y - 45)$ or $(y + 45°)$ (Fig. 4.8c).
4. Once the normal incidence position is set, the prism table is clamped and care is taken not to disturb this position. The telescope is now brought to the original position (Fig. 4.8d).

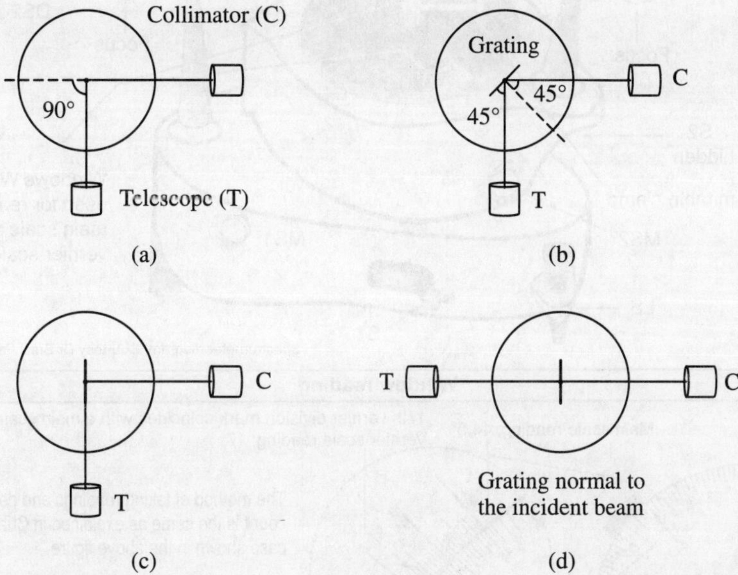

Figure 4.8 Setting of grating for normal incidence

C. Measurement of θ_n

1. The telescope is rotated slowly to the right till the image of the slit due to first order diffraction is observed. A closer examination of the image will reveal that it consists of two very closely spaced lines. (If two lines are not observed, some fine adjustment of the slit width and focussing is done to achieve it). These lines are the sodium doublet lines, referred to as D_1 and D_2 lines.
2. The cross wire is now made to coincide with the line D_1 and the readings of both the windows are noted.
3. Using the vernier adjustment of the telescope the cross wire is now shifted on line D_2, and again the two windows are read.
4. The telescope is further rotated to the right to obtain the second order spectrum. The separation between D_1 and D_2 is more in this order compared to that in the first order but the intensities of the lines are weaker.

CHART 4.2 SPECTROMETER

A spectrometer is an optical instrument which is used to measure angles and thus enables us to study phenomena involving the deviation / dispersion of light. A spectrometer mainly consists of a collimator for producing a parallel beam of light, a prism table for placing a prism or a grating and a telescope for observing the spectrum. The prism table and the telescope can be rotated and their angles of rotation can be measured using the circular scale attached to them. The prism table also has a facility for rotating without changing the reading of the scale. Both collimator and telescope have an objective lens, an eye piece and a focussing arrangement.

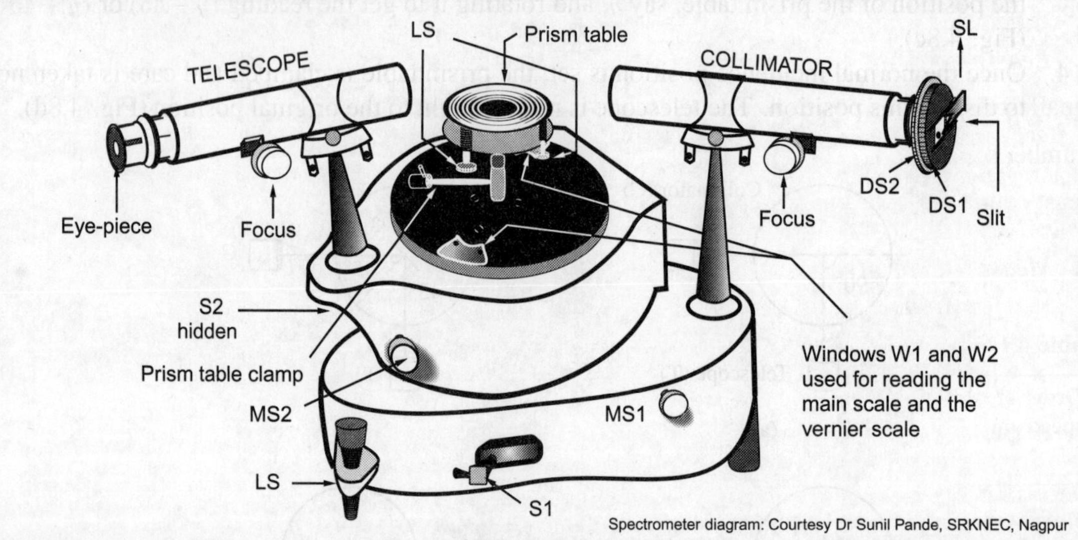

Spectrometer diagram: Courtesy Dr Sunil Pande, SRKNEC, Nagpur

Window reading

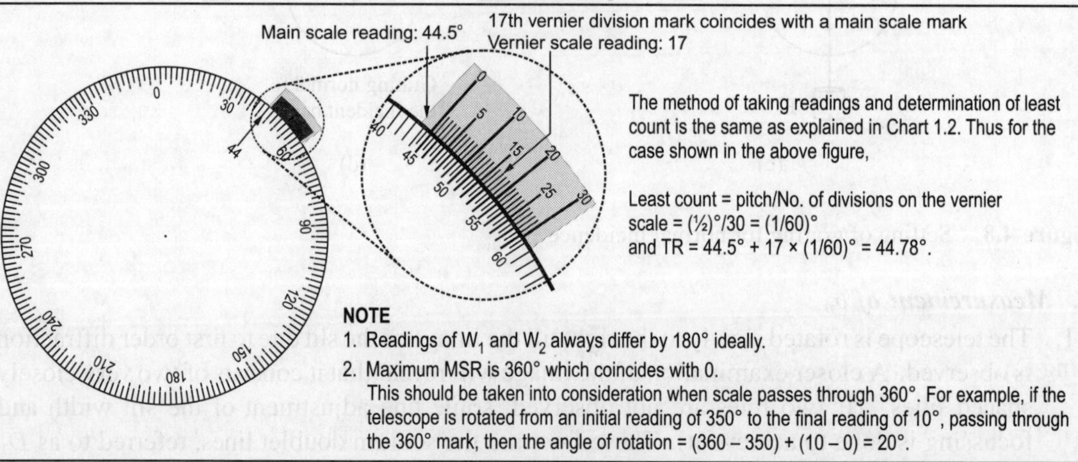

The method of taking readings and determination of least count is the same as explained in Chart 1.2. Thus for the case shown in the above figure,

Least count = pitch/No. of divisions on the vernier scale = (½)°/30 = (1/60)°
and TR = 44.5° + 17 × (1/60)° = 44.78°.

NOTE

1. Readings of W_1 and W_2 always differ by 180° ideally.
2. Maximum MSR is 360° which coincides with 0.
   This should be taken into consideration when scale passes through 360°. For example, if the telescope is rotated from an initial reading of 350° to the final reading of 10°, passing through the 360° mark, then the angle of rotation = (360 – 350) + (10 – 0) = 20°.

Various adjustments

1. The base of the spectrometer and the prism table are properly levelled using a spirit level and the screws LS provided for this purpose.
2. The orientation and the height of the slit are adjusted by the drum screw DS2 and the slide SL, respectively, to get a vertical slit of sufficient height. The slit width is adjusted using the drum screw DS1 to get a thin slit. However the intensity of light coming out of the slit is ensured to be sufficient for observation by positioning the collimator properly in front of the light source.
3. The collimation of a light beam, that is, rendering of an incident beam of light into a parallel beam is achieved by means of a rack and pinion arrangement provided with the collimator. This arrangement adjusts the distance between the slit and the collimating lens to bring the slit into the focal plane of this lens.
4. The collimator and the telescope are properly levelled so that the image of the slit is in the centre of the view.
5. If either the telescope or the windows are to be rotated by small angles, this can be done by locking their positions using S1 and S2 and using micromovement screws MS1 and MS2.

5. Readings of D_1 and D_2 are taken as explained for the first order.
6. The telescope is now rotated to the left of the central line (position of zeroth order principal maxima) and the entire procedure explained above for the right side is repeated.
7. The observations are tabulated in Table 4.1.

Observations and calculations

A. Least count of the spectrometer

Smallest division on main scale = $X°$

Number of divisions on vernier scale = N

Least count of the spectrometer = $(X/N)°$

B. Measurement of θ_n

Table 4.1

Order of spectrum	Line	Window	Spectrum left of the direct image			Spectrum right of the direct image			$P° \sim Q° = 2\theta_n$	Mean θ_n
			MSR	VSD	TR (P°)	MSR	VSD	TR (Q°)		
I	D_1	W_1								θ_1 for D_1
		W_2								
	D_2	W_1								θ_1 for D_2
		W_2								
II	D_1	W_1								θ_2 for D_1
		W_2								
	D_2	W_1								θ_2 for D_2
		W_2								

Using Eq. 4.26 the wavelength λ is calculated for various values of θ_n obtained in Table 4.1.

$$a + b = (2.54/N) \text{ cm}$$

Thus for the first order, $\lambda = (a + b) \sin\theta_1$

and for the second order, $\lambda = \dfrac{(a + b)\sin\theta_2}{2}$

Estimation of error

The error $\Delta\lambda$ in the estimation of λ can be determined by adopting the method outlined in Section 1.4.2.

$$\text{Thus, } \Delta\lambda = 2\left(\frac{a+b}{n}\right)\cos\theta_n \Delta\theta \tag{4.32}$$

where $\Delta\theta$ is the error in the reading of the spectrometer.

For example, for the following measured data,

$$a+b = 16933.3 \text{ Å}, \ \theta_1 = 20.29°, \ \theta_2 = 44.27°$$

Least count of the spectrometer = (1/60)°

$$\Delta\theta = \frac{1}{2}\left(\frac{1}{60}\right)^{\circ}$$

$$\text{Thus } \Delta\lambda = 2 \times 16933.3 \text{ Å} \times \cos 20.29° \times \frac{1}{2} \times \frac{1}{60} \times \frac{\pi}{180} = 4.62 \text{ Å for the first order}$$

$$\text{and } \Delta\lambda = 2 \times \frac{16933.3}{2} \text{ Å} \times \cos 44.27° \times \frac{1}{2} \times \frac{1}{60} \times \frac{\pi}{180} = 1.76 \text{ Å for the second order.}$$

$$\text{Now, } \lambda = (a+b)\sin\theta_1 = 16933.3 \text{ Å} \sin 20.29° = 5872 \text{ Å in the first order}$$

$$\text{and } \lambda = \frac{(a+b)\sin\theta_2}{2} = \frac{16933.3 \text{ Å} \sin(44.27°)}{2} = 5910 \text{ Å in the second order.}$$

Thus, in the first order, the relative error in $\lambda = \Delta\lambda/\lambda = 4.62 \text{ Å}/5872 \text{ Å} = 7.87 \times 10^{-4}$ and the percentage error is 0.08%;

in the second order the relative error in $\lambda = \Delta\lambda/\lambda = 1.76 \text{ Å}/5910 \text{ Å} = 3.0 \times 10^{-4}$ and the percentage error is 0.03%.

Results

The wavelengths of sodium lines are determined as follows.

Line	**Value of λ (in Å) determined in the present experiment**				**Standard value (in Å)**
	First order	% error	Second order	% error	
D_1	$\ldots \pm \ldots$		$\ldots \pm \ldots$		5890.0
D_2	$\ldots \pm \ldots$		$\ldots \pm \ldots$		5895.9

Discussion

This experiment demonstrates how the phenomenon of diffraction can be used for the determination of wavelength of light. The study of diffraction due to multiple slits (Section 4.3.3) forms the theoretical basis of this experiment.

In the following table we compare the theory and the experiment.

Theory	Experiment
Principal maxima should be obtained at angles $\theta = \sin^{-1}[n\lambda/(a+b)]$.	First order and second order maxima are observed at angles θ_1 and θ_2.
Equations 4.28 and 4.31 predict that dispersion and resolving power should be more for higher orders.	Spacing between D_1 and D_2 is more in second order than in first order.

It can be seen from this table that there is a qualitative agreement between the theory and the experiment. Other predictions of the theory—dispersion should increase with decrease in $(a+b)$, and resolving power should increase with increase in N—may also be verified in this experiment if one uses gratings with different values of $(a+b)$ and N. We observe that intensities of D_1 and D_2 are smaller in second order compared to that in first order.

The values used above in the estimation of error result in an error that is quite small—0.08% for first order and 0.03% for second order. However, major errors can creep in if the rulings on grating are non-uniform. For example, if the number of lines etched on the grating departs from 15,000 by ± 500, then the error which creeps into the calculation of λ is

$$\Delta\lambda = \frac{2.54\sin\theta_n}{n} \times \frac{1}{N^2} \times \Delta N = \frac{\Delta N}{N} \times \lambda = \frac{500}{15000} \times \lambda = \frac{\lambda}{30} = 196.3\ \text{Å} = 3.3\%$$

This error is much larger than the error due to instrumental limitations of the spectrometer.

EXAMPLE 4.6
Angles of diffraction in a grating

A diffraction grating has 15,000 lines uniformly spaced over one inch. It is illuminated at normal incidence by yellow light from a sodium vapour lamp. At what angles will the first and second order maxima be observed for the wavelengths of the sodium doublet, 5890 Å and 5895.9 Å? Is the third order maximum visible?

Solution

The grating element, $a + b = \dfrac{2.54\ \text{cm}}{15,000} = 16933\ \text{Å}$.

Using the grating equation, $(a+b)\sin\theta_n = n\lambda$, we get,

$$\theta_n = \sin^{-1}\left(\frac{n\lambda}{(a+b)}\right)$$

Thus, for $D_1(5890\ \text{Å})$, $\theta_n = \sin^{-1}\left(\dfrac{5890}{16933} \times n\right) = \sin^{-1}(0.34784 \times n)$

and for $D_2(5895.9\ \text{Å})$, $\theta_n = \sin^{-1}\left(\dfrac{5895.9}{16933} \times n\right) = \sin^{-1}(0.34819 \times n)$

Hence

$$\theta_1 = 20.355 \text{ and } \theta_2 = 44.081 \text{ for } D_1$$

$$\theta_1 = 20.377 \text{ and } \theta_2 = 44.137 \text{ for } D_2$$

$$\text{For } n = 3, \sin\theta = \left(\frac{3\lambda}{a+b}\right) > 1.$$

As sine of an angle cannot exceed 1, the third order maximum is not visible.

EXAMPLE 4.7

To show that $\Delta\theta_n = \lambda/[N(a+b)\cos\theta_n]$

Show that the angular distance between a principal maxima of order n and either adjacent minima is given by $\Delta\theta_n = \lambda/[N(a+b)\cos\theta_n]$.

Solution

The position of the nth principal maximum is given by

$$(a+b)\sin\theta_n = n\lambda$$

The position of the first minimum adjacent to this maximum is obtained when the path difference is increased by λ/N (see Eq. 4.25). Let this increase in path difference be due to change of angle $\Delta\theta_n$ so that we can write

$$(a+b)\sin(\theta_n + \Delta\theta_n) - (a+b)\sin\theta_n = \lambda/N$$

Thus,

$$(a+b)(\sin\theta_n\cos\Delta\theta_n + \cos\theta_n\sin\Delta\theta_n - \sin\theta_n) = \lambda/N$$

For small $\Delta\theta_n$, $\sin\Delta\theta_n \to \Delta\theta_n$ and $\cos\Delta\theta_n \to 1$

Hence, $\quad (a+b)(\Delta\theta_n\cos\theta_n) = \lambda/N$

or $\quad \Delta\theta_n = \dfrac{\lambda}{N(a+b)\cos\theta_n}$

EXAMPLE 4.8

Comparison of dispersion and resolving power of a grating

For the data in Ex. 4.6 find the angular separation between D_1 and D_2 in both orders. Compare the dispersion and resolving power of the gratings with the following specifications for D_1 in first order. (a) 15,000 lines per inch, (b) 15,000 lines per cm and (c) 5906 lines per cm. What is the dispersion and resolving power for the zeroth order?

Solution

The angular separation is given by

$$\Delta\theta_n = \frac{n\Delta\lambda}{[(a+b)\cos\theta_n]}$$

$$\text{Thus, } \Delta\theta_1 = \frac{1 \times (5895.9 - 5890)}{16933 \times \cos(20.355)} = 0.021°$$

$$\text{and} \quad \Delta\theta_2 = \frac{2 \times (5895.9 - 5890)}{16933 \times \cos(44.081)} = 0.056°$$

These values can also be obtained by taking the difference of θ_1 for D_1 and D_2 and of θ_2 for D_1 and D_2 obtained in Ex. 4.6.

Dispersion and resolving power are obtained using Eqs. 4.28 and 4.31 respectively, and are tabulated below.

N	$(a+b)$ (Å)	θ_1 (degrees)	$\Delta\theta_1/\Delta\lambda$ ($\times 10^{-3}$ degrees/Å) $= n/(a+b)\cos\theta_1$	$R = nN$
(a) 15000	16933	20.35	3.61	15000
(b) 15000	6667	62.06	18.34	15000
(c) 5906	16932	20.36	3.61	5906

It can be seen that the gratings (a) and (b) have the same resolving power while the gratings (a) and (c) have the same dispersion.

In the zeroth order spectrum the spectral lines are neither resolved nor dispersed as can be seen by setting $n = 0$ in the expressions for dispersion and resolving power.

4.4 APPLICATIONS

Diffraction, like interference, finds numerous applications in engineering and technology. Most of these applications exploit the fact that the *diffraction pattern produced by an obstacle contains information about the characteristics of the light used and of the diffracting obstacle*. For example, a grating experiment can be performed to determine the wavelengths present in a given light source by using a grating of known period. In this way a grating can be used as a wavelength separator or a monochromator (Eq. 4.26, that is, $(a+b)\sin\theta = n\lambda$, tells us that a grating bends different wavelengths by different amounts). Alternatively the geometrical characteristics of the grating—grating spacing or the number of grooves present—can be inferred by performing the experiment using light of known wavelength. By analysing the intensity distribution of the diffraction pattern, the diffracting characteristics of the obstacle (such as the shape of the profile of grooves in the grating) can be revealed.

Diffraction occurs only when the wavelength of light is comparable to the size of an obstacle. This is clear from the above equation because for $\lambda > (a+b), \theta$ is not defined, or for $\lambda << (a+b)$, θ, that is, bending is not appreciable. Only for $\lambda \sim (a+b)$ is θ appreciable and diffraction is observed. Study of diffraction thus brought to the fore an interesting rule of nature. Thus nature's secrets are revealed merely by the comparison of two lengths—the length of the probing wave and the length (size) of the object under study! If one wants to measure anything of the order of a few thousand angstroms, light wave is recommended as a probe (see Ex. 4.9).

If details of the order of a few angstroms (for example, the unit cell of a crystalline solid) are to be studied, x-rays ($\lambda \sim 1$ to 100 Å) are more suitable (see Ex. 4.10). This idea first occurred to Max von Laue, in 1912, and led to the emergence of 'x-ray diffraction' as one of the most powerful techniques for characterisation of materials (see Section 9.6). This technique has been used to map many complex structures in the last century. The structure of a DNA molecule—one of the most notable among them—was discovered by Watson and Crick in 1953. With the discovery of matter waves by de Broglie in 1924 (see Section 6.6), a variety of probing waves became available, thus making it possible to explore the microworld. Advanced research techniques—such as electron or neutron diffraction that are available today—mainly use the phenomenon of diffraction of matter waves.

EXAMPLE 4.9

Application of diffraction for determination of separation of lines on an IC

The following results were obtained in an experiment to study the diffraction of different wavelengths of cadmium and mercury light by the lines on an integrated circuit. Use thise data to determine the mean spacing of the lines on the circuit.

Wavelength (Å)	Angle of first order diffraction (degrees)
4680	28.0
4800	28.7
5770	35.5

Solution

Using the equation $(a+b) = \lambda/\sin\theta_1$ values of $(a+b)$ for various wavelengths are calculated as follows.

Wavelength (Å)	$(a+b)$ (Å)
4680	9968.6
4800	9995.3
5770	9936.2

Thus the mean spacing of the lines on the circuit is 9966.7 Å.

EXAMPLE 4.10

Diffraction of x–rays

If diffraction of x-rays ($\lambda = 1$ Å) is to be studied, can the normal grating (with 15,000 lines per inch) be used? If not, which is a more suitable grating?

Solution

Grating period, $a + b = \dfrac{2.54\ \text{cm}}{15000} = 16933$ Å

Using the grating equation,

$$(a + b)\sin\theta_n = n\lambda$$

the first order maximum will be obtained at

$$\theta_1 = \sin^{-1}\left[\frac{\lambda}{a+b}\right] = \sin^{-1}\left(\frac{1}{16933}\right) = 0.003$$

This angle is very close to the central maximum and thus cannot be observed. X-rays will pass through an ordinary optical grating without any appreciable bending.

For studying diffraction of x-rays, $(a + b) \sim 1$ Å is needed. Such a small spacing cannot be achieved mechanically. However a crystalline solid in which atoms are periodically arranged can serve as a natural grating for x-ray diffraction because the interatomic spacing is of the order of a few angstroms in these solids.

Diffraction effects are also important for sound waves and microwaves. The diffraction of sound waves around a rigid sphere was studied first by Rayleigh and then by Ballantine (1928). Equation 4.13 tells us that the directionality of sound waves can be increased by increasing the diameter of the speakers. The arrangement of loudspeaker arrays in a public address system is often guided by diffraction considerations. A parabolic reflector radiating microwaves shows single slit diffraction effects. Diffraction considerations are thus important in the design of these reflectors.

As the amount of diffraction depends on the wavelength, diffraction effects also play an important role in wireless communication. Radio waves having wavelengths ranging from 180 to 6000 m are often employed in standard AM (amplitude modulation) broadcast band. In FM (frequency modulation) band, the waves ranging from 2.7 to 3.6 m are used. AM waves thus suffer more diffraction due to their large wavelength compared to the FM waves. Hence FM reception is poor in places where AM reception is loud and clear.

Diffraction by multiple slits gives well-defined directions in which sharp diffraction maxima occur. Such a grating can be simulated by placing the sources themselves at these multiple slits. One has to only ensure a known phase relationship between waves produced by these sources.

This is essentially the principle behind arrays of radiating antennas used for directional communication. An example would be an array of parabolic antennas for radio communication. When used in reverse, as receiving antenna arrays, such arrays can be used to probe deep space regions. The Giant Metrewave Radio Telescope (GMRT) facility near Pune (India), belonging to the National Centre for Radio Astronomy, a part of the Tata Institute of Fundamental Research, is an example of this. Antenna theory is a subject of specialised study in electronic and communication engineering.

Diffraction studies are also useful in astronomy for measurement of stellar data. Techniques such as correlation interferometry and methods of wide-angle interference are used for collection of these data. Mathematically the diffraction pattern is the two-dimensional Fourier transform of a two-dimensional object. The Fourier transform of the diffraction pattern is the real image of the object. These concepts form the basis of advanced application areas like spatial filtering, optical computation and Fourier optics.

One of the interesting applications of diffraction is in the determination of elastic properties of liquids. The elastic constants are related to the velocity of ultrasonic waves in the liquid. The liquid under study is compressed by sending ultrasonic waves through it using a piezoelectric transducer (Fig. 4.9). These waves produce regions of lower and higher density (that is, with lower and higher refractive index) in a liquid much like ruled and unruled parts of an optical grating. The liquid thus acts as an 'acoustical grating'. When light is passed through such a grating it follows the grating equation $(a + b)\sin\theta = n\lambda$ where $a + b$ is the distance between the centres of the compressed part and the rarefied part next to it, which is just equal to half the wavelength λ' of the ultrasonic waves in the liquid. Thus λ' can be calculated using the equation

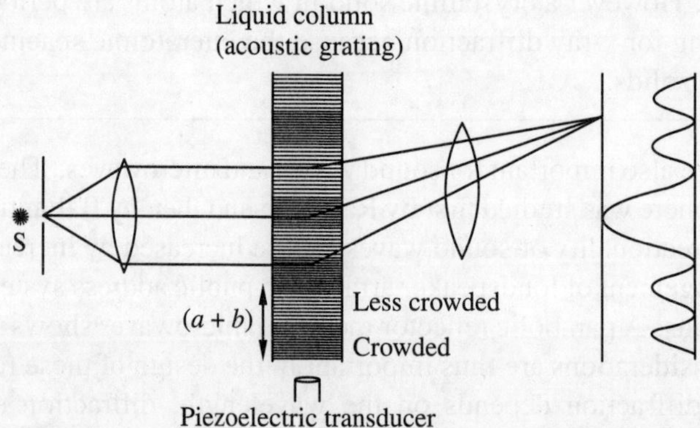

Figure 4.9 Experimental set up for determination of ultrasomic velocity using an acoustic grating

$$\frac{\lambda'}{2}\sin\theta = n\lambda \quad \text{or} \quad \lambda' = \frac{2n\lambda}{\sin\theta} \tag{4.33}$$

The velocity is determined using the relation

$$v = f\lambda' \tag{4.34}$$

where f is the frequency of the ultrasonic waves.

An important consequence of diffraction is that it limits the resolving power of an optical instrument. If the diffraction patterns surrounding the images of the two nearby objects coalesce into each other, the two objects are not perceived as separate. In this case the optical instrument fails to resolve (show as separate) the two objects. If the diffraction patterns do not merge into each other, the objects are well resolved (Fig. 4.10). Diffraction studies provided a deeper insight into the concept of resolution. Rayleigh gave the criterion for resolution: two point sources are just resolved if the angular separation between them is such that the maxima of the diffraction pattern of one source falls on the first minima of the diffraction pattern of the other.

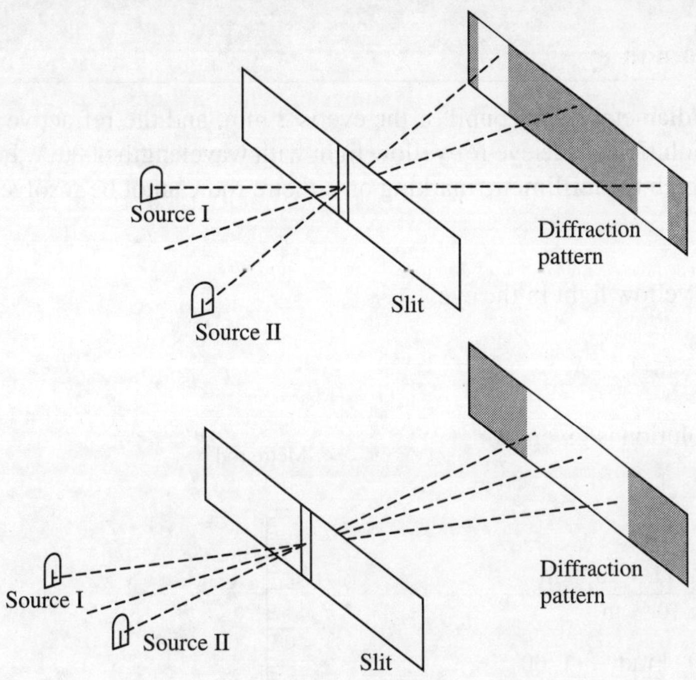

Figure 4.10 Diffraction patterns of (a) well-resolved and (b) unresolved sources

Applying Rayleigh's criterion to the case of a single slit (Eq. 4.13), two sources can be just resolved if their angular separation θ is greater than θ_c where

$$\sin\theta_c = \frac{\lambda}{d} \tag{4.35}$$

The problem of diffraction for a circular slit was solved by Airy, a British astronomer, in 1835. It was shown that

$$\sin\theta_c = \frac{1.22\lambda}{D} \tag{4.36}$$

is the required expression for the limiting angle θ_c for the resolution of two objects by a circular aperture of diameter D. Equation 4.36 also defines the resolution for a lens of diameter D and thus holds good for a telescope objective. In Ex. 4.11 the resolution of the human eye is shown to be equal to 1 minute. For a microscope the limit of resolution (minimum distance that can be just resolved) is given by

$$\Delta = \frac{\lambda}{2n \sin i} \tag{4.37}$$

where i is the angle subtended by the object at the objective and n is the refractive index of the liquid with which the objective is filled.

EXAMPLE 4.11

Resolution of human eye

Assuming that the diameter of the pupil of the eye is 2 mm, and the refractive index to be 1.33, find the angular resolution of the eye for yellow light with wavelength 6000 Å. Find the maximum distance beyond which the millimetre marking on a metre rod cannot be resolved by the eye.

Solution

The wavelength of yellow light in the eye,

$$\lambda = \frac{6000 \text{ Å}}{1.33} = 4511.28 \text{ Å}$$

The angular resolution is given by

$$\theta_c = \frac{1.22\lambda}{D}$$

$$= \frac{1.22 \times 4.511 \times 10^{-7} \text{ m}}{2 \times 10^{-3} \text{ m}}$$

$$= 2.75 \times 10^{-4} \text{ rad} \cong (1/60)^\circ$$

$$= 1 \text{ minute}$$

For the millimetre marking to be resolved (see Fig. 4.11),

L should not exceed

$$\frac{1 \text{ mm}}{\tan \theta_c} \cong \frac{1 \text{ mm}}{\theta_c} = \frac{1 \text{ mm}}{2.75 \times 10^{-4} \text{ rad}}$$

$$= 3636.4 \text{ mm} = 3.64 \text{ m}$$

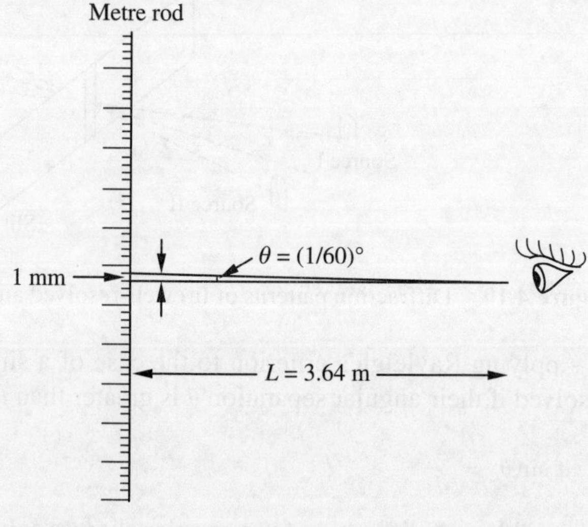

Figure 4.11 Resolution of the human eye

Equations 4.36 and 4.37 indicate that lenses with larger diameter provide higher resolution than those with a smaller diameter. These equations also bring forth the fundamental restrictions imposed by diffraction on the limit of resolution achievable by optical instruments. These equations tell us that no matter how perfect the lenses used are, and how ingeniously the optical instruments are designed, diffraction effects will let us perceive the two objects as separate only if we use an appropriate wavelength. This knowledge provided a breakthrough in the understanding of resolution. It provided a clue that the resolving power of an instrument can be enhanced by reducing the wavelength of the probing waves. This gave a lot of impetus to progress in the field of microscopy. The resolving powers of microscopes have been vastly increased with the discovery of matter waves, as the wavelengths of these waves can be greatly reduced (see Section 6.6). An electron microscope, for example, can resolve details to the order of less than an angstrom. *Had diffraction not been discovered, and subsequently researched into, we would not have known about anything smaller than the wavelength of light!*

5

Polarisation

5.1 THE STORY OF POLARISATION

In the last two chapters it was seen how physicists studied light and found the explanation for why two (or more) light beams produce bright and dark regions and why light bends round sharp obstacles, exhibiting the phenomena of interference and diffraction respectively. Another interesting phenomenon exhibited by light, known as *polarisation*, has attracted the attention of physicists since the second half of the seventeenth century.

In 1669, Erasmus Bartholin, Doctor of Medicine and Professor of Mathematics at the University of Copenhagen, discovered a remarkable optical phenomenon. He observed that crystals of Iceland spar (calcite) produce two images of an object. It was known through SnellŠs law (1621) that change in medium results in bending of light. But with calcite the bending was rather unusual. Two sets of light rays were produced and the waves moving parallel to one plane were bent by a different amount than those moving along a normal to this plane. In 1678, Huygens used this phenomenon of double refraction to demonstrate the polarisation of light. In 1690 he also offered a successful but somewhat incomplete explanation of double refraction using the idea of wavefronts.

In 1808, Etienne Louis Malus (1775–1812), a military engineer and captain in Napoleon's army, discovered the phenomenon of polarisation accidentally. He was looking at light reflected from the windows of Luxembourg palace through a calcite crystal and found that one of the images disappeared. He deduced from these observations that reflected light is linearly polarised at right angles to the plane of incidence. Malus also discovered an experimental relation in 1809 which governs the intensity of light transmitted by a polariser.

An important feature of polarisation, which sets it apart from other properties like interference and diffraction, is that polarisation is possible only for those waves for which apart from the propagation direction, another direction—that of the disturbance—is also important. In 1817, Young suggested the transverse nature of light waves on the basis of the experimental evidence gathered by his contemporaries. Notable among them were Dominique-Francois Arago (1786–1853), French statesman and a scientist, and Augustin Fresnel. They studied double refraction by calcite and were puzzled as to why the refracted rays, though coherent, did not produce any interference pattern.

Young proposed a novel idea to unfold this mystery. He suggested that light beams must have a transverse character and that the two beams must have their planes of vibration at right angles to each other. Interestingly these predictions were found theoretically justifiable by Maxwell in 1873. He identified the vibrations associated with light waves as those of the electric field (and the magnetic field), and propounded that these vibrations are perpendicular to the direction of propagation of the waves (transverse character).

Polarisation experiments also brought forth the transverse character of other electromagnetic waves. For example, when x-rays were discovered in 1898, there was much speculation whether they were waves or particles. In 1906, Charles Glover Barkla (1877–1944) established the transverse wave nature of x-rays on the basis of polarisation experiments.

One of the major components in the research on polarisation was the production of light of desired polarisation. Many innovative and elegant polarisers were designed by physicists and engineers working in this area. The Polaroid Corp. made millions of people aware of the phenomenon of polarisation through their products such as sunglasses and camera filters. Arago constructed one of the first polaroids in 1812 using a stack of glass plates. The Nicol prism—one of the most renowned birefringent polarisers—was built in 1828 by the Scottish physicist, William Nicol (1768–1851). A polariser consisting of parallel conducting wires, called the wire grid polariser, requires the spacing between the adjacent grids to be less than the wavelength of the waves. So it could be easily developed for microwaves. But for light waves ($\lambda < 7000$ Å) this tough task was only recently accomplished when George R Bird and Maxfield Parrish Jr succeeded in placing about 30,000 wires in one inch. On the other hand polaroids have also been developed more easily using conducting polymers. In 1928, Edwin Herbert Land, then a 19-year-old undergraduate at Harvard college, developed a commercial polaroid known as J-sheet. He also invented H-sheet, the molecular analogue of wire grid, in 1938. Many other polaroids such as K-sheet, HR-sheet and others have been developed since then for the production of polarised light.

Research in polarisation also triggered investigations into whether desired polarisation/double refraction can be achieved artificially. These pursuits led to the discovery of a large number of phenomena such as optical activity, electrooptic effect, magnetooptic effect, and photoelastic effect. Faraday, Kerr, Pockels, Biot and Brewster made some pioneering contributions to this effort.

Mathematical formulations concerning theory of polarisation were developed by GG Stokes in 1852 and were refined by R Clarke Jones in 1941. Hans Mueller, then a professor of physics at the Massachusetts Institute of Technology, developed the matrix method for dealing with these formulations.

This chapter introduces the subject of polarisation. It also discusses the different types of polarisation (Section 5.2), the unpolarised character of natural light (Section 5.3) and various methods of producing polarised light (Sections 5.4 and 5.5), Huygen's model of double refraction (Section 5.5) and the analysis of polarised light (Section 5.6). In Section 5.7 applications in engineering and technology that developed as a result of polarisation studies are discussed.

5.2 TYPES OF POLARISATION

In Section 2.3 it was shown that the concept of polarisation arises because there are two natural directions associated with any wave: the direction of propagation of the wave and the 'direction of disturbance' constituting the wave. When the latter coincides with the former we say that the wave has longitudinal polarisation. When the latter is perpendicular to the former, we say the wave has transverse polarisation.

According to MaxwellŠs theory (Section 2.4), light is an electromagnetic wave. There are two fields—the electric and the magnetic—which oscillate with the same frequency at any point in a light wave (Fig. 5.1). For defining polarisation of the light wave, the electric field is chosen by convention. Polarisation is the direction of the electric field. The simplest light wave is one in which the electric field is perpendicular to the direction of propagation. Since, there are two independent mutually perpendicular directions, which are both perpendicular to any given direction of propagation, there are two independent transverse polarisations possible. The plane formed by the direction of the electric field, **E**, and the direction of propagation of the wave (xy-plane in Fig. 5.1) is called the *plane of vibration*. A plane perpendicular to the plane of vibration [containing the magnetic field vectors (xz-plane in Fig. 5.1)] is called the *plane of polarisation* for historical reasons.

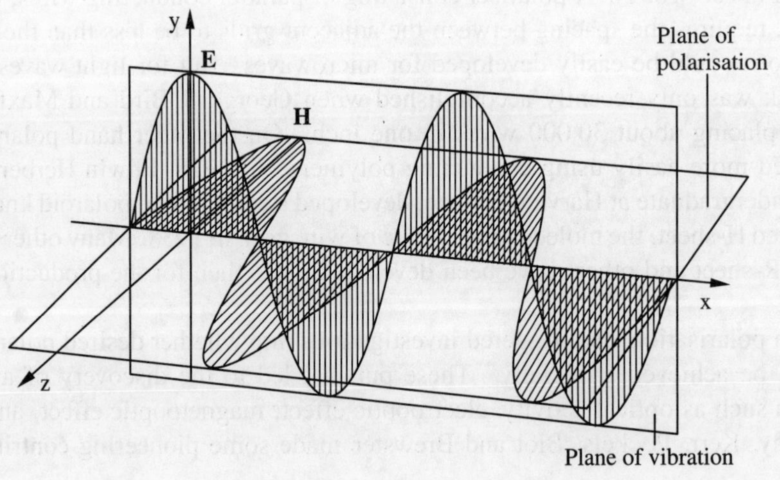

Figure 5.1 Plane polarised wave showing plane of polarisation and plane of vibration

In Section 3.2 which dealt with the superposition of waves, it was implicitly assumed that the participating waves are polarised in the same direction. Here the effect of incorporating the polarisation of individual waves on superposition is discussed. Thus Eqs. 3.4 and 3.5 can be rewritten as,

$$\mathbf{E}_1 = \mathbf{j}A_1 \sin(\alpha_1 - \omega t) \tag{5.1}$$

and $$\mathbf{E}_2 = \mathbf{j}A_2 \sin(\alpha_2 - \omega t) \tag{5.2}$$

where $\mathbf{E}_1$ and $\mathbf{E}_2$ are the instantaneous electric fields of the two interfering light waves, both propagating along the x-direction. The inclusion of **j**, the unit vector along the y-axis, in these equations denotes that both waves are polarised along the y-direction. The resultant electric field will be given by

$$\mathbf{E} = \mathbf{E}_1 + \mathbf{E}_2 = \mathbf{j}\,[A_1 \sin(\alpha_1 - \omega t) + A_2 \sin(\alpha_2 - \omega t)] = \mathbf{j}\,A \sin(\phi - \omega t) \tag{5.3}$$

where,

$$A^2 = A_1^2 + A_2^2 + 2A_1A_2\cos(\alpha_1 - \alpha_2) \text{ and } \tan\phi = \frac{A_1 \sin\alpha_1 + A_2 \sin\alpha_2}{A_1 \cos\alpha_1 + A_2 \cos\alpha_2} \tag{5.4}$$

Thus we see that the resultant is also a plane-polarised wave with its electric vector oscillating along the same direction.

Consider the superposition of two waves polarised along mutually perpendicular directions. Thus the equations for waves polarised along the y- and z-directions, respectively, are

$$\mathbf{E}_y = \mathbf{j}A_1 \sin(\alpha_1 - \omega t) \tag{5.5}$$

and $$\mathbf{E}_z = \mathbf{k}A_2 \sin(\alpha_2 - \omega t) \tag{5.6}$$

Equations 5.5 and 5.6 show that their addition is a vector addition, and the result is not expressible in the form of Eqs. 5.3 and 5.4. Physically this means that *superposition of two waves polarised perpendicular to each other does not lead to interference pattern* (see Ex. 5.1).

EXAMPLE 5.1

Waves polarised perpendicular to each other cannot produce an interference pattern.

Prove that two plane polarised light waves polarised at right angles to each other cannot produce interference effects.

Solution

Let the electric field vectors of two plane polarised waves be given by

$$\mathbf{E}_1 = \mathbf{j}A_1 \sin(\alpha_1 - \omega t)$$

and $$\mathbf{E}_2 = \mathbf{k}A_2 \sin(\alpha_2 - \omega t),$$

where **j** and **k** are unit vectors along the y- and z-axes respectively.

Thus the resultant field is $\mathbf{E} = \mathbf{E}_1 + \mathbf{E}_2$.

The resultant intensity, $I = \langle \mathbf{E}^2 \rangle = \langle \mathbf{E} \cdot \mathbf{E} \rangle$

$$= \langle (\mathbf{E}_1 + \mathbf{E}_2) \cdot (\mathbf{E}_1 + \mathbf{E}_2) \rangle$$

$$= \langle E_1^2 \rangle + \langle E_2^2 \rangle + 2 \langle \mathbf{E}_1 \cdot \mathbf{E}_2 \rangle$$

Thus we have, $I = I_1 + I_2 + I_{12}$ where, I_{12} is called the interference term.

Now $\mathbf{E}_1$ and $\mathbf{E}_2$ are perpendicular to each other. So the dot product $E_1 \cdot E_2 = E_1 E_2 \cos 90° = 0$.

Thus, $I = I_1 + I_2 = \langle E_1^2 \rangle + \langle E_2^2 \rangle$

Now, $\langle E_1^2 \rangle = \left(\dfrac{A_1^2}{T}\right) \displaystyle\int_0^T \sin^2(\alpha_1 - \omega t) dt = \dfrac{A_1^2}{2}$

Similarly $\langle E_2^2 \rangle = \dfrac{A_2^2}{2}$

Thus $I = \dfrac{(A_1^2 + A_2^2)}{2}$

As the intensity is independent of the phase difference, it remains constant and the interference effects cannot be produced.

What is the effect of superposition then? The answer can be sought by relating E_y and E_z. We do this by eliminating t from Eqs. 5.5 and 5.6 by the following steps.

Expanding these equations we get

$$\left(\frac{E_y}{A_1}\right) = \sin \alpha_1 \cos \omega t - \cos \alpha_1 \sin \omega t \tag{5.7}$$

and $$\left(\frac{E_z}{A_2}\right) = \sin \alpha_2 \cos \omega t - \cos \alpha_2 \sin \omega t \tag{5.8}$$

Multiplying Eq. 5.7 by $\cos \alpha_2$ and Eq. 5.8 by $\cos \alpha_1$ and subtracting, we get,

$$\left(\frac{E_y}{A_1}\right) \cos \alpha_2 - \left(\frac{E_z}{A_2}\right) \cos \alpha_1 = \sin(\alpha_1 - \alpha_2) \cos \omega t \tag{5.9}$$

Repeating this using $\sin \alpha_2$ and $\sin \alpha_1$ as multipliers, we get,

$$\left(\frac{E_y}{A_1}\right) \sin \alpha_2 - \left(\frac{E_z}{A_2}\right) \sin \alpha_1 = \sin(\alpha_1 - \alpha_2) \sin \omega t \tag{5.10}$$

Squaring and adding Eqs 5.9 and 5.10 we get

$$\frac{E_y^2}{A_1^2} + \frac{E_z^2}{A_2^2} - \left(\frac{2E_yE_z}{A_1A_2}\right)\cos(\alpha_1 - \alpha_2) = \sin^2(\alpha_1 - \alpha_2) \tag{5.11}$$

If the phase difference $\alpha_1 - \alpha_2 = \alpha$, then Eq. 5.11 reduces to

$$\frac{E_y^2}{A_1^2} + \frac{E_z^2}{A_2^2} - \left(\frac{2E_yE_z}{A_1A_2}\right)\cos\alpha = \sin^2\alpha \tag{5.12}$$

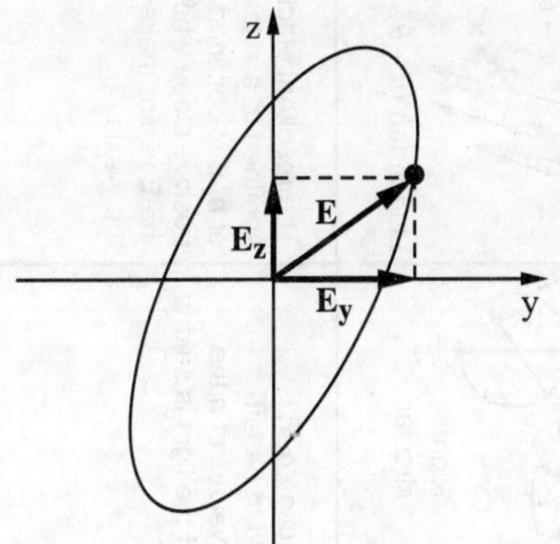

Figure 5.2 Superposition of two mutually perpendicular linearly polarised waves results in an eliptically polarised wave

This equation represents an inclined ellipse, α being its angle of inclination with the y-axis. This means that the resultant electric field which can be expressed as

$$\mathbf{E} = \mathbf{j}E_y + \mathbf{k}E_z \tag{5.13}$$

traces an ellipse in the yz-plane (Fig. 5.2). In this case the light is said to be elliptically polarised. This represents a general case in which the resultant **E** vector varies both in magnitude and direction around the direction of the propagation of the waves.

For different values of α and (A_1/A_2) this general case specialises into different curves. For $\alpha = n\pi$, $n = 0, 1, 2, \ldots$ we get linearly polarised or plane polarised light. For $\alpha = (2n+1)\pi/2$ the light is elliptically polarised with the axes of the ellipse being along the y- and z-axes, and for $\alpha = (2n+1)$ $\pi/2$ and $A_1 = A_2$, we get circularly polarised light. Further we have left- or right-polarised light depending on whether n is even or odd. These different cases are illustrated in a comparative manner in Chart 5.1. Example 5.2 also illustrates how different states of polarisation result due to superposition of component waves.

Thus we note that in interference, since the superposing waves have the same polarisation, the phase difference determines the resultant intensity. Whereas *in superposition of two waves of different polarisations, the phase difference between them determines the resultant polarisation.*

EXAMPLE 5.2
To find the resultant state of polarisation

Find the state of polarisation of the electric field whose y- and z-components are given by the following equations.

(i) $E_y = A\cos(kx + \omega t)$ and $E_z = A\cos(kx + \omega t + \pi/4)$

CHART 5.1 DIFFERENT TYPES OF POLARISED LIGHT

Type	Elliptically polarised	Circularly polarised	Plane polarised
	(i) Following from Eq. 5.12 using different value of α		
Mathematical representation	For $\alpha = \pi/2, 3\pi/2, \ldots (2n+1)\pi/2$, $\sin\alpha = 1$ and $\cos\alpha = 0$ Thus, $E_y^2/A_1^2 + E_z^2/A_2^2 = 1$ (1) Equation (1) represents an ellipse with its minor and major axes along the y- and z-axes.	For $\alpha = (2n+1)\pi/2$, and $A_1 = A_2 = A$ $E_y^2 + E_z^2 = A^2$ (4) Equation (4) represents a circle (circle is an ellipse whose minor and major axes are equal).	For $\alpha = n\pi$, $\sin\alpha = 0$, $\cos\alpha = -1$ for odd n and $\cos\alpha = 1$ for even n Thus, $(E_y/A_1) = (-1)^n (E_z/A_2)$ (7) Equation (7) represents a straight line.
	(ii) Use of Eqs. 5.5 and 5.6		
	$E_y = A_1 \sin(\alpha - \omega t)$ (2) and $E_z = A_2 \sin[\alpha - (2n+1)\pi/2 - \omega t)]$ or $E_z = (-1)^{n+1} A_2 \cos(\alpha - \omega t)$ (3)	$E_y = A \sin(\alpha - \omega t)$ (5) and $E_z = A \sin[\alpha - (2n+1)\pi/2 - \omega t]$ or $E_z = (-1)^{n+1} A \cos(\alpha - \omega t)$ (6)	$E_y = A_1 \sin(\alpha - \omega t)$ (8) and $E_z = A_2 \sin(\alpha - n\pi - \omega t)$ or $E_z = (-1)^n A_2 \sin(\alpha - \omega t)$ (9)
Diagrammatic representation	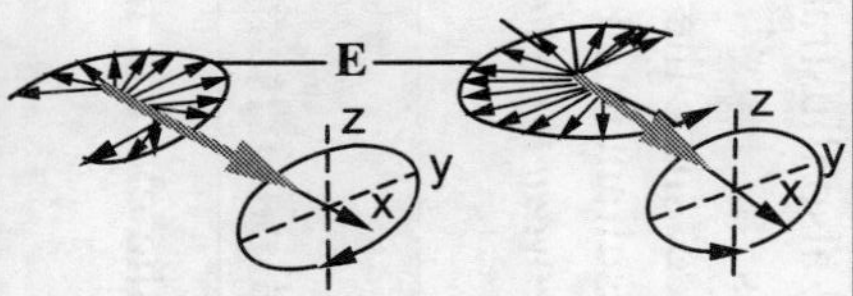 E, z, y, x Left elliptical Right elliptical	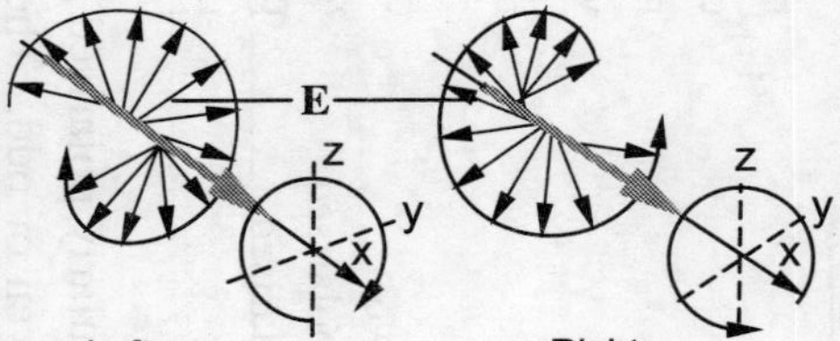 E, z, y, x Left circular Right circular	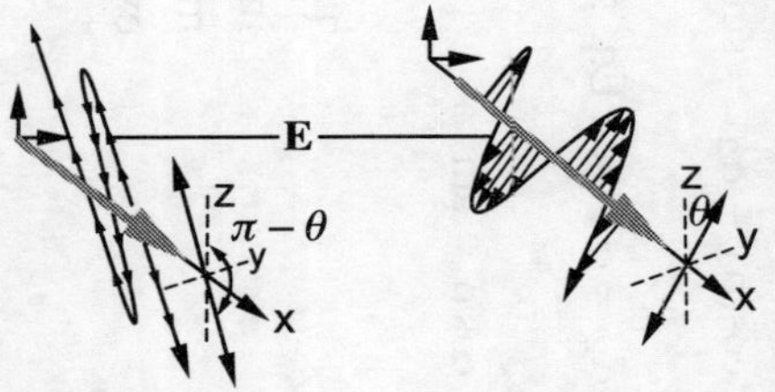 E, z, y, x, $\pi - \theta$, θ Linear, $\tan(\pi - \theta)$ Linear $\tan\theta$
Description	For even values of n the resultant vector **E** rotates in a clockwise direction, when you look into the beam of light propagating to your eye (x-axis in the present case). Light is said to be left-elliptically polarised in this case. For odd valves of n, **E** rotates in an anti-clockwise direction. Such a light is said to be right-elliptically polarised.	For even values of n the direction of **E** is clockwise and the light is said to be left-circularly polarised. For odd values of n the rotation is anti-clockwise and the light is said to be right circularly polarised.	The resultant electric field vector **E** is confined to a single plane. For even valves of n, **E** is confined to a direction with a positive slope whereas for odd values of n the **E** vector makes a negative slope with the y-axis.

(ii) $E_y = A_1 \sin(kx - \omega t)$ and $E_z = A_2 \sin(kx - \omega t - \pi/2)$

(iii) $E_y = A \sin(kx - \omega t)$ and $E_z = A \sin(kx - \omega t - \pi/2)$

(iv) $E_y = A \sin(kx - \omega t)$ and $E_z = A \sin(kx - \omega t)$

The first pair of these equations is a general case and the next three pairs are the equations considered in Chart 5.1 for $n = 0$.

Discuss what will happen if the following changes are made in case (ii) considered above.

(a) Equations in Chart 5.1 are considered for $n = -1$, $n = 1$ and $n = 2$.

(b) If $(kx - \omega t)$ is replaced by (i) $kx + \omega t$, (ii) $-kx - \omega t$ and (iii) $kx + \omega t$.

Solution

(i) $E_y = A\cos(kx + \omega t)$ and $E_z = A\cos(kx + \omega t + \pi/4)$.
Combining these two equations we get,

$$E_y^2 + E_z^2 - \sqrt{2}E_yE_z = A^2/2$$

which is the equation of an ellipse that is tilted at an angle of 45° to the y-axis. Thus the electric field is elliptically polarised.

Table 5.1 E at various time instants

t	E_y	E_z	$\mathbf{E}$
0	A	$A/\sqrt{2}$	$\mathbf{j}A + \mathbf{k}A/\sqrt{2}$
$T/4 = \pi/2\omega$	0	$-A/\sqrt{2}$	$-\mathbf{k}A/\sqrt{2}$
$T/2 = \pi/\omega$	$-A$	$-A/\sqrt{2}$	$\mathbf{j}A - \mathbf{k}A/\sqrt{2}$
$3T/4 = 3\pi/2\omega$	0	$A/\sqrt{2}$	$\mathbf{k}A/\sqrt{2}$
$T = 2\pi/\omega$	A	$A/\sqrt{2}$	$\mathbf{j}A + \mathbf{k}A/\sqrt{2}$

To ascertain whether the polarisation is left or right, we find E_y and E_z for $x = 0$ at various instants of time as shown in Table 5.1. The resultant electric field E rotates in the clockwise direction as shown in Fig. 5.3a. However, as the wave is propagating in the negative x-direction, the rotation of **E** is anticlockwise looking into the source. Thus the light is right elliptically polarised.

It can be noted here that the state of polarisation can also be ascertained if one finds E_y and E_z for just two instants, $t = 0$ and $t = T/4$ and noting the sense of rotation as shown in Fig. 5.3b. The state of polarisation is found in this way for (ii) and (iii), as illustrated in Table 5.2.

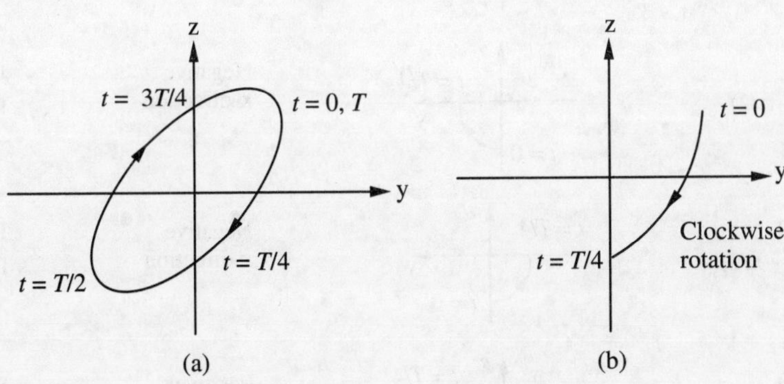

Figure 5.3 (a) **E** for $t = 0$ to $t = T$ and (b) **E** for $t = 0$ and $t = T/4$

Table 5.2 Finding state of polarisation from $\mathbf{E}(0, 0)$ and $\mathbf{E}(0, T/4)$

	$x = 0, t = 0$			$x = 0, t = T/4$			Figure	Polarisation
	E_y	E_z	$\mathbf{E}$	E_y	E_z	$\mathbf{E}$	Direction of propagation: $+x$	
(ii) $E_y = A_1 \sin(kx - \omega t)$, $E_z = A_2 \sin(kx - \omega t - \pi/2)$	0	$-A_2$	$-A_2\mathbf{k}$	$-A_1$	0	$-A_1\mathbf{j}$	z, y, $t = T/4$, $t = 0$	Left elliptically polarised
(iii) $E_y = A \sin(kx - \omega t)$, $E_z = A \sin(kx - \omega t - \pi/2)$	0	$-A$	$-A\mathbf{k}$	$-A$	0	$-A\mathbf{j}$	z, y, $t = T/4$, $t = 0$	Left circularlly polarised

(iv) $E_y = A \sin(kx - \omega t)$, and $E_z = A \sin(kx - \omega t)$

Thus $E_z = E_y$, which is the equation of a straight line.

Hence the electric field is plane polarised at an angle of 45° with the y-axis (because slope $= \tan 45° = 1$).

(a) Equations for $n = -1$ in Chart 5.1 are $E_y = A_1 \sin(kx - \omega t)$ and $E_2 = A_2 \sin(kx - \omega t + \pi/2)$. Extending the method illustrated in Table 5.2 we get right elliptically polarised light for this case. For $n = 2$ we get the same results as that for $n = 0$ and for $n = 1$ the results are the same as that for $n = -1$. This is because the phase shift of 2π does not change the resultant **E**.

Table 5.3 State of polarisation on changing relative signs of kx and ωt

Equations for E_y and E_z	Figure	Direction of propagation	State of polarisation
$E_y = A_1 \sin(kx - \omega t)$ $E_z = A_2 \sin(kx - \omega t - \pi/2)$	z, y, $t = T/4$, $t = 0$	Positive x-direction	Left elliptically polarised
$E_y = A_1 \sin(kx + \omega t)$ $E_z = A_2 \sin(kx + \omega t - \pi/2)$	z, y, $t = T/4$, $t = 0$	Negative x-direction	Left elliptically polarised
$E_y = A_1 \sin(-kx - \omega t)$ $E_z = A_2 \sin(-kx - \omega t - \pi/2)$	z, y, $t = T/4$, $t = 0$	Negative x-direction	Right elliptically polarised
$E_y = A_1 \sin(-kx + \omega t)$ $E_z = A_2 \sin(-kx + \omega t - \pi/2)$	z, y, $t = T/4$, $t = 0$	Positive x-direction	Right elliptically polarised

(b) If $kx - \omega t$ is replaced by $kx + \omega t$, the result remains unchanged. For $-kx - \omega t$ and $-kx + \omega t$ we get right polarised light. These cases are illustrated in Table 5.3.

5.3 WHY NATURAL LIGHT IS UNPOLARISED

The reason why natural light is 'unpolarised' is the same as why it is incoherent (see Section 3.3). A single atom emits, during 10^{-8}s, a wave of certain polarisation. Another atom emits a wave of different polarisation. In this way natural light, comprising of emissions from numerous atoms, contains all possible polarisations (Fig. 5.4a). As the plane of polarisation keeps changing in very short intervals of times ($\sim 10^{-9}$s), the state of polarisation cannot be detected by any known detector, including the eye. The averaging out of the effects of polarisation makes us feel that natural light is unpolarised. By some means if we are able to restrict the electric field vibrations along a particular direction, we say that we have polarised the light (Fig. 5.4b). Radio waves and microwaves are generally plane polarised due to the fixed direction of current oscillations on the antenna.

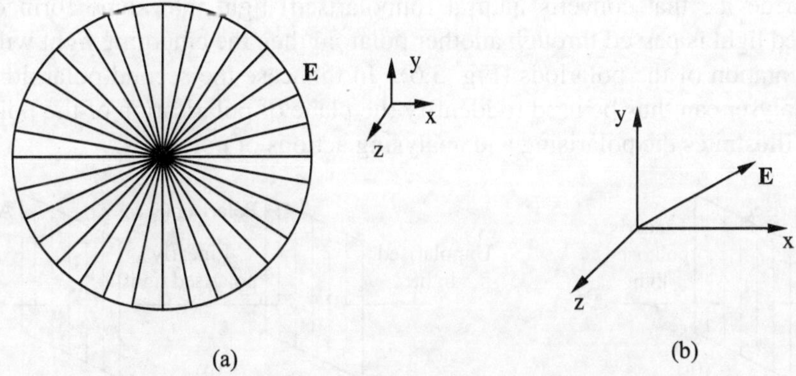

Figure 5.4 (a) Electric field vibrations in natural light keep changing direction in a random manner; (b) The electric field vector for a plane polarised wave oscillates along a single direction

A simple experiment, illustrated in Fig. 5.5, demonstrates how polarisation brings out the transverse character of a wave. A transverse wave on a string, linearly polarised with vibrations

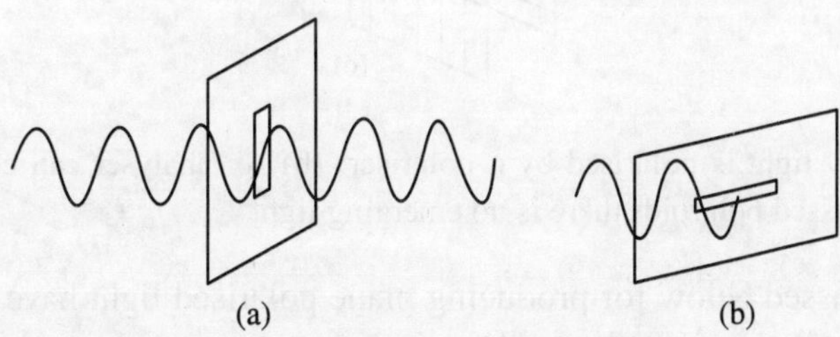

Figure 5.5 (a) Transverse wave polarised along the orientation of the slit can pass through it; (b) Vibrations die out for perpendicular orientation of the slit

along the vertical direction, can be made to pass through a slit only for its vertical orientation (Fig. 5.5a). For a horizontal orientation of the slit the vibrations will completely die out (Fig. 5.5b). But a longitudinal wave on a string will transmit through the slit no matter what its orientation is. That light is a transverse wave can be proved by performing a similar experiment by letting a polarised beam of light fall on a polarising crystal (called as polaroid). The emerging light has maximum intensity (that is the same as the incident intensity) for only a particular orientation of the crystal. It is interesting to note that for a transverse wave there are infinite possible directions of **E**, perpendicular to the direction of propagation. Polarising light means selecting one of these directions.

With this preliminary idea of polarisation we try to understand how polarised light of a desired polarisation can be produced in the subsequent sections.

5.4 PRODUCTION OF PLANE POLARISED LIGHT

The commercial devices used for the production of polarised light are called *polaroids*. A *polariser* is a device that converts natural (unpolarised) light into some form of polarised light. If polarised light is passed through another polaroid then the emerging light will depend on the relative orientation of the polariods (Fig. 5.6). In this case the second polaroid is called an *analyser*. An analyser can thus be used to identify the plane of polarisation of the polarised light. Figure 5.6 also illustrates the polarising and analysing actions of polaroids.

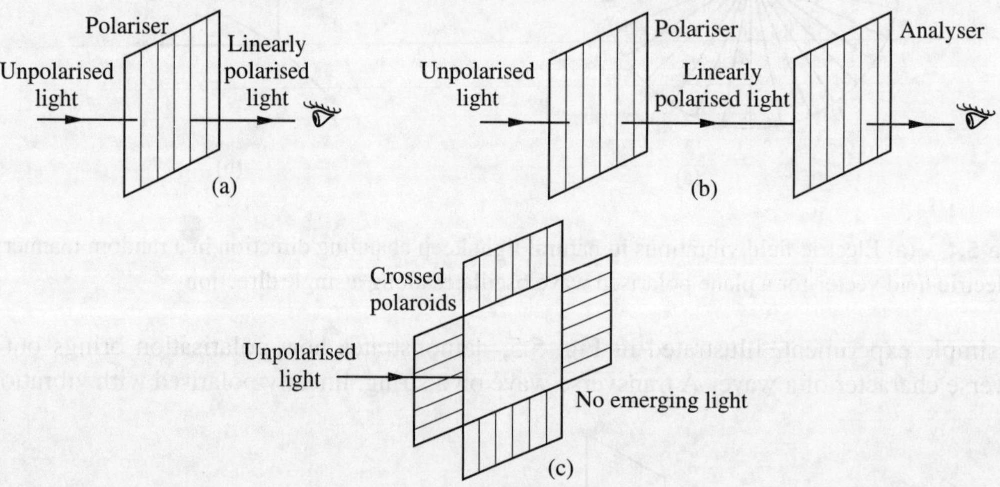

Figure 5.6 (a) Ordinary light is polarised by a polariser; (b) An analyser can control the intensity of polarised light; (c) For crossed polaroids there is no emerging light

The techniques discussed below for producing plane polarised light have a common unifying requirement. They are all based on the exploitation of some asymmetry in the polariser which helps select some particular state from among all states of polarisation. These techniques are discussed in the following pages and are summarised in Chart 5.2.

CHART 5.2 TECHNIQUES USED FOR PRODUCTION OF PLANE POLARISED LIGHT

Technique	Selective absorption	Reflection	Scattering	Double refraction
Principle of the technique	Electric field vibrations in unpolarised light are preferentially absorbed along a particular direction and transmitted along a direction perpendicular to this direction.	Reflection involves re-radiation by the electrons in the reflector of the incident unpolarised light. The component of **E** in the plane of incidence is absorbed (electrons cannot re-radiate this component in the direction of reflected beam). The reflected beam thus contains component of **E** perpendicular to the plane of incidence and is thus plane polarised.	Scattering involves re-radiation of the incident unpolarised light by the free electrons in the scatterer. An electron acts as a dipole and re-radiates maximum energy perpendicular to the axis of the dipole and does not radiate any light along lits axis. Thus the beam perpendicular to the incident direction is devoid of component of **E** in the plane of incidence and is thus plane polarised.	Double refraction or birefringence involves splitting of unpolarised light into two refracted beams, the e-ray and the o-ray, by an optically anisotropic transparent crystal. Splitting is due to the anisotropy of refractive index. Both these rays are plane polarised. Thus plane polarised light can be obtained by using a birefringent crystal and suppressing one of these two rays. This is achieved by a construction known as the Nicol prism.
Figure illustrating the technique	Wire grid Unpolarised light Polarised light y z x **Wire grid polariser** (Absorption of **E** along the y-direction and transmission along the z-direction) y unpolarised light z Tourmaline crystal Linearly polarised light x **Dichroic crystal like tourmaline** (Absorption of **E** along the z-axis and transmission along the y-axis)	Unpolarised light θ_P Plane polarised light Partially polarised light θ_r **Polarisation by reflection** Light polarised perpendiculum to the page Light almost polarised in the plane of the page **Polarisation by a stack of glass plates** (Successive reflections rid the incident beam of the dot component and the final emerging beam is almost plane polarised with **E** in the plane of incidence.)	S_1 S A Plane polarised light S_2 (The scattered beams S_1 and S_2 are partially polarised whereas the beam S is plane polarised.)	e-ray o-ray Crystal (calcite or quartz) Unpolarised beam **Phenomenon of double refraction** (Note $i = 0$ and $r = 0$ for the o-ray which thus follows Snell's law. It is not so for the e-ray that does not obey Snell's law.) calcite 90° optic axis e-ray 48° Unpolarised beam 68° 71° o-ray Canada balsam **The Nicol prism** (The dashed outline corresponds to the natural crystal. It is polished to facilitate the total internal reflection of the o-ray so that the plane polarised e-ray only emerges out of the crystal.)

5.4.1 Selective Absorption

Selective absorption, also known as *dichroism*, is based on the property of the polariser to preferentially absorb electric field vibrations along a particular direction and transmit those perpendicular to this direction. The direction parallel to the transmitted component is called the *transmission axis* or the *pass axis* of the polariser. If A is the amplitude of the plane polarised light falling on the polariser at an angle θ with the transmission axis, then the component $A\cos\theta$ is transmitted, whereas the component $A\sin\theta$ is absorbed (see Fig. 5.7). As intensity is proportional to the square of the amplitude, the transmitted intensity is given by

$$I = I_0 \cos^2\theta \tag{5.14}$$

where I_0 is the maximum value of transmitted intensity occurring at $\theta = 0$ or π. Equation 5.14 is known as *Malus' law*, named after its discoverer (see Ex. 5.3).

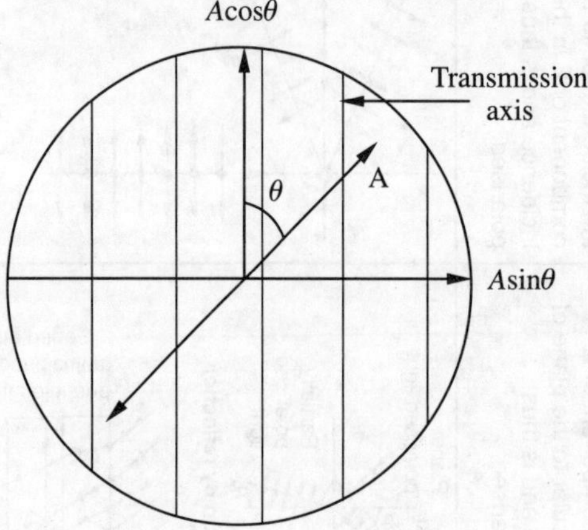

Figure 5.7 A polariser transmits the component parallel to its transmission axis and absorbs the component perpendicular to this axis

EXAMPLE 5.3

Application of Malus' law

An unpolarised beam of light is incident on a group of three polarising sheets which are positioned such that the pass axis of each is rotated by 30° clockwise with respect to the preceding sheet. What fraction of the incident intensity is transmitted?

Solution

As the incident light is unpolarised, the fraction of intensity transmitted by the first sheet, I_1/I_0, is equal to the average value of $\cos^2\theta$, that is,

$$\frac{I_1}{I_0} = <\cos^2\theta>$$

$$= \frac{1}{2\pi}\int_0^{2\pi}\cos^2\theta d\theta$$

$$= \frac{1}{2\pi}\int_0^{2\pi}\left[\frac{(1+\cos 2\theta)}{2}\right]d\theta = \frac{1}{2}$$

Thus, $I_1 = \frac{I_0}{2}$

Let I_2 and I_3 be the intensity transmitted by the second and the third sheet respectively.

Thus, $I_2 = I_1\cos^2\theta$

$$= \left(\frac{I_0}{2}\right)\cos^2 30° = \frac{3I_0}{8}$$

and $I_3 = I_2\cos^2\theta$

$$= \left(\frac{3I_0}{8}\right)\cos^2 30° = \frac{9I_0}{32}$$

Thus 9/32 of the incident intensity is transmitted.

A wire grid polariser is the simplest polariser based on the phenomenon of selective absorption. It consists of a set of parallel conducting wires (see Chart 5.2). When unpolarised light falls on it, the component of electric field parallel to the wires gets absorbed, because it drives electrons in the wires resulting in a current and Joule heat loss. The component perpendicular to the wires is transmitted. The transmission axis of the wire grid polariser is thus perpendicular to the wires.

Some crystalline materials are inherently dichroic due to the anisotropy in their structure. Natural tourmaline is one of the best known dichroic crystals. It absorbs the electric field vibrations perpendicular to the principal axis or the pass axis, a specific direction in a crystal determined by its structure, and transmits the field components parallel to it (see Chart 5.2).

5.4.2 Polarisation by Reflection

Natural light is found to get partially polarised upon reflection. The reflected beam contains predominantly the electric field component perpendicular to the plane of incidence (called as the σ component), whereas the transmitted beam contains the vibrations in the plane of incidence (called as the π component). The σ component and the π component are represented by dots and arrows respectively in Chart 5.2.

It was found experimentally by Brewster that the reflected beam is completely plane polarised in the plane of incidence if the incident beam makes an angle θ_p with the normal, called the *polarising angle* or *Brewster's angle*, at which the reflected beam and the refracted beam are at right angles, that is,

$$\theta_p + \theta_r = 90° \tag{5.15}$$

where θ_r is the angle of refraction.

Using Snell's law,

$$\mu_1 \sin\theta_p = \mu_2 \sin\theta_r \tag{5.16}$$

μ_1 and μ_2 being refractive indices of the media containing incident and refracted beams respectively. Combining Eqs. 5.15 and 5.16 we get,

$$\tan\theta_p = \frac{\mu_2}{\mu_1} \tag{5.17}$$

For the transition of light beams from air to the medium of refractive index μ, Eq. 5.17 becomes

$$\tan\theta_p = \mu \tag{5.18}$$

Equation 5.17/5.18 is known as *Brewster's law* (see Ex. 5.4).

The reason why the electric field component vibrating in the plane of incidence is not reflected at the BrewsterŠs angle is the following. The incident beam sets the electrons in the material into oscillations, which then re-radiate the beam. The component in the plane of incidence cannot radiate sideways and is thus absent in the reflected beam. The dot component radiating perpendicular to the direction of its vibration can thus be observed at 90° to the refracted beam. By allowing the light to suffer reflections at several reflecting surfaces using a pile of plates (see Chart 5.2) the σ component can be entirely eliminated from the transmitted beam and the transmitted beam can be made completely π polarised.

EXAMPLE 5.4

Application of Brewster's law

Sunlight is reflected from a pool of water standing in a roadway. An observer finds that the reflected sunlight is completely polarised. What is the angle of the sun above the horizon?

Solution

Refractive index of water $= \mu = 1.33$

Brewster's angle, $\theta_p = \tan^{-1}\mu = \tan^{-1}(1.33) = 53.1°$

Thus, angle of the sun above the horizon is (see Fig. 5.8)

$\theta_s = 90° - 53.1° = 36.9°$

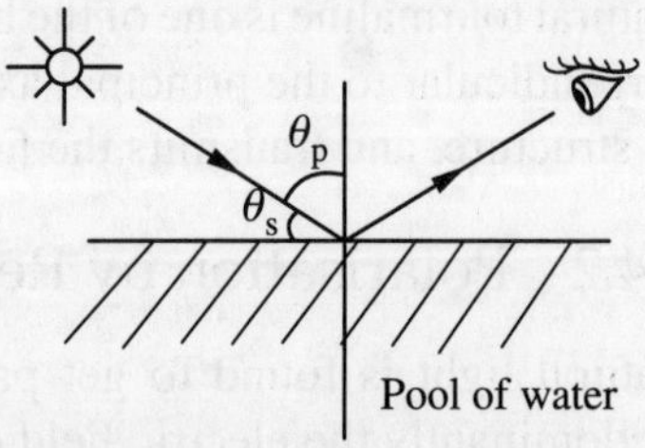

Figure 5.8 Polarisation of sunlight due to reflection

5.4.3 Polarisation by Scattering

Light can be polarised by scattering it sideways by gaseous media. An unpolarised light beam striking a gaseous atom A (see Chart 5.2) will set the electrons in it into oscillations. Since light beams are always transverse, these oscillations can be considered to be due to two oscillating dipoles represented by the arrow and dot at A. As the dipole radiates maximum energy perpendicular to its axis and no energy along it, completely polarised light with its **E** vector perpendicular to the plane of the paper will be observed at right angles to the direction of the unpolarised beam. This is because none of the components in the original beam has **E** field in the plane and also perpendicular to this direction. The scattered beams along other directions will be partially polarised.

5.4.4 Polarisation by Double Refraction

Natural light can also be polarised using some optically anisotropic (optical properties are different in different directions) transparent crystals such as calcite or quartz. These crystals offer different refractive indices to light polarised along different directions. Experiment 5.1 describes the determination of these indices. The underlying physics can be explained as follows.

The incident light wave forces electrons in the crystals into oscillations. The electrons re-radiate the secondary wavelets, which recombine to form the refracted wave. The speed of the wave, and hence its refractive index ($\mu = c/v$), is decided by the difference between the frequency of the **E** vector of the incident beam and the natural frequency of the electrons. This in turn depends on the force constant between atoms. In isotropic materials the force constant is the same in all directions, thus indicating a single index of refraction. In case of a double refracting crystal, an anisotropy in the force constant is manifested in the anisotropy of the refractive index.

The phenomenon of double refraction, also known as *birefringence*, involves splitting of an incident ray into two rays by the crystal (see Chart 5.2). These rays are called as the *ordinary ray* or o-ray and the *extraordinary ray* or e-ray. Both these rays are linearly polarised. Electric field vibrations of the o-ray are perpendicular to the direction of the ray as well as to the optic axis, whereas those of e-ray are perpendicular to the direction of the ray but lie in the plane containing the optic axis and the ray direction. Such a plane is also known as *principal plane* or *principal section* of the crystal. The vibrations of the e-ray thus occur in the principal plane and those of o-ray, occur perpendicular to the principal plane.

The o-rays travel in the crystal with the same speed (v_o) in all directions and thus have a single index of refraction. The crystal thus acts as an isotropic medium for the o-rays. The speed of e-rays, v_e, and the refractive index, μ_e, however vary with the direction. The o-rays follow Snell's law of refraction but the e-rays do not. This is clear from the figure in Chart 5.2 showing $r = 0$ for $i = 0$ for the o-ray, in conformity with Snell's law, but $r \neq 0$ for $i = 0$ for the e-ray, thus contradicting Snell's law.

The principle of production of polarised light using a birefringent crystal is simply to eliminate one of the two rays. This is achieved by sandwiching a layer of material having refractive index between μ_o and μ_e. The geometry of transmission is so chosen that the ray for which the sandwiched material is optically rarer falls on it at an angle of incidence greater than the critical angle. It is thus eliminated by total internal reflection. The construction of one such crystal—named the *Nicol prism* after its inventor—is shown in Chart 5.2. It is made by grinding the ends of a calcite rhombohedron from 71° to 68°, cutting the rhomb diagonally and then polishing and rejoining the two pieces with a layer of Canada balsam sandwiched between the two pieces. The Canada balsam material is transparent and its refractive index is 1.55, a value less than μ_o (1.658) and greater than μ_e (1.486). The o-ray will thus suffer total internal reflection at the Canada balsam layer if it is incident at an angle greater than the critical angle. It can be shown that the critical angle is 69° for the o-ray (see Ex. 5.5). The explanation of double refraction using HuygenŠs model and its use in the production of different types of polarised light will be discussed in the next section.

EXAMPLE 5.5
Critical angle for total internal reflection of the o-ray

Show that the critical angle required for the total internal reflection of the o-ray in a Nicol prism is 69°.

Solution

The critical angle θ is the minimum angle of incidence for which there is no refraction.

Thus for $i = \theta, r = 90°$. For $i > \theta$, the ray suffers total internal reflection.

Using Snell's law, $\mu_1 \sin i = \mu_2 \sin r$, we have

$$1.658 \times \sin\theta = 1.55 \text{ or } \theta = \sin^{-1}\left(\frac{1.55}{1.658}\right) = 69°$$

EXPERIMENT 5.1 BIREFRINGENCE

Aim

To determine the refractive indices of the ordinary- and the extraordinary-ray in a birefringent crystal, reading lens, spirit

Apparatus

Spectrometer, sodium vapour lamp, birefringent crystal, reading lens, spirit level

Formula

$$\mu_e = \frac{\sin[(A+\delta_{me})/2]}{\sin(A/2)} \tag{5.19}$$

$$\mu_o = \frac{\sin[(A+\delta_{mo})/2]}{\sin(A/2)} \tag{5.20}$$

where μ_e and μ_o are the refractive indices for the extraordinary-ray and the ordinary-ray respectively, δ_{me} and δ_{mo} are their respective angles of minimum deviation, and A is the angle of the prism.

Theory

This experiment is based on the phenomenon of birefringence. The experimental arrangement is shown in Fig. 5.9a. When a ray of light is incident on a prism made of a birefringent material it splits into the o-ray and the e-ray. In case of a positive crystal like quartz, e-ray bends more compared to o-ray, whereas for a negative crystal such as calcite, o-ray deviates more than e-ray. However, this splitting does not take place if the ray of light is incident on the prism in the direction of its optic axis. The splitting is maximum in a direction perpendicular to the optic axis. The refractive indices of o-ray and e-ray are defined in terms of their velocities as

$$\mu_0 = \frac{c}{v_0} \tag{5.21}$$

$$\mu_e = \frac{c}{v_{\text{emin}}} \text{ for positive crystal} \tag{5.22}$$

$$\text{and } \mu_e = \frac{c}{v_{\text{emax}}} \text{ for negative crystal} \tag{5.23}$$

where the suffix emax and emin denotes, respectively, the maximum and the minimum velocity of the e-ray inside the crystal.

To ensure maximum splitting, the prism cut with its optic axis parallel to the refracting edge is used. It should be noted that optic axis denotes a direction (axis of symmetry) and any line parallel to the optic axis also denotes an optic axis. The refractive indices for the o-ray and the e-ray can be determined with the help of a spectrometer by measuring the minimum angles of their deviation. The difference $\Delta\mu = \mu_c - \mu_o$ is called the *birefringence.*

Procedure

A. Measurement of δ_{me} and δ_{mo}

1. The planes of spectrometer chassis and prism table are properly adjusted using a spirit level.
2. The source of light is switched on and the slit width of the collimator is adjusted to get an intense beam of light through a narrow slit (see Chart 4.2).
3. The prism is placed on the prism table with one of its glass faces facing the light coming from the collimator. In this position of the prism the light rays are refracted by the prism.
4. The telescope is properly adjusted to view the refracted beams. Two well-defined images of the slit (corresponding to the o-ray and the e-ray) should be visible through the eyepiece (Fig. 5.9b).
5. The telescope and prism table are rotated simultaneously along the same direction so that the images also move along this direction. This is continued till the images start moving in the opposite direction. The position of the telescope is then adjusted such that one of the images (say o-image) retraces its path after touching the centre of the cross wire. The telescope is fixed in this position. This is the minimum deviation position of the prism for the o-ray.
6. Now the collimator is focussed for parallel rays by Schuster's method. For this the prism table is rotated slightly in both directions around the minimum deviation position in

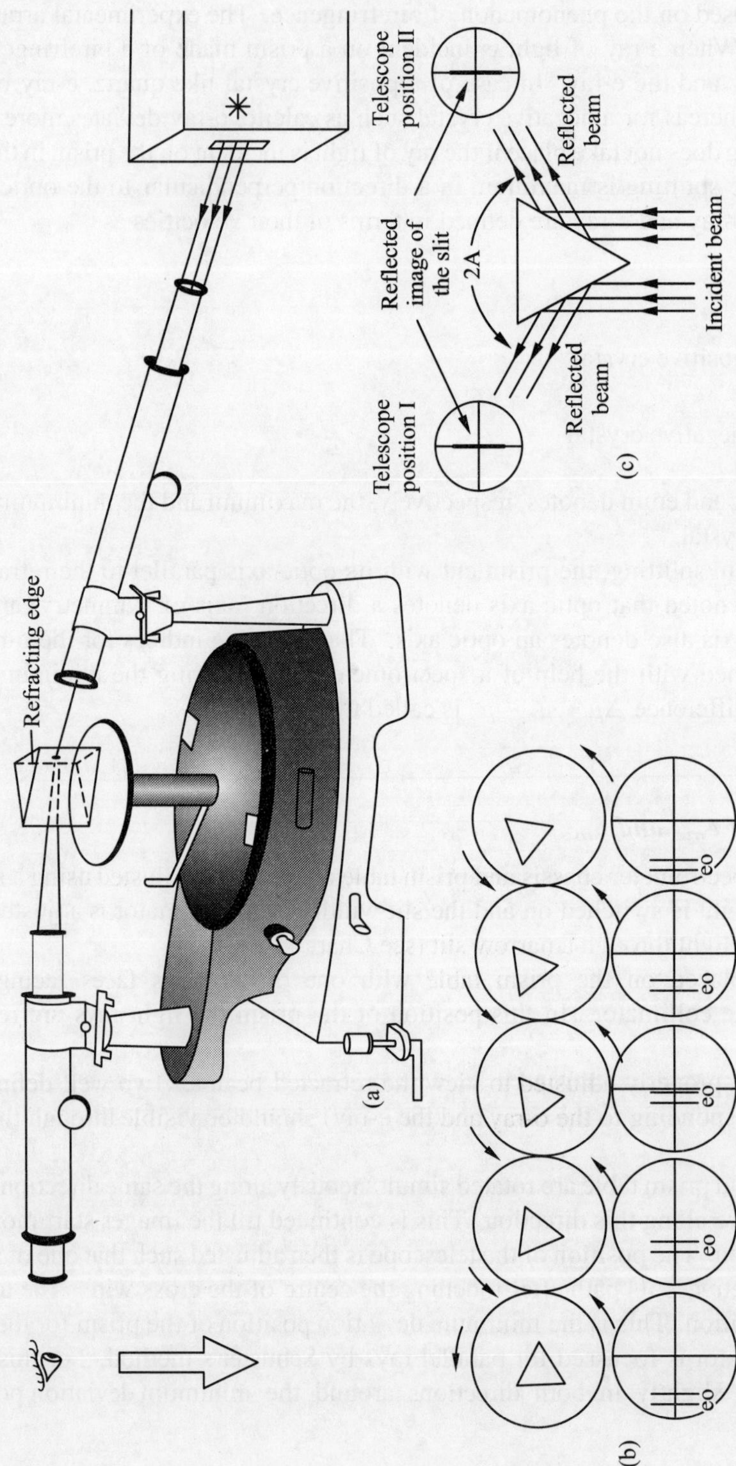

Figure 5.9 (a) Experimental arrangement; (b) Angle of minimum deviation for the e-ray. Minimum deviation for the o-ray means that the e-ray should retrace after touching the cross wire; (c) Angle of prism.

succession so that the image is away from the centre of the cross wire (see Fig. 5.9b). The image gets blurred by this adjustment. The collimator and the telescope are now adjusted so that the image remains as sharp as possible in the complete field of view.

7. The readings of both windows of the spectrometer are recorded.
8. Steps 5 and 6 mentioned above are repeated for the e-ray and the readings of both the windows corresponding to the minimum deviation position of the prism for the e-ray are recorded.
9. The prism is removed and readings of both windows (direct reading) are noted when the telescope is in line with the collimator and the image of the slit coincides with the vertical cross wire.
10. The readings are tabulated in Table 5.4.

B. *Measurement of A*

1. The prism is placed on the prism table so that light falls equally on the two glass faces of the prism (Fig. 5.9c).
2. The telescope is adjusted to view the reflected image of the slit on both sides as shown in the figure.
3. The readings of both the windows are recorded in these two positions of the telescope and are tabulated in Table 5.5.

Observations and calculations

A. *For least count of the spectrometer*

Smallest division on main scale: $X°$

Number of divisions on vernier scale: Y

Least count = $(X/Y)°$

B. *For measurement of angle of minimum deviation*

Table 5.4 Measurement of angle of minimum deviation

Direct reading (degrees)				Reading for minimum deviation (degrees)						δ_m (degrees)	Mean (degrees)
				o-ray			e-ray				
Window No.	MSR	VSD	TR (x)	MSR	VSD	TR (y)	MSR	VSR	TR (z)		
W_1										$\delta_{mo} = x \sim y = \ldots$	
										$\delta_{me} = x \sim z = \ldots$	$\delta_{mo} = \ldots$
W_2										$\delta_{mo} = x \sim y = \ldots$	
										$\delta_{me} = x \sim z = \ldots$	$\delta_{me} = \ldots$

C. For measurement of angle of prism

Table 5.5 Measurement of angle of prism

Window No.	Telescope position 1 (degrees)	Telescope position II (degrees)	Difference (degrees)	Mean (degrees)
	MSR VSR TR (x)	MSR VSD TR (y)	$x \sim y = 2A$	
W_1				
W_2				

μ_o and μ_e are calculated using Eqs. (5.19) and (5.20) by substituting the observed values of $\delta_{\rm mo}$, $\delta_{\rm me}$ and A given in Tables 5.4 and 5.5.

Estimation of error

The error $\Delta\mu$ in the estimation of μ can be determined by adopting the method outlined in Section 1.4.2. Thus,

$$\Delta\mu = \left|\frac{\partial\mu}{\partial A}\right|\Delta A + \left|\frac{\partial u}{\partial\delta\mu}\right|\Delta\delta m$$

$$= \left\{\left|\frac{\sin(A/2)\times\cos[(A+\delta m)/2] - \cos(A/2)\times\sin[(a+\delta m)/2]}{\sin^2(A/2)}\right|\right.$$

$$\left. + \left|\frac{\cos[(A+\delta m)/2]}{\sin(A/2)}\right|\right\}\Delta\theta \tag{5.24}$$

where $\Delta\theta$ is the error in the reading of the spectrometer.

For example, for the following measured data,

$$A = 59.77^\circ,\ \delta_{\rm mo} = 41.06^\circ,\ \delta_{\rm me} = 42.0^\circ$$

and least count of the spectrometer = $(1/60)^\circ$, we have,

$$\Delta\theta = \frac{1}{2}\left(\frac{1}{60}\right)^\circ$$

Thus

$$\Delta\mu_o = \left\{\left|\frac{\sin(59.77^\circ/2)\times\cos[(59.77^\circ+41.06^\circ)/2] - \cos(59.77^\circ/2)\times\sin[(59.77^\circ+41.06^\circ)/2]}{\sin^2(59.77^\circ/2)}\right|\right.$$

$$\left. + \left|\frac{\cos(59.77^\circ+41.06^\circ)/2}{\sin(59.77^\circ/2)}\right|\right\}\times\frac{1}{2}\times\frac{1}{60}\times\frac{\pi}{180}$$

$$= 2.05 \times 10^{-4} + 1.86 \times 10^{-4}$$

$$\Delta\mu_0 = 3.91 \times 10^{-4} = 0.0004$$

and $$\Delta\mu_e = \left\{ \left| \frac{\sin(59.77^\circ/2) \times \cos[(59.77^\circ + 42^\circ)/2] - \cos[59.77^\circ/2] \times \sin[(59.77^\circ + 42^\circ)/2]}{\sin^2(59.77^\circ/2)} \right| \right.$$

$$\left. + \left| \frac{\cos[(59.77^\circ + 42.0^\circ)/2]}{\sin(59.77^\circ/2)} \right| \right\} \times \frac{1}{2} \times \frac{1}{60} \times \frac{\pi}{180}$$

$$= 2.10 \times 10^{-4} + 1.84 \times 10^{-4}$$

$$= 3.94 \times 10^{-4} = 0.0004$$

Now, $$\mu_e = \frac{\sin[(A + \delta_{\rm me})/2]}{\sin(A/2)}$$

$$= \frac{\sin[(59.77^\circ + 42.0^\circ)/2]}{\sin(59.77^\circ/2)} = 1.557$$

Thus the relative error in $\mu_e = \Delta\mu_e/\mu_e = 0.0004/1.557 = 2.57 \times 10^{-4}$ and the percentage error is 0.026%.

$$\mu_o = \frac{\sin[(A + \delta_{mo})/2]}{\sin(A/2)}$$

$$= \frac{\sin[(59.77^\circ + 41.06^\circ)/2]}{\sin(59.77^\circ/2)} = 1.547$$

Thus the relative error in $\mu_o = \Delta\mu_o/\mu_o = 0.0004/1.547 = 2.59 \times 10^{-4}$ and the percentage error is 0.026%.

Result

The refractive indices for the ordinary- and the extraordinary-ray for the given birefringent prism are found to be — ± — and — ± , respectively. The errors in the determination of μ_0 and μ_e are —% and —% respectively.

Discussion

This experiment demonstrates how the phenomenon of double refraction can be studied using a birefringent prism and a spectrometer. The Huygen's model forms the theoretical basis of this experiment.

In the following table we compare the experiment and the theory.

Experiment	Theory
Two images of a single slit are observed when light emerging out of a birefringent prism is seen through the telescope of a spectrometer.	When light is incident on a birefringent crystal, it refracts by different amounts due to anisotropy of the refractive index. Hence two images of a single object should be formed.

It can be seen from this table that there is a qualitative agreement between the theory and the experiment.

In the prisms used in the laboratory the optic axis is normally parallel to the refracting edge, that is, normal to the equilateral top and bottom faces of the prism (Fig. 5.10a). However, the prism with its optic axis perpendicular to the refracting edge, and also perpendicular to the base BC (Fig. 5.10b), can also be used. But if a prism is cut with its optic axis parallel to the base and perpendicular to the refracting edge (Fig. 5.10c), then double refraction is not observed because as the light travels nearly along the optic axis there is no splitting.

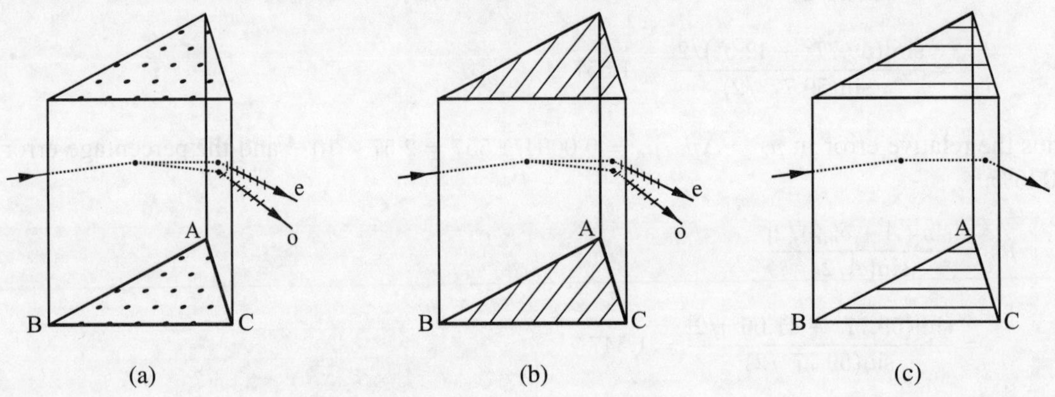

Figure 5.10 Refraction in a negative crystal

An interesting demonstration of the phenomenon of polarisation can be accomplished in this experiment if the emerging rays are viewed through a polaroid. As one rotates the polaroid the rays get extinguished at two different angles when the electric field vectors of the rays become perpendicular to the transmission axis of the polaroid.

5.5 HUYGEN'S MODEL OF DOUBLE REFRACTION AND PRODUCTION OF ELLIPTICALLY AND CIRCULARLY POLARISED LIGHT

The phenomenon of double refraction can be explained using Huygen's wave theory. Imagine a point source embedded in a birefringent crystal. It will give rise to two sets of wavefronts. For the o-ray the wavefront will be spherical as it moves with the same speed in all directions; for the e-ray the wavefront will be an ellipsoid of revolution as its speed is different along different directions.

As the speeds of the o-ray and the e-ray are the same along the optic axis, the ellipsoid of revolution will either contain the spherical wavefront (Fig. 5.11a) or will be contained in it (Fig. 5.11b). The former case corresponds to a negative crystal for which $v_e > v_o$ and $\mu_e < \mu_o$ and for which the ellipsoid of revolution is generated by the rotation of ellipse about its minor axis. In the latter case, corresponding to a positive crystal, $v_e < v_o$ and $\mu_e > \mu_o$ and the ellipsoid of revolution is generated by rotating an ellipse about its major axis. Calcite and quartz are examples of negative and positive crystals respectively. A crystal can either be a uniaxial crystal or a biaxial one depending on whether it has a single optic axis or two optic axes. These distinguishing features of birefringent crystals are summarised in Table 5.6.

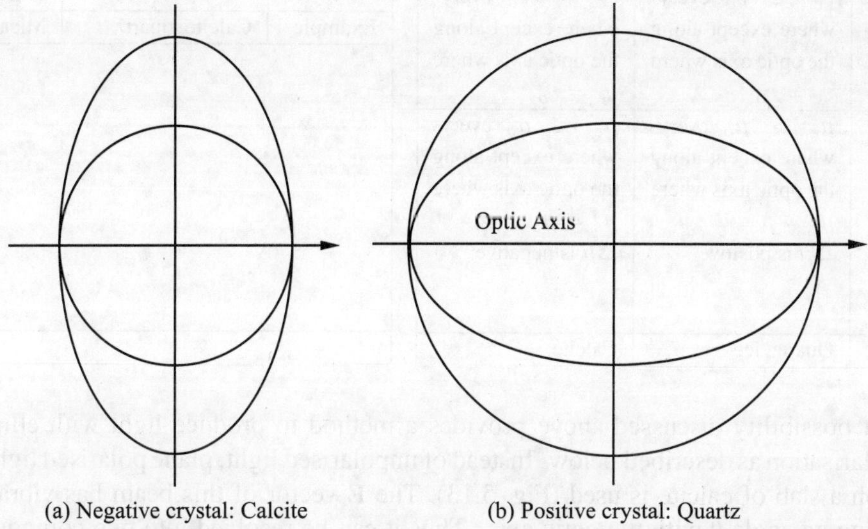

Figure 5.11 Huygen's wavefronts for the o-ray and the e-ray. The ellipsoid of revolution corresponds to the e-ray and the sphere corresponds to the o-ray

Consider the propagation of a plane wave incident normally on a negative crystal. A similar treatment is applicable to a positive crystal. The propagation is determined by the orientation of the optic axis and the following possibilities arise.

1. When the optic axis is parallel to the refracting surface (the surface on which the plane wave is incident), the wavefronts are separated as shown in Fig. 5.12a, but the direction of propagation is the same for both the rays. Thus double refraction does not take place in this case (see also Fig. 5.12b).
2. When the optic axis is normal to the refracting surface, the two rays propagate along the optic axis (Fig. 5.12c) and hence have the same velocity. Thus there is no splitting and double refraction is not observed.
3. When the optic axis is inclined with respect to the refracting face, the Huygen's wavefronts can be constructed as shown in Fig. 5.12d. Along the optic axis the two rays travel with the same velocity and hence cover the same distance. But along other directions $v_e > v_o$ and so the e-ray wavefront contains the wavefront for the o-ray.

Table 5.6 Distinguishing features of birefringent crystals

Type →	**Positive crystal**	**Negative crystal**
Huygen's wavefronts	Ellipsoid of revolution (wavefront corresponding to e-ray) lies completely inside the sphere (wavefront corresponding to o-ray)	Ellipsoid of revolution lies completely outside the sphere
Velocity	$v_e < v_o$ everywhere except along the optic axis where $v_e = v_o$	$v_e > v_o$ everywhere except along the optic axis where $v_e = v_o$
Refractive index	$\mu_e > \mu_o$ everywhere except along the optic axis where $\mu_e = \mu_o$	$\mu_e < \mu_o$ everywhere except along the optic axis where $\mu_e = \mu_o$
$\Delta\mu$ (Birefringence) $= \mu_e - \mu_o$	$\Delta\mu$ is positive	$\Delta\mu$ is negative
Example	Quartz, ice	Calcite

	Uniaxial crystal	**Biaxial crystal**
Definition	The o-ray and the e-ray have the same speeds only along one direction, that is, along the optic axis	The o-ray and the e-ray have the same speeds along two directions, that is, the crystal has three principal refractive indices and two optic axes
Example	Calcite, quartz	Mica, topaz

The first possibility discussed above provides a method to produce light with elliptical and circular polarisation as described below. Instead of unpolarised light, plane polarised light incident normally on a slab of calcite is used (Fig. 5.13). The **E** vector of this beam has vibrations in a plane making an angle θ with the optic axis. Thus it can be resolved into two components with amplitudes, $E\cos\theta$ and $E\sin\theta$, parallel and perpendicular to the optic axis, respectively. Thus the incident beam can be looked upon as a superposition of two linearly polarised waves with their electric field vectors vibrating along and perpendicular to the optic axis. These components are to be identified as the e-ray and the o-ray, respectively, and propagate along the same direction with different speeds as discussed in the first case above. This means that the two rays will cover different paths and will emerge out of the crystal with some path difference. Assuming the crystal thickness to be t the path difference is $(\mu_e - \mu_o)t$ and the corresponding phase difference is

$$\alpha = \frac{2\pi(\mu_e - \mu_o)t}{\lambda}. \tag{5.25}$$

The superposition of these two rays will result in Eq. 5.12 and the polarisation of the resultant wave will be decided by the phase difference α. If the thickness of the crystal is such that the phase difference of $\pi/2$ (path difference of $\lambda/4$) is introduced, then the resultant wave will be elliptically polarised. Such a crystal is known as the *quarter wave plate* (QWP). Thus for a QWP,

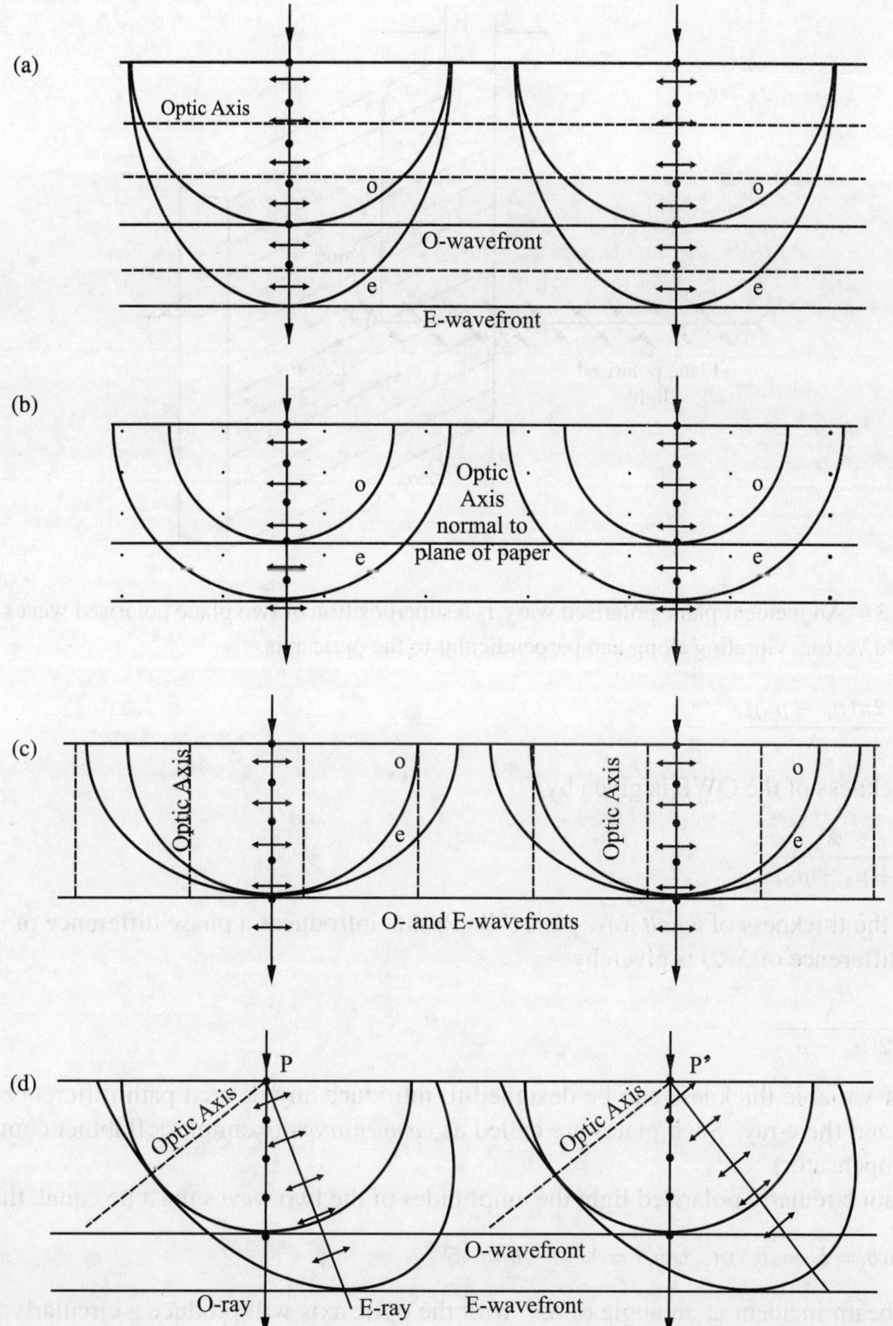

Figure 5.12 Propagation of a plane wave incident normally on a negative crystal: (a) and (b) optic axis parallel, (c) optic axis normal and (d) optic axis inclined to the refracting surface. Case (d) shows double refraction. In this case, for o-ray, $i = 0$ and $r = 0$; for e-ray, $i = 0, r \neq 0$, showing that the e-ray does not satisfy Snell's Law.

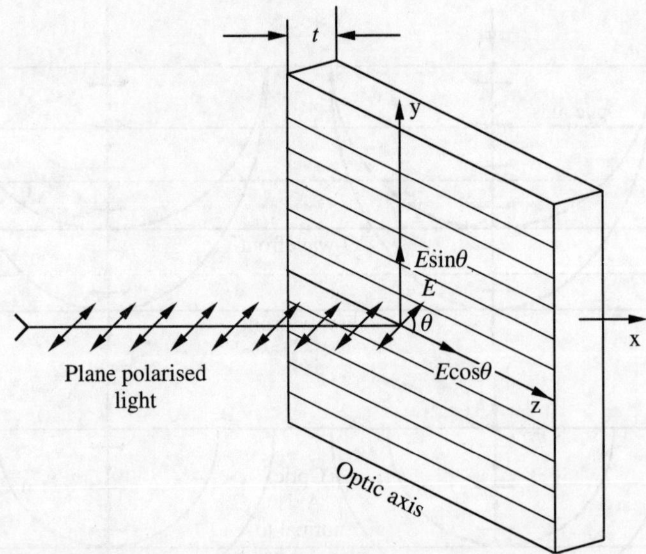

Figure 5.13 An incident plane polarised wave is a superposition of two plane polarised waves with their electric field vectors vibrating along and perpendicular to the optic axis.

$$\alpha = \frac{2\pi(\mu_e - \mu_o)t}{\lambda} = \frac{\pi}{2}$$

or, the thickness of the QWP is given by

$$t = \frac{\lambda}{4(\mu_e - \mu_o)} \tag{5.26}$$

Similarly the thickness of a *half wave plate* (HWP) that introduces a phase difference of π (that is, the path difference of $\lambda/2$) is given by

$$t = \frac{\lambda}{2(\mu_e - \mu_o)} \tag{5.27}$$

A plate of variable thickness can be designed to introduce any desired path difference between the o-ray and the e-ray. Such plates are called as *compensators* (examples: Babinet compensator, Soleil compensator).

To obtain circularly polarised light the amplitudes of the two waves must be equal, that is,

$$E\sin\theta = E\cos\theta \quad \text{or} \quad \tan\theta = 1 \quad \text{or} \quad \theta = 45^\circ$$

Thus the beam incident at an angle of 45° with the optic axis will produce a circularly polarised light.

If $\alpha = 2\pi, 4\pi, \ldots$, then the emergent beam will be linearly polarised with the same plane of polarisation as the incident beam. However, for $\alpha = \pi, 3\pi, \ldots$, the emergent beam will again be linearly polarised with its plane of vibration tilted through an angle 2θ with respect to the plane of vibration of the incident light.

Except for the cases discussed above the emergent beam will be elliptically polarised for any other value of α. In Table 5.7 we summarise the action of wave plates. In this table we make use of Chart 5.1 and study the action of wave plates in the following way.

We look upon the incident wave as a superposition of two components E_y and E_z along the y- and z-axes (see Fig. 5.13). These components can be represented by the equations (see Eqs. 5.5 and 5.6) $E_y = E \sin\theta \sin(kx - \omega t)$ and $E_z = E \cos\theta \sin(kx - \omega t - \alpha)$ so that α is the phase difference between them. These components travel as the o-ray and the e-ray, respectively, inside the wave plate. The wave plate is assumed to be made of a positive crystal so that the e-ray travels slower than the o-ray. It is assumed that the wave plate absorbs no energy but only changes the value of α to α'. For example, with HWP, $\alpha' = \alpha - \pi$ and with QWP, $\alpha' = \alpha - \pi/2$. This new phase difference decides the state of polarisation of the emergent beam.

Examples 5.6 and 5.7 illustrate the role of wave plates in producing light of desired polarisation.

EXAMPLE 5.6
Minimum thickness of wave plate to obtain a desired polarisation

Plane polarised light is incident on a quartz crystal cut parallel to its optic axis. Mention the wave plate that should be used and find its minimum thickness for which the ordinary and the extraordinary ray combine to produce (a) plane polarised light and (b) circularly polarised light. (Given $\mu_o = 1.544, \mu_e = 1.553, \lambda = 5890$ Å).

Solution

We should use a HWP to produce plane polarised light and a QWP with its optic axis inclined at an angle of 45° to the incident E to produce circularly polarised light (see Table 5.7). Minimum thickness of the wave plates in these two cases is given by

(a)
$$t = \frac{\lambda}{2(\mu_e - \mu_o)}$$
$$= \frac{5890 \times 10^{-8}\text{cm}}{2(1.553 - 1.544)}$$
$$= 3.27 \times 10^{-3}\text{cm}$$

(b)
$$t = \frac{\lambda}{4(\mu_e - \mu_o)}$$
$$= \frac{5890 \times 10^{-8}\text{cm}}{4(1.553 - 1.544)}$$
$$= 1.64 \times 10^{-3}\text{cm}$$

Table 5.7 Action of wave plates

A wave with components $E_y = E\sin\theta \sin(kx - \omega t)$ and $E_z = E\cos\theta \sin(kx - \omega t - \alpha)$, whose polarisation is decided by the phase difference α, is incident on a wave plate. Inside the wave plate E_y propagates as the o-ray and E_z, as the e-ray. For a wave plate made of positive crystal $\mu_e > \mu_o$ and $v_e < v_o$. Thus $k_e = 2\pi/\lambda_e = (2\pi/\lambda)\mu_e$ and $k_o = 2\pi/\lambda_o = (2\pi/\lambda)\mu_o$. Hence e-ray leads the o-ray in phase by $k_e x - k_o x = (2\pi/\lambda)x(\mu_e - \mu_o)$. Thus the polarisation of the wave is

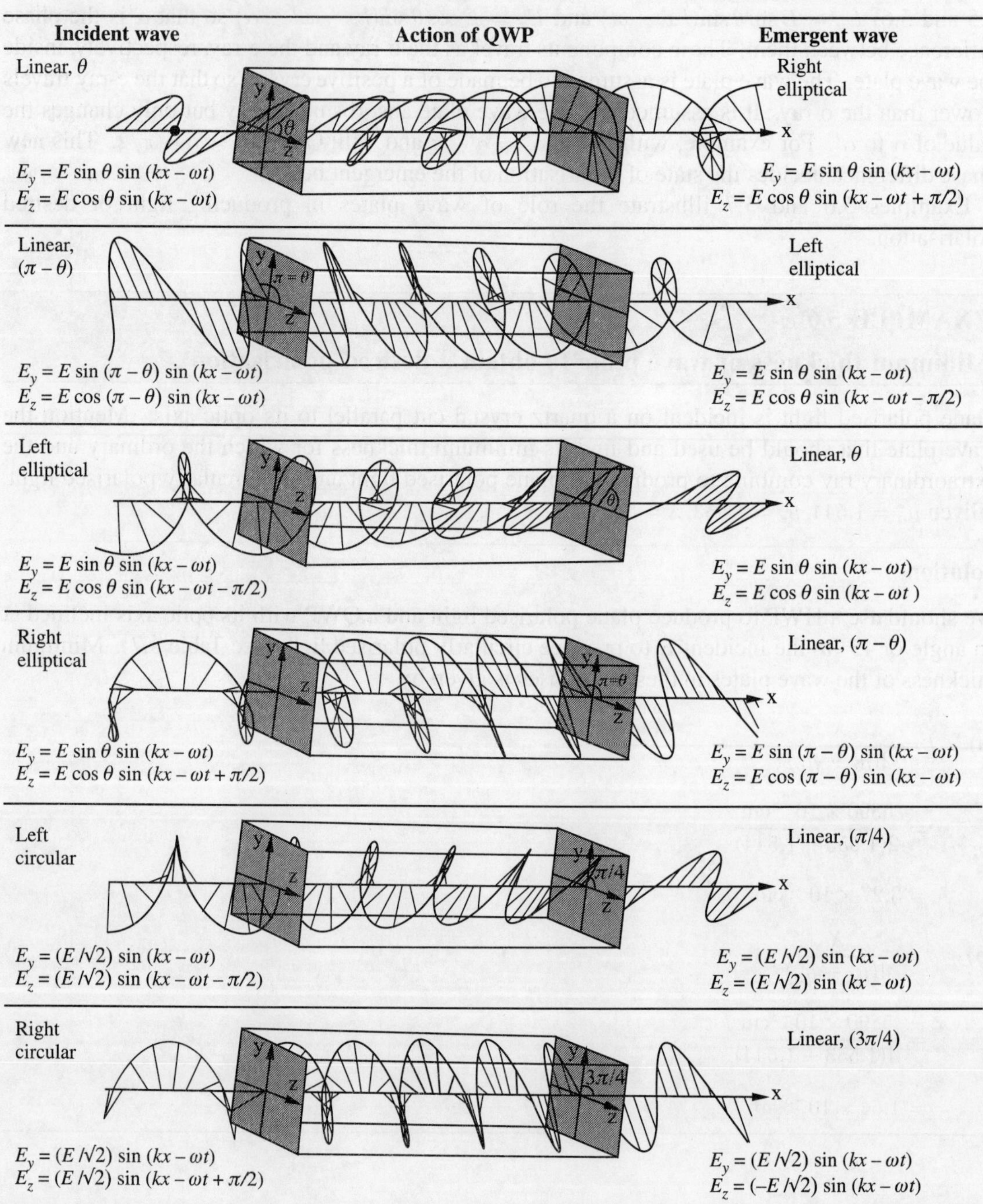

continuously changing inside the plate. For illustrative purposes in the figures below, the difference $k_e x - k_o x$ is exaggerated. When the wave emerges out of the wave plate the phase difference between E_y and E_z is α' with $\alpha' = \alpha - \pi$ for HWP and $\alpha' = \alpha - \pi/2$ for QWP so that α' decides the state of polarisation of the emergent beam.

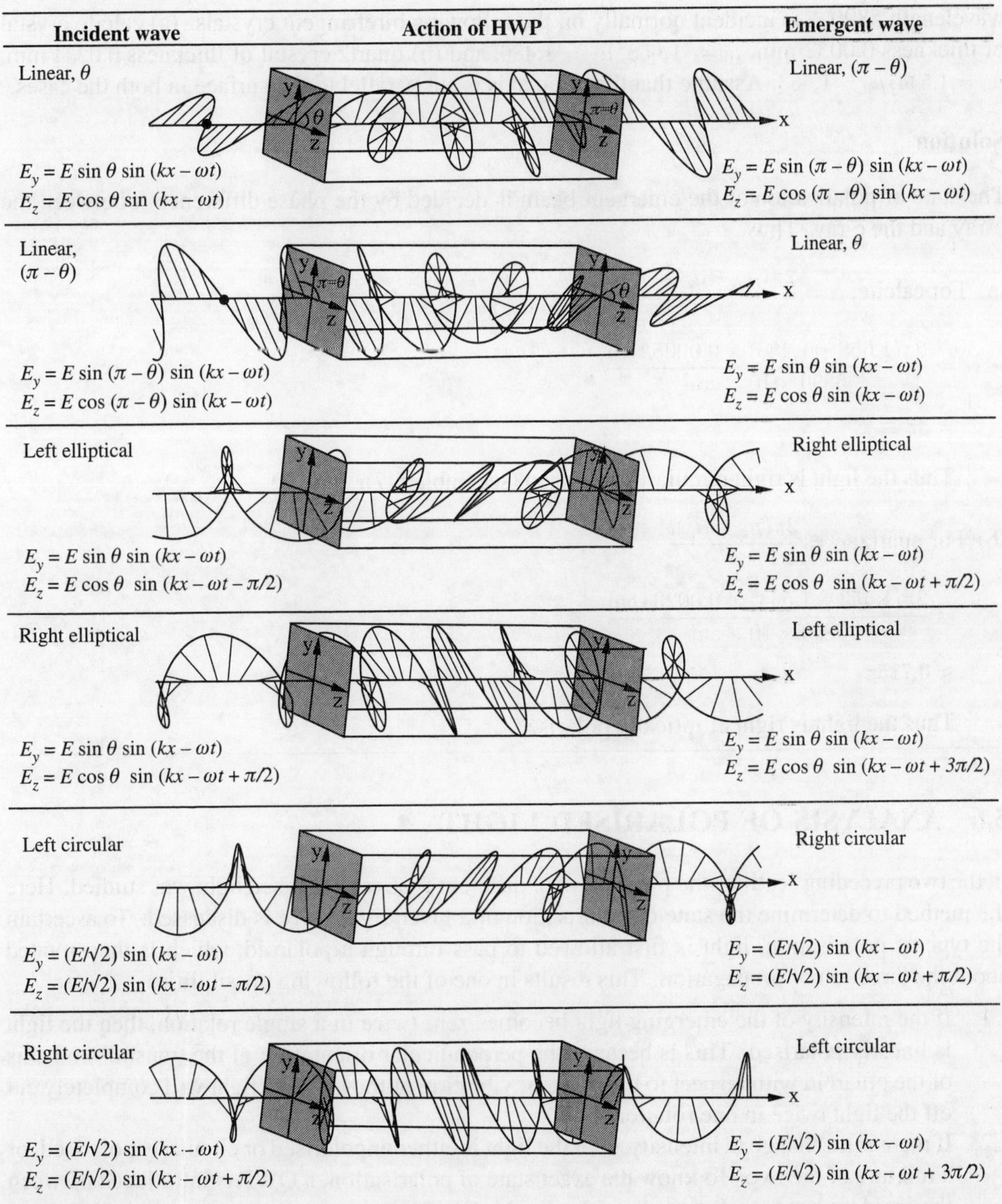

EXAMPLE 5.7

State of polarisation of emergent beam in a birefringent crystal

Find the state of polarisation of the emergent beam when a left circularly polarised beam of wavelength 5890 Å is incident normally on the following birefringent crystals. (a) calcite crystal of thickness 0.0052 mm, $\mu_0 = 1.658$, $\mu_e = 1.486$, and (b) quartz crystal of thickness 0.0234 mm, $\mu_o = 1.544$, $\mu_e = 1.553$. Assume that the optic axis is cut parallel to the surface in both the cases.

Solution

The state of polarisation of the emergent beam is decided by the phase difference α between the o-ray and the e-ray. Thus,

(a) For calcite, $\alpha = \dfrac{2\pi(\mu_o - \mu_e)t}{\lambda}$

$$= \frac{2\pi(1.658 - 1.486) \times 0.00052\,\text{cm}}{[5890 \times 10^{-8}\,\text{cm}]}$$

$$= 3\pi \equiv \pi$$

Thus the light is right circularly polarised (see Table 5.7).

(b) For quartz, $\alpha = \dfrac{2\pi(\mu_e - \mu_o)t}{\lambda}$

$$= \frac{2\pi(1.553 - 1.544) \times 0.00234\,\text{cm}}{5890 \times 10^{-8}\,\text{cm}}$$

$$\approx 0.715\pi$$

Thus the light is right elliptically polarised.

5.6 ANALYSIS OF POLARISED LIGHT

In the two preceding sections the production of different types of polarised light was studied. Here the method to determine the state of polarisation of a given light wave is discussed. To ascertain the type of polarisation, light is first allowed to pass through a polaroid, which is then rotated about the direction of propagation. This results in one of the following possibilities.

1. If the intensity of the emerging light becomes zero twice in a single rotation, then the light is linearly polarised. This is because the perpendicular orientation of the transmission axis of the polaroid with respect to the plane of vibration of the incident light will completely cut off the light twice in one rotation.
2. If there is no change in intensity, then the light is either unpolarised or circularly polarised, or a mixture of the two. To know the exact state of polarisation, a QWP is introduced between the incident beam and the polaroid. As the QWP converts a circularly polarised light to

a linearly polarised one, complete extinction of light intensity twice in one rotation of an analyser will ensure that the incident beam is circularly polarised. If there is no variation in intensity, then the incident beam is unpolarised. If the intensity is minimum but not zero twice in a single rotation of an analyser, then the light is a mixture of unpolarised and circularly polarised light.

3. If the intensity goes to minimum but does not become zero, then the light is either elliptically polarised or a mixture of unpolarised and plane polarised light, or a mixture of elliptically polarised and plane polarised light. To ascertain the state of polarisation, a QWP is introduced in the path of light with its optic axis parallel to the transmission axis of the polaroid, that is, at the position of maximum intensity. As the QWP will convert elliptically polarised light into plane polarised light, the incident beam is elliptically polarised if the rotation of the polaroid results in zero intensity twice in one rotation. If rotation does not result in zero intensity and the same analyser setting as before gives the maximum intensity, then the light is a mixture of unpolarised and plane polarised light. If the maximum intensity is observed at some other analyser setting than before, then the incident beam is a mixture of elliptically polarised and plane polarised light. The analysis of polarised light is summarised in Chart 5.3.

5.7 APPLICATIONS

In this chapter polarisation has been studied as a fundamental phenomenon associated with light waves, and various techniques for production and analysis of polarised light have been discussed. Early work on polarisation was directed mainly towards the scientific investigation of the nature of light, but later studies provided powerful research techniques for investigating the structural details of matter. The spectroscopic investigation of matter based on polarisation relies on the fact that the state of polarisation of the light coming from a particular object is decided by the nature of that object. Such investigations have revealed structures ranging from a galaxy (10^{20} m) to a nucleus (10^{-14} m). Polarisation studies have shown that Saturn rings are made of ice crystals. These studies have also helped us to deduce the size and shapes of microscopic objects like viruses, and interpret spectral lines emitted by various atoms and molecules.

Some of nature's most beautiful phenomena are due to polarisation. The blue colour of the sky is due to the fact that the scattering of light by air molecules in the earth's atmosphere is inversely proportional to the fourth power of the wavelength. Thus violet and blue colours are scattered most and the mixture of scattered colours results in the light blue colour of the sky. At the time of sunset, as the sun's rays traverse the longest distance, the light reaching the earth is depleted of blue colour and therefore appears red. Though normal sunlight is partially polarised, light from the rainbow is completely plane polarised. Scattering also accounts for why most babies have blue or blue-grey eyes at birth. Larger scattering due to smaller particles or shorter molecules in the irises of babies results in blue colour of the eye. As the child grows, the molecules come together and the eye colour turns from blue or blue-grey to brown. Though the normal human eye cannot detect the state of polarisation of light, certain species of bees, ants and beetles are found to possess polarisation-sensing devices built in their eyes.

CHART 5.3 ANALYSIS OF POLARISED LIGHT

Place the polariser in the path of the given light of unknown polarisation. Now rotate the polariser about the direction of propagation of the incident beam through one complete rotation and observe the intensity of the emergent beam.

Light of unknown polarisation

Polariser

Intensity of light	Intensity changes and becomes zero twice	There is no change in intensity	Intensity changes and becomes minimum but not zero twice
Type of light	Linearly polarised	Either unpolarised, circularly polarised, or a mixture of the two	Either elliptically polarised, a mixture of unpolarised and plane polarised light, or a mixture of elliptically polarised and plane polarised light

Now a QWP is introduced before the polariser and the polariser is again rotated.

The light can be

(a) unpolarised or circularly polarised
(b) a mixture of the two
(c) elliptically polarised
(d) a mixture of unpolarised and plane polarised
(e) a mixture of elliptically polarised and plane polarised light

QWP

Polariser

	(a)	(b)	(c)	(d)	(e)	
Intensity of light	There is no change in intensity	Intensity changes and becomes zero twice	Intensity changes and becomes minimum but not zero twice	Intensity changes and becomes zero twice	Intensity changes, becomes minimum but not zero and the same setting of polariser as before gives the maximum intensity	Maximum intensity is observed at some other setting of the polariser than before
Type of light	Unpolarised	Circularly polarised	A mixture of unpolarised and circularly polarised light	Elliptically polarised	A mixture of unpolarised and plane polarised light	A mixture of elliptically polarised and plane polarised light

The breakthroughs in polarisation studies have also contributed to various branches of engineering and technology to make new practical applications of existing knowledge possible. Some of these applications are discussed below.

5.7.1 Applications of Polarising Devices

Many practical devices based on the phenomenon of polarisation employ either a single polaroid or a combination of polaroids. Polaroid sunglasses are made of lenses with a vertical pass axis. They reduce the glare (see Ex. 5.8) due to reflection from horizontal surfaces to a large extent (Fig. 5.14). Polaroids are also used in applications such as aeroplane windows, automobile headlights, visors, camera filters and laser windows. A polariscope, which uses two polarisers in tandem, has numerous uses, especially related to viewing of colourless objects (like white hair, fish scales, silk fibres and transparent crystals during crystal growth).

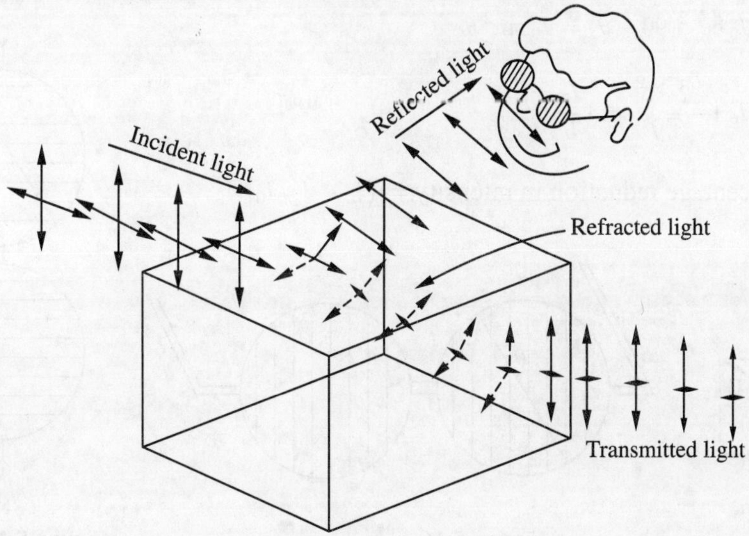

Figure 5.14 Reduction in glare due to reflection using polaroid sunglasses

EXAMPLE 5.8
Glare reduction due to polaroid sunglasses

The amplitude of the horizontal component of the electric field of light on a beach is twice that of the vertical component. Find the percentage reduction in light intensity that reaches a woman's eye sporting polaroid sunglasses in (a) a standing position and (b) a position lying on her side.

Solution

As $E_{\text{horizontal}} = 2E_{\text{vertical}}$, the resultant electric field amplitude is

$E = \sqrt{2^2 + 1^2} = \sqrt{5}$ (Fig. 5.15)

(a) In the standing position the pass axis of the polaroid sunglasses sported by a woman makes an angle θ with **E** as shown in Fig. 5.16.

Thus the transmitted intensity

$$I = I_0 \cos^2\theta = I_0 \left(\frac{1}{\sqrt{5}}\right)^2 = \frac{I_0}{5}$$

or the percentage reduction in intensity = $[(I_0 - I)/I_0] \times 100 = 80\%$.

(b) When lying on her side, the pass axis of the polaroid sunglasses makes an angle $90 - \theta$ with **E** as shown in Fig. 5.17. Thus the transmitted intensity

$$I = I_0 \cos^2(90 - \theta) = I_0 \sin^2\theta$$

$$= I_0 \left(\frac{2}{\sqrt{5}}\right)^2 = 4\frac{I_0}{5}$$

or the percentage reduction in intensity = $[(I_0 - I)/I_0] \times$ 100 = 20%.

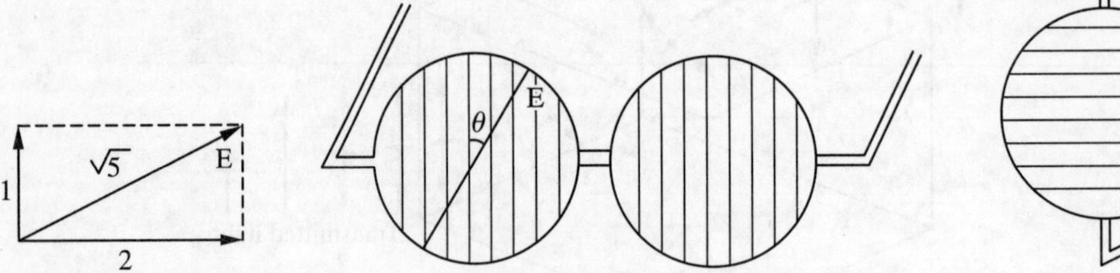

Figure 5.15 **Figure 5.16** **Figure 5.17**

The application of polaroids in stereoscopic (three-dimensional) viewing is explained as follows. The perception of depth by our optic faculties is a highly complicated and only partially understood process. It is believed to be due to the simultaneous impression of both the eyes, each eye viewing the object from a slightly different angle. This can be perceived by holding a finger upright in front of the eyes and viewing it while rapidly opening and closing the two eyes alternately. The finger appears to switch back and forth. In a 3D slide show, a person wears polaroid glasses with pass axes identical to those of the polaroids used in front of the projectors (Fig. 5.18: left eye and left projector are covered by polaroids with horizontal pass axes, and those on right with vertical axes). On viewing the two images simultaneously, depth perception is achieved and a 3D image is seen. If polaroids are not used, then the interference of light coming from the two projectors will result in a blurred image. In 3D photography, a commercial material called *polaroid vectograph* is often used. It is a film consisting of two sheets of polyvinyl alcohol arranged such that their pass axes are perpendicular to each other. Paintings are made using iodine solution on two sides of the film so that they overlap each other. The conduction electrons in iodine align along the pass axes so that the painted regions form the crossed polaroids. Using a linear polariser, either of the two paintings or both of them can be viewed by the proper orientation of the polariser.

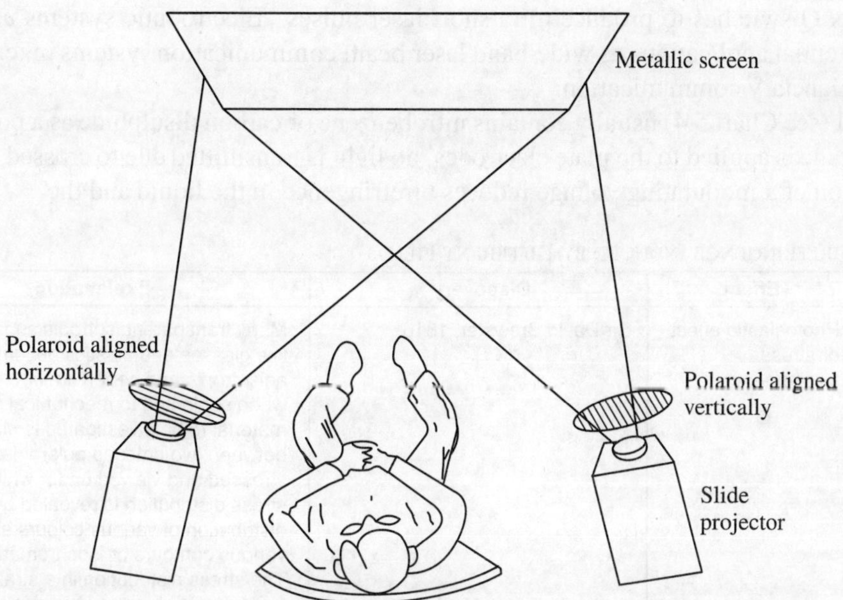

Figure 5.18 3D slide show using polaroids

One of the interesting aspects of research studies in polarisation is that several transparent mineral crystals were explored for their polarising and double refracting properties. These studies helped in the development of artificial gemstones. Since 1961, industrial laboratories (for example, General Electric Research Laboratory) have been able to synthesise diamonds that are better and different from those found naturally. Tiger's eye, cat's eye, yellow diamond and pale yellow diamond are some examples of synthetic diamonds. Mineralogists often use the phenomenon of interference of polarised light for identification of mineral specimens.

5.7.2 Applications of Birefringence

Most of the practical applications of birefringence are based on the principle that birefringence can be induced in an isotropic substance by subjecting it to an electric, magnetic or a mechanical field. Various effects arising out of the interaction of these fields with matter are explained in Chart 5.4.

Photoelasticity (stress induced birefringence) is a field of paramount importance in civil and mechanical engineering. Engineers simulate the stress distribution by application of load to transparent plastic models of structures to check and optimise their design. Photoelastic technique enables engineers to 'see' stress in commonly used engineering components such as gears, turbine blades, hooks, beams and windshields.

Kerr effect and Pockels effect (electric field induced birefringence) are widely used in applications such as electrooptic modulation, high speed photography (shutters) and high speed switching devices. Kerr cells and Pockels cells are used as light beam choppers to replace rotating toothed wheels and as Q-switches to produce ultra-short laser pulses. Electrooptic systems using these cells have potential applications as wide-band laser beam communication systems in terrestrial as well as interplanetary communication.

A Kerr cell (see Chart 5.4) usually contains nitrobenzene or carbon disulphide as a polar liquid. When no voltage is applied to the plate electrodes, no light is transmitted due to crossed polaroids. The application of a modulating voltage induces birefringence in the liquid and the

CHART 5.4 BIREFRINGENCE INDUCED BY DIFFERENT FIELDS

Field	Effect	Discovery	Explanation
Mechanical	Photoelastic effect or stress birefringence	Sir David Brewster, 1816	Many transparent isotropic substances like glasses and plastics become optically anisotropic and hence double refracting when subjected to mechanical stress. The material to be investigated is placed between two crossed polaroids. When stressed and viewed under white light, stress distribution is revealed by the distribution of various colours along various contours or isochromatic regions. The stress distribution in a strained transparent object is shown in the figure.
Electric	Electrooptic effect	J Kerr in glasses, liquids, 1875; F Pockels in piezoelectric crystals,1893	The induction of birefringence due to electric field is known as Kerr effect or Pockels effect if observed in glasses/liquids or piezoelectric crystals, respectively. The experimental arrangement consists of a Kerr cell or a Pockels cell placed between two crossed polaroids as shown in the figures. The applied field is along the beam direction in case of Pockels effect and is at an angle of ± 45° with the pass axes of polarisers in case of Kerr effect. These cells act as electrooptic modulators or shutters.
Magnetic	Magnetooptic effect	W Voigt (in vapours), 1902; Cotton and Mouton (in liquids), 1907	When a constant magnetic field is applied to a transparent medium (vapour in case of Voigt effect and liquid in case of Cotton–Mouton effect) perpendicular to the direction of the incident beam, birefringence is induced, similar to that of a uniaxial crystal with optic axis in the direction of the applied field. This is due to lining up of the magnetically and optically anisotropic molecules in the applied field direction.

cell acts as a variable wave plate. This results in proportional restoration of light and the cell acts as a shutter. Kerr cells can respond to frequencies as high as 10^{10} Hz.

A Pockels cell (see Chart 5.4) is simply an appropriate single crystal of piezoelectric material mounted with its optic axis along the beam direction. Its opposite ends are coated with transparent electrodes usually made of metallic oxides. A Pockels cell can modulate the transmission of light, as explained above, at frequencies well above 10^{10} Hz and can act as a shutter with a response time less than 1ns.

Magnetooptic effects such as Voigt effect or Cotton–Mouton effect find applications in modulation techniques. The signal to be transmitted is applied to the coil as a modulating voltage (see Chart 5.4) and the emerging beam of light carries this information (variations in the modulating voltage) in the form of intensity variations.

CHART 5.4 (Continued)

Figures	**Induced birefringence, $\Delta\mu$**
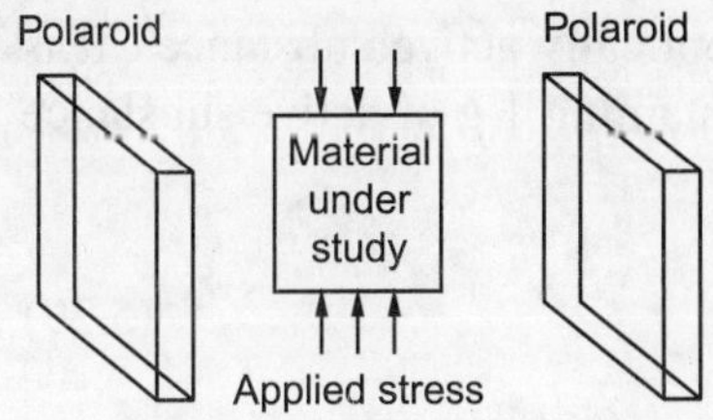 Polaroid Polaroid Material under study Applied stress	*Ks*. *K* is the constant of proportionality depending on the material and s is the amount of stress applied at a given point
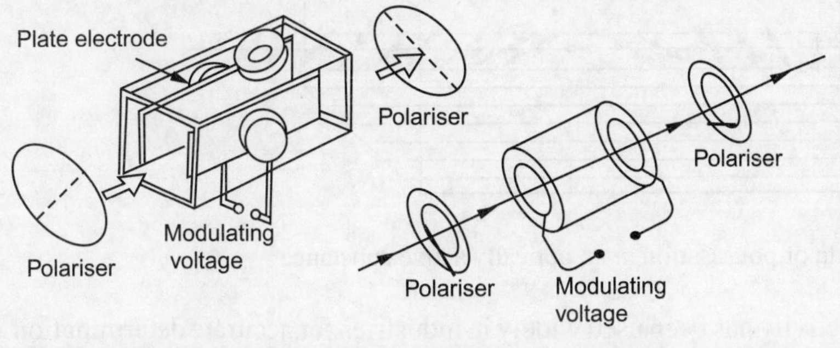 Plate electrode Polariser Modulating voltage Polariser Polariser Polariser Modulating voltage	$K\lambda E^2$. *K* for Kerr effect, is known as the Kerr constant, λ is the wavelength of light used and *E* is the applied electric field. Kerr effect is known as quadratic electrooptic effect as $\Delta\mu \propto E^2$. *KE*, for Pockels effect, where *K* is the constant characteristic of material.
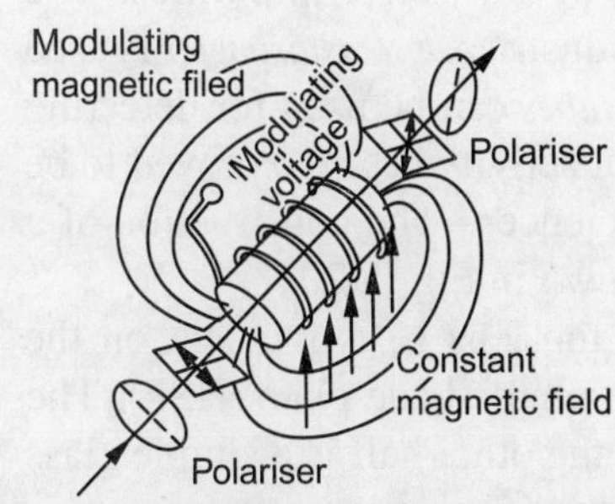 Modulating magnetic filed Modulating voltage Polariser Constant magnetic field Polariser	$K\lambda E^2$. *K* is the constant characteristic of the material, λ is the wavelength of light and *B* is the strength of the applied magnetic field.

5.7.3 Applications of Optical Activity

If a plane polarised beam of light is passed through a quartz crystal, along its optic axis, then the emergent beam is found to have its plane of polarisation tilted through some angle (Fig. 5.19). This phenomenon of the rotation of the plane of polarisation is called *optical activity* and substances exhibiting it are said to be *optically active*. Cinnabar, turpentine, sodium chlorate, sugar solution, and quartz are some well known optically active substances. It is found that different wavelengths suffer different amounts of rotation in their planes of polarisation (rotatory dispersion effect). For a particular wavelength the amount of rotation is proportional to the length, l, traversed by the light beam in the optically active medium. In case of solutions, the rotation is also proportional to the concentration of the optically active material. The rotation produced by d g/cm^3 of the active substance in an inactive solvent is thus given by

$$\theta = \rho l d \tag{5.28}$$

where ρ is called the specific rotation or rotatory power of the optically active substance. It is defined as the rotation produced by a 10 cm column of liquid containing 1 g of active substance for every cubic centimetre of solution,

$$\rho = \frac{10\theta}{ld} \tag{5.29}$$

Figure 5.19 Rotation of the plane of polarisation in an optically active substance

The phenomenon of optical activity has been used widely in industries for accurate determination of the concentration of optically active substances in the presence of non-active impurities. The instrument used for this purpose is called the *polarimeter*. In sugar industries, a *saccharimeter* is used for the accurate determination of sugar concentrations. *Urinalysis tubes* can be used for detecting the concentration of sugar in urine specimens (see Ex. 5.9). Optical activity has also proved to be a useful technique for studying organic molecules in biological sciences. The construction of a typical polarimeter, known as *Laurent's half shade polarimeter*, is shown in Fig. 5.20.

Monochromatic light from the source S, rendered parallel by the lens L, is incident on the polariser N_1. The polarised light coming out of N_1 is incident on a half shade plate (HSP). The HSP is constructed such that one half of it is a half wave plate and the other half is a simple glass plate. The light coming out of these two halves of the HSP will have different intensities.

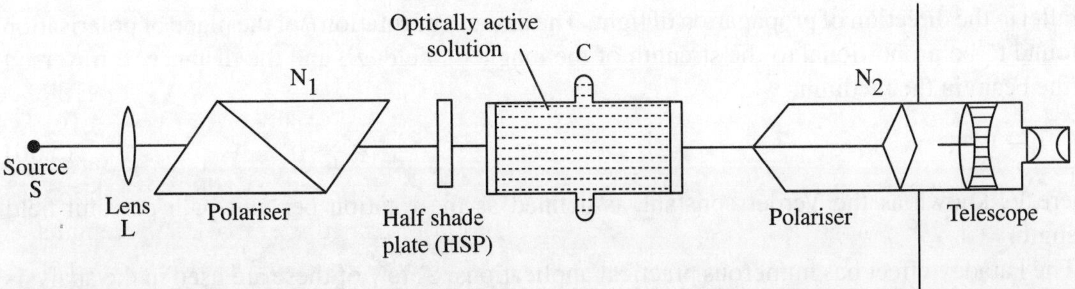

Figure 5.20 Laurent's half shade polarimeter

However, if the polariser N_2 is adjusted such that its principal plane is parallel to the optic axis of the HSP, then the two halves appear to be equally illuminated. In this position of N_2 the glass tube C is filled with the optically active solution. This will result in unequal brightness of the field of view. N_2 is rotated to regain the position of equal brightness. The difference in the two positions of N_2 is the angle of rotation, θ. The concentration of active substance is evaluated with the help of Eq. 5.28 using known values of ρ and l (ρ can be determined using the solution for which the concentration of the active substance is known).

EXAMPLE 5.9
Application of optical activity

A urinalysis tube uses the phenomenon of optical activity to detect the amount of sugar in a specimen of urine. Assuming that the instrument can read accurately to 0.1°, will a 20 cm long tube be sensitive enough to detect a difference in sugar concentration of 1mg/cm^3. (Specific rotation of sugar = 6.6° g^{-1} cm^2)

Solution

The rotation produced due to optical activity is given by

$$\theta = \rho l d$$

Thus the change in θ produced due to change Δd in d is

$$\begin{aligned}\Delta\theta &= \rho l \Delta d \\ &= 6.6^\circ \mathrm{g}^{-1}\,\mathrm{cm}^2 \times 20\,\mathrm{cm} \times 0.001 \mathrm{g/cm}^3 \\ &= 0.132^\circ\end{aligned}$$

which is above the 0.1° least count. Thus the concentration of 1 mg/cm^3 will be detected by the given urinalysis tube.

Optical activity can also be induced in inactive substances with the help of an external field. Michael Faraday discovered this in glass in 1845, using a magnetic field applied in a direction parallel to the direction of propagation of light. The amount of rotation θ of the plane of polarisation is found to be proportional to the strength of the magnetic field, B, and the distance, l, traversed by the beam in the medium,

$$\theta = VBl, \tag{5.30}$$

where V, known as the Verdet constant, is defined as the rotation per unit path per unit field strength.

The Faraday effect has numerous practical applications. A few of these are used in the analysis of mixtures of hydrocarbons, in spectroscopy for studying the properties of energy levels above ground level, and in communication engineering for modulation, coupling and matching purposes.

One of the most remarkable applications of optical activity is in the *liquid crystal display* (LCD). A liquid crystal state is characterised by long-range orientational order of molecules. In the OFF state of LCD the liquid crystals rotate the polarised light through 90° enabling it to pass through the rear polariser (see Fig. 5.21a). A mirror reflects light back through the crystals and then through the front polariser to form the white background of the display. When a key is pressed (ON state) an electric field is applied to specific areas of the crystals which destroys the optical activity of crystals in these areas so that the state of polarisation of light passing through them is unaffected and it is blocked by the rear polariser. No light reaches the mirror and thus the affected areas of the display look black, displaying a particular character (Fig. 5.21b).

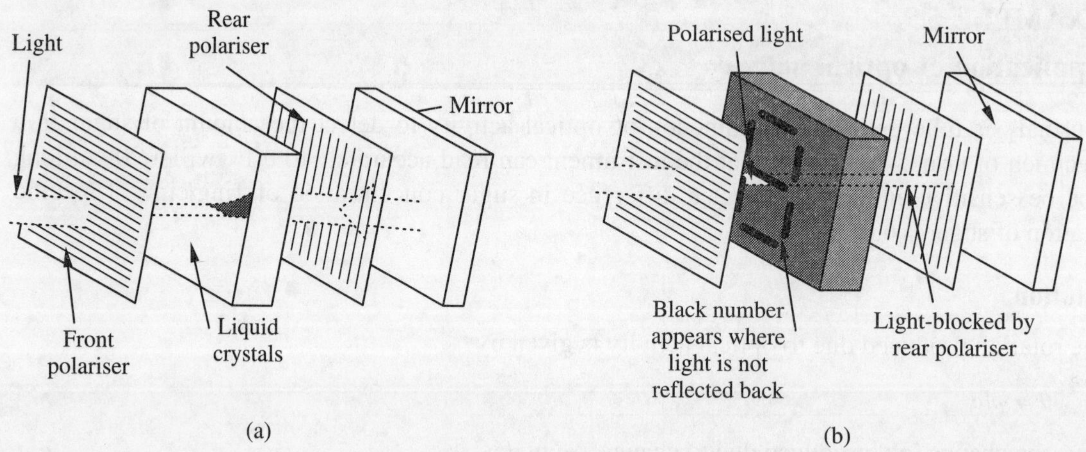

Figure 5.21 A liquid crystal display in (a) OFF state and (b) ON state

LCD devices find wide application in the electronic and computer industries. Applications of liquid crystals are not restricted to display only. Some temperature sensitive liquid crystals are used in medical diagnostics for early detection of tumours, malignant growths and leprosy. Research studies on liquid crystals have also provided some useful links between the physical and biological sciences.

EXERCISES

1. What is meant by plane polarised, elliptically polarised and circularly polarised light?
2. How will you experimentally distinguish between plane polarised, circularly polarised and elliptically polarised light with the help of a polariser (or analyser) alone?
3. State Brewster's law. How can it be used to produce plane polarised light?
4. Describe production of plane polarised light by reflection. Hence, obtain an expression for Brewster law.
5. Describe how elliptically and circularly polarised light can be produced.
6. Explain the terms
   - (i) double refraction
   - (ii) optic axis
   - (iii) principal plane
7. Differentiate between
   - (i) ordinary and extraordinary rays
   - (ii) positive and negative crystals
   - (iii) biaxial and uniaxial crystals

PROBLEMS

1. Discuss the state of polarisation when the x- and y-components of the electric field are given by the following equations:
   - (i) $E_x = E_o \cos(\omega t + kz), E_y = E_o/\sqrt{2}\cos(\omega t + kz + \pi)$
   - (ii) $E_x = E_o \sin(\omega t + kz), E_y = E_o \cos(\omega t + kz)$.
2. Light of intensity I_o is incident on a polariser. What is the intensity of the resultant beam if:
   - (i) incident light is unpolarised?
   - (ii) incident light is plane unpolarised with its electric field making an angle of 30° with the axis of the polariser?
3. Unpolarised light falls on a polarising sheet placed on top of another. What should be the angle between characteristic directions of the sheets if the intensity of transmitted light is one third of the intensity of the incident beam.
4. A half wave plate is fabricated for a wavelength of 3800 Å. For what wavelength does it work as a quarter wave plate?
5. Consider a positive crystal with refractive indices 1.31 and 1.309 for the e-ray and o-ray, respectively. What should be the minimum thickness of that crystal so that it can act as a quarter wave plate for light of wavelength 6000 Å?

6

Quantum Physics

6.1 THE STORY OF QUANTUM PHYSICS

So far we have seen how the various observed properties of light could be explained assuming light to be a wave. By the end of the nineteenth century this seemed to have finally settled the matter in favour of the wave nature of light. However, the citadel of wave theory started developing cracks at the dawn of the twentieth century.

Historically, one of the earliest experiments that indicated that something was wrong with the wave theory was the study of what is known as *blackbody radiation*. All bodies emit radiation. But these radiations are not noticeable at room temperature due to their low intensity and the fact that these radiations are mainly in the infrared (IR) range. An ideal blackbody absorbs all radiations incident upon it. Thus the radiation spectrum—plot of intensity of radiation versus wavelength—emitted by an ideal blackbody consists of all wavelengths. The experimental spectrum observed by Lummer and Pringsheim in 1900 is shown in Fig. 6.1. A point of fundamental importance noticed in this study was that the general characteristics of the spectrum are independent of the material and dependent only on the temperature.

In 1896, assuming the wave nature of light and using general thermodynamic principles, Wien made one of the first attempts to explain the spectrum of blackbody radiation. His predictions agreed well for lower wavelengths but broke down for longer wavelengths. In 1900, Rayleigh and Jeans advanced a theory which agreed with experimental results for higher wavelengths. But their theory predicted infinite intensity for low wavelength radiation ($\lambda \to 0$). This failure of Rayleigh–Jeans theory is often described as the *ultraviolet catastrophe*.

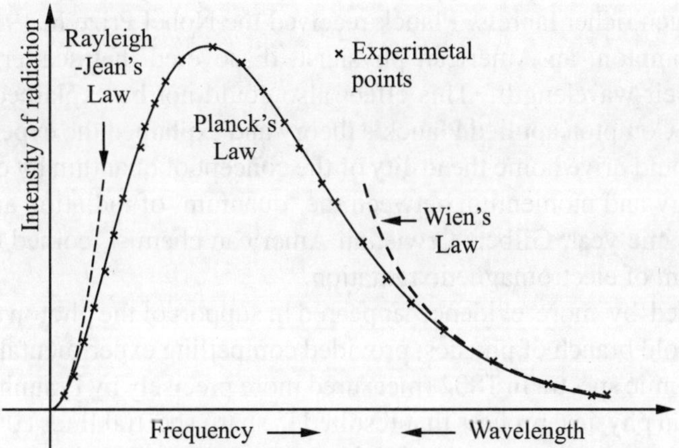

Figure 6.1 Comparison of experiment with theory in blackbody radiation

Max Planck, a German physicist, was curious to find the correct explanation of blackbody radiation and tenaciously devoted six years of his life before he hit upon the solution in 1900. However Planck, a conventional man, was quite uncomfortable with the assumptions he had to make in his model.

1. The energy of the atomic oscillators should be proportional to their frequency.
2. The emission and absorption of energy at any given frequency should be in steps of some basic unit called *quantum*.

Planck struggled to come to terms with these strange and unorthodox ideas, which were a radical departure from the ideas of conventional physics (often called classical physics). Planck obtained the best fit with the experimental data by considering the constant of proportionality to be 6.6×10^{-34} Js. Planck's quantum hypothesis was looked upon by many scientists of that time—including Planck himself—as a mere calculational trick rather than a new fundamental principle. However, the quantum theory gained much more credence in later years as the experimental evidence kept mounting in favour of this theory.

Prominent among the experiments that supported Planck's quantum theory of radiation were the photoelectric effect, Compton effect and line spectra. The photoelectric effect was discovered by Hertz in 1887 when he observed the emission of charge from a metal surface when light was incident on it. It is interesting to note that Hertz discovered this effect accidentally when he was conducting some experiments to prove the electromagnetic wave nature of radiation proposed by Maxwell about a decade ago. Ironically the experiment performed to establish the wave theory of light contained the seeds of quantum theory of light. Hertz was clueless about this new effect. In 1897, JJ Thomson discovered the electron and in 1900, Lenard, who was Hertz's student, showed that the emitted charge consists of electrons.

When the wave theory failed to explain the photoelectric effect, it was Einstein again who provided the crucial insight into cracking this riddle. Einstein was convinced by Planck's theory and extended it to offer a correct theoretical explanation of the photoelectric effect in 1905. It should be noted that Einstein received the Nobel prize in 1921 for his explanation of the photoelectric effect and not for the theory of relativity which he proposed in the same year (1905) and brought him much richer laurels. Planck received the Nobel Prize in 1918.

In 1926, AH Compton, an American physicist, discovered that scattering of x-rays results in an increase in their wavelength. This effect also could not be explained assuming the wave nature of radiation. Compton applied Planck's theory and explained the experimental results quite convincingly. He could drive home the utility of the concept of quantum by explicitly considering the transfer of energy and momentum between the 'quantum' of radiation and the electron in his derivation. In the same year, Gilbert Lewis, an American chemist, coined the word 'photon' to describe the quantum of electromagnetic radiation.

As the years passed by more evidence appeared in support of the photon theory of light. Spectroscopy, a century-old branch of physics, provided compelling experimental evidence. Wollastan recorded the first atomic spectra in 1802 (measured more precisely by Fraunhoffer in 1814). However, the fundamental physical process that resulted in sharp spectral lines could not be understood on the basis of the classical wave theory of light. In 1913, Niels Bohr, a physicist from Denmark, offered the first successful account of spectral lines for hydrogen using Planck's quantum theory. Planck's quantum theory also received strong support through the work of experimentalists who found the value of Planck's constant. For example, the value obtained by Millikan, determined photoelectrically in 1916, agreed within experimental limits with that used by Planck to fit the experimental results of blackbody radiation. The most accurate known value of Planck's constant is $h = (6.62606876 \pm 0.0000052) \times 10^{-34}$ Js.

Physicists were confronted with the interesting situation arising out of these developments in the first quarter of the twentieth century. Properties such as interference, diffraction and polarisation could be explained assuming light to be a wave, whereas others like the photoelectric effect, Compton effect, line spectra and blackbody radiation could be explained by considering light to be made of particles. Thus, what is the true nature of light? Is it a wave? Or is it a particle? It was not before long that the physicists realised that these contradictions are only apparent. They realised that it is like asking the question: What is a coin? Is it heads? Or is it tails? The answer is, it is both. Bohr proposed a complementarity principle according to which the wave and the particle aspects of light complement rather than contradict each other. This reconciliation also came to be known as the *wave–particle dualism* of radiation.

The wave–particle dualism of radiation aroused the curiosity of a young and brilliant French physicist—Prince Louis de Broglie. De Boglie who was a student of history was attracted to physics since his elder brother Maurice was pursuing a career in it. A thought haunted de Broglie: If light, which had been assumed to be a wave for the last two centuries can be assumed to be a particle, then why cannot matter which had hitherto been seen as a particle be assumed to have wave nature? Taking a cue from the inherent symmetry and simplicity of nature, de Broglie extended the idea of wave–particle dualism to matter. He proposed a revolutionary idea of 'matter waves' in 1924 in his doctoral thesis.

The experimental proof for the existence of matter waves was provided almost accidentally by CJ Davisson and LH Germer of USA in 1927. Along with GP Thomson of Scotland, they demonstrated the wave nature of electrons through diffraction experiments. De Broglie was awarded the Nobel Prize in 1929 after his theory was experimentally validated by these scientists. The wave character of many other microparticles—including protons, neutrons and positrons—has been studied in great detail since de Broglie's discovery.

The wave description of a particle was not a simple problem because a wave spreads, whereas a particle is localised in space. In 1927, Werner Heisenberg, a German theoretical physicist, gave deeper thought to these problems and enunciated one of the most fundamental principles in physics known as the 'Uncertainty principle'. Heisenberg's principle uses the concept of a wave packet, rather than a simple wave to describe a particle. Heisenberg's principle asserts that it is not possible for us to view both the particle and wave nature of radiation (or matter) simultaneously with arbitrary accuracy. The experiments that bring out the particle character suppress the wave character and vice versa. Which of the two facets (particle or wave) is revealed to the experimentalist is determined by the experiment itself.

Heisenberg's principle beautifully protects the wave–particle dualism of radiation and matter. Nature's harmony captured by Heisenberg in his principle is illustrated in an interesting manner by JB Rajam (Atomic Physics, 1971): *Reality, whether matter or radiation, is made up of a subtle and almost indefinable fusion of two antagonistic but complementary factors, the continuous wave and the discontinuous particle; it is a discontinuous continuity or a continuous discontinuity and hence not a simple but complex unity*!

The revolutionary research breakthroughs in the first quarter of the twentieth century gave physicists an entirely fresh perspective of the phenomena in the microworld. These developments gave birth to a new branch of physics, which came to be known as *quantum mechanics*. This branch formed the foundation of modern physics. It was named so in the background of the old mechanics, which was referred to as classical mechanics or Newtonian mechanics. The emergence of quantum mechanics led to the fundamental question: What is the most fundamental equation of quantum mechanics, just as Newton's is the most fundamental equation in classical mechanics?

At the University of Zurich, Erwin Schrodinger (1867–1961), a young Austrian physicist, was studying de Broglie's theory for presentation in a seminar as per the suggestions of the eminent physical chemist, Peter Debye. During the course of his study Schrodinger took the above question to heart. In 1926 he discovered the equation and used it to solve the problem of the hydrogen atom. Schrodinger's equation includes all the essential elements of the discoveries of Planck, de Broglie and Heisenberg.

However, it should be noted here that the results of the quantum theory reduce to those obtained by classical physics when macroscopic systems are under study. For such systems the solution of Schrodinger's equation would give results which match those of Newtonian mechanics. Macroscopic systems turn out to be those for which it would be possible to consider h to be zero.

In this chapter the basic quantum concepts such as Planck's theory, wave–particle duality, uncertainty principle and Schrodinger equation are introduced (Sections 6.2 and 6.6–6.9). These concepts are used to study the photoelectric effect and the Compton effect in Sections 6.3–6.5. Application of these concepts to engineering and technology is discussed in Section 6.10.

6.2 PLANCK'S QUANTUM THEORY

Planck made the following assumptions in his theoretical explanation of the observed blackbody radiation spectrum. These assumptions later became the fundamental postulates of the quantum theory of radiation.

1. An atomic oscillator cannot have an arbitrary amount of energy but can only be in one of the discrete energy levels given by

$$E_n = nh\nu \qquad (6.1)$$

where $n = 1, 2, 3, \ldots$ and ν is the frequency of radiation; h is known as the *Planck's constant* and its approximate value is 6.625×10^{-34} Js.

2. An oscillator can emit or absorb energy only by making a transition from one allowed energy state to another defined through Eq. 6.1. Thus the change in oscillator energy is given by

$$\Delta E = h\nu \qquad (6.2)$$

Example 6.1 illustrates why quantisation is not noticed in a large scale oscillator but assumes significance at the atomic level.

These assumptions by Planck were limited to the explanation of blackbody radiation spectrum, that is, the radiant energy was assumed to be quantised only during absorption or emission. Planck assumed that radiation behaves as an electromagnetic wave during propagation.

EXAMPLE 6.1

Quantised energy levels for microscopic and macroscopic systems

Find the first three energy levels of (a) a one second simple pendulum and (b) the hydrogen electron, assuming they are quantised systems. If the pendulum is taken to be that of a typical wall clock with mass, $m = 10$ g and amplitude $a = 1$ cm, find the quantum number n. Find the percentage change in energy for transition of (i) pendulum from nth state to $(n+1)$th state and (ii) hydrogen electron from ground state to the next quantum state.

Solution

(a) For a 1s simple pendulum, $\nu = 1$ Hz.

Assuming Planck's quantisation principle,

$$E_n = nh\nu = n \times 6.625 \times 10^{-34} \text{ Js} \times 1 \text{ s}^{-1} = 6.625 \times 10^{-34} \times n \text{ J}$$

or $E_1 = 6.625 \times 10^{-34}$ J

$$E_2 = 13.25 \times 10^{-34} \text{ J}$$

$$E_3 = 19.87 \times 10^{-34} \text{ J}$$

(b) For a hydrogen electron,

$$E_n = \left(\frac{-13.6}{n^2}\right) \text{eV}$$

Thus, $E_1 = -13.6$ eV, $E_2 = -3.4$ eV, and $E_3 = -1.5$ eV

Now for a given simple pendulum,

$$E = \frac{1}{2}\,\text{m}\,\omega^2 a^2$$

$$= \frac{1}{2}(10 \times 10^{-3}\ \text{kg})(2\pi \times 1)^2 (1 \times 10^{-2})^2$$

$$= 1.97 \times 10^{-5}\ \text{J}$$

$$\text{Thus}\quad n = \frac{E}{h\nu}$$

$$= \frac{1.97 \times 10^{-5}}{(6.625 \times 10^{-34} \times 1)} = 2.97 \times 10^{28}$$

(i) Percentage change in energy of pendulum for transition from the nth state to the $(n+1)$th state

$$= \frac{(E_{n+1} - E_n)}{E_n} \times 100 = \frac{[(n+1)h\nu - nh\nu]}{nh\nu} \times 100$$

$$= \frac{1}{n} \times 100 = \frac{100}{2.97 \times 10^{28}} \cong 10^{-27}$$

(ii) Percentage change in energy of hydrogen electron due to transition from the ground state to the next quantum state

$$= \frac{|(-3.4 + 13.6)| \times 100}{|-13.6|} = 75\%$$

See Ex. 6.18.

Einstein extended Planck's assumption and stated that the electromagnetic energy is quantised not only during absorption or emission but also during propagation. The quantum of electromagnetic radiation is called a photon. The energy of each photon is given by

$$E = h\nu \tag{6.3}$$

Equation 6.3 is also expressed as

$$E = \hbar\omega \tag{6.4}$$

where $\hbar = h/2\pi = 1.05 \times 10^{-34}$ Js and $\omega = 2\pi\nu$ is the angular frequency of radiation. Example 6.2 illustrates how the handy form of Eq. 6.3 is useful for solving numerical problems.

EXAMPLE 6.2

Handy formula to find photon energy

(a) Show that the energy of a photon of wavelength λ is given by E (eV) $= 12422/\lambda$ (Å).
(b) Find the range of photon energies in the visible portion of the electromagnetic spectrum.

Solution

(a) Energy of a photon

$$E = h\nu = \frac{hc}{\lambda}$$

$$= \frac{(6.625 \times 10^{-34}\ \text{Js})(3 \times 10^{8}\ \text{m/s})}{(\lambda \times 10^{-10}\ \text{m})} = \frac{(19.875 \times 10^{-16})\ \text{J}}{\lambda\ \text{Å}}$$

$$\text{or } E(eV) = \frac{12422}{\lambda\ \text{Å}}$$

(b) The wavelength range for visible light is 4000–7000 Å. Hence using the above expression, we get the range of energies for visible photons as 1.8 eV to 3.1 eV.

It is emphasised here that a photon should be understood as a particle of light and not a particle of matter, as assumed in Newton's corpuscular theory. A photon is a concentrated bundle of energy. It has no rest mass meaning that a photon can never exist at rest. It always travels at the speed of light. The mass of a photon is only due to its energy and is given by $m = E/c^2$. This equation can be obtained using the relativistic energy–momentum relation of Einstein

$$E = \sqrt{m_0^2 c^4 + p^2 c^2} \tag{6.5}$$

and using $m_0 = 0$ and $p = mc$ for a photon. The momentum of a photon is given by

$$p = \frac{E}{c} = \frac{h\nu}{c} = \frac{h}{\lambda} \tag{6.6}$$

If a source of radiation is emitting N photons per second, each photon carrying energy $h\nu$, then the intensity of radiation is given by

$$I = Nh\nu \tag{6.7}$$

(Intensity is defined as the rate at which energy is delivered by the source.)

6.3 PHOTOELECTRIC EFFECT

The photoelectric effect, as the name suggests, means generation of electrons using light. When light of suitable frequency is incident on a photosensitive material, electrons are observed to be ejected from its surface. This phenomenon is known as photoelectric effect. The electrons so emitted are called the *photoelectrons* and the current constituted by them is known as *photoelectric current*. The experimental details of the study of the photoelectric effect are dealt with in Expt. 6.1. These details are essential for the further study of the photoelectric effect.

EXPERIMENT 6.1 PHOTOELECTRIC EFFECT

Aim

To study the photoelectric effect

Apparatus

Photocell, battery with potentiometric arrangement, voltmeter, suspended coil galvanometer, lens with a variable diaphragm, mercury discharge lamp, three filters for the transmission of 4358 Å, 5461 Å and 5780 Å lines of mercury

Experimental arrangement

Figure 6.2 shows the experimental arrangement required for the study of photoelectric effect. Light from the source S of mercury vapour lamp is focussed by lens L. D is the variable diaphragm that controls the intensity of light falling on the photocathode, C. F is a filter, which transmits one of the wavelengths of mercury (Hg) light. The frequency of light can thus be varied by using different filters. A photocell is normally biased such that the cathode C is at the negative potential with respect to the anode A. In the reverse mode the polarities of the battery are interchanged so that the cathode is made positive with respect to the anode. A voltmeter measures the voltage across the two electrodes and the suspended coil galvanometer measures the photoelectric current, i_p.

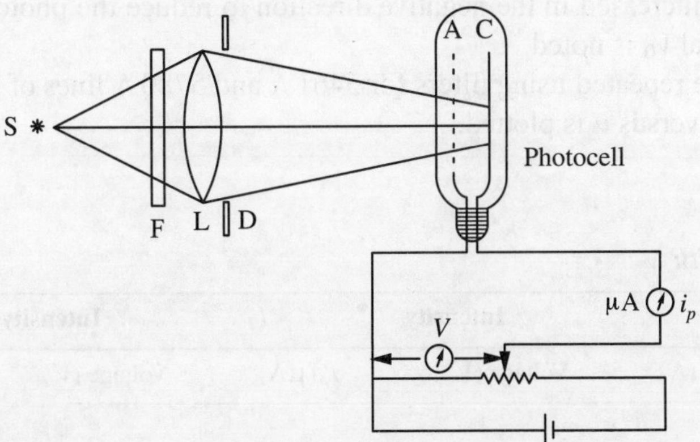

Figure 6.2 Experimental arrangement to study the photoelectric effect

Theory

Light incident on a photocathode results in the ejection of photoelectrons. When these photoelectrons flow through the external circuit, they generate a photocurrent, which can be measured by a galvanometer. If the photocell is connected in a reverse mode, then the negative bias of anode repels the photoelectrons emitted from the cathode. The minimum reverse voltage that makes the photocurrent zero is known as the *stopping potential*, V_0. Evidently this voltage stops the fastest moving photoelectron or the most energetic photoelectron. Hence

$$eV_0 = K_{\max} = \frac{1}{2}mv_{\max}^2 \tag{6.8}$$

Electrons in the cathode material are bound with an energy called the *work function*, ϕ, of the material. Thus the work function of a material can be defined as the *minimum energy* that should be supplied to just free an electron (to just come to the surface of the material without any kinetic energy) and make photoemission happen.

Procedure

A. For variation of i_p with V

1. The connections are made as shown in Fig. 6.2.
2. For a particular orientation of D (corresponding to minimum intensity, I_1), the maximum positive voltage is applied to A and the current, i_p, is noted.
3. The voltage is reduced at some fixed intervals and the corresponding values of i_p are measured. This is continued for negative V till i_p becomes zero.
4. Steps 2 and 3 are repeated for higher intensities I_2 and I_3 by adjusting the diaphragm.
5. The graph of i_p versus V is plotted for I_1, I_2 and I_3.

B. For variation of V_0 with the frequency ν of the incident light

1. The filter for the 4358 Å line of Hg is placed before the lens L.
2. The potential is increased in the negative direction to reduce the photocurrent to zero. The stopping potential V_0 is noted.
3. Steps 1 and 2 are repeated using filters for 5461 Å and 5780 Å lines of Hg.
4. The graph of V_0 versus ν is plotted.

Observations

A. Variation of i_p with V

Intensity I_1		**Intensity** I_2		**Intensity** I_3	
Voltage (V)	i_p(µA)	Voltage (V)	i_p(µA)	Voltage (V)	i_p(µA)

B. Variation of V_0 with ν

Frequency, ν (Hz)	Stopping potential, V_0 (volts)

Graph See Fig. 6.3.

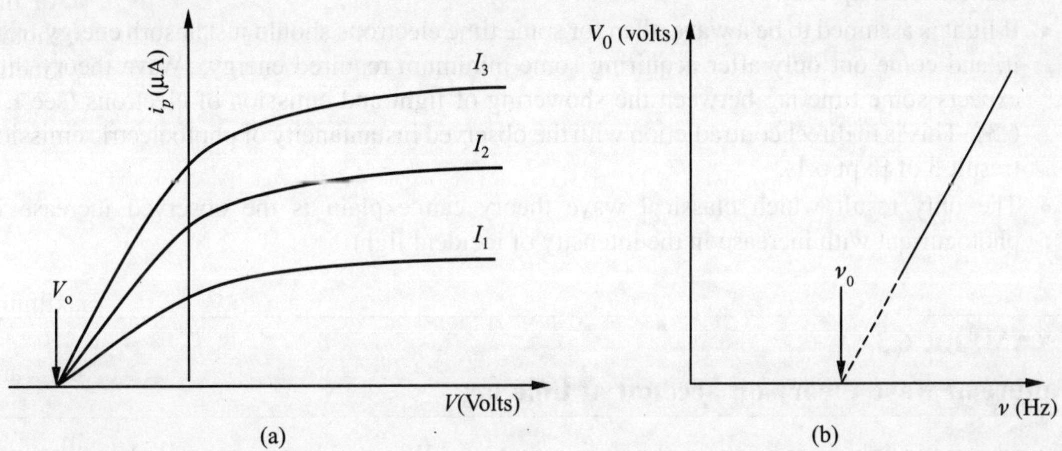

Figure 6.3 (a) Variation of photocurrent with voltage and intensity; (b) Variation of stopping potential with frequency

Results

1. It can be seen from Fig. 6.3a that the photocurrent increases with both the positive voltage and intensity and attains a saturation value. However, the stopping potential (that is, $K_{\max}$ for electrons) is independent of the intensity of incident light.
2. It can be seen from Fig. 6.3b that the stopping potential (that is, $K_{\max}$) increases linearly with the frequency of incident light, indicating that photoemission will not occur for frequencies below a certain minimum frequency, called the *threshold frequency*, ν_0.
3. No measurable time lag is observed between the incidence of light and emission of electrons, or, the photoelectric effect is an instantaneous process.

Discussion

The features of the photoelectric effect are summarised in Fig. 6.3 and the results given above. In attempting to offer a theoretical explanation for these features it is seen that the wave theory of light fails on all three counts (see Chart 6.1). On the other hand, all experimental observations are in good agreement with the quantum theory of light. *This experiment clearly demonstrates how the failure of certain well-established ideas and concepts can pave the way for radically new ideas in physics.*

The wave theory fails to account for the results obtained in the photoelectric effect for the following reasons:

- Wave theory implies that increase in the intensity of incident light should result in increase in the kinetic energy, $K_{\max}$, of the emitted photoelectrons. This is because larger intensity of a light wave implies larger electric field amplitude that will thrust electrons out with higher $K_{\max}$. This contradicts result 1 obtained in Expt. 6.1.
- The wave theory implies a threshold intensity (minimum intensity below which photoemission should not take place), contradicting result 2 of Expt. 6.1 on the existence of a threshold frequency.
- If light is assumed to be a wave, then for some time electrons should just absorb energy from it, and come out only after acquiring some minimum required energy. Wave theory thus expects some time lag between the showering of light and emission of electrons (see Ex. 6.3). This is in direct contradiction with the observed instantaneity of photoelectric emission (result 3 of Expt 6.1).
- The only result which classical wave theory can explain is the observed increase in photocurrent with increase in the intensity of incident light.

EXAMPLE 6.3

Failure of wave theory on account of time lag

Monochromatic light from a source of power output 10^{-5} W is incident on a metal plate 5 m away from the source. Suppose that light is a wave that supplies energy continuously to an electron inside the metal. This to-be-ejected electron collects its energy from a circular area of the plate 10^{-9} m in radius (approximately 10 atomic diametres). If the work function of the metal is 5 eV, how long would it take for the electron to acquire this much energy from the circular area and come to the metal surface?

Solution

Power of the source = 10^{-5} W = 10^{-5} J/s

The circular area from which the electron collects energy is

$$= \pi(10^{-9}\,\text{m})^2 = \pi \times 10^{-18}\,\text{m}^2$$

The area of the 5-metre sphere centred on the light source is

$$4\pi(5\,\text{m})^2 = \pi \times 100\,\text{m}^2$$

Thus, if the light source radiates uniformly in all directions, the rate P at which energy falls on the circular area is given by

$$P = (10^{-5}\ \text{J/s})\,\frac{\pi \times 10^{-18}\ \text{m}^2}{\pi \times 100\ \text{m}^2} = 10^{-25}\ \text{J/s}$$

The work function is the minimum energy required by an electron to overcome the surface barrier. Assuming that all this power is absorbed, the required time is given by

$$t = \frac{5\,\text{eV}}{10^{-25}\,\text{J/s}} \times \frac{1.6 \times 10^{-19}\,\text{J}}{1\,\text{eV}} = 8 \times 10^{6}\,\text{s} \approx 93 \text{ days!}$$

However, photoemission is observed to be an instantaneous effect.

This example thus brings to the fore the striking failure of the wave theory.

Einstein applied Planck's quantum theory of radiation and successfully accounted for the observed features of the photoelectric effect in the following way. He visualised the photoelectric effect as arising due to a collision process in which an incident photon transfers all its energy to an electron bound in a cathode material. The electron uses a part of this energy in overcoming binding forces in the material. The remaining energy appears as the kinetic energy of the electron. Thus Einstein suggested a photoelectric equation,

$$h\nu = \phi + K_{\text{max}} \tag{6.9}$$

where $h\nu$ is the energy of the incident photon, and ϕ or the work function is the energy lost in overcoming binding forces in the material.

When the intensity of incident light is increased, keeping its frequency constant, it merely increases the number of photons in the beam without increasing energy of each photon as per Eq. 6.7. This leaves the kinetic energy of emitted photoelectrons unaffected in agreement with result 1 of Expt. 6.1.

Using Eq. 6.9, if $h\nu < \phi$, then $K_{\text{max}} < 0$, meaning that photoemission will not take place. For photoemission to take place, $h\nu > \phi$; or writing $\phi = h\nu_0$, the condition for photoelectric effect to occur is that $\nu > \nu_0$ where ν_0 is the threshold frequency. The existence of the threshold frequency (result 2 of Expt. 6.1) is thus accounted for by the quantum theory.

The collision process visualised in the photoelectric effect involves an instantaneous transfer of energy from a photon to a bound electron. Thus the quantum theory expects the photoelectric effect to be an instantaneous effect; an explanation, which is in complete agreement with the experiment (result 3 of Expt. 6.1).

Using Eq. 6.8, Eq. 6.9 can be rewritten as,

$$V_0 = \frac{h\nu}{e} - \frac{\phi}{e}$$

This is the equation of a straight line with slope h/e and intercept on y-axis equal to ϕ/e. The agreement of this equation with the experimental graph (Fig. 6.3b) brings out the quantitative success of the quantum theory. From the values of the slope and the intercept on the y-axis, the values of Planck's constant and work function can be determined (see Ex. 6.4). Millikan proposed the experimental determination of h and ϕ in this way for the first time.

EXAMPLE 6.4
Experimental determination of h and ϕ

Using the following data recorded in the photoelectric effect experiment, determine the value of Planck's constant h and the work function ϕ of the photosensitive material used.

Frequency of incident light, ν (Hz)	Stopping potential, V_0 (volts)
10×10^{14}	2.37
6×10^{14}	0.72

Solution

Einstein's photoelectric equation can be written as

$$h\nu_1 = \phi + eV_{01} \tag{6.10}$$

$$h\nu_2 = \phi + eV_{02} \tag{6.11}$$

Solving Eqs. 6.10 and 6.11

$$h = \frac{e(V_{02} - V_{01})}{(\nu_2 - \nu_1)}$$

$$= \frac{(1.6 \times 10^{-19}\ \text{C})(2.37\ \text{V} - 0.72\ \text{V})}{(10 \times 10^{14}\ \text{Hz} - 6 \times 10^{14}\ \text{Hz})} = 6.6 \times 10^{-34}\ \text{J/s}$$

Substituting in Eq. 6.10

$$\phi = h\nu_1 - eV_{01}$$

$$= (6.6 \times 10^{-34}\ \text{Js})(10 \times 10^{14}\ \text{Hz})/(1.6 \times 10^{-19}\ \text{J/eV}) - 2.37\ \text{eV} = 1.75\ \text{eV}$$

6.4 COMPTON EFFECT

When a high frequency radiation, such as x-rays, is incident on a scatterer, the beam scattered in any direction shows two wavelengths, the original wavelength λ and a new higher wavelength λ'. This phenomenon of change in wavelength due to scattering is known as the Compton effect. The original wavelength is referred to as the *unmodified wavelength* and the higher one as the *modified wavelength*. The difference $\lambda' - \lambda$ is called the *Compton shift*.

The experimental arrangement to study the Compton effect is shown in Fig. 6.4. Radiation emanating from a source of monochromatic x-rays is collimated by the collimating slits and is incident on a scatterer. The intensity of scattered x-rays is measured as a function of the scattering angle θ using the x-ray spectrometer. The results obtained by Compton using carbon as a scatterer are shown in Fig. 6.5. The results show that the Compton shift, $\Delta\lambda$, increases with the increase in scattering angle.

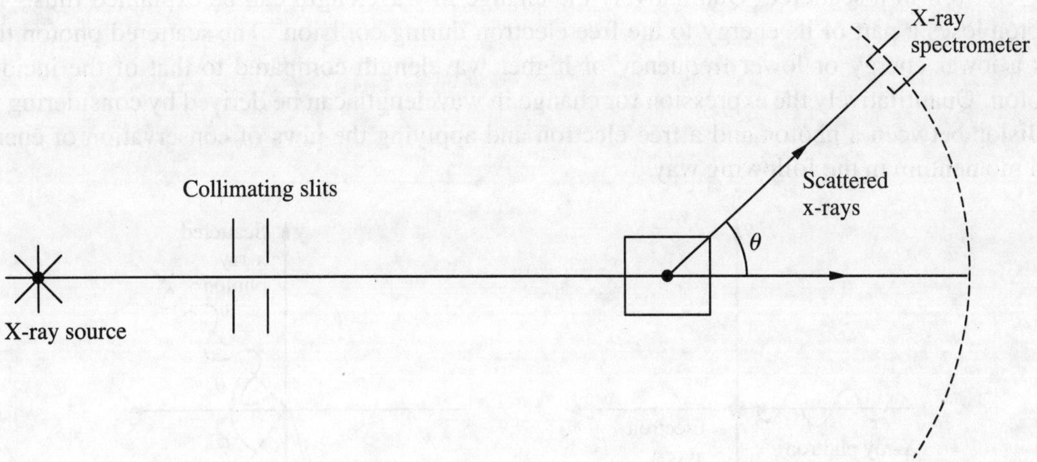

Fig. 6.4 Experimental arrangement to study the Compton effect

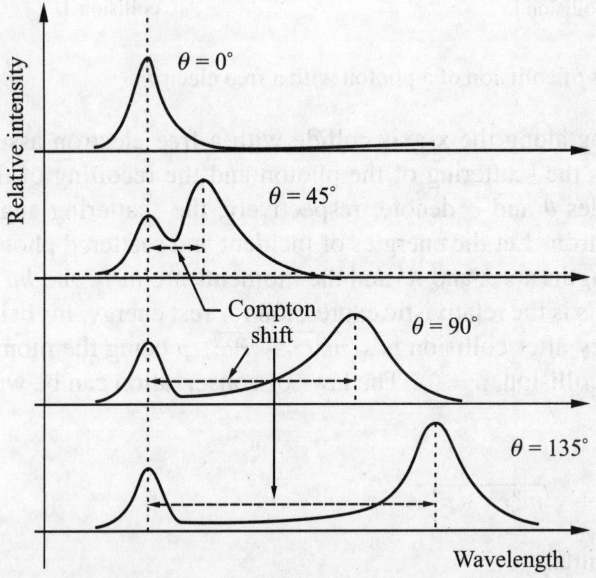

Figure 6.5 Measured intensities of scattered x-rays

Wave theory fails to account for the observed change in wavelength. According to wave theory, scattering is a result of the incident electromagnetic wave setting the free electrons in a scatterer into oscillations. These electrons then re-radiate the waves of the same frequency like a tiny transmitting antenna. Thus wave theory does not expect any change in wavelength. It is in complete disagreement with the experimental observations.

Compton accounted for the experimental observations by assuming the quantum nature of x-rays. He conceptualised the scattering as due to the collision of the incident photon with the free electron in a scatterer. Qualitatively the change in wavelength can be explained thus. The photon loses a part of its energy to the free electron during collision. The scattered photon thus has a lower energy or lower frequency or higher wavelength compared to that of the incident photon. Quantitatively the expression for change in wavelength can be derived by considering the collision between a photon and a free electron and applying the laws of conservation of energy and momentum in the following way.

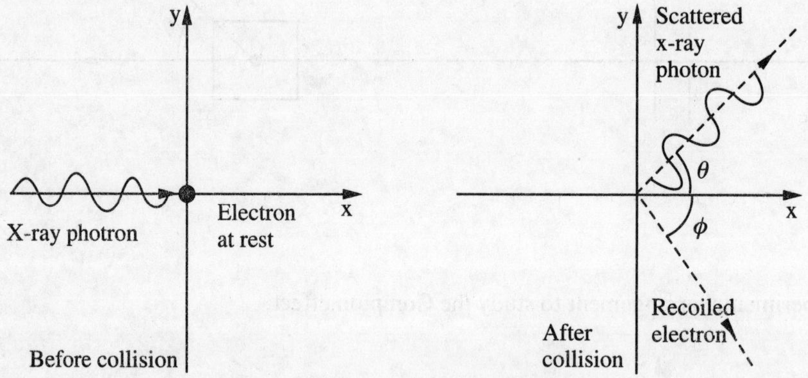

Figure 6.6 Kinematics of collision of a photon with a free electron

Let a photon moving along the x-axis collide with a free electron assumed to be initially at rest. Figure 6.6 shows the scattering of the photon and the recoiling of the electron as a result of collision. The angles θ and ϕ denote, respectively, the scattering angle for photon and the angle of recoil for electron. Let the energies of incident and scattered photon be $h\nu$ and $h\nu'$. The corresponding wavelengths are λ and λ' and the momenta are $h\nu/c$ and $h\nu'/c$. The rest energy of the electron is m_0c^2 (this is the relativistic expression for rest energy, m_0 being the rest mass of the electron), and its energy after collision is $\sqrt{m_0^2c^4 + p^2c^2}$; p being the momentum of the electron after collision (before collision, $p = 0$). The laws of conservation can be written as follows.

Conservation of energy

$$h\nu + m_0c^2 = h\nu' + \sqrt{m_o^2c^4 + p^2c^2} \tag{6.12}$$

Conservation of momentum

$$\text{along x-axis} \quad \frac{h\nu}{c} + 0 = \left(\frac{h\nu'}{c}\right)\cos\theta + p\cos\phi \tag{6.13}$$

$$\text{along y-axis} \quad 0 + 0 = \left(\frac{h\nu'}{c}\right)\sin\theta - p\sin\phi \tag{6.14}$$

To derive an expression for $\Delta\lambda$ in terms of θ, ϕ is first eliminated using Eqs. 6.13 and 6.14. This is achieved by first getting $p\cos\phi$ and $p\sin\phi$ from Eqs. 6.13 and 6.14 and then squaring and adding these expressions. The resulting expression for p^2 is then equated with that obtained from Eq. 6.12. This step eliminates p and we get the expression for $\Delta\lambda$ after a little simplification as follows.

$$\left[\frac{h\nu}{c} - \left(\frac{h\nu'}{c}\right)\cos\theta\right]^2 + \left[-\left(\frac{h\nu'}{c}\right)\sin\theta\right]^2 = \frac{1}{c^2}\left[(h\nu + m_0c^2 - h\nu')^2 - m_0^2c^4\right]$$

Replacing ν by c/λ and ν' by c/λ' throughout, we get

$$\frac{h^2}{\lambda^2} + \frac{h^2}{\lambda'^2} - \left(\frac{2h^2}{\lambda\lambda'}\right)\cos\theta = \frac{h^2}{\lambda^2} + \frac{h^2}{\lambda'^2} + \frac{2hm_0c}{\lambda} - \frac{2hm_0c}{\lambda'} - \frac{2h^2}{\lambda\lambda'}$$

or $$\left(\frac{2h^2}{\lambda\lambda'}\right)(1-\cos\theta) = 2hm_0c\left(\frac{1}{\lambda} - \frac{1}{\lambda'}\right)$$

or $$\Delta\lambda = \left(\frac{h}{m_0c}\right)(1-\cos\theta) \tag{6.15}$$

This is the expression for the Compton shift, as derived by Compton. The expression predicts that as the scattering angle increases, $\Delta\lambda$ should increase. This prediction is in good agreement with the experiment (see Fig. 6.5). In this way the quantitative explanation of the observed $\Delta\lambda$ as a function of θ was offered by Compton using Planck's quantum theory of radiation. It is interesting to note that the Compton effect is purely a quantum phenomenon. In the limit $h \to 0$, we get $\lambda' = \lambda$, a result in agreement with the predictions of classical physics.

The constant $\lambda_c = h/m_0c$ is known as the *Compton wavelength* of a particle. Substituting the numerical constants for an electron we get $\lambda_c = 0.024$ Å. The maximum Compton shift occurs for $\theta = 180°$ for which $\Delta\lambda = 2h/m_0c = 0.048$ Å (see Ex. 6.5).

EXAMPLE 6.5

Incident wavelength in Compton scattering

The maximum energy given to an electron during Compton scattering is 35 keV. Find the wavelength of the incident photon.

Solution

The maximum energy is given to an electron when the photon is backscattered, that is when, $\theta = \pi$.

$$\Delta\lambda = \frac{h}{m_0c}(1-\cos\theta)$$

$$\Delta\lambda = \frac{2h}{m_0c} = \frac{2\times 6.625\times 10^{-34}\,\text{Js}}{(9.1\times 10^{-31}\text{kg})(3\times 10^{8}\,\text{m/s})} = 0.048\ \text{Å}$$

The energy gained by the electron = energy of incident photon – energy of scattered photon

$$\frac{hc}{\lambda} - \frac{hc}{\lambda'} = 35 \text{ keV}$$

or $$\frac{\Delta\lambda}{\lambda\lambda'} = \frac{35 \text{ keV}}{hc} = \frac{35000 \text{ eV}}{12422 \text{ eVÅ}} = \frac{2.82}{\text{Å}}$$

or $$\frac{0.048 \text{ Å}}{\lambda(\lambda + 0.048 \text{ Å})} = \frac{2.82}{\text{Å}}$$

or $$\frac{0.048}{2.82} = \lambda^2 + 0.048\lambda$$

or $$\lambda^2 + 0.048\lambda - 0.017 = 0$$

or $$\lambda = \frac{1}{2}(-0.048 + \sqrt{23.04 \times 10^{-4} + 680 \times 10^{-4}}) = 0.11 \text{ Å}$$

Thus the wavelength of the incident photon is 0.11 Å.

The presence of the original wavelength λ in the scattered beam was explained by Compton in the following way. When x-rays are scattered by free electrons in the scatterer the modified component results as explained above. But if the incident photon collides with the bound electron, then the mass of the whole atom should be taken into account in place of the mass of the electron. As the atomic mass is much larger in comparison to that of an electron (for example, for a carbon atom the mass is about 22000 m_0). Equation 6.15 suggests that the corresponding $\Delta\lambda$ will be negligibly small. Such collisions thus result in $\Delta\lambda \sim 0$ or $\lambda' = \lambda$, hence explaining the presence of the original wavelength in the scattered beam. Example 6.6 illustrates the importance of the Compton effect in the various ranges of the electromagnetic spectrum.

EXAMPLE 6.6
Observability of the Compton effect

Calculate the per cent change in photon energy for a Compton collision with scattering angle equal to 90° for radiation in (a) the microwave range, with $\lambda = 3.0$ cm, (b) the visible range, with $\lambda = 5000$ Å, (c) the x-ray range, with $\lambda = 1$ Å and (d) the γ-ray range, with $\lambda = 0.012$ Å.

Solution

$$\Delta\lambda = \frac{h}{m_0 c}(1 - \cos 90°) = 0.024 \text{ Å}$$

$$\text{Fractional loss of energy} = \frac{E - E'}{E} = \frac{h\nu - h\nu'}{h\nu}$$

$$= \frac{c/\lambda - c/\lambda'}{c/\lambda} = \frac{\lambda' - \lambda}{\lambda'}$$

Thus, the per cent change in photon energy $= \dfrac{\Delta\lambda}{\lambda + \Delta\lambda} \times 100$

We get the following results.

Radiation	λ	**Change in energy** %
Microwave	3.0 cm	8×10^{-9}
Visible	5000 Å	5×10^{-4}
X-ray	1 Å	2.3
Gamma ray	0.012 Å	66.7

It can be seen that the per cent change in energy is too small for the microwave and visible ranges but is appreciable in x-ray and γ-ray ranges. Hence the increase in wavelength due to Compton scattering is appreciable only for x-rays and γ-rays. Compton effect is thus observable only in this part of the electromagnetic spectrum and cannot be observed with visible or other low frequency radiations.

6.5 COMPARISON OF PHOTOELECTRIC EFFECT AND COMPTON EFFECT

The photoelectric effect and Compton effect can be looked upon as two special cases of the collision of a photon and an electron. The photon colliding with a bound electron and transferring all its energy to it results in the photoelectric effect. The photon ceases to exist and an electron is ejected in the process. The photon colliding with a free electron and transferring a part of its energy results in the Compton effect. Here a photon with reduced energy is observed and the electron recoils.

It is impossible for a photon to give all its energy and momentum to a free electron. This can be proved by a contradiction. Assume that the photon is absorbed. The equations of conservation of energy and momentum are thus given by

$$h\nu + m_0c^2 = 0 + \sqrt{m_0^2c^4 + p^2c^2} \tag{6.16}$$

and $$\frac{h\nu}{c} + 0 = 0 + p \tag{6.17}$$

Substituting for pc in Eq. 6.16 we get

$$h\nu + m_0c^2 = \sqrt{m_0^2c^4 + h^2\nu^2}$$

Squaring both sides we get $2h\nu m_0c^2 = 0$.

This is absurd as neither $h\nu$ nor m_0c^2 can be zero. Thus the assumption is wrong and it is impossible for a photon to transfer all its energy and momentum to a free electron. Complete transfer is possible in a photoelectric effect when a photon strikes a bound electron. In the

Compton effect only partial transfer of energy takes place in the collision of a photon with a free electron.

Several features that are common to both the photoelectric effect and the Compton effect allow us to make a comparative study of these two effects. Chart 6.1 illustrates this comparison.

EXAMPLE 6.7
Momentum transfer in the photoelectric effect and the Compton effect

Find p_e/p_p, the ratio of electron momentum to photon momentum, in the photoelectric effect and the Compton effect, using the following data. What conclusion can be drawn from the results obtained?

	Photoelectric effect	Compton effect
Material	Sodium	Carbon
Range and wavelength of incident photon	UV, 2000 Å	X-rays, 1 Å
	Work function = 2.3 eV	Compton shift = 0.048 Å

Solution

The electron momentum $p_e = \sqrt{2mE_e}$, E_e being the energy of electron. $E_e = h\nu - \phi$ for photoelectric effect and $E_e = h\nu - h\nu'$ for Compton effect. The calculations are tabulated as follows.

	Photoelectric effect	Compton effect
Electron energy, E_e	$h\nu - \phi$ $= (12422/2000)\text{ eV} - 2.3\text{ eV}$ $= 3.9\text{ eV}$	$h\nu - h\nu'$ $= (12422/1)\text{ eV} - (12422/1.048)\text{ eV}$ $= 568.9\text{ eV}$
Electron momentum, p_e	$\sqrt{2 \times 9.1 \times 10^{-31} kg \times 3.9 \times 1.6 \times 10^{-19}\text{J}}$ $= 1.07 \times 10^{-24}\text{kg m/s}$	$\sqrt{2 \times 9.1 \times 10^{-31}\text{kg} \times 568.9 \times 1.6 \times 10^{-19}\text{ J}}$ $= 1.3 \times 10^{-23}$ kg m/s*
Momentum of incident photon, $p_p = (E/c)$	$6.2 \times 1.6 \times 10^{-19}\text{J}/(3 \times 10^8\text{m/s})$ $= 3.31 \times 10^{-27}\text{kg m/s}$	$12422 \times 1.6 \times 10^{-19}\text{J}/(3 \times 10^8\text{m/s})$ $= 6.6 \times 10^{-24}\text{kg m/s}$
(p_e/p_p)	323	1.97

*In case of Compton effect p_e can also be obtained easily by noting that

$$\frac{h}{\lambda} + 0 = \frac{-h}{\lambda'} + p_e$$

(Because the Compton shift equal to 0.048 Å corresponds to backscattering, that is, $\theta = 180°$ and $\phi = 0°$. Hence we get the above equation by substituting θ and ϕ in Eqs. 6.13.)

Thus, $p_e = 6.6 \times 10^{-34}\text{Js}\dfrac{[1 + 1.048] \times 10^{10}}{1.048} = 1.3 \times 10^{-23}$ kg m/s

Chart 6.1 Comparative Study of Photoelectric Effect and Compton Effect

	Photoelectric effect	**Compton effect**
Discovery	Experiment: Hertz, 1887, Theory: Einstein, 1905	Compton, 1923 (both experiment and theory)
What it is	Emission of electrons due to incidence of light of suitable frequency on a photosensitive metal	Reduction in frequency of x-rays due to scattering by a suitable target
Experimental arrangement	Photocathode Light beam Evacuated glass bulb Plate μA Photoelectron V V_0: Stopping potential	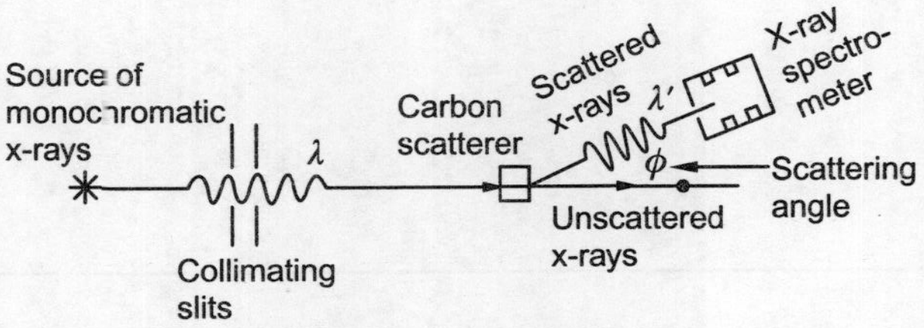 Source of monochromatic x-rays λ Collimating slits Carbon scatterer Scattered x-rays λ' X-ray spectro-meter ϕ Scattering angle Unscattered x-rays
Experimental results	1. Maximum kinetic energy of emitted photoelectrons ($K_{max} = eV_0$) is independent of intensity but depends linearly on the frequency of incident light. 2. ν_0 is the threshold frequency below which photoemission does not occur. 3. Photoelectric effect is instantaneous. See Fig. 6.3.	Scattered radiation contains x-rays of original (unmodified) wavelength, λ, and of a higher (modified) wavelength, λ' $\Delta\lambda$ ($= \lambda' - \lambda$) increases with θ See Fig. 6.5
Why the wave theory fails	Wave theory expects that 1. K_{max} should increase with the intensity of incident light and should be independent of its frequency; 2. There should exist a threshold intensity below which photoemission should not occur; 3. There should be a measurable time lag between incidence of light and emission of electrons. All these expectations disagree with the experimental results.	Wave theory expects that the scattered x-rays should have the same frequencies as that of the incident x-rays, because the free electrons absorb the incident x-ray radiation and re-radiate the same frequency. This expectation disagrees with the experimental results.

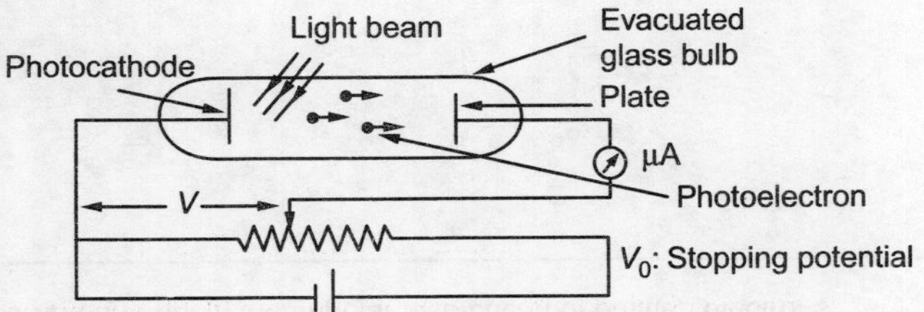

How the quantum theory succeeds	Einstein proposed $h\nu = \phi + K_{max}$ ($h\nu$ is the energy of the incident photon and ϕ is the work function). Thus quantum theory predicts 1. K_{max} should depend linearly on freqency but should be independent of intensity because increase in I (= $Nh\nu$) merely increases the number, (N) of photons without increasing the energy of individual photons. 2. Linear dependence of K_{max} on ν and existence of threshold frequency (for $h\nu < \phi$, $K_{max} < 0$) implies that no photoemission occurs. 3. Photoemission should be instantaneous as energy transfer between photon and electron is instantaneous. All these predictions agree with the experimental results.	Compton proposed a *semiclassical model* based on quantum theory, in which x-rays are regarded as photons and their scattering as a photon–electron collision. The model predicts that the photon should lose energy to the electron and hence the scattered photon should have less energy, hence less frequency or more wavelength. Quantitatively, the expression for change in wavelength derived by Compton using this model, $\Delta\lambda = (h/m_0c)(1 - \cos\theta)$, predicts that $\Delta\lambda$ should increase with the scattering angle. These predictions agree with the experimental results.
Collision mechanism	A photon collides with a bound electron, transfers all its energy, gets absorbed (ceases to exist) and ejects an electron. Conservation of momentum is generally ignored in the mathematical treatment. The net difference between the momenta of photon and photoelectron is transferred to the whole lattice. This corresponds to vanishingly small kinetic energy of the lattice which can be ignored in the energy conservation (see Ex. 6.7). We are interested only in the electron current and not in the angular distribution of emitted electrons.	A photon collides with a free electron, undergoes partial transfer of energy, gets scattered with less energy and the electron recoils. Both conservation of energy and conservation of momentum are taken into account using a simple two-body collision model. This is essential to study the angular distribution of scattered photons.

We know that in a two-body collision model, the momentum transferred by one body to an other body is of the order of its own momentum. The maximum transfer of momentum takes place when the body is backscattered [because, $(\Delta P)_{\max} = p - (-p) = 2p$]. The result obtained above for the Compton effect shows that the two-body collision model is successful. However, in the case of the photoelectric effect, the result, $p_e = 323p_p$, cannot be explained with a simple two-body model.

The release of a photoelectron can thus be compared to a process like a spark (a burning carbon particle of very small momentum) triggering an explosion. Since the photon momentum is very small compared to the electron momentum, the sodium sample will experience recoil with momentum similar to that of the electron. However, the sample will not be thrown back due to its large mass. The photoelectric effect cannot be viewed as a simple two-body collision process and momentum conservation loses significance.

6.6 WAVE–PARTICLE DUALITY OF RADIATION AND CONCEPT OF MATTER WAVES

Phenomena like interference, diffraction and polarisation follow from the wave theory of light but cannot be explained on the basis of the quantum theory of light. On the other hand the photoelectric effect, Compton effect, blackbody radiation and line spectra are consequences of the quantum theory of light but cannot be explained using the wave theory. These apparently contradictory viewpoints were reconciled by Bohr in his *principle of complementarity* which says that the *particle and wave aspects are complementary descriptions of light*. The 'true nature' of light is a dual nature; it includes both wave and particle aspects. This is often labelled as *wave–particle duality*. However, these subtleties cannot be understood on the basis of any experience from everyday life, which is necessarily macroscopic and therefore classical.

Whether light behaves as a wave or a particle is determined not only by the experiment but also by the energies and momenta involved. For example, for longer wavelength waves such as radio waves and microwaves, the photon energies and momenta are too small and the wave characteristics are dominant. On the other hand, for the shorter wavelength γ-rays and x-rays, the momenta and energies of photons increase and the photon description dominates over the wave description. For intermediate visible radiation both wave and particle characteristics are observed. Example 6.8 clarifies this point.

EXAMPLE 6.8

Wave–particle characteristics in various ranges of the electromagnetic spectrum

Find the energy and momentum of a photon in the following ranges of the electromagnetic spectrum. γ-rays, x-rays (λ = 0.01 Å), UV (100 Å), IR (10^{-4} m), microwave (1 m), radio waves (100 m). Remark on the wave–particle characteristics in these ranges.

Solution

Wave type	Wavelength, λ (Å)	Energy E (eV) = $12422/\lambda$ (Å)	Momentum $p = h\nu/c = E/c$ (kg m/s)	Remarks
γ-rays, x-rays	0.01	1.24×10^6	6.6×10^{-22}	↑ Photon energy and momentum increases—particle characteristics become important
UV	100	124.2	6.6×10^{-26}	
IR	10^6	0.0124	6.6×10^{-30}	
Microwave	10^{10}	1.24×10^{-6}	6.6×10^{-34}	$\lambda = h/p$ increases—wave characteristics become important ↓
Radio wave	10^{12}	1.24×10^{-8}	6.6×10^{-36}	

The wave particle correlation can be examined closely through the expressions of intensity in the two models. For a wave $I \propto E_0^2$ (where E_0 is the amplitude of the wave), and in quantum theory, $I \propto N$ (see Eq. 6.7), the number of photons emitted per second. Thus N is proportional to E_0^2. For a large value of N, the number of photons will be distributed in proportion to the squares of the amplitudes at different points. But for a small value of N also it can be concluded that the probability of finding a photon at a particular point and time depends on the value of E_0^2 there.

De Broglie extended the wave–particle duality of radiation to matter. According to de Broglie's hypothesis, each moving body has a wave associated with it. These waves are called *matter waves*. De Broglie reasoned that the expression for the wavelength of a photon, $\lambda = h/p$ should also hold good for matter. Thus the de Broglie equation for wavelength of matter waves is

$$\lambda = \frac{h}{p} \tag{6.18}$$

For a particle with kinetic energy E, the de Broglie wavelength is given by

$$\lambda = \frac{h}{\sqrt{2mE}} \left(\text{as } E = \frac{p^2}{2m} \right)$$

For an electron accelerated through a potential difference of V volts, a handy relation for wavelength is (see Ex. 6.9)

$$\lambda(\text{Å}) = \frac{12.28}{\sqrt{V(\text{volts})}}$$

EXAMPLE 6.9

Handy expression for the de Broglie wavelength of an electron

Show that the de Broglie wavelength of an electron accelerated through a potential difference of V volts is given by $\lambda\,(\text{Å}) = 12.28/\sqrt{V(\text{volts})}$.

Solution

An electron accelerated through a potential difference of V volts gains a momentum equal to

$\sqrt{2meV}$ (because $p^2/2m = eV$).

Thus the de Broglie wavelength,

$$\lambda = \frac{h}{\sqrt{2meV}}$$

$$= \frac{6.625 \times 10^{-34}\ \text{Js}}{\sqrt{2 \times 9.1 \times 10^{-31}\ \text{kg} \times 1.6 \times 10^{-19}\ \text{C} \times V\,(\text{volts})}}$$

$$= \frac{12.28 \times 10^{-10}\ \text{m}}{\sqrt{V\,(\text{volts})}}$$

$$\text{or} \quad \lambda(\text{Å}) = \frac{12.28}{\sqrt{V\,(\text{volts})}}$$

The matter waves postulated by de Broglie were different from the various types of waves known at that time—mechanical and electromagnetic waves. They are also called *probability waves*: their amplitude is related to the likelihood of the particle's presence in some region of space at some time. Table 6.1 shows the comparison of matter waves with mechanical and electromagnetic waves.

Table 6.1 Comparison of matter waves with mechanical and electromagnetic waves

	Governed by	**Oscillating variable**		**Significance**	**Examples**
Mechanical waves	Newton's equation	Displacement, y	Real valued	Energy $\propto \langle y^2 \rangle$	Waves on string, sound waves
Electromagnetic waves	Maxwell's equation	Electric/ magnetic vector, **E/H**	Real valued	Intensity $\propto \langle E^2 \rangle$	Light waves radio waves, x-rays
Matter waves	Schrodinger's equation	Wave function, ψ	Complex valued	Probability density $\propto \lvert\psi\rvert^2$	Waves associated with electrons, neutrons

As in the case of wave–particle duality of radiation, the wave–particle duality of matter forbids the simultaneous presence of these two aspects in the same experiment. Whether matter shows up as particles or as a wave is decided by the relative magnitude of the de Broglie wavelength in comparison with its dimensions and the dimensions of the environment with which it interacts. Example 6.10 provides the numerical appreciation of this point.

De Broglie had no experimental evidence to support his theory. However, he was able to validate Bohr's ideas of stationary orbits and quantisation of angular momentum, considering the wave character of electrons. Example 6.11 demonstrates de Broglie's interpretation of Bohr's theory using his own theory.

EXAMPLE 6.10

De Broglie wavelength for macroscopic and microscopic systems

Calculate the de Broglie wavelength of (a) a rock of mass 50 g thrown with a speed of 40 m/s, and (b) an electron accelerated through 50 V.

Solution

(a) For a rock,

$$\lambda = \frac{h}{mv} = \frac{6.625 \times 10^{-34}\ \text{Js}}{50 \times 10^{-3}\ \text{kg} \times 40\ \text{m/s}} = 3.31 \times 10^{-34}\ \text{m}$$

(b) For an electron,

$$\lambda = \frac{12.28}{\sqrt{V\,(\text{volts})}} = \frac{12.28}{\sqrt{50}} = 1.74\ \text{Å}$$

See Ex. 6.18

EXAMPLE 6.11

Bohr's expression from de Broglie's hypothesis

Obtain Bohr's condition of quantisation of angular momentum using de Broglie's idea of matter waves.

Solution

According to de Broglie's hypothesis an electron orbiting round the nucleus is associated with a wave. As the electron is constrained to move in an orbit, the circumference of the circular orbit should be equal to the integral multiple of the wavelength of the matter wave associated with the electron (Fig. 6.7).

Thus $2\pi r = n\lambda$

Using de Broglie's equation, $\lambda = h/p$, we get

$$pr = mvr = \frac{nh}{2\pi} = n\hbar$$

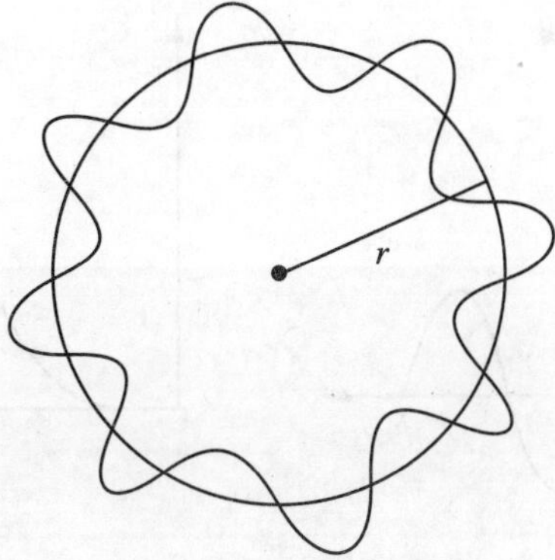

Figure 6.7 Electron wavelengths fitting into Bohr's orbit

which is Bohr's condition for quantisation of angular momentum. Thus the quantum number, n, assumes a new significance. It is the number of de Broglie wavelengths that fit into the circumference of Bohr's stationary orbits.

Davisson and Germer offered the first experimental proof of existence of matter waves by demonstrating wave behaviour of electrons. Chart 6.2 displays how de Broglie's theory is validated by Davisson and Germer's experiment. Figure 6.8 shows the diffraction pattern observed by exploiting the wave character of particles. Since the earliest discovery of Davisson and Germer, the wave character of several microparticles has been studied experimentally, establishing the universality of matter waves.

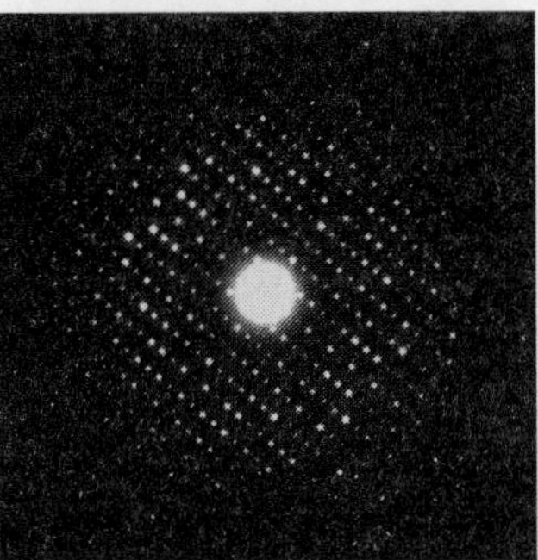

Figure 6.8 The electron diffraction pattern of an inorganic crystal, a tantalum oxide (Source: http://en.wikipedia.org/wiki/File:Tant-ED.jpg)

The hypothesis of matter waves by de Broglie raises the question: What kind of wave represents a particle? Evidently a wave like that shown in Fig. 6.9a, given by the equation $y = A\sin(kx - \omega t)$ hardly looks like a particle which is localised in space. The *wave packet* also called the *wave group*, shown in Fig. 6.9b, looks much more like a particle because it is localised and its amplitude is large only in a finite region.

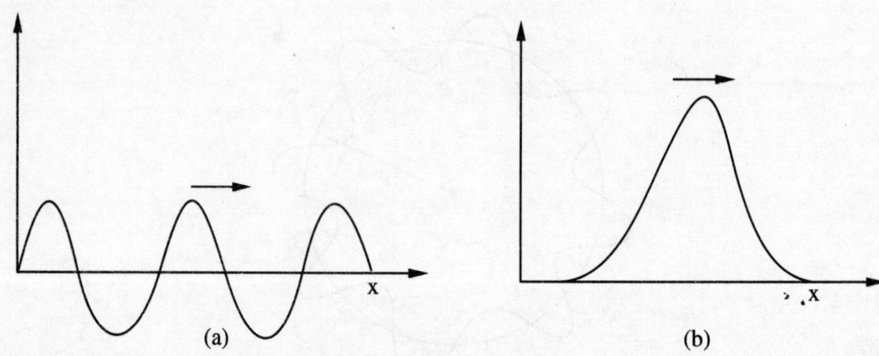

Figure 6.9 Wave representation of a particle: A wave group shown in (b) represents a particle better than the simple wave shown in (a).

Mathematical insight into the construction of a wave packet can be gained using the phenomenon of beats discussed in Ex. 3.2 (ii). The phenomenon is a result of interference of two waves.

$$y_1 = A\sin(kx - \omega t) \text{ and } y_2 = A\sin[(k + dk)x - (\omega + d\omega)t]$$

The equation of the resultant is given by

$$y = 2A\sin[(k + \tfrac{1}{2}dk)x - (\omega + \tfrac{1}{2}d\omega)t]\cos[\tfrac{1}{2}d\omega t - \tfrac{1}{2}dkx]$$

As $d\omega << 2\omega$ and $dk << 2k$

$$y = 2A\cos\left[\frac{d\omega t - dkx}{2}\right]\sin(kx - \omega t) \tag{6.19}$$

Figure 6.10a shows the resultant in which the constituent waves interfere constructively and destructively, producing periodic maxima and minima in space. If more than two waves are added, the resultant contains larger maxima separated by wider regions of minima (Figs 6.10b and 6.10c). The addition of infinite number of waves produces a single maximum resulting in a wave packet of zero width (Fig. 6.10d).

It should be noted that the group consists of waves represented by $\sin(kx - \omega t)$ of Eq. 6.19, and are enveloped by the variations proportional to $\cos[(d\omega t - dkx)/2]$. The velocity of the constituent waves is called the *wave velocity* or the *phase velocity*, v_p, and the velocity of the peak of the envelope is called the *group velocity*, v_g. The two velocities are thus defined, respectively, through the equations

$$kx - \omega t = \text{constant and } d\omega t - dkx = \text{constant}$$

Differentiating these equations with respect to time we get

$$v_p = \frac{\omega}{k} \tag{6.20}$$

$$v_g = \frac{d\omega}{dk} \tag{6.21}$$

CHART 6.2 CONFIRMATION OF DE BROGLIE'S THEORY BY DAVISSON AND GERMER'S EXPERIMENT

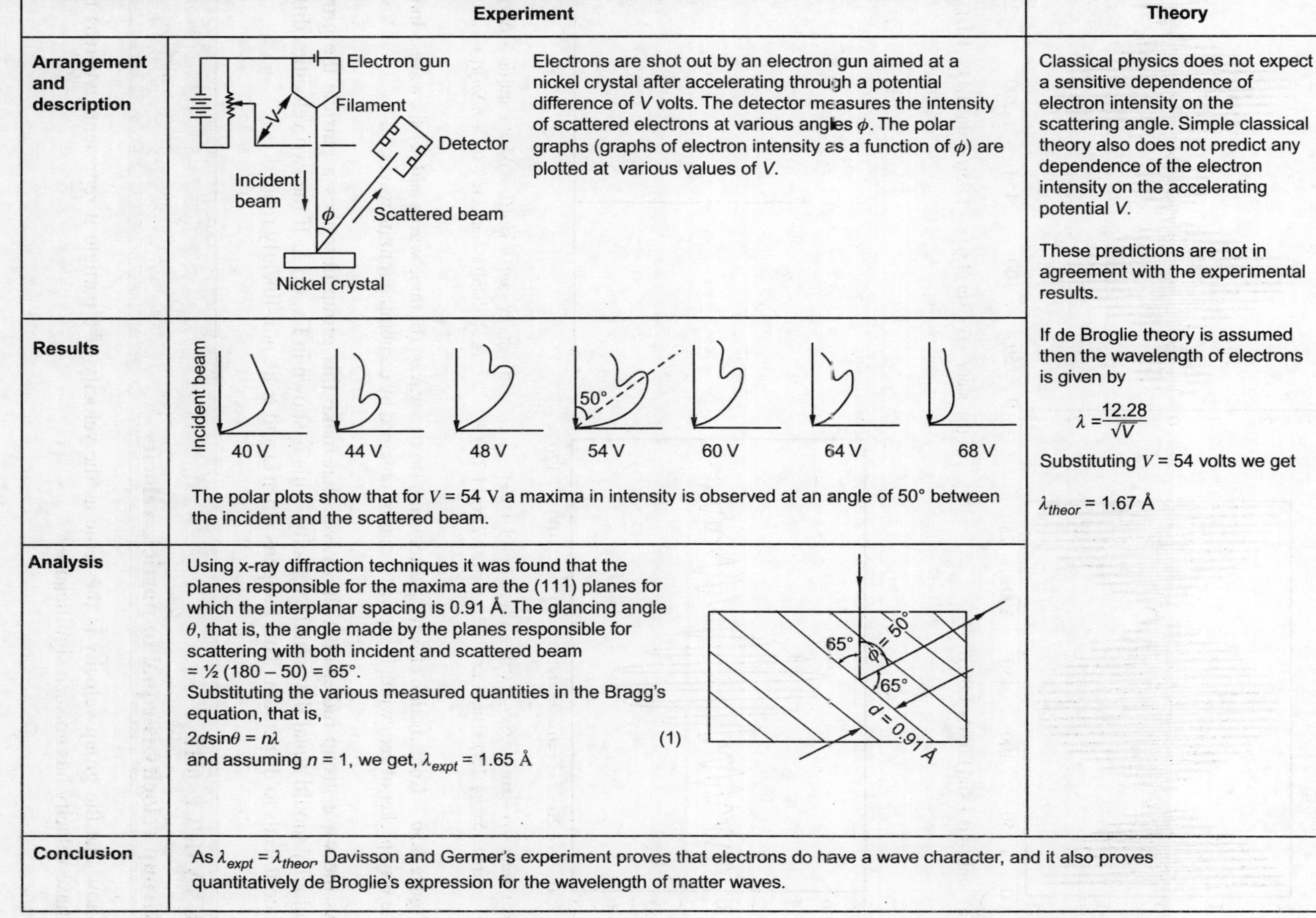

	Experiment	Theory
Arrangement and description	Electrons are shot out by an electron gun aimed at a nickel crystal after accelerating through a potential difference of V volts. The detector measures the intensity of scattered electrons at various angles ϕ. The polar graphs (graphs of electron intensity as a function of ϕ) are plotted at various values of V.	Classical physics does not expect a sensitive dependence of electron intensity on the scattering angle. Simple classical theory also does not predict any dependence of the electron intensity on the accelerating potential V. These predictions are not in agreement with the experimental results.
Results	The polar plots show that for $V = 54$ V a maxima in intensity is observed at an angle of 50° between the incident and the scattered beam.	If de Broglie theory is assumed then the wavelength of electrons is given by $\lambda = \frac{12.28}{\sqrt{V}}$ Substituting $V = 54$ volts we get $\lambda_{theor} = 1.67$ Å
Analysis	Using x-ray diffraction techniques it was found that the planes responsible for the maxima are the (111) planes for which the interplanar spacing is 0.91 Å. The glancing angle θ, that is, the angle made by the planes responsible for scattering with both incident and scattered beam $= \frac{1}{2}(180 - 50) = 65°$. Substituting the various measured quantities in the Bragg's equation, that is, $2d\sin\theta = n\lambda$ (1) and assuming $n = 1$, we get, $\lambda_{expt} = 1.65$ Å	
Conclusion	As $\lambda_{expt} = \lambda_{theor}$, Davisson and Germer's experiment proves that electrons do have a wave character, and it also proves quantitatively de Broglie's expression for the wavelength of matter waves.	

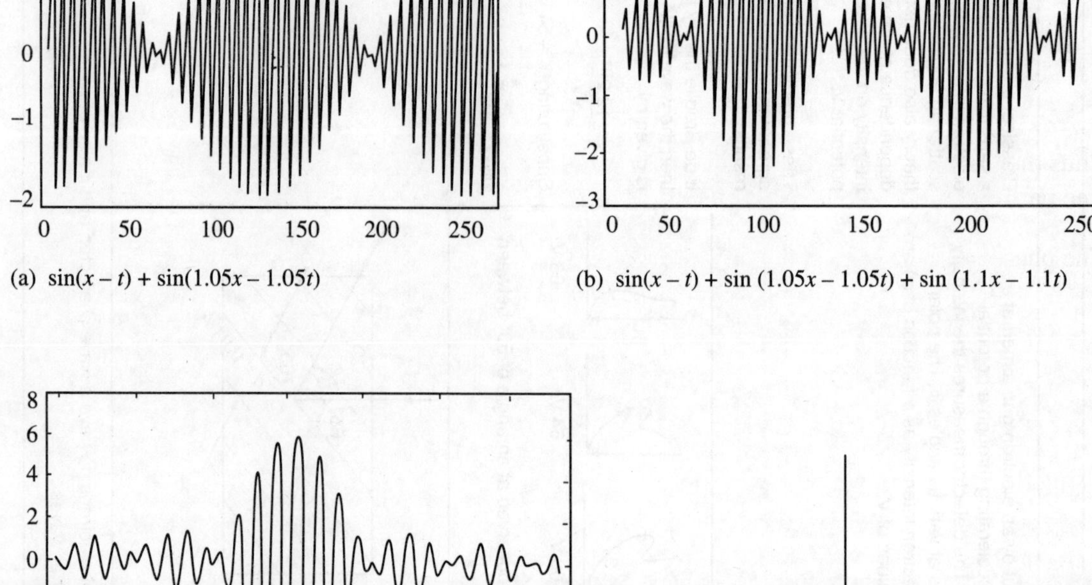

(a) $\sin(x - t) + \sin(1.05x - 1.05t)$

(b) $\sin(x - t) + \sin(1.05x - 1.05t) + \sin(1.1x - 1.1t)$

(c) $\sin(x - t) + \sin(1.05x - 1.05t) + \sin(1.1x - 1.1t) + \sin(1.15x - 1.15t) + \sin(1.2x - 1.2t) + \sin(1.25x - 1.25t)$

(d) $\sin(kx - \omega t) + \sin[(k + \Delta k)x + \sin[(\omega + \Delta\omega)t] + \ldots \sin[(k + n\Delta k)x + (\omega + n\Delta\omega)t] + \ldots$

Figure 6.10 Construction of wave packet using (a) two waves, (b) three waves and (c) six waves. In (d) the wavepacket of zero width is shown. It can be obtained by combining infinite waves.

As the wave group rather than the waves that make this group describes a particle the group velocity should equal the particle velocity. This is shown in Ex. 6.12. If the two velocities differ then it leads to dispersion. This happens when ω and k are non-linearly related.

EXAMPLE 6.12
Group velocity is equal to particle velocity

Show that the group velocity is the same as the velocity of a particle it represents whereas the phase velocity has no such significance.

Solution

The group velocity is given by

$$v_g = \frac{d\omega}{dk} = \frac{dE}{dp}, \text{ because } \omega = \frac{E}{\hbar} \text{ and } k = \frac{p}{\hbar}$$

$$= \frac{d}{dp} \times \frac{p^2}{2\text{m}} = \frac{p}{m} = \frac{mv}{m} = v$$

Thus the group velocity is the same as the particle velocity. (This is true for the matter wave of a free particle as $E = p^2/2m$ for a free particle.)

The phase velocity is given by

$$v_p = \frac{\omega}{k} = \frac{E/\hbar}{p/\hbar} = \frac{E}{p}$$

$$= \frac{p}{2\text{m}} = \frac{v}{2}$$

Using the relativistic equation, $E = mc^2$

$$v_p = \frac{mc^2}{p} = \frac{c^2}{v}$$

As $v < c$, $v_p > c$. Since the phase velocity exceeds the velocity of light, it cannot be the velocity of any transfer of matter or information.

6.7 HEISENBERG'S UNCERTAINTY PRINCIPLE

The description of a particle by a wave packet, as discussed in the previous section, tells us that a better representation of a particle is possible through a narrower wave packet. However a narrow wave packet requires a large number of waves for its construction. It can be seen that the width of the wave packet, Δx, and the width of the propagation constants, Δk, of the waves required to construct this wave packet, are inversely proportional to each other. Figure 6.11 illustrates this point. Using Fourier analysis it can be shown quantitatively that

$$\Delta x \, \Delta k \geq \frac{1}{2} \tag{6.22}$$

The detailed mathematical treatment is beyond the scope of this book.

Using the relation $p = \hbar k$ and hence $\Delta p = \hbar \Delta k$, we get

$$\Delta x \, \Delta p \geq \frac{\hbar}{2} \tag{6.23}$$

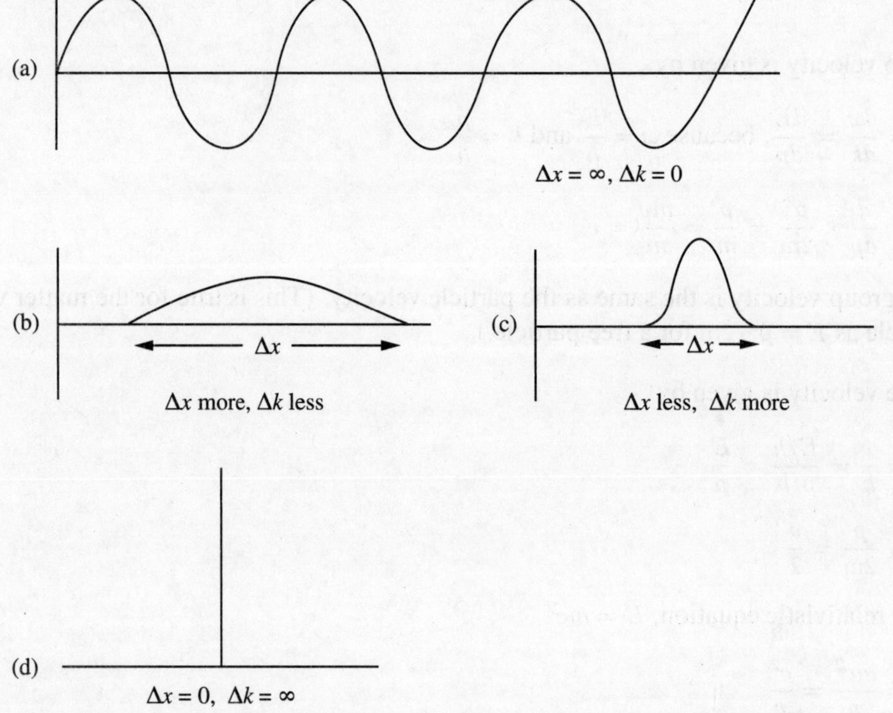

Figure 6.11 The relation between Δx and Δk for wave packets of various widths

Equation 6.23 is known as Heisenberg's uncertainty principle where Δx and Δp are identified, respectively, as the uncertainty in measured position and the uncertainty in measured momentum. For brevity, we call Δx, the position uncertainity, and Δy, the momentum uncertainity. This is the strict relation according to quantum mechanics. Sometimes h, $h/2$ or $\hbar$ is used instead of $\hbar/2$ on the right-hand side of Eq. 6.23. This does not make any pertinent difference as the uncertainty principle is often used only to estimate the quantities involved.

The uncertainty principle is stated formally in the following various ways.

1. The product of the uncertainty (Δx) in the position of an object at some instant and the uncertainty (Δp_x) in its momentum component in the x-direction at the same instant is equal to or greater than $\hbar/2$.
2. It is fundamentally impossible to make simultaneous measurements of a particle's position and momentum with infinite accuracy.
3. The simultaneous measurement of a particle's position and momentum with unlimited precision is impossible.

The uncertainty principle holds for all the components of position and momentum ($x, p_x; y, p_y; z, p_z$) and also for other pairs of variables such as energy and time (E, t) and angular position and angular momentum (ϕ, L). These pairs of variables are known as *canonically conjugate variables*. Thus we have

$$\Delta x\, \Delta p_x \geq \hbar/2$$

$$\Delta y\, \Delta p_y \geq \hbar/2 \tag{6.24}$$

$$\Delta z\, \Delta p_z \geq \hbar/2$$

$$\Delta E\, \Delta t \geq \hbar/2 \tag{6.25}$$

$$\Delta \phi\, \Delta L \geq \hbar/2 \tag{6.26}$$

See Ex. 6.13 for obtaining relations 6.25 and 6.26.

EXAMPLE 6.13
Uncertainty relations for (E, t) and (L, ϕ)

Obtain the uncertainty relations $\Delta E \Delta t > h$ and $\Delta L \Delta \phi > h$ for (a) energy and time and (b) angular momentum and angular position respectively.

Solution

(a) If Δt, the time interval in an atomic process, is the uncertainty in the measurement of time and $\Delta \nu$ the corresponding uncertainty in frequency, then $\Delta \nu > 1/\Delta t$ (as the frequency is the number of waves divided by the time interval). As $E = h\nu$, $\Delta E = h\Delta \nu$.

Hence $\Delta E \Delta t \geq h$

(b) From Fig. 6.12, we have $x = r\phi$, and $L = rp$.

Thus, $\Delta x = r\Delta \phi$ and $\Delta L = r\Delta p$

Hence, $\Delta L \Delta \phi = \Delta p \Delta x \geq h$

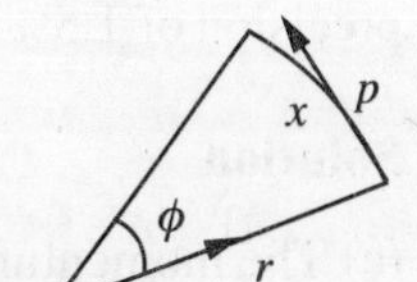

Figure 6.12 $x = r\phi$

Here it should be noted that it is important to develop a clear understanding of the subtle concepts hidden in the uncertainty principle.

First, the uncertainty principle places a fundamental restriction only on the product of the two quantities. Either of the quantities can be determined individually with unlimited precision. For example, it is possible to have a single particle system in which the particle has a completely definite position (or a completely definite momentum). What the uncertainty principle says is that, for a particle, no measurement can give a completely definite position and a completely definite momentum *at the same time*. In case the system has a definite particle position, then its momentum is totally uncertain: $\Delta x = 0$ and $\Delta p = \infty$. At the other extreme, the system has a definite momentum: $\Delta p = 0$, and a totally uncertain position: $\Delta x = \infty$. For the intermediate cases the relative magnitudes of Δx and Δp are limited by $\Delta p \Delta x \geq \hbar/2$.

Secondly, the uncertainty principle provided a revolutionary picture of matter: it was in striking contrast to the classical Newtonian concept. In the classical concept the particle's exact position and momentum are both determinable simultaneously. In fact the state of a single particle is specified in classical mechanics by both position and momentum. In the uncertainty principle the very possibility of such a state is denied!

The uncertainty principle allows us to understand why it is not possible to observe both the particle and wave characteristics in a single experiment. The experiments in which particle characteristics predominate (Δx small, Δp large) suppress wave character, and those in which the wave character shows up more strongly (Δp small, Δx large) the particle character is suppressed.

Lastly, it should be noted that the uncertainties involved in the uncertainty principle are not of the experimental or statistical origins. These are *the inherent, irreducible uncertainties coming into play due to the wave description of a particle.* Any inaccuracy in the measurement will only add to the amount allowed by the uncertainty principle.

Example 6.14 illustrates how the uncertainty principle which assumes significance in the microworld is of little practical significance in everyday life. The uncertainty principle cannot be validated by a laboratory experiment but can be proved with the help of some thought experiments. Two such experiments—Heisenberg's microscope and diffraction of electrons—are illustrated in Chart 6.3.

EXAMPLE 6.14

Application of the uncertainty principle for macroscopic and microscopic systems

Find the uncertainty in position for (a) a ball of mass 45 g moving with a speed of 40 m/s measured to a precision of 1.5% and (b) an electron moving with a speed of 2×10^6 m/s, measured to a precision of 1.5%.

Solution

(a) The momentum of a ball is

$$p = mv = 45 \times 10^{-3} \text{ kg} \times 40 \text{ m/s} = 1.8 \text{ kg m/s}$$

As $(\Delta p/p) \times 100 = 1.5, \Delta p = 0.015p = 0.027$ kg m/s

$$\text{So, } \Delta x = \frac{h}{\Delta p} = \frac{(6.625 \times 10^{-34} \text{ Js})}{(0.027 \text{ kg m/s})}$$

$$= 2.45 \times 10^{-32} \text{ m}$$

(b) For an electron,

$$p = 9.1 \times 10^{-31} \text{ kg} \times 2 \times 10^6 \text{ m/s} = 1.82 \times 10^{-24} \text{ kg m/s}$$

$$\Delta p = 0.015 \times 1.82 \times 10^{-24} \text{ kg m/s} = 2.73 \times 10^{-26} \text{ kg m/s}$$

$$\text{Hence } \quad \Delta x = \frac{6.625 \times 10^{-34} \text{Js}}{2.73 \times 10^{-26} \text{ kg m/s}} = 243 \text{ Å}$$

See Ex. 6.18.

Chart 6.3 Two Thought Experiments to Illustrate the Uncertainty Principle

	Heisenberg microscope	Electron diffraction
Description	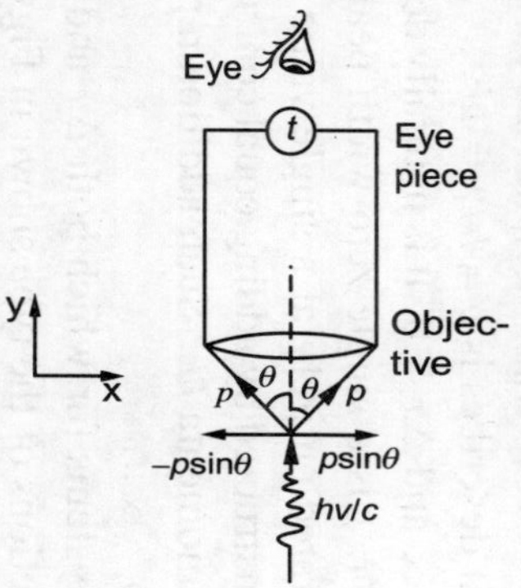 An attempt is made to 'see' an electron. For this at least one photon scattered by an electron should enter within an angle 2θ (apex angle of the cone of light entering the optical system). However, the process of collision results in transfer of momentum. Thus the process of measurement disturbs the system under measurement.	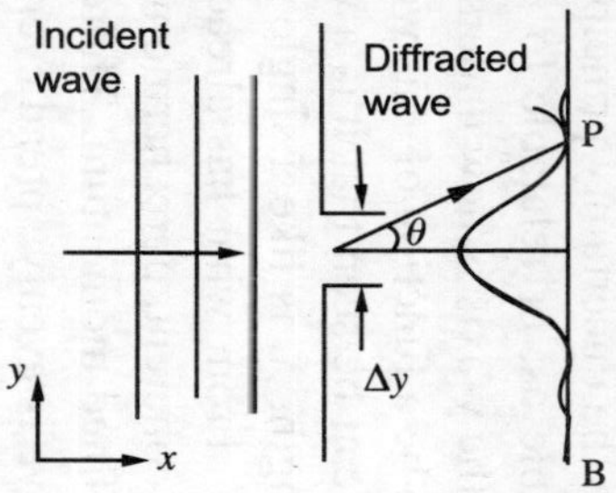 An electron represented by an incident matter wave enters a slit of width Δy. As the electron passes through the slit, the matter wave associated with it diffracts. The narrower the slit width, more is the diffraction. Thus the electron coming out of the slit acquires the y-component of the momentum (initial y-component is zero). The diffracted wave forms a diffraction pattern on the screen B.
Uncertainty in position	The uncertainly in the determination of the position of the electron along the x-axis is given by the limit of resolution of the microscope, that is, $\Delta x = \frac{\lambda}{(2\sin\theta)}$ (1) where λ is the wavelength of light used.	As an electron can pass anywhere through the slit the uncertainly in its location along the y-axis is equal to the slit width, Δy.
Uncertainty in momentum	The x-component of the momentum of the scattered photon will lie between + $(h/\lambda)\sin\theta$ and – $(h/\lambda)\sin\theta$. Thus the uncertainty in momentum, Δp_x, for the photon will be $(2h/\lambda)\sin\theta$. Due to the conservation of momentum, for an electron, $\Delta p_x = (2h/\lambda)\sin\theta$ (2)	For the first minimum at P we can write $\lambda = \Delta y \sin\theta$ (3) For small values of θ, $\sin\theta \approx \theta$, so that $\theta = \lambda/\Delta y = h/(p_x \Delta y)$, where p_x is the x-component of momentum. If Δp_y is the y-component of electron momentum striking at point P, then we can write, $\tan\theta \approx \theta = \Delta p_y/p_x$. Δp_y is assumed to be the measure of the uncertainty in the y-component of the momentum. Thus, $\Delta p_y = p_x\,\theta = h/\Delta y$ (4)
Result	From Eqs. (1) and (2), $\Delta x\, \Delta p_x = h$	From Eq. (4), $\Delta y\, \Delta p_y = h$

6.8 WAVE FUNCTION ψ

A wave implies a variation of some quantity with space and time. For example, a sound wave is characterised by the pressure or density variations in the medium, and a light wave implies the variation of electric and magnetic field vectors. A pertinent question thrown up here is what varies when a particle is described by a wave, or what constitutes the variation in case of matter waves. The quantity whose variations make up matter waves is called a wave function and is denoted by ψ. Though a wave function ψ assigned to every quantum mechanical system provides a complete description of that system, its physical significance is not readily apparent. The following discussion attempts to bring out this significance.

As ψ describes a quantum mechanical system, it should imbibe the uncertainty principle. The wave function should be such that its spread may be interpretable as, or relatable to Δx, the positional uncertainty. Figure 6.11 makes this clear. The label of the y-axis is now understood as the wave function $\psi(x)$. However, to allow for dynamics, ψ has to be a function of x as well as t.

Applying de Broglie's hypothesis to the wave in Fig. 6.11a, it can be seen that it is a wave of momentum $p = h/\lambda$. In other words, a wave of definite wavelength, λ, is like a single particle system with definite momentum $p = h/\lambda$. $\Delta p = 0$ as $\Delta\lambda = 0$. From what has already been studied in earlier chapters about waves, the following emerges: A coherent, pure, harmonic, fixed wavelength wave is a single particle system with a completely defined momentum. Therefore, it is also a system with completely undefined position. That the wave is evenly spread everywhere goes well with the above idea of relating spread to uncertainty in position. Thus the wave function for a single particle with definite momentum can be represented by the equations

$$\psi = A\sin(kx - \omega t) \text{ or } \psi = A\cos(kx - \omega t)$$

or combinations like $\psi = A\exp[\pm\, i(kx - \omega t)] = A[\cos(kx - \omega t) \pm i\sin(kx - \omega t)]$.

These equations are consistent with the uncertainty principle. They are the equations of a wave. But the waves in question are neither mechanical waves, nor electromagnetic waves. They are matter waves. The matter wave with wave function $\psi(x,t) = A\exp[i(kx - \omega t)]$ has the wavelength $\lambda = 2\pi/k$ and the momentum of the single particle which it describes is $p = h/\lambda = \hbar k$.

The other extreme is a single particle system with $\Delta p = \infty$ and $\Delta x = 0$. It is naturally described by a wave function, $\psi(x)$, which has no spread, that is $\psi(x)$ is a single zero-width peak at the particle position. It is the extreme wave packet concentrated entirely at a single value of x as indicated in Fig. 6.8d. Mathematically this $\psi(x)$ can be constructed by adding equal contributions from definite momentum states—$\exp(ikx)$ of all possible momenta $\hbar k$. Such addition represents the fact that the system has all possible values of p and $\Delta p = \infty$.

Between these two extremes there are single particle systems for which both Δx and Δp are non-zero and finite, and are thus represented by wave functions of the type shown in Figs. 6.11b and 6.11c.

From the foregoing discussion it appears that $\psi(x,t)$ can be related to the probability of finding a particle at position x at time t. But $\psi(x,t)$ cannot directly mean the probability because it can be complex as well as negative. Max Born overcame this difficulty in 1926 stating that ψ does not have any physical significance. It is $|\psi|^2$ or $\psi\psi^*$ (ψ^* being the complex conjugate of ψ) which is always positive and has the physical meaning of a probability density: It is the probability per unit volume that the particle will be found around any given point. The probability of finding the particle in volume element dV is given by $|\psi|^2 dV$. This interpretation of $|\psi|^2$ demands the following conditions on ψ.

1. ψ must be continuous and single valued everywhere,
2. Partial derivatives of ψ must be continuous and single valued everywhere,
3. ψ should be normalised, that is,

$$\int |\psi|^2 dV = 1 \tag{6.27}$$

These conditions ensure that the probability density is unambiguously defined everywhere and the probability of finding the particle somewhere in space is always unity.

6.9 THE EQUATION FOR ψ: SCHRODINGER'S EQUATION

Having developed some familiarity with the new picture of matter given by the wave function ψ, the question that arises is how does $\psi(x,t)$ change with time or what is the dynamical equation for $\psi(x,t)$. The situation is similar to that in classical physics where we describe single particles by position and ask how the position changes with time. The answer is: as governed by Newton's equations of motion. Thus the answer to a similar question about $\psi(x,t)$ has to do with something as fundamental as Newton's equations. In other words, such an enquiry will lead us to the most fundamental equation in quantum physics—the quantum dynamical equation. Schrodinger developed this equation and thus it bears his name. It should be carefully noted, however, that *Schrodinger's equation cannot be derived from more basic principles.* It is the basic principle! Schrodinger's research towards developing this equation was guided by the following main considerations.

1. The quantum dynamical equation should be a differential equation like all other basic equations of physics.
2. For macroscopic systems the results of the equation should match with those obtained using classical mechanics. To ensure this the equation should imbibe the classical laws of conservation of energy and conservation of momentum as well as the relation between energy and momentum.
3. The equation should contain all the elements of the new concepts of matter: de Broglie wave–particle dualism, Heisenberg's uncertainty relation and the concept of energy quantum.

The function $y = A\exp[i(kx - \omega t)]$ (6.28)

is a solution of the classical wave equation,

$$\frac{\partial^2 y}{\partial x^2} = \left(\frac{1}{v^2}\right)\frac{\partial^2 y}{\partial t^2} \tag{6.29}$$

where $v = \dfrac{\omega}{k}$ is the wave velocity. This equation can be verified easily by substitution (see Ex. 2.5). From the discussion in the previous section it is known that an identical equation,

$$\psi(x,t) = A\exp[i(kx - \omega t)] \tag{6.30}$$

describes a free particle system. For such a system there are no forces on a particle and hence its momentum has a fixed definite value. Using the quantum relations $E = \hbar\omega$ and $p = \hbar k$, Ex. 6.30 can be written as,

$$\psi(x,t) = A\exp[(-i/\hbar)(Et - px)] \tag{6.31}$$

Differentiating this expression the following equations are obtained,

$$\frac{\partial\psi(x,t)}{\partial t} = -\frac{iE}{\hbar}[\psi(x,t)] \tag{6.32}$$

and $$\frac{\partial\psi(x,t)}{\partial x} = \frac{ip}{\hbar}[\psi(x,t)] \tag{6.33}$$

or $$E\psi(x,t) = i\hbar\frac{\partial\psi(x,t)}{\partial t} \tag{6.34}$$

and $$p\psi(x,t) = -i\hbar\frac{\partial\psi(x,t)}{dx} \tag{6.35}$$

and $$p^2\psi(x,t) = -\hbar^2\frac{\partial^2\psi(x,t)}{\partial x^2} \tag{6.36}$$

Now using the free particle energy–momentum relationship,

$$E = \frac{p^2}{2m} \tag{6.37}$$

From Eqs 6.32–6.36 it can be seen that the following equation will support it.

$$i\hbar\frac{\partial\psi(x,t)}{\partial t} = \frac{-\hbar^2}{2m}\times\frac{\partial^2\psi(x,t)}{\partial x^2} \tag{6.38}$$

For a particle, which is not free but is constrained to move in a region where its potential energy is V,

$$E = \frac{p^2(x)}{2m} + V(x) \tag{6.39}$$

where V as well as kinetic energy and momentum are dependent on x.

Comparison of Eq. 6.39 with Eq. 6.37 motivates generalising Eq. 6.38 to the following equation for any single particle moving in one dimension.

$$i\hbar\frac{\partial\psi(x,t)}{\partial t} = \frac{-\hbar^2}{2m}\times\frac{\partial^2\psi(x,t)}{\partial x^2} + V(x)\psi(x,t) \tag{6.40}$$

Equation 6.40 is known as *Schrodinger's time-dependent equation*. The equation,

$$E\psi(x,t) = \frac{-\hbar^2}{2m}\times\frac{\partial^2\psi(x,t)}{\partial x^2} + V(x)\psi(x,t)$$

$$\text{or} \quad \frac{\partial^2 \psi(x,t)}{\partial x^2} + \frac{2m}{\hbar^2} \times [E - V(x)]\psi(x,t) = 0 \tag{6.41}$$

is known as *Schrodinger's time-independent equation.* These equations contain all the elements of quantum physics and also maintain classical mechanics on the macro-scale as they explicitly contain the *E–p* relation of Newton's theory.

Schrodinger's equation is also expressed in the operator form as

$$H\psi = E\psi \tag{6.42}$$

where H is called the *Hamiltonian operator* or the *total energy operator*

$$H \equiv \frac{p^2}{2m} + V \tag{6.43}$$

where $p^2/2m$ is called the *kinetic energy operator*, and V is called the *potential energy operator*. p^2 is referred to as the *momentum square operator* and is defined through Eq. 6.36 as

$$p^2 \equiv -\hbar^2 \frac{\partial^2}{\partial x^2} \tag{6.44}$$

E is known as the *energy operator* and is defined through Eq. 6.34 as

$$E \equiv i\hbar \frac{\partial}{\partial t} \tag{6.45}$$

Equation 6.42 is also known as an *eigenvalue equation.* ψ is known as the *eigenfunction* and E is known as the *eigenvalue* associated with ψ. An eigen value equation is one in which the operation of a function by a certain operator results in the same function multiplied by some constant. Example 6.15 gives a better idea of an eigenvalue equation.

EXAMPLE 6.15

Eigenfunctions and eigenvalues of d/dx and d^2/dx^2

Which of the following are eigenfunctions of the operators d/dx and d^2/dx^2? Give the eigenvalues when appropriate. (a) $\sin x$, (b) $\cos x$, (c) $\sin^2 x$, (d) $\exp(2x)$, (e) $\exp(ix)$ and (f) $\sin nx$.

Solution

(a) $\dfrac{d}{dx}(\sin x) = \cos x$, not an eigenfunction

$$\frac{d^2}{dx^2}(\sin x) = -\sin x = (-1)\sin x; \text{ the eigenvalue is } -1$$

(b) $\dfrac{d}{dx}(\cos x) = -\sin x$, not an eigenfunction

$$\frac{d^2}{dx^2}(\cos x) = -\cos x = (-1)\cos x; \text{ the eigenvalue is } -1$$

(c) $\dfrac{d}{dx}(\sin^2 x) = 2\sin x \cos x$; not an eigenfunction

$\dfrac{d^2}{dx^2}(\sin^2 x) = 2 - 4\sin 2x$; not an eigenfunction

(d) $\dfrac{d}{dx}[\exp(2x)] = [2\exp(2x)] = (2)[\exp(2x)]$; the eigenvalue is 2

$\dfrac{d^2}{dx^2}[\exp(2x)] = [4\exp(2x)] = (4)[\exp(2x)]$; the eigenvalue is 4

(e) $\dfrac{d}{dx}[\exp(ix)] = [i\exp(ix)] = (i)[\exp(ix)]$; the eigenvalue is i

$\dfrac{d^2}{dx^2}[\exp(ix)] = [-\exp(ix)] = (-1)[\exp(ix)]$; the eigenvalue is -1

(f) $\dfrac{d}{dx}[\sin(nx)] = n\cos(nx)$; not an eigenfunction

$\dfrac{d^2}{dx^2}[(\sin nx)] = (-n^2)[\sin(nx)]$; the eigenvalue is $-n^2$

Schrodinger's equation occupies the same fundamental place in quantum mechanics as Newton's equation in classical mechanics and is used in the same spirit. Newton's law, $F = md^2x/dt^2$, is used with the specific force in any given situation to solve for position x as a function of time. Schordinger's equation is used in a similar way with the specific potential energy in any given situation to solve for the wave function ψ as a function of x and t.

In Chart 6.4 the application of Schrodinger's equation to three different cases is discussed: (i) a free particle, (ii) particle in a one-dimensional infinite potential well and (iii) a harmonic oscillator. It should be noted that the results obtained using the Schrodinger's equation are consistent with de Broglie's principle and Heisenberg's uncertainty principle. Example 6.16 illustrates this point. Schrodinger's equation, like other quantum concepts, assumes importance only in the microworld and is of little significance in the macroworld (everyday life). Example 6.17 elaborates this point.

Example 6.18 illustrates why the different quantum concepts are not of much importance in our day-to-day life but assume a lot of significance in the microworld. The extremely small value of Planck's constant is the underlying reason for this common observation regarding all quantum concepts.

CHART 6.4 APPLICATIONS OF SCHRODINGER EQUATION

	Free particle	Particle confined to one-dimensional well A particle is confined to travel along the x axis between $x = 0$ and $x = L$		Harmonic oscillator
		Infinite potential well	**Finite potential well**	
Description of problem	A free particle is one on which no net force acts.	$V = \infty$, $V = \infty$, $V = 0$, $x = 0$, $x = L$ The walls of the well are infinitely hard. Hence the particle cannot exist outside the well due to the infinite potential barrier, which it cannot surmount.	V, I, II, III, $E < V$, $x = 0$, $x = L$ The height of the potential barrier is finite so that the particle with energy less than the height of the potential barrier is trapped inside the well. The particle now has a certain probability of penetrating into regions I and III.	A harmonic oscillator is one which is acted upon by a force which tries to restore the oscillator to its equilibrium position. This force is given by Hook's law as $F = -kx$ where k is the force constant and x is the oscillator's displacement from the equilibrium position. Angular frequency of oscillator, $\omega = \sqrt{k/m}$
Examples	Particle in an ideal gas	The infinite potential well has no physical counterpart.	Potential wells with barriers of finite height exist, for example, an electron in a one-dimensional molecular chain.	Mass spring system and atom in a crystal lattice.
Potential energy	As $F = -\dfrac{dV}{dx}$ the potential energy is constant and assumed to be zero. Thus, $V = 0$	The potential energy V of the particle is constant inside the well. We assume it to be zero for convenience in both the cases, that is, $V = 0$ inside the well, that is, for $0 < x < L$. The potential is infinite on both sides of the well.	The potential is finite on both the sides of the well.	Energy; $V = \frac{1}{2} kx^2$; x $V = \int -\mathbf{F} . d\mathbf{x} = \frac{1}{2} kx^2$. Thus the potential energy is a parabolic function of x. Here it should be noted that this holds only for small values of x.
Schrodinger's equation	$\dfrac{d^2\psi}{dx^2} + \dfrac{2mE\psi}{\hbar^2} = 0$ (1)	$\dfrac{d^2\psi}{dx^2} + \dfrac{2mE\psi}{\hbar^2} = 0$ (5) inside the well	In region II, $\dfrac{d^2\psi_{II}}{dx^2} + \dfrac{2mE\psi_{II}}{\hbar^2} = 0$ (9) In regions I and III $\dfrac{d^2\psi}{dx^2} + \dfrac{2m(E-V)\psi}{\hbar^2} = 0$ or $\dfrac{d^2\psi}{dx^2} - \alpha^2\psi = 0$ (10) where $\alpha = \dfrac{\sqrt{2m(E-V)}}{\hbar}$	$d^2\psi/dx^2 + \dfrac{2m(E - \frac{1}{2} kx^2)}{\hbar^2}\psi = 0$ (16) Substituting $\alpha = \dfrac{\omega m}{\hbar}$ and $\beta = \dfrac{2mE}{\hbar^2}$, we get $\dfrac{d^2\psi}{dx^2} + (\beta - \alpha^2 x^2)\psi = 0.$ (17)

Solution	$\psi = A \exp[(i\sqrt{2mE}/\hbar)x] +$ $B\exp[-(i\sqrt{2mE}/\hbar)x]$ if the particle moves only in the positive x direction, $B = 0$. Thus, $B = 0$. Thus, $\psi = A \exp\left(\frac{i\sqrt{2mE}}{\hbar}\right)x$ (2)	As the particle cannot exist outside the well, $\psi = 0$ at $x = 0$ and $\psi = 0$ at $x = L$. Therefore $\psi = 0$ for $0 > x > L$. Inside the well, $\psi = A_1 \sin(\sqrt{2mE}/\hbar)x + B_1 \cos(\sqrt{2mE}/\hbar)x]$. Using $\psi = 0$ at $x = 0$ we get $B_1 = 0$ Thus, $\psi = A_1 \sin\left(\frac{i\sqrt{2mE}}{\hbar}\right)x$ (6)	$\psi_{II} = A_2 \sin(\sqrt{2mE}/\hbar)x +$ $B_2 \cos(\sqrt{2mE}/\hbar)x]$ (11) $\psi_I = A_3 \exp(\alpha x) + B_3 \exp(-\alpha x)$ (12) $\psi_{III} = A_4 \exp(\alpha x) + B_4 \exp(-\alpha x)$ (13) We know that ψ_I and ψ_{III} must be finite everywhere. Now $\exp(-\alpha x) \to \infty$ as $x \to -\infty$ and $\exp(\alpha x) \to \infty$ as $x \to \infty$. Thus B_3 and A_4 must be zero. Hence $\psi_1 = A_3 \exp(\alpha x)$ and $\psi_{III} = B_4 \exp(-\alpha x)$ (14)	$\psi = A \exp(-Cx^2)$. Substituting this equation in Eq. (16), we get $4c^2x^2\psi - 2c\psi + \beta\psi - a^2x^2\psi = 0$. Thus, $C = \frac{\alpha}{2} = \frac{\beta}{2} = \frac{\omega m}{2h}$ (18)
Energy eigen value	As the particle is free all energy states are possible and are given by $E = p^2/2m = \hbar^2 k^2/2m$ or $E = \frac{h^2k^2}{8\pi^2 m}$ (3)	Using $\psi = 0$ at $x = L$, we get $\left(\frac{i\sqrt{2mE}}{\hbar}\right)L = n\pi$ or $E_n = \frac{n^2h^2}{8mL^2}$, $n = 1, 2, 3, \ldots$ (7)	The conditions on ψ demand that the wave functions inside and outside the well must match up their values and slopes where they join. This occurs only for certain values of energy that are smaller than that given by Eq. (7).	Since $\alpha = \beta$, $E = ½\hbar\omega$. This represents the ground state energy. Energy of nth state is given by $E_n = (n + ½)\hbar\omega$; $n = 0, 1, 2, 3, \ldots$ (19)
Wave function ψ	$\psi = A \exp(ikx)$ (4)	Substituting $(\sqrt{2mE}/\hbar) = n\pi/L$, we get $\psi_n = A_1 \sin(n\pi/L)x$. To find the value of A_1, the normalisation condition $\int_0^1 \lvert\psi_n^2\rvert dx = 1$ is used. Substituting for ψ_n and solving, we get, $A = \sqrt{2/L}$. Thus $\psi_n = \sqrt{2/L} \sin(n\pi/L)x$. (8)	Here $\psi_{II} = A_3$ at $x = 0$ and $\psi_{II} = B_4$ at $x = L$. The wave function contains both the sine and cosine functions. The complete wave function is $\psi_{II} = A_2 \sin(\sqrt{2mE}/\hbar)x +$ $B_2 \cos(\sqrt{2mE}/\hbar)x]$ (15)	To find the value of A, we use the normalisation condition, $\int_{-\infty}^{\infty} \lvert\psi_n^2\rvert dx = 1$. Substituting for ψ_n and solving, we get, $A = (\omega m/\pi h)^4$. Thus, $\psi_n = (\omega m/\pi h)^4 \exp(-\omega m x^2/2\hbar)$ (20)
Important results	(i) $\lvert\psi^2\rvert = \psi\psi^* = A \exp(ikx) \times A \exp(-ikx) = A^2$. Thus a free particle has the same probability of being found at all points along the x-axis. (ii) Energy levels are continuous.	(i) The ground state energy is not zero as that would make ψ zero, indicating the particle's absence. (ii) Energy levels are quantised. (iii) Wave functions terminate at the boundaries in case of infinite potential well indicating the particle's absence outside the well. The wave functions for finite potential well are similar to that for an infinite potential well except that they extend a little beyond the boundaries of the well indicating that the particle has now a small but finite probability of existing outside the well. **Tunnel effect**: In case of the finite potential well the particle that lacks the energy to surmount the barrier has a small but finite probability of leaking out. This is know as the *tunnel effect*. The tunnel effect occurs in case of the emission of α particles by certain radioactive nuclei and in the operation of tunnel diodes (see Chart 12.3.)		(i) The ground state energy is not zero but it is $E_0 = ½\hbar\omega$, called as the zero point energy. (ii) Energy levels are quantised.

EXAMPLE 6.16

Eigenfunctions for a particle in a one-dimensional infinite potential well

Sketch the first three eigenfunctions for a particle in a one-dimensional infinite potential well of width L. Obtain the expression, $E_n = n^2h^2/8mL^2$ using de Broglie's hypothesis. Show that the uncertainty principle provides a reasonable estimate of the ground state energy.

Solution

The wave function for a particle in an infinite potential well is given by (see Chart 6.4)

$$\psi_n = \left(\sqrt{\frac{2}{L}}\right) \sin\left(\frac{n\pi}{L}\right) x$$

Figure 6.13 shows the three eigenfunctions ψ_1, ψ_2, and ψ_3.

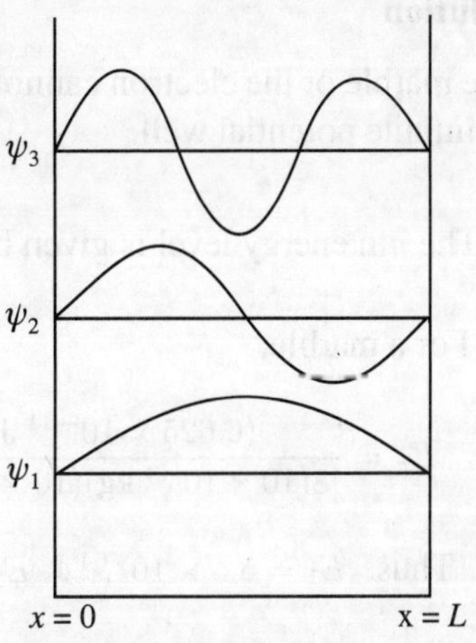

Figure 6.13 Eigenfunctions for a particle in a one-dimensional infinite potential well

These plots show that

$$n\lambda = 2L \tag{6.46}$$

where the wavelength λ is to be identified as a de Broglie wavelength,

$$\lambda = \frac{h}{p} \tag{6.47}$$

The energy of such a particle is entirely kinetic (as the potential energy is constant and is assumed to be zero) and is given by

$$E_n = \frac{p^2}{2m} \tag{6.48}$$

Combining Eqs. 6.46, 6.47 and 6.48, $E_n = \dfrac{n^2h^2}{8mL^2}$

Uncertainty in the position of a particle, $\Delta x = L$.

Thus, $\Delta p_x = \dfrac{h}{2\Delta x} = \dfrac{h}{2L}$

For the momentum to have an uncertainty of $h/2L$ it must be at least as large as $h/2L$. The minimum kinetic energy of the particle must be $E_1 = p^2/2m = h^2/8mL^2$

EXAMPLE 6.17

Application of Schrodinger equation for microscopic and macroscopic systems

Find the first three energy levels of (a) a marble of mass 10 g in a box of width 10 cm and (b) an electron in a box of width 1Å.

Solution

The marble or the electron cannot leave the box. Therefore the problem is like that of a particle in an infinite potential well.

The nth energy level is given by $E_n = \dfrac{n^2h^2}{8mL^2}$

(a) For a marble,

$$E_n = \frac{(6.625 \times 10^{-34}\ \text{Js})^2 n^2}{[8(10 \times 10^{-3}\ \text{kg})(10 \times 10^{-2}\ \text{m})^2]} = 5.5 \times 10^{-64} n^2\ \text{J}$$

Thus, $E_1 = 5.5 \times 10^{-64}$ J, $E_2 = 22 \times 10^{-64}$ J, and $E_3 = 49.5 \times 10^{-64}$ J

(b) For an electron,

$$E_n = \frac{(6.625 \times 10^{-34}\ \text{Js})^2 n^2}{8(9.1 \times 10^{-31}\ \text{kg})(1 \times 10^{-10}\ \text{m})^2}$$

$$= 6.03 \times 10^{-18} n^2\ \text{J} = 37.7 n^2\ \text{eV}$$

Thus, $E_1 = 37.7$ eV, $E_2 = 151$ eV, $E_3 = 339$ eV

See Ex. 6.18

EXAMPLE 6.18

Comparison of results for macroscopic and microscopic systems

Tabulate the results obtained in Exs. 6.1, 6.10, 6.14 and 6.17. What conclusion can be drawn from the comparison of these results?

Solution

	Macroscopic system	Microscopic system
Example 6.1 Energy quantisation, $E_n = nh\nu, E_n = -13.6/n^2$ eV	per cent change in energy $= 10^{-27}\%$	per cent change in energy $= 75\%$
Example 6.10 de Broglie's hypothesis, $\lambda = h/p$	rock, $\lambda = 3.31 \times 10^{-34}$ m	electron, $\lambda = 1.74$ Å
Example 6.14 Heisenberg's uncertainty principle, $\Delta x \Delta p > h$	ball, $\Delta x = 2.45 \times 10^{-32}$ m	electron, $\Delta x = 243$Å
Example 6.17 Application of Schrodinger equation for a particle in a one-dimensional infinite potential well, $E_n = n^2h^2/8mL^2$	marble, $E_1 = 5.5 \times 10^{-64}$ J $E_2 = 22 \times 10^{-64}$ J, $E_3 = 49.5 \times 10^{-64}$ J,	electron, $E_1 = 37.7$ eV, $E_2 = 151$ eV, $E_3 = 339$ eV

It can be seen from these results that

1. The per cent change in energy due to transition from one quantum state to the next quantum state is negligible for a macroscopic system but is appreciable for a microscopic system.
2. The de Broglie wavelength is negligible for a macroscopic system but is appreciable for a microscopic system.
3. The uncertainty in position is negligible for a macroscopic system but is appreciable for a microscopic system.
4. The difference between the consecutive energy levels is negligible for a macroscopic system but is appreciable for a microscopic system.

Thus it can be seen *that the application of quantum concepts does not lead to any observable effects in the macroworld but assumes significance in the the microworld. The underlying reason for this difference is that the magnitude of Planck's constant is extremely small.*

6.10 APPLICATIONS

The basic elements of quantum mechanics have been studied in this chapter. The quantum theory brought about a revolution in physics and is regarded as the most successful theory till date. It can be appreciated that quantum concepts are outside our everyday experience and assume significance only in the subatomic world. What does an engineering student gain through the study of these concepts? Careful thought reveals that much of the progress in engineering and technology in the last century has been made through the exploitation of discoveries in quantum physics (made during 1900–1932). Quantum theory has contributed immensely to widely varying fields such as astrophysics, microelectronics, optics, acoustics, and various areas in chemistry and life sciences. A discipline developed through utilitarian applications based on quantum mechanics has come to be known as *quantum engineering*.

The chapter began with Planck's quantum theory. Planck's ideas were used by several scientists to tackle the problems confronted by them at the beginning of the twentieth century. This resulted in new knowledge of phenomena that subsequently led to new applications in engineering and technology. For example, in 1917 Einstein applied Planck's theory to predict the mechanism of stimulated emission, a key to the laser operation. Einstein's theoretical ideas bore fruit when the first experimental laser was built by Maiman in 1960. In later years, Planck's idea of quantisation of energy was found to possess universal validity. Several quantum particles like phonon, magnon, polaron, graviton and others were discovered for the quantisation of different fields, in analogy with the photon which represents the quantisation of the electromagnetic field.

The study of the photoelectric effect and Compton effect (Sections 6.3–6.5) helped crystallisation of ideas of quantum nature of radiation. The photoelectric effect has been used to construct a large number of devices which are classified as photoemissive, photoconductive or photovoltaic devices. Chart 6.5 describes a few devices based on the photoelectric effect.

In Section 6.6 we studied the wave–particle duality of matter and radiation and the representation of particles by wave packets. The wave–particle duality helps to interpret experimental results in terms of either of these facets as per what suits a particular experiment. The eye provides an excellent example where the gathering and focussing of light is understood in terms of waves, and the absorption of light by rods and cones is understood in terms of photons.

De Broglie's discovery of matter waves opened up new vistas in the fields of microscopy and materials characterisation. It was known that the resolving power of a microscope is inversely proportional to the wavelength of light used. Thus the maximum resolving power of an optical microscope was limited by the minimum wavelength of light, towards the blue-violet end of the visible spectrum. An electron microscope was developed in 1932 that exploits the wave nature of electrons.

The equation, $\lambda = 12.28/\sqrt{V(\text{volts})}$, tells us that the resolving power of an electron microscope can be increased by increasing the potential difference through which the electrons are accelerated. Resolutions as minute as 1 Å and magnifications as high as $\times\ 10^6$ times have been achieved using electron microscopes. This allows the study of minute cellular constituents and molecules. Figure 6.14 shows the high resolution images of a pollen and an algae obtained using a scanning electron microscope. In materials characterisation, newer techniques were developed exploiting the wave properties of microparticles. Techniques such as electron diffraction, neutron diffraction, electron spectroscopy and neutron spectroscopy were developed as refined tools for material characterisation.

The study of wave packets aided the development of techniques of transmission of signals such as amplitude modulation, frequency modulation and chopping (keying). This knowledge further contributed to various developments in the fields of communication engineering and space technology.

Section 6.7 explained the Heisenberg's uncertainty principle, which allows us to develop a better insight into the study of microlevel phenomena. Chart 6.6 displays four of the many important applications of the uncertainty principle.

Sections 6.8–6.9 gave the interpretation of wave function and the development of Schrodinger's equation. Schrodinger's equation occupies a cardinal place in quantum mechanics; just as Newton's equation does in classical mechanics. Thus it has been used with astonishing success in solving a vast number of problems in atomic physics. Schrodinger even extended his

Chart 6.5 Applications of Photoelectric Effect

Effect	Explanation		Applications
Photoemissive effect	Ejection of electrons from the surface of a photosensitive metal. Photoemissive devices use vacuum tubes and require large bias voltage and high frequency radiation for their operation.	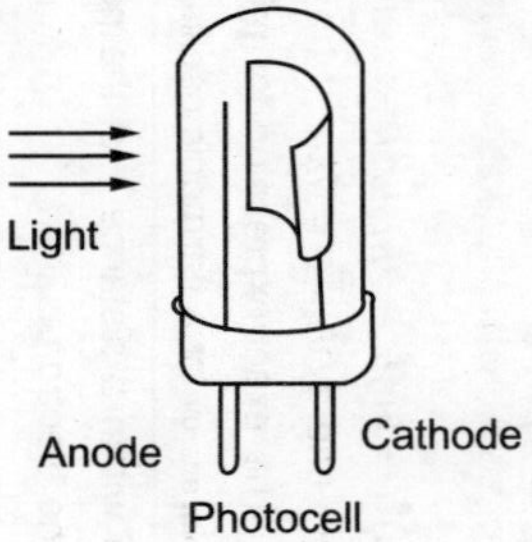 Light Anode Cathode Photocell	• *Photocell:* The anode is usually made of platinum or nickel and the cathode, of potassium or cesium. As the current in vacuum cell is very small, gas-filled cells are also used. Uses of photocell are (i) to measure the amount of illumination, as in photometry and colorimetry; (ii) to reproduce rapid variations in light intensity, as in sound track of a motion picture film or in scanning in TV tube and (iii) to detect a definite large change in light intensity, as in On-Off switches, electric eye. • *Photo multiplier tube:* It consists of several electrodes which are maintained at successively increasing potentials. Thus the number of electrons keep on multiplying and a large photo current is obtained. Photo multiplier tube is used to measure very weak light intensities as in nuclear-radiation, television pickup devices, colourimetry, astronomy and many industrial processes.
Photocon-ductive effect	Generation of free charge carriers inside a photosensitive semiconductor. The conductivity of a semiconductor increases in this effect. Photo-conductive devices use semiconductor chips and require relatively small bias voltage and low frequency radiation for their operation.	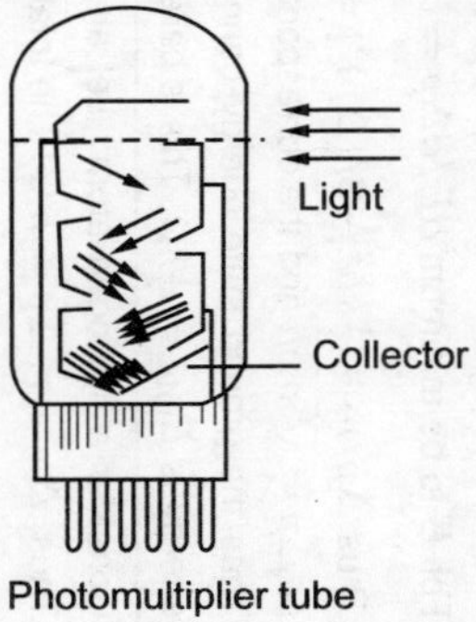 Light Collector Photomultiplier tube	• *Photolectric cells:* A current is generated in a photoconductor when light is shone on it. Moving objects between the light source and the photoconductor interrupts the current and thus get detected. Photo-electric cells are used in systems such as perimeter guard systems to sense and control the movement of objects or intruders. They are also used in xerography and systems that read coded or printed information on data cards and packages. Due to the fact that resistance of photo- conductors changes with the extent of their illumination they are also reffered to as photoresistors or light dependent resistors (LDR). • *Photoconductivity* is also used to examine the perfection of the interiors of large crystals of silicon. Silicon absorbs the visible light but transmits the IR radiation. Thus radiation of variable frequency is incident on a sample and transmitted radiation detected by a detector provides information about the interior of the crystal.
Photovoltaic effect	Generation of voltage due to absorption of light near a *p-n* junction. This voltage is called *photo emf*.	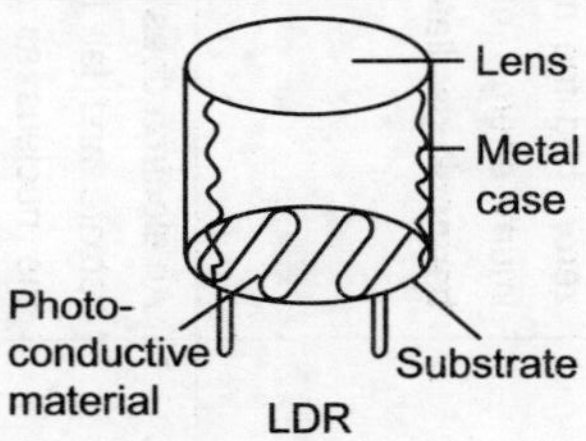 Lens Metal case Photo-conductive material Substrate LDR	• *Solar cell:* Solar cells are used as voltage sources. A large number of cells connected in series or parallel form a solar panel. Solar panels are used as sources of power in satellites, spacecrafts and in large number of solar devices. • *Photodiodes* and *phototransistors* are photovoltaic devices that are widely used in optoelectronic devices. For example, arrays of photo-diodes, fabricated using modern IC techniques can be used to read printed information.

CHART 6.6 APPLICATIONS OF UNCERTAINTY PRINCIPLE

S.No	Problem	Solution
1.	Can the electrons be present inside the nucleus, with protons and neutrons?	If an electron is assumed to be in the nucleus then the uncertainty in its position, Δx, is of the order of the nuclear diameter. $\Delta x = 10^{-14}$ m. Thus, $\Delta p = \hbar/\Delta x = 1.054 \times 10^{-34}\ \text{Js}/10^{-14}\ \text{m} \geq$ 1.054×10^{-20} kg m/s. If this is the uncertainty in a nuclear electron's momentum, the momentum p itself must be at least comparable in magnitude. Now $E = \sqrt{m_0^2c^4 + p^2c^2}$. But with the above value of p, $p^2c^2 >> m_0^2c^4$. Hence $m_0^2c^4$ can be neglected and $E = pc = (1.054 \times 10^{-20}\text{kg m/s})(3.0 \times 10^8\text{m/s}) \geq 3.2 \times 10^{-12}\text{J} \geq 20$ MeV. Thus the kinetic energy of the electron must exceed 20 MeV if it is to be a nuclear constituent. But experiments show that β-rays emitted by radioactive nuclei, comprising of electrons, do not have energy more than 4 MeV. Thus electrons cannot reside in the nucleus.
2.	Show that the zero point energy (ground state energy) of a harmonic oscillator cannot be zero. find the minimum energy of a harmonic oscillator.	Energy of a one-dimensional harmonic oscillator is, $E = p^2/2m + 1/2kx^2 = p^2/2m + 1/2\ m\omega^2x^2$ (as $\omega = \sqrt{k/m}$). Approximating x and p to be uncertainties in these quantities, that is, $x \approx \Delta x$ and $p \approx \Delta p$, we get $E = \frac{1}{2}(\Delta p)^2/m + \frac{1}{2}k\hbar^2/(\Delta p)^2$ (since $\Delta x \Delta p = \hbar$). For E to be minimum $\partial E/\partial \Delta p = 0$ Thus $\Delta p/m + \frac{1}{2}k\hbar^2[-2/(\Delta p)^3] = 0$ or $(\Delta p)^4 = \text{km}\hbar^2 = m^2\omega^2\hbar^2$ or $\Delta p = \sqrt{m\omega\hbar}$ and the corresponding miniumm energy $E_{\min} = \hbar\omega$. Thus the ground state energy cannot be zero. The exact expression for minimum energy is $E_{\min} = \frac{1}{2}\hbar\omega$. This is called the zero point energy for a harmonic oscillator.
3.	An electron does not radiate and fall into the nucleus as expected in Maxwell's theory.	Consider a hydrogen atom. Let an electron be within a distance r of the nucleus. Thus, $\Delta x = r, \Delta p = h/r$. The total energy of the electron is then $E = p^2/2m - (1/4\pi\epsilon_0)e^2/r = h^2/2mr^2 - (1/4\pi\epsilon_0)e^2/r$ If an electron approaches the nucleus, then r decreases and the potential energy becomes more negative but the kinetic energy increases. The energy is minimum at $r = a_0 = h^2(4\pi\epsilon_0)/me^2$ which is the Bohr radius. The lowest energy state is not the one with an electron in the nucleus.
4.	The average time between the excitation of an atom and the radiation from the atom is 10^{-8}s. Find the inherent uncertainty in the frequency of the photon.	$\Delta E \geq \hbar/2\Delta t \geq 1.054 \times 10^{-34}\text{Js}/2(1 \times 10^{-8}\text{s}) \geq 5.3 \times 10^{-27}\text{J}$. The corresponding uncertainty in frequency is $\Delta\nu = \Delta E/h \geq 8 \times 10^6$ Hz This spread results in the irreducible broadening of spectral lines. In practice however, the broadening due to other effects such as Doppler effect dominates and overshadows the broadening due to uncertainty.

(a)

(b)

Figure 6.14 SEM images of (a) pollen and (b) algae
(Source: http://remf.dartmouth.edu/imagesindex.html)

ideas to life sciences and authored a book *What is life?* in 1944, which is said to have inspired Watson's research on genes. This research led him to probe the structure of the DNA molecule, along with Crick, in 1953.

The applications of Schrodinger's equation to the problem of the finite potential well opened up a host of practical applications. Modern diode lasers using fabrication marvels like quantum wells are based on the same simple idea of energy quantisation for electrons trapped in the well. The semiconductor tunnel diodes and a number of superconductor devices such as Josephson junction and SQUID (superconducting quantum interference devices) are based on the quantum mechanical tunnelling phenomenon. The radioactive alpha decay of nuclei has been understood on the basis of tunnelling. The knowledge of tunnelling was most ingeniously used by Gert Binning and Heinrich Rohrer in 1981 in their invention of a scanning tunnelling microscope (STM). The STM uses tunnelling to create images of surfaces with a resolution comparable to atomic size.

Nanocrystallites, Q dots and Q corals are some frontier areas of research based on the quantum mechanical study of finite potential wells. This research is pregnant with a vast potential for applications to various branches of engineering and technology. The quantum dots, for example, act like artificial atoms and have a future in electron optics and computer technology.

The principles of quantum mechanics allow an altogether novel concept of computation. The classical concept of computation is based on finite state machines. The basic concept is that the machine—the computer—is in a definite state. This state changes to another definite state when the machine is given a particular input. The corresponding quantum concept is similar except that the state is a quantum state. The quantum state is of a variety known as *entangled states*. When this state evolves under an input, it is like many classical states evolving simultaneously or parallely. In principle, the consequences of this are far reaching. An example is Shor's algorithm. By using it the problem of prime number factorisation reduces from an exponential time algorithm to a polynomial time one, that is, it can be done in a reasonable time. The quantum computers that employ quantum evolution for performing a computation, are not yet a reality. Development of such machines is one of the most promising frontline research areas in science and technology.

In this chapter the basic aspects of quantum physics have been studied. A brief review of how all these aspects shaped the developments in quantum engineering has also been given. In the next chapter we study LASERS—one of the most important inventions in quantum optics that has contributed immensely to the development of quantum engineering.

EXERCISES

1. What is Compton effect? Why is it observed in light elements? Write an expression for the Compton shift and explain the existence of modified and unmodified radiations in the effect.
2. Explain Compton effect on the basis of quantum theory. Why did the classical theory fail to explain the Compton effect?
3. In the Compton effect what happens: (i) When a photon collides with a free electron in the scattering block? (ii) When a photon collides with a bound electron in the scattering block?
4. Calculate the wavelengths of electrons by (i) Bragg's law and (ii) de Broglie hypothesis, using the data obtained in the Davission and Germer experiment. Explain the importance of the results obtained.
5. Explain the de Broglie concept of matter waves. Explain how de Broglie's hypothesis regarding wave nature of matter leads to Bohr's quantisation of angular momentum of an element.
6. Show that the de Broglie wavelength of an electron accelerated by an electric field V volts/m is $12.28/\sqrt{V}$ Å
7. Explain why a single monochromatic wave cannot represent a particle.
8. What is Heisenberg's uncertainty principle? Explain the significance of this principle. Describe a thought experiment to arrive at this principle.
9. Explain Heisenberg's uncertainty principle and give its applications. Why is it significant only for subatomic particles and not for heavy bodies?
10. State Heisenberg's uncertainty principle and prove that electrons are not the constituents of nucleus.
11. Explain the physical significance of the wave function ψ.
12. Show that the solution of Schrödinger's equation for a particle in an infinite potential well leads to the concept of quantisation of energy. Is the particle allowed to take zero energy? If not, why?
13. Write Schrödinger's time-independent and time-dependent wave equations.
14. Explain in short the phenomenon of tunnelling that occurs when a beam of particles are incident on a potential barrier of finite width.
15. Show that the wave function for a particle confined in a one-dimensional potential well of length L and infinite depth is given by $\psi_n(x) = A\sin(n\pi x/L)$. Hence, using the normalisation condition on ψ show that A is given by $\sqrt{2/L}$. Show the necessary waveforms.
16. Explain the tunnelling effect.

PROBLEMS

1. According to the special theory of relativity, the effective mass of a particle moving with a large velocity is $m = m_0/\sqrt{[1-(v^2/c^2)]}$. Show that the energy of the particle is equal to $\sqrt{(m^2v^2c^2 + m_0^2c^4)}$, where the letters have their usual meaning.
2. Incident x-rays ($\lambda = 1.37$ Å) are scattered by a suitable scatterer at an angle of 30°. Calculate the wavelength of the scattered radiation and velocity of the recoil electron.
3. X-rays of wavelength 22×10^{-12} m (photon energy = 56 keV) are scattered from a carbon target, the scattered radiation being viewed at 85° to the incident beam. (i) What is the Compton shift? (ii) What percentage of its initial energy does an incident photon lose?
4. Calculate the de Broglie wavelength of an electron moving with a velocity of 10^7 m/s.
5. Calculate the wavelength associated with a proton of energy 1 MeV.
6. Calculate de Broglie wavelength of an α-particle accelerated by a potential difference of 6.68×10^{-27} kg.
7. 10-keV electrons are passed through a thin film of metal for which the atomic spacing is 5.5×10^{-11} m. What is the angle of deviation of the first order diffraction maxima?
8. A photon of frequency ν is scattered by an electron. The scattered photon of frequency ν' is observed to travel in a direction 90° to the incident beam. Using the principle of conservation of momentum, show that the de Broglie wavelength of the recoil electron is given by $\lambda = c/\sqrt{(\nu^2 + \nu'^2)}$
9. Show that the uncertainty in the measurement of electron momentum is equal to its momentum itself, if the uncertainty in measurement of the location of the electron is equal to its de Broglie wavelength.
10. A proton is confined to a nucleus of radius 10^{-14} m. Calculate the minimum uncertainty in its momentum. Also calculate the minimum uncertainty in the velocity of the proton.
11. A particle is moving in a one-dimensional infinite potential box of width 25 Å. Calculate the probability of finding the particle within a small interval of 0.05 Å at the centre of the box when it is in its state of least energy.

1. Calculate the following:

   (i) De Broglie wavelengths of (a) 150 gm baseball moving with 35 m/s, and (b) electron accelerated through 120 V;

   (ii) Uncertainty in the position of (a) a 50-gm bullet moving with velocity 30 m/s accurate to 0.01 %, and (b) an electron with velocity 30 m/s accurate to 0.01 %;

   (iii) First three energy levels for: (a) a μg dust particle moving in a one-dimensional box of width 0.1 mm, and (b) an electron confined to 10 Å.

   What conclusions can be drawn from the results obtained?

7

Atomic Physics

7.1 THE STORY OF ATOMIC PHYSICS

In the last chapter we studied quantum physics—the new physics that emerged in the first quarter of the twentieth century to explain phenomena that could not be explained by the classical physics of the nineteenth century. Quantum physics—the physics of the microworld—opened up vast vistas of knowledge giving birth to several new disciplines such as molecular physics, atomic physics, nuclear physics and elementary particle physics. Quantum physics dominated the last century due to its deep philosophical implications and wide technological applications.

The story of physics of the twentieth century is mostly about the pursuits and accomplishments in understanding the microworld through quantum physics. After the initial success of quantum theory in explaining phenomena such as blackbody radiation, photoelectric effect and Compton effect, quantum ideas were applied with a lot of enthusiasm to many research problems—the foremost among them being concerned with the structure of atoms and atomic phenomena.

The word 'atom' makes one curious and fires the imagination. For physicists the atom has been a fascinating domain of enquiry and insight since antiquity. There were curious attempts to know whether matter, though appearing to be continuous, has a definite structure on a microscopic level beyond the direct reach of our senses. Leucippus (500 BC) and later his follower, Democritus (420 BC) speculated that all matter is made of atoms (a word derived from the Greek word *atomos*, meaning indivisible) and that there are open spaces between the atoms. However, Plato and Aristotle (384–472 BC) disagreed with Democritus and opined matter to be continuous. The ideas of these philosophers remained unquestioned for the next twenty centuries, until in 1600, Gassendi, an Italian physicist, revived the atomic theory once again. The atomic theory was substantiated through the research of Dalton (1808) and Avogadro (1811) who suggested models to explain how atoms combine to form a molecule.

More experimental evidence came forth—but only accidentally—in favour of the atomic theory in 1827. Robert Brown, the Scottish botanist, observed the zigzag motion of pollen grains suspended in water. This motion, which came to be known as Brownian motion, was attributed to the random collision of pollen with water molecules. In 1905—now christened his miracle year—Einstein produced three seminal papers, one of which treated the Brownian motion. Using the experimental data on Brownian motion and applying the atomic model, he estimated the size of an atom to be 1×10^{-8} cm, a value close to the size of hydrogen—the smallest atom known today (1.06×10^{-8} cm).

How do we access atoms? The answer is through spectra. Atomic spectra are said to be signatures of atoms as each atom has a unique spectrum. The first atomic spectra were recorded by Wollastan in 1802 (measured more precisely by Fraunhoffer in 1814). However, the fundamental physical process that resulted in sharp spectral lines could not be understood. Research in spectroscopy was aimed mainly at correlating the observed spectra with the atomic structures that produced them.

By the end of the nineteenth century, the atomic theory was largely accepted but very little was known about the atoms themselves. In 1898, the British physicist, JJ Thomson, proposed that atoms are positively charged lumps of atoms with electrons embedded in them (see Fig. 7.1a). Thomson's model got serious attention because he had played an important role in discovering the electron only a year before. However, the pathbreaking research by Ernest Rutherford and his co-workers proved the real atom to be quite different.

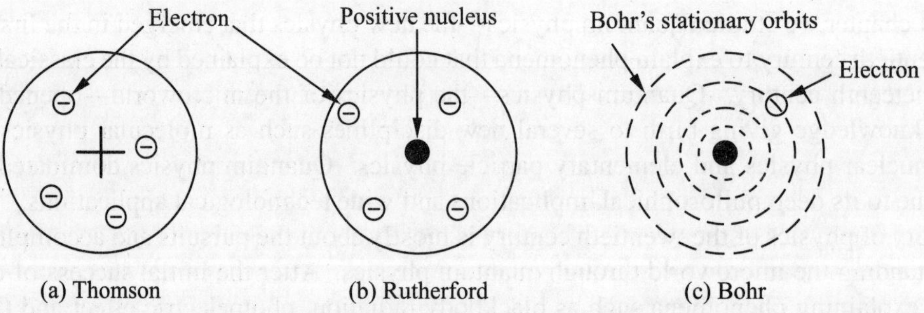

Figure 7.1 Various atomic models

Rutherford, a native of New Zealand, worked as a research student under JJ Thomson at Cambridge. In 1911, he suggested that his co-workers, Hans Geiger and Ernest Marsden, should conduct experiments on scattering of alpha rays from thin metal foils. At that time Marsden was a twenty-year-old student who had not yet earned his bachelor's degree. Their experiments involved shooting high energy alpha particles at thin metal foils and observing the change in their paths. Using the apparatus shown in Fig. 7.2a, they got surprising results (see Fig. 7.2b). Some of the particles were scattered through very large angles, a few of them getting scattered even in the reverse direction. Rutherford wondered at the results: *It was as incredible as you fired a 15-inch shell at a piece of tissue paper and it came back and hit you.*

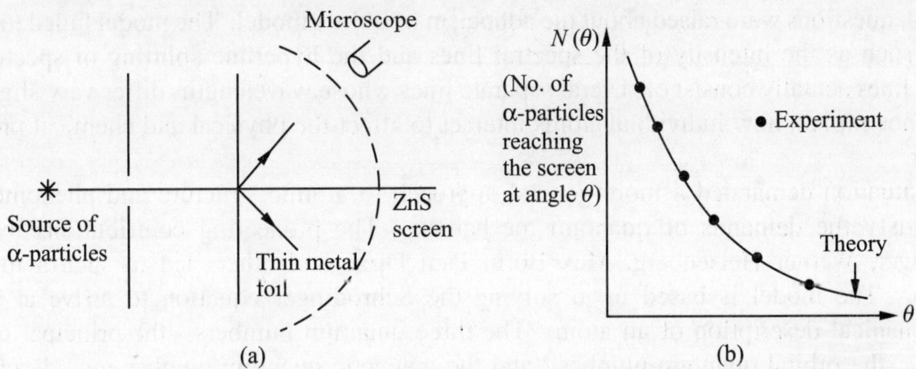

Figure 7.2 Rutherford scattering experiment: (a) Experimental arrangement; (b) Experimental results

Rutherford advanced the planetary model of the atom on the basis of the results obtained in the scattering experiments. According to this model, an atom consists of a tiny nucleus that has a positive charge and in which almost the entire mass of the atom is concentrated. The electrons are in motion at some distance from the nucleus (see Fig. 7.1b). The empty space between atoms allows most of the alpha-particles to pass through the foil; those particles which approach the nuclei get scattered through large angles because of the intense electric field near the nuclei. Experimental results obtained by Rutherford showed good agreement with the theoretical predictions of the model advanced by him. However, the planetary atom was inconsistent with the celebrated electromagnetic theory as it predicted that the accelerated electronic charge should radiate energy and ultimately collapse into the nucleus, contrary to the observed stability of atoms. In 1913, Niels Bohr, the Danish physicist, proposed a model of the hydrogen atom that cleverly tied together the nuclear atom of Rutherford and light quanta of Einstein. He proposed that electrons revolve around the nucleus in certain circular orbits in which they do not radiate energy (Fig. 7.1c). These orbits are called Bohr's stationary orbits and are characterised by the quantisation of angular momentum. Arnold Sommerfeld generalised these to elliptical orbits. The absorption and emission of photons result when electrons make quantum transitions from one stationary orbit to another.

There had been several attempts to explain the spectrum of the hydrogen atom on an empirical basis. They included contributions by Lyman, Paschen, Brackett and Pfund. However, the empirical relation discovered by JJ Balmer, a Swiss schoolteacher, in 1885 was the most successful among the various explanations put forth. Bohr was attracted to Balmer's work in early 1913 and developed his model in the same year.

Bohr's model offered the first successful account of spectral lines for hydrogen. The agreement between the experimental value of Rydberg's constant (constant of proportionality in Balmer's formula) and the theoretical value based on Bohr's model is one of the great triumphs of Bohr's theory. Bohr further explored the domains of classical and quantum physics and propounded the correspondence principle that elucidates how classical physics can be subsumed within quantum physics for large quantum numbers. However, Bohr's model aroused a great controversy as it was a hybrid mixture of classical and quantum concepts. Otto Stern and Max von Laue even threatened to quit physics if Bohr were right. (However, they changed their minds later.) As quantum physics advanced, questions were raised about the adhocism in Bohr's model. The model failed to explain features such as the intensity of the spectral lines and the hyperfine splitting of spectral lines (spectral lines actually consist of several separate lines whose wavelengths differ very slightly). It also did not explain how individual atoms interact to affect the physical and chemical properties of matter.

This situation demanded a more general approach to atomic structure and phenomena that could satisfy the demands of quantum mechanics. The pioneering contributions by Erwin Schrodinger, Werner Heisenberg, Max Born, Paul Dirac and others led to such a model by 1925–26. The model is based upon solving the Schrodinger equation to arrive at a quantum mechanical description of an atom. The three quantum numbers—the principal quantum number n, the orbital quantum number l and the magnetic quantum number m_l—described an atom in this model. The fourth quantum number, the spin quantum number m_s, was added in 1926 for giving a more complete description of an atom. Two young Dutch physicists, Uhlenbeck and Goudsmit, introduced the concept of spin to account for the multiple lines within the spectra and for the so-called 'anomalous Zeeman effect'. Though the spin quantum number was an arbitrary addition, in 1928, Dirac showed by invoking the relativistically correct form of the Schrodinger equation, that it was an intrinsic part of the quantum theory.

In 1925, Wolfgang Pauli (1900–58), an Austrian physicist, proposed his famous exclusion principle: No two electrons can have the same set of four quantum numbers. This principle turned out to be the missing link in understanding the arrangement of electrons in an atom. Interestingly, a few months before Uhlenbeck and Goudsmit introduced the concept of spin, the American physicist Ralph Kronig had conceived of this idea and had told Pauli about it. However, because Pauli ridiculed the idea, Kronig did not publish his work.

The rigorous quantum mechanical model based on a set of four quantum numbers, (n, l, m_l, m_s) could explain almost all the available experimental results and is thus considered the most successful model of the atom even today. Foremost among these successes was the understanding of different basic atoms in nature in a unified manner.

How many different basic atoms are there in nature? Dmitri Mendeleev (1834–1907), the Russian chemist, made the first comprehensive attempt to tabulate the elements. His Periodic Table consisted of 63 elements. Other elements were discovered subsequently and today we know that there are about 92 naturally occurring elements, and 16 artificial elements that can be synthesised in the laboratory using high-energy accelerators and a nuclear reactor. Some of these elements are named after prominent scientists or places: Einsteinium (Es, $Z = 99$), Fermium (Fm, $Z = 100$) and Berkelium (Bk, $Z = 97$). The element of atomic number 101 is called Mendelevium in honour of Mendeleev. The quantum model of the atom helped to unfold the mystery and bring order to the periodic table of elements.

In 1895, an interesting discovery became a part of the story of atomic physics. This was the discovery of x-rays by the German physicist, Wilhem K Roentgen (1845–1923). Roentgen, completed a diploma in mechanical engineering in 1868 and received his doctorate in 1869. He published his work in the *Journal of Physical and Medical Society* and received the first Nobel Prize in 1901, instituted by the Nobel Foundation in Sweden.

One day, Roentgen, working in a darkened room with a gas discharge tube, observed that a sheet of paper coated with barium platinocynide kept nearby was glowing. He noted that the sheet glowed only when high tension was applied to the discharge tube and the glow disappeared when the potential was switched off. Roentgen suspected that this was due to some kind of radiation from the discharge tube. He repeated the experiment several times and found that the radiation could pass through thick sheets of paper and even through a thin sheet of metal. He named the radiation as 'x-rays' to denote its unknown nature.

In 1913, the young British physicist HJG Moseley, who worked under Rutherford, studied characteristic x-rays for as many as 38 elements using his own x-ray tube. Owing to Moseley's work, the characteristic x-ray spectrum became the universally accepted signature of an element, permitting the solution of a number of periodic table puzzles. Moseley's results showed that the spectral lines emitted by these elements belonged to two different series, which were identified with the K- and L-type of characteristic radiation previously observed by Charles Barkla.

This chapter will be mainly concerned with the structure of the atom, since it is this structure that is responsible for nearly all the properties of matter that shape our world. We begin with an overview of the atomic spectra and Bohr's theory (Sections 7.2 and 7.3). Section 7.4 deals with the application of quantum mechanics to the problem of the hydrogen atom. The resulting quantum mechanical description of an atom based on quantum numbers and its application in understanding the periodic table is discussed in Section 7.5, while Section 7.6 deals with the x-ray spectra of elements. Various applications of the knowledge of atomic physics are discussed in the last section.

7.2 ATOMIC SPECTRA

In Expt. 4.1 we saw that light coming through the collimator in a spectrometer and incident on a grating deviates through an angle that depends on the wavelength of the light. Thus one observes an image of the slit at this angle, when viewed through the telescope of the spectrometer. If the slit is vertical, a set of vertical lines, corresponding to the wavelengths present in the source of light, can be seen. This set of images is known as *line* or *emission spectrum* and is produced when light from a gas, through which electric discharge is passing, or from a flame to which a volatile salt is added, is dispersed by a spectrometer. An *absorption spectrum* is obtained when light with a continuous spectrum, for example, light from a filament lamp, is passed through an absorbing layer of the element and then analysed with a spectrograph. The absorption lines that appear as dark lines on a bright background coincide exactly with the corresponding emission lines (Fig. 7.3). The Fraunhoffer lines in the solar spectrum are an example of absorption spectrum; the cooler gases surrounding the sun absorb light of certain wavelengths emitted by the sun.

Line spectra are also called atomic spectra, as the wavelengths of the lines are characteristic of the atoms of the element emitting the light. Each element has its unique atomic spectra, which are therefore known as signatures of atoms. For example, the line spectra for hydrogen, helium and mercury are shown in Fig. 7.4. The wavelengths in the spectrum of an element are found to fall into series called the spectral series. Table 7.1 shows the different spectral series for hydrogen.

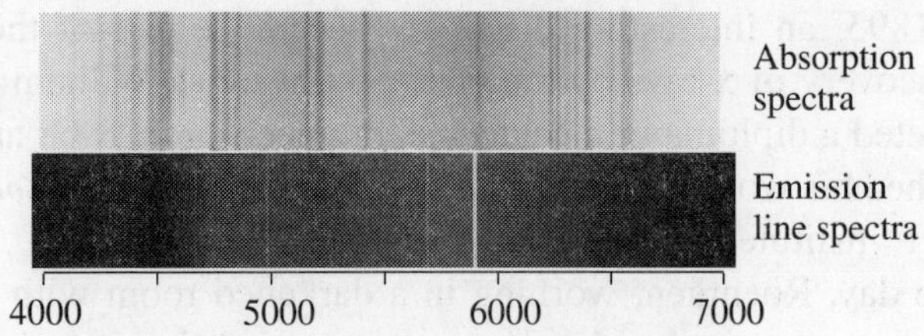

Figure 7.3 Spectra of sodium vapor (Source: http://en.wikipedia. org/wiki/File: Sodium_spectrum.jpg)

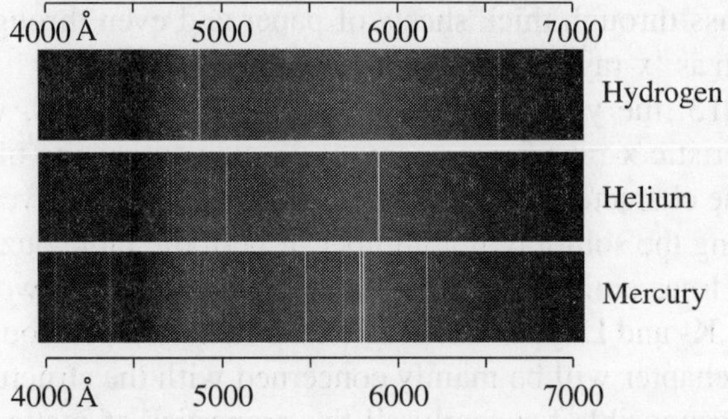

Figure 7.4 Emission line spectra of hydrogen, helium and mercury (Adapted from: http://astro.u-strasbg.fr/~koppen/discharge/discharge.html)

Table 7.1 Spectral series for hydrogen

Series	**Expression for wavelength, λ (R is the Rydberg constant)**	**Range of electromagnetic spectrum**
Lyman	$\frac{1}{\lambda} = R\left(\frac{1}{1^2} - \frac{1}{n^2}\right), \ n = 2, 3, \ldots$	Ultraviolet
Balmer	$\frac{1}{\lambda} = R\left(\frac{1}{2^2} - \frac{1}{n^2}\right), \ n = 3, 4, \ldots$	Visible
Paschen	$\frac{1}{\lambda} = R\left(\frac{1}{3^2} - \frac{1}{n^2}\right), \ n = 4, 5, \ldots$	Infrared
Brackett	$\frac{1}{\lambda} = R\left(\frac{1}{4^2} - \frac{1}{n^2}\right), \ n = 5, 6, \ldots$	Infrared
Pfund	$\frac{1}{\lambda} = R\left(\frac{1}{5^2} - \frac{1}{n^2}\right), \ n = 6, 7, \ldots$	Infrared

7.3 BOHR'S THEORY

In 1913, Bohr applied the ideas of the quantum theory of radiation to the Rutherford nuclear atom to develop a theory that could account for many of the properties of atomic spectra. His theory is based on the following postulates.

1. An atomic system possesses a number of states which are radiatively stable. Contrary to classical electromagnetism, there is no radiation of energy, even when the motion of charged particles in these states is an accelerated motion. Such states are called the *stationary states* of the system.
2. The different possible stationary states of an electron revolving round the positive nucleus are those for which the angular momentum is quantised.
3. Any emission or absorption of radiation will correspond to a transition between two stationary states, and the frequency, ν, of the radiation emitted or absorbed, is given by the equation $h\nu = E_2 - E_1$, where E_1 and E_2 are the energies of the corresponding stationary states.

Bohr's theory could account for the hydrogen atom spectrum in the following way. According to the second postulate stated above, the angular momentum, L, of an electron in a hydrogen atom revolving round the nuclear charge, $+\,e$, in a circular orbit of radius r is

$$L = mvr = n\hbar, \; n = 1, 2, 3, \ldots \tag{7.1}$$

m being the mass and v being the linear velocity of the electron.

The electrostatic force of attraction on the electron due to the nucleus provides the necessary centripetal force so that

$$\frac{mv^2}{r} = \frac{1}{4\pi\epsilon_0} \times \frac{e^2}{r^2} \tag{7.2}$$

Using Eqs. 7.1 and 7.2 we get

$$r_n = \frac{n^2h^2\epsilon_0}{\pi me^2} \tag{7.3}$$

and $$v_n = \frac{e^2}{2h\epsilon_0 n} \tag{7.4}$$

where the suffix n is used to denote the nth Bohr orbit. The total energy E_n of the electron in this orbit is the sum of its kinetic and potential energies given by

$$E_n = \frac{1}{2}mv_n^2 - \frac{1}{4\pi\epsilon_0} \times \frac{e^2}{r_n} \tag{7.5}$$

Using Eqs. 7.3 and 7.4 in Eq. 7.5, we get

$$E_n = \frac{-me^4}{8n^2h^2\epsilon_0^2} \tag{7.6}$$

In Ex. 7.1 we obtain handy expressions for radius, velocity and energy of electrons in Bohr orbits by substituting the numerical values of the fundamental numerical constants in Eqs. 7.3, 7.4 and 7.6.

EXAMPLE 7.1

Handy formulae for r_n, v_n and E_n for the hydrogen atom

Show that the radius, the velocity and the energy of an electron in the nth Bohr orbit are given respectively by $r_n = 0.53n^2$ Å, $v_n = 2.2 \times 10^6/n$ m/s and $E_n = -13.6/n^2$ eV. Find these parameters for the electron in the ground state. Use these results to develop a quantitative perspective of Bohr's model of the Hydrogen atom.

Solution

Using Eqs. 7.3, 7.4 and 7.6 and substituting the numerical values of constants we get

$$r_n = \frac{n^2h^2\epsilon_0}{\pi me^2} = \frac{(6.625 \times 10^{-34}\ \text{Js})^2 \times (8.85 \times 10^{-12}\ \text{F/m}) \times n^2}{3.14 \times 9.1 \times 10^{-31}\ \text{kg} \times (1.6 \times 10^{-19}\ \text{C})^2}$$

$$v_n = \frac{e^2}{2h\epsilon_0 n} = \frac{(1.6 \times 10^{-19}\ \text{C})^2}{(2 \times 6.625 \times 10^{-34}\ \text{Js}) \times (8.85 \times 10^{-12}\ \text{F/m}) \times n}$$

$$E_n = \frac{-me^4}{8n^2h^2\epsilon_0^2} = \frac{-[9.1 \times 10^{-31}\text{kg} \times (1.6 \times 10^{-19}\ \text{C})^4]}{8 \times (6.625 \times 10^{-34}\ \text{Js})^2 \times (8.85 \times 10^{-12}\ \text{F/m})^2 \times n^2}$$

Simplifying, we get

$$r_n = 0.53n^2\ \text{Å},\ v_n = \frac{2.2 \times 10^6}{n}\ \text{m/s and}\ E_n = \frac{-13.6}{n^2}\ \text{eV for the ground state},\ n = 1.$$

Thus $r_1 = 0.53$ Å, $v_1 = 2.2 \times 10^6$ m/s and $E_1 = -13.6$ eV.

These results provide a feel of the magnitudes involved at the atomic level. The radii increase beyond 0.53 Å, the velocities decrease below 2.2×10^6 m/s and the total energies increase above -13.6 eV as we move to the higher Bohr orbits. The negative total energy implies that the electron is bound to the nucleus and for a large value of n, velocity and energy approach zero.

According to Bohr's third postulate, the frequency of the quantum of radiation emitted or absorbed when an electron in a hydrogen atom makes a transition from one stationary state E_n to another stationary state E_k is given by

$$\nu = \frac{E_n - E_k}{h} \tag{7.7}$$

The corresponding wave number, that is, the number of waves in unit length of light path in vacuum is given by

$$\nu = \frac{1}{\lambda} = \frac{\nu}{c} = \frac{E_n - E_k}{hc} \tag{7.8}$$

Using Eq. 7.6

$$\nu = \frac{me^4}{8h^3\epsilon_0^2 c}\left[\frac{1}{k^2} - \frac{1}{n^2}\right] = R\left[\frac{1}{k^2} - \frac{1}{n^2}\right] \tag{7.9}$$

where the constant, $R = \dfrac{me^4}{8h^3\epsilon_0^2 c}$ (7.10)

is known as the *Rydberg constant* and Eq. 7.9 is the *Rydberg formula*. Thus the Rydberg constant is known in terms of fundamental constants and its theoretical value can be compared against the experimental value to verify Bohr's theory. However, the above derivation of the Rydberg formula is based on the assumption that the nucleus is stationary with respect to the electron. This is a reasonable assumption as the hydrogen nucleus consisting of a single proton is about 1840 times heavier than the electron. However, the precision of spectroscopic measurements is so high that the motion of the nucleus cannot be neglected for the purpose of testing agreement between theory and experiment. If the motion of the nucleus is taken into consideration, then the mass m of the electron should be replaced by its reduced mass given by

$$m' = \frac{mM}{m + M} \tag{7.11}$$

where M is the mass of the nucleus. Thus the refined expression for the Rydberg constant becomes

$$R = \frac{m'e^4}{8h^3\,\epsilon_0^2 c} = \frac{e^4}{8h^3\,\epsilon_0^2 c} \times \frac{mM}{m + M} \tag{7.12}$$

In Ex. 7.2 we examine the agreement between the theoretical and the experimental values of R for the hydrogen atom. Through this problem we can appreciate how the incorporation of reduced mass in the expression for R leads to improvement in agreement with the experimental result. Example 7.3 illustrates the agreement between theory and experiment for singly ionised helium. These problems thus bring out the validity of Bohr's theory for hydrogen and helium.

EXAMPLE 7.2

Incorporation of reduced mass of electron in Bohr's theory

The experimental value of the Rydberg constant for hydrogen is $R_H = 109677.58\ \text{cm}^{-1}$. Compare this with the theoretical results obtained using Bohr's theory considering (a) rest mass and (b) reduced mass of an electron. What do the results signify?

Solution

(a) The Rydberg constant for hydrogen, considering the rest mass of electron, is

$$R = \frac{me^4}{8h^3\epsilon_0^2 c}$$

$$= \frac{9.109382 \times 10^{-31}\ \text{kg} \times (1.602176 \times 10^{-19}\ \text{C})^4}{8 \times (6.626069 \times 10^{-34}\ \text{Js})^3 \times (8.854188 \times 10^{-12}\ \text{F/m})^2 \times 2.997925 \times 10^8\ \text{m/s}}$$

$$= 109737.16\ \text{cm}^{-1}$$

(b) Incorporation of reduced mass leads to $R = \dfrac{me^4}{8h^3\epsilon_0^2 c} \times \dfrac{1}{1 + (m/M)}$

$$= \frac{109737.16\ \text{cm}^{-1}}{1 + (9.109382 \times 10^{-31}\ \text{kg}/1.672622 \times 10^{-27}\text{kg})}$$

$$= 109677.43\ \text{cm}^{-1}$$

This result is in closer agreement with the experimental value (109677.58 cm^{-1}) than the one obtained above by neglecting the reduced mass.

This example tells us that the precision of spectroscopic measurements is so great that in testing the theory, the motion of the nucleus cannot be neglected, even though the nucleus is much heavier than the electron.

EXAMPLE 7.3

Seeking agreement between experiment and Bohr's theory for helium

Using the following experimental data compare the results obtained using Bohr theory with the experimental value of $M/m = 1836.12$, obtained using Millikan's oil drop experiment.

The Rydberg constant for singly ionised helium, $R_{He} = 109722.269\ \text{cm}^{-1}$ and that for hydrogen, $R_H = 109677.58\ \text{cm}^{-1}$.

Solution

For hydrogen,

$$R_H = \frac{me^4}{8h^3\epsilon_0^2 c}\ \frac{1}{1 + (m/M)} \tag{7.13}$$

and for helium,

$$R_{He} = \frac{me^4}{8h^3\epsilon_0^2 c}\ \frac{1}{1 + (m/4M)} \tag{7.14}$$

where we have assumed that the mass of the helium atom is four times the mass of a proton.

From Eqs. 7.13 and 7.14 we get, $\dfrac{R_H}{R_{He}} = \dfrac{4M + m}{4(m + M)}$.

Rearranging, we get

$$\frac{M}{m} = \frac{(4R_H - R_{He})}{4(R_{He} - R_H)} = \frac{4 \times 109677.58\,\text{cm}^{-1} - 109722.269\,\text{cm}^{-1}}{4(109722.269\,\text{cm}^{-1} - 109677.58\,\text{cm}^{-1})} = 1840.4$$

We see that the theoretical value M/m is in excellent agreement with the experimental value given above, thus validating Bohr's theory for the singly ionised helium.

More refined spectroscopic analysis shows that spectral lines consist of a number of component lines lying close together corresponding to close energy levels. Sommerfeld succeeded in part in accounting for the fine structure of the spectral lines of hydrogen and He^+ by postulating the existence of elliptic orbits as well as circular orbits, and by taking into account the relativistic variation of the electron mass with velocity. However, Bohr's theory could not succeed any further due to its inherent limitation that it was a hybrid combination of classical and quantum ideas. Stationary states in Bohr's model are characterised by the quantisation of angular momentum on one hand and by the dynamic equilibrium governed by classical mechanics on the other.

Bohr's theory showed serious disagreement with the experimental observations in the case of spectra of many electron atoms, for example, two electron atoms such as neutral helium and singly ionised lithium, and the multiplet structures of spectral lines such as sodium doublet or magnesium triplet. It could not further account for the splitting of spectral lines in a magnetic field because it necessitated introduction of a third quantum number, the magnetic quantum number, that was related to three-dimensional or spatial quantisation of the orientation of the elliptic orbits in space. In the next section we will see how the application of quantum mechanics overcame these limitations of Bohr's theory.

7.4 APPLICATION OF QUANTUM MECHANICS TO HYDROGEN ATOM

The study of the hydrogen atom using quantum mechanics invokes the application of various ideas of quantum physics that were studied in the last chapter. De Broglie's theory of wave mechanics is first applied to the electron in the hydrogen atom. Secondly, the Heisenberg uncertainty principle is applied to reconcile different ideas and develop a perspective of atomic structure. Lastly, the Schrodinger equation is applied to study the electron in the field of the hydrogen nucleus.

7.4.1 Application of de Broglie's Theory

According to de Broglie's theory the electron revolving round the nucleus in a hydrogen atom is associated with a matter wave, that is, an electron wave. As the electron is constrained to move in a circular orbit, the circumference of the circular orbit should be equal to the integral multiple of the wavelength of the electron wave (see Fig. 7.5a). Thus, $2\pi r = n\lambda$, where, λ, the wavelength of the electron wave is given by the de Broglie equation as $\lambda = h/mv$. Thus, we have $mvr = nh/2\pi = n\hbar$, which is Bohr's condition for quantisation of angular momentum.

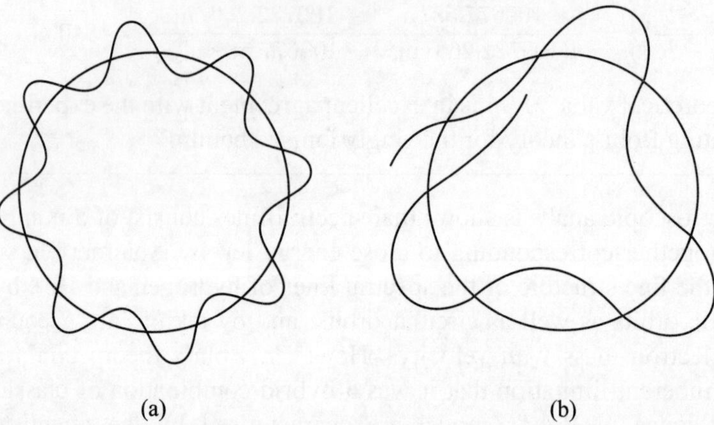

Figure 7.5 (a) Circumference of electron orbit is equal to integral multiple of the wavelength of the electron wave; (b) Destructive interference will occur if fractional number of wavelengths fit the circumference

The behaviour of electron waves in the hydrogen atom is analogous to the standing wave vibrations of a wire loop. Thus we may postulate that *an electron circles a nucleus only if its orbit contains an integral number of de Broglie wavelengths.* A fractional number of wavelengths is not possible because destructive interference will occur (Fig. 7.5b). Thus the quantum number, n, assumes a new significance. It is the number of de Broglie wavelengths that fit into the circumference of Bohr's stationary orbits. In Table 7.2 the electron waves for the first eight orbits are displayed.

7.4.2 Application of Uncertainty Principle

The postulate stated above contains both the particle and wave characters of the electron since the electron wavelength is derived from the orbital velocity required to balance the attraction of the nucleus. That the antithetical wave and particle characters of an electron can never be observed simultaneously, though they are inseparable in nature, is brought out in Heisenberg's uncertainty principle.

According to the uncertainty principle the position and momentum, or energy and time, of an electron cannot be measured simultaneously with unlimited precision. Thus Bohr's concept of stationary orbits contradicts the uncertainty principle, because, for Bohr's orbits the momenta and energies are sharply defined so that $\Delta p = 0$ and $\Delta E = 0$. The uncertainty relations, $\Delta x \Delta p \geq \hbar$ and $\Delta E \Delta t \geq \hbar$, imply that $\Delta x = \infty$ and $\Delta t = \infty$, contrary to the observation that electrons lie in the vicinity of the nuclei and have an average de-excitation time of 10^{-8} s.

Thus, the concept of sharply defined orbits no longer holds in quantum mechanics. However, the uncertainty principle can be used to show that the uncertainty in the position of an electron in the ground state is the same as Bohr's first radius (see Ex. 7.4).

Table 7.2 Electron waves in the hydrogen atom

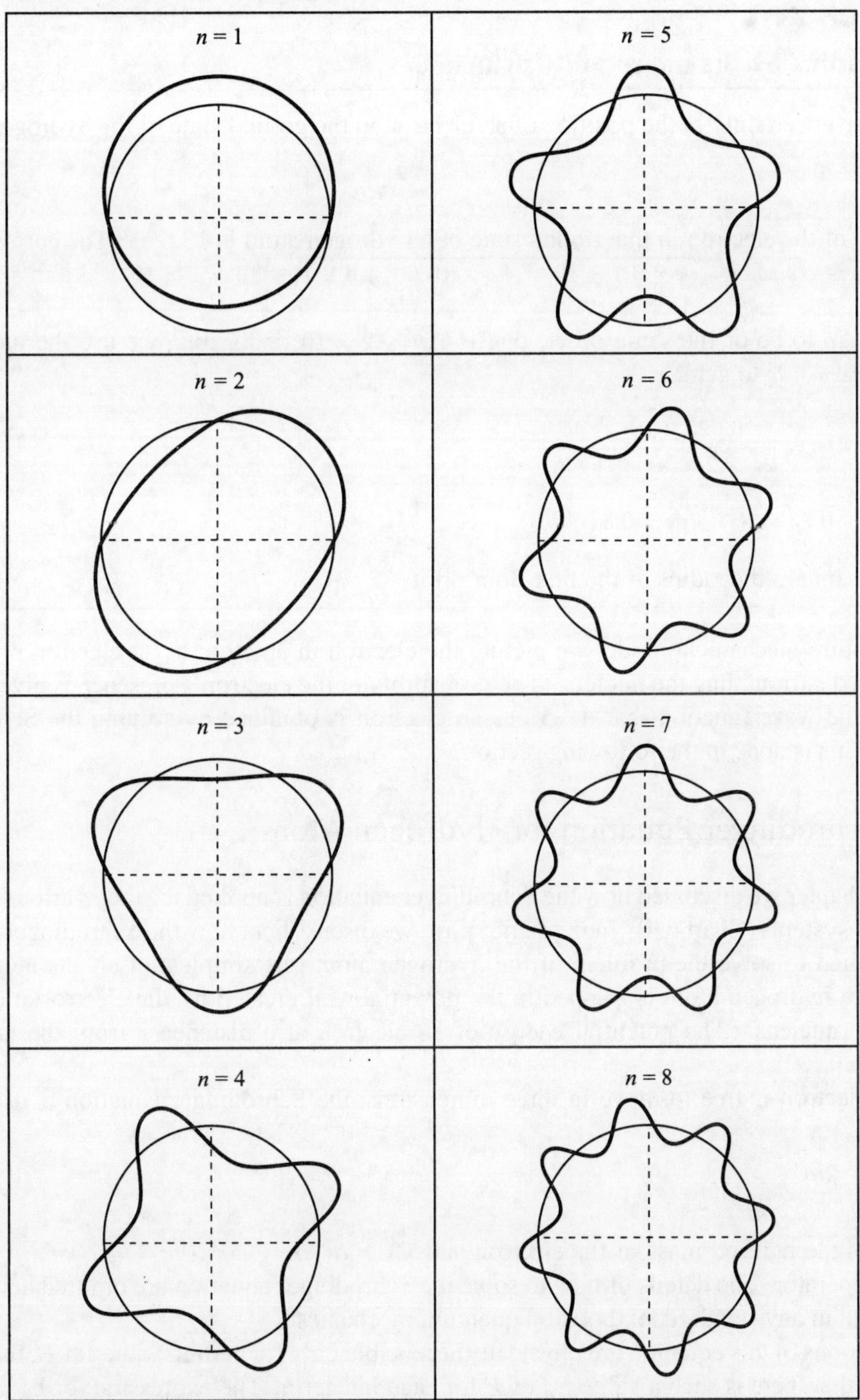

EXAMPLE 7.4

Bohr's radius by the uncertainty principle

Estimate the uncertainty in the position of an electron in the ground state of the hydrogen atom.

Solution

The energy of the electron in the ground state of a hydrogen atom is 13.6 eV. The corresponding momentum is $\sqrt{2mE} = \sqrt{2 \times 9.1 \times 10^{-31}\ \text{kg} \times 13.6 \times 1.6 \times 10^{-19}\,\text{J}} = 2 \times 10^{-24}$ kg m/s.

Assuming Δp to be of the same order, that is $\Delta p \sim 2 \times 10^{-24}$ kg m/s, we use the uncertainty relation $\Delta x \Delta p \geq \hbar$, to get Δx.

$$\Delta x \sim \hbar/p \sim \frac{6.625 \times 10^{-34}\ \text{Js}}{2\pi \times 2 \times 10^{-24}\ \text{kg m/s}}$$

$$\sim 0.53 \times 10^{-10}\ \text{m} = 0.53\ \text{Å}$$

This is the same as the radius of the first Bohr orbit.

In the quantum mechanical model we picture the electron in an atom as an electron probability density cloud surrounding the nucleus. The probability of the electron's presence is given by $|\psi|^2$ where, ψ, the wave function that describes an electron is obtained by solving the Schrodinger equation. This is done in the following section.

7.4.3 Schrodinger Equation for Hydrogen Atom

In the last chapter we discussed how the Schrodinger equation is applied to study various quantum mechanical systems (Chart 6.4). In the same spirit, we discuss here how the Schrodinger equation can be applied to solve the problem of the hydrogen atom, the simplest of all the atoms. The electron in a hydrogen atom is trapped in the potential well created by the electrostatic field of the positive nucleus. The potential energy of an electron at a distance r from the nucleus is $V = -e^2/4\pi\epsilon_0 r$.

As the electron is free to move in three dimensions, the Schrodinger equation is used in the form

$$\nabla^2\psi + \frac{2m'}{\hbar^2}(E - V)\psi = 0 \tag{7.15}$$

where m' is the reduced mass of the electron and $\nabla^2 = \partial^2/\partial x^2 + \partial^2/\partial y^2 + \partial^2/\partial z^2$ is called the Laplacian operator. The details of how to solve the Schrodinger equation are omitted here as they can be found in any standard textbook on quantum mechanics.

The solutions of this equation are physically acceptable only for certain values of E, the energy. It turns out that there is such a value E_n of E for each integer n. The expression for E_n coincides with Eq. 7.6. For $E = E_n$, there are n^2 solutions of the equation. These solutions are called states of electron with energy E_n. These states can be classified by two more integers: l which varies from 0 to $n - 1$ and m which, for each value of l, varies from $-l$ to $+l$. The integers n, l and m are called *quantum numbers*. For an acceptable set of n, l and m, the solution in spherical polar coordinates r, θ, and ϕ of the electron is given by

$$\psi_{nlm}(r, \theta, \phi) = R_{nl}(r) Y_{lm}(\theta, \phi) \tag{7.16}$$

where, $R_{nl}(r)$ and $Y_{lm}(\theta, \phi)$ are special functions of r, and of θ and ϕ.

Figure 7.6 shows plots of $R_{nl}(r)$ for $n = 1, 2, 3, 4$.

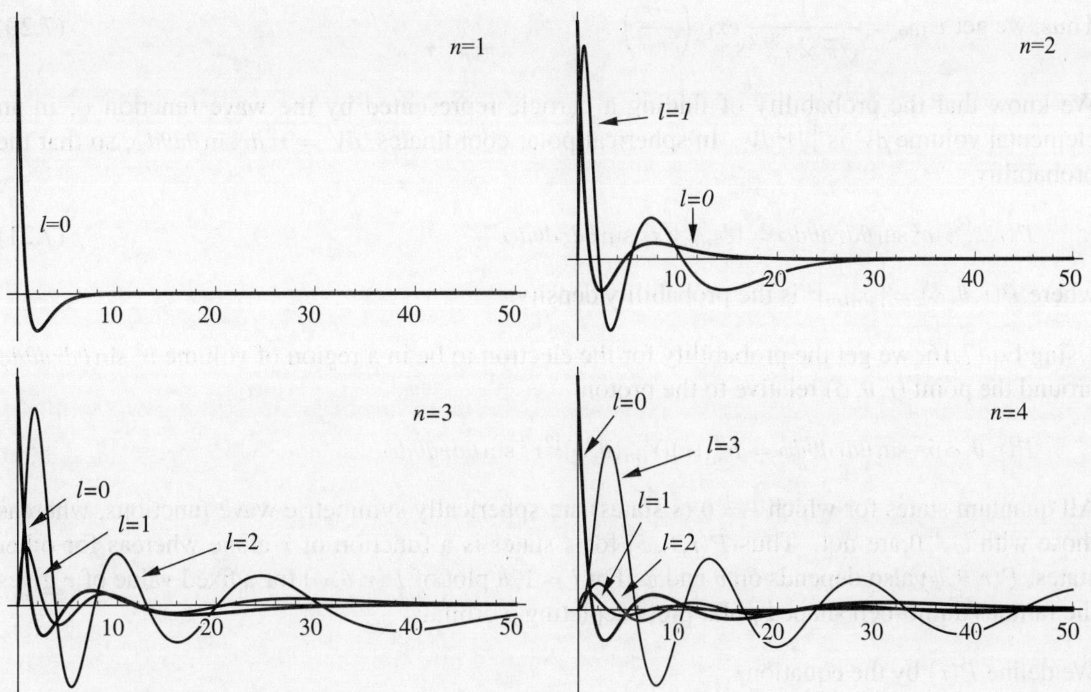

Figure 7.6 Plots of radial parts of hydrogen electron wavefunctions for $n = 1$, 2, 3 and 4. For each n, the functions are plotted for $l = 0, 1, \ldots (n-1)$.

The complete wave function is thus quite complicated. However, the simplified ground state wave function can be written as

$$\psi_{100} = A \exp\left(\frac{-r}{r_1}\right) \tag{7.17}$$

where r_1 is the radius of the first Bohr orbit. The constant A can be determined using the normalisation condition,

$$\int |\psi_{100}|^2 dV = 1$$

Thus, $\int\limits_0^\infty \left[A \exp\left(\frac{-r}{r_1}\right)\right]^2 4\pi r^2 dr = 1$ (7.18)

Solving, we get $A = \frac{1}{\sqrt{\pi} \times r_1^{3/2}}$ (7.19)

where we have used the result $\int\limits_0^\infty x^n \exp(-ax)dx = \frac{n!}{a^{n+1}}$

Thus, we get $\psi_{100} = \frac{1}{\sqrt{\pi} \times r_1^{3/2}} \exp\left(\frac{-r}{r_1}\right)$ (7.20)

We know that the probability of finding a particle represented by the wave function ψ, in an elemental volume dV is $|\psi|^2 dV$. In spherical polar coordinates, $dV = r^2 dr \sin\theta d\theta d\phi$, so that the probability,

$$P(r, \theta, \phi) r^2 \sin\theta dr d\theta d\phi = |\psi_{nlm}|^2 r^2 \sin\theta dr d\theta d\phi \quad (7.21)$$

where $P(r, \theta, \phi) = |\psi_{nlm}|^2$ is the probability density.

Using Eq. 7.16, we get the probability for the electron to be in a region of volume $r^2 \sin\theta dr d\theta d\phi$ around the point (r, θ, ϕ) relative to the proton

$$P(r, \theta, \phi) r^2 \sin\theta dr d\theta d\phi = R_{nl}^2(r) |Y_{lm}(\theta, \phi)|^2 r^2 \sin\theta dr d\theta d\phi$$

All quantum states for which $l = 0$ (s states) are spherically symmetric wave functions, whereas those with $l \neq 0$ are not. Thus $P(r, \theta, \phi)$ for s states is a function of r alone whereas for other states, $P(r, \theta, \phi)$ also depends on θ and ϕ. For $l = 1$, a plot of $P(r, \theta, \phi)$ for a fixed value of r gives the famous dumb-bell shaped polar plots depicting p orbitals.

We define $P(r)$ by the equation,

$$P(r)dr = \left[\int_0^\pi d\theta \sin\theta \int_0^{2\pi} d\phi P(r, \theta, \phi)\right] r^2 dr = R_{nl}^2(r) r^2 dr \quad (7.22)$$

as the radial probability density, that is, the probability that the electron will be found in the spherical shell bounded by spherical surfaces of radii r and $r + dr$. The most probable radius for the nth orbit corresponds to the peak of the plot of $P(r)$ versus r. It can be shown that the most probable radius for the electron in the ground state of the hydrogen atom is the same as Bohr's radius (see Ex. 7.5).

The quantum mechanical treatment of the hydrogen atom discussed above succeeded in solving many of the problems that could not be handled by Bohr's theory. In addition, it also yielded all the correct ideas and results of Bohr's theory. However, for elementry purposes, the simplicity and elegance of Bohr's theory sometimes outweighs the quantum mechanical method due to the mathematical complexities involved in the latter. It is possible to go quite a long way with Bohr's model by using the quantum numbers of the quantum mechanical theory and making corresponding corrections in the conclusions of Bohr's theory. In this sense Bohr's theory may be considered a first approximation to quantum mechanics.

EXAMPLE 7.5

Bohr radius as the most probable radius

Using the ground state wave function given by Eq. 7.20, show that the most probable radius for the electron in the ground state of the hydrogen atom is the same as the radius of the first Bohr orbit.

Solution

The ground state wave function is $\psi_{100} = \dfrac{\exp(-r/r_1)}{(\sqrt{\pi} \times r_1^{3/2})}$

The corresponding probability density is given by

$$P(r) = |R_{nl}|^2 r^2 = \left(\frac{1}{\pi r_1^3}\right) \exp\left(\frac{-2r}{r_1}\right) r^2$$

The maximum of $P(r)$ as a function of r occurs at $\dfrac{dP(r)}{dr} = 0$.

Thus, $r^2 \exp\left(\dfrac{-2r}{r_1}\right)\left(\dfrac{-2}{r_1}\right) + \exp\left(\dfrac{-2r}{r_1}\right) 2r = 0$

or $\quad r = r_1$

The most probable radius of the electron in the ground state of the hydrogen atom is therefore the same as the radius of the first Bohr orbit.

7.5 QUANTUM NUMBERS AND THE PERIODIC TABLE

In the last section we saw that ψ, the solution of the Schrodinger equation for the electron in the hydrogen atom, is characterised by three quantum numbers, n, l and m (denoted as m_l hereafter). The fourth quantum number, m_s, representing the spin of an electron was added in an ad hoc fashion in the Schrodinger treatment. However, it emerged as a natural consequence of the relativistic quantum theory of Pauli and Dirac. Thus, the set of four quantum numbers (n, l, m_l, m_s), completely defines the wave function or the state of a given electron. These numbers are usually quoted instead of the wave function as they are less cumbersome.

The principal quantum number, $n = 1, 2, 3, \ldots$ that arises from the solution of the radial part of the Schrodinger equation fulfils the same role as the quantum number, n, in Bohr's theory. It governs the total energy of the electron, and the energy levels obtained using quantum mechanics are the same as those given by Bohr's formula.

The orbital quantum number, $l = 0, 1, 2, \ldots, n - 1$, that arises from the solution of the polar part governs the magnitude of the electron's angular momentum about the nucleus. It is related to the shape of the electron orbital and defines the suborbitals denoted by the letters s, p, d and f corresponding to $l = 1, 2, 3$ and 4, respectively.

The magnetic quantum number, $m_l = 0, \pm 1, \pm 2, \ldots, \pm l$, that arises from the solution of the azimuthal part governs the direction of angular momentum, that is, the component of the angular momentum along a prescribed direction which may be, for example, an external magnetic field. If an external magnetic field is applied along the z-direction, then the component of angular momentum along the field direction is quantised, that is, $L_z = m_l \hbar$. The phenomenon that the angle made by an angular momentum vector with the field direction can take only discrete values is referred to as *space quantisation* (see Fig. 7.7a).

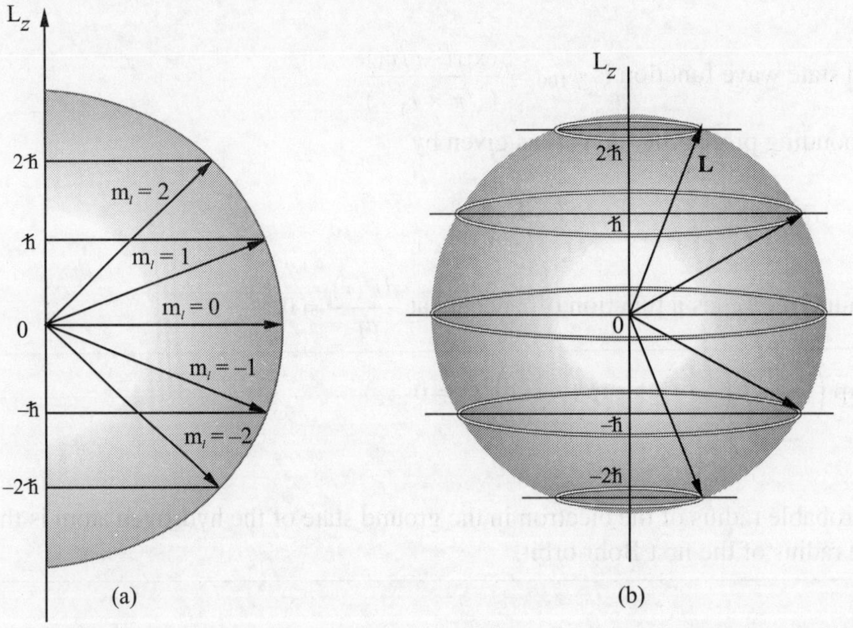

Figure 7.7 (a) Space quantisation of orbital angular momentum for $l = 2$; (b) The angular momentum vector precesses about the z-axis tracing out a cone in space

It should be noted here that only one component of the angular momentum is quantised. This can be understood using the uncertainty principle. If all the three components L_x, L_y and L_z are quantised, or have definite values, say $L_x = 0, L_y = 0$ and $L_z = L$ (that is, the angular momentum is along the z-direction so that the electron is in the xy-plane all the time), then $\Delta z = 0$ and $\Delta p_z \sim \infty$; this is impossible for an electron in a hydrogen atom. Thus the angular momentum vector does not point in a specific direction but precesses about the z-axis, tracing out a cone in space as shown in Fig. 7.7b. The values of L_x and L_y average out to zero, whereas a fixed value of $L_z = m_l \hbar$ is maintained as this cone is traced out.

The spin quantum number, $m_s = \pm \frac{1}{2}$ represents the spin of the electron. The inherent angular momentum is referred to as the spin of an electron and there are only two possible ways in which it can be oriented with respect to an external field. Figure 7.8 shows the two possible orientations of the spin angular momentum. The two possible values of spin $\pm\hbar/2$ are referred to simply as the *spin up* and the *spin down* states.

Spin is an intrinsic property associated with all elementary particles. Particles such as electrons, protons, neutrons, positrons and antiprotons have spin $\pm\hbar/2$ and are called fermi particles or fermions. Particles having zero or integral spin are called bosons, for examples, photons and mesons. Fermions obey Pauli's exclusion principle that states that no two particles can have the same set of quantum numbers. In other words, only one fermion can occupy a given quantum state. There is no limit on the number of bosons that can occupy the same quantum state at the same time.

Both the orbital and spin motion of an electron give rise to the magnetic moment. The magnetic moment associated with a current loop has magnitude equal to the product of the current and the area enclosed by the loop, and is perpendicular to the plane of the loop (see Fig. 7.9). Thus, for a circular Bohr orbit of radius r, an electron circulating with frequency ν has a magnetic moment

$$\mu_l = iA = -e\nu\pi r^2 \tag{7.23}$$

Using Eq. 7.1, that is, $L = mvr = m_l\hbar$ and that $v = 2\pi r\nu$ in the above equation we get,

$$\mu_l = -\frac{e}{2m}\mathbf{L} = -\frac{e\hbar}{2m}m_l \tag{7.24}$$

The quantity $-e/2m$, that is the ratio of the magnetic moment to the orbital angular momentum, is called the *gyromagnetic ratio*. The minus sign tells us that the magnetic moment is directed opposite to the orbital angular momentum because of the negative charge of the electron. The quantity $\mu_B = e\hbar/2m$ is called the *Bohr magneton*. It has a numerical value approximately equal to 9.27×10^{-24} Am^2(or J/T).

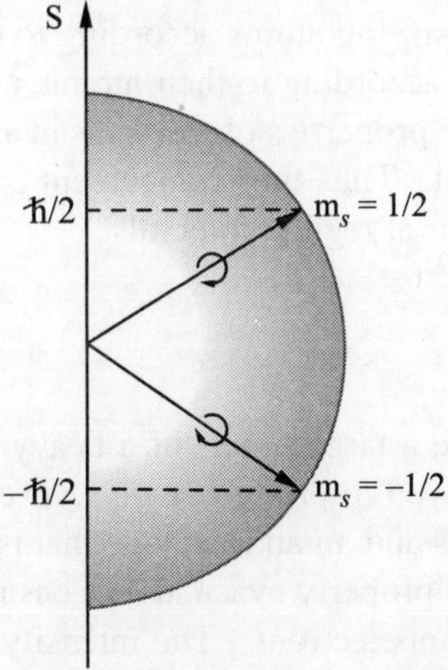

Figure 7.8 The two possible orientations of the spin angular momentum are spin up and spin down

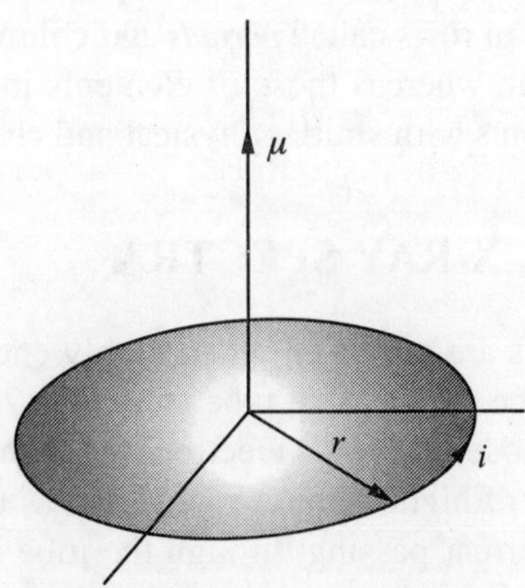

Figure 7.9 Magnetic moment due to a circular current

As shown in the Pauli–Dirac theory, the spin is twice as effective as the orbital momentum in terms of the magnetic moment that arises. Thus, the gyromagnetic ratio for spin is

$$\gamma_s = \frac{\mu_s}{S} = \frac{e}{m} \tag{7.25}$$

However, the magnetic moment due to electron spin is also just one Bohr magneton since the spin is $\hbar/2$ instead of $\hbar$. That is,

$$\mu_s = -\frac{e\hbar}{2m} = -\mu_B \tag{7.26}$$

It is emphasised here that the energy of a state may not uniquely identify the state. The energy is determined by n whereas the electronic state is determined by all the four quantum numbers. Thus, the energy level, E_n, corresponds to

$$2 \times \sum_{l=0}^{n-1} (2l+1) = 2 \times [1 + 3 + \ldots + 2(n-1) + 1] = 2n^2 \text{ states}$$

We say that the energy level E_n is *degenerate* and its *degeneracy* is $2n^2$. In Dirac's relativistic theory, the energy of an electron is also determined to a small extent by the quantum number l, but the energy difference due to different l values is negligible compared to that due to different n values.

When atoms are placed in a magnetic field, the interaction of the magnetic moment of the electron with the field shifts the electronic energy levels leading to the Zeeman effect and the fine structure of atomic spectra. In heavier atoms there is an additional interaction between the spin magnetic moment and the magnetic field of the revolving nucleus as is seen by the electron. This gives the spin–orbit energy contribution to energies of electronic states. The spin–orbit energy is considerable only for outer electrons in heavier atoms. Thus for these atoms only the total angular momentum J—the sum of the orbital and the spin angular momenta—of the electron is important so that the relevant quantum numbers are j and m_j rather than l, m_l, S and m_s.

The Pauli exclusion principle provides a method of classifying atoms according to their electronic structures. Elements are listed in the Periodic Table according to their atomic numbers in series of rows called *periods* and columns called *groups*. The properties of elements in a group are similar, whereas those of elements in a period are different. Thus the arrangement is such that elements with similar physical and chemical properties recur at regular intervals.

7.6 X-RAY SPECTRA

X-rays are produced when highly energetic electrons strike a target made of a heavy metal. In practice, a Coolidge tube (see Fig. 7.10) is usually employed to produce x-rays. It consists of a cathode that emits electrons through thermionic emissions and an anti-cathode that is usually a metal of high atomic weight such as tungsten. The tube is properly evacuated to ensure that all the current passing through the tube is carried by the thermoelectrons. The intensity of x-rays depends on the number of electrons striking the target per second or the heater current, and the wavelengths of x-rays depend on the voltage across the tube.

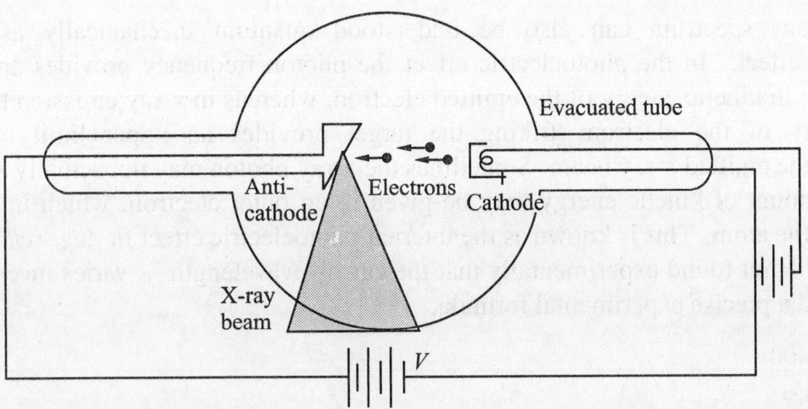

Figure 7.10 Coolidge tube for production of x-rays

The spectrum of x-ray radiation comprises of a continuous spectrum called *bremsstrahlung* (meaning breaking radiation), characterised by a sharp cut-off at the short wavelength end and a characteristic spectrum superimposed upon the continuous spectrum consisting of a few discrete peaks (Fig. 7.11). The continuous spectrum can be accounted for by the classical theory of electromagnetic radiation. According to classical electrodynamics, accelerated charges emit electromagnetic radiations. The stoppage of fast moving electrons by a target results in their deceleration, thus causing the emission of x-rays. Electrons lose energy as they are scattered by the atoms in the target material. Thus, the photons that are emitted in successive collisions of electrons with target atoms give the continuous spectrum.

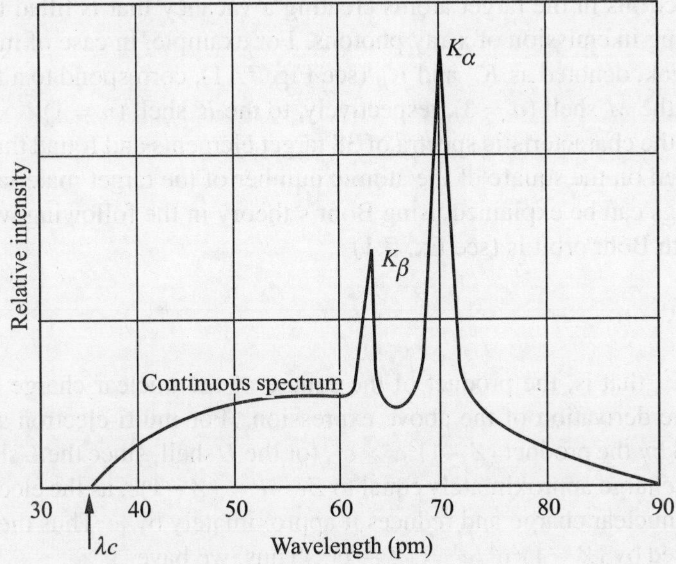

Figure 7.11 Continuous and characteristic x-ray spectrum for the molybdenum target

The continuous spectrum can also be understood quantum mechanically as an inverse photoelectric effect. In the photoelectric effect the photon frequency provides an upper limit on the change in kinetic energy of the emitted electron, whereas in x-ray emission the change in kinetic energy of the electron striking the target provides an upper limit on the photon frequency in the emitted x-ray beam. Sometimes the x-ray photon may not actually appear but an equivalent amount of kinetic energy may be given to an outer electron, which in turn may be ejected from the atom. This is known as the internal photoelectric effect or *Auger effect*.

Duane and Hunt found experimentally that the cut-off wavelength λ_c varies inversely with V. They obtained a precise experimental formula,

$$\lambda_c = \frac{12400}{V} \tag{7.27}$$

Since the work functions that are a few electron volts are much smaller compared to the accelerating potential in the x-ray tube (that is in kV), we can neglect the work function and write $eV = hc/\lambda_c$. Solving (see Ex. 6.2), we get

$$\lambda_c = \frac{hc}{eV} = \left(\frac{12422}{V}\right) \text{Å} \tag{7.28}$$

which is approximately the same as the Duane–Hunt formula. Thus it is appropriate to regard x-ray production as an inverse photoelectric effect.

Thus the cut-off wavelength of the continuous spectrum is independent of the target material and depends only on the voltage of the x-ray tube. However, the wavelengths corresponding to the characteristic x-ray spectrum are characteristic of the target material used in the x-ray tube. The discrete peaks in the characteristic spectrum arise because the incident electron knocks out one of the low orbit electrons in the target atoms creating a vacancy that is filled by the electrons in higher orbits resulting in emission of x-ray photons. For example, in case of molybdenum target, the characteristic peak, denoted as K_α and K_β (see Fig. 7.11), correspond to a transition from the L shell ($n = 2$) and the M shell ($n = 3$), respectively, to the K shell ($n = 1$).

Moseley studied the characteristic spectra of 38 target elements and found that the frequency of a given line depended on the square of the atomic number of the target material used. Moseley's experimental findings can be explained using Bohr's theory in the following way. The energy of an electron in the nth Bohr orbit is (see Ex. 7.1)

$$E_n = -\frac{13.6}{n^2} \text{ eV}$$

The product, $e^2 \times e^2$, that is, the product of the square of the nuclear charge and the electronic charge is used in the derivation of the above expression. For multi-electron atoms this product needs to be replaced by the product $(Z-1)^2 e^2 \times e^2$, for the L shell, since the L shell electron 'sees' an effective nuclear charge approximately equal to $Ze - e = (Z-1)e$, as the electron present in the K shell screens the nuclear charge and reduces it approximately by e. Thus the above expression needs to be multiplied by $(Z-1)^2 e^4/e^4 = (Z-1)^2$. Thus, we have

$$E_n = -\left[\frac{13.6(Z-1)^2}{n^2}\right] \text{eV} \tag{7.29}$$

For K_α transition, we have $\Delta E = E_2 - E_1 = \left[\frac{-13.6(Z-1)^2}{2^2}\right] - \left[\frac{-13.6(Z-1)^2}{1^2}\right]$ eV

or $$\Delta E = 10.2(Z-1)^2 \text{ eV} \tag{7.30}$$

and the corresponding frequency of the emitted photon is

$$\nu = \frac{\Delta E}{h} = 2.46 \times 10^{15}(Z-1)^2 \text{ Hz} \tag{7.31}$$

The same line of argument with $(Z-1)$ replaced by $(Z-2)$ holds for the M shell.

Thus, the frequency is proportional to the square of the atomic number in agreement with the experimental findings of Moseley.

7.7 APPLICATIONS

In this chapter we studied the physics of atoms in the light of the quantum theory. This knowledge had myriad practical applications and thus proved to be of immense importance in engineering and technology. The physics of atoms assumed great importance.

In Section 7.2 we studied about the atomic spectra. The branch of physics dealing with the study of various spectra is known as spectroscopy. It is concerned with the systematisation and classification of atomic energy levels and has played a central role in developing our understanding of atomic structure. Spectroscopy has proved to be a useful means of analysis and has diverse applications in industry and research for the identification and determination of elements that help in solving many complex problems.

In Sections 7.3 and 7.4 we studied the theory of the hydrogen atom using Bohr's model and the more rigorous quantum mechanical model. The refinements in various models helped in getting a better interpretation of the spectrum of the hydrogen atom. They also facilitated a better understanding of various atoms in nature on the basis of four quantum numbers.

In Section 7.5 we studied the significance of quantum numbers and their use in understanding the periodic table. Such studies provided insights into how the atomic structure of elements decides their physical and chemical properties. Several research studies were undertaken to look into aspects such as electronic configurations, various allowed and disallowed energy levels and the probabilities of various transitions between these levels in a vast number of materials. The knowledge gained through such studies led to new applications in engineering and technology.

The discovery of the spin or the intrinsic angular momentum as the fourth quantum number turned out to be of immense importance as spin plays a crucial role in atomic and molecular structure. The concept of spin extends to other sub-atomic particles too. There it plays a role in explaining many atomic properties. For example, the two isotopes of helium, $^3_2$He, and $^4_2$He, owe their very different properties at low temperature to spin. In $^4_2$He, both the protons and the neutrons have equal but oppositely directed spins resulting in zero net spin angular momentum. Helium exists in a liquid state at atmospheric pressure even at the lowest temperature so far achieved. It can become a superfluid below about 2 K to exhibit extraordinary properties. For example it can flow through openings that no other liquid or gas can negotiate.

Spin also plays a crucial role in the behaviour of collection of particles. For example, the free electrons in metals at low temperature, each having a spin of one half, can match up to form pairs with a net spin of zero. This state results in a complete loss of resistivity and the metal turns into a superconductor. This is a special case of the fact mentioned earlier that particles with half integral spin, follow Fermi–Dirac statistics, while those with integral spin follow Bose–Einstein statistics. Magnetic materials such as ferrites are widely used as engineering materials. Their peculiar properties are due to spin.

As magnetic moments are associated with the orbital and spin motion of electrons, the study of effect of magnetic field on the spectra of elements assumed particular importance. In 1896, Pieter Zeeman, the Dutch physicist, observed that spectral lines split in the presence of magnetic field. This effect, known as the *Zeeman effect* confirmed space quantisation and also helped the two Dutch graduate students, Samuel Goudsmit and George Uhlenbeck, to discover spin. The strong magnetic fields associated with sunspots were detected by means of the Zeeman effect. Sunspots appear dark because they are cooler than the rest of the solar surface and follow an 11-year cycle. Many terrestrial phenomena follow this cycle.

In Section 7.6 we studied about the x-ray spectra of elements. Roentgen's discovery of x-rays in 1895 stimulated a lot of scientific research and virtually hundreds of papers on x-rays were published in the year 1896 alone. In the following years a large number of Nobel prizes have been awarded for research investigations pertaining to x-rays. The research findings have found wide application in almost all the fields of human endeavour in modern life. Applications in the field of medical diagnostics led to the emergence of radiology as a separate branch of medicine.

The first radiograph ever taken is that of the hand of Roentgen's wife on 22 December 1895 (Fig. 7.12). One late evening when Roentgen did not return home for dinner, Mrs Roentgen went to his laboratory. Roentgen asked his wife to keep her hand between the x-ray tube and the photographic film. When the film was developed it clearly showed the bones of the fingers of Mrs Roentgen's hand and the ring worn by her. The application potential of x-rays in imaging was realised very soon and six months later FH Williams gave a live demonstration of skeletal imaging in Boston Medical School. The differential absorption of x-rays by various human organs helps radiologists in imaging the internal organs of human body.

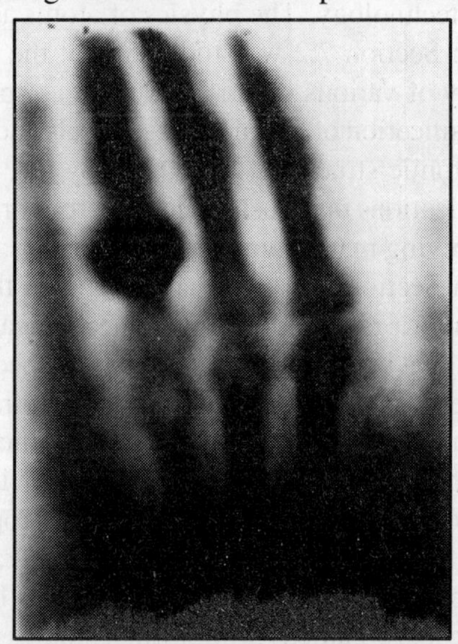

Figure 7.12 The first radiograph ever taken (Source: http://en.wikipedia.org/wiki/File: Anna_Berthe_Roentgen.gif)

The field of radiology witnessed another seminal discovery in 1971 when Allan Cormack and Godfrey Hounsfield developed the CAT (computer aided tomography) scanner or simply the CT-scanner. Cormack, born in South Africa, was a professor of physics at Tufts University,

Massachusetts, USA, and Godfrey Hounsfield was a British engineer who headed the project undertaken by EMI Corporation in Great Britain to construct the scanner. Cormack got the idea of the CT-scanner when he was hospitalised for a long time in a Cape Town hospital. He realised that the conventional two-dimensional x-ray technique had a number of limitations that needed to be overcome. Hounsfield got the idea to develop a scanner when he was working in the field of radars and computers. Cormack and Hounsfield shared the Noble Prize for medicine in 1979 for their seminal work.

In computer aided tomography, the structural details are obtained using computerised techniques. In this technique a narrow beam of x-rays is used to scan a thin slice of the body structure (Fig. 7.13). The beam is attenuated by various tissue elements in the path as it passes through the slice. The relative attenuation is calculated by computational methods on the basis of which a final image is displayed on the video monitor. With CAT the information of the internal structure of the organ can be obtained to a precision of a few millimetres. The CT-scanner has thus gained immense recognition as a powerful tool in non-invasive diagnosis of various diseases.

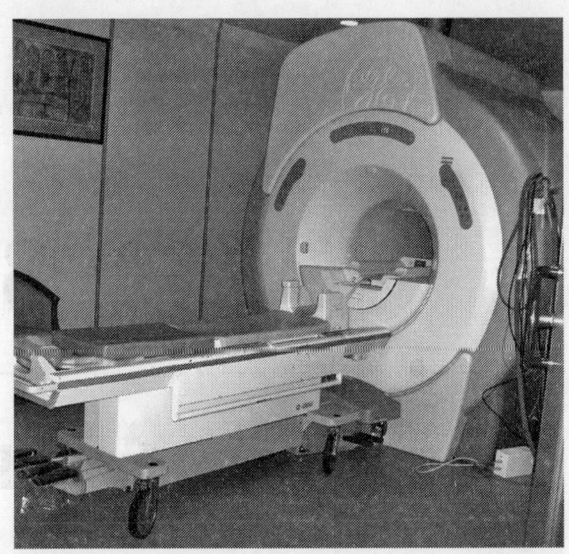

Figure 7.13 In a CT-scanner, a thin x-ray beam rotates around an area of the body, generating a three-dimensional image of the internal structure

Moseley's research on the characteristic x-ray spectra of 38 elements provided fresh insights into the classification of elements in the periodic table. Moseley showed that it is the nuclear charge (that is, the atomic number and not the atomic mass), which is the real basis for numbering the elements. The characteristic x-ray spectrum shows more impressive regularity from element to element compared to that seen in optical or IR spectrum. This is because in the x-ray spectrum the *K* electrons play an important role, whereas optical and IR spectra arise due to transitions of the outermost electrons. The *K* electrons lie very close to the nucleus and are thus more sensitive probes of the nuclear charge than the outermost electrons that are heavily screened from the nucleus by the electrons in the inner orbits.

As the tiny nucleus arouses our curiosity due to its 'core' importance, the physics of the nucleus will be studied in the next chapter.

8

Nuclear Physics

8.1 THE STORY OF NUCLEAR PHYSICS

In the last chapter we studied how deeper enquiry into the structure of an atom led physicists to a wealth of knowledge in atomic physics. As physicists looked deeper into the atom using the insight provided by quantum theory, more was revealed about the core of the atom—the nucleus. The nucleus contains almost the entire mass of the atom but occupies only a quadrillionth ($1/10^{15}$) of its volume. It has also been thoroughly explored in the last century leading to the vast knowledge of nuclear physics that is available today.

It is worth appreciating that in the quest for knowledge a small part of a large domain itself becomes large (see Fig. 8.1). *Physics thus embodies an enquiry into the folded logic of nature.* As more layers are unfolded by reason, more layers emerge, confronting us with the unending reserves of natureŠs deep rooted knowledge. The exploration becomes a bottomless search; the more we know the more we come to know that there is much more to know.

In the last chapter we appreciated that most of the properties of atoms, and hence of the molecules and the different states of matter that they make, are attributable to the electrons and not to the nucleus. The only role that was ascribed to the nucleus was to provide the atoms with most of its mass and the requisite charge to hold its electrons in bondage. The electronic structure of atoms was understood well before anything was known about the nucleus, because the forces that hold the nucleus together are much stronger than the electric forces that hold the electrons to the nucleus. Thus it is harder to break the nucleus apart than to break an atom, and find out what is inside it. However, many brilliant minds researched the nucleus to unravel its mysteries. Their efforts contributed to the development of nuclear physics as a distinct branch of physics during the first three decades of the last century. Physicists mainly investigated what the nucleus consists of, what forces and energies hold it together, and how to understand nuclear phenomena using theoretical models.

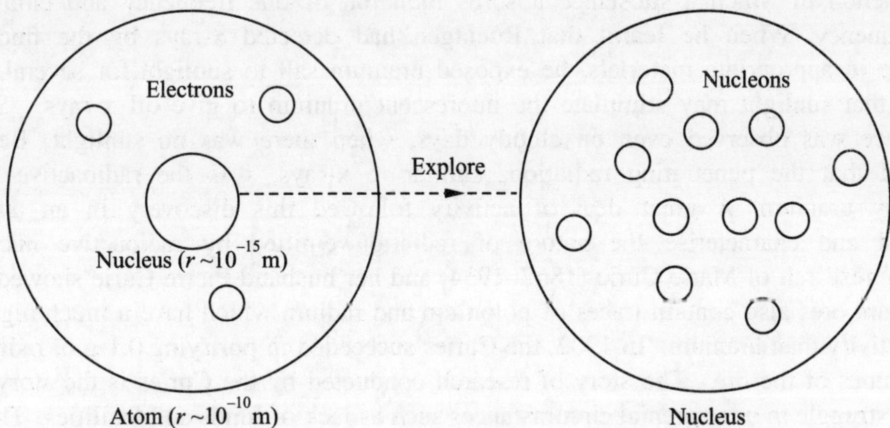

Figure 8.1 A small part (nucleus) of a large domain (atom) itself becomes large!

The atomic nucleus was first discovered in 1911 by Rutherford, though the year 1896 in which Becquerel discovered radioactivity in uranium is considered the year that marks the birth of nuclear physics. Initial research by Rutherford and his contemporaries established that the dimensions of the nucleus are of the order of 10^{-15} m, and energies of the order of several million electronvolts. Thus an atom came to be considered as concentrated nuclear mass and a lot of empty space. It was understood that ordinary matter is mostly empty space. If all the empty space within atoms could be removed, the earth would fit into a large apartment house and our bodies would reduce to specks that could be visible only with a microscope.

Nuclear force —the enigmatic force that keeps so many positively charged protons confined to the nuclear region—aroused the interest of many physicists. In 1932, Werner Heisenberg proposed the neutron–proton model of the nucleus based on the discovery of the neutron by the English physicist, James Chadwick, and the experimental results of Cockroft and Walton. This model explained the short-range nature of nuclear force. In 1935, Hideki Yukawa (1907–81) proposed another model of the nucleus.

Yukawa—the first to receive the Nobel Prize from Japan—explored the enigma of the nuclear force in the early 1930s. In his model he proposed that nuclear force results from the virtual exchange of particles called the π *mesons* or *pions*. In 1947, CF Powell, the British physicist, discovered the pion that decays rapidly to a comparatively longer-lived particle called *muon*. In the 1930s Maria Mayor (1906–72), who studied at Gottingen under Max Born and the German physicist, JHD Jensen, independently proposed the shell model of the nucleus. In 1936, Bohr proposed the liquid drop model of the nucleus.

In 1896, Antoine Becquerel (1852–1908), the French physicist, discovered radioactivity—a purely nuclear phenomenon. This was 15 years before Rutherford inferred the existence of the nucleus. BecquerelŠs grandfather, father and son were also physicists, and all of them were professors at the Paris Museum of Natural History. Many of the ideas and techniques of nuclear physics are based on radioactivity and hence the study of radioactivity is essential to nuclear physics.

Becquerel discovered radioactivity by accident. His research interest was fluorescence, a phenomenon in which a substance absorbs radiation of one frequency and emits one of other frequency. When he learnt that Roentgen had detected x-rays by the fluorescence they cause in appropriate materials, he exposed uranium salt to sunlight for several days, in the hope that sunlight may stimulate the fluorescent uranium to give off x-rays. Since the fluorescence was observed even on cloudy days, when there was no sunlight, Becquerel discovered that the penetrating radiation, similar to x-rays, was the radioactive radiation emitted by uranium. A great deal of activity followed this discovery in an attempt to understand and characterise the nature of radiation emitted by radioactive nuclei. The pioneering research of Marie Curie (1867–1934) and her husband Pierre Curie showed in 1898 that uranium ores also contain traces of polonium and radium which have a much higher level of radioactivity than uranium. In 1902, the Curies succeeded in purifying 0.1 g of radium from several tonnes of the ore. The story of research conducted by the Curies is the story of their tenacious struggle in uncongenial circumstances such as lack of funds and facilities. The Curies shared the 1903 Nobel Prize in physics with Becquerel. Many other radioactive nuclides were discovered in subsequent years.

The fact that certain atoms undergo spontaneous disintegration aroused the interest of physicists in the hope of causing the disintegration of ordinary inactive nuclides by bombarding them with energetic particles. The first disintegration based on these ideas was made by Rutherford (1919) who showed that the nuclei of nitrogen atoms emit swift protons when bombarded with α-particles from radium C. In 1934, Joliot and Curie discovered artificial or induced radioactivity. This started a new line of research that resulted in the discovery of hundreds of radioactive nuclides.

The investigation of the radiations from natural and artificial radionuclides showed that the nucleus has energy levels analogous to atomic energy levels. Nuclear spectroscopy that deals with the identification and classification of these levels is also a source of information about the structure of the nucleus. Many of the observed energy levels, and other properties of nuclei, can be understood using the nuclear shell model proposed by Mayor and Jensen.

In 1936, the Dutch physicist, CT Gorterin, advanced the concept of nuclear magnetic resonance (NMR). Felix Bloch and Edward Purcell, both from the USA, demonstrated the phenomenon of NMR for protons in 1946. Their discovery motivated NMR studies in condensed matter. Bloch, who did his undergraduate work in engineering, also did pioneering work in the development of the band theory of solids (Chapter 11).

In this chapter we study the different aspects of the atomic nucleus such as its structure, binding energy, force, spin and magnetic moment (Section 8.2). Section 8.3 deals with the study of the radioactivity that includes the study of the radioactive decay law, nuclear reactions and detection of nuclear radiation. In Section 8.4 the models of a nucleus and the basic elements of nuclear spectroscopy are studied.

Nuclear physics is unique. It merges the most fundamental and the most applied topics. It has wide applicability in engineering, technology and medicine. Various applications based on the topics covered in this chapter are discussed in the last section.

8.2 ATOMIC NUCLEUS

8.2.1 Nuclear Structure

The nucleus is basically composed of protons and neutrons, together referred to as *nucleons*. The *mass number* A is the total number of nucleons and the *atomic number* Z is the number of protons in the nucleus; Z is also the number of electrons as the total charge on an atom is zero. The chemical properties of an element are determined by Z. The number of neutrons, N, is equal to $A - Z$. The symbol used for a nuclear species (also called a nuclide) is ${}^{A}_{Z}X$, where X is the chemical symbol of the atom.

Nuclides with the same value of Z but different values of N and A are called *isotopes*; those with the same value of A but different values of N and Z are called *isobars* and those with the same value of N but different values of Z and A are called *isotones*. Isobars having an odd number as the value of A, such that Z and N differ by one, are called *mirror nuclei*. Examples of these various nuclides are given in Table 8.1.

Table 8.1 Isotopes, isobars and isotones

	Same value of	Different values of	Example
Isotopes	Z	N and A	Hydrogen (${}^{1}_{1}H$), deuterium (${}^{2}_{1}H$) and tritium (${}^{3}_{1}H$)
Isobars	A	Z and N	${}^{14}_{6}C$ and ${}^{14}_{7}N$, ${}^{15}_{8}O$ and ${}^{15}_{7}N$ are mirror nuclei
Isotones	N	Z and A	${}^{198}_{80}Hg$ and ${}^{197}_{79}Au$

It is seen that for very light nuclei N and Z are almost equal, but as the mass number increases, N soon exceeds Z (Fig. 8.2), and for the heaviest stable nuclides such as ${}^{209}_{83}Bi$, N exceeds Z by more than 1.5 times. The numbers Z or N when equal to 2, 8, 20, 28, 50, 82 or 126 correspond to nuclei that seem to be unusually stable. These numbers are called the *magic numbers*.

The nuclear radius is given by an empirical relation,

$$r = r_0 A^{1/3} \tag{8.1}$$

where, the constant r_0 is experimentally found to be $\sim 1.2 \times 10^{-15}$ m. Nuclear distances are generally expressed in terms of Fermi, equal to one femtometer.

$$1 \text{ fermi} = 1 \text{ femtometer (fm)} = 10^{-15} \text{ m}$$

Thus the radius of ${}^{27}Al$ is $1.2 \times (27)^{1/3} = 3.6$ fm and that of ${}^{64}Zn$ is $1.2 \times (64)^{1/3} = 4.8$ fm. Example 8.1 illustrates an important consequence of Eq. 8.1—all nuclear matter has nearly the same density and is about 10^{13} times that of the atomic density. Equation 8.1 tells us that the volume of the nucleus is proportional to A, an important experimental fact.

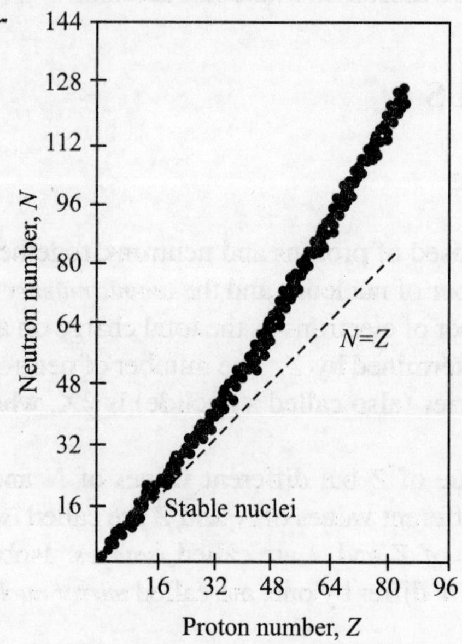

Figure 8.2 Plot of N versus Z: $N = Z$ for light nuclei but N exceeds Z as the mass number increases

EXAMPLE 8.1
Nuclear density and atomic density

Show that all nuclear matter has nearly the same density of the order of 10^{17} kg/m^3 and is 10^{13} times that of atomic density. Use this result to understand a white dwarf and neutron star.

Solution

Nuclear density is given by

$$\rho_{\text{nucleus}} = \frac{\text{nuclear mass}}{\text{nuclear volume}}$$

$$= \frac{m_p A}{(4\pi r^3/3)} = \frac{m_p A}{[4\pi(r_0 A^{1/3})^3/3]} = \frac{3m_p}{4\pi r_0^3}$$

$$= \frac{3 \times 1.67 \times 10^{-27}\ \text{kg}}{4 \times 3.14 \times (1.2 \times 10^{-15}\ \text{m})^3}$$

$$= 2.3 \times 10^{17}\ \text{kg/m}^3$$

For an atom of radius a and mass number A, the density is given by

$$\rho_{\text{atom}} = \frac{(m_p A + m_e Z)}{[(4/3)\pi a^3]} \simeq \frac{m_p A}{[(4/3)\pi a^3]}, \text{ as electronic masses are negligible.}$$

As atomic dimensions are given approximately by a_0, the Bohr radius, we have

$$\frac{\rho_{\text{nucleus}}}{\rho_{\text{atom}}} \sim \left(\frac{a_0}{r_0}\right)^3 \sim \left[\frac{0.5 \times 10^{-10}\text{ m}}{1.2 \times 10^{-15}\text{ m}}\right]^3 \sim 7 \times 10^{13}$$

Thus nuclear density is greater than atomic density by 13 orders of magnitude. Hence, if the atomic space is squeezed then there would be a 10^{13}-fold increase in the density of matter. Such enormous densities are encountered in stars that convert into a white dwarf or a neutron star in the final stages of their evolution. A white dwarf is a star whose radius is much smaller compared to that of the Sun, for example, Sirius. A neutron star is produced when the electrons are forced into the nucleus and convert the protons into neutrons. A neutron star is one or two suns packed into a sphere of only about 10 km radius.

8.2.2 Nuclear Force

The nearly constant value of the binding energy per nucleon suggests that the nuclear force is an extremely short-range force. Each nucleon interacts only with its adjacent neighbours. This is referred to as the saturation of nuclear force. According to the neutron–proton model of Heisenberg, all nucleons contribute to the attractive forces that hold the nucleons together, whereas the repulsive forces between the protons is the cause of nuclear instability. Thus, as Z increases, an excess of N over Z is required to achieve stability. There is a limit to the ability of neutrons to prevent the disruption of large nuclei due to the Coulombic forces of repulsion among protons. This limit is represented by ${}^{209}_{83}\text{Bi}$, the heaviest stable nuclide.

Nuclear force is the strongest of the four fundamental forces of nature. The force becomes negligibly small if nucleons are separated by distances greater than about 3 fm. This feature sharply contradicts the long-range nature of gravitational and electrostatic forces that can in principle extend to infinity. In 1935, Yukawa proposed that the nuclear force can be represented by the potential of the form, $V \sim [\exp(-r/b)]/r$, where, b, the range of the nuclear force is about 1 Fermi. For a proton, Coulomb potential is present in addition to Yukawa potential, whereas a neutron is affected by only Yukawa potential. Yukawa proposed that the strong nuclear force results from the virtual exchange of particles called the π mesons or pions. Example 8.2 demonstrates how Yukawa used the uncertainty principle to show that the rest mass of a pion is about 220 times that of the electron. It is interesting to note that the range of the force is inversely proportional to the mass of the exchange particle. Thus a very short-range nuclear force requires a massive exchange particle. The Coulomb force having infinite range requires an exchange particle, the photon, having zero rest mass.

EXAMPLE 8.2

Determination of the rest mass of a pion using the uncertainty principle

Use the uncertainty principle to show that the rest mass of a pion is 220 times that of the electron.

Solution

The energy required to create a pion of mass m_π, $\Delta E = m_\pi c^2$. For this to be a virtual particle, it must be created and re-absorbed within the time Δt given by the uncertainty principle, $\Delta E \Delta t \sim \hbar$.

In the Yukawa model, the range of nuclear force is b. Assuming that the pion travels within this range at a speed comparable to that of light, the transit time is at least, $\Delta t \sim b/c$.

Thus, $\Delta E = \dfrac{\hbar}{\Delta t} = \dfrac{\hbar c}{b}$

or $\quad m_\pi c^2 = \dfrac{\hbar c}{b}$

or $\quad m_\pi = \dfrac{\hbar}{bc}$

Considering the maximum range of nuclear force of about 1.7 fm, we get

$$m_\pi = \frac{1.05 \times 10^{-34}\ \text{Js}}{1.7 \times 10^{-15}\ \text{m} \times 3 \times 10^{8}\ \text{m/s}}$$

$$\sim 2 \times 10^{-28}\ \text{kg}$$

This is about 220 times the rest mass of an electron.

8.2.3 Nuclear Binding Energy

The binding energy E_B of a nucleus is the difference between the total energy of the bound system of nucleons and the total energy of the separated nucleons. It is found that the atomic mass is less than the sum of the masses of its constituent particles in the free state. If Δm is the decrease in mass when these particles combine to form an atom, the relativistic relation $E = mc^2$ that expresses the mass–energy equivalence can be used to write

$$E_B = \Delta m c^2 \tag{8.2}$$

where the difference Δm, known as the *mass defect*, is given by

$$\Delta m = [Zm_p + (A - Z)m_n] - \text{mass of } {}^A_Z\text{X} \tag{8.3}$$

Thus the mass defect is the amount of mass that would be converted to energy if a particular atom were to be assembled from its constituent particles. The same amount of energy is needed to break the atom into its constituent particles. The masses of atomic particles are generally expressed in atomic mass units (amu), denoted by u. We have

$$1\,\text{u} = 1.66 \times 10^{-27}\,\text{kg} \times (3 \times 10^{8}\,\text{m/s})^2 = 931.5\,\text{MeV} \tag{8.4}$$

Example 8.3 provides an appreciation of the magnitude of binding energies of nuclei as compared to that of electrons. Example 8.4 shows that the average binding energy per nucleon for an α-particle is 7.07 MeV. It is found that for the majority of nucleides, except the lighter ones, the binding energy per nucleon is in the range of 7–8 MeV.

EXAMPLE 8.3

Nuclear binding energy and electronic binding energy

Generally the electrons are bound to the atom by energies between 0 and 10 eV. Compare this with the binding energy of the deuteron. Given: mass of proton, $m_p = 1.007276470$ u, mass of neutron, $m_n = 1.008665012$ u and mass of deuteron, $m_d = 2.013553215$ u.

Solution

Binding energy,

$$E = (m_p + m_n - m_d)c^2$$

$$= (1.007276470\,\text{u} + 1.008665012\,\text{u} - 2.013553215\,\text{u})c^2$$

$$= 0.002388267\,\text{u} \times c^2 = 0.002388267\,\text{u} \times 931.5\,\text{MeV}$$

$$= 2.225\,\text{MeV}$$

Thus we see that the binding energy of a simple two-particle nucleus of the deuteron is about 10^6 times the binding energy of an electron in a hydrogen atom. It is true in general that the nuclear binding energy is a million times greater than the electronic binding energy.

EXAMPLE 8.4

Average binding energy per nucleon

Find the average binding energy per nucleon for an α-particle. Given: the mass of an α-particle, $m_\alpha = 4.001506106$ u.

Solution

The binding energy of an α-particle is given by

$$E = (2m_p + 2m_n - m_\alpha) \times 931.5 \text{ MeV}$$

$$= [2 \times (1.007276470\,\text{u} + 1.008665012\,\text{u}) - 4.001506106\,\text{u}] \times 931.5 \text{ MeV} = 28.30 \text{ MeV}$$

Thus the binding energy per nucleon is $\dfrac{28.30}{4} = 7.075$ MeV per nucleon.

The graph of binding energy per nucleon versus *A* (Fig. 8.3) suggests that a nucleus with a higher mass number may undergo fission to produce a more stable nucleus, that is, one with more binding energy. Similarly, nuclei with low atomic number may undergo fusion to produce more stable nuclei with higher atomic number, and thus a larger binding energy. The peak at $A = 4$ corresponds to the exceptionally stable ${}^{4}_{2}\text{He}$ nucleus, that is, the α-particle. The curve shows a maxima for $A = 56$, indicating the most stable nucleus—that of ${}^{56}_{26}\text{Fe}$, an isotope of iron.

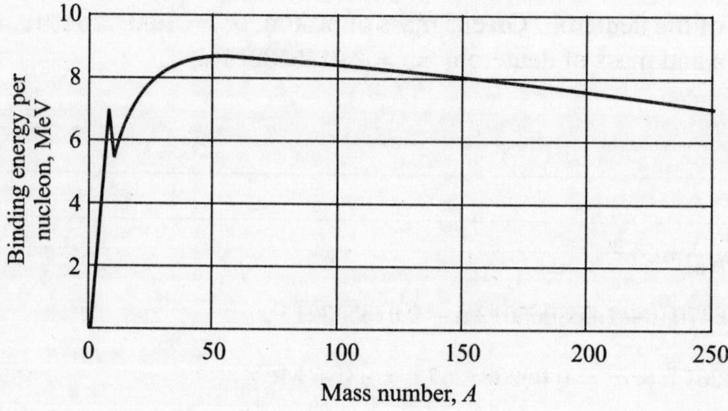

Figure 8.3 Binding energy per nucleon as a function of mass number (The peak at $A = 4$ corresponds to exceptionally stable ${}^{4}_{2}\text{He}$ nucleus: the α-particle.)

8.2.4 Nuclear Spin and Magnetic Moment

In Section 7.5 we discussed the spin of an elementary particle. The spin of a nuclide is the algebraic sum of the spins of its constituent particles. As protons and neutrons are spin-1/2 particles, that is, fermions, nuclides with odd mass number are fermions, whereas those with even mass number are bosons (zero or integer spin).

In Section 7.5 we also discussed the magnetic moment associated with the spin of an electron. Nuclides also have a magnetic moment due to the spins of the constituent particles. Just as the magnetic moment of an electron is measured in units of Bohr magneton, nuclear magnetic moment is measured in *nuclear magnetons.*

Thus the nuclear magneton is defined as $\mu_N = \dfrac{e\hbar}{2m_p}$ (8.5)

which is similar to the expression for the Bohr magneton (see Eq. 7.24), the only difference being that the mass of the electron is replaced by that of the proton. Substituting the values of numerical constants in Eq. 8.5, we get,

$$\mu_N = 5.05 \times 10^{-27} \text{ J/T} \tag{8.6}$$

The magnetic moment of a proton and that of a neutron are given, respectively, by

$$\mu_p = 2.7928\mu_N = 1.41 \times 10^{-26} \text{ J/T}$$

$$\mu_n = -1.9130\mu_N = -9.66 \times 10^{-27} \text{ J/T} \tag{8.7}$$

8.3 RADIOACTIVITY

Radioactivity is the process by which an unstable nuclide emits particles spontaneously to transform itself into a different nuclide. The emitted radioactive rays consist of α- or β-particles that are sometimes accompanied by γ-particles. The emission of α-particles by a parent nucleus results in a daughter nucleus (*decay product*) with mass number (A) four units less and atomic number (Z) two units less than that of the parent nucleus. The emission of β-particles leaves A unchanged but increases Z by 1. The emission of γ-rays leaves A and Z unchanged as γ-rays consist of photons. Radioactive decay can also occur through two other processes: positron emission and electron capture. Table 8.2 gives the different nuclear transformations that accompany various decay processes.

Table 8.2 Radioactive decay processes*

Decay process	Nuclear transformation			Example		
α-decay	${}^{A}_{Z}\text{X}$	$\longrightarrow$	${}^{A-4}_{Z-2}\text{Y} + {}^{4}_{2}\text{He}$	${}^{238}_{92}\text{U}$	$\longrightarrow$	${}^{234}_{90}\text{Th} + {}^{4}_{2}\text{He}$
β-decay	${}^{A}_{Z}\text{X}$	$\longrightarrow$	${}_{Z+1}^{A}\text{Y} + e^-$	${}^{14}_{6}\text{C}$	$\longrightarrow$	${}^{14}_{7}\text{N} + e^-$
γ-decay	${}^{A}_{Z}\text{X}^*$	$\longrightarrow$	${}^{A}_{Z}\text{X} + \gamma$	${}^{87}_{38}\text{Sr}^*$	$\longrightarrow$	${}^{87}_{38}\text{Sr} + \gamma$
Positron emission	${}^{A}_{Z}\text{X}$	$\longrightarrow$	${}_{Z-1}^{A}\text{Y} + e^+$	${}^{64}_{29}\text{Cu}$	$\longrightarrow$	${}^{64}_{28}\text{Ni} + e^+$
Electron capture	${}^{A}_{Z}\text{X} + e^-$	$\longrightarrow$	${}_{Z-1}^{A}\text{Y}$	${}^{64}_{29}\text{Cu} + e^-$	$\longrightarrow$	${}^{64}_{28}\text{Ni}$

** denotes the nucleus in an excited state*

8.3.1 Radioactive Decay Law

The rate at which a particular decay process occurs is found to be proportional to the number of nuclides present, that is,

$$\frac{dN}{dt} = -\lambda N \tag{8.8}$$

where N is the number of unstable nuclides of a given species present at time t, and λ is called the *decay constant* or the *disintegration constant*. The negative sign indicates that N decreases with increasing t. Rearranging Eq. 8.8 and integrating, we get,

$$\int_{N_0}^{N} \frac{dN}{N} = -\int_0^t \lambda dt \tag{8.9}$$

Solving, we get,

$$\ln\left(\frac{N}{N_0}\right) = -\lambda t$$

$$\text{or} \quad N = N_0 \exp(-\lambda t) \tag{8.10}$$

where N_0 is the number of parent nuclei at $t = 0$.

Equation 8.10 is called the *radioactive decay law*. The number of daughter nuclei at any time t will be given by,

$$N_{\text{daughter}} = N_0 - N = N_0 - N_0 \exp(-\lambda t)$$

$$\text{or} \quad N_{\text{daughter}} = N_0[1 - \exp(-\lambda t)] \tag{8.11}$$

Figure 8.4 shows how the number of parent and daughter nuclei vary with time.

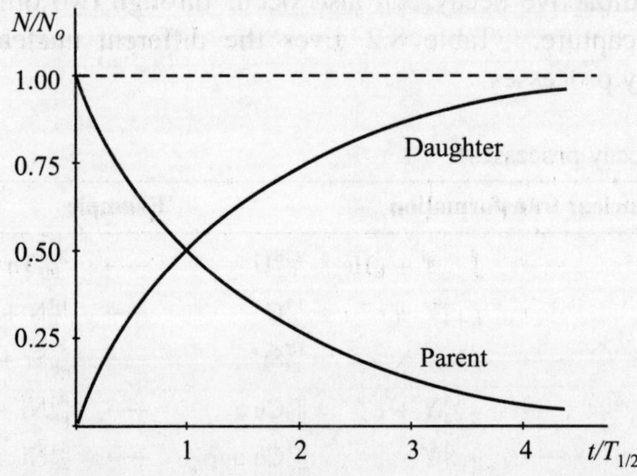

Figure 8.4 Variation of parent nuclei and daughter nuclei with time

The time $T_{1/2}$ required for the parent nuclei to decay to half of its initial population is called the *half-life* of the radioactive decay. Thus,

$$\frac{N}{N_0} = \frac{1}{2} = \exp[-\lambda T_{1/2}] \tag{8.12}$$

$$\text{or} \quad T_{1/2} = \frac{\ln 2}{\lambda} = \frac{0.693}{\lambda} \tag{8.13}$$

The average lifetime, T_{av}, is the average of the lifetimes of all N_0 nuclei. Thus it is given by the sum of the lifetimes of all the atoms divided by the initial number. The number of atoms that decay between time t and $t + dt$ is

$$dN = \lambda N dt \tag{8.14}$$

From Eq. 8.10,

$$dN = \lambda N_0 \exp(-\lambda t) dt \tag{8.15}$$

As the decay process is statistical, the lifetime of an atom can vary from 0 to ∞. Hence the average lifetime is given by

$$T_{av} = \frac{1}{N_0} \int_0^{\infty} t dN \tag{8.16}$$

Using Eq. 8.15,

$$T_{av} = \lambda \int_0^{\infty} \exp(-\lambda t) t dt \quad \text{or} \quad T_{av} = 1/\lambda \tag{8.17}$$

Using Eq. 8.13 we get,

$$T_{av} = \frac{T_{1/2}}{0.693} \quad \text{or} \quad T_{av} = 1.44 T_{1/2} \tag{8.18}$$

The *activity* of a given radioactive specimen is defined as the number of disintegrations per second. Thus, the activity,

$$a = \lambda N = \lambda N_0 \exp(-\lambda t) \tag{8.19}$$

or $$a = a_0 \exp(-\lambda t) \tag{8.20}$$

where a_0 is the initial activity of the specimen.

The SI unit of activity is the becquerel (Bq).

1 Bq = 1 decay per second

However, it is generally measured in the units of Curie (Ci) or Rutherford (Rd).

1 Ci = 3×10^{10} disintegrations per second

and 1 Rd = 1×10^{6} disintegrations per second

8.3.2 Nuclear Reactions

Apart from the unstable nuclei found in nature (natural radioactivity), it is also possible to induce disintegration in certain nuclides artificially (artificial radioactivity) by bombarding them with high energy particles. The bombardment disrupts the stable nucleus and transforms it to one that is radioactive. These induced transmutations of nuclides are called *artificial nuclear reactions* in analogy with the chemical reactions. Thus a nuclear reaction can be represented by the equation

$$x + {}^{A}_{Z}\mathrm{X} \longrightarrow {}^{A}_{Z}\mathrm{Y} + y$$

Here the symbols on the left are the reacting nuclides and those on the right, the product nuclides. The target nucleus X is bombarded by the particle x resulting in the product nucleus Y and the product particle y. The reaction may be represented in an abbreviated notation as ${}^{A}_{Z}\mathrm{X}\ (x, y){}^{A}_{Z}\mathrm{Y}$. The reaction is called (x, y) reaction. For example, the transmutation of nitrogen to oxygen due to the capture of an α-particle can be represented by the equation

$${}^{14}_{7}\mathrm{N} + {}^{4}_{2}\mathrm{He} \longrightarrow [{}^{18}_{9}\mathrm{F}] \longrightarrow {}^{17}_{8}\mathrm{O} + {}^{1}_{1}\mathrm{H}$$

where ${}^{18}_{9}\mathrm{F}$ is an unstable nucleus formed by the capture of an α-particle by the nitrogen nucleus. This kind of nucleus is called a *compound nucleus*. A few examples of nuclear reactions are mentioned in Chart 8.1. In these examples Cn denotes the compound nucleus.

CHART 8.1 NUCLEAR REACTIONS

Reaction	Equation	Example
$(\alpha, \mathbf{p})$	${}^{A}_{Z}\mathrm{X} + {}^{4}_{2}\mathrm{He} \longrightarrow [{}^{A+4}_{Z+2}\mathrm{Cn}] \longrightarrow {}^{A+3}_{Z+1}\mathrm{Y} + {}^{1}_{1}\mathrm{H}$	${}^{10}_{5}\mathrm{B} + {}^{4}_{2}\mathrm{He} \longrightarrow [{}^{14}_{7}\mathrm{N}] \longrightarrow {}^{13}_{6}\mathrm{C} + {}^{1}_{1}\mathrm{H}$
$(\alpha, \mathbf{n})$	${}^{A}_{Z}\mathrm{X} + {}^{4}_{2}\mathrm{He} \longrightarrow [{}^{A+4}_{Z+2}\mathrm{Cn}] \longrightarrow {}^{A+3}_{Z+2}\mathrm{Y} + {}^{1}_{0}\mathrm{n}$	${}^{9}_{4}\mathrm{Be} + {}^{4}_{2}\mathrm{He} \longrightarrow [{}^{13}_{6}\mathrm{C}] \longrightarrow {}^{12}_{6}\mathrm{C} + {}^{1}_{0}\mathrm{n}$
$(\mathbf{p}, \alpha)$	${}^{A}_{Z}\mathrm{X} + {}^{1}_{1}\mathrm{H} \longrightarrow [{}^{A+1}_{Z+1}\mathrm{Cn}] \longrightarrow {}^{A-3}_{Z-1}\mathrm{Y} + {}^{4}_{2}\mathrm{He}$	${}^{6}_{3}\mathrm{Li} + {}^{1}_{1}\mathrm{H} \longrightarrow [{}^{7}_{4}\mathrm{Be}] \longrightarrow {}^{3}_{2}\mathrm{He} + {}^{4}_{2}\mathrm{He}$
$(\mathbf{p}, \gamma)$	${}^{A}_{Z}\mathrm{X} + {}^{1}_{1}\mathrm{H} \longrightarrow [{}^{A+1}_{Z+1}\mathrm{Cn}] \longrightarrow {}^{A+1}_{Z+1}\mathrm{Y} + \gamma$	${}^{26}_{13}\mathrm{Al} + {}^{1}_{1}\mathrm{He} \longrightarrow [{}^{28}_{14}\mathrm{Si}] \longrightarrow {}^{28}_{14}\mathrm{Si} + \gamma$
$(\mathbf{d}, \alpha)$	${}^{A}_{Z}\mathrm{X} + {}^{2}_{1}\mathrm{H} \longrightarrow [{}^{A+2}_{Z+1}\mathrm{Cn}] \longrightarrow {}^{A-2}_{Z-1}\mathrm{Y} + {}^{4}_{2}\mathrm{He}$	${}^{16}_{8}\mathrm{O} + {}^{2}_{1}\mathrm{H} \longrightarrow [{}^{18}_{9}\mathrm{F}] \longrightarrow {}^{14}_{7}\mathrm{N} + {}^{4}_{2}\mathrm{He}$
$(\mathbf{d}, \mathbf{p})$	${}^{A}_{Z}\mathrm{X} + {}^{2}_{1}\mathrm{H} \longrightarrow [{}^{A+2}_{Z+1}\mathrm{Cn}] \longrightarrow {}^{A+1}_{Z}\mathrm{Y} + {}^{1}_{1}\mathrm{H}$	${}^{23}_{11}\mathrm{Na} + {}^{2}_{1}\mathrm{H} \longrightarrow [{}^{25}_{12}\mathrm{Mg}] \longrightarrow {}^{24}_{11}\mathrm{Na} + {}^{1}_{1}\mathrm{H}$
$(\mathbf{d}, \mathbf{n})$	${}^{A}_{Z}\mathrm{X} + {}^{2}_{1}\mathrm{H} \longrightarrow [{}^{A+2}_{Z+1}\mathrm{Cn}] \longrightarrow {}^{A+1}_{Z+1}\mathrm{Y} + {}^{1}_{0}\mathrm{n}$	${}^{9}_{4}\mathrm{Be} + {}^{2}_{1}\mathrm{H} \longrightarrow [{}^{11}_{5}\mathrm{B}] \longrightarrow {}^{10}_{5}\mathrm{B} + {}^{1}_{0}\mathrm{n}$
$(\mathbf{n}, \alpha)$	${}^{A}_{Z}\mathrm{X} + {}^{1}_{0}\mathrm{n} \longrightarrow [{}^{A+1}_{Z}\mathrm{Cn}] \longrightarrow {}^{A-3}_{Z-2}\mathrm{Y} + {}^{4}_{2}\mathrm{He}$	${}^{27}_{13}\mathrm{Al} + {}^{1}_{0}\mathrm{n} \longrightarrow [{}^{28}_{13}\mathrm{Al}] \longrightarrow {}^{24}_{11}\mathrm{Na} + {}^{4}_{2}\mathrm{He}$
$(\mathbf{n}, \mathbf{p})$	${}^{A}_{Z}\mathrm{X} + {}^{1}_{0}\mathrm{n} \longrightarrow [{}^{A+1}_{Z}\mathrm{Cn}] \longrightarrow {}^{A}_{Z-1}\mathrm{Y} + {}^{1}_{1}\mathrm{H}$	${}^{14}_{7}\mathrm{Al} + {}^{1}_{0}\mathrm{n} \longrightarrow [{}^{15}_{7}\mathrm{N}] \longrightarrow {}^{14}_{6}\mathrm{C} + {}^{1}_{1}\mathrm{H}$
$(\mathbf{n}, \mathbf{2n})$	${}^{A}_{Z}\mathrm{X} + {}^{1}_{0}\mathrm{n} \longrightarrow [{}^{A+1}_{Z}\mathrm{Cn}] \longrightarrow {}^{A-1}_{Z}\mathrm{Y} + {}^{1}_{0}\mathrm{n} + {}^{1}_{0}\mathrm{n}$	${}^{27}_{12}\mathrm{Al} + {}^{1}_{0}\mathrm{n} \longrightarrow [{}^{28}_{13}\mathrm{Al}] \longrightarrow {}^{26}_{13}\mathrm{Al} + {}^{1}_{0}\mathrm{n} + {}^{1}_{0}\mathrm{n}$
$(\mathbf{n}, \gamma)$	${}^{A}_{Z}\mathrm{X} + {}^{1}_{0}\mathrm{n} \longrightarrow [{}^{A+1}_{Z}\mathrm{Cn}] \longrightarrow {}^{A+1}_{Z}\mathrm{Y} + \gamma$	${}^{9}_{4}\mathrm{Be} + {}^{1}_{0}\mathrm{n} \longrightarrow [{}^{10}_{4}\mathrm{Be}] \longrightarrow {}^{10}_{4}\mathrm{Be} + \gamma$
$(\gamma, \mathbf{n})$	${}^{A}_{Z}\mathrm{X} + \gamma \longrightarrow [{}^{A}_{Z}\mathrm{Cn}] \longrightarrow {}^{A-1}_{Z}\mathrm{Y} + {}^{1}_{0}\mathrm{n}$	${}^{2}_{1}\mathrm{H} + \gamma \longrightarrow [{}^{2}_{1}\mathrm{H}] \longrightarrow {}^{1}_{1}\mathrm{H} + {}^{1}_{0}\mathrm{n}$

The conservation of energy in a nuclear reaction can be expressed as

$$E_x + m_x c^2 + M_\mathrm{X} c^2 = E_\mathrm{Y} + M_\mathrm{Y} c^2 + E_y + m_y c^2 \tag{8.21}$$

where the masses $m_x, M_\mathrm{X}, M_\mathrm{Y}$ and m_y denote the masses of the incident particle, target nucleus, product nucleus and the product particle, respectively, and $E_x, E_\mathrm{X}, E_\mathrm{Y}$ and E_y

denote the corresponding kinetic energies. The target nucleus is assumed to be at rest initially so that $E_{\mathrm{X}} = 0$.

The difference between the total kinetic energy of the products and that of the incident particle is called the *energy balance* or the *Q value* of the nuclear reaction. For $Q > 0$, total kinetic energy of the products is greater than that of the reactants and the reaction is said to be *exothermic* or *exoergic*. For $Q < 0$, the reaction is *endothermic* or *endoergic*. Thus, using Eq. 8.21, the Q value is given by

$$Q = E_{\mathrm{Y}} + E_y - E_x = (m_x + M_{\mathrm{X}} - M_{\mathrm{Y}} - m_y)c^2 \tag{8.22}$$

The recoil kinetic energy E_{Y} of the product nucleus is hard to measure. However, conservation of linear momentum can be used to eliminate it. Considering the collision shown in Fig. 8.5 and the momenta as $p_x, P_{\mathrm{X}}, P_{\mathrm{Y}}$ and p_y, the conservation of momentum along the x-axis and the y-axis

$$p_x = p_y \cos\theta + P_{\mathrm{Y}} \cos\phi \tag{8.23}$$

$$0 = p_y \sin\theta - P_{\mathrm{Y}} \sin\phi \tag{8.24}$$

Squaring and adding Eq. 8.23 and 8.24, we get

$$P_{\mathrm{Y}}^2 = p_x^2 + p_y^2 - 2p_x p_y \cos\theta \tag{8.25}$$

Using $p^2 = 2mE$ and rearranging, we get

$$E_{\mathrm{Y}} = \left(\frac{m_x}{M_{\mathrm{Y}}}\right) E_x + \left(\frac{m_y}{M_{\mathrm{Y}}}\right) E_y - \left(\frac{2}{M_{\mathrm{Y}}}\right) \sqrt{m_x E_x m_y E_y}\,(\cos\theta) \tag{8.26}$$

Using Eq. 8.22, we have

$$Q = E_y \left[1 + \left(\frac{m_y}{M_{\mathrm{Y}}}\right)\right] - E_x \left[1 - \left(\frac{m_x}{M_{\mathrm{Y}}}\right)\right] - \left(\frac{2}{M_{\mathrm{Y}}}\right) \sqrt{m_x E_x m_y E_y}\,(\cos\theta) \tag{8.27}$$

Examples 8.5–8.7 illustrate the application of Eq. 8.22 and 8.27.

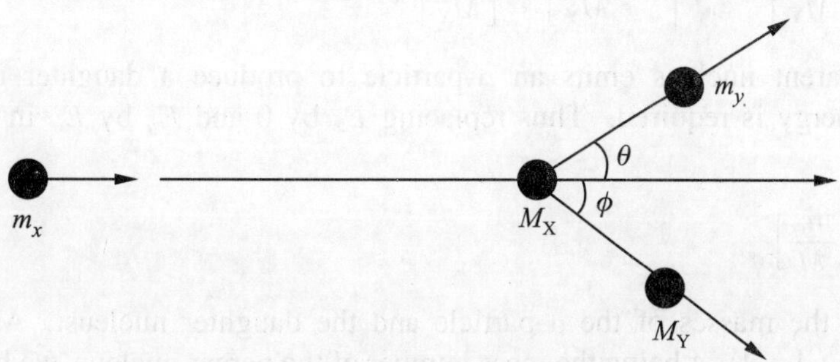

Figure 8.5 Collision of incident particle and target nucleus resulting in product nucleus and product particle

EXAMPLE 8.5

Q value of a nuclear reaction

Find the Q value of a nuclear reaction, ${}^{14}_{7}\mathrm{N}(\alpha, p){}^{17}_{8}\mathrm{O}$. The masses are given as $M_{\mathrm{N}} = 14.00753$ u, $m_\alpha = 4.00387$ u, $M_{\mathrm{O}} = 17.0045$ u and $m_p = 1.00184$ u.

Solution

The given nuclear reaction is ${}^{14}_{7}\mathrm{N} + {}^{4}_{2}\mathrm{He} \longrightarrow {}^{17}_{8}\mathrm{O} + {}^{1}_{1}\mathrm{H}$.

The Q value is given by

$$Q = (m_\alpha + M_{\mathrm{N}} - M_{\mathrm{O}} - m_p) \times 931.5 \text{ MeV}$$

$$= (4.00387\,\mathrm{u} + 14.00753\,\mathrm{u} - 17.0045\,\mathrm{u} - 1.00184\,\mathrm{u}) \times 931.5 \text{ MeV}$$

$$= 4.7 \text{ MeV}$$

EXAMPLE 8.6

Q value for α-decay

Show that the Q value for α-decay is given by $Q = E_\alpha \left[\dfrac{A}{A-4}\right]$.

Solution

The Q value of a nuclear reaction is given by

$$Q = E_y \left[1 + \frac{m_y}{M_{\mathrm{Y}}}\right] - E_x \left[1 - \frac{m_x}{M_{\mathrm{Y}}}\right] - \left[\frac{2}{M_{\mathrm{Y}}}\right] \sqrt{m_x E_x m_y E_y}\,(\cos\theta)$$

In α-decay the parent nucleus emits an α-particle to produce a daughter nucleus, that is, no bombarding energy is required. Thus replacing E_x by 0 and E_y by E_α in the above equation, we get

$$Q = E_\alpha \left[1 + \frac{m_\alpha}{M}\right]$$

m_α and M being the masses of the α-particle and the daughter nucleus. As m_α/M can be approximated to $4/(A-4)$, A being the mass number of the parent nucleus, we have

$$Q = E_\alpha \left[\frac{A}{A-4}\right]$$

EXAMPLE 8.7

Angle of ejection of a product particle in a nuclear reaction

Find the angle at which the proton is ejected in the reaction

$$^{4}_{2}\text{He} + ^{10}_{5}\text{B} \longrightarrow ^{13}_{6}\text{C} + ^{1}_{1}\text{H}$$

if the energy of the incident α-particle is 2 MeV, that of the proton is 5 MeV and the Q value of the reaction is + 4 MeV.

Solution

Here $Q = +\,4$ MeV, $E_x = 2$ MeV, $E_y = 5$ MeV, $m_x = 4$ u, $m_y = 1$ u, $M_\text{Y} = 13$ u.

Substituting the given data in the equation

$$Q = E_y \left[1 + \frac{m_y}{M_\text{Y}}\right] - E_x \left[1 - \frac{m_x}{M_\text{Y}}\right] - \left[\frac{2}{M_\text{Y}}\right] \sqrt{m_x E_x m_y E_y}\,(\cos\theta)$$

We get,

$$4\text{ MeV} = 5\text{ MeV}\left[1 + \frac{1}{13}\right] - 2\text{ MeV}\left[1 - \frac{4}{13}\right] - \left[\frac{2}{13\text{ u}}\right] \sqrt{4\text{ u} \times 1\text{ u} \times 2\text{ MeV} \times 5\text{ MeV}}\,(\cos\theta)$$

Solving, we get $\cos\theta = 0$ or $\theta = 90°$.

Thus the proton is ejected at right angles to the incident direction.

8.3.3 Detection of Nuclear Radiation

The detection of nuclear radiation is based on the ionisation caused in the material through which they pass. Various nuclear detectors differ in the material within which the ionisation is produced and the way in which it is detected or measured. The commonly used detectors are discussed in Chart 8.2 as they share the same fundamental basis.

8.4 NUCLEAR MODELS AND SPECTROSCOPY

Nuclei are more complicated than atoms. Nuclear force is more complicated than the simple CoulombŠs law that explains forces between atoms. Hence a comprehensive nuclear theory is not available. However, two of the various models of the nucleus, which are more successful than others, are discussed in the following.

8.4.1 Liquid Drop Model

This model, suggested by Bohr, is based on the assumption that nucleons move around within the nucleus and interact strongly with each other like the molecules in a drop of a liquid. The liquid drop model is instructive as it brings out the following similarities between the liquid and the nucleus. The nucleus has surface tension like a liquid drop and the surface energy reduces the total binding energy. For both the nucleus as well as a liquid, density is independent of volume. In both cases volume is directly proportional to the number of nucleons or molecules. The total binding energy of the nuclei, as well as the liquids (the heat of vapourisation), is nearly proportional to their mass.

The liquid drop model, however, cannot account for important properties of nuclides such as the magic numbers. The model is refined on lines similar to the free electron model for metals. In this refined model, known as the *fermi gas model*, it is assumed that each nucleon moves in an average potential due to all the nucleons that can be represented by a three-dimensional square well. Thus the analysis presented in Chart 6.4 for a particle confined to an infinite potential well can be applied to obtain a rough estimate of the energy levels of a nucleon in the finite potential well of a nucleus. In Ex. 8.8 we apply Eq. 7 in Chart 6.4 to estimate energy levels of a nucleon in a typical nucleus and compare them with the electronic energy levels.

EXAMPLE 8.8
Comparison of electronic and nuclear energy levels

Assuming infinite potential well, compare the order of energies for (a) an electron confined to atomic dimension of the order of angstroms, and (b) a nucleon confined to nuclear dimension of the order of fermi.

Solution

The expression for the energy of a particle confined to an infinite potential well of width L is $E_n = n^2h^2/8mL^2$. Thus,

(a) for an electron confined to 1 Å,

$$E_n = \frac{n^2 \times (6.625 \times 10^{-34}\ \text{Js})^2}{[8 \times 9.1 \times 10^{-31}\ \text{kg} \times (10^{-10}\ \text{m})^2]}$$

$$= 0.6 \times 10^{-17} n^2 \text{J} = 37.5\ n^2\ \text{eV}$$

(b) for a nucleon confined to 1 fm,

$$E_n = \frac{(6.625 \times 10^{-34}\ \text{Js})^2}{[8 \times 1.67 \times 10^{-27}\ \text{kg} \times (10^{-15}\ \text{m})^2]}$$

$$= 3.28 \times 10^{-11} n^2\ \text{J} = 205 n^2\ \text{MeV}$$

Thus, we find that the electronic energy levels are of the order of electron volts, whereas the nuclear energy levels are of the order of million electron volts.

CHART 8.2 NUCLEAR DETECTORS

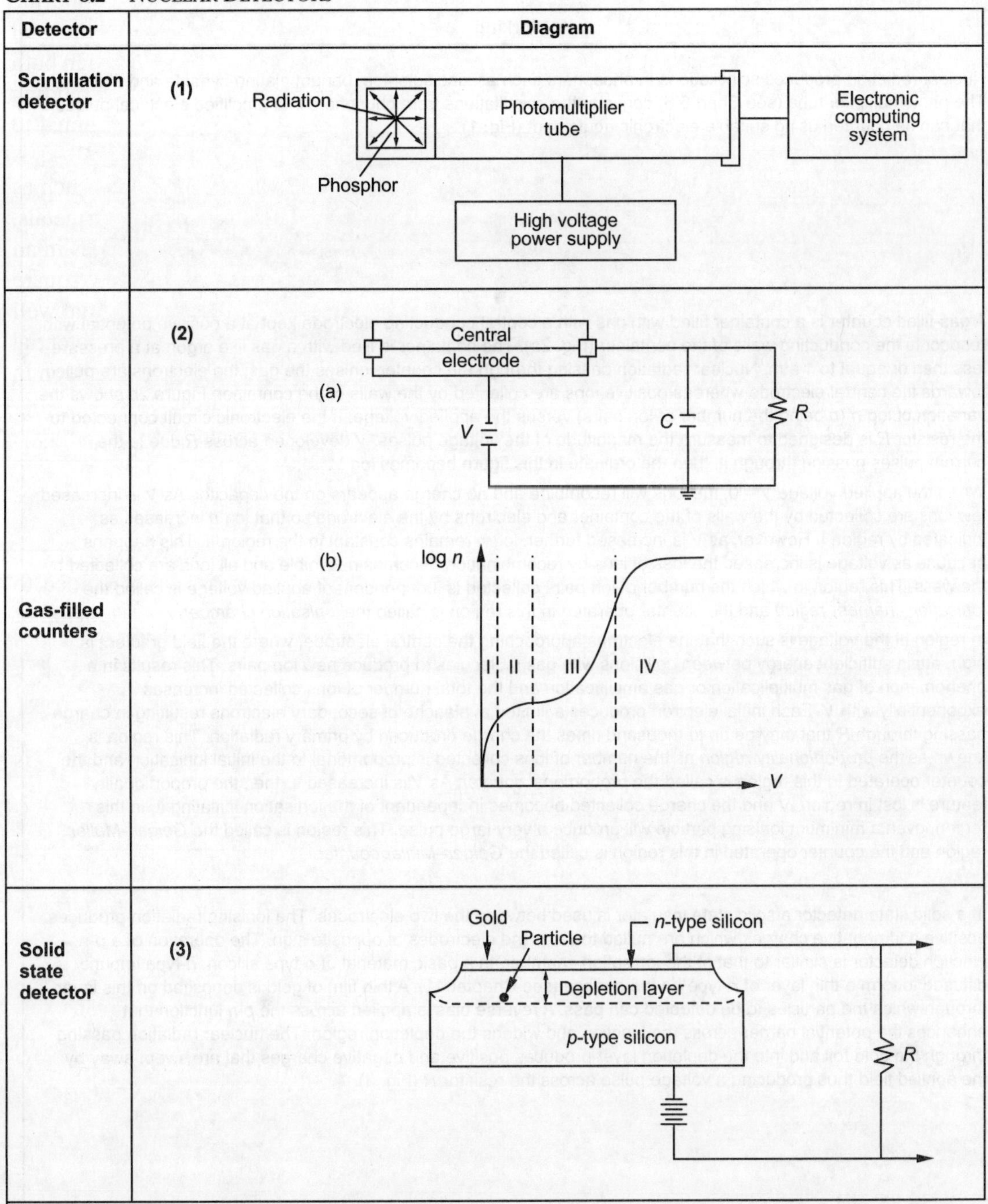

Apart from the above detectors, there are devices that can be used to view the tracks of charged particles directly. In *the photographic emulsion,* a charged particle ionises atoms in emulsion layers. The track of the particle corresponding to chemical changes in the emulsion can be seen on developing the emulsion. In a *cloud chamber,* a super cooled gas is used just below its condensation point. The charged particle ionises the gas in its path and the

Working
Nuclear radiation produces scintillations in phosphors such as zinc sulphide, barium platino-cynaide and diamond. The photomultiplier tube (see Chart 6.5) converts the scintillations from phosphors into amplified electrical pulses that can be counted using suitable electronic equipment (Fig. 1).
A gas-filled counter is a container filled with gas with a central conducting electrode kept at a positive potential with respect to the conducting walls of the container (Fig. 2a). The container is filed with a gas like argon at a pressure less than or equal to 1 atm. Nuclear radiation passing through the counter ionises the gas; the electrons are pulled towards the central electrode whereas positive ions are collected by the walls of the container. Figure 2b shows the variation of log n (n being the number of ion pairs) versus the applied voltage. If the electronic circuit connected to the resistor R is designed to measure the magnitude of the voltage pulses, V developed across R due to the current pulses passing through it, then the ordinate in this figure becomes log V. When the applied voltage $V = 0$, the ions will recombine and no charge appears on the capacitor. As V is increased few ions are collected by the walls of the container and electrons by the electrode so that log n increases as indicated by region I. However, as V is increased further, log n remains constant in the region II. This happens because as voltage is increased the loss of ions by recombination becomes negligible and all ions are collected by the walls. This region in which the number of ion pairs collected is independent of applied voltage is called the *ionisation chamber region* and the counter operated in this region is called the *ionisation chamber*. In region III the voltage is such that the electrons approaching the central electrode, where the field gradient is high, attain sufficient energy between collisions with gas molecules to produce new ion pairs. This results in a phenomenon of gas multiplication or gas amplification and the total number of ions collected increases exponentially with V. Each initial electron produces a limited avalanche of secondary electrons resulting in charge passing through R that may be up to thousand times the charge produced by primary radiation. This region is known as the *proportionality region* as the number of ions collected is proportional to the initial ionisation and the counter operated in this region is called the *proportional counter*. As V is increased further, the proportionality feature is lost in region IV and the charge collected becomes independent of the ionisation initiating it. In this region, even a minimum ionising particle will produce a very large pulse. This region is called the *Geiger–Muller* region and the counter operated in this region is called the *Geiger–Muller counter*.
In a solid state detector a solid state insulator is used between the two electrodes. The ionising radiation produces positive and negative charges which are pulled towards the electrodes of opposite sign. The operation of a p-n junction detector is similar to that of the ionisation chamber. In a basic material of *p*-type silicon, *n*-type impurity is diffused to form a thin layer of *n*-type on the surface (see Chapter 11). A thin film of gold is deposited on this layer through which the particles to be detected can pass. A reverse bias is applied across the *p-n* junction that enhances the potential barrier across the junction and widens the depletion region. The nuclear radiation passing through the gold foil and into the depletion layer produces positive and negative charges that are swept away by the applied field thus producing a voltage pulse across the resistor R (Fig. 3).
ions serve as centres for condensation of the supercooled gas, resulting in the track that can be seen as well as photographed. In a bubble chamber invented in 1952 by Glaser, a liquid maintained near its boiling point is used. Bubble tracks are produced due to ionisation by the charged particles that can be photographed.

8.4.2 Shell Model

In the Fermi gas model the Schrodinger equation can be solved to get the energy states of nucleons in a manner similar to that used for obtaining the energy states of an electron in an atom. This is known as the *shell model* of the nucleus. The state of a nucleon is represented by a principal quantum number and a total angular momentum quantum number.

The shell model successfully accounts for the magic numbers. The nuclei with magic nucleon numbers (Z or N) 2, 8, 20, 28, 50, 82, 126, ... are the most stable as they form close shells analogous to the close electronic shells exhibited by the noble gases that are also the most stable. In this sense the atomic numbers of noble gases, namely 2, 10, 18, 36, 54, 86, ... may be called the magic electron numbers.

The nuclei $^{40}_{20}\text{Ca}$ and $^{208}_{82}\text{Pb}$ are extremely stable as both Z and N are magic numbers: both protons and neutrons form closed shells. The α-particle exhibits exceptional stability due to double magic numbers, $Z = N = 2$. Its binding energy is so high (see Fig. 8.3) that it has little or no attraction for another nucleon. Hence there is no stable nuclide with $A = 5$. The shell model also predicts the correct ground state angular momentum for almost all nuclides. However, it ignores factors such as saturation of nuclear force, nearly constant nuclear density and the variation of binding energy with mass number. A collective model of the nucleus that combines both the shell model and the liquid drop model has been evolved and is considered a remarkably successful model of the nucleus.

8.4.3 Spectroscopy

In the last chapter we appreciated that optical spectroscopy corresponds to transitions of valence electrons in atoms. The characteristic x-rays correspond to transition of electron in the inner electronic shells. The emission of γ-rays by radioactive nuclei corresponds to the difference between nuclear energy levels. The study of γ-rays is thus referred to as *γ-ray spectroscopy* or *nuclear spectroscopy*. Table 8.3 summarises how spectroscopy helps us to explore atoms and nuclei.

Table 8.3 Spectroscopy of atoms and nuclei

Emission	Energy range	Transition between	Spectroscopy
Visible photons	eV	Valence (outermost) electronic energy levels	Atomic spectroscopy
X-rays	keV	Inner electronic energy levels	X-ray spectroscopy
γ-rays	MeV	Nuclear energy levels	γ-ray/nuclear spectroscopy

When a nucleus undergoes transition from one of its excited energy levels to a lower level, resulting in the emission of a γ-ray photon, the nucleus recoils to conserve linear momentum. The recoil kinetic energy reduces the energy of the photon. Thus the energy and momentum conservation equations can be written as

$$\Delta E = h\nu + \frac{p^2}{2M} \tag{8.28}$$

and $$\frac{h\nu}{c} = p \tag{8.29}$$

where, ΔE is the actual difference between the energy levels, $h\nu$ is the energy of the photon, and p is the recoil momentum of the nucleus. Substituting p from Eq. 8.29 in Eq. 8.28 we get,

$$\Delta E = h\nu\left(1 + \frac{h\nu}{2Mc^2}\right) \tag{8.30}$$

or $$h\nu = \Delta E\left(1 + \frac{h\nu}{2Mc^2}\right)^{-1} \tag{8.31}$$

For $h\nu << Mc^2$, $h\nu$ does not differ much from ΔE, so $h\nu$ can be replaced on the right-hand side of Eq. 8.31 by ΔE. Expanding the parenthesis on the right using the binomial theorem and neglecting the higher powers of $h\nu/2Mc^2$ we get

$$h\nu = \Delta E\left(1 - \frac{\Delta E}{2Mc^2}\right) = \Delta E - \frac{(\Delta E)^2}{2Mc^2} \tag{8.32}$$

Thus the reduction in the energy of the photon can be as much as $(\Delta E)^2/2Mc^2$ due to the recoil of the nucleus.

The effect of recoil of the nucleus is a broader spectral line, that is, instead of a single precise energy there is a small spread of energy. Though this spread is small, it is many times that due to natural broadening (see Chart 6.6). The emission of γ-rays without recoil of the nucleus is possible if the atom is bound to its neighbours in a crystal with binding energy sufficient to prevent its recoil. If this condition is satisfied, the crystal recoils as a whole with momentum equal and opposite to that of the γ-ray but with very little energy. This is like the collision between a light particle and a very heavy particle, where the latter takes away very little energy. The result is γ-rays with sharply defined energy. For this the crystal needs to be cooled to a very low temperature. The phenomenon of recoilless γ-emission by a nucleus is known as *Mossbauer effect.*

8.4.4 Nuclear Magnetic Resonance

The magnetic potential energy of the nucleus having the z-component of the magnetic moment μ_z in a constant magnetic field B parallel to the z-axis, is

$$U_m = -\mu_z B \tag{8.33}$$

Thus the energy is minimum when μ_z is positive and maximum when μ_z is negative. In a magnetic field, splitting of energy levels takes place in a manner similar to that in the Zeeman effect (Section 8.5). Figure 8.6 shows the splitting of energy levels for a proton into the spin-up and the spin-down sublevels. The energy difference between these sublevels is

$$\Delta E = 2\mu_{pz} B \tag{8.34}$$

Thus the absorption of the photon of this energy will lead to raising the proton in the lower state to the upper state. Similarly, a proton in the upper state, flipping its spin to fall to the lower state, will result in the emission of a photon of the same energy. The frequency of the photon, ν_L, corresponding to ΔE is given by

$$\nu_L = \frac{\Delta E}{h} = \frac{2\mu_{pz}B}{h} \tag{8.35}$$

This is the frequency with which a nuclear magnetic moment μ precesses around an external magnetic field B parallel to the z-axis. It is called the *Larmor frequency* (see Fig. 8.7).

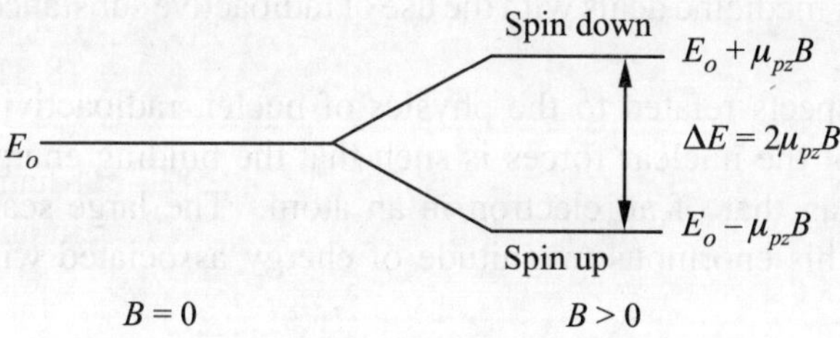

Figure 8.6 Spliting of energy levels for a proton into the spin-up and the spin-down sublevels

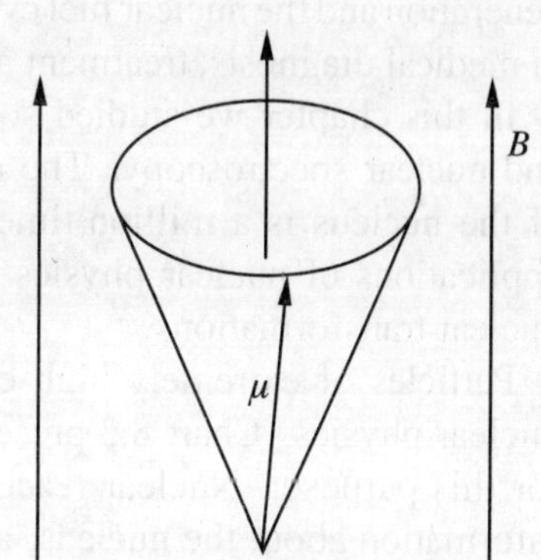

Figure 8.7 A nuclear magnetic moment μ precesses around an external field B with a frequency called the Larmor frequency

If a sample of substance that contains nuclei with spin of 1/2 is placed in a magnetic field **B**, then the spins of most of these nuclei will be aligned parallel to **B** (spin-up state), as this is the lowest energy state. If an electromagnetic radiation of the Larmor frequency ν_L is now applied to the sample, then the nuclei will receive the right amount of energy to flip their spins to the higher state (spin-down). The phenomenon is known as *nuclear magnetic resonance* (NMR).

NMR thus provides an experimental method to determine nuclear magnetic moments. An RF coil is placed around the sample to supply radiation of fixed frequency. **B** is varied until the energy absorbed is a maximum. Alternatively a broad-spectrum RF pulse is applied and the frequency of the radiation the sample gives off as its excited nuclei return to the lower energy state is measured. From the measured values of resonance frequency ν_L, and B, μ can be calculated using Eq. 8.35.

8.5 APPLICATIONS

The term 'nuclear' evokes different responses from different people: People concerned with energy problems see the generation of electricity using nuclear power plants as a solution for the energy crisis; patients undergoing treatment for dreaded diseases may owe their lives to nuclear medicine; physicists engaged in research in nuclear physics find their research investigations more exciting and challenging compared to those in other areas of research; and those engaged in learning nuclear physics marvel at the unimaginable knowledge nature has packed in this tiny entity.

As nuclear physicists explored the various nuclei, the knowledge gained has led to wide-ranging applications that have influenced almost every walk of modern life. Everyday use of terms such as nuclear energy, nuclear power, nuclear devices and nuclear radiation brings home the relevance of nuclear knowledge today. This knowledge has had a deep impact on the technological scene and has led to the development of specialised branches such as nuclear engineering and nuclear medicine.

Nuclear engineering deals with the production and use of nuclear energy. It involves the development, design, construction and operation of power plants that convert nuclear energy to other useful forms such as heat or electrical energy. Nuclear chemical engineering is a branch of chemical engineering that deals with the production and use of radioisotopes, nuclear power generation and the nuclear fuel cycle. Nuclear medicine deals with the use of radioactive substances in medical diagnosis, treatment and research.

In this chapter we studied some basic aspects related to the physics of nuclei, radioactivity and nuclear spectroscopy. The magnitude of the nuclear forces is such that the binding energy of the nucleus is a million times greater than that of an electron in an atom. The large-scale applications of nuclear physics arise from this enormous magnitude of energy associated with nuclear transformations.

Particles of extremely high energy are required in order to pursue much of the research in nuclear physics. Chart 8.3 presents a comparative study of the devices, called *accelerators*, used for this purpose. Nuclear reactions induced by high energy particles are a valuable source of information about the nucleus, and have produced many new nuclear species. These artificially made nuclides yield further information about the nuclei, and find application in other branches of science, such as chemistry, biology and medicine.

8.5.1 Applications of Fission and Fusion

It is interesting to note that until 1939 nuclear physics was considered a scholarly activity of little practical importance. However, the discovery of fission by O Hahn (1879–1968) and F Strassman in Germany changed this scenario drastically. In the late 1930s Lise Meitner, Otto Hann and Fritz Strassmann studied the effect of bombarding uranium with slow (thermal) neutrons. They found a number of radio-nuclides resulted from this bombardment. In 1939, they identified one of the radio-nuclides as barium. This was quite puzzling, because Hahn and Strassmann wondered how this middle mass element ($Z = 56$) come to be produced by bombarding uranium ($Z = 92$) with neutrons. Meitner and her nephew Frisch finally solved the riddle. They suggested a mechanism in which uranium splits into two fragments, one of which is barium. Frisch named this process as *fission*.

The fission of uranium 235 after the capture of a thermal neutron may result in one of the following reactions.

$$n + {}^{235}_{92}\text{U} \longrightarrow {}^{141}_{56}\text{Ba} + {}^{92}_{36}\text{Kr} + 3n$$

$$n + {}^{235}_{92}\text{U} \longrightarrow {}^{140}_{54}\text{Xe} + {}^{94}_{38}\text{Sr} + 2n$$

In fission, that is, splitting of a heavier nucleus into two lighter ones, the greater binding energy of the product nuclei causes energy to be given off. The release of an enormous amount of energy (see Ex. 8.9), and the production of more neutrons than consumed, raised the possibility of a chain reaction with the neutrons produced triggering another fission. Enrico Fermi first suggested such a possibility in 1939.

CHART 8.3 ACCELERATORS

Accelerators are devices used to accelerate charged particles to very high energies that are required to bring about the desired nuclear reactions. Accelerators are quite far removed from ordinary industrial engineering and are generally considered a part of nuclear physics. However, the design and construction of accelerators depend strongly on electrical and mechanical engineering.

Linear accelerator

A linear accelerator consists of a set of coaxial cylindrical tubes of increasing length placed in a vacuum chamber. Alternate tubes are connected together. The odd-numbered tubes are connected to one terminal of the high frequency power supply and the even-numbered to the other terminal. The ions introduced at one end of the chamber along the axis of the tubes get accelerated each time they come to the gap between the tubes. Inside the tube the speed of ions remains constant, as the potential is constant. The tubes are of increasing length because the speed of ions increases after each tube and hence they need to cover longer distances in successive tubes in order to synchronise with the reversal of potential. If the number of gaps is N, then the ions will acquire energy equal to NqV (q is the charge on the ions and V is the applied voltage) when they come out of the final tube. The linear accelerator can accelerate protons and even electrons to several hundred MeV.

The Stanford linear accelerator, which is 3.2 km long produces electrons of energy 20 GeV.

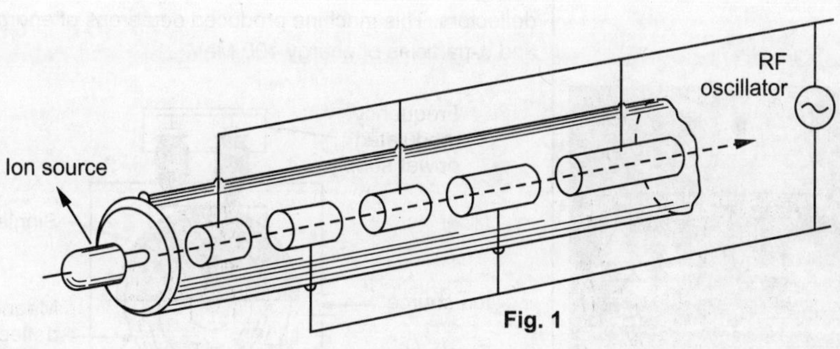

Fig. 1

There are other accelerating devices apart from those described above. The early accelerators like the voltage multiplier of Cockcroft and Walton and the electrostatic generator of van de Graff can produce particles of the order of a few MeV. The betatron is an electron accelerator that uses the principle of a transformer. The cloud of electrons located inside an annular, doughnut-shaped vacuum chamber takes the place of secondary winding. Electrons move in a circular orbit within the chamber and gain energy by induction due to the change of magnetic flux linking the orbit with time. The betatron at the General Electric Research Laboratories produces 100 MeV electrons and the one at the University of Illinois can produce electrons with energies greater than 300 MeV.

Accelerators have grown from devices that fit in an ordinary laboratory producing particles of energies of the order of a few MeV to monstrous machines that require many acres of ground and operate in the GeV range. Three important accelerators are discussed below.

Cyclotron

The Cyclotron, also known as the *magnetic resonance* accelerator, is the most famous of all accelerators. The first cyclotron built by Lawrence and Livingston produced 1.2 MeV protons. It had a magnet with pole faces 28 cm in diameter. A 1.5 m cyclotron can produce protons of about 10 MeV, deuterons of about 20 MeV and α-particles of about 40 MeV. The construction, working and the other details of a cyclotron are included in Chart 13.3. Figure 2 shows the image of 60-inch cyclotron, circa 1939, showing beam of accelerated ions (work of the Lawrence Berkely National Laboratory, taken from http://en.wikipedia.org/wiki/File:Cyclotron_with_glowing_beam.jpg.)

Fig. 2

Synchrotron/Synchrocyclotron

The major limitation of a cyclotron is (see Chart 13.3) that it cannot accelerate electrons because the increase in electronic mass due to the relativistic relation

$$m = \frac{m_0}{\sqrt{1-(v^2/c^2)}}$$

disturbs the resonance condition, $f = Bq/2\pi m$. It is possible to overcome this limitation by adjusting either the frequency of the electric field, f, or the magnetic field, B, or both, so that the resonance condition is maintained. If the electric field frequency is kept fixed and the magnetic field is adjusted, then the machine is called a *synchrotron*. If the magnetic field is kept constant and the frequency of the electric field is varied appropriately, then the machine is called a *frequency modulated (FM) cyclotron* or *synchrocyclotron*. Figure 3 shows the first synchrocyclotron built at the University of California. There is only one dee and the oscillating potential is applied between it and a ground connection. The ion beam is removed with the aid of electrostatic and magnetic deflectors. This machine produced deuterons of energy 200 MeV and α-particles of energy 400 MeV.

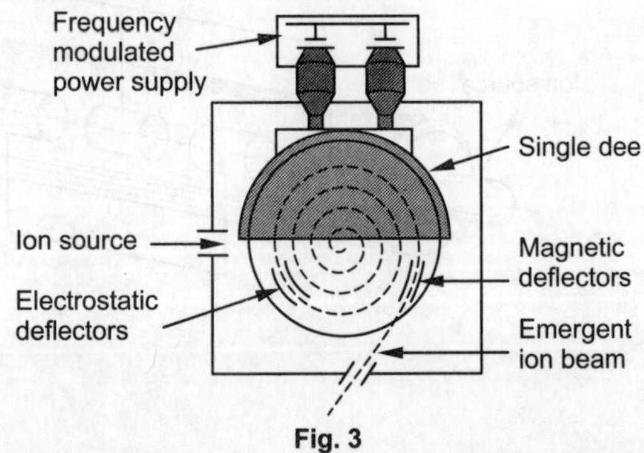

Fig. 3

The electron synchrotron accelerates electrons by the use of a radio frequency cavity and can produce electrons of energy greater than 300 MeV. Proton synchrotron, based on the same principle as electron synchrotron, can produce protons with energies in BeV. The 'Cosmotron' at the Brookhaven National Laboratory produced 3 BeV protons in 1952, the betatron at the University of California produced 6 BeV protons in 1954, and proton energies of 10 BeV were reached in the Russian proton synchrotron at Dubna in 1957.

The alternating gradient synchrotron is based on the principle that alternate focussing and defocussing ion lenses can be arranged so as to have a net focussing effect. The alternate gradient synchrotron at the Brookhaven National Laboratory can accelerate protons to 30 BeV.

If the rate of neutron production exceeds the rate of loss of neutrons, the device becomes an atomic bomb. However, if the chain reaction proceeds in a controlled manner, such that a steady state is attained in which the number of neutrons produced per unit time equals those used up in the same time, then the result is a chain-reacting pile, or a *nuclear reactor*, which can be used as a source of neutrons or energy. Fermi and his collaborators developed the first fission reactor in a squash court at the University of Chicago in 1942. A nuclear reactor is a very efficient source of energy as the fission of 1 kg of U^{235} per day evolves energy at the rate of about 1000 MW per day. A conventional thermal power plant would require 3000 tonnes of coal per day to produce this much power (see Ex. 8.10).

EXAMPLE 8.9

Amount of energy released in fission

Find the amount of energy released due to the fission of 1 gm of U^{235}. Compare it with the amount of energy liberated by burning coal or exploding TNT.

Solution

The fission of one Uranium 235 nucleus liberates 200 MeV energy. Thus the energy released by 1 gm is

$$= \frac{6.023 \times 10^{23}\ \text{atoms/mole}}{235\ \text{gm/mole}} \times 200\ \text{MeV}$$

$$= 5.13 \times 10^{23}\ \text{MeV} = 5.13 \times 10^{29}\ \text{eV} \times 1.6 \times 10^{-19}\ \text{J/eV}$$

$$= 8.21 \times 10^{10}\ \text{J}$$

$$\text{The energy in colonies} = \frac{8.21 \times 10^{10}\ \text{J}}{4.187\ \text{J/cal}} = 2 \times 10^{10}\text{cal}$$

Comparing it with 7000 kcal released in burning 1 kg of coal and with 1000 cal released in exploding 1 kg of TNT, we find that the fission of 1 gm of Uranium 235 is equivalent to

$$\text{(a) burning } \frac{2 \times 10^{10}\ \text{cal}}{7 \times 10^{6}\ \text{cal/kg}} = 3000\ \text{kg} = 3 \text{ tonnes of coal}$$

$$\text{(b) exploding } \frac{2 \times 10^{10}\ \text{cal}}{1000\ \text{cal/kg}} = 20,000 \text{ tonnes of TNT}$$

EXAMPLE 8.10

Power output of a nuclear power plant

Show that the complete fission of 1 kg of Uranium 235 in a day can produce heat power at a rate of 1000 MW. How much coal would be needed in a conventional power plant to produce this much heat power? Find the electric power generated assuming 30% conversion efficiency from heat to electricity.

Solution

Using the result obtained in the last example, the energy released by 1 kg of uranium 235 in one day $= 8.21 \times 10^{13}$ J/day.

Thus the power output $= 8.2 \times 10^{13}$ J/$(24 \times 60 \times 60\,\text{s}) = 9.5 \times 10^{8}$ J/s ≈ 1000 MW.

This is equivalent to burning 3000 tonnes of coal per day in a conventional power plant. For 30% conversion efficiency the electric power generated is 0.30×1000 MW $= 300,000$ kW.

The inverse process of fission is *fusion*, in which a heavy nucleus is formed from two lighter ones. The greater binding energy of the product nucleus also causes energy to be released. For fusion to take place, the nuclei should overcome the mutual repulsion and come within the range of nuclear force so as to 'fuse'. The fusion of hydrogen nuclei into helium nuclei is the basic energy producing process in the sun. This process can take place in several different reaction sequences, the most predominant being the proton–proton cycle. In this cycle four hydrogen nuclei combine to form a helium nucleus releasing about 26.7 MeV energy. These reactions have been taking place in the sun for the last 4.5 billion years and are estimated to continue for another 5 billion years. In stars having a higher core temperature, compared to that of the sun, fusion reactions other than the proton–proton cycle also take place, the carbon cycle being the most predominant. In the carbon cycle, carbon acts as a catalyst but the end result is again the conversion of four protons to an α-particle. The first thermonuclear reaction occurred on earth at Eniwetok Atoll on 1 November 1952, when the USA exploded a fusion device, generating energy equivalent to 10 milllion tonnes of TNT.

The proponents of nuclear power argue that fission and fusion will be our ultimate power choices when we have exhausted all our fossil fuels. They foresee power generation in future by either 'burning rocks' (fission of uranium extracted from ores) or 'burning water' (fusion of deuterium extracted from water). In Chart 8.4 a comparative study of the applications of fission and fusion processes for energy generation is presented.

8.5.2 Applications of Radioactivity

Radioactivity has numerous useful applications in different fields such as archeology, medicine, industry and agriculture.

Archeology

Radioactivity is used in archeology for estimating the age of organic and inorganic substances. A technique called *radiocarbon dating* enables the determination of the age of geological and biological specimens such as seeds, wooden artifacts, and plant and animal remains. This technique was developed by Willard Libby in the 1940s and had a tremendous impact on archeology and related fields.

Radiocarbon is the β-active carbon isotope ${}^{14}_{6}C$ that is produced in the upper atmosphere due to nuclear reaction caused by cosmic rays. It combines with the oxygen molecules to form CO_2

CHART 8.4 THE NUCLEUS AS A SOURCE OF ENERGY: FISSION AND FUSION

The nuclear reactor is a device for producing nuclear energy in a controlled manner. Though the reactor may be based either on a fission process or a fusion process, most of the reactors in the world at present are fission-based because a sustained and controllable source of fusion power is considerably more difficult to achieve. However, the fusion-based reactors promise to become the ultimate source of energy on earth and their development is being pursued vigorously in many countries in the world because they are safe, relatively non-polluting and can make use of the oceans for the supply of limitless fuel.

Fission reactor

The heart of a reactor is a core, which contains the fuel that sustains the chain reaction, thereby producing heat. The reactor is said to be critical when the reactions are sustained, that is, on an average one neutron from a fission process causes another fission. The operating level of a reactor can be described with the help of an effective neutron multiplication factor, k, defined as

$$k = \frac{P}{(A + L)} \quad (1),$$

where, P, A and L denote, respectively, the rates of production, absorption and leakage of neutrons. For $k = 1$, the reactor operates at a constant power output level, that is, at a critical level. When $k < 1$ the power output level of a reactor is dying down and the reactor is said to be subcritical. For $k > 1$ the power output level increases rapidly and the reactor is said to be supercritical.

The neutron multiplication factor can be controlled by the following factors. (i) The loss of neutrons can be minimised by optimising the reactor geometry. A large reactor core for which the ratio of surface to volume is small can minimise the neutron loss by leakage. (ii) Since the fission is induced most effectively by thermal (slow) neutrons, there is a need to moderate the neutron speeds. This is achieved by mixing the uranium fuel with a substance called the *moderator*, for example, heavy water (D_2O) or graphite. The moderator should not be an absorber of neutrons, that is, it should have a low absorption cross section for neutrons. (iii) The finer control of fission rate is achieved by inserting rods made of good neutron absorbers such as boron or cadmium, called the *control rods*. Control rods are inserted in the fissionable material at appropriate depth so as to achieve the desired operating level of a reactor. These rods are instrumental in starting or shutting the reactor. Figure 1 shows a broad outline of the most common type of a nuclear power plant.

Fig. 1

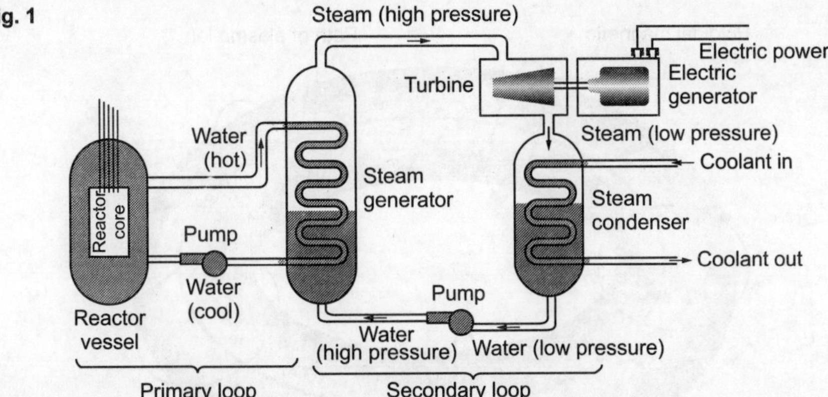

In this reactor, called the pressurised water reactor, water is used as the moderator as well as the heat transfer medium. In the primary loop, water is circulated through the reactor vessel and transfers energy at high temperature and pressure from the hot reactor core to the steam generator. The steam generator is a part of the secondary loop in which the evaporation provides high-pressure steam to operate the turbine. The turbine then drives the electric generator. The secondary loop is completed as the low-pressure steam from the turbine is cooled and condensed to water and forced back into the steam generator by a pump.

A reactor is called a breeder reactor if it can provide electrical power and at the same time produce an equivalent amount of fissionable material, which in turn can be used as a reactor fuel. A reactor that uses natural uranium as its fuel and produces plutonium 239, which is also fissionable is an example of a breeder reactor.

The design and construction of a reactor depend on several engineering inputs apart from the essential requisite of reactor physics. The basic applications of a reactor are production of useful energy for electricity generation, production of weapon grade plutonium for military purposes, scientific research and testing of nuclear technology, and production of radioactive isotopes.

Fusion reactor

For self-sustaining fusion reactions extreme temperature and density is required. High temperature is needed so that some nuclei have the energy needed to overcome the Coulomb repulsion between protons and get within the range of the nuclear forces in order to fuse; high density is needed to increase the frequency of such collisions. Moreover these conditions are to be maintained long enough to ensure the fusion of enough of the fuel. At such large temperatures the atoms are stripped of all their electrons and hence the reactor fuel is in the form of a fully ionised gas or plasma. The confinement of this plasma for sustained fusion reaction is one of the most difficult problems in a fusion reactor. In stars the confinement of plasma is possible due to the enormous gravitational forces. The plasma confinement in a terrestrial reactor can be achieved by the following two schemes.

1. In the Russian-designed Tokamak scheme the plasma is confined in a toroidal region by specially designed magnetic field, in the form of a modified torus or a doughnut as shown in Fig. 2. It uses a polodial field whose field lines are circles around the toroid axis. The poloidal field is produced by a current set up in the plasma itself by the changing field of an electromagnet in the centre of the toroid.
2. In the inertial confinement scheme laser beams are used for confinement of plasma. Intense laser beams are focussed on a small solid pallet containing deuterium and tritium from all angles. These beams vaporise the pellet converting it into plasma and then compress it from all directions due to the large pressure exerted. This results in large increase in density and temperature inducing fusion. The alpha particles generated by the fusion are also forced to remain inside the plasma.

Fig. 2

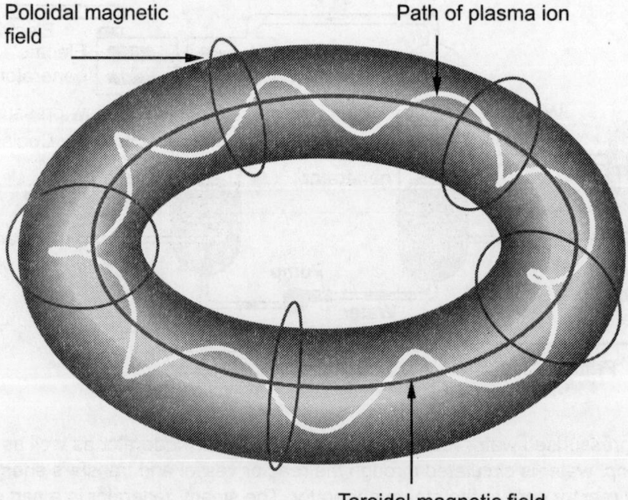

molecules. As plants inhale CO_2 and animals depend on plants, radiocarbon is present in all living organisms. The ratio of ${}^{14}_{6}C$ to ${}^{12}_{6}C$ is the same, about 1.3×10^{-12}, in all living organisms, because they continuously exchange CO_2 with their environment. However, when an organism dies, the input of radiocarbon stops and the ratio of ${}^{14}_{6}C$ to ${}^{12}_{6}C$ decreases steadily due to the β-decay of ${}^{14}_{6}C$. Assuming the ratio of ${}^{14}_{6}C$ to ${}^{12}_{6}C$ in the atmosphere to be constant for thousands of years, the quantity of ${}^{14}_{6}C$ remaining can be used to estimate how long ago the organic material died. Example 8.11 illustrates the use of radiocarbon dating in estimation of the age of a tree.

EXAMPLE 8.11

Radioactive dating of a tree

A piece of charcoal of mass 25 gm, obtained from an archeological site, has a carbon 14 activity of 250 disintegrations per minute. How long ago did the tree from which the sample of charcoal is obtained die?

Solution

As the half life of carbon 14 is 5730 years, the decay constant is given by

$$\lambda = \frac{0.693}{T_{1/2}} = \frac{0.693}{5730 \text{ years}}$$

$$\lambda = 2.3 \times 10^{-10} \text{ disintegrations per minute per nucleus}$$

The number of nuclei in 25 g of carbon is

$$N(^{12}\text{C}) = \left[\frac{(6.023 \times 10^{23} \text{ nuclei/mole})}{(12 \text{ gm/mole})}\right] \times 25 \text{ gm} = 1.25 \times 10^{24} \text{ nuclei}$$

As the ratio of ^{14}C to ^{12}C is about 1.3×10^{-12} in all living organisms, we have,

$$\frac{N_0(^{14}\text{C})}{N_0(^{12}\text{C})} = 1.3 \times 10^{-12}$$

or, $$N_0(^{14}\text{C}) = 1.3 \times 10^{-12} \times 1.25 \times 10^{24} \text{ nuclei}$$

$$= 1.625 \times 10^{12} \text{ nuclei}$$

Thus the initial activity,

$$R_0 = N_0\lambda = 1.625 \times 10^{12} \text{ nuclei} \times 2.3 \times 10^{-10} \text{ disintegrations per minute per nucleus}$$
$$= 374 \text{ disintegrations/minute}$$

Now, $a = a_0 \exp(-\lambda t)$

or, $$t = \frac{\ln(a_0/a)}{\lambda} = \frac{\ln(374/250)}{2.3 \times 10^{-10} \text{ disintegrations/minute}} = 1.751 \times 10^9 \text{ min} = 3331 \text{ years}$$

Thus the tree from which the sample of charcoal is obtained died 3331 years ago.

The estimations based on radiocarbon dating are limited to 60,000–70,000 years whereas the earth itself dates back to 4.5 billion years. Thus the radionuclides with much longer half lives such as rubidium 97 and strontium 87 (half life of 47 billion years) are used in geology. Beryllium 10 is used to date sedimentary rocks and carbon 14 and chlorine 36 are used to study deep ocean layers and underground water reservoirs. In geochronology, the oldest rocks on the surface of the earth are estimated to be 3.3 billion years old and the earth itself is estimated to be about 4.5 billion years old. The youngest lunar rocks were estimated to be about 3 billion years old during the Apollo expedition in 1972.

Medicine

Despite the hazards, the use of ionising radiation in medical research, diagnosis and therapy has been invaluable and has saved many lives. The medicinal uses of radioactivity include the use of nuclear radiation, such as γ-rays and naturally occurring radionuclides like radium and radon, for cancer therapy and other treatments. Nuclear radiation is also used to destroy malignant tumours. Today a large number of radionuclides are produced in reactors and accelerators that are useful in nuclear medicine for the treatment of various diseases. For example, the radioisotope, iodine 131 is used in the treatment of thyroid disorders (see Ex. 8.12). Artificially produced radio isotopes called *tracers* are used in biological and medical research.

EXAMPLE 8.12

Radioactivity of iodine 131

Iodine 131 is a radioactive isotope made in nuclear reactors for use in the treatment of hyperthyroidism. It gets concentrated in the thyroid gland and acts as a radiation source. Its half life is 8 days. If a hospital orders 20 mg of this radioisotope and stores it for 48 days, how much of the original amount will be present?

Solution

Half life of iodine 131, $T_{1/2} = 8$ days.

Hence the decay constant, $\lambda = 0.693/8$ days $= 0.086625$ decays/day

Now, $N = N_0 \exp(-\lambda t)$.

Thus the amount that will be present after 48 days is

$$N = 20 \text{ mg} \exp[(-0.086625 \text{ decays/day}) \times 48 \text{ days}]$$

$$= 20 \times \exp(-4.158) = 0.313 \text{ mg}$$

Thus, only 0.313 mg of the original 20 mg will be present after 48 days.

Industry and agriculture

The variations of the intensity of γ-rays due to absorption in materials are used in industry for the detection of internal defects and measurement of thickness. The ionising power of nuclear radiation is used in factories to avoid accumulation of charge on moving parts due to friction.

In agriculture nuclear radiation is used for the preservation of food. It has been used to eliminate a major agricultural pest, the screwworm fly, without employing pesticides. It is also used to induce plant mutations that can lead to improved varieties of many crops, such as wheat, peas and beans, with higher yields and greater resistance to diseases.

8.5.3 Applications in Medical Diagnostics

Auger camera

In 1957, HO Auger developed an ingenious gamma camera. It measures the radioactivity in a large area, displays the results on an oscilloscope screen, and also records them on a film or a tape. Auger cameras are used in the diagnosis of diseases of the thyroid, liver, brain, kidneys, lungs, spleen, heart and circulatory system.

Positron emission tomography

In 1928, Paul AM Dirac, an English physicist who studied electrical engineering before switching first to mathematics and then to physics, developed his relativistic quantum mechanical equation for the electron. Speculating about the solutions of this equation he predicted the *positron*—a positively charged electron. The positron was detected in 1932 by Carl Anderson in the United States in the stream of secondary particles that result from collisions between cosmic rays and atomic nuclei in the atmosphere.

Positron emission tomography, known as PET, is an imaging system that is useful in metabolic studies. In PET, a suitable positron-emitting radio nuclide is administered to the patient. The positrons travel only a short distance before they encounter electrons and get annihilated, producing two γ-ray photons with energy 511 keV travelling in nearly opposite directions. A ring of detectors placed around the patient detects the location of the annihilation and hence of the emitting nucleus on the basis of these directions. A computer system constructs an image of the distribution of the absorbed radio nuclide in the organ of interest. For example, in a normal brain, metabolic activity produces a similar PET pattern in each hemisphere, whereas irregularities in the patterns indicate degeneration of brain tissue.

Magnetic resonance imaging

Magnetic resonance imaging (MRI) is the latest and the most popular imaging technique. This technique is based on the pioneering discoveries of the British physicist, Peter Mansfield, and the American chemist, Paul Lauterbur, in the early 1970s, concerning the use of NMR to visualise different structures. In 2003 they shared the Nobel Prize in medicine for the crucial importance of their discoveries in the development of applications of medical importance. They suggested the use of a non-uniform magnetic field, which means that the resonance frequency for a particular nucleus depends on the position of the nucleus in the field. The introduction of gradients in the magnetic field makes it possible to create two-dimensional images of structures that cannot be visualised by other techniques.

The proton NMR is usually employed because water constitutes about two thirds of the human body weight. By changing the direction of the field gradient, an image that shows the proton density in a thin slice of the body can then be constructed using a computer. The difference in water content among tissues and organs can thus be brought out in the MR scans. In many diseases the pathological process results in change of water content which is reflected in the scan. The advanced computer processing facilitates the construction of a three-dimensional image that reflects the chemical structure of the tissue, including differences in the water content and movement of the water molecules. This results in a very detailed image of the organ under investigation.

Currently, MRI is used for examining all organs of the body. However, it is more useful in the detailed imaging of the brain and the spinal chord, as nearly all brain disorders lead to alteration in water content. The MRI technique has also been used as an important pre-operative tool. Since MRI yields detailed three-dimensional images, it is possible to get distinct information about the location of a lesion. Such information is very valuable before surgery.

A great advantage with MRI is that in contrast to ordinary x-ray or computer tomography (CT) examinations it does not use ionising radiation. MRI is also safer because RF radiation, unlike x-radiation, has too little quantum energy to disrupt chemical bonds and so cannot harm living tissues. MRI also offers higher resolution than x-ray tomography. For these reasons MRI has replaced several invasive models of examination and reduced risk and discomfort for many patients. In 2002, approximately 22,000 MRI cameras were in use worldwide, and more than 60 million MRI examinations were performed.

8.5.4 Harmful Effects of Radiation

The greater knowledge of the nucleus also brought in its wake the fear of nuclear war and dangers associated with the harmful effects of nuclear radiation. Nuclear weapons brought a gruesome but rapid end to World War II. Currently, the threat of the possible use of nuclear weapons has become a central factor in international politics.

Exposure to nuclear radiation is harmful to humans because of its ionising nature. However, the extent of damage depends on the actual absorption of the dose in a particular organ. The amount of energy actually absorbed is measured in grays (Gy), with 1 Gy corresponding to 1 J/kg. For most applications the unit rad (acronym for radiation absorbed dose) is used. One rad is defined as the amount of radiation that deposits 10^{-2} J of energy into 1 kg of absorbing material. Thus, 1 Gy = 100 rad.

However, the rad is not the best unit for measuring the degree of biological damage because the degree of biological damage depends not only on the dose but also on the type of the radiation. For example, a given dose of alpha particles causes about 10 times more biological damage than an equal dose of x-rays. The relative biological effectiveness (RBE) factor for a given type of radiation is defined as the number of rads of x-radiation or gamma radiation that produce the same biological damage as 1 rad of the radiation used. The dose equivalent is obtained by multiplying the absorbed dose (in rads or Gy) by the RBE factor. Thus the RBE factor is 1 for x-rays and electrons, 10 for α-rays and so on (see Ex. 8.13).

EXAMPLE 8.13

Dose equivalent of Co^{60} gamma rays

A cancer is irradiated with 1000 rad of cobalt 60 gamma rays. Find the biologically equivalent dose in rem if RBE factor for cobalt 60 gamma rays is 0.7.

Solution

The biologically equivalent dose is the product of the RBE factor and the dose in rad. So the dose equivalent = $0.7 \times 1000 = 700$ rem.

The SI unit of dose equivalent is the sievert (Sv). A dose equivalent of 1 Sv causes the same biological effect regardless of the radiation type by which it was acquired. Thus, one Sv is the amount of any radiation that has the same biological effect as that produced when 1 kg of body tissue absorbs 1 joule of x-rays or γ-rays. An earlier unit of dose equivalent that is still in common use is rem, standing for roentgen equivalent in man. Here roentgen is the unit of exposure that is defined only for x-rays and gamma rays as the amount of ionisation produced in unit mass of dry air at standard temperature and pressure (STP). The conversion from sievert to rem is 1 Sv = 100 rem.

All of us are exposed to chronic low-level radiation arising from cosmic rays and from radio nuclides naturally present in the environment. This natural background radiation delivers to each of us a dose of about 0.13 rem/year. However, some geographical locations receive much more hazardous radiation than the average chronic doses from natural sources. For example, some of the soils have high thorium content and the water from some wells and mineral springs contains many times the usual radioactivity. The artificially produced radiation, mostly from medical diagnostic x-rays, also adds a considerable amount to this background radiation. The upper limit of radiation dose recommended by the US government (apart from background radiation) is about 0.5 rem/year. As many occupations involve much higher radiation exposures, an upper limit of 5 rem/year has been set for combined whole body exposure. A dose of 400–500 rem results in a mortality rate of about 50%. The most dangerous form of exposure is ingestion or inhalation of radioactive isotopes, especially those elements the body retains and concentrates, such as strontium 90.

It is found that fetuses and infants are much more vulnerable to radiation than are adults. For example, in one study it was found that children whose mothers received pelvic x-rays while pregnant had a 30%–40% increase in the incidence of cancer. We have a limited knowledge of the immediate effects of large radiation doses on humans; it comes mainly from the studies of victims of atomic bomb explosions and of occasional accidents.

The last two chapters have provided us the necessary background to understand how various atoms combine to produce a myriad variety of solids around us. We study this in the next chapter.

9

Structure and Properties of Matter

9.1 THE STORY OF MATTER

In Chapters 3–5 we studied light, a manifestation of energy. Here we study matter. The study of interaction of matter and energy is interspersed throughout this book. Since ancient times human beings have been enquiring into the nature of matter. Matter can be defined as something that occupies space and thus brings in its ambit everything that is found everywhere in the universe—from the farthest star to the smallest speck of dust.

Ancient Greek philosophers held different views about matter. Thales (about 600 BC) believed that all matter was made of water; Empedocles (500 BC) opined that all matter was made of earth, air and fire. With the revival of the atomic theory in 1600, and its wider acceptance by the nineteenth century, a great deal of effort was directed towards understanding how atoms come together to form molecules or solids. The problem was one of finding out those arrangements of atoms which have lower energy compared to the sum of the energies of their non-interacting constituents. Our knowledge of various types of bonding in solids and structures of many molecules has been enriched by the works of Madelung (1910), van der Waal (1837–1923) and many other scientists engaged in this pursuit.

Both scientists and others were fascinated by the periodicity of crystalline solids. Upto the eighteenth century, however, knowledge of the structure of solids remained restricted to mere visual inspection under low magnification. In the nineteenth century, metals were etched and examined with the recently discovered optical microscopes, to reveal their microstructural features. At the same time mineralogists studied the periodic features and symmetries in crystals on the basis of their shapes. However, all these investigations could not reveal anything smaller than the wavelength of light (4000–7000 Å). With the invention of the electron microscope, and the field ion microscope, the resolution increased to the level of substructure, revealing information about fine particles and imperfections in crystals.

A real breakthrough in this research was provided in 1912 by the German physicist Max von Laue, when he suggested that x-rays with wavelength in the range 1–100 Å (discovered in 1895 by Roentgen; see Chapter 7) should be used to reveal more minute details of crystal structure. In the same year Friedrich and Knippling, following the brilliant suggestion of Laue, demonstrated the diffraction of x-rays by a crystal, using the crystal as a three-dimensional grating, in the same way as an optical grating diffracts visible light (see Chapter 4).

Structure determination at deeper levels was possible using the spectroscopic techniques, which revealed the electronic structure. Nuclear magnetic resonance (NMR) and Mossbauer techniques revealed the nuclear structure of solids. The vast experimental data obtained on a variety of solids using these powerful tools of characterisation helped scientists to study the dependence of various solid state properties on their internal structure and invent theories which would explain this dependence. These theories could explain, for example, why steel is strong, why lead bends easily, why quartz is brittle and why talc is soft. However, the most important advantage of these studies was that it gave scientists the *power of prediction!* They could tell beforehand what properties a particular material will have!

The area of physics incorporating systematic studies in this field came to be called *solid state physics*, subsumed subsequently into *condensed matter physics* and later into its modern incarnation—*materials science*. These studies, aimed at improvement in the properties of solids, gave an impetus to the development of new materials having desired properties. The story of development of many useful materials is similar to that of developing better (more intense and coherent) light that culminated in the development of the laser (Chapter 14).

The bonding and structure of solids (Sections 9.2–9.6) are studied in this chapter. This is followed by a discussion of how the bonding and structure of solids decide their properties (Section 9.7). Research in the physics of materials brought forth new knowledge and provided new insights into the structure–property correlation. This helped in the development of novel materials and their applications (Section 9.8).

9.2 BONDING

Though atoms are the basic constituents of all matter, single isolated atoms seldom exist in nature. (Only at very high altitudes where the density of gases is too low do individual atoms exist.) Atoms have the property of joining together to form molecules. This bonding comes about by a rearrangement of motion of the outermost electrons of the combining atoms in such a manner as to lower energy. Electrons thus *glue* the atoms together. Bonding can result in the formation of a molecule as simple as O_2 (diatomic) or as complex as the double helix of deoxyribonucleic acid (DNA)—the basic unit of life—consisting of millions of atoms.

As mentioned above the principle governing the formation of a molecule is that the energy of the molecule should be lower than the total energy of the participating atoms before interaction. If the interaction increases the energy, the atoms will repel one another instead of forming a molecule. When atoms come together, their electron clouds repel each other, whereas nuclei and electron clouds attract each other. For a two-atom system the net potential energy can be written as

$$V(r) = \frac{-\alpha}{r^n} + \frac{\beta}{r^m} \tag{9.1}$$

where r is the distance between the centres of atoms, α and β are the positive constants denoting the strength of attractive and repulsive forces respectively, and n and m are positive powers. The force between two atoms can be written as

$$F(r) = \frac{-dV(r)}{dr} = \frac{-n\alpha}{r^{n+1}} + \frac{m\beta}{r^{m+1}} \tag{9.2}$$

The variation of $F(r)$ and $V(r)$ are shown in Figs. 9.1a and 9.1b respectively. The molecule becomes stable at that value of r for which $V(r)$ is minimum and $F(r)$ is zero. This value (r_0) is called the *bond length* and the corresponding energy V_0, the *bond energy*.

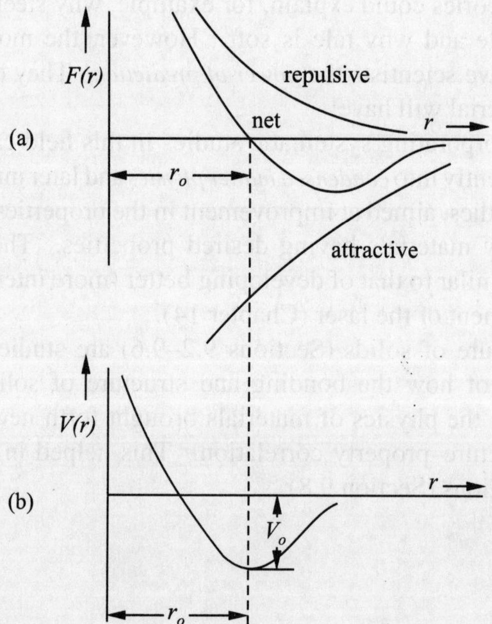

Figure 9.1 Potential energy and force between two atoms as functions of the separation between them

Atoms can join together in one of the following ways resulting in five known types of bonding. But in all these cases the bond formation is dictated by the principle of reduction in the energy of the molecule.

Ionic bond

In the *ionic bond* one or more electrons from one atom may transfer to the other, and the resulting positive and negative ions attract each other. The ionic bond is readily formed between elements that lie at the extreme left and right of the periodic table. Thus alkali halides formed between the alkali metals and halogens have ionic bonds. The ionic bond is non-directional because the transfer of electron(s) results in an inert gas configuration around both the nuclei, meaning that the bonding force is uniform in all directions.

Covalent bond

When one or more pairs of electrons are shared by neighbouring atoms, it is known as the *covalent bond*. It is also known as an *electron pair bond* or a *homopolar bond*. Covalent bonding is common in non-metallic elements, which have half-filled outer shells. Carbon–carbon bonds in living systems are covalent. The covalent bond is generally directional as the overlapping orbitals corresponding to the bonding electrons are directionally oriented.

Metallic bond

In the *metallic bond* individual atoms donate their outermost electrons thus forming an electron gas in which positive nuclei are embedded. It is characteristic of elements left of the fourth column in the periodic table, which have a small number of valence electrons. Metallic bonds are stronger in transition metals because even the *d* electrons add to the electron gas. As the 'electron glue' in this case is mobile, all pervasive and delocalised, the metallic bond is non-directional.

Van der Waals bond

There is neither a transfer nor sharing of electrons. There is only a distortion of the electron clouds of the participating atoms and a weak short-range bond known as *van der Waals bond* is formed. It is also known as a *fluctuating dipole bond* as it results due to momentary charge fluctuations in the charge distribution around an atom, giving rise to a weak temporary dipole within an atom. This dipole–dipole attraction is non-directional. Inert gases, which are chemically inactive solidify at low temperatures due to van der Waals bonding.

Hydrogen bond

If hydrogen is one of the constituent atoms, then, due to the asymmetry of charge distribution the molecule has a permanent dipole moment. The bonding in this case is known as *hydrogen bonding*. The polar character of the hydrogen bonds gives it directionality.

The ionic, covalent and metallic bonds are strong bonds and are called *primary bonds*. About 1–10 eV of energy is required to break such bands. Van der Waals bond and hydrogen bonds, have energies in the range of 0.01–0.5 eV and are examples of *secondary bonds*.

9.3 BONDING IN SOLIDS

In the last section we studied how atoms bond together to form a molecule. In this section we study the role of bonding in solids. A solid state, also called the condensed matter state, is characterised by closely packed aggregates of atoms or molecules which are held together by strong mutual forces of electrostatic attraction. Thus solids have a fixed shape. In the gaseous state atoms do not have any appreciable forces of attraction. Thus gases lack shape, a fixed volume as well as a free surface. The liquid state is an intermediate of these two and has no fixed shape. Liquid molecules can slide over each other to fill any available volume. However a liquid has a free surface.

Solids can be broadly classified as *crystalline* and *amorphous*, depending on the range of the order in the arrangement of the constituent atoms or molecules. If the order extends over only a short-range, the solid is called an amorphous ('without form') solid. A crystalline solid exhibits a long-range order. A crystal has a sharp melting point as the bonds that break during melting have nearly the same strength. An amorphous solid melts over a range of temperatures as the weaker bonds rupture at a lower temperature, softening the solid before melting. Crystals can be classified into the following types depending upon the bonding they exhibit.

9.3.1 Ionic Crystals

Ionic bonds in crystals, like those in molecules, are formed by the transfer of an electron from one atom to the other. As ionic bonds are non-directional and ionic interaction is long-range, each ion tends to surround itself with as many ions of the opposite sign as possible to reduce the potential energy. This tendency promotes the formation of close-packed structures and high coordination number. Two common types of ionic crystal structures—NaCl structure and CsCl structure—are shown in Fig. 9.2. In the first, each atom has six nearest neighbours while in the second this number is eight. These structures are discussed in greater detail in the next section.

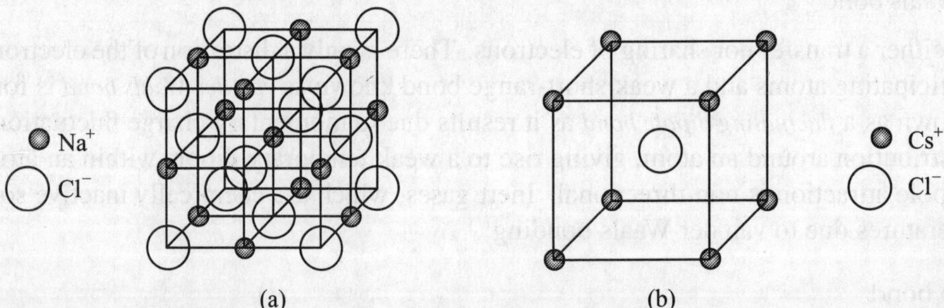

Figure 9.2 (a) NaCl and (b) CsCl structure

The energy required to form an ionic crystal from individual neutral atoms is called the *cohesive energy* of the crystal. It is interesting to note that the cohesive energy of the NaCl crystal is much greater than the bond energy in molecular NaCl, so that the formation of NaCl crystal—and not NaCl molecule—is energetically favoured. Molecules of NaCl can be obtained only by vaporising a NaCl crystal. Ionic solids are hard and have high melting and boiling temperatures and low thermal expansion coefficients due to their high bond strength. They are good thermal and electrical insulators due to the inert gas configurations around both the cations and anions. Ionic solids are brittle, as atoms cannot slip past one another when stressed.

9.3.2 Covalent Crystals

In covalent crystals the sharing of electrons by neutral adjacent atoms causes the cohesion of atoms. This requirement of electron sharing by adjacent atoms naturally restricts the distance between covalently coordinated atoms making the covalent interaction short-range. This combined with the directionality of covalent bonds restricts the number of atoms that can surround any atom. Therefore, the number of nearest neighbours in covalent crystals is low. Solids having covalent bonding usually crystallise in the diamond cubic (DC) structure shown in Fig. 9.3 (discussed in the next section).

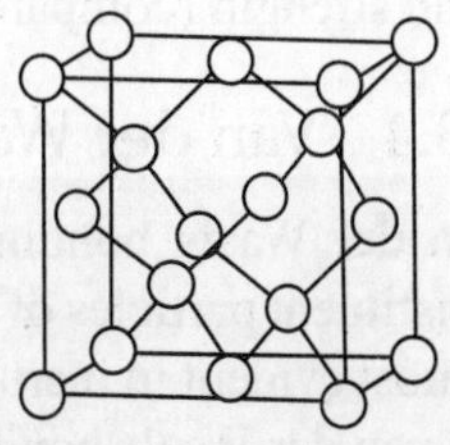

Figure 9.3 DC structure

In DC solids the number of atoms surrounding a particular atom is restricted to four. Owing to this, covalent crystals have open structures as opposed to the close-packed structures in ionic crystals. This also explains the relatively low densities observed in covalent crystals. Covalent crystals are hard, have high melting and boiling temperatures and low thermal expansion coefficients due to their high bond strength. They are good thermal and electrical insulators as electrons stay near the atoms rather than take part in conduction. Their brittleness arises due to the inability of atoms to slip past one another when stressed.

9.3.3 Metallic Crystals

In metallic crystals the interaction between the electron gas and the positive metal ions accounts for the strong cohesive force. As metallic bonds are non-directional, each metal ion tries to surround itself with as many neighbours as possible, for minimising the potential energy. Thus metals crystallise mostly in the hexagonal close-packed structure or the face-centred cubic structure, as shown in Figs 9.4a and 9.4b, for both of which there are 12 surrounding atoms for each atom, or in the body-centred cubic structure (Fig. 9.4c), for which the number of nearest neighbours is 8 (more discussion in the next section).

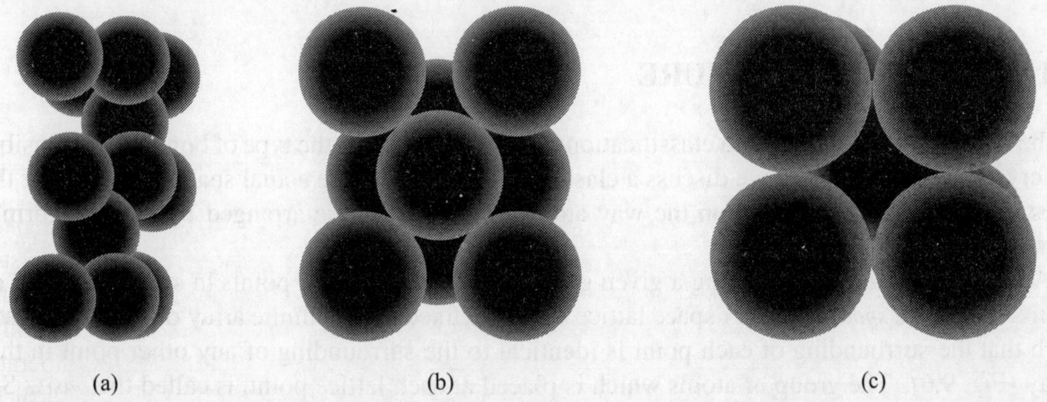

Figure 9.4 Close packed structures: (a) hexagonal close packing, (b) face-centred cubic and (c) body-centred cubic

The presence of copious free electrons in metals makes them good thermal and electrical conductors, and also opaque and lustrous. Metals are soft, ductile and malleable due to their low bond strength (compared to ionic and covalent) and non-directionality.

9.3.4 Van der Waals Crystals

Van der Waals bonding, arising due to temporary fluctuating dipoles, is present between the constituent particles of all crystals but is not appreciable due to its low strength and short-range. It is most evident in inert gas crystals and those molecular solids in which the intermolecular bond is the van der Waals bond (the atoms within the molecule may be held together by a primary bond). Van der Waals crystals are characterised by very low melting and boiling temperatures and low mechanical strength due to the low bond strength.

9.3.5 Hydrogen Bonded Crystals

In some crystals, particularly ice, hydrogen bonding arises due to the permanent dipole that the hydrogen atom forms with one of the other participating atoms. Hydrogen bonds are stronger than van der Waals bonds but weaker than the primary bonds. This accounts for higher boiling temperature of water than other non-metallic hydrides such as H_2Se (–42°C) and H_2S (–60°C).

It is not necessary that only one type of bond should exist in a particular solid. Many solids exhibit mixed bonding. Graphite, for example, has covalent bonding between its constituent carbon atoms (primary, intramolecular bonds) forming planar sheets as shown in Fig. 9.5, but the intersheet bonds (secondary intermolecular bonds) are van der waals bonds. The weak intersheet bonding accounts for its softness and for its suitability as a lubricant.

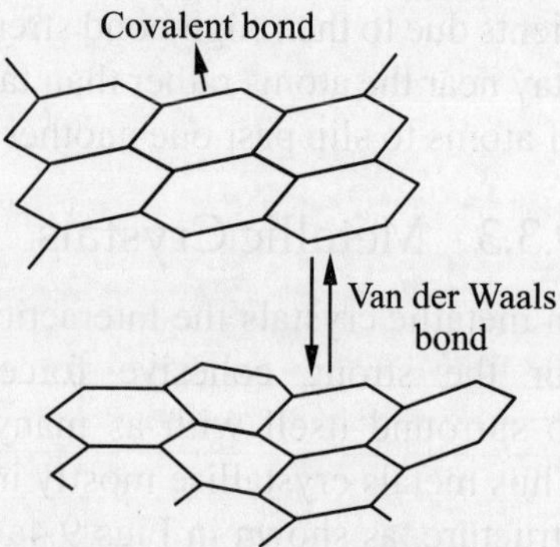

Figure 9.5 Intermolecular and intramolecular bonds in graphite

9.4 CRYSTAL STRUCTURE

In the last section we discussed a classification of crystals based on the type of bonding responsible for crystal formation. Here we discuss a classification based on the actual spatial structure of the crystal. This means we focus on the way atoms or molecules are arranged in space to form a crystal.

A crystal is obtained by placing a given group of atoms at a set of points in space. This set of points is called a *space lattice*. A space lattice can be defined as an infinite array of points in space such that the surrounding of each point is identical to the surrounding of any other point in that array (Fig. 9.6). The group of atoms which is placed at each lattice point, is called the *basis*. So, the equation is:

$$\boxed{\text{a crystal}} = \boxed{\text{a lattice}} + \boxed{\text{a basis}}$$

The central idea is that a crystal displays spatial symmetry. This symmetry is reflected in the lattice of the crystal. The question arises as to how many lattices are possible in all. To answer this question we note that there are two distinct types of symmetries possessed by a lattice: (i) translation symmetry, and (ii) point symmetry. Translation symmetry describes the fact that if all points of a lattice are translated by, that is, shifted by certain vectors, the lattice does not change. In other words translation of each lattice point by any such vector takes it to another lattice point.

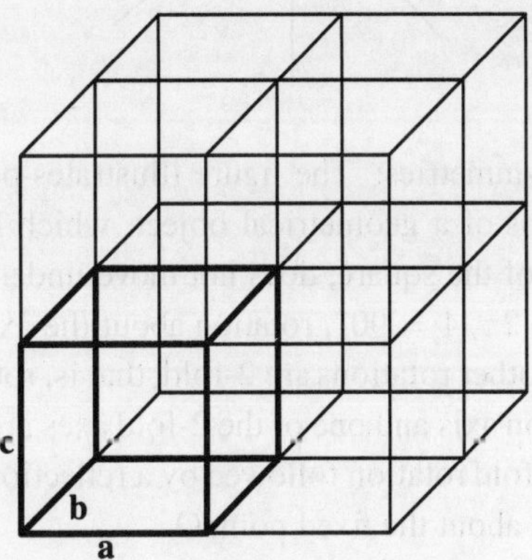

Figure 9.6 A space lattice: **a**, **b** and **c** are basis vectors

The vectors by which the entire lattice can be translated without changing it are of the form

$$\mathbf{r} = l\mathbf{a} + m\mathbf{b} + n\mathbf{c} \tag{9.3}$$

where, $\mathbf{a}, \mathbf{b}, \mathbf{c}$ are called *basis vectors* of the lattice and l, m, n are integers. In fact, the lattice is generated by starting with any one lattice point and translating it by such vectors given by all integer sets (l, m, n).

A *point symmetry* is the invariance of the lattice under operations such as a rotation about an axis, or a reflection in a plane, or a combination of two such operations. Point symmetries are rotation–reflection symmetries. These are illustrated in Fig. 9.7. To grasp the difference between point symmetries and translation symmetries, we note that a molecule possesses point symmetries but no translation symmetry, whereas a crystal possesses both.

The type of lattice is restricted by the requirement that point symmetries allowed should be consistent with translation symmetry. Bravais proved that this means only 7 basic lattices are allowed. These are the *primitive* or *P-lattices* given in Table 9.1. New lattices can be constructed by the process of centring these P-lattices. This is the process of adding more lattice points in highly symmetric positions in such a manner that the point symmetry originally present is not destroyed. This process gives the face-centred, body-centred and base-centred lattices of Table 9.1. Finally we have seven crystal symmetries—the 7 basic P-lattices—possessing a total of 14 distinct lattices. These 14 lattices are called *Bravais lattices*. A crystal can be built by choosing any group of atoms, that is, any basis, and placing it in identical orientation at each of the lattice points of any of these 14 Bravais lattices.

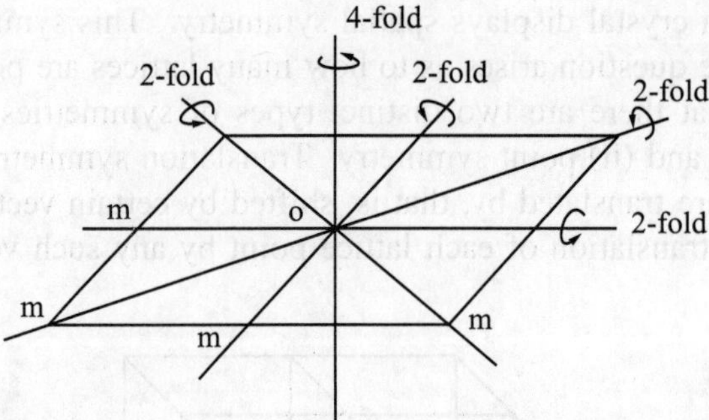

Figure 9.7 Rotation–reflection symmetries: The figure illustrates point symmetries of a square. Point symmetries are symmetry operations of a geometrical object, which leave one point unchanged, that is, one point, in this case the centre O of the square, does not move under any of the operations. The highest symmetry here is the 4-fold, that is, $2\pi/4 = 90°$, rotation about the axis passing through the centre of the square and normal to its plane. The other rotations are 2-fold, that is, rotation by 180°. Mirror reflections in mirrors, containing the 4-fold rotation axis and one of the 2-fold axes are other symmetry operations. These are indicated by the symbol m. A 2-fold rotation followed by a reflection in the mirrors normal to the 2-fold axis gives the operation of inversion about the fixed point O.

Thus, purely from a geometric/mathematical view, an infinite number of crystals is possible. But, very few of them actually exist in nature. This is because, for a particular crystal—containing a particular basis in a particular lattice—to exist, it should also be thermodynamically stable. We find that any given basis can occupy lattice points of only one lattice, or, at most a few lattices, to form a real crystal. Thus NaCl crystallises only in a face-centred cubic structure. Most metals crystallise in either face-centred cubic structure (for example, Ag, Al, Cu, Ni and Pt) or hexagonal structure (for example, Cd, Co, Mg, Zn and Ti). Barium titanate crystallises only in a cubic- or a tetragonal-structure.

We have seen that a space lattice can be constructed if the primitive basis vectors **a**, **b**, and **c**, are known. The relative magnitudes of **a**, **b**, **c** and angles α, β, γ between them fix the Bravais lattice generated (see Table 9.1). For example, if $|\mathbf{a}| = |\mathbf{b}| \neq |\mathbf{c}|$ and $\alpha = \beta = \gamma = \pi/2$, a tetragonal P-lattice is generated; if centring is specified, a centred lattice is generated. A triclinic system is the least symmetric (and most complex), and the cubic system is the most symmetric (and most simple) of all the crystal systems. We will be dealing only with the cubic crystals in the rest of this chapter.

The *unit cell* is a useful concept in understanding the structure of a lattice. A unit cell is a *parallelepiped* generated by the basis vectors **a**, **b** and **c**. For a P-lattice, that is, one without any centring, such a unit cell is called a *primitive unit cell*. The entire P-lattice can be generated by repeating the primitive unit cell by shifting it parallel to **a**, **b**, **c** an infinite number of times so as to fill the whole space. Many times this operation is described as tiling the whole space with the primitive unit cell. The P-lattice gets its letter P from the adjective ‘primitive’ qualifying the unit cell using which we can generate it. Since each of the 8 lattice points in a primitive unit cell is shared by 8 primitive cells meeting at that point, there is one lattice point per primitive unit cell.

Table 9.1 The Bravais lattices

Crystal systems	Primitive basis vectors and angle between them	Unit cell
Cubic 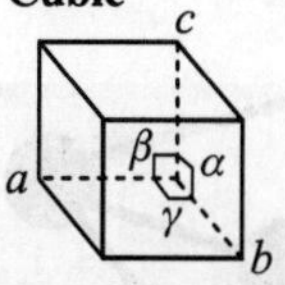	$a = b = c$ $\alpha = \beta = \gamma = 90°$	1. Simple: Lattice points at the eight corners of the unit cell 2. Body-centred: Lattice points at the eight corners and at the body-centre of the unit cell 3. Face-centred: Lattice points at the eight corners and at the six face centres of the unit cell
Tetragonal 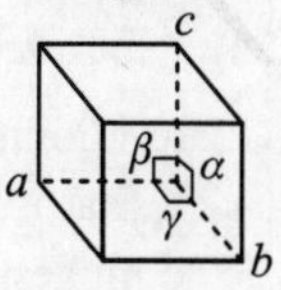	$a = b \neq c$ $\alpha = \beta = \gamma = 90°$	4. Simple: Lattice points at the eight corners of the unit cell 5. Body-centred: Lattice points at the eight corners and at the body-centre of the unit cell
Orthorhombic 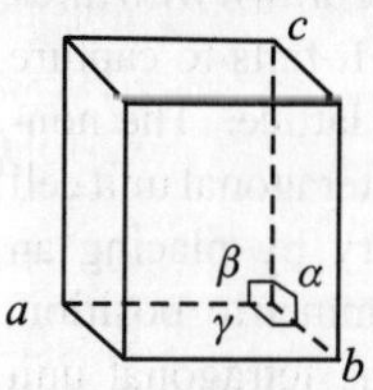	$a \neq b \neq c$ $\alpha = \beta = \gamma = 90°$	6. Simple: Lattice points at the eight corners of the unit cell 7. End-centred (or side-centred or base-centred): Lattice point at the eight corners and at two face centres opposite to each other 8. Body-centred: Lattice points at the eight corners and at the body-centre of the unit cell 9. Face-centred: Lattice points at the eight corners and at the six face centres of the unit cell
Rhombohedral 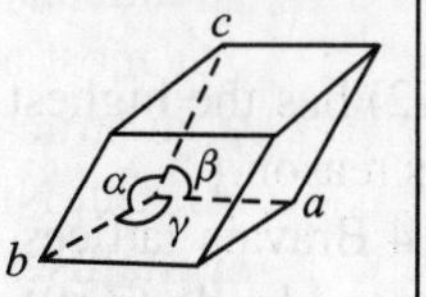	$a = b = c$ $\alpha = \beta = \gamma \neq 90°$	10. Simple: Lattice points at the eight corners of the unit cell
Hexagonal 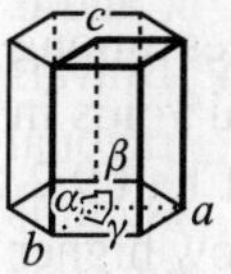	$a = b \neq c$ $\alpha = \beta = 90°$ $\gamma = 120°$	11. Simple: Lattice points at the eight corners of the unit cell outlined by thick lines or points at the twelve corners of the hexagonal prism and at the centres of the two hexagonal prisms and at the centres of the two hexagonal faces
Monoclinic 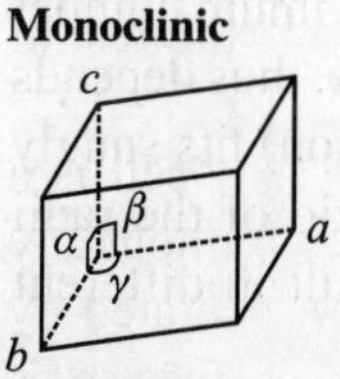	$a \neq b \neq c$ $\alpha = \beta = 90° \neq \gamma$	12. Simple: Lattice points at the eight corners of the unit cell 13. End-centred: Lattice points at the eight corners and at two face centres opposite to each other
Triclinic 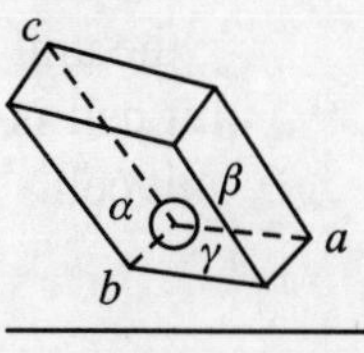	$a \neq b \neq c$ $\alpha \neq \beta \neq \gamma \neq 90°$	14. Simple: Lattice points at the eight corners of the unit cell

This process to generate a lattice can be extended to centred lattices. The only difference is that the unit cell is not a primitive cell. The unit cell of such a lattice, which will give the entire lattice upon repetition, contains more than one lattice point. For example, in the non-primitive unit cell of a body-centred lattice, there are two lattice points per unit cell. This follows because the lattice point at the body-centre belongs entirely to one such cell. The whole idea behind working with non-primitive unit cells is that the full symmetry of the crystal can be captured at the small cost of having to handle a unit cell which is more complex than a primitive unit cell. Figure 9.8 illustrates this idea.

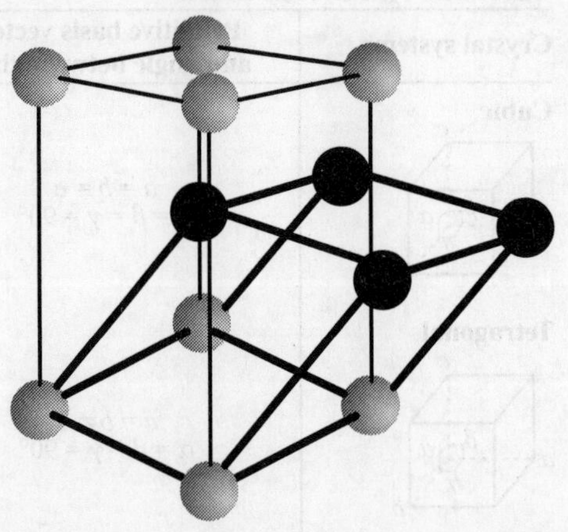

Figure 9.8 The unit cell drawn with thick lines is a primitive cell. It fails to capture the full symmetry of the lattice. The non-primitive body-centred teteragonal unit cell captures higher symmetry by placing an additional atom at a symmetric position, namely, the centre of the tetragonal unit cell

A unit cell is characterised by parameters such as number of atoms per unit cell (Z), coordination number (CN), atomic radius (r), packing fraction (PF), percentage void (VS) and density (ρ). These parameters are defined and calculated for three cubic unit cells, namely, simple cubic (SC), body-centred cubic (BCC) and face-centred cubic (FCC) in Chart 9.1. This chart brings out the correlation between the coordination number and the packing efficiency in different unit cells. FCC with maximum coordination number (12) has the highest packing (74%). [FCC is also referred to as cubic closest packing (CCP) for this reason.]

Though the structure of a particular crystal is identifiable with one of the 14 Bravais lattices, some other structures have also been popularly used. Diamond cubic (DC), zinc blende (ZnS), rocksalt (NaCl) and cesium chloride (CsCl) are four such structures. These structures can be understood better in terms of what are known as *interstitial voids*. They are the void spaces among the atoms or ions, which are packed into a crystal lattice. Two main types of interstitial voids in close-packed structures are *tetrahedral voids* and *octahedral voids*. They are discussed in Chart 9.2.

As noted in the last section, NaCl and CsCl structures which exhibit ionic bonding show higher coordination and better packing as each cation (anion) surrounds itself with the maximum number of anions (cations). The coordination number in ionic crystals, also known as *ligancy*, thus depends on the relative sizes of ions. If the sizes are such that the smaller ion (usually a cation) fits snugly in the voids between the closely packed bigger ions (usually anions) then the ratio of the radii $r_{\text{cation}}/r_{\text{anion}}$ is called the *critical radius ratio*. The ranges of radius ratio that result in different configurations are determined in Chart 9.3.

The determination of the various parameters defined in Chart 9.1 for unit cells of DC, ZnS, NaCl and CsCl structures is given in Chart 9.4.

CHART 9.1 UNIT CELL PARAMETERS FOR CUBIC SYSTEM

Type → Parameter with its definition/description ↓	**Simple cubic (SC)** Atoms at the eight corners of the cube
Number of atoms per unit cell (Z) Considering a 3D structure, contribution of an atom to a unit cell is calculated and multiplied by the number of all such atoms. **Z** →	As the corner is shared by eight unit cells in contact, contribution of corner atom = 1/8 Number of corner atoms = 8 Thus, $Z = 1$ 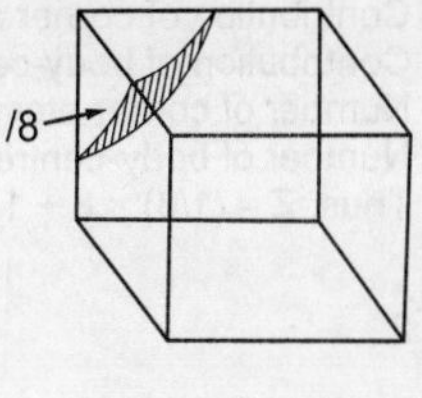 **1**
Coordination number (CN) It is defined as the number of nearest neighbours. A particular atom (lattice point) is considered and its nearest neighbours are counted. The number will be the same for any atom considered as reference. **CN** →	Considering O as the starting atom and counting the nearest neighbours, we get CN = 6 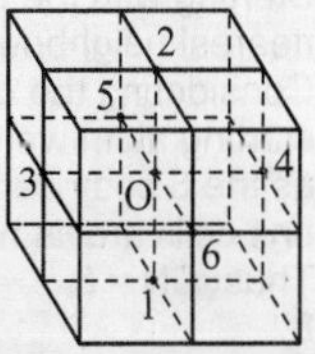 **6**
Atomic radius The radius of an atom is calculated by assuming that the atoms are spherical and that the nearest neighbouring atoms touch each other. r is obtained in terms of the cube edge a. r →	$2r = a$ Thus, $r = a/2$ 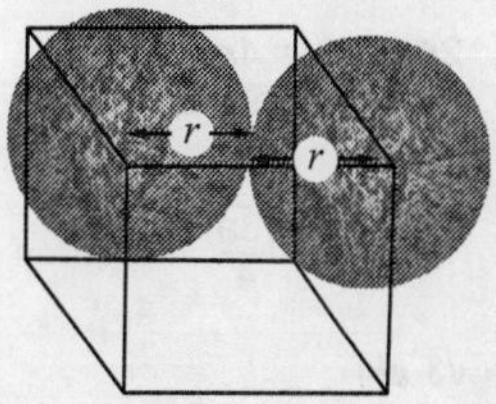 $a/2$
Packing fraction (PF) It is defined as the ratio of volume occupied by the atoms in the unit cell to the volume of the unit cell. It is a measure of how well the atoms are packed in a unit cell. $PF = \dfrac{Z \times (4/3)\,\pi r^3}{a^3}$ (1) **PF** →	Substituting $Z = 1$ and $r = a/2$ in Eq.(1), PF = $\pi/6$ = 0.52 = 52% **52%**
Percentage of void space (VS) The space in the unit cell not utilised by atoms is the void space (VS). VS = (1 – PF) × 100 **VS** →	(1 – 0.52) × 100 = 48 **48%**
Density (ρ) It is the ratio of the mass of atoms in the unit cell to the volume of the unit cell. Thus, $\rho = ZM/N_A a^3$ (2) M = molecular/atomic weight of the material and N_A is the Avogadro number. ρ →	As $Z = 1$ $\rho = \dfrac{M}{N_A\, a^3}$ $\dfrac{M}{N_A\, a^3}$

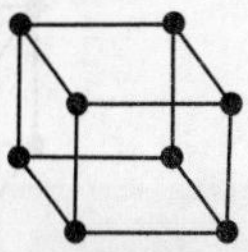

Body-centred cubic (BCC)	**Face-centred cubic (FCC)**
Atoms at the 8 corners + one atom at the body centre	Atoms at 8 corners + atoms at 6 face centres
Contribution of corner atom = 1/8 Contribution of body-centred atom = 1 Number of corner atoms = 8 Number of body-centred atoms = 1 Thus, $Z = (1/8) \times 8 + 1 \times 1 = 2$ **2**	Contribution of corner atom = 1/8 Contribution of face-centred atom = 1/2 Number of corner atoms = 8 Number of face-centred atoms = 6 Thus, $Z = (1/8) \times 8 + 1/2 \times 6 = 4$ **4**
Starting with the body-centred atom the nearest neighbours are 8 corner atoms. Considering the corner atom as the starting atom we get the same result, as the 8 body-centred atoms of adjacent unit cells are its nearest neightbours. Thus CN = 8 **8**	Considering O as the starting atom,four face centred atoms are nearest in one plane and there are three such mutually perpendicular planes. Thus, CN = 12 **12**
$2r + r + r = 4r$ = body diagonal $= \sqrt{3}\, a$ Thus, $r = \frac{\sqrt{3a}}{4}$ $\sqrt{3}\, a/4$	$2r + r + r = 4r$ = face diagonal $= \sqrt{2}\, a$ Thus, $r = \frac{\sqrt{2}a}{4} = \frac{a}{2\sqrt{2}}$ $a/2\sqrt{2}$
Substituting $Z = 2$ and $r = \sqrt{3}\, a/4$ in Eq. (1), $\text{PF} = \frac{\pi\sqrt{3}}{8} = 0.68 = 68\%$ **68%**	Substituting $Z = 4$ and $r = a/2\sqrt{2}$ in Eq. (1), $\text{PF} = p/3\sqrt{2} = 0.74 = 74\%$ **74%**
$(1 - 0.68) \times 100 = 32$ **32%**	$(1 - 0.74) \times 100 = 26$ **26%**
As $Z = 2$ $\rho = 2M/N_A\, a^3$ $\mathbf{2\, M/N_A\, a^3}$	As $Z = 4$ $\rho = 4M/N_A\, a^3$ $\mathbf{4M/N_A\, a^3}$

CHART 9.2 TETRAHEDRAL AND OCTAHEDRAL VOIDS

	Tetrahedral void	Octahedral void
How it is formed	It is the space between three spheres on a close packed plane and a fourth sphere either on the top (Fig. 1) or at the bottom (Fig. 2). It is the cavity between the four spheres. Fig. 1 Fig. 2	It is the space between three spheres on a close packed plane and three more spheres on the top (Fig. 3) [or at the bottom (Fig. 4)] such that the centres of the three spheres are directly over (below) the three triangular valleys, with no sphere over (below) the central valley. Fig. 3 Fig. 4
Why the name	The void gets the name tetrahedral because a regular tetrahedron is formed by joining the centres of the four spheres.	The void gets the name octahedral because a regular octahedron is formed by joining the centres of the six spheres.
Number of void sites in a 3D array of spheres	In a three dimensional array there are two tetrahedral voids and one octahedral void for every sphere (Fig. 5). Fig. 5	
Number of void sites in a FCC unit cell	In the FCC unit cell, the tetrahedral voids are located at quarter of the way and three quarters of the way along the four body diagonals. Thus there are a total of 8 tetrahedral void sites, (that is, 2 sites per atom) in the FCC unit cell (Fig. 6). Fig. 6 1: (1/4, 1/4, 1/4), 2: (1/4, 1/4, 3/4), 3: (1/4, 3/4, 1/4), 4: (3/4, 1/4, 1/4), 5: (1/4, 3/4, 3/4), 6: (3/4, 1/4, 3/4), 7: (3/4, 3/4, 1/4), 8: (3/4, 3/4, 3/4),	In the FCC unit cell, the octrahedral voids are located at the centre of the unit cell and at the centres of the cube edges. Thus there are a total of (¼ × 12) + 1 = 4 octahedral void sites, (that is, one site per atom) in the FCC unit cell (Fig. 7). Fig. 7 1: (1/2, 0, 0), 2: (0, 1/2, 0), 3: (0, 0, 1/2) 4: (1/2, 1, 0), 5: (1/2, 0, 1), 6: (0, 1/2, 1) 7: (1, 1/2, 0), 8: (1, 0, 1/2), 9: (0, 1, 1/2) 10: (1/2, 1, 1), 11: (1, 1/2, 1), 12: (1, 1, 1/2) 13: (1/2, 1/2, 1/2)

9.5 MILLER INDICES

The fact that atoms are periodically arranged in a crystal makes it convenient to use geometrical concepts to designate planes and directions in a crystal. A system devised by Miller is widely used for this purpose. A crystal direction is a vector connecting two lattice points. Thus the position vector (that is, the vector passing through the origin to the lattice point) defined in Eq. 9.3 is a crystal direction. The components l, m and n along the three axes, when reduced to the smallest proportional integers, are called the *indices* and are denoted as $[lmn]$ in the Miller indices notation (the numbers are enclosed in square brackets with no commas between them; however commas are used to separate two digit numbers). Any negative index, say $-l$, is denoted using a bar, that is, as $\bar{l}$. A family of equivalent crystal directions is denoted by $< lmn >$; for example, in a cubic crystal,

$$< 100 >= \{[100], [010], [001], [\bar{1}00], [0\bar{1}0], [00\bar{1}]\}.$$

Examples 9.1 and 9.2 deal with the Miller indices of directions in a cubic lattice.

Miller indices for a crystal plane are determined using the following procedure. The determination for a plane is shown in Fig. 9.9.

1. Find the intercepts made by the plane along the three crystallographic axes: $2a$, $3b$, c Measure these intercepts as multiples of the fundamental vectors: 2, 3, 1.
2. Take reciprocals of these numbers: $\frac{1}{2}, \frac{1}{3}, 1$.
3. Convert to the smallest integers in the same ratio by multiplying each fraction by the LCM of the denominators: 3, 2, 6.
4. Enclose these numbers in parentheses without commas: (326).

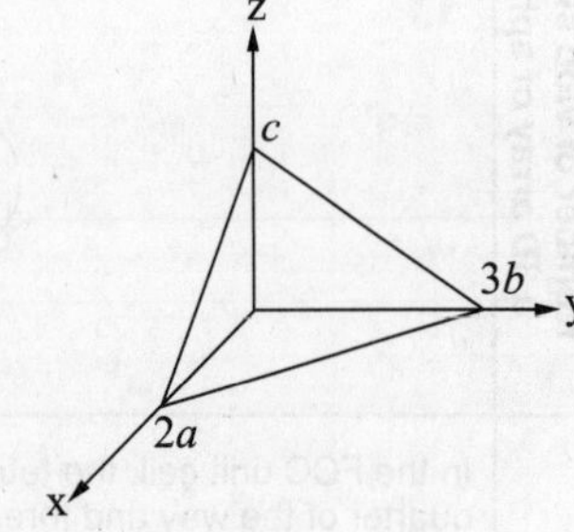

Figure 9.9 Miller indices for crystal planes

The family of planes is denoted by $\{hkl\}$. For example, in a cubic crystal,

$$\{110\} = \{(110), (101), (011), (\bar{1}10), (\bar{1}01), (0\bar{1}1), (1\bar{1}0), (10\bar{1}), (01\bar{1}), (\bar{1}\bar{1}0), (\bar{1}0\bar{1}), (0\bar{1}\bar{1})\}.$$

Examples 9.3 and 9.4 deal with the Miller indices of planes in a cubic lattice.

The above procedure can be summarised to define Miller indices as the *reciprocals of the intercepts made by a plane on the crystallographic axes when reduced to the smallest integers.* It should be understood here that (hkl) does not denote a single plane but a set of *equidistant parallel planes*. Moreover, these are crystal planes or lattice planes, that is, each plane contains lattice points. For a primitive cell there are no lattice points between any two adjacent planes in any one (hkl) set.

It can be shown that the plane nearest to the origin belonging to the (hkl) set makes intercepts a/h, b/k and c/l, respectively, on the a, b and c axes. The planes of the (hkl) set thus cut the a axis into h parts, the b axis into k parts and the c axis into l parts (see Ex 9.5).

CHART 9.3 LIGANCY AND RADIUS RATIO

In finding the critical radius ratio $r_{cation}/r_{anion} = r/R$ for a particular ligancy, we should consider the critical configuration as shown in Fig. 1 where the anions just touch each other and contact the central cation. If the anions do not touch the central cation (Fig. 2) the configuration becomes unstable because the central cation can rattle around in its cage of anions. If all the anions touch the central cation but do not touch one another (Fig. 3), the configuration is stable but not critical.

Fig. 1

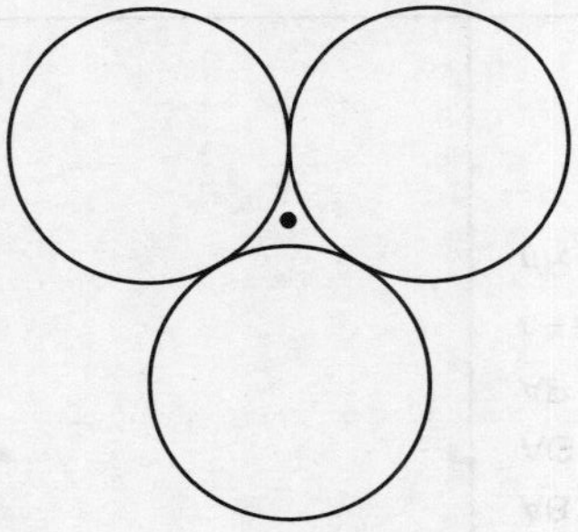

Fig. 2

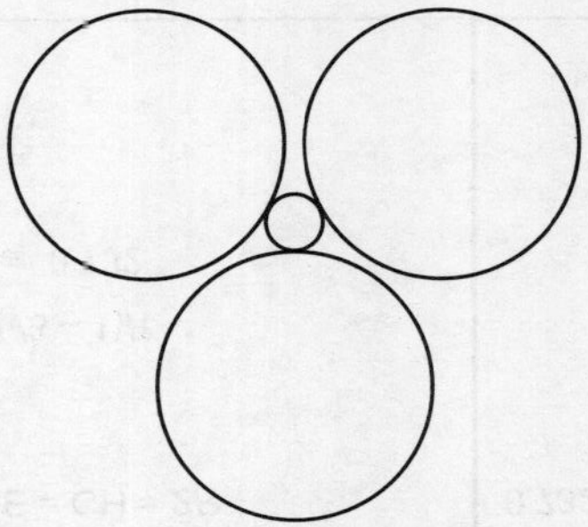

Fig. 3

Disposition of anions about central cation	Ligancy	Figure for critical configuration	Derivation of critical radius ratio	Range of radius ratio	Configuration
Corners of triangle	3	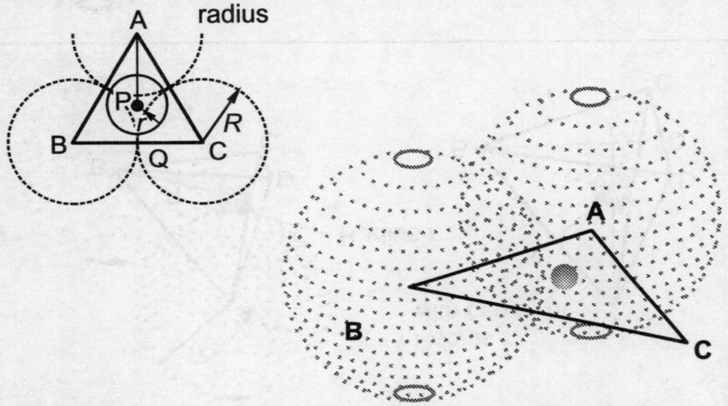	AB = BC = CA = 2BQ = 2R AQ = $\sqrt{3}R$ AP = 2 $\sqrt{3}R/3$ r = AP – R = $R(2-\sqrt{3})/\sqrt{3}$ $r/R = (2-\sqrt{3})/\sqrt{3} \equiv 0.155$	0.155 – 0.225	Triangular

Corners of tetrahedron	4	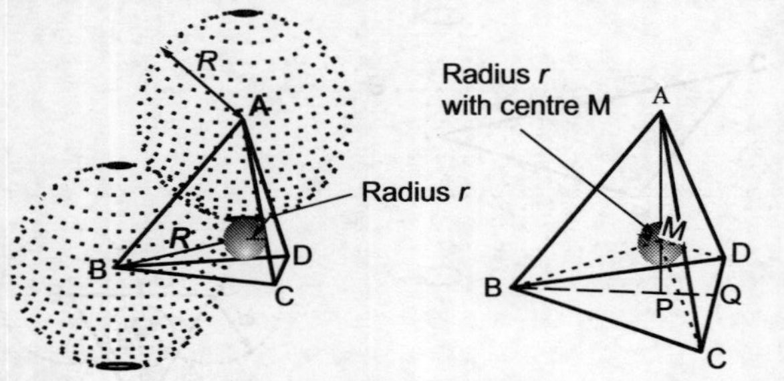 Radius r with centre M; Radius r	$AB = BC = CD = DA = BD = CA = 2R$ $CQ = DQ = R$ $AM^2 = BM^2 = BP^2 + MP^2$ $AP = \sqrt{(4R^2 - 4R^2/3)} = 2\sqrt{2}R/\sqrt{3}$ $= BP^2 + (AP - AM)^2$ $AM = BM = CM = DM$ $AM = (BP^2 + AP^2)/2AB$ $= (4R^2/3 + 8R^2/3)/(22\sqrt{2}R/\sqrt{3}) = 3R/\sqrt{6}$ $r = AM - R = 3R/\sqrt{6} - R$ $r/R = (\sqrt{3/2}) - 1 \cong 0.225$	0.225 – 0.414	Tetrahedral
Corners of octahedron	6	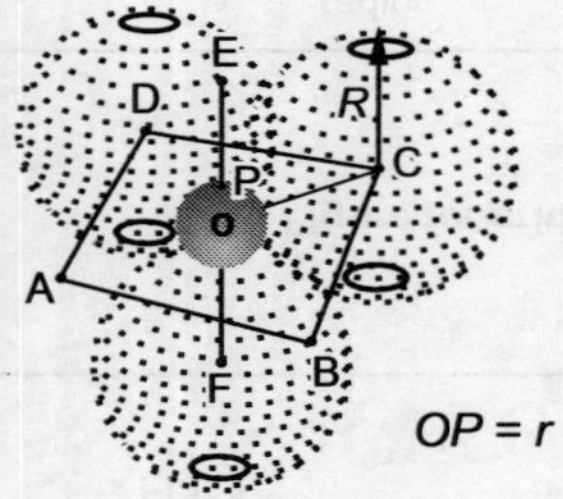 $OP = r$	Octahedral sites: A, B, C, D, E, F $AB = BC = EA = FA = 2R$ $AO = DO = R + R = AC/2$ $= \sqrt{(8R^2)/2} = \sqrt{2}R$ $r/R = \sqrt{2} - 1 \cong 0.414$	0.414 – 0.732	Octahedral
Corners of cube	8	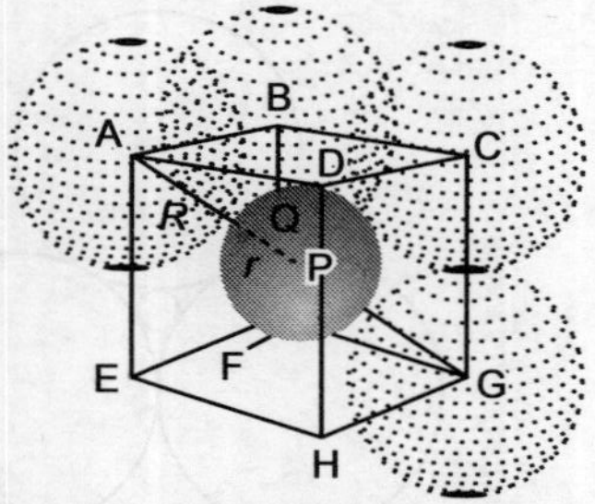	$AB = BC = AE = GH = 2R$ $AG = 2\sqrt{3}R$ $AP = \sqrt{3}R$ $r = AP - R = (\sqrt{3} - 1)R$ $r/R = \sqrt{3} - 1 \cong 0.732$	0.732 – 1.0	Cubic

The correlation between ligancy and radius ratio derived in the above table agrees well in case of ionic crystals. For example, NaCl, with $r_{Na}/R_{Cl} = 0.55$ and MgO, with $r_{Mg}/R_O = 0.59$ exhibit octahedral coordination, whereas CsCl, with $r_{Cs}/R_{Cl} = 0.94$, exhibits cubic coordination. However, in case of covalent crystals the above considerations may not lead to correct prediction of ligancy. For example, ZnS having $r_{Zn}^{2+}/R_S^{2-} = 0.48$, is predicted to be octahedral from the above table but shows tetrahedral coordination. Silica or silicates that exhibit Si–O bond with 50% ionic and 50% covalent character show tetrahedral geometry. This observation agrees with the prediction of the above table because r/R for Si–O is 0.29.

CHART 9.4 SOME COMMONLY USED CRYSTAL STRUCTURES

Structure	Unit cell		
	Spece lattice	**Basis**	**Configuration**
Diamond cubic (DC) ○ C atom	**FCC:** The structure can also be looked upon as two interpenetrating sublattices, one sublattice having its origin at (O, O, O) and the other at (¼, ¼, ¼), that is, one quarter of the way along the body diagonal.	**Two carbon atoms:** The basis has two atoms: one at a lattice site, say at one corner of the unit cell and the other atom at a site ¼ of the way along the body diagonal which is not a lattice site.	**Tetrahedral:** Half of the tetrahedral void sites are occupied by the carbon atoms.
Zinc blende (ZnS) ○ S atom ● Zn atom	**FCC:** The structure can also be looked upon as two interpenetrating FCC sublattices, one sublattice having its origin at (0, 0, 0) and the other at (¼, ¼, ¼), that is, one quarter of the way along the body diagonal.	**ZnS molecule:** The basis has two atoms: one at a lattice site, say at one corner of the unit cell and the other atom (either S or Zn) at a site ¼ of the way along the body diagonal which is not a lattice site. In the unit cell shown in the figure sulphur atoms occupy the lattice sites, whereas zinc atoms are not at lattice sites.	**Tetrahedral:** Half of the tetrahedral void sites are occupied by the Zn atoms.
Rocksalt (NaCl) ○ Cl^- ● Na^+	**FCC:** The structure can also be looked upon as two interpenetrating FCC sublattices, one sublattice having its origin at (0, 0, 0) and the other at (½, 0, 0) [or (0, ½,0) or (0, 0, ½)], that is, half of the way along the cube edge.	**NaCl molecule:** The basis has two ions: one (either Na^+ or Cl^-) at a lattice site, say at one corner of the unit cell and the other ion (either Na^+ or Cl^-) at a site ½ the way along the cube edge which is not a lattice site. In the unit cell shown in the figure Cl^- ions occupy the lattice sites whereas Na^+ ions are not at lattice sites.	**Octahedral:** All the octahedral void sites are occupied by the Na^+ ions.
○ Cl^- ● Cs^+	**SC:** The structure can be looked upon as two interpenetrating SC sublattices, one sublattice having its origin at (0, 0, 0) and the other at (½, ½, ½), that is, half of the way along the body diagonal.	**CsCl molecule:** The basis has two ions: one (either Cs^+ or Cl^-) at a lattice site say at one corner of the unit cell and the other ion (either Cs^+ or Cl^-) at a site ½ the way along the body diagonal which is not a lattice site. In the unit cell shown in the figure Cl^- ions occupy the lattice sites whereas Cs^+ ion is not at a lattice site.	**Cubic:** A cubic void site is occupied by the Cs^+ ion

Unit cell		
Number of atoms per unit cell (Z)		**Coordination number (CN)**
Contribution of corner atom = 1/8, contribution of face-centred atom = 1/2, contribution of atoms at 1/4 of the body diagonal = 1, number of corner atoms = 8, number of face-centred atoms = 6, number of atoms along body diagonals = 4. Thus Z = (1/8) × 8 + (1/2) × 6 + 4 = 8. Thus there are 8 atoms per unit cell.		CN = 4, as each C atom is surrounded by 4 nearest C atoms as shown in the figure. Distance between nearest neighbours $= \sqrt{3}\frac{a}{4}$
Contribution of corner S atom = 1/8, contribution of face-centred S atom = 1/2, contribution of Zn atom at 1/4 of the body diagonal = 1. Thus number of S atoms = (1/8) × 8 + (1/2) × 6 = 4 and number of Zn atoms 1 × 4 = 4. Thus number of ZnS molecules per unit cell = 4.		CN = 4, as each Zn atom is surrounded by 4 nearest S atoms as shown in the figure. Similarly each S atom is surrounded by 4 nearest Zn atoms. Distance between nearest neighbours $= \sqrt{3}\frac{a}{4}$
Contribution of corner Cl^- ion = 1/8, contribution of face-centred Cl^- ion = 1/2, contribution of Na^+ ion at the centre of the cube edge = 1/4, contribution of Na^+ ion located at the centre of the unit cell = 1. Thus number of Cl^- ions = (1/8) × 8 + (1/2) × 6 = 4 and number of Na^+ ions = (1/4) × 12 + 1 × 1 = 4. Thus number of NaCl molecules per unit cell = 4.		CN = 6, as each Na^+ ion is surrounded by 6 nearest Cl^- ions and each Cl^- ion is surrounded by 6 nearest Na^+ ions as shown in the figure. Distance between nearest neighbours $= \frac{a}{\sqrt{2}}$
Contribution of corner Cl^- ion = 1/8, contribution of Cs^+ ion at the body centre of the unit cell = 1. Thus number of Cl^- ions = (1/8) × 8 = 1 and number of Cs^+ ion = 1 × 1 = 1. Thus number of CsCl molecules per unit cell = 1.		CN = 8, as each Cs^+ ion is surrounded by 8 nearest Cl^- ions and each Cl^- ion is surrounded by 8 nearest Cs^+ ions as shown in the figure. Distance between nearest neighbours $= \sqrt{3}\frac{a}{2}$

CHART 9.4 (Continued)

Structure	Unit cell: Atomic/Ionic radii, packing fraction, percentage Void	Density (ρ)
Diamond cubic (DC) ○ C atom	If a = side of cube, $2r = \sqrt{3a/4}$ Thus, $r = \sqrt{3a/8}$ (1) As Z = 8 and $r = \sqrt{3a/8}$, Packing Fraction, PF = $[Z.(4/3)\pi r^3]/a^3 = \sqrt{3\pi/16} = 0.34 = 34\%$. Percentage of void space, VS = (100 – 34) % = 66 %.	As Z = 8, $\rho = ZM/N_A a^3$. $= 8\,M/N_A\,a^3$ where M is the atomic weight of carbon.
Zinc blende (ZnS) ○ S atom ● Zn atom	$r_S + r_{Zn} = \sqrt{3a/4}$ (2) The packing fraction and percentage of void space depends on the relative sizes of Zn and S atoms. PF = $[4 \times (4/3)\pi r^3_{Zn^{2+}} + 4 \times (4/3)\pi r^3_{S_2^-}]/a^3$ (3) Using the ionic radii of Zn^{2+} and S^{2-} as 0.6 Å and 1.74 Å respectively, we get, a = 5.4 Å. Thus, PF = 58.4 % and VS = 41.6 %	$r = (4M_S + 4M_{Zn})/N_A a^3$ $= 4M_{ZnS}/N_A a^3$ where M_S is the atomic weight of sulphur, M_{Zn} is the atomic weight of zinc and M_{ZnS} is the molecular weight of ZnS
Rocksalt (NaCl) ○ Cl^- ● Na^+	$r_{Cl} + r_{Na} = a/2$ (4) The packing fraction and percentage of void space depends on the relative sizes of Na^+ and Cl^- ions. PF = $[4 \times (4/3)\pi r^3_{Cl^-} + 4 \times (4/3)\pi r^3_{Na^+}]/a^3$ (5) Using the ionic radii of Na^+ and Cl^- as 1.02 Å and 1.81 Å respectively, we get, a = 5.66 Å. Thus, PF = 64.6 % and VS = 35.4 % For critical configuration, $r_{Na} = 0.414\ r_{Cl}$ and hence, a = 5.12 Å, PF = 79.3 % and VS = 20.7 %.	$= (4M_{Cl} + 4M_{Na})/N_A a^3$ $= 4M_{NaCl}/N_A a^3$, where, M_{Cl} is the atomic weight of Cl, M_{Na} is the atomic weight of Na and M_{NaCl} is the molecular weight of NaCl.
○ Cl^- ● Cs^+	$r_{Cl} + r_{Cs} = \sqrt{3a/2}$ (6) The packing fraction and percentage of void space depends on the relative sizes of Cs^+ and Cl^- ions. $PF = [1 \times (4/3)r^3_{Cl^-} + 1 \times (4/3)r^3_{Cs^+}]/a^3$ (7) Using the ionic radii of Cs^+ and Cl^- as 1.7 Å and 1.81 Å respectively, we get, a = 4.05 Å. Thus, PF = 68.4% and void = 31.6% For critical configuration, $r_{Cs} = 0.732\ r_{Cl}$ and hence, a = 3.62 Å, PF = 72.9 % and VS = 27.1 %.	$r = (4M_{Cl} + 4M_{Cs})/N_A a^3$ $= 4\,M_{CsCl}/N_A a^3$, where, M_{Cl} is the atomic weight of Cl, M_{Cs} is the atomic weight of Cs and M_{CsCl} is the molecular weight of CsCl.

If the signs of all indices in a crystal direction are changed, we get the antiparallel or opposite direction. Doing this for a plane results in the indices of a plane located at the same distance on the other side of the origin. Thus (hkl) denotes a set of planes with intercepts: ... $(-3a/h, -3b/k, -3c/l)$, $(-2a/h, -2b/k, -2c/l)$, $(-a/h, -b/k, -c/l)$, $(0, 0, 0)$, $(a/h, b/k, c/l)$, $(2a/h, 2b/k, 2c/l)$, $(3a/h, 3b/k, 3c/l)$, ..., that is, $n(a/h, b/k, c/l)$, n being an integer.

For a cubic crystal it can be shown that $[hkl]$ is perpendicular to (hkl) (see Ex. 9.6).

EXAMPLE 9.1
Miller indices of directions

(a) Find the Miller indices of the direction shown in the unit cube beside.

(b) Draw the direction [213] in a cubic unit cell.

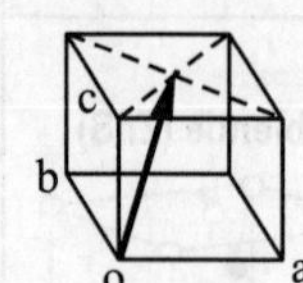

Solution

(a) The components of the direction shown are 1/2, 1/2, 1. Multiplying by 2, we get the smallest proportional integers as 1, 1, 2.

Thus the miller indices of the direction are [112].

(b) The direction [213] has the components 1/3, 1/6, 1/2 along the three axes.

Thus the direction can be shown as seen beside.

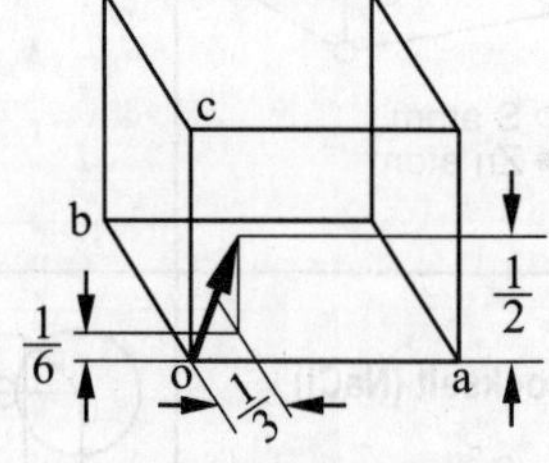

EXAMPLE 9.2
Miller indices for edges, face diagonals and body diagonals in a cubic unit cell

Show the family of crystal directions represented by cube edges, face diagonals and body diagonals in a cubic unit cell.

Solution

Cube edge: $\langle 100 \rangle = \{[100], [010], [001], [\bar{1}00], [0\bar{1}0], [00\bar{1}]\}$

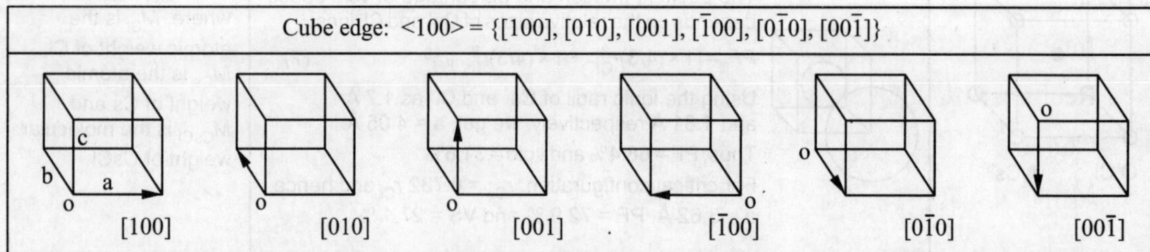

Face diagonal: $\langle 110 \rangle = \{[110], [101], [011], [\bar{1}10], [\bar{1}01], [0\bar{1}1], [1\bar{1}0], [10\bar{1}], [01\bar{1}], [\bar{1}\bar{1}0], [\bar{1}0\bar{1}], [0\bar{1}\bar{1}]\}$

[110] [101] [011] $[\bar{1}10]$ $[\bar{1}01]$ $[0\bar{1}1]$

$[1\bar{1}0]$ $[10\bar{1}]$ $[01\bar{1}]$ $[\bar{1}\bar{1}0]$ $[\bar{1}0\bar{1}]$ $[0\bar{1}\bar{1}]$

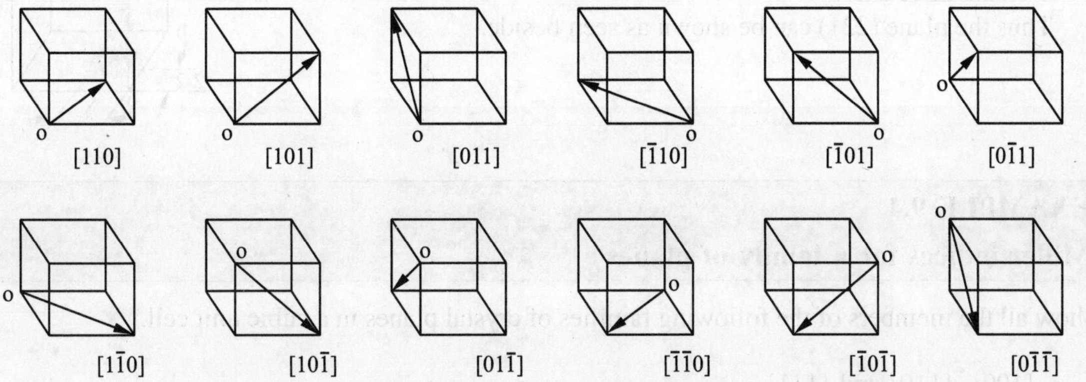

Body diagonal: $\langle 111 \rangle = \{[111], [\bar{1}11], [1\bar{1}1], [11\bar{1}], [\bar{1}\bar{1}1], [\bar{1}1\bar{1}], [1\bar{1}\bar{1}]\ [\bar{1}\bar{1}\bar{1}]\}$

[111] $[\bar{1}11]$ $[1\bar{1}1]$ $[11\bar{1}]$

$[\bar{1}\bar{1}1]$ $[\bar{1}1\bar{1}]$ $[1\bar{1}\bar{1}]$ $[\bar{1}\bar{1}\bar{1}]$

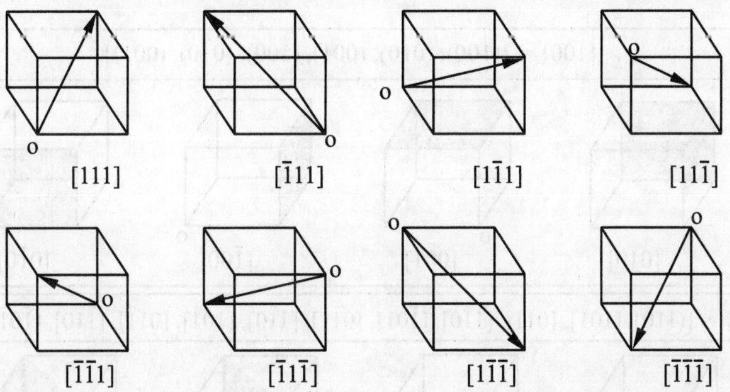

EXAMPLE 9.3

Miller indices of planes

(a) Find the Miller indices of the plane shown in the unit cube.

(b) Draw the plane (221) in a cubic unit cell.

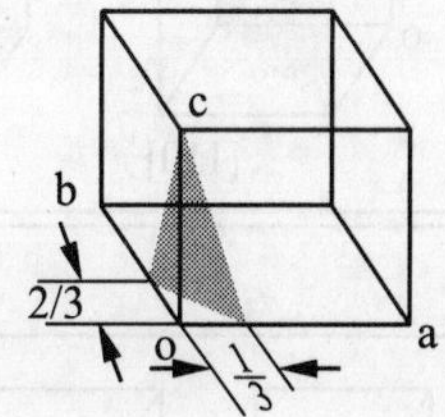

Solution

(a) Using the method explained in Fig. 9.9, we proceed as follows.

Intercepts : 1/3, 2/3, 1

Reciprocals : 3, 3/2, 1

Conversion into smallest integers : 6, 3, 2

Miller indices (632).

(b) The plane (221) makes intercepts 1/2, 1/2, 1 on the three axes.
Thus the plane (221) can be shown as seen beside.

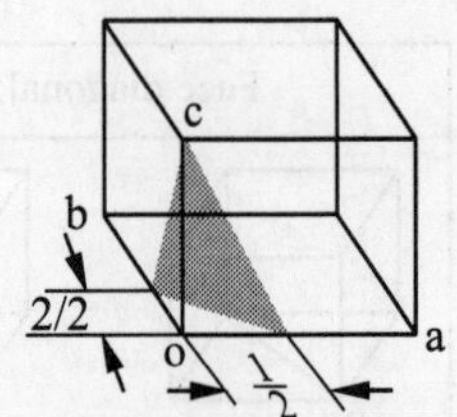

EXAMPLE 9.4
Miller indices for a family of planes

Show all the members of the following families of crystal planes in a cubic unit cell.

{100}, {110} and {111}

Solution

$\{100\} = \{(100), (010), (001), (\bar{1}00), (0\bar{1}0), (00\bar{1})\}$

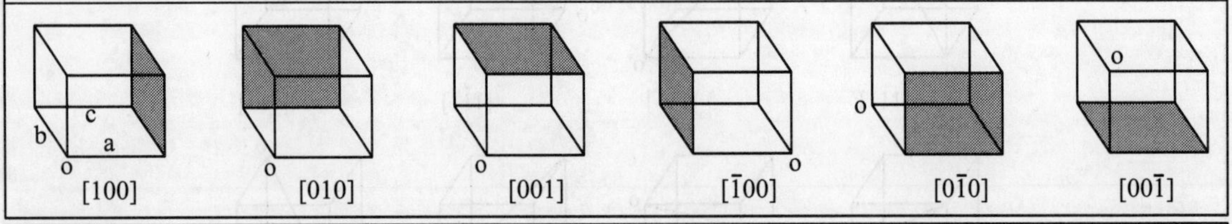

[100] [010] [001] $[\bar{1}00]$ $[0\bar{1}0]$ $[00\bar{1}]$

$\{110\} = \{(110), [101], [011], [\bar{1}10], [\bar{1}01], [0\bar{1}1], [1\bar{1}0], [10\bar{1}], [01\bar{1}], [\bar{1}\bar{1}0], [\bar{1}0\bar{1}], [0\bar{1}\bar{1}]\}$

[110] [101] [011] $[\bar{1}10]$ $[\bar{1}01]$ $[0\bar{1}1]$

$[1\bar{1}0]$ $[10\bar{1}]$ $[01\bar{1}]$ $[\bar{1}\bar{1}0]$ $[\bar{1}0\bar{1}]$ $[0\bar{1}\bar{1}]$

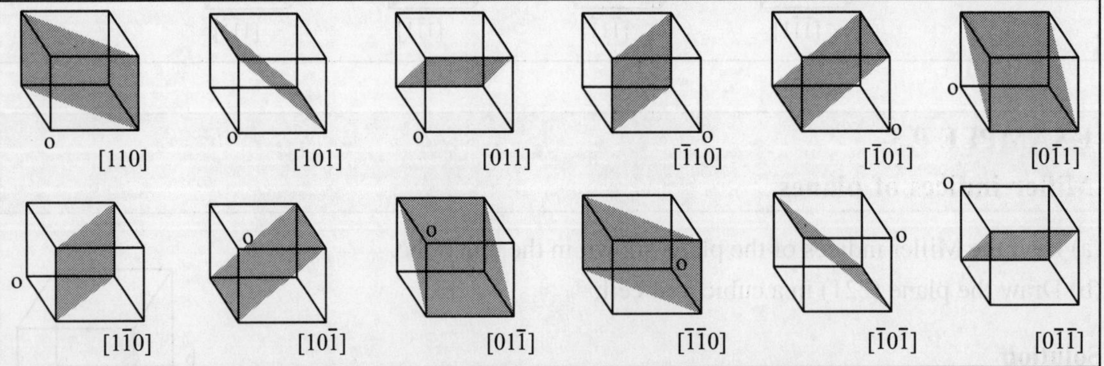

$\{111\} = \{(111), (\bar{1}11), (1\bar{1}1), (11\bar{1}), (\bar{1}\bar{1}1), (\bar{1}1\bar{1}), (1\bar{1}\bar{1}), (\bar{1}\bar{1}\bar{1})\}$

[111] $[\bar{1}11]$ $[1\bar{1}1]$ $[11\bar{1}]$ $[\bar{1}\bar{1}1]$ $[\bar{1}1\bar{1}]$ $[1\bar{1}\bar{1}]$ $[\bar{1}\bar{1}\bar{1}]$

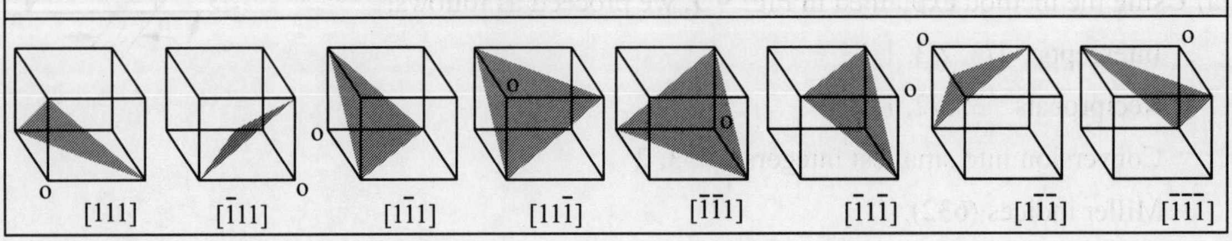

EXAMPLE 9.5
Plane nearest to the origin in (hkl) set

Show that the smallest intercepts made by a crystal plane of the (*hkl*) set on the a, b and c axes are $a/h, b/k$ and c/l respectively. Illustrate using the example of Fig. 9.9.

Solution

Consider any one plane belonging to the (*hkl*) set. Suppose it makes intercepts pa, qb and rc, on the a, b and c axes respectively. Then, we can write

$$h = \frac{\text{LCM}}{p},\ k = \frac{\text{LCM}}{q},\ l = \frac{\text{LCM}}{r} \tag{9.4}$$

where LCM is the lowest common multiple of p, q and r. By definition h, k, l are coprime, that is, they do not have a common integer factor.

Any other plane of the (*hkl*) set makes intercepts on the a, b and c axes bearing the same proportion as $pa : qb : rc$. To find the smallest intercepts we consider the intercepts αpa, αqb and αrc for some value of α and find the smallest value of α. The equation of such a plane is

$$\frac{x}{\alpha pa} + \frac{y}{\alpha qb} + \frac{z}{\alpha rc} = 1$$

Now, for this plane to be a crystal plane of a primitive cell, (da, eb, fc) should satisfy the above equation for some integers d, e, f. Thus we want the smallest value of α such that the equation

$$\frac{d}{\alpha p} + \frac{e}{\alpha q} + \frac{f}{\alpha r} = 1$$

does have some integer solutions for d, e, f.

Using Eq. 9.4,

$$\frac{dh}{\alpha \times \text{LCM}} + \frac{ek}{\alpha \times \text{LCM}} + \frac{fl}{\alpha \times \text{LCM}} = 1$$

or $\quad dh + ek + fl = \alpha \times \text{LCM}$

As the left-hand side of the above equation is necessarily an integer, it is obvious that the smallest value of α which will satisfy this equation is (1/LCM). For any value of α smaller than (1/LCM), there is no integer solution for d, e and f.

Thus, the smallest positive intercepts are $\alpha pa = \dfrac{pa}{\text{LCM}} = \dfrac{a}{h}$, $\alpha qb = \dfrac{qb}{\text{LCM}} = \dfrac{b}{k}$

and $\alpha rc = \dfrac{rc}{\text{LCM}} = \dfrac{c}{l}$

For example, in Fig. 9.9, (*hkl*) = (326). Thus the smallest intercepts are $a/3, b/2$, and $c/6$. Figure 9.10 shows the plane nearest to the origin with these intercepts and the planes with higher

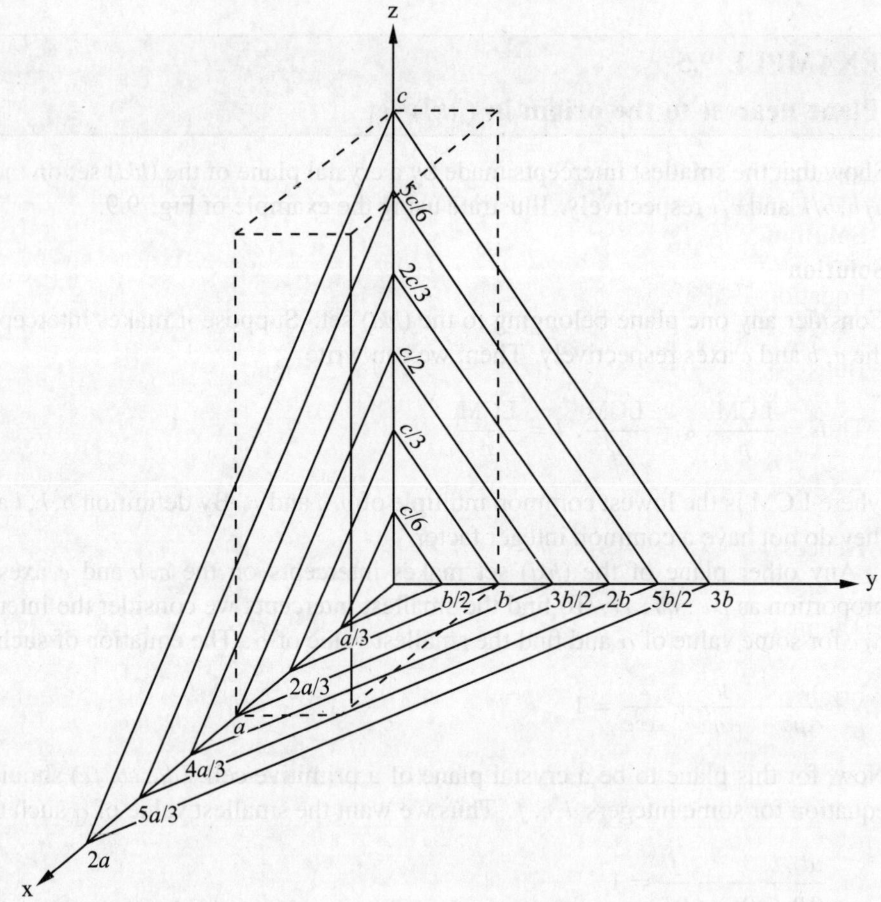

Figure 9.10 Different planes belonging to the (326) set. Note that all these are crystal planes. In this figure some of the planes do not seem to contain lattice points. These planes do not contain lattice points in the unit cell shown but contain infinite number of lattice points if the whole 3D lattice is considered. For example, for the first plane we have $x/(a/3) + y/(b/2) + z/(c/6) = 1$ or $3x/a + 2y/b + 6z/c = 1$ which is satisfied by points $(-a, -b, c)$, $(-3a, -b, 2c)$, $(5a, 2b, -3c)$, and so on, which are all lattice points.

intercepts. In Fig. 9.9 we started with intercepts $2a$, $3b$ and c and obtained the indices (326). Here we find that (326) set means an infinite number of planes on both sides of the origin corresponding to intercepts $n(a/3, b/2, c/6)$, where $n = \cdots, -3, -2, -1, 0, 1, 2, 3, \cdots$.

It can be seen that the plane nearest to the origin has intercepts $a/3$, $b/2$ and $c/6$ on the a, b, c axes respectively. Also, it can be seen that the planes of the (326) set cut the a a-xis into 3 parts, b-axis into 2 parts and c-axis into 6 parts. Thus the planes in the (hkl) set cut the a-axis into h parts, b-axis into k parts and c-axis into l parts. It can be seen from Fig. 9.10 that the number of planes in the (326) set between the origin and the first plane with all integer intercepts (2a, 3b and c) is equal to 6, the LCM of 2, 3 and 6. Thus, the number of planes in the (hkl) set between the origin and the first plane with all integer intercepts is equal to the LCM of h, k and l.

EXAMPLE 9.6
[hkl] is perpendicular to (hkl)

Show that $[hkl]$ is perpendicular to (hkl) in a cubic lattice.

Solution

Equation of a (hkl) plane is given by $\frac{x}{a/h} + \frac{y}{b/k} + \frac{z}{c/l} = 1$

But $\quad a = b = c$ in a cubic unit cell

Thus, $hx + ky + lz - a = 0$

A vector perpendicular to this plane is $r = ha + kb + lc$, which has the indices $[hkl]$.

Hence, $[hkl]$ is perpendicular to (hkl).

The interplanar spacing, d_{hkl} is the distance between two adjacent (hkl) planes. It is the distance between the origin and the nearest (hkl) plane. It is just the length of the normal drawn from the origin to the nearest (hkl) plane. Now the equation of the plane nearest to the origin can be written as

$$\frac{x}{a/h} + \frac{y}{b/k} + \frac{z}{c/l} = 1$$

or $\left(\frac{h}{a}\right)x + \left(\frac{k}{b}\right)y + \left(\frac{l}{c}\right)z - 1 = 0$

The length of the normal from the origin to this plane is given by

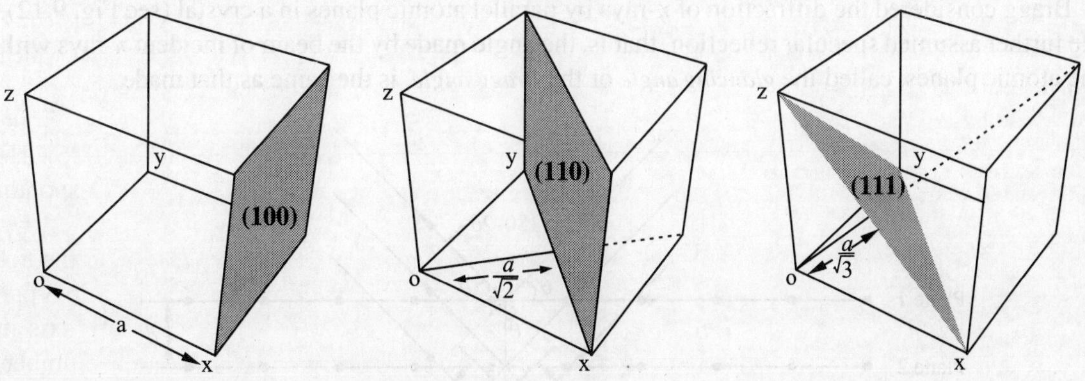

Figure 9.11 (100), (110) and (111) planes and their interplanar spacings. $d_{100} = a$, $d_{110} = a/\sqrt{2}$, $d_{111} = a/\sqrt{3}$

$$\frac{|(h/a)\times 0+(k/b)\times 0+(l/c)\times 0-1|}{\sqrt{(h/a)^2+(k/b)^2+(l/c)^2}}$$

Thus for a cubic crystal, for which $a = b = c$,

$$d_{hkl} = \frac{a}{\sqrt{h^2+k^2+l^2}}. \tag{9.5}$$

Hence we have $d_{100} = a$, $d_{110} = a/\sqrt{2}$ and $d_{111} = a/\sqrt{3}$. These relations are easily verified from Fig. 9.11.

9.6 DETERMINATION OF CRYSTAL STRUCTURE BY X-RAY DIFFRACTION

We have studied the geometry of a crystal structure so far. The techniques to determine the internal structure (the details of the unit cell) of a crystal are discussed here. X-ray diffraction (XRD) is the most widely used technique for this purpose. In Chapter 4 the diffraction patterns produced by different obstacles when illuminated by visible light were studied. The crux of this study is that the directions of the diffracted beams are decided by the geometry of the obstacle (for diffraction by grating, for a given λ, θ is decided by the grating element, $a + b$; see Eq. 4.26), and the intensities of the diffracted beams are decided by the diffracting characteristics of an obstacle. (For a grating it is decided by the detailed shape of the grating rulings.) Thus a closer study of the diffraction pattern reveals the structure and diffracting characteristics of an obstacle.

Thus, x-rays with a wavelength of the order of a few angstroms, can be diffracted by the unit cell of a crystal with dimensions of the same order, revealing details of the structure of the crystal. *The directions of diffracted x-ray beams throw light on the basic symmetry of the crystal, whereas the intensities tell us about the distribution of electrons in the unit cell.* (X-rays are diffracted mainly by electrons; diffraction by nuclei is negligible in most cases.)

Bragg considered the diffraction of x-rays by parallel atomic planes in a crystal (see Fig. 9.12). He further assumed specular reflection, that is, the angle made by the beam of incident x-rays with the atomic planes, called the *glancing angle* or the *Bragg angle*, is the same as that made

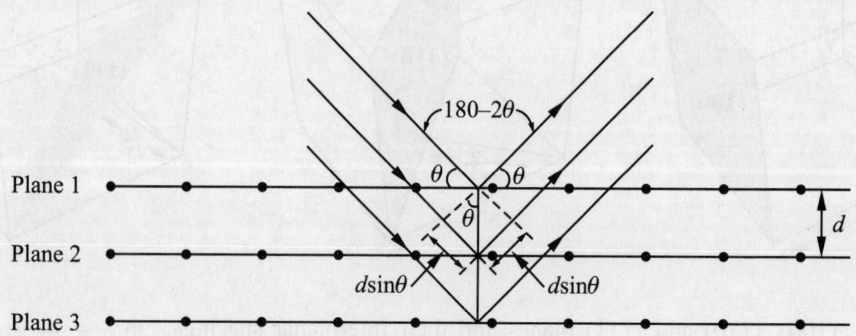

Figure 9.12 Diffraction of x-rays by atomic planes: Bragg's law

by the reflected beam with these planes. The beams reflected from plane 1 and plane 2 will interfere constructively only if the path difference between them is $n\lambda$, n being an integer and λ, the wavelength of x-rays. Thus, if d is the interplanar spacing, then the Bragg condition for reflection is

$$2d\sin\theta = n\lambda \tag{9.6}$$

Bragg's law is used to determine crystal structure by employing one of the following experimental techniques.

Laue technique

In this technique a single crystal specimen is mounted in the path of the incident x-ray beam, which is a mixture of several wavelengths (called as continuous x-rays or 'white' x-rays). This means that in Eq. 9.6, θ is constant whereas λ is variable. Thus a particular $\lambda - d$ combination will satisfy Bragg's law and produce a diffraction maximum. On a photographic plate these diffracted beams form an assembly of 'Laue spots'. The atomic arrangement in the crystal can be deduced from a careful study of the positions and intensities of the Laue spots. However, the Laue technique found restricted usage due to the precision it demands in mounting the crystal.

Rotating crystal method

In this method a single crystal specimen is mounted in the path of a monochromatic beam of x-rays and is slowly rotated. Thus λ is constant but θ is varied. A particular $\theta - d$ combination will produce diffraction maxima. This method provides more accuracy in the measurement of the diffracted intensities. Sophisticated x-ray diffractometers use this technique for both material characterisation and quantitative analysis.

Powder method

In this method a sample in the form of a powder is placed in the path of a monochromatic beam of x-rays. Thus λ is constant here and a particular combination of θ and d will produce a diffraction maxima. This is a widely used technique for crystal structure analysis.

Examples 9.7 and 9.8 illustrate the application of Bragg's law for the determination of crystal structure and density, respectively.

EXAMPLE 9.7

Determination of crystal structure using Bragg's law

A first order diffraction maximum is observed for a chromium crystal at a glancing angle of 30° using a beam of x-rays of wavelength 1.67 Å. If the diffraction is due to the (111) plane, find the crystal structure of chromium. The atomic radius of chromium atoms is 1.25 Å.

Solution

Using Bragg's law, $2d\sin\theta = n\lambda$, and the given data: $n = 1, \theta = 30°, \lambda = 1.67$ Å and $d = d_{111}$, we have,

$$d_{111} = \frac{1.67\text{Å}}{2\sin 30°} = 1.67 \text{ Å}$$

Now, $d_{hkl} = \dfrac{a}{\sqrt{h^2 + k^2 + l^2}}$

Thus, $d_{111} = \dfrac{a}{\sqrt{3}}$ or $a = \sqrt{3} \times 1.67$ Å

Now the atomic radius, $r = 1.25$ Å

Thus, $\dfrac{r}{a} = \dfrac{1.25}{\sqrt{3} \times 1.67} = 0.4321 \approx \dfrac{\sqrt{3}}{4}$

As the relation $r = \sqrt{3}a/4$ holds good for BCC, chromium crystallises in BCC structure.

EXAMPLE 9.8
Determination of density using Bragg's law

For silver, a first order diffraction maximum is observed at the glancing angle of 30°. The wavelength of x-rays used is 2.88 Å and the plane responsible for diffraction is (110). Determine the density of silver. The unit cell of silver is FCC and its atomic weight is 108.

Solution

Using the given data, that is, $n = 1, \theta = 30°, \lambda = 2.88$ Å, and $d = d_{110}$, in the Bragg equation, that is, $2d\sin\theta = n\lambda$, we have,

$$2d_{110}\sin 30° = 2.88 \text{ Å}$$

or $\quad d_{110} = 2.88$ Å

or $\quad \dfrac{a}{\sqrt{2}} = 2.88$ Å

Thus, $a = 4.073$ Å

Now, density, $\rho = ZM/N_A a^3$

$$= \frac{4 \times 108 \text{ kg}}{(6.023 \times 10^{26}/\text{kmole}) \times (4.073 \times 10^{-10} \text{ m})^3}$$

$$= 10615.2 \text{ kg/m}^3$$

9.7 MATERIALS AND THEIR PROPERTIES

We have studied various types of bonding and structures exhibited by the crystalline solids so far. These two factors decide the behaviour or properties of a solid. The experimental study of various properties of a material tells us the macroscopic behaviour of that material. This behaviour is then interpreted using theoretical models developed by taking into account the microscopic details of the materials. These atomic models unify our understanding of the various properties of materials. If the structure–property correlation of solids is studied exclusively from the point of view of the usefulness of these solids, then the solids under study are regarded as *materials*, and the study as *materials science*.

It is interesting to know that all properties of materials are ultimately dependent on the participating atoms and their electron clouds, and thus can be finally expressed and understood in terms of electrical quantities. However, detailed studies involving interaction among atoms and their electron clouds are complicated and are usually undertaken as research problems. For general purposes, the models are based on the appropriate basic principles of physics. For example, mechanical properties are explained on the basis of the principles of mechanics, and principles of thermodynamics form the base for thermal properties of materials. Such studies have developed into independent subjects like *mechanics of solids* and *thermodynamics of solids*.

Characterisation is the process of studying the properties of a material using some tool or technique—usually called a *probe*. Though several properties of a material can be studied using different probes, the unifying principle involved is shown in Fig. 9.13. A property is the result of interaction of matter with some form of energy. Thus the material to be probed is subjected to some energy field and the result of this interaction is studied by means of a detector. A detector may be a scope (such as a spectroscope, oscilloscope or eye), a meter (such as a spectrometer or thermometer), or a graph/recorder (such as a spectrograph or photograph). For example, when a material is illuminated with light, that is, subjected to an optical radiation field, its colour can be noted with the eye. The wavelengths emitted by it can be measured with a spectrometer and its spectra can be recorded using a spectrograph.

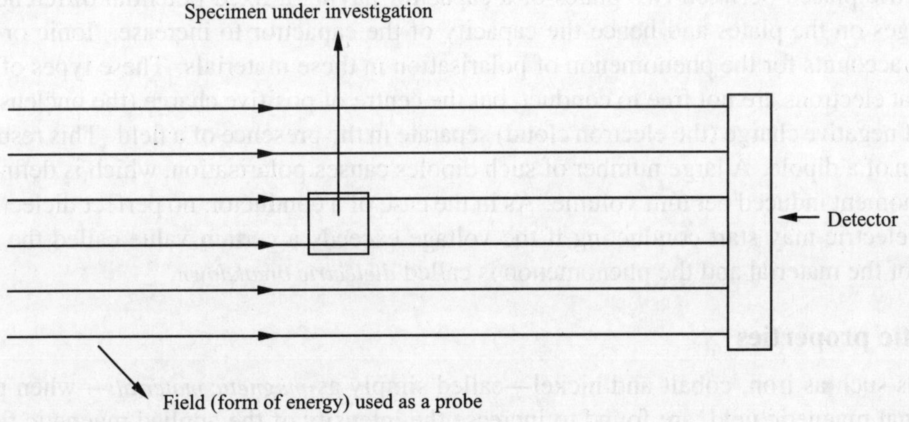

Figure 9.13 Principle of characterisation of materials

Some important material properties arising out of the interaction of matter with different fields—electric, magnetic, optical, thermal, acoustical and mechanical—are listed in Table 9.2. The correlation of the observed characteristics of materials with their internal structure using atomic models is discused briefly in the following. The electric and magnetic properties are studied in more detail in the next chapter.

Table 9.2 Some important properties of solids

Field applied	Properties
Electrical	conductivity, resistivity, dielectric constant, dielectric strength
Magnetic	magnetic permeability, susceptibility, coercive field, remnant polarisation
Optical	colour, refractive index, reflectivity, transmittivity
Thermal	specific heat, thermal expansion, thermal conductivity, Debye temperature
Acoustical	sound velocity, acoustic attenuation, reflection, transmission
Mechanical	elasticity, strength, hardness, ductility, malleability, plasticity

Electrical properties

Materials such as metals and alloys conduct electricity when subjected to an electric field. The ability of a material to conduct electricity is measured in terms of its *conductivity*. The conduction of electricity through metals is due to the flow of free electrons—that are present in them in large numbers due to the metallic bond—from the low potential end to the high potential end. However, the flow of electrons is not smooth but is impeded by the thermal motion of atoms and the presence of imperfections in the metal. This makes any metal, an imperfect conductor, and the current through a metal is limited by its resistivity.

However, in some materials such as alkali halides, carbon and diamond, the application of an electric field does not result in a current. On the other hand, these materials, called *insulators* or *dielectrics*, show *polarisation*, that is, separation of charges in the presence of an electric field. A dielectric placed between two plates of a capacitor having a fixed potential difference causes the charges on the plates and hence the capacity of the capacitor to increase. Ionic or covalent bonding accounts for the phenomenon of polarisation in these materials. These types of bonding mean that electrons are not free to conduct, but the centre of positive charge (the nucleus) and the centre of negative charge (the electron cloud) separate in the presence of a field. This results in the induction of a dipole. A large number of such dipoles causes polarisation, which is defined as the dipole moment induced per unit volume. As in the case of a conductor, no perfect dielectric exists and a dielectric may start conducting if the voltage exceeds a certain value called the *dielectric strength* of the material and the phenomenon is called *dielectric breakdown*.

Magnetic properties

Materials such as iron, cobalt and nickel—called simply as *magnetic materials*—when placed in an external magnetic field, are found to increase the intensity of the applied magnetic field. The magnetic behaviour of a solid is explained on the basis of the motion of its electrons! An electron orbiting round the nucleus (and also spinning about itself) is like a tiny current loop and is thus equivalent to a tiny magnet. Materials having many such magnetic dipoles, normally randomly oriented, show temporary magnetism in the presence of the magnetic field because these dipoles are aligned in the direction of the field. Such materials are known as *paramagnetic* materials. *Ferromagnetic* materials (such as Fe, Co and Ni) contain already aligned groups of dipoles (called *domains*) in the absence of a magnetic field. The alignment and growth of these domains accounts for a large increase in the intensity of the magnetic field when these materials are placed in an external field.

Optical properties

Optical radiation is ubiquitous. So all materials possess a characteristic called colour arising out of its interaction with the optical field. A material has the colour which it reflects, that is, it absorbs wavelengths other than the one corresponding to its colour. An atomic model explains colour in the following way. An optical beam, which is an electromagnetic radiation field with frequency of the order of 10^{14} Hz, sets electrons in forced vibrations; electrons re-radiate this wave with altered energy. The amount of energy absorbed or radiated depends, in a sensitive way, on the web of forces that hold the electrons in place. In most metals a large number of electrons re-radiate all the wavelengths incident upon them making them opaque and highly reflective. In non-conducting solids, the electric field in the light beam induces a dipole as explained in the case of dielectrics. However, as the field here is alternating, the dipole also oscillates in response to the field, a process leading to absorption and transmission of light. It is interesting to note that the dielectric constant, ϵ_r, of a material is equal to the square of its refractive index—an example of the unification achieved through an atomic model. Refraction of light and polarisation of charges are understood as two special cases of the response of an atom to the applied field through the use of the atomic model.

Thermal properties

When a thermal field is applied, some materials show quicker heat transport than others. Thermal conductivity is a measure of how fast a material conducts heat. It is observed that metals are better thermal conductors than non-metals. This is because, the atomic model suggests, the transport of heat is by lattice vibrations in non-metals, whereas it is by both lattice vibrations and free electrons in metals. The thermal expansion coefficient of a solid is a measure of the expansion it undergoes when placed in a thermal field. Ionic and covalent crystals, owing to the high bond strength, resist the increase in their interatomic distances and consequently have lower thermal expansion coefficients compared to metals.

Acoustic properties

Characterisation of materials using acoustic field involves sending sound waves (usually of high frequencies, called ultrasonic waves) through the solid and measuring how fast they travel (velocity) and how fast they decay (attenuation) in the solid. These measured properties are generally explained on the basis of the *elastic constants*, which in turn depend on the interatomic bonding and forces. This study is quite complicated but interesting for crystalline solids for which the elastic constants are different for different directions—a property called as *elastic anisotropy*. Thus we need 21 *second order elastic constants* (SOEC) for the least symmetric (triclinic) crystal but only 3

SOEC for the most symmetric (cubic) crystal. The second order elasticity refers to the assumption of harmonic oscillations, that is, the presence of harmonic forces (forces linearly proportional to the displacement of an atom from its equilibrium position) between the atoms. Departure from linearity, that is, the anharmonicity of lattice forces is measured by the *third order elastic constants* (TOEC) of the crystal. Again we have 56 TOEC for a triclinic crystal, but only 6 TOEC for a cubic crystal. The velocity and attenuation of sound waves inside a crystal is decided in a complex manner by, respectively, the SOEC and the TOEC of that crystal.

Mechanical properties

Mechanical properties include a variety of characteristics that depend on the response of a material to stress and strain. The deformation produced in materials due to an externally applied stress is a measure of its elastic modulus. Covalent crystals, due to the large values of their bond strengths, resist deformation when stressed. The characteristics such as ductility and brittleness are also decided by atomic bonding. Ionic crystals are brittle because the bonds between their atoms are inflexible. Metals, on the other hand, are ductile because the flexible and non-directional metallic bonding allows the atoms to slide over each other, and thus change shape without breaking.

In addition to the properties discussed above, the behaviour of materials is also studied by applying more than one field at a time. Thus a material also shows properties such as electrooptic effect, photoacoustic effect, piezoelectric effect and so on.

9.8 APPLICATIONS

In this chapter we studied crystalline solids as an important state of matter on one hand and as an important class of materials on the other. That matter and materials have played a pervasive role in human civilisation is evident through the names of the various ages: the *Bronze age*, the *Iron age* and the *Atomic age*. Up to the nineteenth century, the application of materials in technology was limited by the knowledge of materials to the macrostructural level. Metalsmiths in many parts of the world knew of the treatment of metals (such as differential polishing, scraping and etching with natural acids) in order to reveal the distribution of their hard and soft constituents. The beautiful and remarkable patterns on the Japanese and Damascus swords bear testimony to this knowledge. The quality of special types of steels produced by the Agarias of Central India matches steels produced today.

In Sections 9.2 to 9.5 the bonding and structure in crystals was studied. Research in these aspects has enriched our knowledge of the various atoms and their aggregates. Today we know that the incredible variety of matter around us is not because of a myriad number of atoms but because of the countless ways in which atoms can combine together to form aggregates, just as a colour printer can produce any conceivable colour by combining just three colours. This helped us to appreciate the vastness of nature on one hand, but on the other it boosted the enthusiasm of scientists to synthesise newer materials. Chemists are leaders in this trade and the branch known as *designer chemistry* has emerged out of these efforts. A million chemists could each make a new sample every day for a million years and still not come close to running out of things to try.

In Section 9.6 we studied the XRD technique for the determination of crystal structures. The use of this technique was effective in bringing out the details of the smallest building blocks of crystals. It was used with great success by the father and son team of WH Bragg and WL Bragg in the characterisation of a large number of unit cells. X-rays have been one of the most powerful tools for characterisation of solids in the last century. The structure of the chemical unit of life—DNA—was studied and reported by James Watson and Francis Crick in 1953 using x-ray analysis.

In Section 9.7 the structure–property correlation in materials was studied through unified models. The results of research on materials have shown rapid augmentation during the last two centuries, with improvement in both the level of structure determination and the accuracy and precision of the characterisation techniques. This spurt in knowledge led to the improvement of existing materials and to the discovery of many new materials, which in turn contributed greatly to the developments in all branches of engineering and technology. As it is a gigantic task to review this entire development, an attempt is made here to trace it through the following representative examples. The applications of dielectric and magnetic materials, conductors, semiconductors and superconductors are discussed in the next two chapters.

Polymeric (Plastic) materials

The use of plastic as a basic material has grown rapidly. *Polymers* are macromolecular substances, that is, substances with large molecules made of many *monomers* or single units. Plastics are polymers which flow under heat and pressure and can mould into various shapes. Wood, paper, silk and leather are examples of natural polymers. Some synthetic polymers have been developed as copies of natural polymers, for example, synthetic rubber is copied from natural rubber. Plastics perform an extremely broad range of functions. They are used as electrical insulators in wall sockets and light switches, as biomaterials in heart valves and artificial kidneys, as adhesives and in many other domestic and industrial appliances. Despite the success of plastic as a basic material, its indiscriminate use has been considered a potential threat to the ecosystem due to its non-biodegradibility.

Metal alloys

Alloys are blends of metals (also called *solid solutions*) which have properties that are both different and desirable from their individual constituents. For example, bronze, which is made of copper and tin, is stronger and harder than both. Some alloys such as brass (Cu and Zn), cast iron (Fe and C), stainless steel (Fe, C, Cr and Ni) are widely used in automobiles, power generating equipment and in electrical gadgets.

Composites

Composites are blends of more than one material type, to accomplish a combination of the best characteristics of the component materials. For example, if we want to increase the stiffness of a solid without the accompanying disadvantage of brittleness, we can use an AB (Aluminium boron) composite. The stiffness is increased due to boron (its elastic modulus is one of the highest among elements.), but brittleness is avoided due to the cushioning effect of the ductile matrix of aluminium. Fibre reinforced composites are lightweight materials of unusual strength. Fibreglass has the strength of glass and the flexibility of a polymer. Fibre composites are also used in aircraft, tennis rackets, and elsewhere where strong but light-weight materials are desired. Shatterproof windshields, integrated circuits and steel-reinforced concrete are a few of the many examples of novel applications of composite materials.

Glass, liquid crystals and ceramics

Glasses are formed when a substance in the molten state is cooled suddenly—a technique called *quenching*—so that atoms do not get sufficient time to arrange themselves in an orderly manner, that is, to crystallise. Theoretically every solid can be made into glass, but the quenching rate required may not make glass formation practical in many cases. *Metallic glasses*, one of the interesting outcomes of this research, have a potential for wide ranging applications. Glasses can be coloured by the addition of small amounts of various substances such as chromium oxide (green glass) and manganese dioxide (violet glass). *Photochromic glasses* get a temporary colour due to interaction of the UV component of sunlight with silver chloride crystals dispersed in these glasses.

Liquid crystals represent a state of matter that is not as ordered as a solid crystal but also not as disordered as a liquid. The crystalline nature of liquid crystals is very weak and can be randomised by applied fields such as an optical or magnetic field. *Liquid crystal displays* (LCD) used in calculators, digital watches, panel meters, video monitors and various other displays exploit the double refraction property of liquid crystals.

Ceramics are non-metallic solids made of inorganic compounds like oxides, nitrides, borides, and others. The bonding in ceramics is predominantly ionic. Rocks, minerals and fire clays are some examples of ceramics. Ceramics are widely used in electrical and electronic industries in insulating dielectrics, and in piezoelectric and ferroelectric applications. Ceramic materials are also used in diverse applications such as refractories for industrial furnaces, in internal combustion engines, in armour systems to protect military personnel and vehicles from ballistic projectiles, and in packaging of integrated circuits. Ceramic tiles are used in the space shuttle to protect the inner aluminium structure of the shuttle during its ascent out of and re-entry into the earth's atmosphere. Ceramic tiles withstand large changes in temperature ($>1000°C$) during these plunges and are able to absorb thermal shocks without cracking.

Besides those mentioned above, a large number of advanced materials have been developed and are in the process of being developed to suit various applications. The remarkable properties accomplished with these materials include, among others, a 100-fold increase in strength-to-density ratio in composites, reduction in size of ICs to less than a micron, 3-fold increase in working temperature of ceramics, a more than 200-fold increase in magnetic strength of hard magnetic materials, 100-fold increase in the speed of cutting tools due to great increase in the hardness and chemical resistance of thin film coatings, increase in transition length of glass by four orders of magnitude, and a 10-fold increase in transition temperature of superconductors. These are only a few examples of the research findings of materials scientists and engineers in their quest for new and improved materials. Several such amazing and dramatic advances remind us of the words of Goddard: *It's difficult to say what is impossible, for the dream of yesterday is the hope of today and the reality of tomorrow.*

Just as the internal circuit of an electrical device dictates its capabilities, the internal structure of materials dictates its capabilities. Scientists are specialists in the study of the internal structure (circuit!) of materials. They take pleasure in mending the circuitry of atoms to make the material behave in a desired way. *Scientists know the art of engineering the atoms.* They are artisans working at the atomic level. The pursuits of materials science hold promise of several amazing discoveries in the future that will change our world in an unimaginable way.

EXERCISES

1. Define: (i) Space lattice, (ii) unit cell, (iii) coordination number, (iv) Miller indices and (v) void space.
2. The edge of the unit cell of a cubic lattice is a and the atomic radius is r for SC and FCC lattices. Find (i) the number of atoms per unit cell, (ii) atomic radius in terms of a and (iii) the packing fraction.
3. Determine the packing fractions of FCC and BCC lattices.
4. For BCC and FCC lattices, calculate (i) the number of atoms per unit cell, (ii) the atomic radius and (iii) the atomic packing fraction.
5. Define packing fraction. Show that the packing fraction of FCC structure is more than that of BCC structure.
6. Derive the relation between atomic radius and cube edge for the SC, BCC and FCC lattices.
7. Draw separate planes representing (100), (210) and (111) in a cubic crystal.
8. Draw the following planes in a cubic unit cell: (112) and (120)
9. Find the Miller indices of atomic planes with intercepts a, $2a$ and ∞ on x-, y- and z-axes respectively. Find the distance of the plane from the origin.
10. Deduce the relation between interplanar distance and Miller indices of the planes for a cubic system.
11. What are Miller indices? Draw crystal planes with Miller indices (210), (101) and (111) in a cubic unit cell.
12. Draw crystal planes for (212), (101) and (210) in a cubic unit cell.
13. Draw separate planes for (100), (020) and (111) in a cubic unit cell.
14. Deduce the relationship between interplanar distance and Miller indices of the planes for a cubic system. Draw crystal planes (120) and (001) in a cubic unit cell.
15. Explain and deduce Bragg's law for x-ray diffraction.
16. What are tetrahedral and octahedral voids?

PROBLEMS

1. Sodium crystallises in a cubic lattice. The edge of the unit cell is 4.3 Å. The density of sodium is 963 kg/m^3, its atomic weight being 23. What type of unit cell does sodium form?
2. The density of copper is 8980 kg/m^3 and the unit cell dimension is 3.61 Å. The atomic weight of copper is 63.54. (i) Determine the crystal structure. (ii) Calculate atomic radius and interplanar spacing of (110) plane.
3. Obtain the Miller indices for a plane making intercepts of 1 Å, 2 Å, and 4 Å along the a, b, and c crystal axes with 2 Å, 3 Å, and 6 Å as unit cell parameters respectively.
4. A beam of x-rays with $\alpha = 0.842$ Å is incident on a crystal at a glancing angle of $8°35'$ when the first order Bragg's diffraction occurs. Calculate the glancing angle for 3rd order diffraction.
5. For silver, a first order diffraction maximum is observed at a glancing angle of 30°. The wavelength of the x-rays used is 2.88 Å and the plane responsible for diffraction is (110). Determine the density of silver. The unit cell of silver is FCC and its atomic weight is 108.

6. Sodium chloride crystallises in FCC structure. The density of sodium chloride is 2180 kg/m^3. If the atomic weight of sodium is 23 and that of chlorine is 35.5, calculate the distance between two adjacent atoms.
7. Iron has BCC structure and the atomic radius is 1.24 Å. Calculate the density of iron crystal. Atomic weight of iron is 55.85.
8. X-rays of wavelength 1.5 Å are diffracted in the first order by (110) planes of a nickel crystal at the angle of 27°6′. Find the lattice constant of the unit cell of nickel which has FCC structure.

10

Dielectric and Magnetic Materials

10.1 THE STORY OF DIELECTRIC AND MAGNETIC MATERIALS

In the last chapter we saw how the microscopic behaviour of materials—their bonding, structure and phenomenology on the microscopic scale—governs their macroscopic behaviour, that is, various observed properties of these materials. In this chapter we use this approach to study the dielectric and magnetic properties of materials. The response of materials to an applied electric/magnetic field is decided by their electrical/magnetic properties. As electromagnetism unifies the study of electricity and magnetism, these properties can be aptly studied as *electromagnetism in matter*.

The story of dielectric and magnetic materials weaves around a common theme. Both these materials show pronounced effect on in the flux density of the field in which they are placed. Dipoles are the basic cause of this macroscopic observation—electric dipoles in dielectric materials and magnetic dipoles in magnetic materials. Investigation into formation and behaviour of these dipoles constituted the research aimed at improving these materials. The story of these materials is thus an account of the progress towards the accomplishment of this aim.

The early work of Andre Ampere (1775–1836), Carl Gauss (1777–1855), Charles Coulomb (1784–1806), Michael Faraday (1791–1867), Hans Oersted (1820) and others established the basic laws of electromagnetism. It was known through these laws that the charge on the plates of a parallel plate capacitor could be increased by increasing the area of plates and decreasing the separation between them. Similarly, increasing the number of turns in a current-carrying solenoid increased the intensity of the magnetic field produced by it. However, practical constraints limited the maximum value of stored charge or magnetic field intensity that could be achieved in this way.

It was found that a further increase in charge storage for a given geometry of capacitor could be achieved if free air between the plates of a capacitor was replaced by a dielectric material of constant ϵ, which is greater than ϵ_0, the dielectric constant or permittivity of free space (that is, the relative dielectric constant, $\epsilon_r = \epsilon/\epsilon_0 > 1$). Similarly, a further enhancement of the magnetic field could be achieved for a given solenoid if the space inside the solenoid was filled by a magnetic material of permeability constant, μ, which is greater than μ_0, the permeability of free space (that is, the relative permeability, $\mu_r = \mu/\mu_0 > 1$). Thus, the aim of improving the dielectric and magnetic materials brought the material properties of permittivity and permeability into focus. Hence, the research was actually about understanding what causes ϵ and μ, and how to increase their values.

Polarisation, the microscopic phenomenon responsible for charge storage in a material medium, was probed further and an atomic model of dielectric materials was developed that provided the insight that ϵ depended linearly on the electric dipole moment per unit volume of the dielectric. On similar lines the atomic model of *magnetisation*, the microscopic phenomenon responsible for field enhancement, showed that μ depended linearly on the magnetic dipole moment per unit volume of the material. Thus, it was understood that materials having larger dipole moments per unit volume could be better dielectric and magnetic materials. In 1912, Peter Debye, the American physical chemist, postulated the existence of permanent dipoles. The unit of dipole moment is named after him in honour of his pioneering contribution in the study of dielectrics using dipoles. Thus, materials could be placed in order of their increasing response to fields as

1. Elemental dielectrics and diamagnetic materials, having no permanent dipoles but dipoles induced due to an applied field;
2. Polar dielectrics and paramagnetic materials, having permanent dipoles that align due to an applied field;
3. Ferroelectrics and ferromagnetic materials, having permanent dipoles already aligned in the absence of a field into regions called domains, that align along an applied field, giving a very large polarisation or magnetisation.

In 1895, Pierre Curie discovered experimentally that the magnetisation of a paramagnetic sample is directly proportional to the external magnetic field. In the subsequent years Curie and Weiss did pioneering experimental work in the study of properties of magnetic materials. In 1905, the French physicist, Paul Langevin, developed the classical theory of paramagnetism that attempted to explain the experimental observations of Curie and Weiss. In 1907, Weiss proposed the molecular field theory to explain the phenomenon of ferromagnetism. In 1880, the Curies first identified the phenomenon of *piezoelectricity* (pressure electricity) in which electric charges appear on the surface of some crystals upon compression and the charges disappear when the pressure is removed.

It is interesting to note that L Neel and Bitter theoretically predicted antiferromagnetism and some properties of antiferromagnetic materials in the 1930s, before the experimental discovery of antiferromagnetic MnO by Bizette, Squire and Tsai in 1938. A large number of antiferromagnetic materials have been discovered since then. In 1948, Neel also explained many of the observed properties of ferrites. It was Neel who coined the term *ferrimagnetism*, a phenomenon exhibited by ferrites. Ferroelectricity was first discovered in Rochelle salt in 1921. There has been keen interest in the study of ferroelectrics since then and more than 3000 ferroelectrics have been discovered or developed.

In this chapter we begin with the application of the basic concepts of electromagnetism to the study of dielectric and magnetic behaviour of materials (Section 10.2). The microscopic models of polarisation and magnetisation are discussed in Section 10.3. The concept of internal field in solids is discussed in Section 10.4. In Section 10.5 we deal with ferroelectricity, ferromagnetism and related phenomena. The various types of dielectric and magnetic materials are discussed in Section 10.6 and the various applications of this knowledge are discussed in Section 10.7.

10.2 ELECTROMAGNETISM IN MATERIALS

Materials are made of atoms and molecules, which contain electric charges but are neutral as a whole. The response of any material to an electric or magnetic field is determined by the type of molecules making up the material. In case of solid materials, the response is also modified by its structure, that is, the spatial arrangement of molecules. It is important to note that the response of materials to an electric field is governed by charge displacements as well as molecular reorientation. The magnetic field largely involves reorientation of molecules.

When an electric field is applied, conductors facilitate the flow of charges whereas insulators obstruct their flow. However, in some materials such as alkali halides, carbon and diamond, the application of an electric field results in holding the charges and creation of surface charge density. Such materials are called *dielectric materials*. As in the case of a conductor and insulator, no perfect dielectric exists and a dielectric may start conducting if the voltage exceeds a certain value called the *dielectric strength* of the material; the phenomenon is called *dielectric breakdown*. When a magnetic field is applied to materials such as iron, cobalt and nickel, the intensity of the magnetic field is found to increase. Such materials are called *magnetic materials*.

This behaviour can be explained on the basis of the phenomenon of polarisation in dielectric materials and magnetisation in magnetic materials. Dipoles are the cause of both phenomena. Polarisation occurs due to electric dipoles, and magnetisation, due to magnetic dipoles. *Permittivity* (or dielectric constant) of a dielectric material measures its ability to store electricity (that is, electric charge in the form of electric dipoles), and *permeability* of a magnetic material measures its ability to store magnetism (that is, magnetic dipoles).

In Chart 10.1 we review the basic concepts required for the study of electromagnetism in materials.

The basic relations derived in Chart 10.1 are given below.

Dielectric		**Magnetic**	
$D = \epsilon E = \epsilon_0 E + P$		$B = \mu H = \mu_0 (H + M)$	
$\epsilon = \epsilon_0 \epsilon_r$	(10.1)	$\mu = \mu_o \mu_r$	(10.3)
$\chi = \dfrac{P}{E} = \epsilon_0 (\epsilon_r - 1)$	(10.2)	$\chi = \dfrac{M}{H} = \mu_r - 1$	(10.4)

CHART 10.1 APPLICATION OF BASIC CONCEPTS OF ELECTROMAGNETISM TO THE STUDY OF MATERIALS

	Dielectric	**Magnetic**
Basic concept	According to Gauss' law, the electric field **E** satisfies $\int \mathbf{E.ds} = q_0/\varepsilon_0$, where ε_0 is the permittivity of the free space and q_0 is the charge enclosed by the Gaussian surface. For a parallel plate capacitor with charge q_0 and potential difference V_0 volts, the capacitance is $C_0 = q_0/V_0 = q_0/E_0 d$. For the Gaussian surface I shown in Fig. 1, $E_0 = q_0/A\varepsilon_0$. Thus, $C_0 = \varepsilon_0 A/d$, where A is the area of the plates and d is the separation between the plates of a capacitor.	According to Ampere's law, the magnetic field **B** satisfies $\int \mathbf{B.dl} = \mu_0 i$, where μ_0 is the permeability of the free space and i is the current enclosed by the Amperean loop. Using Ampere's law it can be shown that the magnetic field produced by a solenoid of n turns per unit length is $B = \mu_0 ni$. The magnetic field strength **H** is given by $\int \mathbf{H.dl} = i$, or as $H = B/\mu_0$, so that $H = ni$ for a solenoid (see Fig. 3).
Figure	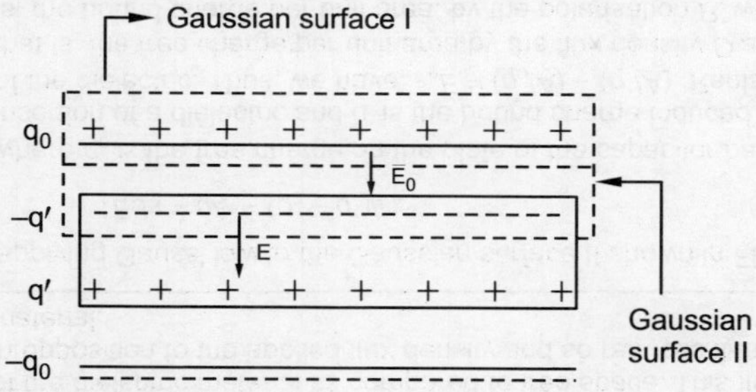 Gaussian surface I; q_0; E_0; $-q'$; E; q'; Gaussian surface II; $-q_0$ Fig. 1 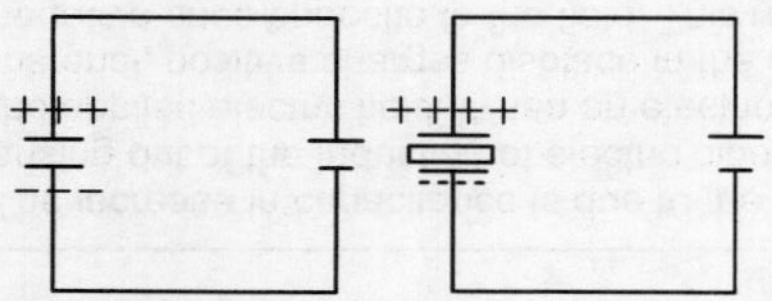 Fig. 2a 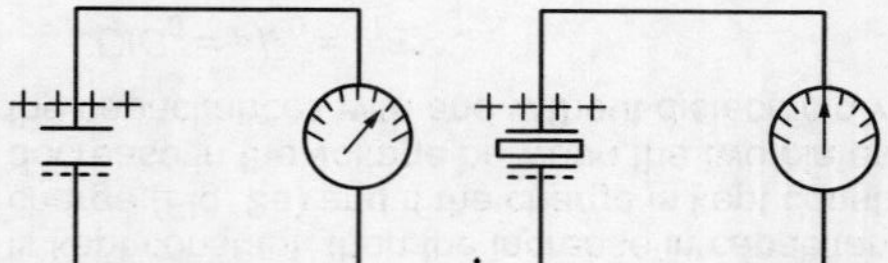 Fig. 2b	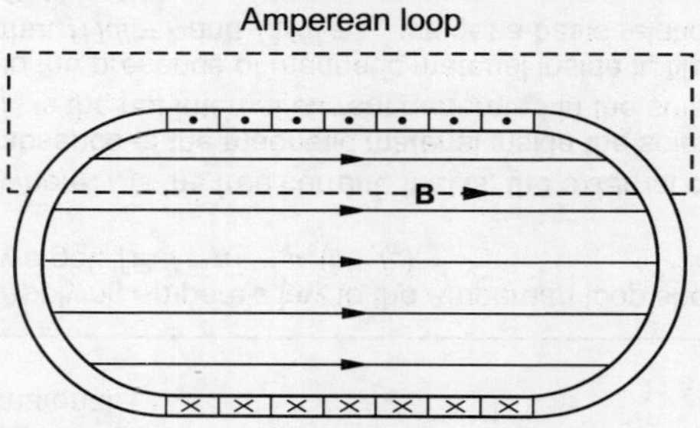 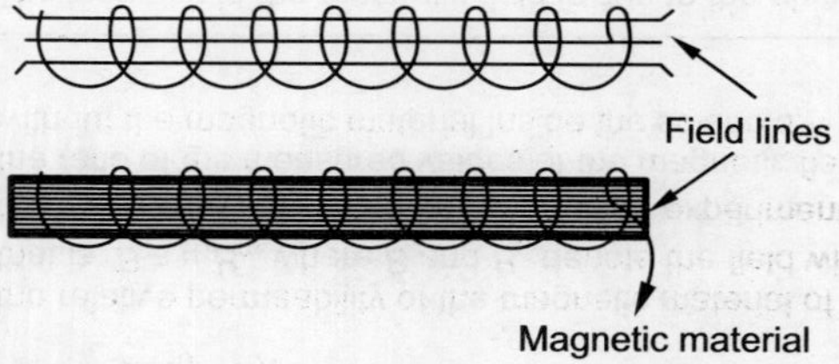 Amperean loop; B Fig. 3 Field lines; Magnetic material Fig. 4

Macroscopic property	If a dielectric material is placed between two plates of a capacitor then its capacitance is found to increase. If the voltage between two plates is kept constant, then the increase in capacitance is due to increase in charge (Fig. 2a) and if the charge is kept constant, then it is due to decrease in the voltage between the two plates (Fig. 2b). The ratio of the capacitances with and without dielectric given by $C/C_0 = \varepsilon/\varepsilon_0 = \varepsilon_r$ (1) defines the relative permittivity or the dielectric constant of the material. It can thus be found experimentally using the measured values of C and C_0.	If a magnetic material is placed inside a solenoid the effective magnetic field is found to increase (Fig. 4). The field increases by a factor of $\mu/\mu_0 = \mu_r$, (6) the relative permeability of the magnetic material of the solenoid, that is, $B = \mu_r B_0$, where B and B_0 denote the field with and without the solenoid. Thus μ_r can be determined experimentally by finding the ratio of the measured values of the magnetic field with and without the magnetic material inside the solenoid.
Microscopic phenomenon	The increase in capacitance is due to the phenomenon of polarisation arising out of the induction of electric dipoles in the dielectric material by the applied electric field. When an electric field is applied to a dielectric material, positive charges displace in the direction of the field and negative ones opposite to the field. This results in the induction of local dipoles throughout the dielectric. Polarisation is defined as the electric dipole moment induced per unit volume or the charge induced per unit area. Polarisation measures the flux density arising from the presence of the dielectric material as compared to free space. This flux density is in opposition to the applied flux density and so reduces it inside the material.	The increase in the magnetic field is due to the phenomenon of magnetisation arising out of the induction of magnetic dipoles in the magnetic material by the applied magnetic field. Magnetic dipole moments arise due to orbital motion and spin (called the intrinsic angular momentum) of electrons. An electron orbiting round the nucleus (and also spinning about itself) is like a tiny current loop and is thus equivalent to a tiny magnet. Magnetisation is defined as the magnetic dipole moment induced per unit volume. It increases the magnetic flux density inside the material.
Basic relations	Applying Gauss' law to the Gaussian surface II shown in Fig. 1, we get, $\int E ds = q/\varepsilon = (q_0 - q')/\varepsilon_0$ (2) where q_0 is the free charge on the plate of the capacitor before the insertion of a dielectric and q' is the bound charge induced on the faces of the dielectric. Thus, we have, $\varepsilon_0 E = (q_0/A) - (q'/A)$. Replacing q_0/A, that is, the free charge per unit area by the flux density D and q'/A, that is, the bound charge per unit area, by the polarisation P, we get a basic relation between D, E and P as $D = \varepsilon E = \varepsilon_0 E + P$ (3) The electric susceptibility χ is defined as $\chi = P/E$. (4) Using Eqs. 1, 3 and 4, we have $\chi = \varepsilon_0(\varepsilon_r - 1)$ (5) If the material is anisotropic then each of the three components of **P** is related to each of the three components of **E** and we need nine numbers to describe polarisation and χ is a tensor.	Applying Ampere's law to the Amperean loop shown in the figure, we get, $\int B dl = \mu i_f = \mu_0(i_f + i_m)$, where, i_f is the free current, that is, the external current in absence of the magnetic material inside the solenoid and i_m is the net internal current generated in the solenoid due to the presence of magnetic material inside it. Noting further that $\int H dl = i_f$ and $\int M dl = i_m$, we get a basic relation between B, H and M as $B = \mu H = \mu_0(H + M)$ (7) The magnetic susceptibility χ is defined as $\chi = M/H$. (8) Thus, Using Eqs. 6, 7 and 8 we have $\chi = \mu_r - 1$ (9) If the material is anisotropic then each of the three components of **M** is related to each of the three components of **H** and we need nine numbers to describe magnetisation and χ is a tensor.

CHART 10.2 DERIVATION OF AN EXPRESSION FOR INDUCED DIPOLE MOMENT

Dielectric

Consider a single atom consisting of a positive nucleus of charge Ze surrounded by a uniformly distributed electronic charge Ze. As the nuclear radius is much smaller than the atomic radius, the nucleus may be considered as a point charge. Let the total negative charge be distributed homogeneously throughout a sphere of radius R where $R \approx 10^{-10}$ m (see Fig.1).

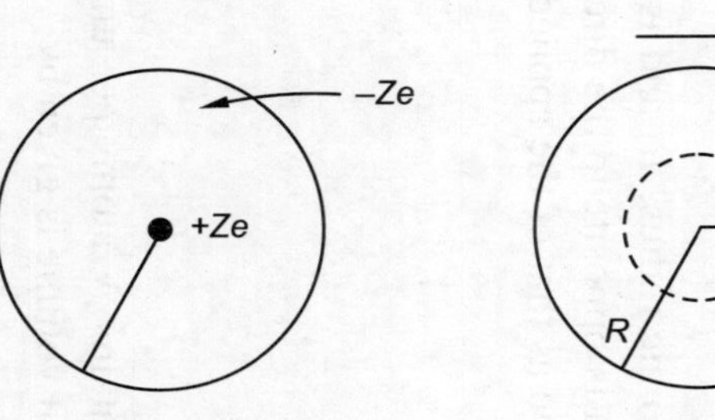

Fig. 1

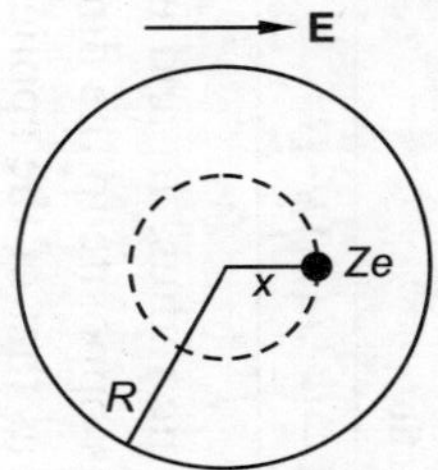

Fig. 2

When an electric field E is applied, the centre of positive charge and the centre of negative charge separate as the positive charge shifts towards the negative potential, and the negative charge towards the positive potential. But as they are pulled apart, a force of attraction between the two centres of charge tries to drive the nucleus back to the original position. Eventually an equilibrium is attained in which the nucleus is displaced by an amount x as shown in Fig. 2.

The force experienced by the nucleus along E is ZeE. The force experienced due to the negative charge contained in the sphere of radius x is

$(1/4\pi\varepsilon_0)(Ze/x^2) \times$ charge in the sphere of radius x

$$= (1/4\pi\varepsilon_0)(Ze/x^2) \times \{[Ze/(4\pi R^3/3)]\ [(4/3)\pi x^3]\}$$

$$= [(Ze)^2 x/R^3].$$

Thus, in equilibrium $ZeE = (1/4\pi\varepsilon_0)[(Ze)^2 x/R^3)(1/4\pi\varepsilon_0)]$ (1)

or $\quad x = (4\pi\varepsilon_0 R^3/Ze)E$ (2)

The resulting dipole moment that is induced in an atom is thus

$$\mu_{\text{d ind}} = Zex = 4\pi\varepsilon_0 R^3 E \quad (3)$$

Diamagnetic

In Section 7.5 we noted that the magnetic moment associated with the orbital motion of an electron in a Bohr orbit is given by (see Eq. 7.23)

$$\mu = -\tfrac{1}{2}\, e\omega R^2 \quad (4)$$

where we have used $\omega = 2\pi v$ for the angular frequency of the electron and R is the radius of the Bohr orbit. Denoting the angular frequency in the absence of field by ω_0, we have (see Eq. 7.2)

$$\omega_0^2 = (1/4\pi\varepsilon_0)\, e^2/mR^3 \quad (5)$$

Fig. 3

Now, if the magnetic field of intensity B is applied into the plane of paper as shown in Fig. 3, then the electron experiences a Lorentz force $-e$ (**v** × **B**) (see Chapter 13). Thus, for the electron orbit to be stable, we should have

$$mR\omega^2 = (1/4\pi\varepsilon_0)e^2/R^2 + evB$$

or $\quad \omega^2 = (1/4\pi\varepsilon_0)e^2/mR^3 + (eB/m)\omega$ (6)

Using Eq. 5, we have $\omega^2 = \omega_0^2 + (eB/m)\omega$. Using iterative solution, we get,

$$\omega = \omega_0 + (eB/2m) \quad (7)$$

where we have used the approximation $eB/m << \omega_0$, which is valid for ordinary fields of the order of 1 Wb/m^2 ($eB/m \approx 10^{11}$ /s $<< \omega_0$, which is of the order of 10^{15}/s. Thus the magnetic moment in the presence of field is

$$\mu = -\tfrac{1}{2}\, e\omega R^2 = -\tfrac{1}{2}\, eR^2[\omega_0 + (eB/2m)].\ \text{Thus,}$$

$$\mu = \mu_0 - (e^2/4m)BR^2 \quad (8)$$

where μ_0 is the moment in the absence of the field. Thus, the dipole moment induced due to the application of field is

$$\mu_{\text{m ind}} = -(e^2/4m)R^2 B \quad (9)$$

In the table on p. 307 E and H denote the applied electric field (measured in V/m) and the applied magnetic field (measured in ampere/m or A/m), respectively. D and B are the symbols for the flux density (coulomb/m^2) and the magnetic induction (weber/m^2 or Wb/m^2 or T) respectively. The polarisation P and the magnetisation M have the same units as D and H, respectively. The dimensionless constant χ denotes electric or magnetic susceptibility. We use the same symbol χ for both types of susceptibility. The type meant in any particular instant is generally clear from the context. The magnetic permeability of a magnetic material changes as it is magnetised, and is usually measured by either its initial permeability or its maximum permeability.

There are microscopic models that help us to understand the various types of polarisation and magnetisation in different materials. We discuss these models in the next section.

10.3 MICROSCOPIC MODELS OF POLARISATION AND MAGNETISATION

10.3.1 Electronic Polarisation and Diamagnetism

Dielectric materials with no permanent electric dipole moment in the absence of an electric field exhibit electronic polarisation. The rare gases such as argon and helium are examples of such dielectric media. Similarly, magnetic materials with no permanent magnetic dipole moments in the absence of a magnetic field exhibit diamagnetism. All materials possess diamagnetism. However, copper, mercury, silicon and sulphur exhibit strong diamagnetic property.

The basic expressions for the dipole moment induced in dielectric and diamagnetic materials due to the applied field are derived in Chart 10.2. These expressions are

Dielectric		**Diamagnetic**	
$\mu_{\text{d ind}} = 4\pi\epsilon_0 R^3 E$	(10.5)	$\mu_{\text{m ind}} = -(e^2/4m)R^2 B$	(10.6)

Thus, the induced dipole moment depends on the atomic radius, R, and is proportional to the applied field in both the cases. However, its direction is opposite to the direction of the applied field in the case of diamagnetic materials and the same as that of the applied field in the case of dielectric materials.

Equation 10.5 can be expressed as

$$\mu_{\text{d ind}} = \alpha_e E \tag{10.7}$$

where $\alpha_e = 4\pi\epsilon_0 R^3$ (10.8)

is called the electronic polarisability. If the gas contains N atoms per unit volume, then the polarisation, that is, the electric dipole moment per unit volume is given by

$$P_e = N\mu_{\text{d ind}} = N\alpha_e E \tag{10.9}$$

Comparing this equation with Eq. 10.2 we get,

$$\epsilon_0(\epsilon_r - 1) = N\alpha_e \tag{10.10}$$

Using Eq. 10.8,

$$\epsilon_r - 1 = 4\pi NR^3 \tag{10.11}$$

Similarly, the magnetisation M, that is, the magnetic dipole moment per unit volume is given by

$$M = N\mu_{\text{m ind}} = -\left(\frac{Ne^2}{4m}\right) R^2 B \tag{10.12}$$

Using Eqs. 10.3 and 10.4, we get,

$$\chi = \mu_r - 1 = -\left(\frac{Ne^2}{4m}\right) R^2 \mu_0 \mu_r \tag{10.13}$$

Thus it can be seen from Eqs. 10.8 and 10.11 that electronic polarisability and dielectric constant for a monoatomic gas are independent of temperature for normal temperatures as both N and R do not depend on temperature. These features are found to be true experimentally for the rare gases. Similarly, the diamagnetic susceptibility and permeability are also independent of temperature (Eq. 10.13), as long as the electronic structure of the material is independent of the temperature.

Examples 10.1 and 10.2 demonstrate how the microscopic models discussed above are in agreement with the experimental data.

EXAMPLE 10.1

Comparison of experimental result and theory of electronic polarisation

The experimental value of the dielectric constant of He measured at 0°C and 1 atm is $\epsilon_r = 1.0000684$. Under these conditions the gas contains 2.7×10^{25} atoms/m^3. Use this data to check the validity of Eq. 10.11.

Solution

Equation 10.11 is $\epsilon_r - 1 = 4\pi NR^3$

Substituting the given values of ϵ_r and N we have,

$$R^3 = \frac{0.0000684}{4\pi \times 2.7 \times 10^{25}/\text{m}^3}$$

or $R = 0.6 \times 10^{-10}$ m

As this is the correct order of magnitude for the radius of an atom, the experimental data is in agreement with the microscopic model discussed in Chart 10.2.

EXAMPLE 10.2

Comparison of experimental result and theory of diamagnetism

The experimental value of the susceptibility of diamagnetic materials is of the order of -10^{-5}. Check its agreement with Eq. 10.13 derived using the microscopic model of diamagnetism discussed in Chart 10.2. Use $R = 1$ Å, $N = 5 \times 10^{28}$ atoms/m^3, $\mu_0 = 4\pi \times 10^{-7}$ H/m and $\mu_r \cong 1$.

Solution

Equation 10.13 is $\chi = -\left(\frac{Ne^2}{4m}\right) R^2 \mu_0 \mu_r$

Substituting the given values of χ, R, N, μ_0 and μ_r, we have

$$\chi = -\frac{5 \times 10^{28}/\text{m}^3 \times (1.6 \times 10^{-19}\ \text{C})^2 \times (10^{-10}\ \text{m})^2 \times (4\pi \times 10^{-7}\ \text{H/m})}{4 \times 9.1 \times 10^{-31}\ \text{kg}}$$

or $\chi \cong -10^{-5}$

Thus the theoretical value obtained using the microscopic model discussed in Chart 10.2 is in agreement with experimental data.

10.3.2 Ionic Polarisation

Ionic polarisation arises due to the relative displacement of the atomic components of a molecule in the presence of an electric field. For example, in the NaCl molecule, the Na^+ ion is bound to the Cl^- ion and so its net dipole moment is $e \times a$ where a is the interatomic separation. In the presence of a field, the Na^+ and Cl^- ions are displaced in opposite directions until ionic bonding forces the process to stop. The dipole moment is thus increased in the presence of a field. Whereas the electronic polarisability measures the shift of the ions relative to the nuclei to which they belong, the ionic polarisability measures the shift of the ions relative to each other (Fig. 10.1). The dipole moment induced by ionic polarisation is again proportional to the applied field.

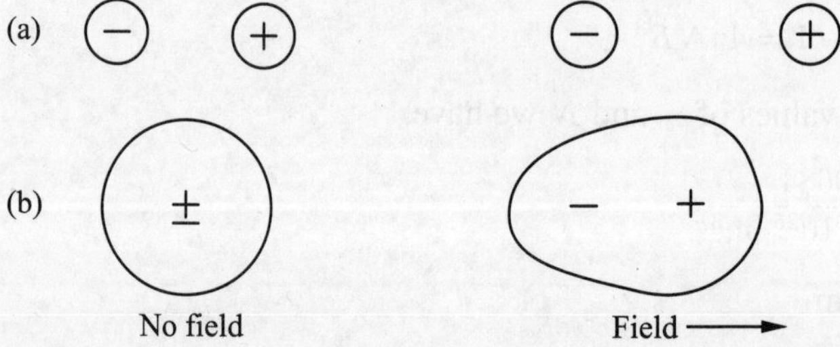

Figure 10.1 (a) In ionic polarisation the ions shift relative to each other; (b) In electronic polarisation they shift relative to the nuclei to which they belong

The ionic polarisation is given by $P_i = N\alpha_i E$ (10.14)

where α_i is the ionic polarisability. In most materials $\alpha_i = 0.1\alpha_e$. Just like α_e, it is largely independent on temperature.

10.3.3 Orientational Polarisation and Paramagnetism

The molecules in which the centres of negative and positive charge do not coincide possess a net electric dipole moment even in the absence of a field. Such molecules are called *polar molecules*. Figure 10.2 illustrates the difference between a non-polar and a polar molecule considering a hypothetical A_2B molecule. For example, methane (CH_4) is non-polar whereas CH_3Cl is polar, because the electronegativity of chlorine is more than that of hydrogen and hence the chlorine atom pulls the bonding electrons closer to itself.

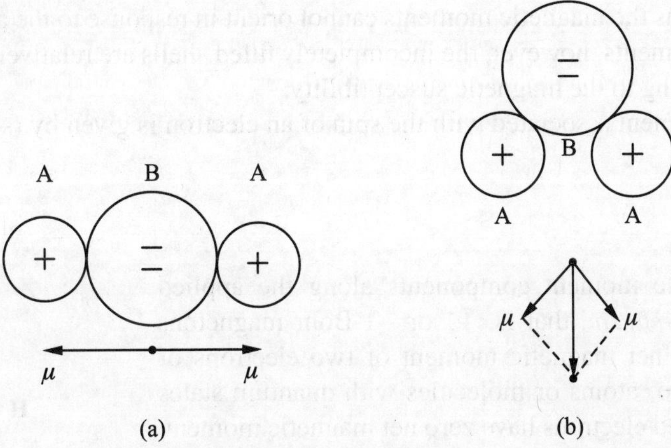

Figure 10.2 For a hypothetical molecule A_2B, the dipole moment is (a) zero for a non-polar configuration and (b) non-zero for a polar configuration

The origin of a permanent dipole moment in magnetic materials can be understood on the basis of magnetic moment, discussed in Section 7.5. The permanent magnetic moment arises due to the angular momentum that, in general has three contributions: the orbital angular momentum of electron, spin angular momentum of electron and spin angular momentum of nucleus.

The angular momentum, $L(= m_l\hbar)$ results in the magnetic moment (see Eq. 7.24),

$$\mu_l = -\left(\frac{e}{2m}\right)L = -\left(\frac{e\hbar}{2m}\right)m_l \tag{10.15}$$

Thus, for example, for the orbital quantum number, $l = 1$, the possible magnetic quantum numbers are $m_l = 1, 0, -1$. The corresponding components of the orbital angular momentum along the field direction are $\hbar, 0, -\hbar$, resulting in the possible magnetic dipole moment components

$-(e/2m)\hbar, 0, (e/2m)\hbar$ or $-1, 0, 1$ Bohr magnetons (as 1 Bohr magneton = $e\hbar/2m = 9.27 \times 10^{-24}$ Am^2). Thus, the magnetic quantum number m_l corresponds to magnetic moment components equal to $-m_l$ Bohr magnetons.

Thus, the completely filled electronic shells contribute nothing to the orbital permanent dipole moment. For example, for the completely filled L shell ($n = 2$), the possible l values are 0 and 1 that correspond to m_l values of $1, 0, -1$, giving a net zero magnetic moment. The incompletely filled shells may or may not result in zero magnetic moment depending on the electronic configuration. The transition elements with incompletely filled inner shells have non-zero orbital magnetic moment. These elements are the iron group elements (atomic numbers: 21–28), the rare earths (atomic numbers: 39–45 and 58–71) and elements with atomic numbers, 89–92. Among these the iron group is of utmost engineering importance.

The orbital magnetic moment of individual atoms makes little contribution to the net magnetic moment in the solid state. This is because the incompletely filled electronic shell in iron group atoms lies near the outside of atoms and is thus highly susceptible to interaction with neighbouring atoms in a solid. Thus the magnetic moments cannot orient in response to the applied field. In the case of rare earth elements, however, the incompletely filled shells are relatively deeper inside the atom, thus contributing to the magnetic susceptibility.

The magnetic moment associated with the spin of an electron is given by (see Eq. 7.26)

$$\mu_s = -\left(\frac{e\hbar}{2m}\right) \tag{10.16}$$

Thus the spin dipole moment components along the applied field are $e\hbar/2m$ or $-e\hbar/2m$, that is, +1 or −1 Bohr magnetons (Fig. 10.3). As the net magnetic moment of two electrons of opposite spins is zero, atoms or molecules with quantum states containing only paired electrons have zero net magnetic moment. Thus, as in the case of orbital moment, completely filled shells do not contribute to the total spin moment.

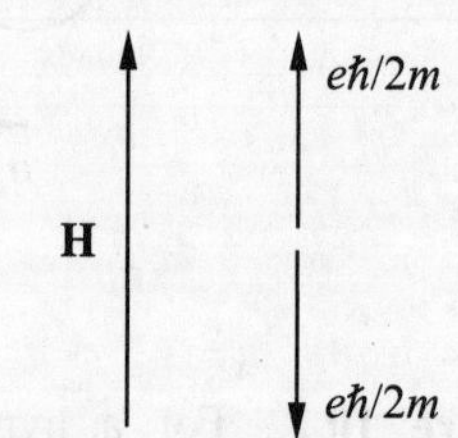

Figure 10.3 The two possible spin dipole moment components along the applied field direction

The net spin magnetic moment of an atom is thus decided by Hund's rule, which governs the order of filling of electron orbitals in atoms. For instance, the alkali metal atom having only one unpaired electron in the outermost s orbital has a spin moment of one Bohr magneton. An atom with three electrons in the p-orbital has a net magnetic moment of three units as all the spins are aligned in the same direction. However, an atom with four electrons in the p-orbital has a net moment of only two units as the spin of the fourth electron is opposite to that of the first three. Similarly, an atom with five electrons in the d-orbital has a net magnetic moment of five units, whereas that with nine electrons in the d-orbital has a net moment of only one unit.

The elements of the iron group that are important from the engineering point of view have incompletely filled $3d$ shell resulting in a net spin magnetic moment. Table 10.1 gives the spin configurations of electrons of the iron group elements. The spin configurations for free atoms of this group are the same as those for divalent ions.

Table 10.1 Spin configurations of electrons in free atoms or divalent ions of the iron group of elements

Element	Atomic number	Electrons in $3d$ shell	Spin configuration	Resultant spin in Bohr magneton
Scandium	21	1	[↑][ ][ ][ ][ ]	1
Titanium	22	2	[↑][↑][ ][ ][ ]	2
Vanadium	23	3	[↑][↑][↑][ ][ ]	3
Chromium	24	4	[↑][↑][↑][↑][ ]	4
Manganese	25	5	[↑][↑][↑][↑][↑]	5
Iron	26	6	[↑↓][↑][↑][↑][↑]	4
Cobalt	27	7	[↑↓][↑↓][↑][↑][↑]	3
Nickel	28	8	[↑↓][↑↓][↑↓][↑][↑]	2

The simple deduction of the net magnetic moment for a free atom discussed above has to be modified in the more complicated case of solid state, where the electron energy levels of the outermost orbitals interact to form energy bands (see Chapter 11). In the transition metals, the energy bands corresponding to the $4s$ orbital overlap that of the $3d$ orbital to some extent. The $4s$ electrons thus spend some time in the $3d$ band. For five or more $3d$ electrons, the spins of the overlapping $4s$ electrons must be opposite to that of the unpaired $3d$ electrons thus reducing the net magnetic moment. For instance, iron atom with a moment of 4 units as a free atom has a moment of only 2.2 units as a part of the crystal.

The third contribution to the total angular momentum is due to the nuclear spin. The angular momentum due to nuclear spin is of the same order of magnitude as that due to the orbital and the spin motion of electrons. However, the large mass of the nucleus ($\sim 10^3$ times that of an electron), results in a magnetic moment that is negligible in comparison to the magnetic moments of electrons. As the major contribution to the total magnetic moment comes from the spin magnetic moment, we neglect the other two contributions in further discussion.

In the presence of an applied electric field the permanent electric dipole moments of polar molecules get aligned in the direction of the field. This contribution is called the *orientation polarisation*. Similarly, the permanent magnetic dipoles get aligned in the presence of an external magnetic field. This is known as *paramagnetism*, and the materials as paramagnetic materials. In both the cases the net dipole moment is zero in the absence of an external field as the dipole moments are randomly oriented due to thermal effect. On application of a field the alignment of dipole moments gives rise to a net dipole moment in the field direction.

The orientation of electric dipoles in an electric field and magnetic dipoles in a magnetic field can be studied using Langevin's theory. This study is presented in a comparative manner in Chart 10.3.

We obtain the following expressions for the orientational polarisation and magnetisation for a paramagnetic material.

$$P_0 = \frac{N\mu_p^2 E}{3kT} \tag{10.17}$$

$$M = \frac{N\mu_0\beta^2 H}{kT} \tag{10.18}$$

Thus we find that both P_0 and M are inversely proportional to the temperature. This is to be expected as greater the thermal agitation, lower the alignment of dipoles and vice-a-versa. Thus, orientation polarisability is given by (see Eq. 10.9)

$$\alpha_o = \frac{\mu_p^2}{3kT} \tag{10.19}$$

In general polar molecules have all the three contributions and the total polarisability α is given by

$$\alpha = \alpha_i + \alpha_e + \alpha_o \tag{10.20}$$

$$\text{or, } P = N\alpha E = N\left(\alpha_e + \alpha_i + \frac{\mu_p^2}{3kT}\right)E \tag{10.21}$$

Using Eq. 10.2 we have

$$\chi = \epsilon_0(\epsilon_r - 1) = N\left(\alpha_e + \alpha_i + \frac{\mu_p^2}{3kT}\right) \tag{10.22}$$

Similarly, for paramagnetic materials the magnetic susceptibility is obtained using Eq. 10.18.

$$\chi = \frac{N\mu_0\beta^2}{kT} = \frac{C}{T} \tag{10.23}$$

$$\text{where, } C = \frac{N\mu_0\beta^2}{k} \tag{10.24}$$

is called the *Curie constant* and Eq. 10.23 is known as the *Curie law* of paramagnetism. For magnetic susceptibility there are no counterparts of ionic and electronic polarisability.

Equations 10.22 and 10.23 indicate that the susceptibility is proportional to $1/T$ in both the cases. Figure 10.4 shows the plots of χ versus $1/T$ for these cases. The permanent dipole

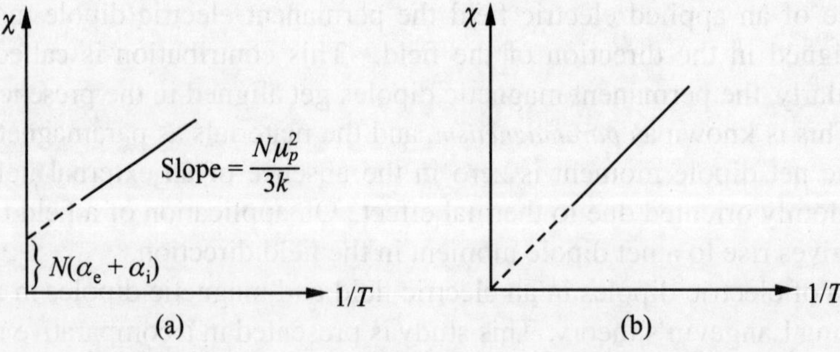

Figure 10.4 Variation of susceptibility with temperature for (a) dielectric material and (b) paramagnetic material

CHART 10.3 ORIENTATIONAL POLARISATION AND PARAMAGNETISM

We consider a system of N permanent dipoles/m^3 of the material in the absence of a field at temperature T. For a dielectric material the individual dipole has a moment of magnitude μ_p and for a magnetic material the dipole has a spin magnetic moment equal to a Bohr magneton, denoted by β. As the dipoles are randomly oriented, the vector sum of all the individual μ_ps is zero and there is no polarisation in the absence of an electric field. Similarly, the magnetisation M is zero in the absence of a magnetic field, as the number of 'up' spins equals the number of 'down' spins. When the external field (**E** or **H**) is applied, the dipoles align in the direction of the field. The applied field tries to orient the dipoles along it whereas the thermal energy tries to randomise its motion.

Orientational polarisation

The orientational polarisation, P_0 is the sum of the components of the individual μ_p along the field, that is,

$$P_0 = \int_0^{\pi} N(\theta)d\theta\mu_p \cos\theta \qquad (1)$$

where $\mu_p \cos\theta$ is the component of the dipole moment μ_p that makes an angle θ with E along the field direction and $N(\theta)d\theta$ is the number of molecules for which the direction of μ_p at a given instant lies within an angle between θ and $(\theta + d\theta)$ with the field direction. This number is thus proportional to the solid angle $2\pi \sin\theta\, d\theta$ as indicated in Fig. 1.

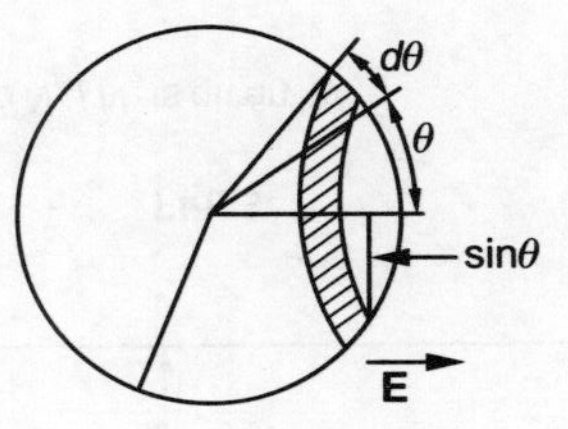

Fig. 1

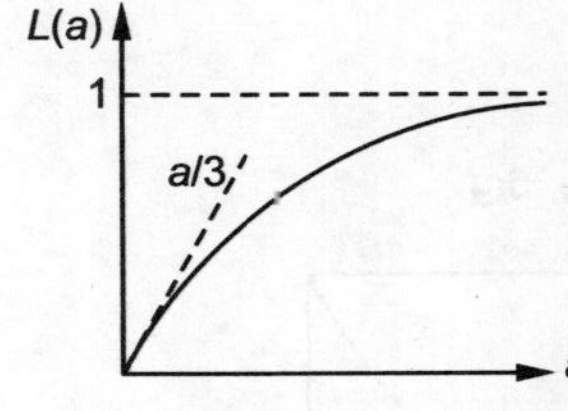

Fig. 2

However, in the presence of an external field $N(\theta)d\theta$ is no longer proportional to $2\pi\sin\theta\, d\theta$, as the external field along x-axis makes the x-axis a preferred direction. Thus, the above number should be weighted by the Boltzman's factor, $\exp[-W(\theta)/kT]$ to get the expression for $N(\theta)d\theta$, where $W(\theta)$ is the energy of a dipole making an angle θ with the field direction, and is given by the equation, $W(\theta) = -\mu_p E \cos\theta$. Thus,

$$N(\theta)\, d\theta = A\, 2\pi\sin\theta d\theta\exp[\mu_p E \cos\theta/kT] \qquad (2)$$

where k is Boltzman's constant, and A is the constant of proportionality that can be determined from the total number of dipoles N, given by

$$N = \int_0^{\pi} N(\theta)d\theta \qquad (3)$$

Using Eqs. (1), (2) and (3), we get $P_0 = N\int_0^{\pi}\mu_p\cos\theta\exp[\mu_p E\cos\theta/kT]\sin\theta d\theta \,/ \int_0^{\pi}\exp[\mu_p E\cos\theta/kT]\sin\theta d\theta$.

Substituting $t = \mu_p E\cos\theta/kT$ and $a = \mu_p E/kT$, we get

$$P_0 = (NkT/E)\int_{-a}^{a} te^t\,dt \Big/ \int_{-a}^{a} e^t\,dt = N\mu_p[\coth(a) - (1/a)] = N\mu_p L(a) \qquad (4)$$

where the function $L(a)$ is called the Langevin function (Fig. 2).

For large values of a, that is, for large fields and small temperatures, $L(a)$ approaches unity and P_0 approaches $N\mu_p$, implying alignment of all the dipoles along the applied field. For small values of a, that is, for small fields and large temperatures, $L(a)$ approaches $a/3$ so that P_0 becomes $P_0 = N\mu_p^2 E/3kT$ (5)

Thus, the orientational polarisation is proportional to the square of the permanent dipole moment and inversely proportional to the temperature.

Paramagnetism

The magnetisation is given by the net spin magnetic moment along the field direction, that is,

$$M = (N_p - N_a)\beta \tag{6}$$

where N_p and N_a denote, respectively, the number of dipoles/m^3, parallel and antiparallel to the field.

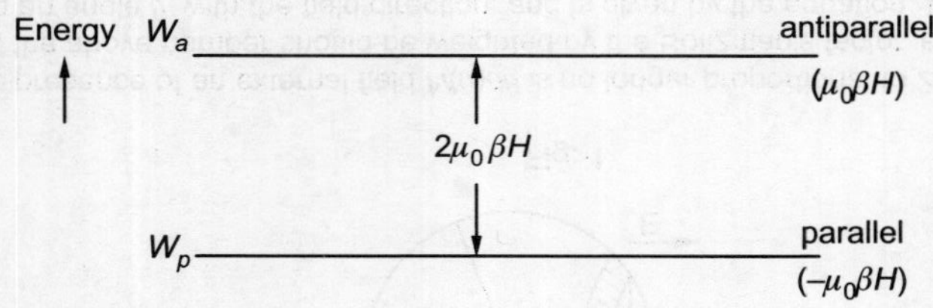

Energy W_a antiparallel $(\mu_0\beta H)$; $2\mu_0\beta H$; W_p parallel $(-\mu_0\beta H)$

Fig. 3

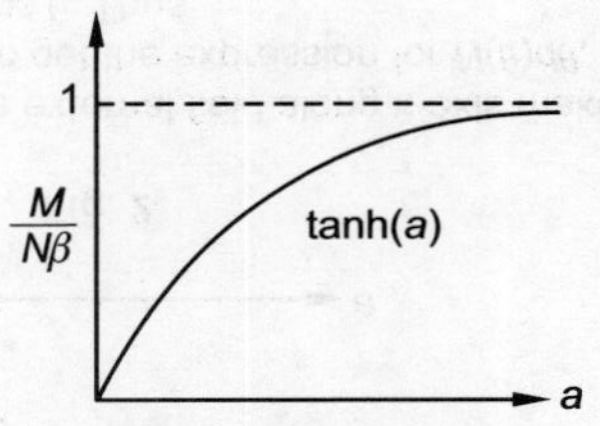

$\frac{M}{N\beta}$; 1; tanh(a); a

Fig. 4

According to Boltzman's statistics the ratio N_p / N_a is given by

$$N_p/N_a = \exp[-(W_p - W_a)/kT] \tag{7}$$

where, W_p and W_a are the energies corresponding to the parallel and antiparallel orientations respectively. The energy for antiparallel orientation is larger than that for parallel orientation and the energy difference is $2\mu_0\beta H$ (Fig. 3). Thus,

$$N_a/N_p = \exp[-2\mu_0\beta H/kT] \tag{8}$$

Now, $N = N_a + N_p$ (9)

Using Eqs. 8 and 9, we get $N_p = N/[1 + \exp(-2\mu_0\beta H/kT)]$ (10)

and $N_a = N\exp[(-2\mu_0\beta H/kT)]/[1 + \exp(-2\mu_0\beta H/kT)]$ (11)

Substituting Eqs. 10 and 11 in Eq. 6 we get $M = N\beta\tanh(\mu_0\beta H/kT)$ (12)

Fig. 4 shows the variation of $M/N\beta$ as a function of $a = \mu_0\beta H/kT$. For large values of a, that is, for large fields and small temperatures, tanh(a) approaches unity and M approaches $N\beta$, implying alignment of all the dipoles along the applied field, that is, saturation magnetisation. For small values of a, that is, for small fields and large temperatures, tanh(a) approaches a so that M becomes $M = N\mu_0\beta^2 H/kT$ (13)

Thus the magnetisation is proportional to the square of the spin magnetic moment and inversely proportional to the temperature.

moment can be obtained from the slope of the line and the value of $\alpha_i + \alpha_e$ can be obtained from the intercept on the y-axis as shown in Fig. 10.4a. Examples 10.3–10.5 provide the numerical appreciation of the above equations.

EXAMPLE 10.3

Determination of dipole moment and polarisabilities

The dielectric constant of a gas at 1 atm and 27°C is found to be $\epsilon_r = 1.006715$; at 177°C, $\epsilon_r = 1.005970$. The number of molecules of the gas per cubic metre is 2.44 $\times 10^{25}$. Calculate the dipole moment of the molecules and the sum of the electronic and ionic polarisabilities.

Solution

Using Eq. 10.22, we have

$$\epsilon_0(\epsilon_{rT} - 1) = N\left(\alpha_e + \alpha_i + \frac{\mu_p^2}{3kT}\right) \tag{10.25}$$

and $$\epsilon_0(\epsilon_{rT'} - 1) = N\left(\alpha_e + \alpha_i + \frac{\mu_p^2}{3kT'}\right) \tag{10.26}$$

where $T = 27\,°\text{C} = 300$ K and $T' = 177°\text{C} = 450$ K, $\epsilon_{rT} = 1.006715$ and $\epsilon_{rT'} = 1.005970$.

From Eqs. 10.25 and 10.26 we get

$$\epsilon_0(\epsilon_{rT} - \epsilon_{rT'}) = \left(\frac{N\mu_p^2}{3k}\right)\left[\left(\frac{1}{T}\right) - \left(\frac{1}{T'}\right)\right]$$

Substituting the given values we get

$$\epsilon_0(\epsilon_r T - \epsilon_r T') = (8.85 \times 10^{-12}\text{farad/m})(1.006715 - 1.005970)$$

$$= \left(\frac{2.44 \times 10^{25}\ \text{molecules/m}^3 \times \mu_p^2}{3 \times 1.38 \times 10^{-23}\ \text{J/K}}\right) \times \left(\frac{1}{300\ \text{K}} - \frac{1}{450\ \text{K}}\right)$$

Thus, $\mu_p = 3.173 \times 10^{-30}\ \text{Cm} = \dfrac{3.173 \times 10^{-30}\ \text{Cm}}{3.33 \times 10^{-30}\ \text{Cm/debye}} = 0.95$ debye

Substituting in Eq. 10.25 we get,

$$\epsilon_0(\epsilon_r T - 1) = (8.85 \times 10^{-12}\text{farad/m})(1.006715 - 1)$$

$$= (2.44 \times 10^{25}\ \text{molecules/m}^3) \times \left[\alpha_e + \alpha_i + \frac{(3.173 \times 10^{-30}\ \text{Cm})^2}{3 \times 1.38 \times 10^{-23}\ \text{J/K} \times 300\ \text{K}}\right]$$

Thus, $\alpha_e + \alpha_i = 1.6 \times 10^{-39}$ farad m^2.

EXAMPLE 10.4
Field required to cause the desired orientational polarisation

The molecules of a polar gas have a permanent dipole moment of 1.2 debye units. What should be the magnitude of the externally applied field to cause an orientational polarisation of 0.5% of the saturation polarisation at 27°C?

Solution

The orientational polarisation is given by $P_0 = \dfrac{N\mu_p^2 E}{3kT}$

where, the saturation polarisation is $(P_0)_{\text{sat}} = N\mu_p$

Thus, we have $\quad 0.005N\mu_p = \dfrac{N\mu_p^2 E}{3kT}$

or, $\quad E = \dfrac{3 \times 1.38 \times 10^{-23}\ \text{J/K} \times 300\ \text{K} \times 0.005}{1.2 \times 3.33 \times 10^{-30}\ \text{Cm}} = 1.55 \times 10^7\ \text{V/m}$

EXAMPLE 10.5
Estimation of susceptibility of paramagnetic materials at room temperature

Estimate the susceptibility of paramagnetic materials at room temperature using the following data: $N = 5 \times 10^{28}/\text{m}^3$ and $\beta = 1$ Bohr magneton.

Solution

For a paramagnetic material, $\chi = N\mu_0\beta^2/kT$

$$= \frac{5 \times 10^{28}/\text{m}^3 \times 4\pi \times 10^{-7}\ \text{H/m} \times (9.27 \times 10^{-24}\ \text{Am}^2)^2}{1.38 \times 10^{-23}\ \text{J/K} \times 300\ \text{K}}$$

$$\approx 10^{-3}$$

This estimated order of magnitude is in close agreement with the experimental value of susceptibility for paramagnetic materials.

10.4 INTERNAL FIELD

In case of gases, the atoms or molecules experience the same electric field that is externally applied. This is because the atoms are non-interacting, provided the density of the gas is reasonably small. However, in the case of liquids or solids, the atoms are so close together that the field created by surrounding atoms also affects the field of a particular atom. The actual field acting at the location of a given atom thus differs from the applied field and is called the *local field* or the *internal field*. It is the sum of the applied field and the field created by the surrounding atoms.

To find the expression for the internal field consider an infinite string of similar equidistant atoms of polarisability α_e as shown in Fig. 10.5. Let $\mathbf{E_i}$ be the internal field seen by an atom A. Due to symmetry, $\mathbf{E}_i$ will be parallel to $\mathbf{E}$ and will be the same for all atoms.

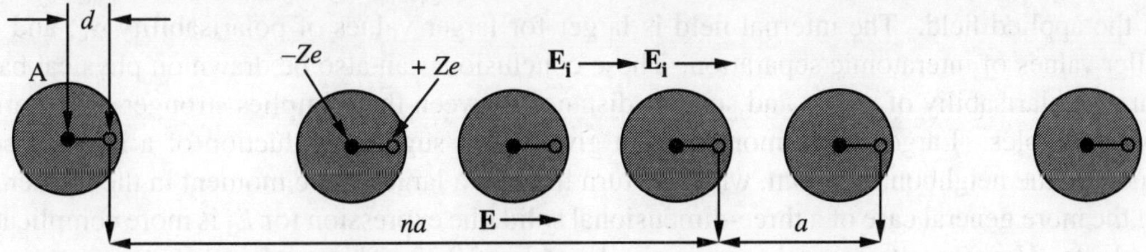

Figure 10.5 String of atoms of polarisability α_e in an external field $\mathbf{E}$ parallel to the string

The dipole moment induced in each atom of the string is thus

$$\mu_{\text{ind}} = \alpha_e \mathbf{E}_i \tag{10.27}$$

The field at A, due to an atom situated at a distance *na* (*a* being the interatomic distance), is

$$E_n = \frac{Ze}{4\pi\epsilon_0}\left[\frac{1}{(na)^2} - \frac{1}{(na+d)^2}\right] \tag{10.28}$$

where d is the distance between the centres of positive and negative charges in a particular atomic dipole, or,

$$E_n = \frac{Ze}{4\pi\epsilon_0 (na)^2}\left[1 - \left(1 + \frac{d}{na}\right)^{-2}\right]$$

Assuming $d \ll a$ and using binomial expansion, we get

$$E_n = \frac{\mu_{\text{ind}}}{2\pi\epsilon_0 (na)^3} \text{ where } \mu_{\text{ind}} = Zed \tag{10.29}$$

Taking into account atoms on both sides of A, we get the total field at A as

$$E_i = E + \frac{\mu_{\text{ind}}}{2\pi\epsilon_0}\left[2\sum_{n=1}^{\infty}\frac{1}{(na)^3}\right] \tag{10.30}$$

Thus, $$E_i = E + \left(\frac{\mu_{\text{ind}}}{\pi\epsilon_0 a^3}\right)\sum_{n=1}^{\infty}\frac{1}{n^3} \tag{10.31}$$

The sum in Eq. 10.31 is approximately equal to 1.2 and so

$$E_i = E + \frac{1.2\mu_{\text{ind}}}{\pi\epsilon_0 a^3} \tag{10.32}$$

Substituting for μ_{ind} from Eq. 10.27 and simplifying, we get,

$$E_i = \frac{E}{1 - (1.2\alpha_e/\pi\epsilon_0 a^3)} \tag{10.33}$$

From Eq. 10.33 it can be concluded that the internal field seen by an atom in a string is larger than the applied field. The internal field is larger for larger values of polarisability α_e, and for smaller values of interatomic separation. These conclusions can also be drawn on physical basis as larger polarisabiliy of atoms and smaller distance between them implies stronger cooperation between dipoles. Large dipole moment of a given atom suggests induction of a larger dipole moment in the neighbouring atom, which in turn induces a large dipole moment in the former.

In the more general case of a three-dimensional solid, the expression for E_i is more complicated to evaluate. However, its general form is similar to Eq. 10.33. The number density of atoms N, replaces $1/a^3$ and as $N\mu_{\text{ind}} = P$, the polarisation, from Eq. 10.32 we get,

$$E_i = E + \frac{\gamma P}{\epsilon_0} \tag{10.34}$$

Here γ is called the *internal field constant*. It is a dimensionless constant depending on the internal arrangement of atoms in the material. For example, it is $1.2/\pi$ for a one-dimensional string of atoms considered above. For cubic solids, γ can be shown to be equal to 1/3, and the expression for E_i becomes

$$E_i = E + \left(\frac{P}{3\epsilon_0}\right) \tag{10.35}$$

Equation 10.35 is also called the expression for *Lorentz field*. Now, using Eqs. 10.2 and 10.9, we get

$$P = \epsilon_0(\epsilon_r - 1)E = N\alpha_e E_i$$

Substituting for E_i from Eq. 10.34 and simplifying, we get

$$\frac{\epsilon_r - 1}{1 + \gamma(\epsilon_r - 1)} = \frac{N\alpha_e}{\epsilon_0} \tag{10.36}$$

For the special case of cubic solids, $\gamma = 1/3$ so that

$$\frac{\epsilon_r - 1}{\epsilon_r + 2} = \frac{N\alpha_e}{3\epsilon_0} \tag{10.37}$$

This equation is called the *Clausius–Mosotti equation* (see Ex. 10.6).

Again we note that in the case of magnetic materials we use concepts analogous to those discussed above. Thus, the internal field seen by a given magnetic dipole is equal to the sum of the applied field and the contribution from the neighbouring dipoles. This contribution tends to align the given dipole in the same direction as its neighbours. According to Weiss hypothesis, $\mathbf{H}_i$ is defined, in analogy with Eq. 10.34, as

$$\mathbf{H}_i = \mathbf{H} + \gamma\mathbf{M} \tag{10.38}$$

where **H** is the applied field and γ is the internal field constant, also called the *molecular field constant* or the *Weiss constant*, that determines the strength of the interaction between the dipoles. Thus, γ**M** measures the tendency of neighbouring dipoles to align a given dipole in the direction of **M**.

EXAMPLE 10.6

Determination of relative dielectric constant

Find the relative dielectric constant of sulphur using the following data. Atomic weight = 32, density = 2.08 gm/cm^3, electronic polarisability = 3.28 $\times 10^{-40}$ farad/m^2.

Solution

Using Clausius Mosotti equation,

$$\frac{\epsilon_r - 1}{\epsilon_r + 2} = \frac{N\alpha_e}{3\epsilon_0}$$

where N, the number of atoms per unit volume is given by

$$N = \frac{N_A\rho}{M} = \frac{(6.023\times 10^{26}\ \text{atoms/kmole})\times(2.08\times 10^{3}\ \text{kg/m}^3)}{32\ \text{kg/kmole}}$$

$$= 3.915\times 10^{28}\ \text{atoms/m}^3$$

Thus, we have

$$\frac{\epsilon_r - 1}{\epsilon_r + 2} = \frac{(3.915\times 10^{28}\ \text{atoms/m}^3)\times(3.28\times 10^{-40}\ \text{farad/m}^2)}{3\times 8.85\times 10^{-12}\ \text{farad/m}}$$

or, $\epsilon_r = 3.81$

10.5 FERROELECTRICITY, FERROMAGNETISM AND RELATED PHENOMENA

The presence of already aligned dipoles, called *domains*, in the absence of any external field results in the phenomena of *ferroelectricity* and *ferromagnetism* in materials. However, in the absence of any external field, the domains are randomly oriented and the net moment or the bulk polarisation/magnetisation is zero. When the field is applied, the domains grow as well as orient in the field direction, thus augmenting the flux density by many orders of magnitude. The ferromagnetic properties of transition elements are due to the way spins of the inner unpaired electrons are aligned in their crystal lattices, as discussed earlier in Section 10.3.3.

10.5.1 Hysteresis

In the mechanisms discussed in Section 10.3, the polarisation or magnetisation depended linearly on the applied field. However, in the case of ferroelectric and ferromagnetic materials, they depend nonlinearly on the history of the material. The variation of P versus E and that of M versus B follows what is known as the *hysteresis curve* (Fig. 10.6).

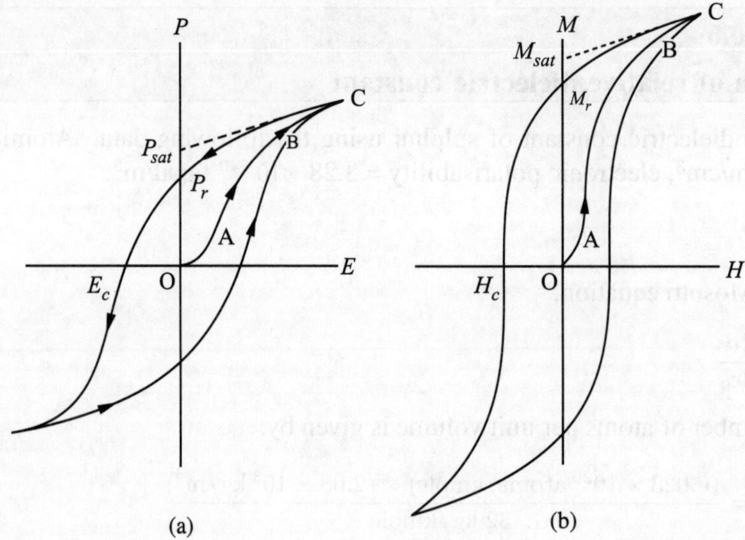

Figure 10.6 Hysteresis curve for (a) ferroelectric material and (b) ferromagnetic material

When a virgin specimen of a material is placed in a field, polarisation (magnetisation) increases nonlinearly along the curve OABC as shown in the figure and approaches the saturation value P_{sat} (M_{sat}). However, upon reversal of the field, $P(M)$ does not become zero, even though $E(H)$ is made zero. The polarisation (magnetisation) $P_r(M_r)$ present in the material for $E = 0$ ($H = 0$) is known as the *remnant polarisation* (*remnant magnetisation*). Thus the material is said to be spontaneously polarised (magnetised) and the field applied in the opposite direction that is required to reduce this polarisation (magnetisation) to zero is known as the *coercive field*, $E_c(H_c)$. Along the curve OABC, the *differential permittivity/differential permeability* are defined as

$$\epsilon_r = 1 + \left(\frac{1}{\epsilon_0}\right)\left(\frac{dP}{dE}\right) \tag{10.39}$$

and $$\mu_r = 1 + \frac{dM}{dH} \tag{10.40}$$

The internal area of the hysteresis loop is a measure of energy lost or the work done in the magnetising and demagnetising cycle (polarising and depolarising cycle). Each time the loop is traversed, energy equal to the area of the loop is dissipated as heat. Example 10.7 illustrates the calculation of power loss per second due to hysteresis in a transformer core.

EXAMPLE 10.7

Calculation of power loss due to hysteresis in a transformer core

Calculate the power loss due to hysteresis in a transformer core that is wound with a coil carrying an alternating current at a frequency of 50 Hz. The magnetisation is uniform throughout the core volume of 0.01 m^3 and the area of the hysteresis loop is 50,000 units on a graph drawn with the units of scales for the two axes as 10^{-4} Wb/m^2 and 10^2 A/m.

Solution

Energy lost during each cycle = area of the hysteresis loop

$$= 50,000 \times 10^{-4}\ \text{Wb/m}^2 \times 10^2\ \text{A/m.} = 500\ \text{J/m}^3$$

Energy lost in the core in each cycle = 500 J/m^3 $\times$ 0.01 m^3 = 5 J

Thus the power loss due to hysteresis = energy lost per cycle $\times$ number of cycles/s

$$= 5\ \text{J} \times 50\ \text{Hz} = 250\ \text{W}$$

10.5.2 Curie–Weiss Law

The phenomenon of ferroelectricity and that of ferromagnetism both vanish above a certain temperature θ_f called the *Curie temperature*. At 0 K the domains are perfectly aligned in the field direction. As temperature increases, thermal energy causes the domains to deviate from the perfect parallel alignment. Finally, as the temperature reaches the Curie temperature, ferroelectricity/ferromagnetism in the material completely disappears and the material becomes paraelectric/paramagnetic. When the material is cooled from a temperature above its Curie temperature, the domains reform and the material once again attains its ferroelectric/ferromagnetic character.

Above the Curie temperature, the permittivity/permeability follows a *Curie–Weiss law*,

$$\epsilon_r = \frac{C}{T-\theta},\ T > \theta_f \tag{10.41}$$

$$\mu_r = 1 + \frac{C}{T-\theta} \quad \text{or} \quad \chi = \frac{C}{T-\theta},\ T > \theta_f \tag{10.42}$$

where C is called the *Curie constant* and θ is a characteristic temperature that is usually a few degrees smaller than the ferroelectric Curie temperature. However, in case of ferromagnetic materials, θ is usually somewhat higher than the ferromagnetic Curie temperature (Fig. 10.7). Thus the material that has become paramagnetic from the initial ferromagnetic state differs from the truly paramagnetic material in that $\theta = 0$ K for the latter (compare Fig. 10.7 with Fig. 10.4b).

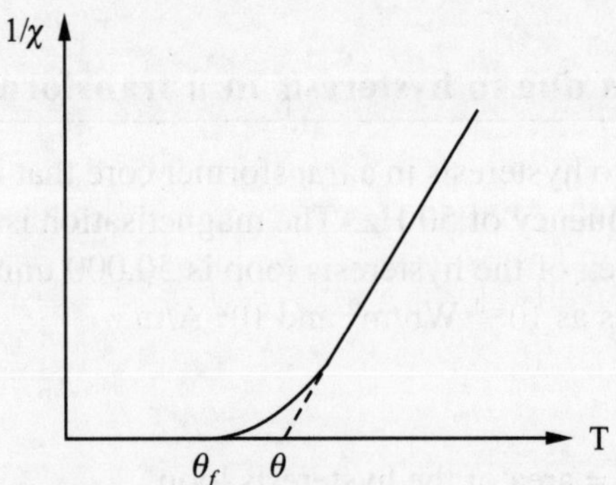

Figure 10.7 The reciprocal of susceptibility as a function of temperature for a ferromagnetic material showing θ_f and θ, the ferromagnetic and the paramagnetic Curie temperature, respectively.

The Curie–Weiss law can be easily obtained if we extend Eq. 12 of Chart 10.3 to the present case by replacing H by the internal field, H_i. Thus,

$$M = N\beta \tanh\left(\frac{\mu_0 \beta H_i}{kT}\right) \tag{10.43}$$

For sufficiently high temperature, $\mu_o \beta H_i \ll kT$ and as $\tanh(a) \cong a$ for $a << 1$, we have,

$$M = \frac{N\mu_0 \beta^2 H_i}{kT} \tag{10.44}$$

Using Eq. 10.38, we get

$$M = \frac{N\mu_0 \beta^2}{kT}(H + \gamma M) \tag{10.45}$$

or $$M = \frac{(N\mu_0\beta^2/kT)H}{1 - (N\mu_o\beta^2\gamma/kT)} \tag{10.46}$$

Substituting, $$C = \frac{N\mu_0\beta^2}{k} \tag{10.47}$$

and $$\theta = \gamma C \tag{10.48}$$

we get $\chi = \dfrac{M}{H} = \dfrac{C}{(T - \theta)}$ which is the same as the Curie–Weiss law.

If the experimental values of C and θ are substituted in the above equation, we get $\gamma \approx 10^3$. This value is not in agreement with the value obtained if one uses the classical model of the internal field as arising from the magnetic interaction of atomic dipoles (see Ex. 10.8). One needs to resort to the quantum concept involving exchange forces to explain this interaction.

EXAMPLE 10.8
Failure of the classical model of internal field

Calculate the energy of a point dipole in the field of another point dipole if the two dipoles are placed parallel to each other and parallel to the line joining their centres. Assume the strength of each dipole to be 1 Bohr magneton and the distance between the dipoles as 2Å.

Solution

The axial magnetic field of a magnetic dipole of strength μ_d at a distance r from the dipole is $\mu_d/2\pi r^3$ and the corresponding magnetic flux density is $B = \mu_0\mu_d/2\pi r^3$.

The energy of another aligned dipole μ at this distance is thus $U = \mu_d B = -\mu_0\mu_d^2/2\pi r^3$

Substituting 1 Bohr magneton for μ_d and 2 Å for r we get,

$$U = \frac{4\pi \times 10^{-7}\,\text{H/m} \times (9.27 \times 10^{-24}\,\text{Am}^2)^2}{2 \times (2 \times 10^{-10}\,\text{m})^3} = 21.5 \times 10^{-25}$$

This is equivalent to kT with $T = 0.15$ K. Thus the ferromagnetic interactions cannot be explained classically because the interaction energy should be of the order of kT where $T = \theta_f = 1000$ K.

10.5.3 Spontaneous Polarisation/Magnetisation

The polarisation of a solid dielectric material is given by $P = N\alpha E_i$, where α is the total polarisability and the internal field, E_i is given by Eq. 10.34. Simplifying we get,

$$P = \frac{N\alpha E}{1 - (N\alpha\gamma/\epsilon_0)} \tag{10.49}$$

The equation indicates the possibility of spontaneous polarisation in ferroelectric materials. For $N\alpha\gamma/\epsilon_0 = 1$, we get a non-zero value of P for zero electric field. Thus $N\alpha\gamma/\epsilon_0 = 1$ is the condition for spontaneous polarisation.

Spontaneous magnetisation can be understood by substituting the internal field H_i given by Eq. 10.38 in Eq. 10.43. Thus,

$$M = N\beta \tanh\left[\frac{\mu_0\beta}{kT}(H + \gamma M)\right] \tag{10.50}$$

The Curie–Weiss law tells us that the susceptibility and hence the magnetisation are non-zero for zero value of H at $T = \theta$, thus indicating the possibility of spontaneous magnetisation. Substituting $H = 0$ in Eq. 10.50 we get.

$$M = N\beta \tanh\left[\frac{\mu_0 \beta\gamma M}{kT}\right] \tag{10.51}$$

The maximum value of M is $N\beta$, when all the dipoles are aligned in the field direction (see Ex. 10.9). Denoting this by M_{sat}, that is, the *saturation magnetisation*, and substituting

$$x = \frac{\mu_0 \beta\gamma M}{kT} \tag{10.52}$$

in Eq. 10.51, we get,

$$\frac{M}{M_{\text{sat}}} = \tanh(x) \tag{10.53}$$

EXAMPLE 10.9
Saturation magnetisation for pure iron

Calculate a theoretical value for the saturation magnetisation of pure iron, assuming all magnetic moments due to the four unpaired $3d$ electrons to be aligned along the magnetic field. Pure iron has a BCC unit cell with a lattice constant equal to 2.87 Å.

Solution

The saturation magnetisation is given by $M_{\text{sat}} = N\mu$
where μ is equal to 4 Bohr magnetons and N, the number of atoms per cubic metre, is given by

$$N = \frac{2 \text{ atoms/unit cell}}{(2.87 \times 10^{-10}\text{ m})^3/\text{unit cell}} = 8.46 \times 10^{28} \text{ atoms/m}^3$$

Thus,

$$M_{\text{sat}} = (8.46 \times 10^{28}\text{ atoms/m}^3) \times (4\text{ Bohr magnetons/atom}) \times$$

$$(9.27 \times 10^{-24}\text{Am}^2/\text{Bohr magneton})$$

or $M_{\text{sat}} = 3.14 \times 10^6\,\text{A/m}$

Figure 10.8 shows the graph of M/M_{sat} as a function of x. Now using Eq. 10.52 we also have

$$\frac{M}{M_{\text{sat}}} = \left(\frac{kT}{N\mu_0\beta^2\gamma}\right)x \tag{10.54}$$

Using Eqs. 10.47 and 10.48 we get,

$$\frac{M}{M_{\text{sat}}} = \left(\frac{T}{\theta}\right) x \tag{10.55}$$

Thus, M/M_{sat} depends linearly on x, T/θ being the slope of the line. Thus, Fig. 10.8 shows the variation of M/M_{sat} with x for different T/θ values, that is, different slopes. The line with a smaller slope for which $T' < \theta$ intersects the $\tanh(x)$ curve. The point of intersection gives the value of M/M_{sat} at $T = T'$. For $T \geq \theta$, the only point of intersection is $x = 0$ indicating no spontaneous magnetisation.

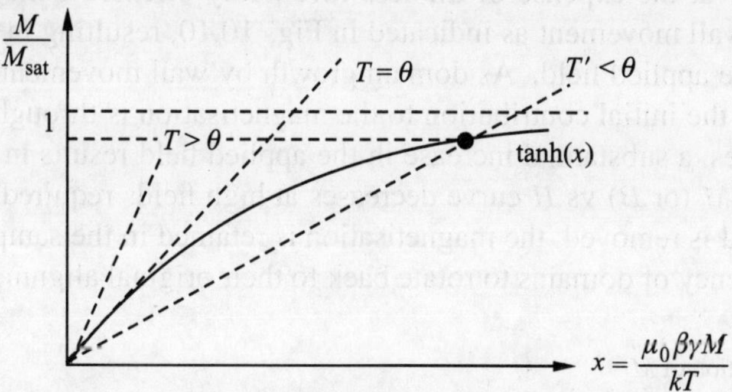

Figure 10.8 Variation of M/M_{sat} with x for different T/θ values

To study the variation of M/M_{sat} with temperature we find the points of intersection for a large number of straight lines with different slopes. Figure 10.9 shows the plot of M/M_{sat} as a function of T/θ. It can be seen from this plot that spontaneous magnetisation (that is, the ferromagnetic behaviour) lasts only up to $T \leq \theta$. For $T > \theta$, the spontaneous magnetisation vanishes and the material shows paramagnetic behaviour. At $T = 0$, the spontaneous magnetisation becomes equal to M_{sat}.

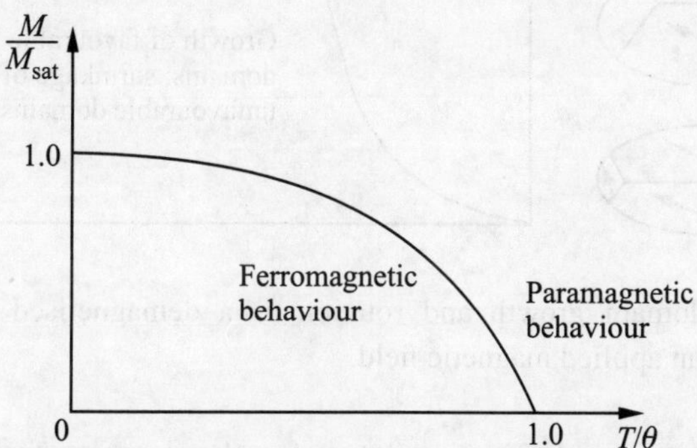

Figure 10.9 M/M_{sat} as a function of T/θ

10.5.4 Ferromagnetic Domains

Ferromagnetic domains are microscopic volume regions in which the magnetic dipole moments of atoms of ferromagnetic materials tend to align themselves in a parallel direction below the Curie temperature. In the absence of an external field the domains are aligned at random so that there is no net magnetic moment for a bulk sample. All the magnetic dipoles in each domain are aligned, but the domains themselves are aligned at random so that there is no net magnetisation. Magnetic domains are formed in a crystal due to the creation of positive exchange energy between the dipoles in the domain. This is possible if the ratio of the atomic spacing to the diameter of the 3d orbit is in the range 1.4–2.7. For this reason iron (Fe), cobalt (Co) and nickel (Ni) are ferromagnetic, but manganese (Mn) and chromium (Cr) are not.

When an external magnetic field is applied, domains whose moments are initially parallel to the applied field grow at the expense of the less favourably oriented domains. These domains grow by the domain wall movement as indicated in Fig. 10.10, resulting in the rapid increase of magnetisation with the applied field. As domain growth by wall movement requires less energy than domain rotation, the initial contribution to the magnetisation is through this process. When domain growth finishes, a substantial increase in the applied field results in rotation of domains. Thus the slope of the M (or B) vs H curve decreases at high fields required for domain rotation. When the applied field is removed, the magnetisation is retained in the sample though some of it is lost due to the tendency of domains to rotate back to their original alignment.

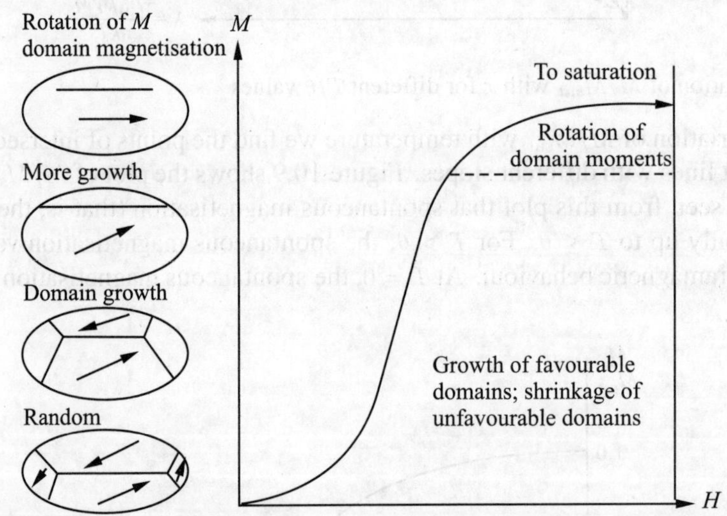

Figure 10.10 Magnetic domain growth and rotation as a demagnetised ferromagnetic material is magnetised to saturation by an applied magnetic field.

The following different types of energies determine the domain structure of a ferromagnetic material. The most stable structure is attained when the net potential energy of the material is minimum.

1. *Exchange energy:* It is the energy associated with the coupling of individual magnetic dipoles into a single magnetic domain.
2. *Magnetostatic energy:* It is the magnetic potential energy due to the external magnetic field surrounding a sample of ferromagnetic material.
3. *Magnetocrystalline anisotropy energy:* It is the energy required during the magnetisation of a

ferromagnetic material to rotate the magnetic domains because of crystalline anisotropy. For example, the magnetising energy required in Fe for the [111] direction of magnetisation is much larger compared to that for the [100] direction. The [111] and the [100] directions are thus referred to as the hard and the easy directions of magnetisation in Fe.

4. *Domain wall energy:* It is the energy associated with the disorder of dipole moments in the wall volume between magnetic domains.
5. *Magnetostrictive energy:* It is the energy due to the mechanical stress caused by magnetostriction (discussed in the next section) in a ferromagnetic material.

10.5.5 Electrostriction, Piezoelectricity and Magnetostriction

The polarisation of a dielectric material results in ionic displacement causing a slight strain in the material. This electromechanical effect, that is, the mechanical distortion due to an electric field is called *electrostriction* and the materials, electrostrictive materials. The reverse effect, that is, a mechanical stress resulting in the polarisation of a material is known as the *piezoelectric effect*. Electrostriction is present in all materials but only those materials, which become polarised due to application of mechanical stress, exhibit piezoelectricity.

Ferroelectric crystals always exhibit the piezoelectric property. In the absence of an external field the centres of the positive and the negative charges have a certain separation in the specimen of such a crystal. When an external voltage is applied this separation increases or decreases, thereby elastically straining the crystal. Conversely, a compressive stress that causes a decrease in the separation produces a potential difference between the two ends of a crystal and a tensile stress that causes an increase in the separation produces a potential difference of opposite sign.

Similarly, when a ferromagnetic material is magnetised, its dimensions undergo a slight change and it either expands or contracts in the direction of magnetisation. This magnetically induced reversible elastic strain is called *magnetostriction*. Magnetostriction is attributed to the change in the bond length between the atoms in a ferromagnetic material when their electron-spin dipole moments rotate to align in the field direction. The fields of the dipoles may attract or repel each other, causing the material to contract or expand.

10.5.6 Antiferromagnetism and Ferrimagnetism

In some materials the exchange forces produce a tendency for antiparallel alignment of electron spins of neighbouring atoms. These materials are known as *antiferromagnetic materials* if the spins of neighbouring atoms are of equal magnitude (Fig. 10.11a). Manganese and chromium are antiferromagnetic in the solid state at room temperature. They have negative exchange energy because the ratio of their atomic spacing to diameter of the $3d$ orbit is less than about 1.4.

In some materials the constituent atoms are antiferromagnetically coupled but due to unequal moments they do not cancel (Fig. 10.11b). Thus, the moments of all couples add to produce a net magnetic dipole moment and hence a net magnetisation. These materials are called *ferrimagnetic materials*. As a group, ferrimagnetic materials are called *ferrites*.

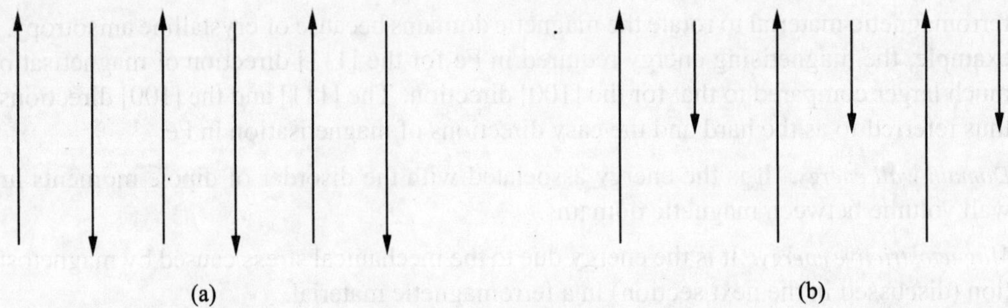

Figure 10.11 Coupling of electron spins: (a) Antiferromagnetic coupling and (b) Ferromagnetic coupling. The length of an arrow is a measure of the magnetic moment of the atom

The antiferromagnetic materials are characterised by a Neel temperature, just as ferromagnetic materials are characterized by the Curie temperature. Below the Neel temperature the spins tend to order. Thus at 0 K, the ordering is such that equal number of spins are oriented parallel and antiparallel to the applied field. This is in contrast to the ferromagnetic material in which all spins are aligned in the field direction at 0 K.

Experimentally it is observed that the susceptibility of antiferromagnetic materials is maximum at the Neel temperature. Above the Neel temperature, T_N, the antiferromagnetic materials are found to follow an equation analogous to the Curie–Weiss law,

$$\chi = \frac{C}{(T+\theta)}, \quad \text{for } T > T_{\text{N}} \tag{10.56}$$

where C is the Curie constant and θ is the paramagnetic Curie temperature. Figure 10.12 shows the variation of χ as a function of $1/T$ for antiferromagnetic materials. The variation for the paramagnetic and the ferromagnetic materials is also included for comparison.

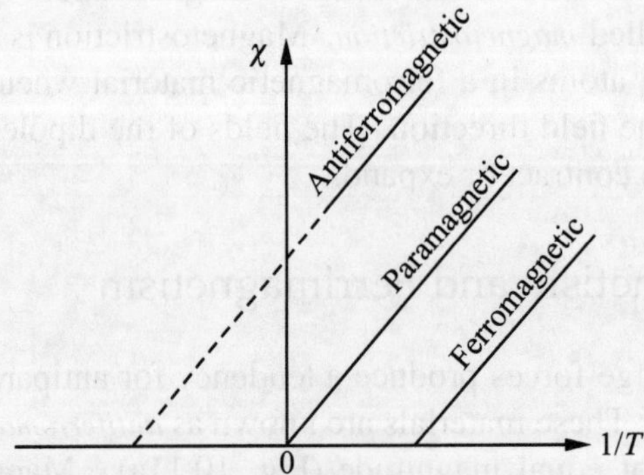

Figure 10.12 Variation of susceptibility with reciprocal temperature for paramagnetic, ferromagnetic and antiferromagnetic material

To understand how Eq. 10.56 describes the situation, we consider an antiferromagnetic material having a BCC structure as shown in Fig. 10.13. All atoms are identical, carrying magnetic dipole moment of 1 Bohr magneton that can either be in 'spin up' or 'spin down' state. The spin of atoms occupying the body centre site is opposite to the spin of atoms occupying corner B sites.

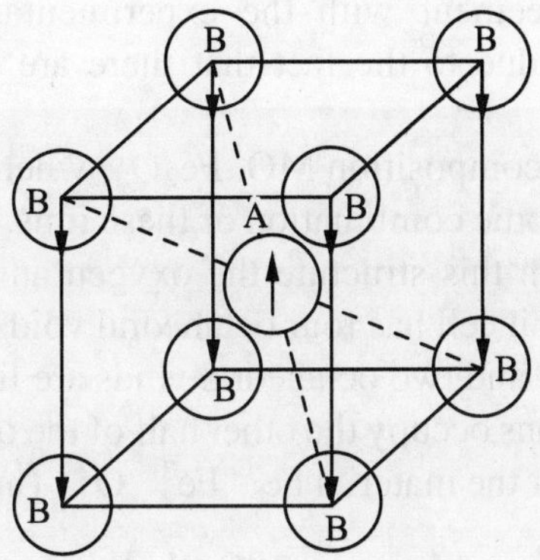

Figure 10.13 Antiferromagnetic BCC crystal showing two sublattices, A and B, having antiparallel spins

Using Eq. 10.38 we can write the internal fields $\mathbf{H_a}$ and $\mathbf{H_b}$ at sites A and B respectively as

$$\mathbf{H_a} = \mathbf{H} - \gamma\mathbf{M_b} \text{ and } \mathbf{H_b} = \mathbf{H} - \gamma\mathbf{M_a} \tag{10.57}$$

where $\mathbf{M_a}$ and $\mathbf{M_b}$ are the magnetisations of the A sublattice and the B sublattice, respectively. The negative sign indicates that an atom at A site tends to align its dipole moment opposite to the direction of the magnetisation of the B sublattice. The internal field constant γ measures the strength of the exchange interaction.

Now, using Eq. 10.43, we have,

$$\mathbf{M_a} = N\beta \tanh\left[\frac{\mu_0\beta}{kT}(\mathbf{H} - \gamma\mathbf{M_b})\right] \text{ and } \mathbf{M_b} = N\beta \tanh\left[\frac{\mu_0\beta}{kT}(\mathbf{H} - \gamma\mathbf{M_a})\right] \tag{10.58}$$

For very high temperature, $\tanh x \cong x$. Thus we get,

$$\mathbf{M_a} = \frac{N\mu_0\beta^2}{kT}(\mathbf{H} - \gamma\mathbf{M_b}) \text{ and } \mathbf{M_b} = \left(\frac{N\mu_0\beta^2}{kT}\right)(\mathbf{H} - \gamma\mathbf{M_a}) \tag{10.59}$$

Hence the total magnetisation is

$$\mathbf{M} = \mathbf{M_a} + \mathbf{M_b} = \left(\frac{N\mu_0\beta^2}{kT}\right)(2\mathbf{H} - \gamma\mathbf{M}) \tag{10.60}$$

As $\mathbf{M}$ and $\mathbf{H}$ are in the same direction we can express Eq. 10.60 in the scalar form as

$$\frac{M}{H} = \frac{N\mu_0\beta^2}{kT}\left[2 - \gamma\left(\frac{M}{H}\right)\right]$$

$$\text{or} \quad \chi = \frac{2C}{(T+\theta)}, \text{ where } C = \frac{N\mu_0\beta^2}{k} \text{ and } \theta = \gamma C \tag{10.61}$$

Thus, this equation shows agreement with the experimentally observed relation given by Eq. 10.56. The factor of 2 is due to the fact that there are two types of sites—A and B—both numbering N.

Most simple ferrites have the composition MO. Fe_2O_3, where M is a divalent metal ion such as Fe^{2+}, Mn^{2+}, Co^{2+}, Mg^{2+} or some combination of these ions. Most of these ferrites crystallise in the inverse spinel structure. In this structure the oxygen anions form the FCC packing. As discussed in Chart 9.2, an FCC unit cell has four octahedral voids and eight tetrahedral voids. Out of these only one tetrahedral void and two octahedral voids are filled. The metal ions occupy half of the octahedral sites and Fe^{3+} ions occupy the other half of the octahedral sites and all tetrahedral sites. For example, the unit cell of the material $Fe^{2+}Fe_2^{3+}O_4^{2-}$ (magnetite) is shown in Fig. 10.14.

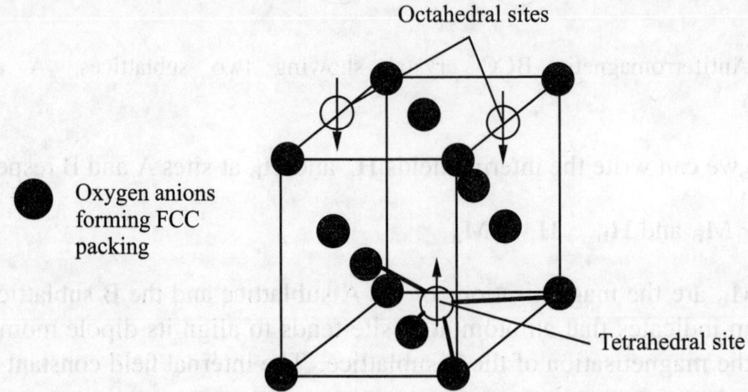

Figure 10.14 Unit cell of $Fe^{2+}Fe_2^{3+}O_4^{2-}$

The Fe^{2+} ion has a magnetic moment of 4 Bohr magnetons as it has four unpaired $3d$ electrons left after the loss of two $4s$ electrons. Similarly, the Fe^{3+} ion has a magnetic moment of 5 Bohr magnetons due to the loss of two $4s$ and one $3d$ electron. The antiferromagnetic coupling between the octahedral and the tetrahedral sites results in the mutual cancellation of moments due to Fe^{3+} ions (Fig. 10.15). Thus, the net magnetic moment is due to Fe^{2+} ions alone (or the appropriate metal ion), that is, 4 Bohr magnetons—a value that is in agreement with the experiment.

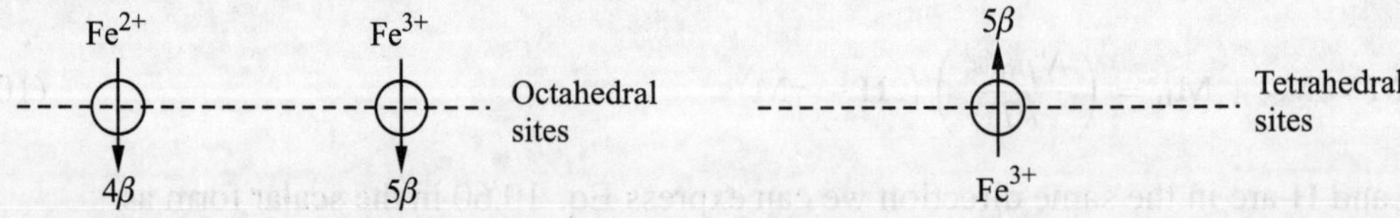

Figure 10.15 Antiferromagnetic coupling between octahedral and tetrahedral sites results in net magnetic moment of 4 Bohr magnetons in a magnetite unit cell

CHART 10.4 CLASSIFICATION OF DIELECTRIC MATERIALS

An electrical field applied to a material can bring about relative displacement between centres of positive and negative charges within molecules. It can also reorient molecules in which these centres are already relatively displaced. Behaviour of materials is determined by the relative importance of these phenomena.

Elemental dielectrics	Ionic dielectrics without permanent dipoles	Polar dielectrics
Dielectric materials consisting of atoms of only one element, for example, diamond and phosphorous, are called elemental dielectrics. As there are no permanent dipoles only the contribution due to electronic polarisation is present. Equations 10.11 and 10.37 are thus valid for elemental dielectrics. As the quanti-ties on the right hand side of these equations are affected only slightly by temperature, the dielectric constant ε_r is almost independent of temperature. $\varepsilon_r - 1$ for most of the solids is of the order of unity or ten, whereas for gases it is of the order of 10^{-3} or 10^{-4}. This difference is due to the number of atoms per unit volume.	In inoic crystals (for example, alkali halides) the total polarisation is the sum of electronic and ionic polarisation. Thus $P = P_e + P_i$ and $\alpha = \alpha_e + \alpha_i$. It should be noted that P_i usually constitutes a considerable fraction of the total polarisation. At optical frequencies the contribution of ionic polarisation to the total polarisation is negligible because the relatively heavy positive and negative ions can no longer follow the rapid field variations and only electronic polarisation is present. Thus, $P_e = \varepsilon_0(\varepsilon_{re} - 1) E,$ (1) where ε_{re} is the dielectric constant measured at optical frequencies. At much lower frequencies the dielectric constant is denoted as ε_{rs}, that is the static dielectric constant, and both P_i and P_e contribute to the total polarisation in this case. Thus, $P_i + P_e = \varepsilon_0(\varepsilon_{rs} - 1) E$ (2) From Eqs. 1 and 2, $P_i = \varepsilon_0(\varepsilon_{rs} - \varepsilon_{re}) E$. The difference between ε_{rs} and ε_{re} is thus a measure of the ionic polarisation. The dielectric constant ε_{re} is often referred to as the index of refraction and is denoted by n^2 (See Example 10.10).	They are materials in which molecules carry a permanent dipole moment, for example, solid nitrobenzene. In the solid state, as the permanent dipoles are essentially frozen, that is, they cannot orient in response to the applied field, only the electronic and ionic polarisations contribute to the total polarisation. However, on melting all the three mechanisms contribute and as the orientation polarisation decreases with increasing temperature, liquids show a decrease in their dielectric constant with increasing temperature. Figure 1 shows the variation of ε_r with temperature for nitrobenzene. 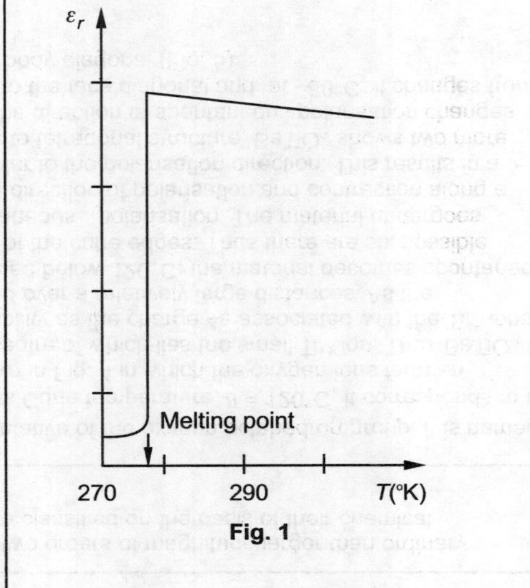 **Fig. 1**

Ferroelectric materials

Ferroelectric naterials are characterised by a large dielectric constant: of the order of thousand or about two orders of magnitude larger than ordinary dielectrics. They exhibit nonlinear dependence of P on E and hysteresis effect. Ferroelectric materials are classified on the basis of their chemical composition and structure as discussed below.

The first ferroelectric material discovered was the Rochelle salt ($NaKC_4H_4O_6.4H_2O$), a representative of the tartarate group. It has two transition temperatures, –18°C and 23°C, that is, it is ferroelectric in the range –18°C – 23°C (Fig. 2).

The ferroelectric material KH_2PO_4, discovered by Busch and Scherrer in 1935, is a representative of the dihydrogen phosphates and arsenates and the alkali metals. Its Curie temperature is 123 K, that is, it is ferroelectric for temperatures below 123 K (Fig. 3).

$BaTiO_3$ is a representative of the oxygen octahedron group. It is named so because above its Curie temperature, θ = 120°C, it corresponds to the cubic structure shown in Fig. 4 in which the oxygen ions form an octahedron, at the centre of which lies the small Ti^{4+} ion. Thus $BaTiO_3$ has a high ionic polarisability as the charge 4e associated with the Ti^{4+} ions that can be displaced over a relatively large distances. As the temperature is reduced below 120°C, the material becomes spontaneous polarised along any of the cube edges. Thus there are six possible directions for spontaneous polarisation. The material undergoes expansion along the direction of polarisation and contraction along a direction perpendicular to the polarisation direction. This results in a transition from cubic to tetragonal structure. $BaTiO_3$ shows two more transitions. At 5°C, the direction of spontaneous polarisation changes from the cube edge to the face diagonal and at –80°C, it changes from face diagonal to the body diagonal (Fig. 5).

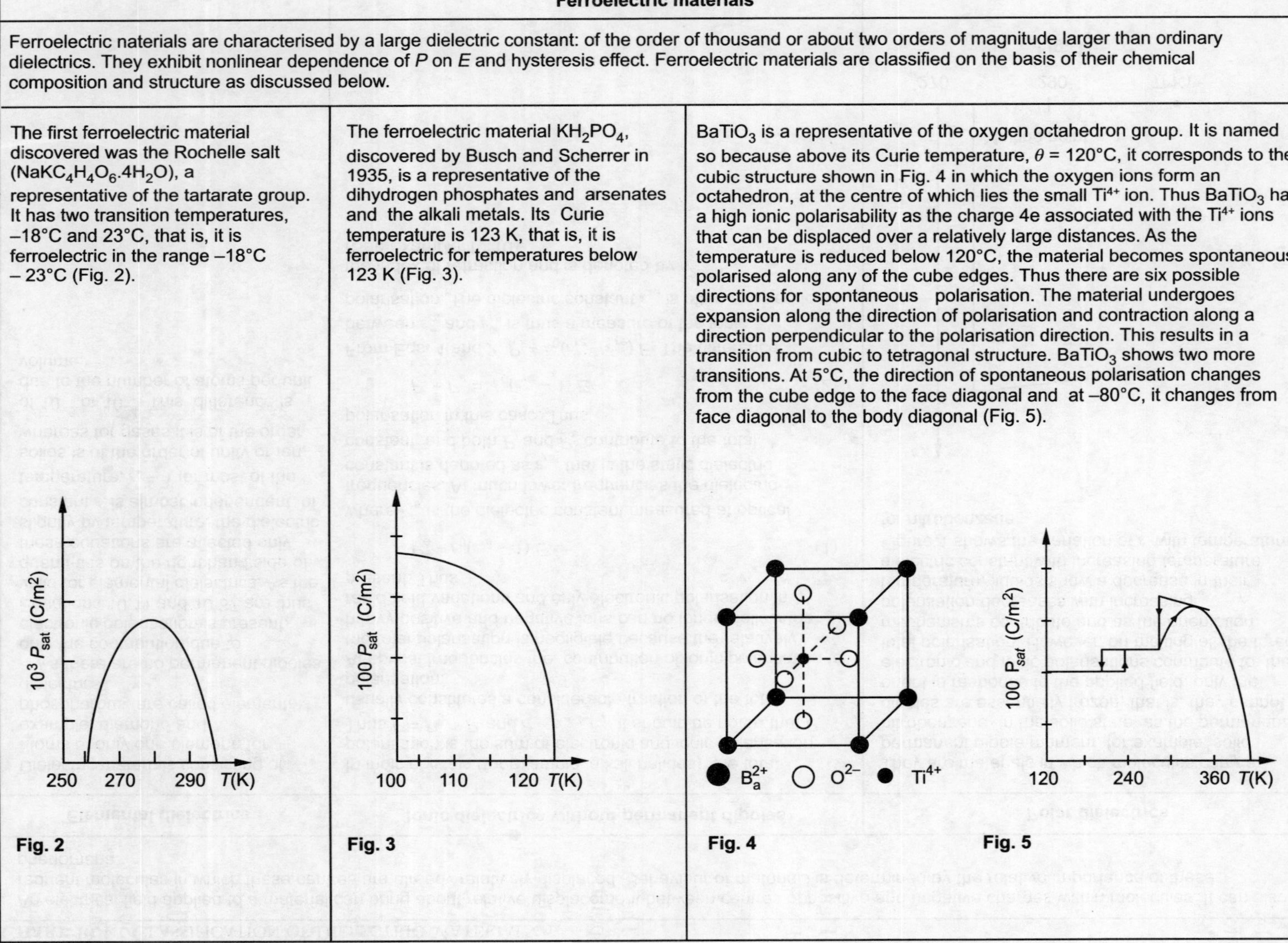

Fig. 2

Fig. 3

Fig. 4

Fig. 5

The magnetic moment of ferrites can be enhanced by replacing some of the magnetic Fe^{2+} ions by non-magnetic ions like Zn^{2+} or Cd^{2+}. This can be understood as follows. Zinc ions prefer tetrahedral sites forcing some of the Fe^{3+} ions from tetrahedral sites to octahedral sites. As zinc has no magnetic moment, the net magnetic moment increases as shown in Fig. 10.16.

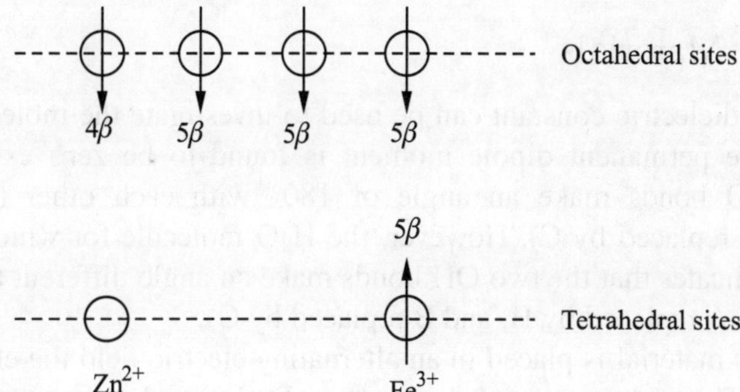

Figure 10.16 Addition of non-magnetic zinc to magnetite increases magnetisation because the Zn^{2+} ions prefer tetrahedral positions and force Fe^{3+} ions to move into octahedral sites

EXAMPLE 10.10
Percentage of electronic and ionic polarisability

The value of ε_r for glass is 6.75 at frequencies 10^9 Hz. What percentage may be attributed to ionic polarisability? Given: refractive index of glass is 1.5.

Solution

At optical frequencies $P_e = \epsilon_0(\epsilon_{re} - 1)E = \epsilon_0(n^2 - 1)E$ and at lower frequency $P = P_i + P_e = \epsilon_0(\epsilon_{rs} - 1)E$.

Thus, $\quad P_i = \epsilon_0(\epsilon_{rs} - n^2)E$

Percentage that may be attributed to ionic polarisation

$$= \left[\frac{P_i}{(P_e + P_i)}\right] \times 100 \quad = \left[\frac{\epsilon_0(\epsilon_{rs} - n^2)E}{\epsilon_0(\epsilon_{rs} - 1)E}\right] \times 100 \quad = \left[\frac{(6.75 - 2.25)}{(6.75 - 1)}\right] \times 100 = 78.3\%$$

10.6 CLASSIFICATION OF MATERIALS

So far we studied the various mechanisms of polarisation and magnetisation and related phenomena. This study forms a basis for classification of materials. Charts 10.4 and 10.5 illustrate, respectively, the classification that facilitates comparative study of various dielectric and magnetic materials.

10.7 APPLICATIONS

In this chapter we were able to appreciate the basic aspects of the physics of dielectric and magnetic materials. These materials serve as useful engineering materials in a vast variety of applications that are discussed briefly in the following.

10.7.1 Dielectric Materials

Experimental data of dielectric constant can be used to investigate the molecular structure. For example, for CO_2 the permanent dipole moment is found to be zero experimentally. This indicates that the CO bonds make an angle of 180° with each other (Fig. 10.2a with A replaced by O, and B replaced by C). However, the H_2O molecule for which μ_p is found to be 1.84 debye, which indicates that the two OH bonds make an angle different than 180° with each other (Fig. 10.2b with A replaced by H, and B replaced by O).

When the dielectric material is placed in an alternating electric field the electrical energy gets converted into heat. This is known as *dielectric loss.* Dielectric heating is widely employed in industry for preheating in the moulding of plastics, for quick heating of thermosetting glues in cabinet and furniture making, for accelerating jelling and drying of foam rubber, in foundry core baking and for drying of paper and textile products. Dielectric materials are also used as insulation for wires, cables and electrical equipment, as polarisable media for capacitors, in devices used for the propagation or refraction of electromagnetic waves, and for a variety of dielectric devices, such as rectifiers and dielectric amplifiers.

The high dielectric constant of ferroelectric crystals makes them useful for storing energy in small-sized capacitors in electrical circuits. The names Seignette–electrics or Rochelle–electrics, that are widely used, are derived from the name of the first substance found to have the property of ferroelectricity, Seignette salt or Rochelle salt.

Transducers using piezoelectric elements are used for converting vibrations into electrical energy in such applications as, crystal microphones, phonograph pickups, vibration pickups, gas lighters, dynamic pressure sensing elements and electromechanical resonators such as frequency controlling quartz crystal. The use of quartz crystal for oscillator stabilisation has made possible the modern radio and TV broadcasting industry and mobile radio communications with aircraft and ground vehicles. Quartz crystal resonators are also used in electric—wave or frequency—separation filters that find applications in telephone systems for carrier frequency separation and in radio communication for selecting a desired signal frequency band while rejecting undesired frequencies. Piezoelectric transducers that convert electrical energy into vibrations are used in such application as ultrasonic generator, underwater sound ranging equipment (sonar devices) and in ultrasonic cleaning devices, which use a liquid medium for washing small- and medium-sized objects.

10.7.2 Magnetic Materials

Among the magnetic materials the diamagnetic and paramagnetic materials are of little importance to engineers because of their weaker magnetic properties. However, superconductors,

Chart 10.5 Classification of Magnetic Materials

Absence of magnetic monopoles in nature means that atomic and molecular magnetism arise from electron currents within them. Materials respond to a magnetic field by modifying these currents and by reorienting them. Generally the former type of response is relatively much less important.

	Diamagnetic	Paramagnetic	Ferromagnetic	Antiferromagnetic	Ferrimagnetic
	Lack permanent dipole moments		Permanent dipole moments are associated with atoms in the material		
Response to the applied field	Filed lines are repelled by magnetic field 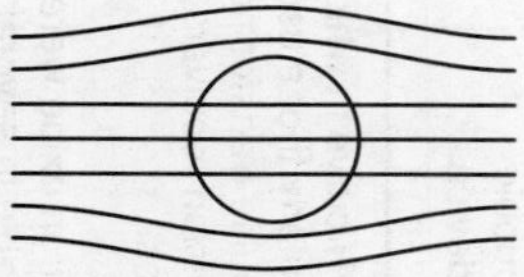	Field lines are pulled towards the material and permeate through it 	When a ferromagnetic material is placed in a magnetic field the field lines crowd into the material 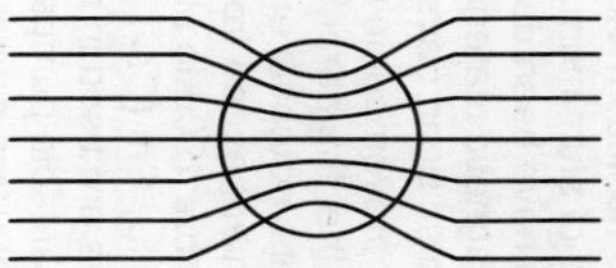		
Interaction between individual dipoles		Zero or negligible 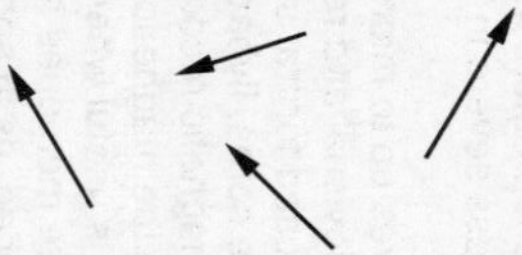	Parallel orientation 	Antiparallel orientation of equal moments 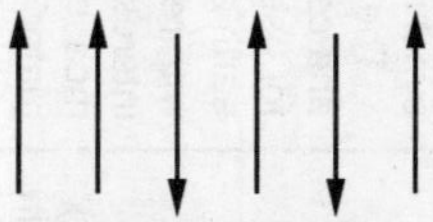	Antiparallel orientation of unequal moments 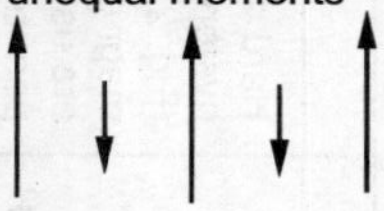
Suscepti-bility, χ	Small and negative, of the order of 10^{-5}; often cancelled by positive magnetic effects;	Of the order of 10^{-3} at room temperature	Of the order of 10^{6}	Inversely proportional to temperature, governed by the Curie–Weiss law $\chi = C/(T+\theta)$, for $T > T_N$.	
Tempera-ture depen-dence	Independent of the temperarure as long as the electronic structure of the material is independent of the temperature	Inversely proportional to temperature, governed by the Curie law $\chi = C/T$	Inversely proportional to temperature, governed by the Curie–Weiss law $\chi = C/(T-\theta)$, for $\theta > \theta_f$		
Magnetic permea-bility	μ_r is slightly smaller than 1, that is, μ is slightly smaller than μ_0	μ_r is slightly greater than 1, that is, μ is slightly greater than μ_0	μ_r is of the order of 10^3, that is, $\mu >> \mu_0$		
Materials	Diamagnetism occurs in all materials; some prominent examples are cadmium, copper, silver, tin and zinc	Aluminium, calcium, oxygen, platinum and titanium	Iron, cobalt, nickel	Manganese, chromium	Magnetite, Fe_3O_4 (loadstone)

Chart 10.6 Soft and Hard Magnetic Materials

Soft magnetic materials

Soft magnetic materials have narrow hysteresis loops with low coercive forces (Fig. 1) , which make them easy to magnetise and demagnetise. Thus, they are useful for applications such as cores for large power transformers, small electronic transformers and stator and rotor materials for motors and generators.

Fig. 1

Iron, which has been mainly used as a soft magnetic material, underwent many modifications in the last hundred years for improvement in its performance. It has been alloyed with 3%–4% silicon for reducing core losses. Further improvement was achieved first by the use of laminated (stacked sheet) structure and then, in 1940, by the production of grain oriented iron–silicon sheet. Iron–cobalt and iron–nickel alloys have also been developed for applications in air–borne high speed generators and motors, servometers, loudspeaker telephone diaphragms, magnetic pole tips and in audio and instrument transformers, instrument relays and for rotor and stator laminations respectively.

Hard magnetic materials

Hard magnetic materials have wide hysteresis loops with high coercive forces (Fig. 2), which make them difficult to magnetise and demagnetise. Thus they are useful for applications requiring permanent magnets such as those in loudspeakers, telephone receivers and synchronous and brushless motors.

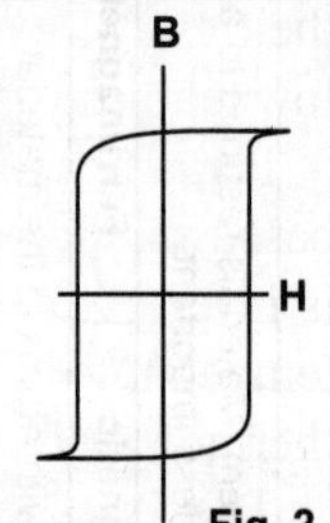

Fig. 2

The alloys made of aluminium, nickel and cobalt (alnico) have been developed for use as hard magnetic materials. Rare earth alloys have also been developed with remarkable improvement in magnetic strength. It is interesting to note that the parameter $(BH)_{max}$ which is the measure of magnetic potential energy of a hard magnetic materials has been enhanced from its value less than 10 kJ/m^3 for steels before 1900 to about 240 kJ/m^3 by 1980 for a commercial alloy Sm $(Co_{0.68}\ Cu_{0.10}Fe_{0.21}\ Zr_{0.01})_{7.4}$. Rare earth magnets are used in medical devices such as thin motors in implantable pumps and valves and in aiding eyelid motion, for electronic wrist watches, travelling wave tubes, and in direct current and synchronous motors and generators.

Soft ferrites

A magnetic material used in high frequency applications should have high resistivity because, if the material is conductive, then eddy current energy losses would be greater at the higher frequencies. The resistivity of soft ferrites is typically 10^6–10^{11} times higher than that of metals. Hence the eddy current problem that prevents penetration of magnetic flux into the material is much less severe in ferrites than in iron.

Soft ferrites can therefore be used in transformer cores up to microwave frequencies. They are widely used for low signal, memory core audiovisual and recording head applications. At low signal levels soft ferrite cores are used for transformers and low energy inductors. They are also used in deflection yoke cores, flyback transformers and convergence coils for television receivers and in magnetic recording heads for various types of magnetic tapes. Some computers use the magnetic core memories based on the 0 and 1 binary logic. The magnetic core is useful where loss of power does not cause loss of information. Since magnetic core memories have no moving parts they are used when high shock resistance is needed, as in some military uses.

Modern ferrite devices stemmed from the contributions of the Japanese and Dutch scientists during World War II. In 1956, the ferromagnetic garnets were discovered in France and the United States. Single crystal garnet films form the basis of bubble domain device technology and bulk garnets find applications in microwave devices.

Hard ferrites

In 1952, the Philips company introduced the magnetically hard ferrite—barium ferrite—in the Netherlands under the trade name feroxdure. Permanent magnets made using the hard ceramic ferrites find wide applications in generators, relays, motors, loudspeakers, telephone ringers, and receivers. They are also used in holding devices for door closers, seals and latches and in many toy designs.

Ferrite devices are sometimes classified on the basis of the region of the *B/H* curve in which the device operates. The most important devices in the linear region are high-quality inductors, particularly those used in filters in frequency-division multiplex communication systems and low power, wide band and pulse transformers. Nonlinear B/H devices find usage in the form of deflecting yokes and flyback transformers for television receivers. Highly nonlinear B/H devices are used as square loop toroidal cores for digital computer memories.

Magnetic Storage: Tape recorders using thin plastic coated with magnetic iron oxide were introduced in 1935 for the German magnetophone. Audio magnetic recorders have become indispensable in several walks of modern life. Sheets of magnetic material with thickness of a few micrometres or less, called as magnetic thin films, are widely used in the electronics industry. Magnetic films can be single crystalline, polycrystalline or amorphous and their applications include magnetic bubble technology. Magnetic bubble memory is a memory storage technology that uses localised magnetised regions to store information. It is the integrated ciruit analogy to rotating magnetic memories such as discs and tapes.

which show perfect diamagnetism, form an important group of engineering materials (see Chapter 11). Paramagnetic salts entered the group of electrical engineering materials by virtue of their essential role in solid state masers. They are the working materials used in obtaining very low temperatures (<1 K) by adiabatic demagnetisation.

Paramagnetic materials also find applications in thermometry. A magnetic thermometer is based on Curie's law, which states that the magnetic susceptibility of a paramagnetic material is inversely proportional to the temperature. Magnetic thermometers have been devised for the temperature range 1mK to 1 K using paramagnetic thermometric materials. Nuclear magnetic thermometers have also been developed for use only in the ultra-low temperature region.

The ferromagnetic and ferrimagnetic materials are of utmost importance in engineering. They are broadly classified as soft and hard magnetic materials depending on their hysteresis characteristics that also decide their corresponding applications. A comparative study of these materials is presented in Chart 10.6.

11

Conductors, Semiconductors and Superconductors

11.1 THE STORY OF CONDUCTORS

In the last chapter we studied the dielectric and magnetic properties of solids. Another important property of solids that attracted the attention of scientists was the electrical property. The questions that fascinated physicists were: What mechanisms contribute to electrical conduction in solids? What accounts for different conductivities in different solids? Why do some solids not conduct at all?

In order to improve conductors scientists were motivated to research into what causes conduction of current in materials and how it can be enhanced. The early work of Coulomb (1785), Ampere (1875) and Ohm (1827) established the basic law of electricity, that is, the current, I, produced in a conductor is proportional to the voltage, V, applied across it with $1/R$ as the constant of proportionality, or $V = IR$. The resistance R of the conductor was found to vary directly with the length and inversely with the area of cross section of the conductor, $R = (1/\sigma)(l/A)$, where σ is the conductivity of the material. Thus a simple, practical method to increase current (for a fixed voltage) is to use a short and thick conductor. However, having reached the limit prescribed by this method the role of conductivity was probed further.

In the beginning of the last century Drude and Lorentz enquired into the microscopic meaning of conductivity and proposed the free electron model. This model assumed the metal to be a framework of positive ions embedded in a sea of free electrons. Drude's model was refined by Hendrik Lorentz who assumed the sea of electrons to be a free electron gas which obeys the Maxwell–Boltzman statistics in thermal equilibrium. Maxwell–Boltzman statistics, the earliest known statistics, governing the behaviour of large assemblies of particles was developed by James Maxwell of Scotland (who formulated the electromagnetic equations) and Ludwig Boltzman of Vienna. Boltzman made pioneering contributions to the field of kinetic theory of gases and to statistical mechanics. The constant, k $= 1.381 \times 10^{-23}$ J/K, in the expression $3kT/2$, for the average energy of an atom in a monoatomic gas, is named after him. In the Drude–Lorentz model the interaction of electrons with lattice defects is ignored.

The Drude–Lorentz model was successful in explaining conductivity in metals on two counts. First, it led to the correct form of Ohm's law. Second, it was in broad agreement with an experimental law discovered by Weidemann and Franz which states that the ratio k$/\sigma$ of thermal and electrical conductivities is the same for all metals and depends only on temperature. It is interesting to note that though the constancy of the ratio k$/\sigma$ was validated by the Drude–Lorentz model, the theoretical values of k and σ obtained using this model departed from their experimental values. It seemed that both the expressions for k and σ contain some common error that cancels in the ratio k/σT. Moreover the theoretical value of k$/\sigma T$ found from the Drude–Lorentz model, 1.11×10^{-8} WΩ/K^2, did not agree with the experimental value, 2.45×10^{-11} WΩ/K^2. The free electron theory also failed to explain other observed properties of solids related to specific heat and paramagnetism.

The expression for conductivity of metals, derived on the basis of free electron theory, is $\sigma = ne^2\lambda/mv$, where, n is the free electron density, e and m are the electronic charge and mass, v is the mean speed and λ is the mean free path of electrons. This relation suggests that the conductivity of a conductor can be increased by increasing the values of n and λ. Both these parameters depend on the structure of the material: The electronic configuration decides the value of n and the periodicity of atoms decides λ. The value of λ is also found to vary inversely with temperature and imperfections in a material. Thus, the clues to improve the conductivity were to choose materials with larger n and λ, use pure materials and reduce the operating temperature. The understanding of conductivity seemed to have reached a plateau with the free electron theory.

In 1911, however, this field got a new lease of life when physicists came to know of a startling breakthrough. H Kammerling Onnes, a Dutch physicist, and his assistant Gilles Holst were studying, not the conductivity but something else (related to the production of low temperature using liquid helium), at the University of Leiden in Netherlands. They cooled some mercury to approximately below 4.2 K and found to their utter surprise that the resistivity (reciprocal of conductivity) of the mercury disappeared completely. They had discovered 'superconductivity'! For many years superconductivity remained a frontier area of research in physics all over the globe. In 1933, Meissner and Ochsenfield discovered the perfect diamagnetic property of superconductors.

Initially the group of HK Onnes discovered many superconducting elements as the liquid helium facility was available only with them. Later, Toronto, Berlin and other places gained prominence due to their reports on superconductivity. Further research in superconductivity was directed towards finding new materials that could show superconductivity at higher temperature, called the high T_C superconductors. (T_C is the transition temperature, the temperature at which a material goes from a normal state into the superconducting state.)

Up to 1973 the known superconducting materials were niobium (1925, 9.2 K), NbC (1940, 10.1 K), Nb_3Sn (1954, 18 K) and Nb_3Ge (1973, 23.2 K). In 1986, Muller and Bednorz reported a remarkable increase in transition temperature in lanthanum barium copper oxide (35 K) that was followed by discoveries of Paul Chu and his associates in yttrium barium copper oxide (91 K). Recently, in January 2001, magnesium diboride, a commonly found material in chemistry laboratories, was found to be superconducting below 39 K.

The theoretical understanding of superconductivity eluded physicists until 1957, when Bardeen, Cooper and Schrieffer explained the subtle mechanism of this phenomenon in terms of conduction by electron pairs using a model named after them—the *BCS theory*. Though no significant increase in T_C has been reported after 1986, theoretical and experimental research aimed at bringing T_C close to the room temperature are still going on. Nature has not yet completely revealed the secret of room temperature superconductivity.

In 1926, Enrico Fermi, an Italian physicist, theorised that electrons in a solid obey statistics which is different from Maxwell–Boltzman statistics. This new theory came to be known as Fermi–Dirac statistics in honour of Fermi and PAM Dirac, an English physicist, who studied electrical engineering before switching first to mathematics and then to physics. In 1928, A Sommerfield refined the free electron model using Fermi Dirac statistics instead of Maxwell–Boltzman statistics. He, however, considered the interior of a metal to be a region of constant potential, neglecting the periodic variations due to the presence of atoms. Sommerfield's model could resolve discrepancies relating to specific heat and paramagnetism but failed to account for the observed differences in conductivities of different solids. The use of Fermi–Dirac statistics also improved agreement with the experimental value of $k/\sigma T$.

Felix Bloch, a Swiss physicist, an undergraduate in engineering, systematically studied the effect of periodic potential on electron motion in 1928. He explained the formation of energy bands in solids by solving Schrodinger's equation for an electron moving in a periodic potential of a solid. The pioneering work of Bloch, Kronig, Penny, Heitler and London led to the evolution of the *band theory* which has been extremely successful in explaining the observed electrical characteristics of a variety of solids.

The band theory expanded the scope of research into the study of electrical properties of solids on the basis of their energy band diagrams and opened up new avenues of knowledge. One of the precious gifts of these investigations was a deeper insight into the novelties of semiconductors. A large number of semiconductors were probed, with silicon and germanium attracting the most attention—due to the very high degree of purity with which they could be obtained. In 1939, AH Wilson developed a general model of semiconductors.

Solids → conductors → semiconductors → pure semiconductors → impure (doped) semiconductors →→→ . . . → the tree expanded.

We will try to get to the roots of this tree in this chapter. In Section 11.2 the free electron theory of metals is discussed. Section 11.3 deals with how energy bands are formed in solids; in Section 11.4 the concept of Fermi level is discussed, and in Sections 11.5 and 11.6, semiconductors and superconductors are discussed respectively. The applications of this study are discussed in the last section.

11.2 FREE ELECTRON THEORY OF METALS

The conduction of electricity through metals is due to the flow of free electrons—that are present in them in large numbers due to the metallic bond—from the low potential end to the high potential end. In the free electron theory it is assumed that conduction electrons in metals are free to move throughout the volume of the sample. They collide with one another and also with atoms of the metal.

If there are n electrons per unit volume in a conductor, then the current density is given by (see Ex. 11.1)

$$J = nev_d \tag{11.1}$$

where v_d is the *drift velocity* of the electrons in a solid. It is the average velocity in the direction of the accelerating force. The drift is superimposed on the random thermal motion of an electron in a crystal lattice (Fig. 11.1).

The corresponding conductivity is given by

$$\sigma = ne\mu \tag{11.2}$$

where $\mu = v_d/E$ is called the *mobility* of free electrons.

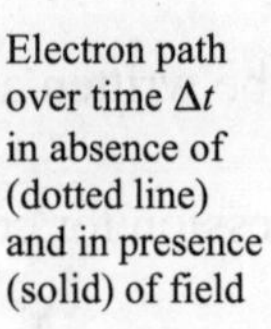
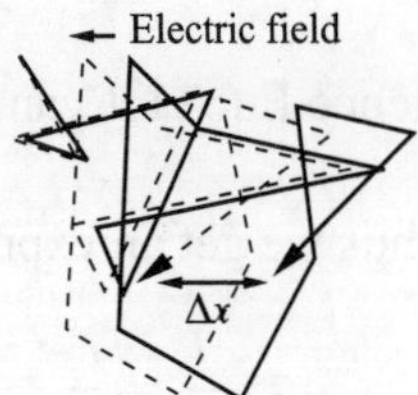

Figure 11.1 Drift velocity of electrons under the action of electric field

EXAMPLE 11.1
Expression for conductivity of a metal

Show that the conductivity of a metal is given by $\sigma = ne\mu$, where n is the density and μ is the mobility of electrons.

Solution

As n is the number of electrons per unit volume, a rectangular block of metal of length l and area of cross section A (Fig. 11.2) contains nAl electrons. Thus the total charge in this block $q = nAle$.

If this charge flows in time t then the current constituted is

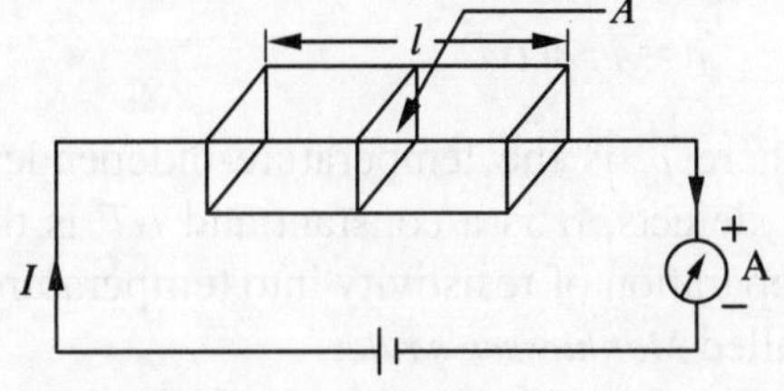

Figure 11.2

$$I = \frac{q}{t} = \frac{nAle}{t} = neAv_d$$

where v_d is the average drift velocity of electrons in the metal. The corresponding current density is $J = I/A = nev_d$.

Using conductivity, $\sigma = J/E$ and mobility, $\mu = v_d/E$, we get, $\sigma = ne\mu$

If τ is the mean time between collisions, then we can write

$$v_d = a\tau \tag{11.3}$$

where, a is the acceleration of an electron due to the applied field E, given by

$$a = \frac{F}{m} = \frac{eE}{m}$$

Thus we have,

$$v_d = \frac{eE}{m}\tau \tag{11.4}$$

Hence Eq. 11.1 can be written as $J = \left(\frac{ne^2\tau}{m}\right)E.$

Thus we get an expression for conductivity and also one for resistivity as

$$\sigma = \frac{J}{E} = \frac{ne^2\tau}{m} \text{ and } \rho = \frac{1}{\sigma} = \frac{m}{ne^2\tau} \tag{11.5}$$

In these expressions τ can also be expressed as $\tau = \lambda/v_F$ where λ is the mean free path between the collisions of a free electron and v_F is the *fermi velocity* corresponding to the fermi energy E_F [that is obtained by the relation $\frac{1}{2}mv_F^2 = E_F$ (see Section 11.4)].

Resistivity in metals is due to the scattering of electrons by collision with other electrons or with atoms. When an electron is scattered it changes its path randomly. It no longer travels in the direction of the field and hindrance to the flow of electrons or resistance to current is produced. Scattering due to atoms vibrating about their mean positions increases with increase in temperature, because the amplitude of vibrations increases with the temperature. On the other hand, scattering due to structural defects, that is, lack of periodicity, does not depend on the temperature but depends on the purity of the metal and its history. The resistivity of a metal is thus expressed as

$$\rho = \rho_i + \alpha T \tag{11.6}$$

where ρ_i is the temperature-independent part of resistivity that depends on the concentration of defects, α is a constant and αT is the temperature-dependent part due to atomic vibrations. Separation of resistivity into temperature-dependent and temperature-independent components is called *Matthiessen's rule.*

The free electron theory of metals, thus suggested, that *resistivity of a perfect, pure, single crystal of a metal decreases to zero as T approaches zero.*

11.3 FORMATION OF ENERGY BANDS IN SOLIDS

We are familiar with electronic energy levels of isolated atoms. What will happen if a large number of such non-interacting atoms, as in a gas, are brought together as in a solid? To understand this let us consider two hydrogen atoms, which are initially separated by a large distance r (Fig. 11.3).

Electrons confined to these atoms are like particles confined to finite potential wells of width δ. We can thus apply the results of Chart 6.4 (particle confined to an infinite potential well) as a first approximation. In Fig. 11.3, the ground state wave functions, ψ, of electrons are shown. It should be noted that ψ does not terminate at the boundaries as in the case of an infinite potential well but extends beyond the well. This means that there is a small but finite probability of an electron existing outside the potential well (that is, outside the atom). The energy eigenvalue corresponding to the ground state is given by

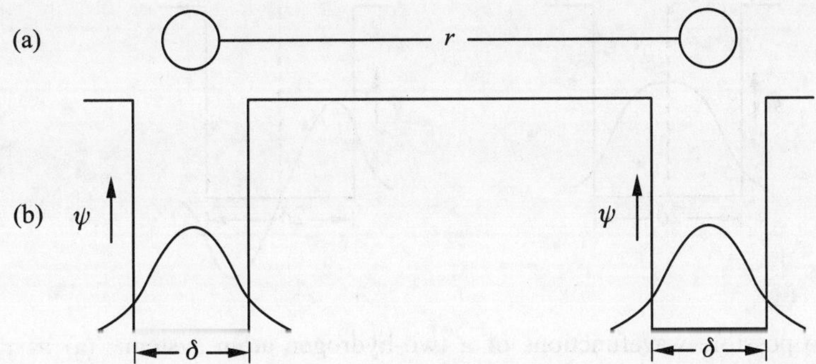

Figure 11.3 Two hydrogen atoms (a) separated by a large distance r and (b) represented by finite potential wells.

$$E_1 = \frac{h^2}{8m\delta^2} \tag{11.7}$$

Imagine that the distance r between atoms is gradually reduced. Any modification in the wavefunctions is governed by the conditions imposed on ψ (see Section 6.8): ψ and its partial derivatives should be continuous. ψ can maintain this continuity by taking either a symmetrical form or an antisymmetrical form as shown in Fig. 11.4a. As atoms are brought even closer, such that $r = \delta$ (Fig. 11.4b) we have a single two-atom system as against two independent systems of non-interacting atoms of Fig. 11.3. The electronic energies corresponding to these two wavefunctions for this two-atom system can be thus expressed as

$$E_{1\text{symmetrical}} = \frac{h^2}{8m(2\delta)^2} = \frac{1}{4}\left(\frac{h^2}{8m\delta^2}\right) \tag{11.8}$$

and $$E_{1\text{antisymmetrical}} = \frac{2^2h^2}{8m(2\delta)^2} = \frac{h^2}{8m\delta^2} \tag{11.9}$$

The splitting of energy levels on reducing r is shown in Fig. 11.5a for a two-atom system, in Fig. 11.5b for a three-atom system, and in Fig. 11.5c for a seven-atom system. Figure 11.5c shows two additional features. The higher energy levels split first, that is at larger value of r than the lower energy levels. Secondly, the higher energy levels split more broadly compared to the lower energy levels. The obvious reason for this is that the low energy core electrons spend most of their time near the nucleus and hence these levels split last (that is, at much smaller r) and split least. From Figs. 11.5a–11.5c it can be concluded that *each level splits into the same number of levels as the number of atoms in a system.*

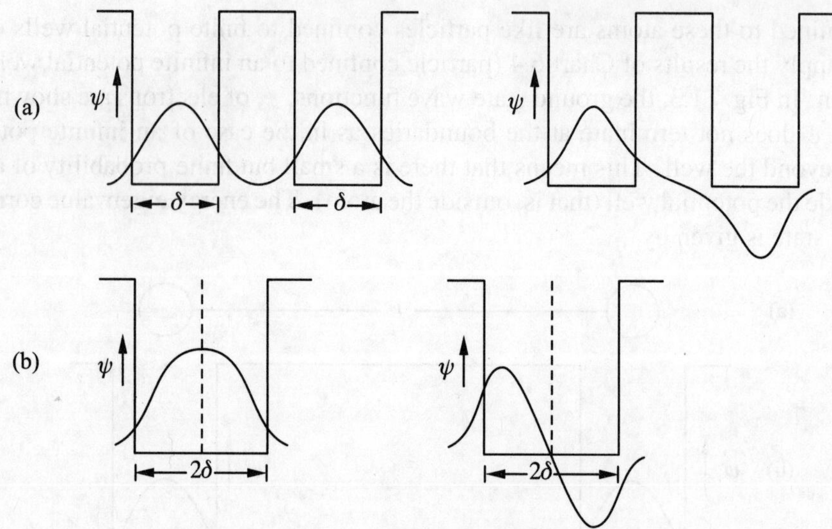

Figure 11.4 Two possible wavefunctions of a two-hydrogen atom system: (a) as r is reduced and (b) when $r = \delta$

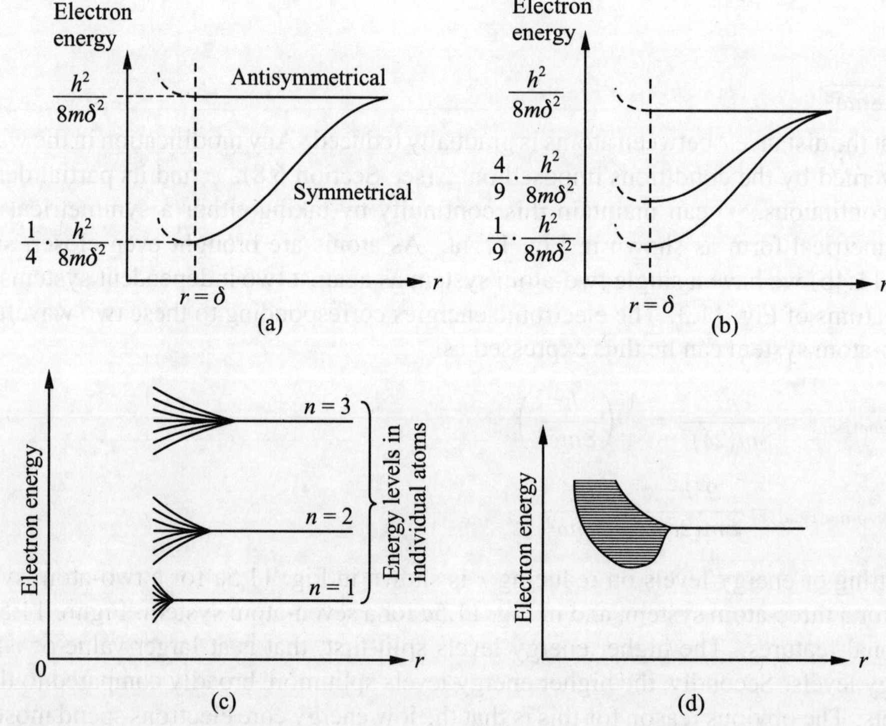

Figure 11.5 Splitting of energy levels in (a) two-atom system, (b) three-atom system, (c) seven-atom system and (d) one mole of a solid

Actually, the situation is a bit more complicated. The individual energy levels of atomic electrons may be degenerate, that is, there may be more than one electron state of that energy. For example, the $2s, 2p_x, 2p_y$ and $2p_z$ orbitals of a hydrogen electron all have the same energy: $-13.6/2^2$ eV = -3.4 eV. So the splitting of individual energy levels may be into even larger number of levels than the number of combining atoms.

In one mole of a solid there are about 10^{24} atoms. The splitting of energy levels into this many levels results in extremely small spacing between the adjacent energy levels (Fig. 11.5d). *Such a collection of levels is called an energy band.* This is how energy bands are formed in a solid.

It should be clearly understood here that band formation does not depend on whether an energy level in an individual atom is actually occupied or not. All energy levels, occupied or unoccupied by electrons, split into bands. The energy levels in these bands are then available for electrons, which can be said to belong to the whole solid rather than to its atomic constituents. These electrons are obviously the ones that occupied the outer shells of the individual atoms participating in the formation of the solid. The electrons in the inner shells of the atoms are more or less unaffected.

11.3.1 Origin of Forbidden Bands in Solids

In isolated atoms there are certain allowed energy levels for electrons. The splitting of these levels produces allowed energy bands in solids. Likewise there are disallowed or *forbidden energy bands* or *gaps* in solids. Splitting of valence levels results in a *valence band* and a *conduction band*. The valence band consists of energy levels for which the corresponding electron wave functions are localised between two atoms. An electron occupying any of the valence wave functions remains localised to a region between two atoms. Conduction band energies correspond to electron wave functions, which extend over the entire body of the solid. An electron occupying any of these wave functions can move all over the solid and can, therefore, lead to electrical conduction. These are the so-called *Bloch wave functions*.

If these two bands overlap, the solid is a *conductor*. If they are separated by a forbidden energy gap, the solid is either an *insulator* or a *semiconductor*, depending on whether the gap is large (5–15 eV) or small (~1 eV) (Fig. 11.6). However it should be noted that only partially

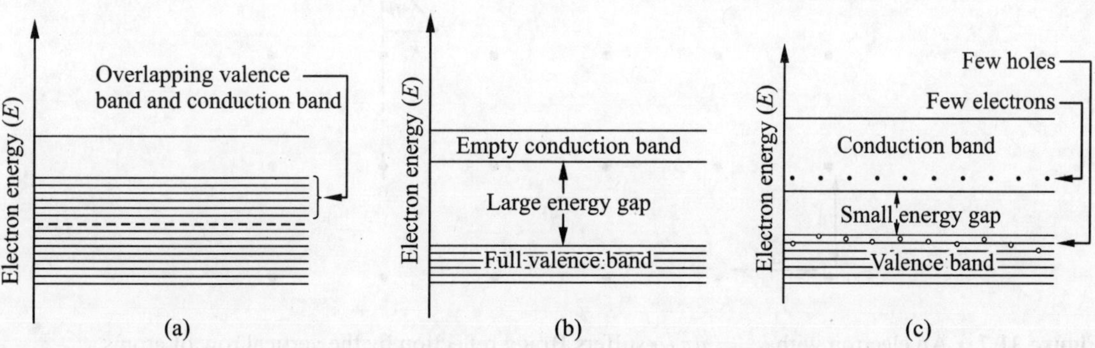

Figure 11.6 Band structure of (a) metal, (b) insulator and (c) semiconductor

filled bands can lead to conduction. Completely filled or completely empty bands do not lead to conduction due to non-availability of levels and non-availability of charges respectively.

Each solid has a unique band diagram, the details of which can be worked out with rigorous quantum mechanical calculations. Chart 11.1 displays the band diagrams of sodium, silicon and carbon. These diagrams show the variation of electron energy with the interatomic separation in these solids. However, it should be borne in mind that the interatomic spacing cannot be reduced gradually as shown in these diagrams to obtain a solid. The solid exists at a particular interatomic spacing. Thus the electrical behaviour of a solid is decided by the vertical slice through the diagram at the value of r at which that solid exists. In the actual band structure of a solid the x-axis thus loses significance and is generally not labelled. Chart 11.1 also includes the actual band structures of these solids and a discussion of their electrical properties.

The origin of forbidden energy gap in solids can also be understood in the following alternative way. In a crystal lattice (see Chapter 9) the electron moves through a region of periodically varying potential. The crystal planes can diffract the electron waves of wavelength λ if the Bragg equation

$$n\lambda = 2d\sin\theta, n = 1, 2, \ldots \tag{11.10}$$

is satisfied. Using the expression for wave number $k = 2\pi/\lambda$, Eq. 11.10 can be written as

$$k = \frac{n\pi}{d\sin\theta} \tag{11.11}$$

According to Eq. 11.11 there is at least one set of values of k corresponding to a given set of crystal planes, that is d, and different integers n and glancing angles θ for which electrons are diffracted away and do not remain confined to the crystal. Figure 11.7 shows the Bragg reflection of electrons by crystal planes in a square lattice.

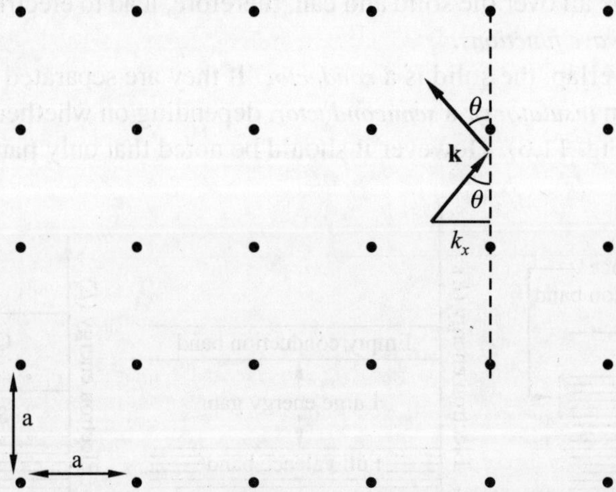

Figure 11.7 An electron with $k_x = n\pi/a$ suffers Bragg reflection by the vertical row of atoms

CHART 11.1 ENERGY BAND DIAGRAMS OF SOLIDS

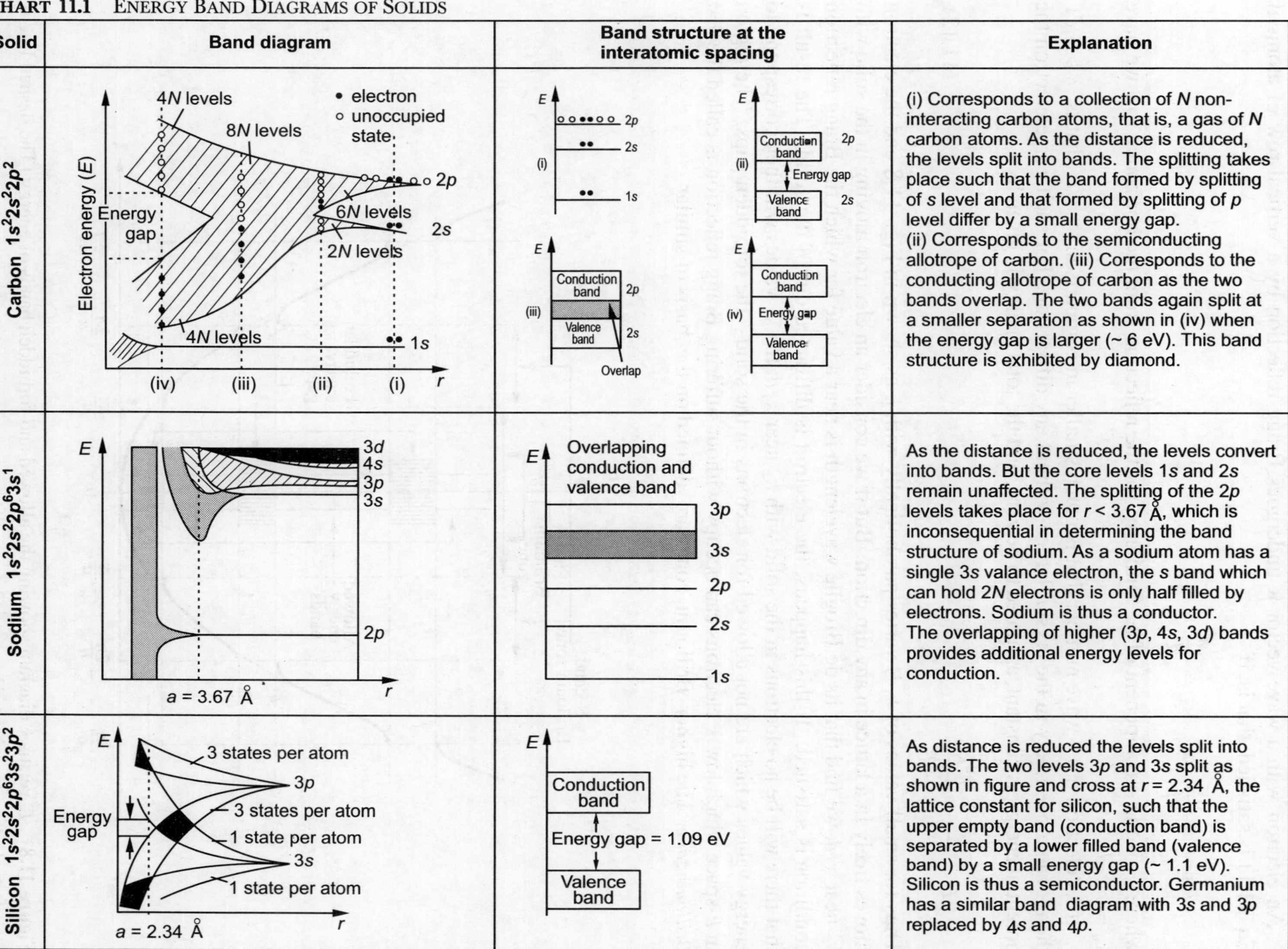

Solid	Band diagram	Band structure at the interatomic spacing	Explanation
Carbon $1s^2 2s^2 2p^2$			(i) Corresponds to a collection of *N* non-interacting carbon atoms, that is, a gas of *N* carbon atoms. As the distance is reduced, the levels split into bands. The splitting takes place such that the band formed by splitting of *s* level and that formed by splitting of *p* level differ by a small energy gap. (ii) Corresponds to the semiconducting allotrope of carbon. (iii) Corresponds to the conducting allotrope of carbon as the two bands overlap. The two bands again split at a smaller separation as shown in (iv) when the energy gap is larger (~ 6 eV). This band structure is exhibited by diamond.
Sodium $1s^2 2s^2 2p^6 3s^1$			As the distance is reduced, the levels convert into bands. But the core levels 1*s* and 2*s* remain unaffected. The splitting of the 2*p* levels takes place for $r < 3.67$ Å, which is inconsequential in determining the band structure of sodium. As a sodium atom has a single 3*s* valance electron, the *s* band which can hold 2*N* electrons is only half filled by electrons. Sodium is thus a conductor. The overlapping of higher (3*p*, 4*s*, 3*d*) bands provides additional energy levels for conduction.
Silicon $1s^2 2s^2 2p^6 3s^2 3p^2$			As distance is reduced the levels split into bands. The two levels 3*p* and 3*s* split as shown in figure and cross at $r = 2.34$ Å, the lattice constant for silicon, such that the upper empty band (conduction band) is separated by a lower filled band (valence band) by a small energy gap (~ 1.1 eV). Silicon is thus a semiconductor. Germanium has a similar band diagram with 3*s* and 3*p* replaced by 4*s* and 4*p*.

An electron with a wave vector **k** undergoes Bragg reflection by a vertical row of atoms if Eq. 11.11 is satisfied, that is, if

$$k \sin\theta = \frac{n\pi}{d} \quad \text{or} \quad \mathbf{k}_x = \frac{n\pi}{d}$$

where k_x is the x component of k. Similarly, for the reflection from the horizontal row we have $k_y = n\pi/d$.

If an electron has a wave number that is much smaller in comparison to π/d (that is, $\lambda >> d$), then it can move freely in the crystal lattice without any diffraction. For such a free electron the potential energy is constant, assumed to be zero, and the total energy is given by

$$E = \frac{p^2}{2m} = \frac{\hbar^2 k^2}{2m} \tag{11.12}$$

Thus for small values of k, E varies parabolically with k as shown in Fig. 11.8 and the electron moves freely in a lattice in any direction. But if we consider an electron moving in the solid with k_x near π/d, we find that its de Broglie wavelength is near a value for which the Bragg reflection condition is satisfied. If this happens, the electron is diffracted out of the solid. The result is that there will be no electrons in the solid with k_x near π/d; the k_x values will then correspond to energy values which are not allowed for electrons in the solid—the forbidden gaps. The region in k-space which low-k electrons can occupy without suffering Bragg reflection is called the *first Brillouin zone*. The higher Brillouin zones are defined on the basis of similar

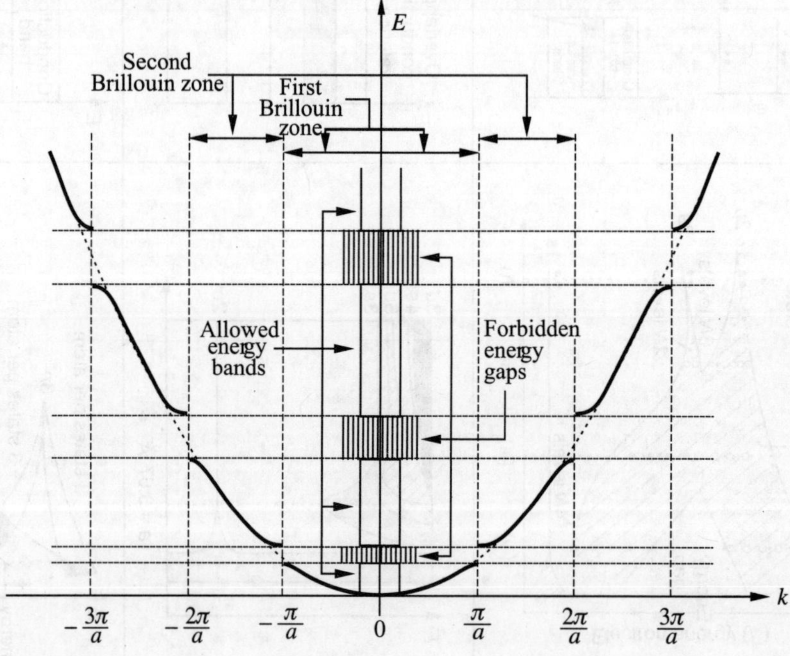

Figure 11.8 E versus k diagram showing the allowed and forbidden bands of energy (The dashed line is for a free electron for which, $E = \hbar^2 k^2/2m$.)

restrictions on k. The gap between the possible energies in the first and second Brillouin zones is the forbidden energy gap.

11.3.2 Effective Mass

As electrons interact with the periodically varying potential of a crystal lattice, their response to the external field is different from that of free electrons. The group velocity of the wave packet associated with an electron is given by (see Eq. 6.21),

$$v_g = \frac{d\omega}{dk} = \frac{d(E/\hbar)}{dk} = \frac{1}{\hbar} \times \frac{dE}{dk} \tag{11.13}$$

where we have used the relation $E = \hbar\omega$, ω being the angular frequency of the de Broglie wave of an electron. Under the action of an externally applied electric field, if the electron covers a distance dx with velocity v_g along the direction of the accelerating force, F, then the energy acquired by it is

$$dE = F dx = F v_g dt = \frac{F}{\hbar}\left(\frac{dE}{dk}\right) dt \tag{11.14}$$

Rearranging we get,

$$\frac{dk}{dt} = \frac{F}{\hbar} \tag{11.15}$$

The acceleration, a, of an electron is given by

$$\begin{aligned} a &= \frac{dv_g}{dt} = \frac{1}{\hbar} \times \frac{d}{dt}\left(\frac{dE}{dk}\right) \\ &= \frac{1}{\hbar} \times \frac{dk}{dt} \times \frac{d}{dk}\left(\frac{dE}{dk}\right) = \frac{1}{\hbar} \times \frac{dk}{dt} \times \frac{d^2E}{dk^2} \end{aligned} \tag{11.16}$$

Substituting from Eq. 11.15 we get

$$a = \frac{F}{\hbar^2} \times \frac{d^2E}{dk^2} \tag{11.17}$$

If one compares this equation with Newton's equation for free electrons, that is, $a = F/m_e$, then we find that the quantity $[(1/\hbar^2)(d^2E/dk^2)]^{-1}$ may be identified as the mass of an electron in the periodic lattice of a solid. It is called the *effective mass* m_e^* of an electron, that is,

$$m_e^* = \left[\frac{1}{\hbar^2} \times \frac{d^2E}{dk^2}\right]^{-1} \tag{11.18}$$

The concept of effective mass should be seen in the context of the free electron theory of certain transport phenomena, like conductivity, in solids. The assumption of the free electron theory that electrons constitute a gas of free particles has to be modified because of the presence of the periodic lattice in a solid in which the electrons move. It is clear from Eqs. 11.17 to 11.18 that such a modification may be done in a simple and transparent manner. All that is to be done is to allow a variable electron mass to take into account the periodic potential seen by the electron as it moves in a lattice. This potential leads to an E–p relation, which is very much like the quadratic E–p relation for a free particle. The only difference is that at certain values of p, or, equivalently k, the energy curve flattens out (see Fig. 11.8). This is what leads to a modification in the free motion of the electron—a modification, which may be described simply by saying that the mass has become variable as a function of p or k. The effective mass is not constant but depends on the second derivative of energy with respect to k. It can be positive or negative. It is equal to m_e only for small values of k for which the energy varies parabolically with k. In conductors the two masses are usually equal but they differ in case of insulators and semiconductors.

11.4 FERMI ENERGY AND FERMI LEVEL

Electrons in a solid follow what is known as *Fermi–Dirac distribution function.* The concept of a distribution function has its origin in statistical mechanics. Statistical mechanics is a branch of physics that deals with the study of collective behaviour of large assemblies of particles in statistical terms. The three main statistics developed for such a study are *Maxwell–Boltzman*, *Bose–Einstein* and *Fermi–Dirac* statistics. We are concerned here with the basic elements of Fermi–Dirac statistics that govern the distribution of electrons in solids. Electrons are also categorised as *fermions* for this reason.

The Fermi–Dirac distribution function is defined through the equation

$$p(E) = \frac{1}{1 + exp[(E - E_F)/kT]} \tag{11.19}$$

where k is Boltzman's constant and E_F is the *Fermi energy.* If we take any small energy interval $[E, E+dE]$, the probability that an electron—randomly picked out from a collection of elecltrons—has energy within this interval is given by $p(E)dE$. As the exponential is always non-negative, Eq. 11.19 says that $p(E) \leq 1$. If $p(E) = 1$, the energy level is occupied; if $p(E) = 0$ the level is empty.

As the concept of Fermi level is of cardinal importance in semiconductor physics it is elaborated in Chart 11.2. The first column in this chart explains the concept mathematically. The second column contains the graphical representation and the third column explains the physical meaning of the concept. The chart also lists various ways of looking at the concept of Fermi level. Equation 11.19 shows that the distribution function is such that the probability of occupation of level ΔE above E_F is the same as the probability of non-occupation of level ΔE below E_F. This is proved in Ex. 11.2. Examples 11.3 and 11.4 deal with some other important applications of the concept of Fermi level in metals.

The discussion so far implies that the physics of semiconductors is centred around two questions: *How are valence band and conduction band positioned (that is, what is the band structure)*

CHART 11.2 FERMI LEVEL AND FERMI ENERGY

	Mathematical (Using Eq. 11.19) $p(E)= 1/\{1 + \exp[(E–E_F)/kT]\}$	**Graphical**	**Physical**
$T = 0$ K	For $E < E_F$, $p(E) = 1/[\exp(-\infty) +1] = 1/(0 + 1) = 1$ For $E > E_F$, $p(E) = 1/[\exp(\infty) + 1] = 1/\infty = 0$	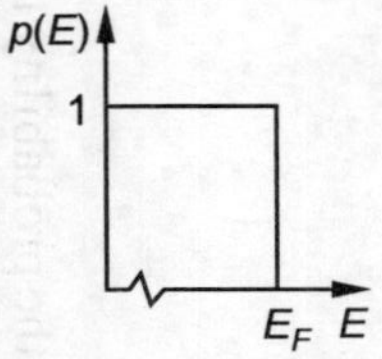 $p(E)$; 1; E_F E	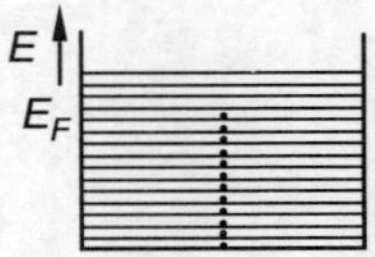 E; E_F All energy levels below E_F are filled with electrons. All energy levels above E_F are empty.
$T > 0$ K	For $E < E_F$, $p(E) = \dfrac{1}{1 + \{1/\exp[(E - E_F)/kT]\}} < 1.$ As the temperature increases $\exp[(E - E_F)/kT]$ decreases, thus the denominator of $p(E)$ increases and $p(E)$ decreases below 1. For $E > E_F$, $p(E) = \left\{1 + \dfrac{1}{\exp[(E - E_F)/kT]}\right\} > 0.$ As the temperature increases $\exp[(E - E_F)/kT]$ decreases, thus denominator of $p(E)$ decreases and $p(E)$ increases above 0. For $E = E_F$, $p(E_F) =1/(1 + 1) = 0.5.$ For $E << E_F$, $(E - E_F)$ is a very small quantity, hence $p(E) \sim 1.$ For $E >> E_F$, $(E - E_F)$ is a very large quantity, hence $p(E) \sim 0.$	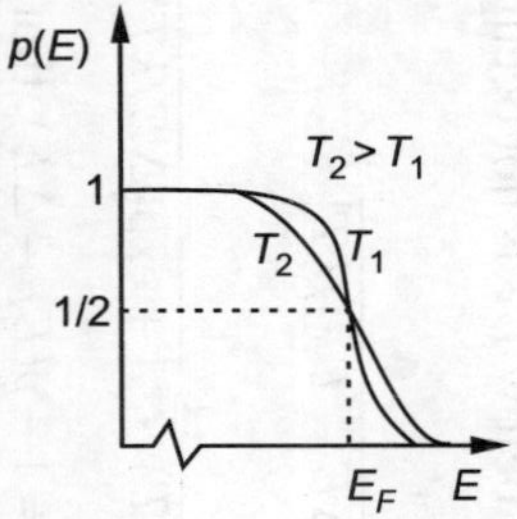 $p(E)$; $T_2 > T_1$; 1; T_2; T_1; 1/2; E_F E	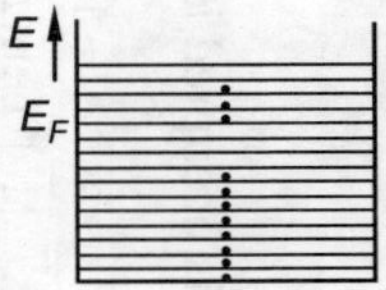 E; E_F As temperature increases, electrons gain thermal energy. Electrons from energy levels just below Fermi level quit their levels and occupy energy levels just above it. However, for energy levels much below or much above the Fermi level there is no change because E_F is typically of the order of a few electron volts, whereas kT is only a fraction of an electron volt at ordinary temperatures (at room temperature, $kT = 0.026$ eV). Secondly electrons occupying levels much below the Fermi level cannot leave their levels and occupy the levels just above them as these levels are already occupied by electrons. (Pauli exclusion principle forbids occupation of a level by more than one electron).
Definitions and other related points	The Fermi level is the highest occupied energy level at 0 K. Alternatively, it is the maximum energy an electron can possess at 0 K. For this reason the Fermi level is also called the *Fermi brim*. The Fermi level is defined as an energy level which has 50% chance of occupation. The Fermi level is a reference energy level and need not always be occupied. In fact there may be no electron state for $E = E_F$! This is the case if the Fermi level lies in the forbidden energy gap. The Fermi level can also be defined as the level corresponding to the weighted average of the energies corresponding to the conduction band and the valence band—density of electrons in the conduction band and density of holes in the valence band being the respective weights. The significance of this definition is brought out in Chart 11.4. Fermi speed is the electron speed corresponding to the Fermi energy, given through the relation $E_F = ½ mv_F^2$. Fermi temperature T is defined through the equation $E_F = kT_F$.		

for a particular case under study and where does the Fermi level lie in the band structure for that case? We pursue these questions in this chapter and the next chapter.

EXAMPLE 11.2
$p(E_F + \Delta E) = 1 - p(E_F - \Delta E)$

Show that the probability of occupation of the level $E_F + \Delta E$ is the same as the probability of non-occupation of the level $E_F - \Delta E$.

Solution

The probability that the energy level ΔE above E_F is occupied is

$$p(E_F + \Delta E) = \frac{1}{1 + \exp[(E_F + \Delta E - E_F)/kT]} = \frac{1}{[1 + \exp(\Delta E/kT)]} \tag{11.20}$$

Now the probability that the energy level ΔE below E_F is not occupied is

$$1 - p(E_F - \Delta E) = 1 - \frac{1}{1 + \exp[(E_F - \Delta E - E_F)/kT]}$$

$$= 1 - \frac{1}{1 + \exp(-\Delta E/kT)} = \frac{1}{1 + \exp(\Delta E/kT)} \tag{11.21}$$

From Eqs. (11.20) and (11.21), $p(E_F + \Delta E) = 1 - p(E_F - \Delta E)$, that is, the probability that a level at a particular energy increment above E_F is occupied equals the probability that a level below E_F by same energy decrement is empty.

EXAMPLE 11.3
Fermi energy in metals

In a metal the density of energy states per unit energy range, $N(E)$ is given by

$$N(E) = \left(\frac{8\sqrt{2}\pi m^{3/2}}{h^3}\right)\sqrt{E}$$

where m is the mass of electron and E is the energy at which $N(E)$ is to be evaluated. Find the expression for Fermi energy E_F in metals. Find the Fermi energy for copper for which the number of conduction electrons is $8.48 \times 10^{28}/\text{m}^3$.

Solution

$N(E)dE$ represents the number of energy states per unit volume in the energy range dE. Thus $N(E)p(E)dE$ represents the number of occupied energy levels per unit volume, where $p(E)$ is the probability of occupation of energy level E. If we integrate $N(E)p(E)dE$ in the limits $E = 0$ and

$E = E_F$, the result should be n, the number of conduction electrons per unit volume of the metal. Thus,

$$n = \int_0^{E_F} N(E)p(E)dE$$

Now E_F is the highest occupied energy level at 0 K. Thus $p(E) = 1$ for all $E < E_F$.

Thus,

$$n = \frac{8\sqrt{2}\pi m^{3/2}}{h^3} \int_0^{E_F} \sqrt{E}dE$$

$$= \frac{8\sqrt{2}\pi m^{3/2}}{h^3} \times \frac{2E_F^{3/2}}{3}$$

Solving for E_F, now leads to,

$$E_F = \left[\frac{3nh^3}{16\sqrt{2}\pi m^{3/2}}\right]^{2/3} = \frac{0.121h^2}{m} \times n^{2/3}$$

$$= \left[\frac{0.121(6.625 \times 10^{-34}\text{ Js})^2}{9.1 \times 10^{-31}\text{ kg}}\right] n^{2/3}$$

Thus, $\quad E_F = 5.84 \times 10^{-38} \times n^{2/3}\text{ J} = 3.65 \times 10^{-19} \times n^{2/3}\text{ eV}.$

For copper,

$$E_F = 3.65 \times 10^{-19} \times (8.48 \times 10^{28})^{2/3}\text{ eV}$$

$$= 7.04\text{ eV}$$

EXAMPLE 11.4

Increase in the fraction of electrons due to heating

Calculate an approximate value for the fraction of electrons that are excited from below E_F to above E_F when copper is heated from 0 K to 300 K.

Solution

The value of kT at room temperature is 0.026 eV. Thus, only those electrons that are within a range of 0.026 eV are affected by the change in temperature from 0 K to 300 K.

The fraction of electrons that are excited from level $E_F - kT$ to E_F is $\dfrac{\int_{E_F-kT}^{E_F} N(E)dE}{\int_0^{E_F} N(E)dE}$

$$= \frac{(8\sqrt{2}\pi m^{3/2}/h^3)\int_{E_F-kT}^{E_F} \sqrt{E}dE}{(8\sqrt{2}\pi m^{3/2}/h^3)\int_0^{E_F} \sqrt{E}dE}$$

$$= \frac{E_F^{3/2} - (E_F - kT)^{3/2}}{E_F^{3/2}} = 1 - \left[1 - \frac{kT}{E_F}\right]^{3/2} = \frac{(3/2)kT}{E_F}$$

(As $kT << E_F$, only the first two terms in the binomial expansion are retained.)

Hence the fraction of electrons excited

$$= \frac{3}{2}\left(\frac{0.026}{7.04}\right) = 0.0055 = 0.55\%$$

Thus we see that only a very small fraction of electrons are affected.

11.5 SEMICONDUCTORS: INTRINSIC AND EXTRINSIC

Semiconductors are solids that are characterised by a small forbidden energy gap between their valence band and conduction band. Semiconductors are classified as *intrinsic* and *extrinsic* depending on their impurity concentration. An intrinsic semiconductor is a pure crystal. Practically, it is one in which impurity concentrations are so low that its electrical properties are not affected by them but are solely determined by the band structure of the pure material. Tetravalent elements like Ge and Si belonging to the IV group are examples of intrinsic semiconductors. As each atom has four vacancies in its outer shell, it forms covalent bonds with its four neighbours, thus completing its outer shell with eight electrons. These materials crystallise in diamond structure (see Chapter 9).

In extrinsic semiconductors impurities are deliberately added in small and controlled quantity through a process known as *doping*. Doping alters the electrical characteristics of materials in a desired manner. Extrinsic semiconductors are classified as either n-type or p-type, depending on whether the dopant is pentavalent (P, As, Sb or Bi) or trivalent (B, Al, Ga or In).

The physics of semiconductors can be understood either through crystal representation or band structure representation. Both these representations convey the same knowledge. Chart 11.3 compares intrinsic, n-type and p-type semiconductors using these two representations. The correlation between these two representations is also brought out in this chart.

The expressions for Fermi energy intrinsic and extrinsic semiconductors can be derived using Eq. 11.19 in conjunction with other known properties of these semiconductors. The initial considerations for these derivations are as follows.

The energy of the bottom of the conduction band is E_C, and E_V is the energy of the top of the valence band. (In Chart 11.3 we assumed $E_g = E_C - E_V$.) The number density of electrons in the conduction band is n and p is the number density of holes in the valence band. N is the total number density of electrons in both the bands and N_D and N_A denote the number density of donor atoms and acceptor atoms respectively. N_D^+ and N_A^- denote the number density of ionised donors and ionised acceptors respectively. (When the fifth electron of a donor atom is stripped off, it gets a net positive charge, and when the acceptor atom accepts the fourth electron it gets a net negative charge as explained in Chart 11.3).

CHART 11.3 CORRELATION BETWEEN CRYSTAL AND BAND REPRESENTATION

	Intrinsic		**Extrinsic** ***n*-type**
	Crystal representation	**Band representation**	**Crystal representation**
$T = 0$ K	Atom (Si or Ge) Electron Covalent bond	E Completely empty conduction band Forbidden energy gap, E_g Completely filled valence band	Weakly bound fifth electron Donor atom
$T > 0$ K	Free electron Broken covalent bond Hole	E Few electrons E_g Few holes	Fifth electron donated to the lattice +
Correlation between two representations	1. Electron confined to the atom (in the covalent bond, that is, a bound electron) 2. Electron outside the influence of atom (that is, free electron) 3. At 0 K all covalent bonds are intact 4. At $T > 0$ K some covalent bonds are broken due to thermal excitation. A broken covalent bond results in a free electron and a vacancy (called as hole). Electron–hole pairs are thus created. 5. The energy required to break a covalent bond, that is, the strength of a covalent bond.	Electron in the valence band. Electron in the conduction band. The valence band is completely filled and the conduction band is completely Electrons jump from the valence band to the conduction band. As a result there are few electrons in the conduction band and an equal number of holes in the valence band. The energy required for an electron in the valence band to jump to the conduction band, that is, the energy gap, $E_C - E_V$ or E_g.	1. Pentavalent donor atoms replace the atoms of host crystal sub-stitutionally, making bonds with four neighbors. 2. The fifth electron is very loosely bound to the donor atom. The energy required to make it free (ionisation energy) is very small: of the order of 0.01 eV. 3. At room temperature nearly all donor atoms donate their 5th electron to the lattice. 4. Ionisation of donor atoms does not create a vacancy as the loosely bound 5th electron is donated to the lattice and a covalent bond is not ruptured for this process.

* Levels are created and not a band as doping means impurities are added in extremely small quantity.

Extrinsic	Extrinsic	
***n*-type**	***p*-type**	
Band representation	**Crystal representation**	**Band representation**
E Completely empty conduction band E_C E_D Donor levels filled Completely filled valence band	Acceptor atom Incomplete covalent band	E Completely empty conduction band Acceptor levels empty E_A Completely filled valence band
E Electrons from donor levels and valence band E_D Donor levels empty	A hole is created Covalent band completed	E Acceptor levels filled E_A Holes produced by the transition of electrons to acceptor levels or conduction band
Donor levels (with energy = E_D) are created just below the conduction band. The difference $E_C - E_D$ is very small ~ 0.01 eV. At room temperature donor levels become empty as electrons jump from donor levels to the conduction band. Jumping of an electron from donor levels to the conduction band does not create hole in the valence band.	1. Trivalent acceptor atoms replace the atoms of the host crystal substitutionally, making bonds with three neighbours, with one bond incomplete. 2. It requires a very small energy ~ 0.01 eV to ionise the acceptor atoms, that is, to complete the covalent bond by an electron from the neighbouring parent atom. 3. At room temperature nearly all acceptor atoms are ionised by electrons in their neighbouring atoms. 4. Ionisation of acceptor atoms does not produce additional electrons as the electrons are used in the completion of covalent band.	Accepter energy levels (with energy = E_A) are created just above the valence band. The difference $E_A - E_V$ is very small ~ 0.01 eV. At room temperature all the acceptor levels are filled by transition of electrons from the valence band. Jumping of electron from valence band to acceptor level does not create electron in the conduction band.

It can be shown that (see Ex. 11.5)

$$n = N_C \exp\left[\frac{-(E_C - E_F)}{kT}\right] \tag{11.22}$$

and $$p = N_V \exp\left[\frac{-(E_F - E_V)}{kT}\right] \tag{11.23}$$

where $$N_C = 2\left(\frac{2\pi m_e^* kT}{h^2}\right)^{3/2} \tag{11.24}$$

and $$N_V = 2\left(\frac{2\pi m_h^* kT}{h^2}\right)^{3/2} \tag{11.25}$$

where m_e^* and m_h^* denote the effective mass of the electron and hole respectively.

EXAMPLE 11.5

To show $n = N_C \exp[-(E_C - E_F)/kT]$ and $p = N_V \exp[-(E_F - E_V)/kT]$

Derive the following expressions for the number density of electrons (n) in the conduction band, and the number density of holes (p) in the valence band.

$$n = N_C \exp\left[\frac{-(E_C - E_F)}{kT}\right]$$

$$p = N_V \exp\left[\frac{-(E_F - E_V)}{kT}\right]$$

where $$N_C = 2\left(\frac{2\pi m_e^* kT}{h^2}\right)^{3/2}$$

and $$N_V = 2\left(\frac{2\pi m_h^* kT}{h^2}\right)^{3/2}$$

Solution

We know that the density of energy states per unit energy range for free electrons in a metal is given by

$$N(E) = \left(\frac{8\sqrt{2}\pi m^{3/2}}{h^3}\right)\sqrt{E}$$

We can extend this expression for electrons moving in the periodic potential of lattice by replacing the electron mass m by its effective mass m_e^* (see Section 11.3.2). Thus, in the present case

$$N(E) = \left(\frac{8\sqrt{2}\pi(m_e^*)^{3/2}}{h^3}\right)\sqrt{E - E_C}$$

where E_c represents energy of the bottom level of the conduction band. Now the total number of electrons per unit volume in the conduction band or the number density of electrons in the conduction band is given by (see Ex. 11.3)

$$n = \int_{E_C}^{\infty} N(E)p(E)dE$$

where the upper limit of integration is assumed to be ∞ to simplify the integration.

Thus,

$$n = c\int_{E_C}^{\infty} \frac{(E - E_C)^{1/2}dE}{1 + \exp[(E - E_F)/kT]}$$

where $\quad c = \dfrac{8\sqrt{2}\pi(m_e^*)^{3/2}}{h^3}$

As $\exp[(E - E_F)/kT] >> 1$ for E in conduction band, we have

$$n = c\int_{E_C}^{\infty} (E - E_C)^{1/2} \exp\left[-\left(\frac{E - E_F}{kT}\right)\right] dE$$

Substituting $x = (E - E_C)/kT$, that is, $dE = kTdx$ and thus, the limits of integration as, $x = 0$ to $x = \infty$ we get,

$$n = c\left\{\exp\left[\frac{-(E_C - E_F)}{kT}\right]\right\}(kT)^{3/2}\int_0^{\infty} x^{1/2}\exp(-x)dx$$

Now, $\displaystyle\int_0^{\infty} x^{1/2}\exp(-x)dx = \sqrt{\pi}/2$

Thus we get,

$$n = N_C \exp\left[-\frac{E_C - E_F}{kT}\right]$$

where $N_C = 2(2\pi m_e^* kT/h^2)^{3/2}$

Now the number density of holes in the valence band is given by

$$p = \int_{-\infty}^{E_V} N(E)[1 - p(E)]dE$$

where the factor $1 - p(E)$ denotes the probability of non-occupation of energy level E, that is, the absence of an electron in level E or the presence of a hole in level E. $N(E)$ in this case is

$$\left[\frac{8\sqrt{2}\pi(m_h^*)^{3/2}}{h^3}\right]\sqrt{E_V - E}$$

where E_V represents the energy of the top of the valance band.

Now, $1 - p(E) = 1 - \dfrac{1}{1 + \exp[(E - E_F)/kT]}$

$$= \frac{1}{1 + \exp[-(E - E_F)/kT]}$$

which can be approximated to $\exp[(E - E_F)/kT]$, as $1 << \exp[-(E - E_F)/kT]$ for E in valence band.

Thus $p = c \displaystyle\int_{-\infty}^{E_V} (E_V - E)^{1/2} \exp\left[\frac{E - E_F}{kT}\right] dE$

where $c = \dfrac{8\sqrt{2}\pi (m_h^*)^{3/2}}{h^3}$

Substituting $(E_V - E)/kT = x$, that is, $dE = -kTdx$ and the limits of integration as, $x = \infty$ to $x = 0$, we get,

$$p = -c\left\{\exp\left[-\frac{(E_F - E_V)}{kT}\right]\right\}(kT)^{3/2} \int_{\infty}^{0} x^{1/2} \exp(-x)dx$$

Using $-\displaystyle\int_{\infty}^{0} x^{1/2} \exp(-x)dx = \int_{0}^{\infty} x^{1/2} \exp(-x)dx = \sqrt{\pi}/2$, we get

$$p = N_V \exp\left[-\frac{(E_F - E_V)}{kT}\right]$$

where $N_V = 2(2\pi m_h^* kT/h^2)^{3/2}$.

Exploiting the common mathematical treatment required in the derivation of the expressions for Fermi level for different types of semiconductors, we include these derivations in Chart 11.4. The expressions derived are given in Table 11.1.

The special cases of these expressions are discussed in Chart 11.4. The chart also displays the positions of Fermi level in the band structures of the respective semiconductors and discusses their physical significance.

Table 11.1 Expressions for Fermi level

Type of semi-conductor	**Expression**	
Intrinsic	$E_F = (E_C + E_V)/2 + \frac{3}{4}kT \log(m_h^*/m_e^*)$	(11.26)
n-type	$\exp(E_F/kT) = \{-1 + [1 + 4(N_D/N_C)\exp[(E_C - E_D)/kT]\}^{1/2}/2\exp(-E_D/kT)$	(11.27)
p-type	$\exp(-E_F/kT) = \{-1 + [1 + 4(N_A/N_V)\exp[(E_A - E_V)/kT]\}^{1/2}/2\exp(E_A/kT)$	(11.28)

The expressions for the density of charge carriers, n and p, can now be obtained by substituting the expressions for Fermi energy [E_F or $\exp(E_F/kT)$] in Eqs. 11.22 and 11.23 respectively. For an intrinsic semiconductor (see Chart 11.3)

$$n = p = n_i \tag{11.29}$$

Hence, $$np = n_i^2 \tag{11.30}$$

where n_i is called the *intrinsic density* of charge carriers. It is found quantitatively that this equation also holds approximately for extrinsic semiconductors. This is because an increase in the number of one type of charge carrier decreases the number of the other type due to recombination; the product np remains constant. For n-type material, an increase in n results in a decrease in p, and for p-type material, an increase in p results in a decrease in n; but np is nearly equal to n_i^2. The symbols n_n, p_n, n_p and p_p are used to denote respectively the number density of electrons in n-type, holes in n-type, electrons in p-type and holes in p-type material. Thus we have

$$n_n p_n \approx n_i^2 \tag{11.31}$$

and $$n_p p_p \approx n_i^2 \tag{11.32}$$

Here $n_n \approx N_D$ and $p_p \approx N_A$. Equations 11.30–11.32 are called the *Law of mass action.*

Exploiting the common mathematical treatment required in the derivation of the expressions for n and p for different types of semiconductors, these derivations are included in Chart 11.5. Using these expressions the expressions for conductivity can be determined and the mechanisms of conduction in these semiconductors can be understood. This is also included in Chart 11.5.

For a semiconductor as electrons and holes both contribute to conductivity, we have (see Eq. 11.2)

$$\sigma = ne\mu_e + pe\mu_h \tag{11.33}$$

where μ_e is the mobility for electrons and μ_h is the mobility for holes. Using Eq. 11.29, the conductivity for an intrinsic semiconductor, known as *intrinsic conductivity*, is given by,

$$\sigma = n_i e(\mu_e + \mu_h) \tag{11.34}$$

See Ex. 11.6.

It should be noted that conventional current due to electrons is in the same direction as the current due to holes. Thus the two currents add up. The conductivity is the sum of electron conductivity $n_i e\mu_e$ and hole-conductivity $n_i e\mu_h$. The extrinsic conductivities are given by

$$\sigma_n = ne\mu_e \text{ for } n\text{-type} \tag{11.35}$$

and $$\sigma_p = pe\mu_h \text{ for } p\text{-type} \tag{11.36}$$

As the conductivity of a semiconductor depends on its energy gap and temperature (see Eq. 4 of Chart 11.5), the temperature dependence of conductivity in semiconductors can be studied experimentally and the value of the forbidden energy gap can be determined. Experiment 11.1 describes the four-probe method for such a study.

CHART 11.4 DERIVATION OF FERMI ENERGY

	Intrinsic	Extrinsic: *n*-type	Extrinsic: p-type
Starting step	As explained in Chart 11.3, the electrons and holes are generated in pairs. Hence density of electrons is equal to density of holes. $n = p$ (1)	The starting point is the charge neutrality equation which says that the total positive charge equals the total negative charge. Thus, $p + N_D^+ = n + N_A^-$ (5)	
		As explained in Chart 11.3 the electrons come to the conduction band from both the donor levels and the valence band. Thus $n >> p$. Also $N_D^+ >> N_A^-$. Thus Eq. (5) becomes $n \sim N_D$ (6)	As explained in Chart 11.3 the holes are created due to transition of electrons from the valence band to the acceptor levels and to the conduction band. Thus, $P >> n$. Also $N_A^- >> N_D^+$. Thus Eq. (5) becomes $p \sim N_A^-$ (11)
Derivation of the expression for E_F	Using Eqs 11.22 and 11.23 in Eq. (1), $N_C \exp[-(E_C - E_F)/kT] = N_V \exp[-(E_F - E_V)/kT]$. Taking log of both the sides and rearranging, we get, $E_F = (E_C + E_V)/2 + ½\, kT \log(N_V/N_C)$. Substituting the expressions for N_C and N_V from Eqs 11.24 and 11.25 we get, $E_F = \left(\frac{E_C + E_V}{2}\right) + ¾\, kT \log\left(\frac{m_h^*}{m_e^*}\right)$ (2)	$N_D^+ = N_D \times$ probability that a donor atom gets ionised $= N_D[1 - p(E_D)] = N_D\{1 - 1/[1 + \exp((E_D - E_F)/kT)]\}$ Thus, Eq. (6) becomes $N_C \exp[-(E_C - E_F)/kT] = N_D\{1 - 1/[1 + \exp((E_D - E_F)/kT)]\}$ or $[\exp(E_F/kT)]^2 \times \exp[-(E_C + E_D)/kT] + \exp(E_F/kT) \times \exp(-E_C/kT) - N_D/N_C = 0$ This equation is quadratic in $\exp(E_F/kT)$. Solving it we get, $\exp(E_F/kT) = \{-1 + [1 + 4(N_D/N_C) \exp((E_C - E_D)/kT)^{1/2}\}/\{2 \exp(-E_D/kT)\}$ (7) where we have ignored the negative root as the exponential is always positive.	$N_A^- = N_A \times$ probability that an acceptor atom gets ionised $= N_A\, p(E_A) = N_A\{1/[1 + \exp(E_A - E_F)/kT)]\}$ Thus, Eq. (11) becomes $N_V \exp[-(E_F - E_V)/kT] = N_A\{1/[1 + \exp((E_A - E_F)/kT)]\}$ or $[\exp(-E_F/kT)]^2 \times \exp[(E_V + E_A)/kT] + \exp(-E_F/kT) \times \exp(E_V/kT) - N_A/N_V = 0$ This equation is quadratic in $\exp(-E_F/kT)$. Solving it we get, $\exp(-E_F/kT) = \{-1 + [1 + 4(N_A/N_V) \exp((E_A - E_V)/kT)]^{1/2}\}/[2 \exp(E_A/kT)]$ (12) where we have ignored the negative root as the exponential is always positive.

Special cases	If one assumes that the widths of the valence band and conduction band are much smaller compared to the width of the forbidden energy gap, then the expression for E_F can be easily obtained by solving the equation, Total density of electrons in both the bands = N = density of electrons in the conduction band + density of electrons in the valence band, $N = Np(E_C) + Np(E_V)$ or $N = N \times 1/\{1 + \exp[(E_C - E_F)/kT]\} + N \times 1/\{1 + \exp[(E_V - E_F)/kT]\}$. Solving this equation, we get, $E_F = (E_C + E_V)/2$ (3) which is the special case of the more rigorous equation (2) derived above for $m_e^* = m_h^*$.	(i) For high doping concentration and low temperature, the second term under the square root is much grater than 1, so 1 can be neglected. (ii) For low doping concentration and high temperature, the second term under the square root sign is much smaller than 1, so only the first two terms of the binomial expansion are retained. Simplifying Eq. (7) we get the following special expressions for E_F. For high N_D and low T, $E_F = (E_C + E_D)/2 + ½\, kT \log(N_D/N_C)$ (8) For low N_D and high T, $E_F = E_C - kT \log(N_C/N_D)$ (9) Thus, at $T = 0$, $E_F = (E_C + E_D)/2$ (10) At very high temperatures, the second term in Eq. (9) approaches $(E_C - E_V)/2$, hence $E_F \sim (E_C + E_V)/2$, that is, the intrinsic value.	Simplifying Eq. (12) we get the following special expressions for E_F. For high N_A and low T, $E_F = (E_A + E_V)/2 + ½\, kT \log(N_V/N_A)$ (13) For low N_A and high T, $E_F = E_V + kT \log(N_V/N_A)$ (14) Thus, at $T = 0$, $E_F = (E_V + E_A)/2$ (15) At very high temperatures, the second term in Eq. (14) approaches $(E_C - E_V)/2$, hence $E_F \sim (E_C + E_V)/2$, that is, the intrinsic value.
Position of Fermi level	E E_C Conduction band E_F E_V Valence band	E E_C Conduction band E_F E_D N_D increasing E_{Fi} E_V Valence band T(K)	E E_C Conduction band E_{Fi} E_A E_F E_V N_A increasing Valence band T(K)
Significance of E_F	Using the definition of E_F in Chart 11.2, $E_F = (n\, E_C + p\, E_V)/(n + p)$ (4) Thus the Fermi level lies at the centre of the energy gap [that is, at $(E_C + E_V)/2$] as $n = p$.	It can be seen from Eq. (4) that the Fermi level shifts toward the conduction band as $n >> p$. However, for both p-type and n-type semiconductors, the Fermi level approaches the intrinsic value, E_{Fi} at very high temperatures because at these temperatures the thermally excited electron–hole pairs far outnumber the carriers due to doping. The intrinsic carriers swamp out the extrinsic effect.	It can be seen from Eq. (4) that the Fermi level shifts toward the valence band as $p >> n$.

CHART 11.5 DERIVATION OF CHARGE CARRIER DENSITY AND CONDUCTIVITY

	Intrinsic	Extrinsic	
		n-type	*p*-type
Density of charge carriers	Using Eqs. 11.22, 11.23 and 11.30, we have,	Since the energy gap does not change with impurity concentration, Eq. 1 is also valid for extrinsic semiconductors.	
	$n_p = n_i^2 = N_C \exp[-(E_C - E_F)/kT] \times N_V \exp[-(E_F - E_V)/kT] = N_C N_V \exp[-(E_C - E_V)/kT]$	(i) At high concentration and low temperature, from Chart 11.4, we have	
	or $n_i = 2(2\pi kT/h^2)^{3/2}(m_e^* m_h^*)^{3/4} \exp[-(E_C - E_V)/2kT]$ (1) Under the assumption that the widths of the conduction band and valance band are much smaller in comparison to the forbidden gap, we have (see Chart 11.4), $n = N\,p(E_C) = N\{1/[1 + \exp((E_C - E_F)/kT)]\}$ Using $E_F = (E_C + E_V)/2$, $n = N\{1/[1 + \exp((E_C - E_V)/2kT)]\}$.	$\exp[E_F/kT] = (N_D/N_C)^{1/2} \exp[(E_C + E_D)/2kT]$ (5) Substituting Eq. (5) in the expression for *n*, we get, $n = N_C \exp[-(E_C - E_F)/kT]$ $= N_C \exp(-E_C/kT)(N_D/N_C)^{1/2} \times \exp[(E_C + E_D)/2kT]$ or $n = (N_D N_C)^{1/2} \exp[(E_D - E_C)/2kT]$ or $n \propto \sqrt{N_D}$. (6)	$\exp[-E_F/kT] = (N_A/N_V)^{1/2} \exp[-(E_V + E_A)/2kT]$ (9) Substituting Eq. (9) in the expression for *p*, we get, $p = N_V \exp[-(E_F - E_V)/kT]$ $= N_V \exp(E_V/kT)(N_A/N_V)^{1/2} \times \exp[-(E_V + E_A)/2kT]$ or $p = (N_A N_V)^{1/2} \exp[(E_V - E_A)/2kT]$ or $p \propto \sqrt{N_A}$ (10)
	Using $E_C - E_V = E_g$, and $1 << \exp(E_g/2kt)$	(ii) At low doping concentration and high temperature, from Chart 11.4, we have,	
	we get $n = N \exp(-E_g/2kT)$ (2) Thus the intrinsic density is mainly controlled by the exponential term and is independent of E_F.	$\exp[E_F/kT] = (N_D/N_C)\exp(E_C/kT)$. (7) Substituting Eq. (7) in the expression for *n*, we get, $n = N_C \exp(-E_C/kT)(N_D/N_C)\exp(E_C/kT)$ or $n = N_D$ (8)	$\exp[-E_F/kT] = (N_A/N_V)\exp(-E_V/kT)$. (11) Substituting Eq. (11) in the expression for *p*, we get, $p = N_V \exp(E_V/kT)(N_A/N_V)\exp[-E_V/kT]$ or $p = N_A$ (12)

Conductivity	$\sigma = ne\mu_e + pe\mu_h = n_i e(\mu_e + \mu_h)$ (3) Putting n_i from Eq. 1, we get $\sigma = 2\left(\frac{2\pi kT}{h^2}\right)^{3/2}(m_e^* m_h^*)^{3/4}$ $\exp[-(E_C - E_V)/2kT]\, e(\mu_e + \mu_h)$ or $\sigma = A \exp[-E_g/2kT]$ where $A = 2\left(\frac{2\pi kT}{h^2}\right)^{3/2}(m_e^* m_h^*)^{3/4}\; e(\mu_e + \mu_h)$. Taking log, $\log \sigma = \log A - (E_g/2kT)$ (4) Thus the conductivity increases with increase in temperature. This is expected as the increase in temperature ruptures more covalent bonds, thus creating more number of electron– hole pairs. 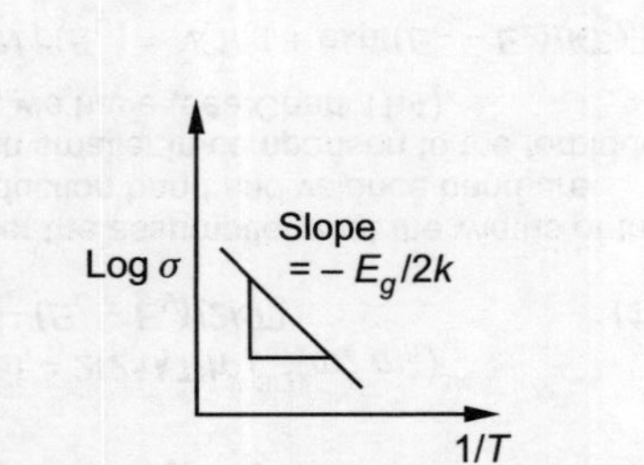 	The electrical conductivity of an extrinsic semiconductor varies as shown in figure. At low temperatures the conductivity is determined by the number density of ionised impurity atoms (donors/acceptors) because a small amount of energy, $E_C - E_D$ in the case of *n*-type and $E_A - E_V$ in case of p-type is required for ionisation. Thus the slope of log σ versus 1 / T in this range, called as *extrinsic range*, is given by $-(E_C - E_D)/k$ for *n*-type and by $-(E_A - E_V)/k$ for *p*-type. After complete ionisation of impurities any further increase in temperature does not affect conductivity. This range is called *exhaustion range* for *n*-type and *saturation range* for *p*-type signifying the complete ionisation of impurity atoms. This range is important for devices as conductivity is practically independent of temperature in this range. At high temperatures electrons from the valence band are promoted to the conduction band as the thermal energy is sufficient to overcome the forbidden energy gap. Thus the semiconductor behaves as an essentially intrinsic one. The slope of log σ versus $1/T$ in this intrinsic range is $-E_g/2k$. 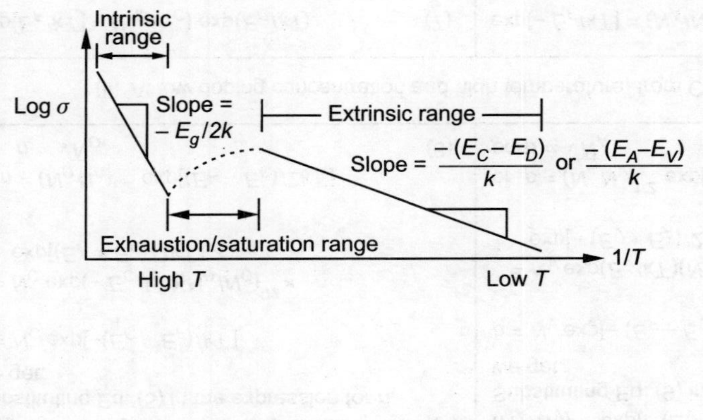
See Experiment 11.1 for determination of *Eg* using resistivity measurements. See Examples 11.7–11.12 for the application of various expressions derived in this chart.		

EXAMPLE 11.6
Intrinsic resistivity for Ge and Si

The mobilities of free electrons and holes in pure germanium are 0.38 m²/Vs and 0.18 m²/Vs. The corresponding values for pure silicon are 0.13 m²/Vs and 0.05 m²/Vs, respectively. Find out the value of intrinsic resistivity for both the materials. Assume $n_i = 2.5 \times 10^{19}$/m³ for Ge and $n_i = 1.5 \times 10^{16}$/m³ for Si.

Solution

The intrinsic conductivity is given by $\sigma = n_i e(\mu_e + \mu_h)$

and the intrinsic resistivity by, $\rho = 1/\sigma$

Thus for pure Ge,

$$\sigma = (2.5 \times 10^{19}/\text{m}^3) \times (1.6 \times 10^{-19}\ \text{C}) \times [(0.38 + 0.18)\ \text{m}^2/\text{Vs}]$$

$$= 2.24\ \mho\text{m}^{-1}$$

and $\rho = \dfrac{1}{2.24} = 0.446\ \Omega\text{m}$

For pure Si,

$$\sigma = (1.5 \times 10^{16}/\text{m}^3) \times (1.6 \times 10^{-19}\ \text{C}) \times [(0.13 + 0.05)\ \text{m}^2/\text{Vs}]$$

$$= 0.432 \times 10^{-3}\ \mho\text{m}^{-1}$$

and $\rho = 2.315 \times 10^3\ \Omega\text{m}$

The resistivity of Si is more than that of Ge due to its higher value of energy gap. As the energy required to break the covalent bonds is larger in the case of silicon, fewer electron–hole pairs are generated.

EXAMPLE 11.7
Variation of n/N with temperature

Find the fraction of electrons in the valence band of intrinsic germanium that can be thermally promoted across the forbidden energy gap of 0.72 eV into the conduction band at (a) 30 K, (b) 300 K and (c) the melting point of Ge, which is = 1210 K.

Solution

The fraction of electrons is given by

$$n/N = \exp(-E_g/2kT)$$

$$= \exp\left[\frac{-0.72\ \text{eV}}{2 \times (0.026/300) \times T\ \text{eV}}\right] \text{ (because } kT = 0.026 \text{ eV at T} = 300 \text{ K)}$$

Thus we get the results shown in the table.

Thus the fraction of electrons that can be promoted across the forbidden energy gap increases with the increase in temperature. This is expected as thermal energy breaks more covalent bonds, thus creating more number of electron–hole pairs.

T (K)	n/N
30	7.4×10^{-61}
300	9.7×10^{-7}
1210	0.032

EXAMPLE 11.8
Variation of n/N with E_g

Estimate the fraction of electrons in the conduction band at a room temperature of (a) germanium ($E_g = 0.72$ eV), (b) silicon ($E_g = 1.10$ eV) and (c) diamond ($E_g = 5.6$ eV).

Solution

$$\frac{n}{N} = \exp\left(\frac{-E_g}{2kT}\right)$$

$$= \exp[-E_g\,\text{eV}/(2 \times 0.026\,\text{eV})]$$

Germanium	0.72	9.7×10^{-7}
Silicon	1.10	6.5×10^{-10}
Diamond	5.6	1.7×10^{-47}

Thus we get the results shown in the table.

The significance of these results is that the equation $(n/N) = \exp(-E_g/2kT)$ provides a defining relation for solids on the basis of their energy gaps. The materials with small energy gap (semiconductors) contain appreciable number of electrons in their conduction band. But materials with large energy gap (insulators) have negligible number of electrons in the conduction band. For metals, $E_g = 0$ hence $(n/N) \sim 1$ meaning that most of the electrons in a metal are conduction electrons. For perfect insulators, $E_g = \infty$. Hence $(n/N) \sim 0$.

EXAMPLE 11.9
Temperature at which $E_F = E_C$

An n-type silicon material has the donor concentration equal to 1 atom per 2×10^8 silicon atoms. Find the value of the temperature at which the Fermi level coincides with the edge of the conduction band. The given density of atoms in silicon = 5×10^{28} atoms/m³. Assume $m_e^* = m_e$.

Solution

For n-type semiconductors

$$n = N_C \exp\left[-\frac{(E_C - E_F)}{kT}\right] \approx N_D$$

Thus, Fermi level coincides with the edge of the conduction band, that is, $E_F = E_C$, provided

$$N_D = N_C$$

$$= 2\left(\frac{2\pi kTm_e^*}{h^2}\right)^{3/2}$$

$$= 2\left(\frac{2\pi kTm_e}{h^2}\right)^{3/2} \quad (\text{as } m_e^* = m_e)$$

$$= 2\left[\frac{2\pi \times 1.38 \times 10^{-23}\ \text{J/K} \times 9.1 \times 10^{-31}\ \text{kg}}{(6.625 \times 10^{-34}\ \text{Js})^2}\right]^{3/2} \times (T\ \text{K})^{3/2}$$

$$= 4.82 \times 10^{21} \times T^{3/2}/\text{m}^3$$

Now density of atoms in silicon = $5 \times 10^{28}/\text{m}^3$.

So, density of donor atoms, $N_D = \dfrac{5 \times 10^{28}\ \text{m}^{-3}}{2.0 \times 10^8} = 2.5 \times 10^{20}/\text{m}^3$

Hence, $2.5 \times 10^{20} = 4.82 \times 10^{21} \times T^{3/2}$

or $\quad T = 0.14$ K

Thus the complete ionisation of donor atoms is achieved at temperatures that are low enough!

EXAMPLE 11.10
Calculation of N_D, N_D^+, n and p

A sample of Si is doped with phosphorus. The donor impurity level lies 0.045 eV below the bottom of the conduction band. At 300 K, E_F is 0.010 eV above the donor level. Calculate (a) the impurity concentration, (b) the number of ionised impurities, (c) the free electron concentration and (d) the hole concentration. For Si, $E_C = 1.100$ eV, $m_e^* = 0.31m_e$, and $m_h^* = 0.38m_e$.

Solution

Using the given data the band structure of the sample can be drawn as shown in Fig.11.9.

Thus,

$$E_C - E_D = 0.045\ \text{eV}$$

$$E_C - E_F = 0.035\ \text{eV}$$

$$E_F - E_D = 0.01\ \text{eV}$$

Also, kT at 300 K = 0.026 eV

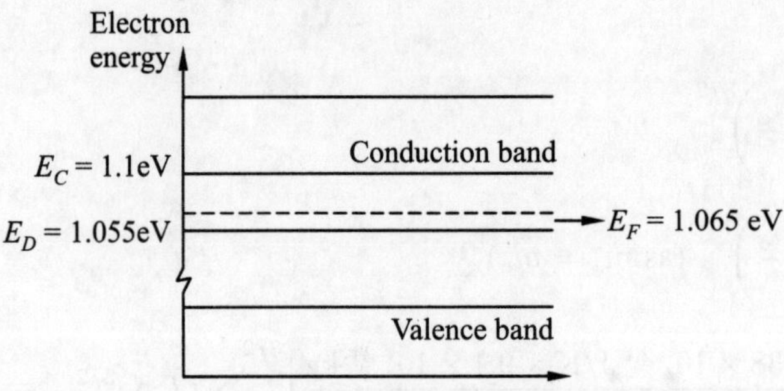

Figure 11.9

The constants N_C and N_V are given by

$$N_C = 2\left[\frac{2\pi m_e^* kT}{h^2}\right]^{3/2}$$

$$= 2\left[\frac{2\pi \times 0.31 \times 9.1 \times 10^{-31}\ \text{kg} \times 0.026 \times 1.6 \times 10^{-19}\ \text{J}}{(6.625 \times 10^{-34}\ \text{Js})^2}\right]^{3/2} = 4.35 \times 10^{24}/\text{m}^3$$

$$N_V = 2\left[\frac{2\pi m_h^* kT}{h^2}\right]^{3/2}$$

$$= 2\left[\frac{2\pi \times 0.38 \times 9.1 \times 10^{-31}\ \text{kg} \times 0.026 \times 1.6 \times 10^{-19}\ \text{J}}{(6.625 \times 10^{-34}\ \text{Js})^2}\right]^{3/2}$$

$$= 5.91 \times 10^{24}/\text{m}^3$$

(a) As the sample is doped with phosphorus, a group V element, the sample is an n-type material. Thus we can write,

$$n \sim N_D^+$$

$$N_C \exp\left[\frac{-(E_C - E_F)}{kT}\right] = N_D[1 - p(E_D)] = N_D\left[1 - \frac{1}{1 + \exp[(E_D - E_F)/kT]}\right]$$

or $\quad 4.35 \times 10^{24}/\text{m}^3 \times \exp\left[\frac{-0.035\ \text{eV}}{0.026\ \text{eV}}\right] = N_D\left[1 - \frac{1}{1 + \exp[(-0.01\ \text{eV}/0.026\ \text{eV})]}\right]$

or $\quad N_D = 2.8 \times 10^{24}/\text{m}^3$

(b) The number of ionised impurities is given by

$$N_D^+ = N_D\left\{1 - \frac{1}{1 + \exp[(E_D - E_F)/kT]}\right\}$$

$$= (2.8 \times 10^{24}/\text{m}^3) \times 0.4050 = 1.13 \times 10^{24}/\text{m}^3$$

(c) The free electron concentration is given by

$$n = N_C \exp\left[\frac{-(E_C - E_F)}{kT}\right] = 1.13 \times 10^{24}/\text{m}^3$$

(d) The hole density is given by

$$p = N_V \exp\left[\frac{E_V - E_F}{kT}\right], \; E_V = 0 \text{ and } E_F = 1.065 \text{ eV}$$

$$= 5.91 \times 10^{24}/\text{m}^3 \exp(-1.065 \text{ eV}/0.026 \text{ eV}) = 9.6 \times 10^6/\text{m}^3$$

EXAMPLE 11.11

Calculation of n_n, p_n and ρ

A silicon wafer is doped with 10^{21} phosphorus atoms/m^3. Calculate (a) the majority carrier concentration, (b) the minority carrier concentration and (c) the electrical resistivity of the doped silicon at room temperature. Assume complete ionisation of the dopant atoms; n_i (Si) $= 1.5 \times 10^{16}$ m^{-3}, $\mu_e = 0.135$ m^2/Vs and $\mu_h = 0.048$ m^2/Vs.

Solution

Since the silicon is doped with phosphorous, a group V element, the doped silicon is n-type.

(a) As complete ionisation is assumed, the majority carrier concentration is

$$n_n = N_d = 10^{21} \text{ electrons/m}^3$$

(b) The minority carrier concentration is

$$p_n = \frac{n_i^2}{N_d} = \frac{(1.5 \times 10^{16} \text{ m}^{-3})^2}{10^{21} \text{ m}^{-3}} = 2.25 \times 10^{11} \text{ holes/m}^3$$

(c) For a doped semiconductor,

$$\sigma = e(n\mu_e + p\mu_h)$$

$$= 1.6 \times 10^{-19} \text{ C}[(10^{21} \text{ m}^{-3} \times 0.135 \text{ m}^2/\text{Vs}) + (2.25 \times 10^{11} \text{ m}^{-3} \times 0.048 \text{ m}^2/\text{Vs})]$$

$$= 21.6 \, \mho \, \text{m}^{-1}$$

$$\text{or } \rho = 0.046 \, \Omega \, \text{m}$$

It should be noticed that the second term is negligible and the conductivity is given mainly by $\sigma = ne\mu_e$, which is true as the given semiconductor is of n-type.

EXAMPLE 11.12
Increase in conductivity with temperature for pure silicon

Show that the electrical conductivity of pure silicon increases 2390 times when the temperature is raised from room temperature to 200°C. Assume that the energy gap of silicon is 1.1 eV.

Solution

The electrical conductivity of a pure semiconductor is given by (see Chart 11.5)

$$\sigma = A \exp\left(\frac{-E_g}{2kT}\right)$$

Thus, $\sigma_{473K} = A \exp\left[\frac{-E_g}{2k \times 473\text{ K}}\right]$ and $\sigma_{300K} = A \exp\left[\frac{-E_g}{2k \times 300\text{ K}}\right]$.

Hence, $\dfrac{\sigma_{473K}}{\sigma_{300K}} = \exp\left[\dfrac{-E_g}{2k}\left(\dfrac{1}{473\text{ K}} - \dfrac{1}{300\text{ K}}\right)\right]$.

$$= \exp\left[\frac{-1.1\text{ eV}}{2 \times 8.62 \times 10^{-5}\text{ eV/K}} \times \left(\frac{1}{473\text{ K}} - \frac{1}{300\text{ K}}\right)\right] = 2390$$

Thus, $\sigma_{473K} = 2390 \times \sigma_{300K}$. The electrical conductivity of the silicon increases 2390 times when the temperature is raised from room temperature to 200°C.

EXPERIMENT 11.1 ENERGY GAP BY FOUR PROBE METHOD

Aim

To determine the forbidden energy gap of a semiconductor using four probe method

Apparatus

Four probe arrangement, thermometer (0–200°C), constant current source, digital voltmeter, oven and semiconductor sample

Formulae

$$\rho = \frac{\rho_0}{G(t/d)} \tag{11.37}$$

$$\rho_0 = \frac{2\pi dV}{I} \tag{11.38}$$

$$E_g = 0.396\times \text{ slope of the linear portion of the curve between } \log_{10}\rho \text{ versus } \frac{10^3}{T} \tag{11.39}$$

where ρ is the resistivity of the sample, ρ_0 is its uncorrected value, $G(t/d)$ is the correction factor, t is the thickness of the sample, d is the distance between adjacent probes, V is the voltage across the inner pair of probes and I is the current through the outer probes.

Theory

We know that the resistance of a wire is proportional to its length l and inversely proportional to its area of cross section A. The constant of proportionality is called the resistivity, ρ, of the material of the wire. Thus,

$$R = \frac{\rho l}{A} \tag{11.40}$$

The reciprocal of resistivity is known as the conductivity, σ. The microscopic equation for σ is

$$\sigma = \frac{J}{E} \tag{11.41}$$

where J is the current density given by I/A and E is the electric field given by V/l. Substituting these expressions in Eq. 11.41, we get Eq. 11.40, through the macroscopic form of Ohm's law: $V = IR$.

Figure 11.10 shows a simple arrangement for the measurement of conductivity. However, this simple method is not straightforward for semiconductors for two reasons.

1. It is not convenient to produce a specimen of accurately known dimensions and uniform cross section.
2. The contacts between metal and semiconductor are not ohmic; they show different rectifying $V - I$ characteristics if the biasing is reversed.

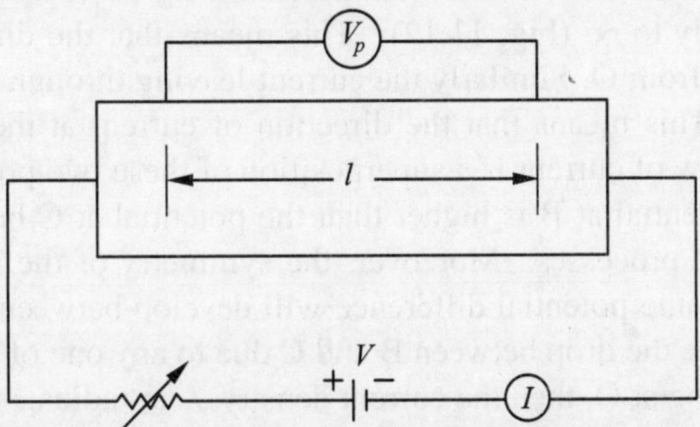

Figure 11.10 Experimental arrangement for measurement of conductivity of a bar-shaped sample of uniform thickness

A four probe arrangement eliminates these difficulties and is generally preferred for conductivity measurements in semiconductors. It consists of four spring-loaded equispaced needles in contact with the specimen, as shown in Fig. 11.11. A constant current is passed through the outer probes,

A and D, and the voltage developed across the inner probes B and C is measured using a digital voltmeter.

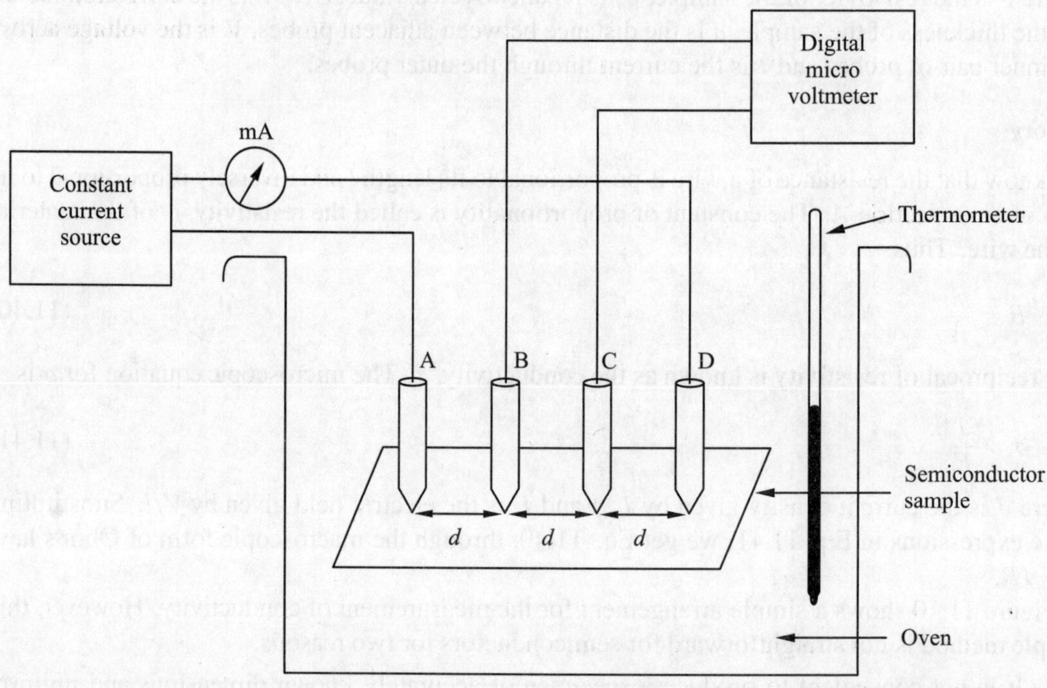

Figure 11.11 Experimental set up for measurement of conductivity using four probe method

The situation can be visualised in terms of the following processes. The current entering at A spreads symmetrically to ∞ (Fig. 11.12). This means that the direction of current at the point P is along OP away from O. Similarly the current leaving through the probe D is collected symmetrically from ∞. This means that the direction of current at the point P′ is along O′P′ towards O′. The actual flow of current is a superposition of these two processes. We note that in both the processes the potential at B is higher than the potential at C because the current flows from B to C in both these processes. Moreover, the symmetry of the situation implies that in both these processes the same potential difference will develop between B and C. So the actual potential difference is twice the drop between B and C due to any one of these processes.

If I is the current entering at O, then the current density J_r at radius r from O is

$$J_r = \frac{I}{2\pi r^2} \tag{11.42}$$

where a factor of 2 appears because current flows only in the bottom half plane. The field E_r is given by

$$E_r = \frac{I}{2\pi \sigma r^2} \tag{11.43}$$

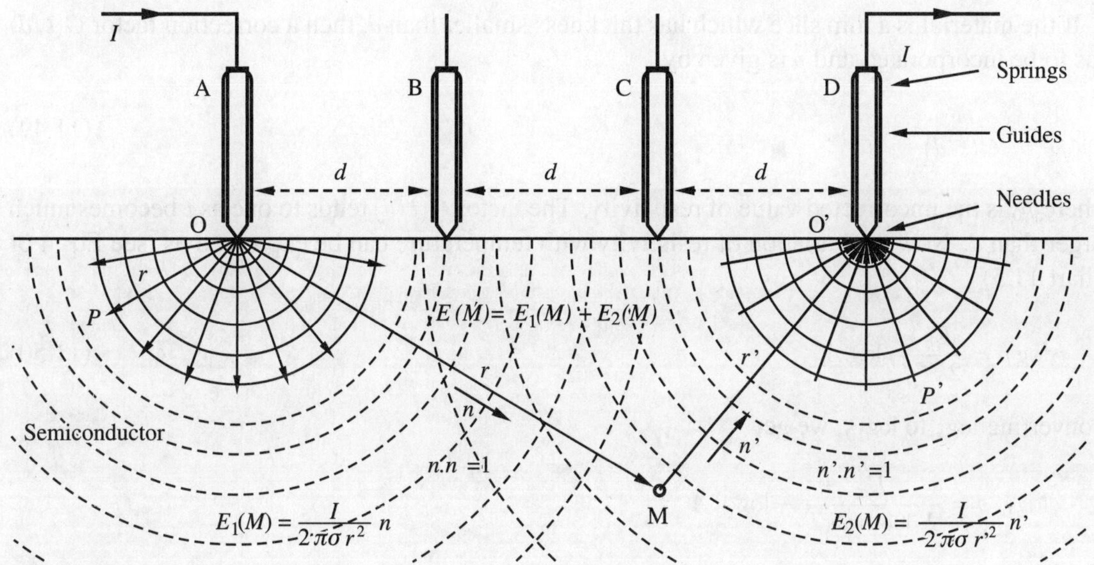

Figure 11.12 Determination of ρ for a semiconductor sample

The potential at some radius a is given by

$$V_a = -\int E_r dr = -\int_0^a \frac{I dr}{2\pi\sigma r^2} = -\frac{I}{2\pi\sigma}\int_0^a \frac{dr}{r^2} = \frac{I}{2\pi\sigma a}$$

Thus the potential at point B due to current entering at A is

$$V_B = \frac{I}{2\pi\sigma d} \tag{11.44}$$

and that at C is

$$V_C = \frac{I}{2\pi\sigma(2d)} \tag{11.45}$$

Hence the potential difference developed between B and C due to current entering at A is

$$V_{B-C} = \frac{I}{4\pi\sigma d} \tag{11.46}$$

and hence the actual voltage V, between B and C, will be twice this voltage, that is,

$$V = \frac{I}{2\pi\sigma d} \tag{11.47}$$

As $\sigma = \dfrac{1}{\rho}$, the resistivity for materials with very large thickness in comparison with the probe spacing, d is given by

$$\rho = \frac{2\pi d V}{I} \tag{11.48}$$

If the material is a thin slice which has thickness smaller than d, then a correction factor $G(t/d)$ has to be incorporated and ρ is given by

$$\rho = \frac{\rho_0}{G(t/d)} \tag{11.49}$$

where ρ_0 is the uncorrected value of resistivity. The factor $G(t/d)$ tends to one as t becomes much larger than d. Now the variation of resistivity with temperature can be expressed as (see Eq. 4 of Chart 11.5)

$$\log \rho = \frac{E_g}{2kT} - \log A \tag{11.50}$$

Converting $\log_e$ to $\log_{10}$, we get

$$\log_{10} \rho = \frac{E_g}{2.303 \times 2kT} - \log_{10} A$$

$$\text{or} \quad \log_{10} \rho = \left[\frac{E_g}{2.303 \times 2k \times 10^3}\right] \frac{10^3}{T} - \log_{10} A \tag{11.51}$$

Thus, if a graph is plotted between $\log_{10} \rho$ and $10^3/T$ we will get a straight line, the slope of which is given by $\dfrac{E_g}{(2.303 \times 2k \times 10^3)}$

Substituting $k = 8.6 \times 10^{-5}$ eV/K, we get the expression for the energy gap as

$$E_g = 0.396 \times \text{ slope of plot between } \log_{10}\rho \text{ and } 10^3/\text{T} \tag{11.52}$$

Procedure

1. The connections are made as shown in Fig. 11.11.
2. The four-probe arrangement is placed in an oven and the thermometer is inserted in the slot provided.
3. The constant current source is switched on. The mode switch of the constant current source is turned to current mode and the current is adjusted to a value of 8 mA. This current is to be maintained constant throughout and should be checked and assured for constancy.
4. Now the mode switch is turned to voltage mode and the oven is switched on.
5. As the temperature of the oven increases the voltage is recorded at intervals of 10°C from 40°C to 150°C.
6. The observations are tabulated and a graph is plotted between $\log_{10}\rho$ and $10^3/T$.

Observations

$I = 8$ mA, $d =$ — cm, $t =$ — cm, $G(t/d) =$ —

S. No.	Temperature (°C)	Voltage, V (mV)	Temperature (K)	Resistivity, $\rho = 2\pi dV/IG(t/d)$ (Ohm-cm)	$10^3/T$ (K^{-1})	$\log_{10}\rho$
	40					
	50					
	–					
	–					
	–					
	–					
	–					
	–					
	–					
	–					
	–					
	150					

Graph

See Fig. 11.13.

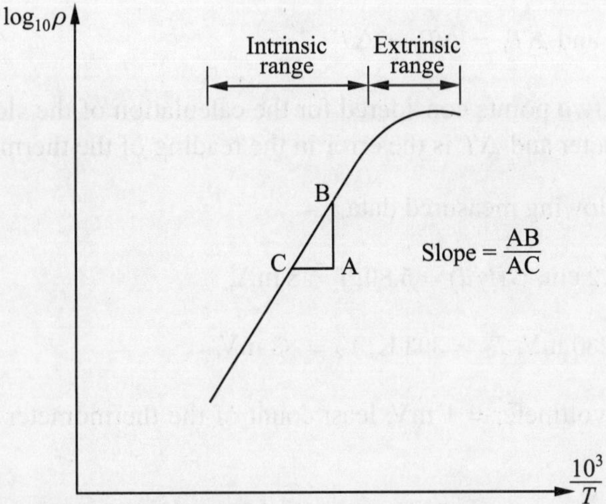

Figure 11.13 Variation of $\log_{10}\rho$ with $10^3/T$

Calculations

From the graph, the slope of the linear portion (intrinsic portion) of the graph = ——
Hence $E_g = 0.396$ (eV/K) × slope (K) = —— (eV)

Estimation of errors

The error ΔE_g in the estimation of E_g can be determined by adopting the method outlined in Section 1.4.2. Thus, using Eq. (11.39),

$$E_g = 0.396 \times \frac{\log_{10}\rho_1 - \log_{10}\rho_2}{(10^3/T_1) - (10^3/T_2)}, \text{ we have,}$$

$$\Delta E_g = \left|\frac{\partial E_g}{\partial \rho_1}\right| \Delta\rho_1 + \left|\frac{\partial E_g}{\partial \rho_2}\right| \Delta\rho_2 + \left|\frac{\partial E_g}{\partial T_1}\right| \Delta T_1 + \left|\frac{\partial E_g}{\partial T_2}\right| \Delta T_2 \tag{11.53}$$

Simplifying we get,

$$\Delta E_g = K\left\{\left[\frac{T_1 T_2}{(T_2 - T_1)}\right]\left[\frac{(V_1 + V_2)}{V_1 V_2}\right]\Delta V + \left[\ln\left(\frac{V_1}{V_2}\right)\right]\left[\frac{T_1^2 + T_2^2}{(T_2 - T_1)^2}\right]\Delta T\right\} \tag{11.54}$$

where we have used

$$K = \frac{0.396}{(2.303 \times 1000)} = 1.7195 \times 10^{-4}$$

$$\rho_1 = \frac{2\pi d V_1}{IG(t/d)}, \ \rho_2 = \frac{2\pi d V_2}{IG(t/d)}$$

$$\Delta\rho_1 = \frac{2\pi d}{IG(t/d)}\Delta V_1, \ \Delta\rho_2 = \frac{2\pi d}{IG(t/d)}\Delta V_2$$

$$\Delta V_1 = \Delta V_2 = \Delta V \text{ and } \Delta T_1 = \Delta T_2 = \Delta T$$

(T_1, V_1) and (T_2, V_2) are two points considered for the calculation of the slope, ΔV is the error in the reading of the voltmeter and ΔT is the error in the reading of the thermometer.

For example, for the following measured data,

$t = 0.05$ cm, $d = 0.2$ cm, $G(t/d) = 5.89$, $I = 8$ mA,

$T_1 = 353$ K, $V_1 = 230$ mV, $T_2 = 393$ K, $V_2 = 88$ mV,

least count of the voltmeter = 1 mV, least count of the thermometer = 1 K, $\Delta V = 0.5$ mV and $\Delta T = 0.5$ K,

$$\Delta E_g = 1.7195 \times 10^{-4}\ \text{eV/K}\left[\frac{353\ \text{K} \times 393\ \text{K}}{393\ \text{K} - 353\ \text{K}}\right]\left[\frac{230\ \text{mV} + 88\ \text{mV}}{230\ \text{mV} \times 88\ \text{mV}}\right] 0.5\ \text{mV}$$

$$+1.7195 \times 10^{-4}\ \text{eV/K}\left[\ln\left(\frac{230\ \text{mV}}{88\ \text{mV}}\right)\right]\left[\frac{(353\ \text{K})^2 + (393\ \text{K})^2)}{(393\ \text{K} - 353\ \text{K})^2}\right] 0.5\ \text{K}$$

$$= 0.0047\ \text{eV} + 0.0144\ \text{eV} = 0.0191 \approx 0.02\ \text{eV}$$

Now, $\rho_1 = \dfrac{2\pi \times 0.2\ \text{cm} \times 230\ \text{mV}}{8\ \text{mA} \times 5.89} = 6.13\ \Omega\text{cm}$

and $\rho_2 = \dfrac{2\pi \times 0.2 \text{ cm} \times 88 \text{ mV}}{8 \text{ mA} \times 5.89} = 2.35\ \Omega\text{cm}$

Hence $E_g = 0.396 \times \dfrac{\log_{10} 6.13 - \log_{10} 2.35}{(10^3/353 \text{ K}) - (10^3/393 \text{ K})} = 0.57 \text{ eV}$

Thus the relative error = 0.02 eV/0.57 eV = 0.035 and the percentage error is 3.5 %.

Results

1. The resistivity of a given semiconductor is found to vary as shown in Fig. 11.13.
2. The energy gap of the semiconductor is found to be —— ± —— eV. The error in the determination of energy gap is —— %.

Discussion

This experiment enables us to understand how a semiconductor behaves as its temperature is changed. We studied the theoretical behaviour, that is, how a semiconductor should behave, in Chart 11.5, and the theory section of this experiment. In the following table we compare the theory and the experiment.

Theory	Experiment
Equation 4 of Chart 11.5, derived using band theory, predicts that resistivity of a semiconductor should decrease with increase in temperature.	Figure 11.13 shows that the measured value of $\log \rho$ increases as $10^3/T$ increases, or as T decreases.

It can be seen from this table that there is a qualitative agreement between the theory and the experiment.

In semiconductors resistivity decreases with the increase in temperature because thermal energy causes more covalent bonds to break, thus generating more electron–hole pairs, that is, more number of electrons are promoted from valence band to conduction band. In metals, on the other hand, resistivity increases with the increase in temperature as thermal energy randomises the motion of electrons, thus increasing the scattering and decreasing the mean free path of these electrons. Metals thus exhibit positive temperature coefficient (PTC) of resistivity whereas semiconductors exhibit negative temperature coefficient (NTC) of resistivity. In this experiment the decrease in resistivity causes the decrease in resistance, and hence the observed decrease in voltage, because the current through the sample is kept constant.

However, the behaviour discussed above pertains to an intrinsic semiconductor. This corresponds to low $10^3/T$, or high T values in Fig. 11.13. For lower temperatures, that is high values of $10^3/T$, the material is essentially extrinsic. It should be noted here that in practice an ideally intrinsic, or the purest material for which $n = p$, can never be obtained. Materials that are considered to be intrinsic (not externally doped) always contain some impurities making them either n- or p-type. It is only at high temperatures that the thermally generated carriers outnumber the carriers due to impurities and extrinsic character of the material is swamped by the intrinsic character.

11.6 SUPERCONDUCTIVITY

Superconductivity was discovered by K Onnes in 1911 when he observed that the resistivity of mercury disappered completely for temperatures below 4.15 K (see Fig. 11.14). The phenomenon of superconductivity is observed in most metals, many alloys and certain chemical compounds. For most of the superconducting metals the temperature at which a transition from normal state to superconducting state takes place, called the *transition temperature*, T_C, lies between 0.1 K to 10 K. It is significant that the elements that are good conductors at room temperature, for example Cu, Au, Ag and the alkali metals do not show superconductivity. It is also interesting that superconducting elements are relatively poor conductors at room temperature.

It is observed that the transition temperature of a superconductor is reduced by the application of a magnetic field. If the magnetic field exceeds a certain critical value H_C, then its superconductivity disappears altogether. The relation between critical field and temperature is observed to be

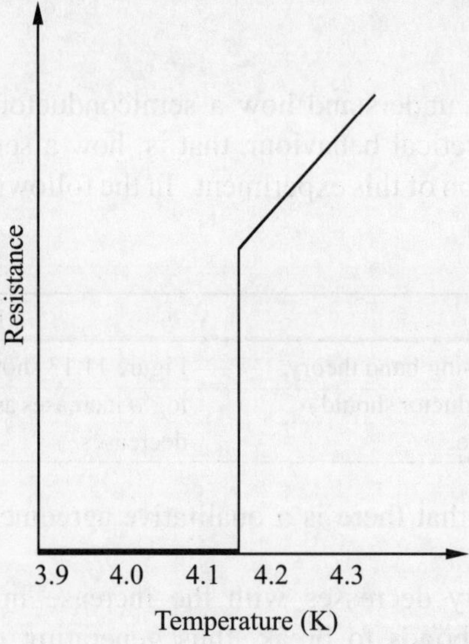

Figure 11.14 Below $T_c = 4.15$ K mercury becomes superconducting, that is, shows no resistance

$$H_C(T) = H_C(0)\left[1 - \frac{T^2}{T_C^2}\right] \tag{11.55}$$

where, $H_C(0)$ is the critical magnetic field at 0 K. From this relation, at $T = 0$, $H_C = H_C(0)$ and at $T = T_C, H_C = 0$. When $T = T_1 < T_C$, as shown in Fig. 11.15, the material is superconducting for $H_C < H_{C1}$. If H_C exceeds H_{C1}, the material loses superconductivity and turns normal. On the other hand when the applied magnetic field is equal to H_{C1}, the material will turn normal if $T > T_1$. The critical magnetic field required to destroy superconductivity goes on decreasing with increasing temperature.

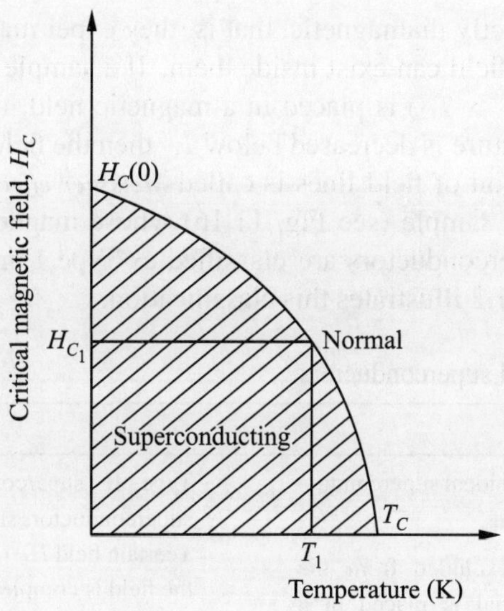

Figure 11.15 Variation of critical magnetic field with temperature for a superconducting material

The magnetic field that destroys superconductivity need not be an external field; it may arise due to current flowing through a superconducting ring. We know that the magnetic field produced by current I is given by $H = I/2\pi r$, r being the radius of the ring. Thus, corresponding to the critical field H_C, there is a critical current I_C given by

$$I_C = 2\pi r H_C \tag{11.56}$$

which will destroy superconductivity. Equation 11.56 is referred to as *Silsbee's rule*.

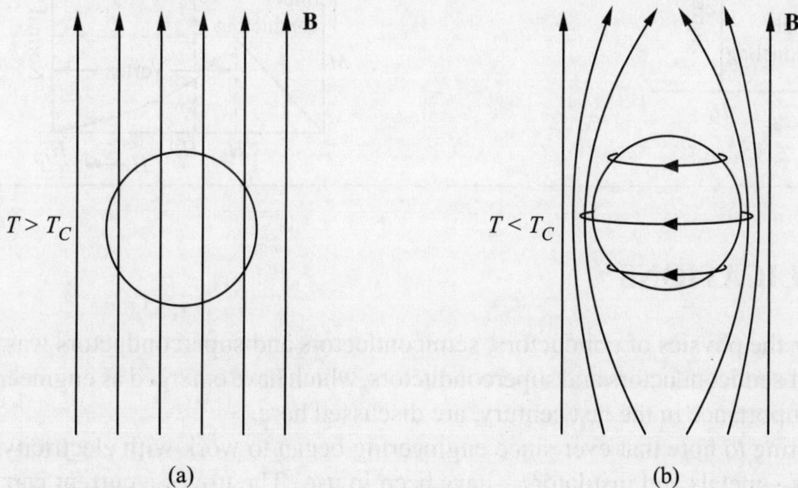

Figure 11.16 The Meissner effect: (a) Field lines pass through a material in normal state and (b) Field lines are expelled completely when material is in superconducting state

Superconductors are perfectly diamagnetic, that is, they expel magnetic field lines completely out of them, or no magnetic field can exist inside them. If a sample of superconducting material in the normal state (that is $T > T_C$) is placed in a magnetic field, then field lines can permeate through it. But if the temperature is decreased below T_C then the field lines are expelled out. This phenomenon of total expulsion of field lines is called *Meissner effect*. It arises because currents appear on the surface of the sample (see Fig. 11.16) whose magnetic field exactly cancels the original field inside it. Superconductors are classified as Type I or Type II depending on their magnetic behaviour. Table 11.2 illustrates this classification.

Table 11.2 Type I and Type II superconductors

Type 1	Type II
Type I superconductor, also called ideal superconductors, show complete diamagnetism All of the magnetic field is excluded from the superconductor when the material is placed in a magnetic field. The Meissner effect is thus complete and we have $B = 0$ inside the material or equivalently, $M = -H$. The magnetisation M is equal to $-H$ up to H_C, where it drops to zero as shown in the figure. 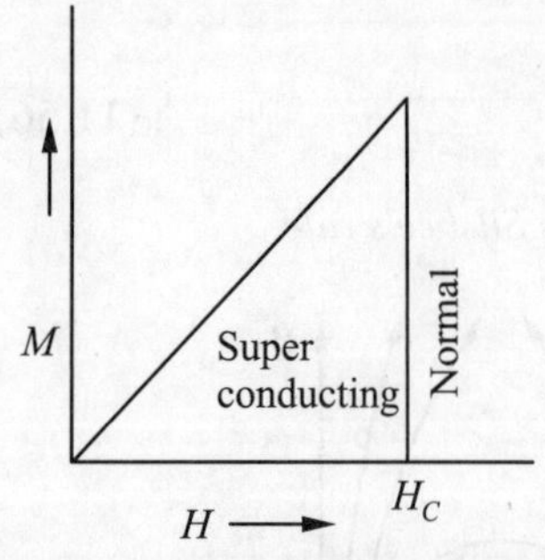 	Type II superconductors, also called the hard superconductors show complete diamagnetism below a certain field H_{C1} called the lower critical field. Thus the field is completely excluded for applied fields below H_{C1}. At H_{C1} the field begins to penetrate the specimen, and the penetration increases until H_{C2} called the upper critical field, is reached. At H_{C2} the magnetisation vanishes and the specimen becomes normal. Type II superconductor is, neverthless, completely superconducting for all fields below H_{C1}. The Meissner effect is thus incomplete in the region between H_{C1} and H_{C1} called the vortex region. 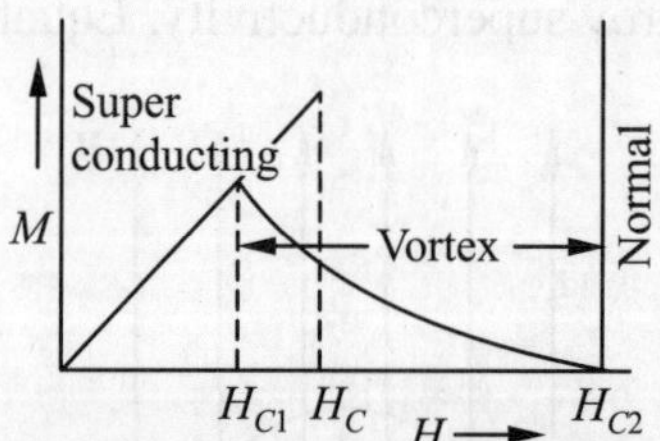

11.7 APPLICATIONS

In this chapter the physics of conductors, semiconductors and superconductors was studied. The applications of semiconductors and superconductors, which have emerged as engineering materials of immense importance in the last century, are discussed here.

It is interesting to note that ever since engineering began to work with electricity, materials of two categories—metals and insulators—have been in use. The use of a current carrying metallic wire clad with an insulating material shows that these materials enjoyed almost equal importance in electricity. Semiconductors had been known for quite some time. Mendeleev had discovered

CHART 11.6 SEMICONDUCTOR DEVICES AND APPLICATIONS

Device	Description	Application
Thermistor	It is a device having resistance that is sensitive to temperature changes, usually decreasing with increase in temperature. Metals usually exhibit positive temperature coefficient of resistivity and resistance increases in most metals at the rate of about 0.4% per degree centigrade rise in temperature. This results due to increased thermal motion of ions and consequent decrease of electron mean free path as temperature is raised. Though Ge and Si exhibit 6% to 8% decrease in resistance per degree centigrade rise in temperature, they are not used as thermistors due to the marked sensitivity of their properties to impurities. Commercial thermistors are usually made of sintered mixtures of Mn_2O_3, NiO_2 and	Thermistors are used in electronic circuits to compensate for change in resistance with temperature of those components where such variation cannot be tolerated. They are also used as temperature sensors in microwave power measuring equipments and electronic thermometers. They can yield power information over the range 10^{-5} mW to 20 mW with a typical burnout level of 400 mW.
Gunn effect device	It is a device based on the property of voltage controlled, bulk negative conductance in GaAs due to a mechanism called as the transferred electron mechanism. The *V–I* characteristics of a Gunn diode are as shown in the figure. The region A-B is the one showing negative differential resistivity (NDR). Current density, *J* / A / B / Electric field, *E*	Gunn effect devices are used as microwave oscillators. Gunn (1963) observed microwave oscillations in a GaAs sample while studying high field effects. The device is named after its inventor.
Photo detector or radiation detector	It is a device based on a photoconductive effect. If a photon of energy $E = h\nu$ is incident on an intrinsic semiconductor material having energy gap E_g, where $E_g < E$, then the photon is absorbed by an electron in the valance band to get promoted to the conduction band (see Ex. 11.13). Thus electron–hole pairs are created by absorption of light leading to conduction of current. This is known as photoconductive effect or the phenomenon of photoconduction. A photodetector or a radiation detector detects the radiation due to change in its bulk resistivity, upon illumination. Out of the three important photodetectors, namely, photoconductor, photodiode and the avalanche photodiode, we discuss the photoconductor here. The other photodetectors are discussed in Chart 12.3. A photoconductor consists of a slab of semiconductor in bulk form or in the form of a thin film deposited on an insulating substrate, with ohmic contacts fixed at opposite ends. Upon illumination a photoconductor drives a photocurrent through the external circuit. The sensitive photoconductors are made of CdS and CdTe. Photoconductors are also known as photoresistors or LDRs (light dependent resistors).	Photoconductor is at the heart of the process of xerography. Generally selenium is used as a photoconductor. It is a good insulator in the dark but conducts in the presence of light. A plate coated with selenium is electrostatically charged and is illuminated with a projected image of the object to be copied. The illuminated portions thus become conducting and leak away. The charge retaining portions of the plate then attract the negatively charged black powder called toner. Now the paper with its reverse positively charged is placed on the plate. It attracts the toner from the plate surface and is fused onto the paper permanently by heat. A copy of the original image is thus produced.
Detector of charged particles	A high energy proton or a particle passing through a semiconductor crystal can excite electrons from valence band to conduction band, thus increasing its conductivity, Thus a semiconductor can be used as a detector of charged particles.	

germanium a long time ago and called it ekasilicon to bring home its closeness with silicon, another semiconducting material known at that time. The name germanium was, however, given by Winkler, the German scientist who studied its properties thoroughly and gave the name in honour of his homeland.

However, semiconductors were known to be materials that were good for nothing: neither good conductors, nor good insulators. Both electrical engineering and electronics paid no attention to them for a long time. It is worth noting here that the history of physics is replete with examples where research that was initially considered moribund has come out with interesting possibilities in the future.

Research in electrical properties of solids culminated in the development of the band theory—the most successful theory for solids—in the first quarter of the last century. After this the capabilities of semiconductors began to be noticed. That the electron can be promoted from the valence band to the conduction band by supplying it with energy became the basis for the development of many new bulk semiconductor devices. Photoconductor and thermistor are two examples of such devices, which employ optical energy and thermal energy, respectively, for this purpose. A brief description and applications of these devices are included in Chart 11.6. The knowledge of band theory also found application in the development of tailor-made materials of desired band gap for specific applications. This came to be known as *band gap engineering*. However, the properties of semiconductors, which proved to be most useful for applications, were associated with the dramatic changes in conductivity due to doping. Example 11.14 gives the quantitative idea of these changes. It should be remembered here that along with physics, the technology of preparing pure crystals also played a major role in this progress. With the discovery of n-type and p-type semiconductors it became clear that a large number of devices could be built using these semiconductors for a variety of applications. These devices are discussed in the next chapter.

EXAMPLE 11.13

Photon energy required for generation of electron–hole pair

Calculate the required photon energy and associated wavelength of the photons in order to create an electron–hole pair in a block of intrinsic Ge, Si and GaAs at 300K (Eg = 0.72 eV, 1.1 eV and 1.32 eV respectively for Ge, Si and GaAs).

Solution

The minimum required energy for a photon to create an electron–hole pair in a semiconductor is just equal to the energy gap of the semiconductor.

$$E_{\text{min}} = E_g = 0.72 \text{ eV for Ge,}$$

$$= 1.1 \text{ eV for Si}$$

$$= 1.32 \text{ eV for GaAs}$$

The corresponding wavelength λ is

$$\lambda = \frac{c}{\nu} = \frac{ch}{E_g}$$

Thus (see Ex. 6.2), $\lambda(\text{Å}) = 12422/E_g$ (eV)

For Ge, $\lambda = 12422/0.72$ eV = 17252.8 Å

For Si, $\lambda = 12422/1.1$ eV = 11292.7 Å

and for GaAs, $\lambda = 12422/1.32$ eV = 9410.61 Å

EXAMPLE 11.14

Increase in conductivity due to doping

The conductivity of a pure silicon sample at room temperature is $4 \times 10^{-4}\ \Omega^{-1}\text{m}^{-1}$. Show that the conductivity increases 27,000 times if the sample is doped with 1 phosphorous atom per 10^8 silicon atoms. (For silicon, atomic weight, $M = 28.1$, density, $d = 2330$ kg/m^3 and $\mu_e = 0.135$ m^2/Vs).

Solution

Number of silicon atoms per m^3,

$$N = \frac{dN_A}{M}$$

$$= \frac{(2330\ \text{kg/m}^3 \times 6.023 \times 10^{26}\ \text{atoms/kmole})}{28.1\ \text{kg/kmole}} = 5 \times 10^{28}\ \text{atoms/m}^3$$

The number of phosphorous atoms doped,

$$N_D = 5 \times 10^{28}/10^8 = 5 \times 10^{20}/\text{m}^3$$

Since phosphorous is a group V element, the doped silicon is of n-type. Assuming complete ionisation, the concentration of electrons is

$$n = n_n \approx N_D = 5 \times 10^{20}/\text{m}^3$$

and the conductivity is

$$\sigma = ne\mu_e$$

$$= 5 \times 10^{20}/\text{m}^3 \times 1.6 \times 10^{-19}\ \text{C} \times 0.135\ \text{m}^2/\text{Vs}$$

$$= 10.8\ \Omega^{-1}\text{m}^{-1}$$

Thus the increase in conductivity is $10.8/4 \times 10^{-4} = 2.7 \times 10^4 = 27000$ times.

Applications of superconductors are based mainly on their two remarkable properties: zero resistivity and perfect diamagnetism. Zero resistivity in superconductors means a large current carrying capacity with no heat loss (I^2R-loss; as $R = 0$, this loss is zero). This makes them useful in electrical power transmission, microelectronics and electrical appliances. The magnetic property of superconductors makes them useful in transformers, motors and in generation of very strong magnetic fields. The superconducting magnets used in the accelerator at the Fermi National Laboratory produce fields, which can accelerate elementary particles to the highest energies.

One of the most novel applications of superconductivity is the development of the magnetically levitated (Maglev) train. This train 'floats' above the rails due to the repulsion between its superconducting magnets and the magnetic field that they induce in the tracks. This overcomes the limitation on the speed imposed by the friction between the rails and the wheels of the train and help in achieving higher speeds. A record speed of 517 km per hour was achieved by the Maglev train developed in Japan in 1979.

Superconductors are also useful in the detection of very small magnetic fields, using the phenomena of *flux quantisation* and *Josephson tunnelling*. The meaning of flux quantisation is that the magnetic flux through a superconducting loop is quantised; the quantum of flux, called the *fluxon*, being given by $\phi = h/2e$. Josephson effect is based on the phenomenon of tunneling a thin insulating barrier by an electron pair. Josephson effect enables the counting of fluxons (and even small fractions of it) facilitating the measurement and detection of very small magnetic fields.

A Josephson junction made of two superconducting films separated by a thin oxide barrier (10 to 20 Å thick) can be used as a switch with switching time of the order of 10^{-11}s. Such ultra-fast switches can be used in computers to enhance the speed of computation. A device employing one or more Josephsons's junctions for flux detection is called a 'superconducting quantum interference device' or SQUID. SQUIDs have been used to detect very small magnetic fields generated by brain activity and to explore the functioning of different parts of the brain.

EXERCISES

1. Explain the formation of energy bands in solids on the basis of the band theory.
2. Draw a graph showing variation of electron energy levels of germanium as a function of its interatomic distance. Explain from the graph why germanium is an insulator at 0 K and semiconductor at 7 K.
3. How are band structures of insulators and semiconductors similar? How are they different?
4. Explain the concept of a hole. Why are holes not generated in metals? Can the electrical conduction of semiconductors be improved to the extent possible in metals? Discuss.
5. Explain in brief the concept of the Fermi level.
6. What is the Fermi–Dirac distribution function? State its significance. Draw a graph showing its variation with energy at different temperatures and discuss it.
7. Show diagrammatically the Fermi level in metals, semiconductors (intrinsic) and insulators at 0 K and higher temperatures.
8. Draw a neat sketch of a band diagram of an intrinsic semiconductor at room temperature and show that the Fermi level lies in the middle of the energy gap.
9. How does the Fermi level change with temperature in extrinsic semiconductor? Discuss the effect of increasing the amount of dopant in extrinsic semiconductors.
10. What is the law of mass action?
11. State the expressions for conductivity in intrinsic, n- and p-type semiconductors.

PROBLEMS

1. Calculate the temperature at which Si (E_g = 1.14 eV) will have the same fraction of electrons in the conduction band as germanium (E_g = 0.72 eV) has at 300 K.
2. The resistivity of intrinsic silicon at 270°C is 3000 Ωm. Calculate the intrinsic carrier density. Assume $\mu_e = 0.17$ m^2/Vs and $\mu_h = 0.035$ m^2/Vs. The resistivity of intrinsic silicon is 2.3 × 10^3 Ωm at 300 K. Calculate its resistivity at 100°C. Assume $E_g = 1$ eV.
3. Calculate the current produced in a Ge plate of 1 cm^2 area and 0.03 mm thickness when a potential difference of 2 V is applied across its faces. Assume a free electron concentration of 2 × 10^{19}/m^3, the electron and hole mobilities being 0.39 m^2/Vs and 0.19 m^2/Vs, respectively.
4. What fraction of the conductivity of intrinsic silicon at room temperature is due to (i) electrons and (ii) holes? The electron and hole mobilities are 0.135 m^2/Vs and 0.048 m^2/Vs respectively.

QUESTIONS

1. In chemical bonding, what is energy level splitting? Why does it lead to a stable molecule?
2. Why do we have a band of available electronic energies in solids?
3. Make a table of concepts in chemical bond formation in molecules and energy band formation in solids, the rows of which show parallel concepts. For example, bonding energy ($E - \Delta$) and VB are in a single row.

4. What do you understand by the concept of density of states?
5. What is Fermi energy for a metal?
6. Where, on the energy band diagram of the metal, is it located? How is it related to W, the work function of the metal?
7. Plot the density of states for a metal with electron energy on X-axis.
8. What do you need to know in order to calculate E_F for a metal?
9. On the scale on which the value calculated by you is numerically correct, what is the value of E_{vacuum}?
10. Suppose we have only five energy levels $E_i, i = 0, 1, 2, 3, 4$ available to 20 electrons in some system. Assume Ei is proportional to i and that the distribution of electrons into quantum states of these five energy levels is $\{8, 4, 4, 3, 1\}$. This is at TK. What is the average energy? Discuss what might happen to this distribution if temperature increases to $1.2TK$. [Hint: Temperature is a measure of $averageenergy$.]
11. What is the Fermi function?
12. Can E_F be in the forbidden energy gap? Why, or why not? If yes, what does it mean to say that half of the states with energies close to E_F are occupied?
13. Define $x = \frac{E-E_F}{kT}$, express $p(E, T)$ as a function of x and find $\frac{dp(E,T)}{dx}$.
14. Plot the Fermi function and its x-derivative $vs.x$.
15. Is $f(E_F + \epsilon, T) = 1 - f(E_F - \epsilon, T)$? Why, or why not?
16. Express equality of the light green and the light blue areas as a mathematical equation.
17. A material has $E_C > E_F$. Give reasns to show that it cannot be a normal semiconductor.
18. The Fermi level for both a conductor as well as an insulator lies between the VB and CB edges. Then why do they have very different electrical properties?
19. What is the concept of effective mass of an electron?
20. Suppose you are studying the motion of a steel ball-bearing of physical mass m in oil. You use the concept of effective mass, m^*. Will m^* be larger than, equal to, or smaller than m? Will it be constant, or variable? Why?
21. What are n and p? Why are they important?
22. What is a hole? Is it a free particle like the CB electron?
23. What is the difference between energy band diagrams at zero and non-zero temperatures?
24. For an intrinsic semiconductor, why should n equal p?
25. Why is it not necessary to draw lower energy bands in the energy band diagram?
26. Why is number density n of electrons in CB equal to the number of occupied quantum states with energies in CB?
27. Look up values of m_e^* and m_h^* for silicon. Estimate n_i at room temperature from the expression above? How does it compare with n for copper? Estimate the temperature at which n_i for Si is double of its value at 10 K.
28. Consider the two processes: (i) motion of a hole and (ii) EHP recombination. In both an electron fills a broken bond. Then in what way are the two processes different?
29. What is law of mass action?

30. In the lecture above we talked about an approximate method to find location of E_F on the energy band diagram. Why is that method approximate?
31. Is your answer to the above question reasonable from the point of view of the exact result for location of E_F Explain.
32. You learned about the band diagram of an intrinsic semiconductor in the last lecture. Is the location of E_F obtained above in line with our idea of the Fermi energy for a conductor?
33. If we increase the temperature of an intrinsic semiconductor there will be more electrons in the CB and more holes in the VB. Why is this true?
34. From the above question, we should expect conductivity to rise with temperature. Does the formula we obtained above for σ_i show this? Is this obvious from the graph of $\ell n(\sigma_i) vs : \frac{1}{T}$ above?
35. Why does the conductivity of copper not increase with temperature, as happens for an intrinsic semiconductor?
36. What is an extrinsic semiconductor?
37. Make a list of elements normally used as donor and acceptor dopants for Si and Ge. Try to find out which are used the most. Look up the web, or a book. Also, try to find how an intrinsic semiconductor like gallium arsenide is doped. What are the dopants to convert intrinsic GaAs to n and p-types?
38. What are impurity energy levels?
39. What is a bound hole?
40. What is the ground to say that the extra electron on donor impurity is *looselybound*?
41. Is it true that even in extrinsic semiconductors the only current carriers are electrons in CB and holes in VB? Explain.
42. Suppose the impurity concentration is very high. What might happen to impurity energy levels?
43. Collect data from the web and make a table of electron and hole effective masses (m_e^*, m_h^*), dielectric constant (ϵ_r), radius and energy of loosely bound electron states for donor impurity atoms ($r_{1,e}, E_{1,e}$), and the same of loosely bound hole states for acceptor impurity atoms in Si, Ge and as many other intrinsic semiconductors as you can manage. [You will have to calculate $r_{1,e}, E_{1,e}, r_{1,h}, E_{1,h}$ from formulae discussed in this lecture.]

12

Diodes and Transistors

12.1 THE STORY OF DIODES AND TRANSISTORS

In the last chapter some basic concepts in semiconductor physics were studied. Basic research in the physics of semiconductors was also accompanied by applied research where the potential of different theoretical concepts in building new devices was explored.

The story had a lot to gain from the progress in electronic devices prior to the advent of semiconductor devices. The story can be traced back to 1883 when Thomas Edison discovered that electric charges emitted by a filament in the bulb are attracted to a positively charged metal plate placed near it. He inferred this on the basis of the observed current through the plate circuit. However, when the plate was negatively charged, no current flowed. This effect, known as 'Edison effect' was patented by Edison, but was unexplained as the electron was not known at that time. With the discovery of the electron by JJ Thomson in 1897, the Edison effect was properly understood and used in the construction of vacuum tube diodes, also called as *valves*.

In 1907, Lee DeForest, an American scientist, developed a triode valve by adding a third electrode, called the *grid*, between the filament and the plate. The vacuum tubes dominated the first half of the twentieth century. However, the limitations of size, requirement of a heating filament, and life of the filament, left much scope for better options.

Applied research in semiconductor physics brought forth newer and better options. One of the first and simplest of the devices developed through this research was a *p-n* junction diode. The discovery of a diode also triggered renewed interest among physicists to develop devices based on *p-n* junction, called the junction devices. The crowning achievement came in 1948 with one of the most remarkable discoveries of the twentieth century, the discovery of the *transistor*. It was credited to the semiconductor research group of Bell Telephone Laboratory, led by William Schokley and including John Bardeen and Walter Brattain. Interestingly, Bardeen, who studied electrical engineering at the University of Wisconsin and solid state physics at Princeton University, later developed the theory of superconductivity in 1957 along with Leon Cooper and J Robert Schrieffer. Bardeen was the first person to receive two Nobel prizes in the same subject.

Transistors had sown the seeds of a revolution in electronics by promising immense prospects for the miniaturisation of technology. The invention of the integrated circuit in 1958, and the microprocessor in the early 1970s bear testimony to this promise.

In this chapter we study the physics of semiconductor devices, *p-n* junction diodes and transistors (Sections 12.2 and 12.3). The application of these devices to engineering and technology are discussed in the last section.

12.2 *p–n* JUNCTION DIODE

We discussed the *p*-type and *n*-type semiconductors in the last chapter. Here the physics of a *p–n* junction diode, an essential component of all electronic circuits is discussed. From the name we may conclude that a *p–n* junction diode is a device made by joining a *p*-type semiconductor with an *n*-type semiconductor. But in practice, a *p–n* junction is obtained by doping an intrinsic semiconductor with both pentavalent and trivalent impurities so that an *n*-region and a *p*-region are formed one after the other to produce a junction. Nevertheless, considering the *p–n* junction as being formed by joining *n*- and *p*-type semiconductors simplifies the study and so we make such an assumption.

The various mechanisms in *n*-type and *p*-type semiconductors using crystal representation and band representation were seen in Chart 11.3. We can understand the various mechanisms in a diode also using these representations. Furthermore, it is useful to make a comparative study of the various features of a diode under three different modes: unbiased mode, forward–biased mode and reverse–biased mode. Chart 12.1 presents this comparative study.

- In the *unbiased mode* no external voltage is applied to a diode. However, there exists an internal voltage drop or a potential barrier across the junction in the steady state of a stand-alone diode.
- In the *forward-biased mode*, the *p*-region is connected to the positive terminal of the external voltage source and the *n*-region to the negative terminal.
- In the *reverse-biased mode* the connections are reversed, that is, the *n*-region is connected to the positive and the *p*-region to the negative terminal.

From Chart 12.1 it is clear that forward-biasing reduces the potential barrier across the junction and reverse-biasing increases this barrier. In all the three modes the behaviour of a diode is decided by how the height of the potential barrier affects the various currents flowing across the junction.

CHART 12.1 *p–n* **JUNCTION DIODE**

	Unbiased		
	Before contact	**Immediately after contact**	**Unbiased steady state**
	Pictorial representation		
Schematic diagram	⊖ Acceptor ion ⊕ Donor ion *p* *n* • Electron ○ Hole	*p* *n*	Exposed acceptor ions V_0 Exposed donor ions Pictorial represen-tation *p* *n*
Band diagram	*E* *E* E_{FN} E_{FP} *E*: Electron energy	*E*	Depletion region *E* Conduction band $V_B = V_0$ eV_0 Conduction band E_F Valence band Valence band
Physical explanation	A *p*-type material contains negative acceptor ions (because they accept electrons) and holes as majority carriers. An *n*-type material contains positive donor ions (because they donate electrons) and electrons as majority carriers. However, overall charge neutrality exists in both the semiconductors, and the charge carriers are said to cover the ions, that is, electrons cover the donor ions and holes cover the acceptor ions. Fermi levels occupy the positions as shown in figure.	After contact electrons and holes flow under the concentration gradients across the junction (electrons flow from *n* to *p*; holes flow from *p* to *n*) much in the same way as heat flows under a thermal gradient or water flows under a pressure gradient. Electrons flow from the conduction band of *n*-region into the empty levels in the conduction band of *p*-region. Holes flow from valance band of *p*-type into that of *n*-type material. Heat flow proceeds towards thermal equilibrium, water flow proceeds towards hydrostatic equilibrium. In the same way in the present case migration of charge carriers proceeds towards electronic equilibrium.	In equilibrium the migration of charge carriers is prevented due to formation of a potential barrier as explained below. Migration of charge carriers result in exposing or uncovering the donor and acceptor ions on two sides of a junction. The avenues of these static ions gives rise to an electric field which prevents further diffusion of charge carriers. Just as the thermal equilibrium and the hydrostatic equilibrium are characterised by the constancy of temperature and pressure, respectively, *the electronic equilibrium is characterised by the constancy of Fermi level*. The constancy of Fermi level across the *p–n* junction demands the raising of energy levels in the con-duction band by an amount eV_o where V_o is called the *junction potential*. The energy hill eV_o acts as a barrier to the further flow of charge carriers. As the narrow region on both sides of a junction is devoid of mobile charge carriers it is called as *depletion region*.
	Mathematical treatment		
Density of charge carriers			$V_B = V_0$. Thus we have, see (Eq.12.3) $n_{n0} = n_{p0} \exp[eV_o/kT]$ (1) which leads to $V_o = (kT/e) \ln (N_D N_A/n_i^2)$ (2) See Eq. 12.5
Diode eq./ Rectifier eq.			$J = J_0 [\exp(eV/kT) - 1]$ (5) or the corresponding current, $I = I_0 [\exp(eV/kT) - 1]$ (6) as $I = JA$ and $I_0 = J_0 A$, *A* being the area of cross section.

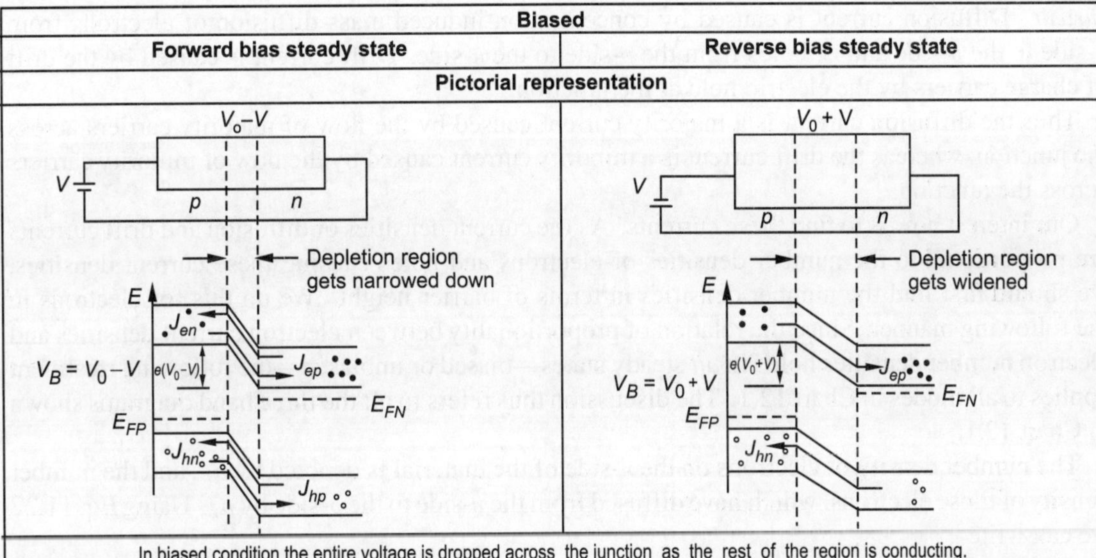

In biased condition the entire voltage is dropped across the junction as the rest of the region is conducting.

Forward bias steady state	Reverse bias steady state
In forward bias the positive terminal of the battery is connected to the *p*-region and negative terminal to the *n*-region. Thus the effective voltage across the junction reduces to $V_0 - V$. *Height of the potential barrier decreases* to $e(V_0 - V)$. The depletion region gets narrowed down. The condition of electronic equilibrium no longer holds and the Fermi level is no longer constant (Fermi levels get separated). Low potential barrier means less opposition to the migration of charge carriers. Thus majority electrons and holes flow across the junction in response to the applied voltage and heavy current flows.	In reverse bias the positive terminal of the battery is connected to the *n*-region and the negative terminal to the *p*-region. Thus the effective voltage across the junction increases to $V_0 + V$. *Height of the potential barrier increases to* $e(V_0 + V)$. The depletion region gets widened. The condition of electronic equilibrium no longer holds and the Fermi level is no longer constant (Fermi levels get separated). High potential barrier means more opposition to the migration of charge carriers. Thus current due to majority carriers is greatly reduced. However the potential hill favours the flow of minority carriers (electrons to *n*-region and holes to *p*-region) and a *small current that is independent of the applied voltage flows.*
Mathematical treatment	
$V_B = V_0 - V$. Thus we have, $n_n = n_p \exp[e(V_0 - V)/kT]$ or $n_p = n_{p0} \exp(eV/kT)$ (3)	$V_B = V_0 + V$. Thus we have, $n_n = n_p \exp[e(V_0 + V)/kT]$ or $n_p = n_{p0} \exp(-eV/kT)$ (4)
Here we have used Eq. 1 and $n_n = n_{n0}$ as this does not depend on V_B.	
Thus the concentration increases by a factor of $\exp(eV/kT)$. As a result heavy current flows.	Thus the concentration decreases by a factor of $\exp(-eV/kT)$. As a result negligible current flows due to majority carriers. But due to a favourable bias a small current due to minority carriers flows.
See Ex. 12.2	
As $\exp[eV/kT] >> 1$, we have $J = J_0 \exp[eV/kT]$ and correspondingly, $I = I_0 \exp[eV/kT]$ (7) Thus a large current flows in forward bias that increases exponentially with the applied voltage.	As $\exp[-eV/kT] << 1$, hence $J = -J_0$ or $I = -I_0$ (8) Thus a small saturation current due to minority carriers flows in reverse bias. We also find that the current is independent of voltage. However this current can be enhanced by increasing the temperature or showering light of suitable frequency on the diode.
See Ex. 12.3	

The total current across a junction is the sum of two components: *diffusion current* and *drift current*. Diffusion current is caused by concentration-induced mass diffusion of electrons from n-side to the p-side and of holes from the p-side to the n-side. Drift current is caused by the drift of charge carriers by the electric field at the junction.

Thus the diffusion current is a majority current caused by the flow of majority carriers across the junction, whereas the drift current is a minority current caused by the flow of minority carriers across the junction.

Our interest now is to find these currents. As the current densities of diffusion and drift currents are proportional to the number densities of electrons and holes causing these current densities, we should first find the number densities in terms of barrier height. We do this for electrons in the following manner. Since the relation of proportionality between electron current densities and electron number densities holds in *all* steady states—biased or unbiased—the following treatment applies to all modes in Chart 12.1. The discussion thus refers to all the three band diagrams shown in Chart 12.1.

The number density of electrons on the n-side of the material is denoted by n_n, and the number density of those electrons, which have diffused from the n-side to the p-side by n_p. Using Eq. 11.22 we can write

$$n_n = N_C \exp\left[-\frac{(E_{CN} - E_{FN})}{kT}\right] \tag{12.1}$$

and $$n_p = n_p(V_B) = N_C \exp\left[-\frac{(E_{CN} + eV_B - E_{FN})}{kT}\right] \tag{12.2}$$

In the above equations, E_{CN} and E_{FN} denote the conduction band and Fermi energies on the n-side of the junction. In the unbiased steady state, Fermi levels are equalised, that is, $E_{FN} = E_{FP}$ (Fermi energy on p-side), as explained in Chart 12.1. In a biased steady state, this is no longer true and the Fermi levels on the two sides of the junction differ, causing a non-zero steady state current in the junction. However, E_{CN} is different from E_{CP} (conduction band energy on p-side) in all the three modes, the difference of the two being equal to eV_B where V_B is the height of the potential barrier across the junction. This is why the expression for n_p contains the energy $E_{CP} = E_{CN} + eV_B$ in the exponential factor. Dividing Eq. 12.1 by Eq. 12.2 we get,

$$V_B = \left(\frac{kT}{e}\right) \ln\left(\frac{n_n}{n_p}\right) \tag{12.3}$$

Now the potential barrier across the junction is related to the externally applied biasing potential V as

$$V_B = V_0 - V \tag{12.4}$$

where $V > 0$ for forward bias, $V < 0$ for reverse bias and $V = 0$ in the unbiased state.

V_0 is thus the potential barrier at the junction in the unbiased steady state. It characterises the dynamic equilibrium of zero net current across the junction in which the diffusion and drift currents cancel each other. We first calculate V_0 by applying Eq. 12.2 to the special case of zero bias or $V_B = V_0$.

If we denote the number densities n_n and n_p when the bias V is zero by n_{n0} and n_{p0}, then V_0 can be written in terms of n_{n0} and n_{p0}. Actually, $n_{n0} = n_n$ as this does not depend on V_B. For the unbiased steady state, we note that $n_{n0} = N_D$ and $p_{p0} = N_A$ at room temperature, where p_{p0} denotes the hole density on the p-side when $V = 0$. But, by the Law of mass action, for the p-side, we have $p_{p0}n_{p0} = n_i^2$, n_i being the intrinsic charge carrier density of thermally generated electron–hole pairs in the host intrinsic semiconductor. Thus, finally we can write,

$$V_0 = \left(\frac{kT}{e}\right) \ln \frac{n_{n0}}{n_{p0}} = \frac{kT}{e} \ln \frac{N_D N_A}{n_i^2} \tag{12.5}$$

Equation 12.5 gives the unbiased steady state junction potential in terms of the doping concentrations on the *n*- and *p*-side of the *p-n* junction. Example 12.1 illustrates an application of Eq. 12.5.

In the biased steady state ($V \neq 0$), using Eq. 12.4 in Eq. 12.2 we get,

$$n_p = N_C \exp\left[-\frac{(E_{CN} + eV_0 - E_{FN})}{kT}\right] \exp\left(\frac{eV}{kT}\right) \tag{12.6}$$

or, $$n_p = n_{p0} \exp\left(\frac{eV}{kT}\right) \tag{12.7}$$

Equation 12.7 thus relates the diffused electron density in the biased state, n_p, with the diffused electron density in the unbiased state, n_{p0}. It can be seen that for $V > 0$, that is, for forward bias, n_p increases by a factor of $\exp(eV/kT)$ and for $V < 0$, that is, for reverse bias, n_p decreases by the same factor. Example 12.2 provides a quantitative idea of these changes.

EXAMPLE 12.1
Determination of V_0

The bulk *n*-type and *p*-type materials of a *p-n* junction diode made of germanium have conductivities of 10^4 ℧/m and 10^2 ℧/m, respectively at 300 K. Find the contact potential difference at 300 K, assuming $n_i = 2.5 \times 10^{19}/\text{m}^3$, $\mu_e = 0.38\ \text{m}^2/\text{Vs}$ and $\mu_h = 0.18\ \text{m}^2/\text{Vs}$.

Solution

Almost the entire current is carried by majority carriers in an impurity semiconductor. The bulk conductivity of the *n*-region and the *p*-region is given, respectively, by

$$\sigma_n = en\mu_e \text{ and } \sigma_p = ep\mu_h$$

or $$n = n_{n0} = \frac{10^4\ \mho/\text{m}}{1.6 \times 10^{-19}\ \text{C} \times 0.38\ \text{m}^2/\text{Vs}} = 1.64 \times 10^{23}/\text{m}^3$$

and $$p = p_{p0} = \frac{10^2\ \mho\,\text{m}}{1.6 \times 10^{-19}\ \text{C} \times 0.18\ \text{m}^2/\text{Vs}} = 3.47 \times 10^{21}/\text{m}^3$$

For p-type the Law of mass action gives

$$n_{p0} = \frac{n_i^2}{p_{p0}} = \frac{(2.5 \times 10^{19}\ \text{m}^{-3})^2}{(3.47 \times 10^{21}\ \text{m}^{-3})}$$

$$= 1.80 \times 10^{17}\ \text{m}^{-3}$$

Thus, $V_0 = \frac{kT}{e} \ln\left(\frac{n_{n0}}{n_{p0}}\right) = (0.026\ \text{eV}/e) \ln\left(\frac{1.64 \times 10^{23}\ \text{m}^{-3}}{1.80 \times 10^{17}\ \text{m}^{-3}}\right) = 0.36\ \text{V}$

EXAMPLE 12.2

Carrier concentration under forward bias and reverse bias

Find the concentration of electrons diffused to the p-side of the p-n junction in terms of the diffused concentration in unbiased state, under the following conditions: (a) forward bias of 0.1 V, (b) reverse bias of 1V.

Solution

In forward bias we have,

$$n_p = n_{p0} \exp\left[\frac{eV}{kT}\right]$$

Thus, $n_p = n_{p0} \exp\left(\frac{0.1\ \text{eV}}{0.026\ \text{eV}}\right) = 47 n_{p0}$

In reverse bias

$$n_p = n_{p0} \exp\left[\frac{-eV}{kT}\right]$$

Thus, $n_p = n_{p0} \exp\left(\frac{-1.00\ \text{eV}}{0.026\ \text{eV}}\right)$

$$= 1.98 \times 10^{-17} n_{p0}$$

Thus a small forward bias of 0.1 V augments the electron concentration 47 times, whereas a large reverse bias of 1V reduces it 10^{-17} times.

Now we proceed to find the current densities. The four current densities, which can be separately recognised (see Fig. 12.1), are

1. Hole current density, J_{hp}, flowing from p-side,
2. Hole current density, J_{hn}, flowing from n-side
3. Electron current density, J_{en}, flowing from n-side
4. Electron current density, J_{ep}, flowing from p-side.

J_{en} and J_{hp} are the diffusion current densities due to the majority carriers and J_{ep} and J_{hn} are drift current densities due to minority carriers.

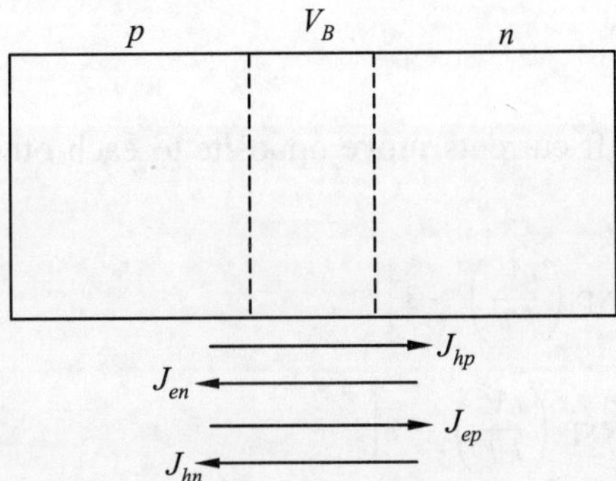

Figure 12.1 Four components of current density in a diode

As the diffusion current density due to electrons is proportional to the number of electrons diffused to the p-side, we have

$$J_{en} = Kn_p = Kn_{p0} \exp\left(\frac{eV}{kT}\right) \tag{12.8}$$

K being the proportionality constant. Equation 12.8 can be written as

$$J_{en} = J_{en0} \exp\left(\frac{eV}{kT}\right) \tag{12.9}$$

Equation 12.9 thus relates the diffusion current density due to electrons, $J_{en} = Kn_p$, in the biased state with the density, $J_{en0} = Kn_{p0}$, in the unbiased state.

Now, in the unbiased steady state ($V = 0$), the diffusion and drift currents for both types of charge carriers cancel each other, that is, we have,

$$J_{en0} = J_{ep0} \text{ and } J_{hp0} = J_{hn0} \tag{12.10}$$

and Eq. 12.9 becomes

$$J_{en} = J_{ep0} \exp\left(\frac{eV}{kT}\right) \tag{12.11}$$

Further, $J_{ep0} = J_{ep}$ and $J_{hn0} = J_{hn}$ (12.12)

because the minority current density is essentially independent of the bias as it is caused by a fall of all the thermally generated minority carriers down the potential hill. Therefore, Eq. 12.11 becomes

$$J_{en} = J_{ep} \exp\left(\frac{eV}{kT}\right) \tag{12.13}$$

Identical considerations apply to majority holes diffusing from the *p*-side to the *n*-side and minority holes drifting from the *n*-side to the *p*-side. Thus, we can also write

$$J_{hp} = J_{hn} \exp\left(\frac{eV}{kT}\right) \tag{12.14}$$

Now, as the diffusion and drift currents move opposite to each other, the net electron and hole currents can be written as,

$$J_e = J_{\text{en}} - J_{ep} = J_{ep0}\left[\exp\left(\frac{eV}{kT}\right) - 1\right] \tag{12.15}$$

$$J_e = J_{hp} - J_{hn} = J_{hn0}\left[\exp\left(\frac{eV}{kT}\right) - 1\right] \tag{12.16}$$

Therefore, the total current density at the *p-n* junction is

$$J = J_e + J_h = (J_{ep0} + J_{hn0})\left[\exp\left(\frac{eV}{kT}\right) - 1\right]$$

or $$J = J_0\left[\exp\left(\frac{eV}{kT}\right) - 1\right] \tag{12.17}$$

where, $J_0 = J_{ep0} + J_{hn0}$ is the current density due to minority carriers, called the *saturation current density*. Equation 12.17 is known as the *diode equation* or the *rectifier equation*. It describes the current in the *p-n* junction in forward bias ($V > 0$) as well as reverse bias ($V < 0$). In forward bias, $\exp(eV/kT)$ is large compared to 1 and

$$J = J_0 \exp\left(\frac{eV}{kT}\right) \tag{12.18}$$

In reverse bias the exponential term with negative V is negligible compared to 1 and

$$J = -J_0 \tag{12.19}$$

Example 12.3 illustrates the application of the diode equation.

The various features of diodes discussed so far and in Chart 12.1 can also be studied experimentally. Experiment 12.1 describes how the static characteristics of a *p-n* junction diode can be obtained. The conclusions of this experimental study are in close agreement with the theoretical conclusions drawn here.

EXAMPLE 12.3
Current through a – junction diode

The current flowing in a *p-n* junction diode at room temperature is 0.1 μA, when a large reverse bias voltage is applied. Calculate the current flowing when forward bias of 0.1 V is applied.

Solution

The volt ampere relation for a junction diode is

$$I = I_0\left[\exp\left(\frac{eV}{kT}\right) - 1\right]$$

where $kT = 0.026$ eV at room temperature.

For large reverse bias $I = I_0 = 0.1\ \mu\text{A}$.

Thus, for forward bias of 0.1V, $I = 0.1\ \mu\text{A} \times \left[\exp\left(\frac{0.1\text{ eV}}{0.026\text{ eV}}\right) - 1\right]\mu\text{A}$

$= 4.58\ \mu\text{A}$

EXPERIMENT 12.1 *p–n* JUNCTION DIODE

Aim

To study the static characteristics of *p-n* junction diodes and to determine their forward and reverse resistances

Apparatus

p-n Junction diodes of Ge and Si, power supply (0–18 V), voltmeter (0–1.5 V and 0–10 V), milliammeter (0–25 mA), and microammeter (0–50 μA)

Formulae

Forward resistance,

$$R_{FB} = \frac{\delta V_{FB}}{\delta I_F} \tag{12.20}$$

Reverse resistance, $R_{RB} = \dfrac{\delta V_{RB}}{\delta I_R}$ (12.21)

where δV_{FB} and δV_{RB} denote the changes in voltage in forward bias and reverse bias respectively and δI_F and δI_R are the respective current changes.

Theory

The behaviour of a diode under different biases was studied using band theory in Chart 12.1. A *p-n* junction diode offers a very small resistance when connected in forward bias (Fig. 12.2a) because the potential barrier is reduced and the charge carriers experience less opposition to their flow across the junction. A diode for this reason, is represented as an arrowhead and a bar as shown in

Fig. 12.3. The arrowhead indicates the direction of conventional current, that is, the direction in which holes flow under forward bias (from *p*-region to *n*-region).

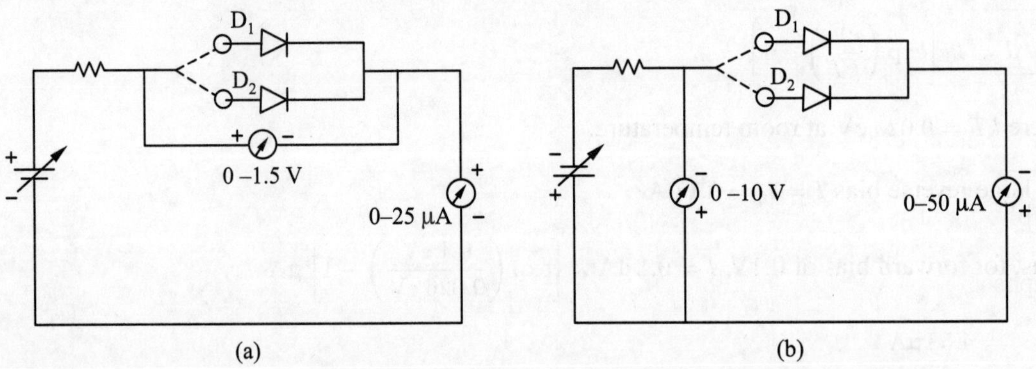

Figure 12.2 Circuit diagrams for the study of diodes in (a) forward bias; (b) reverse bias

In reverse bias (Fig. 12.2b), the potential barrier increases thus opposing the flow of majority carriers across the junction. However, the flow of minority carriers is not opposed and a small current flows in a direction opposite to the arrowhead. A *p-n* junction diode thus offers a large resistance when connected in reserve bias.

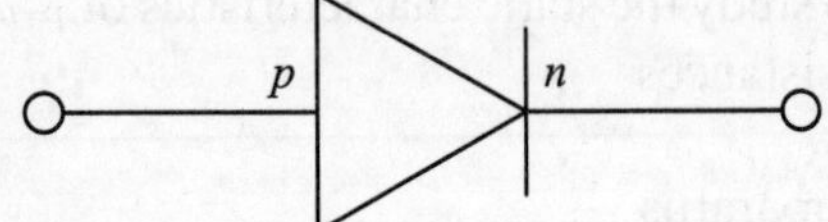

Figure 12.3 Symbol for a *p-n* junction diode

A diode is thus a rectifying device that conducts heavily under forward bias and poorly under reverse bias. The minimum voltage required for the diodes to start conduction is called the *break-point/threshold/offset or cut-in voltage*, V_C. This minimum voltage is required to overcome the internal barrier potential. V_C is approximately 0.2 V for Ge and 0.6 V for Si. V–I characteristics of a diode show that I does not change linearly with V. For this reason a diode is called a *nonlinear device*.

Procedure

A. For forward characteristics

1. The connections are made as shown in Fig. 12.2a for diode D_1.
2. Voltage is increased in steps of 0.1 V and the corresponding current is noted. The readings are recorded in the range 0–1.5 V.
3. Steps 1 and 2 are repeated for diode D_2. The observations are tabulated.

B. For reverse characteristics

1. The connections are made as shown in Fig. 12.2b for diode D_1.
2. Steps 2 and 3 of (A) are repeated by using a 0–10 voltmeter and increasing the voltage in steps of 1 V.

Graphs are plotted between I and V in the first quadrant for forward bias and in the third quadrant for reverse bias.

Observations

Forward bias				**Reverse bias**			
D_1		**D_2**		**D_1**		**D_2**	
V_{FB} (V)	I_F (mA)	V_{FB} (V)	I_F (mA)	V_{RB} (V)	I_R (μA)	V_{RB} (V)	I_R (μA)

Graph

Graphs are constructed as in Fig. 12.4.

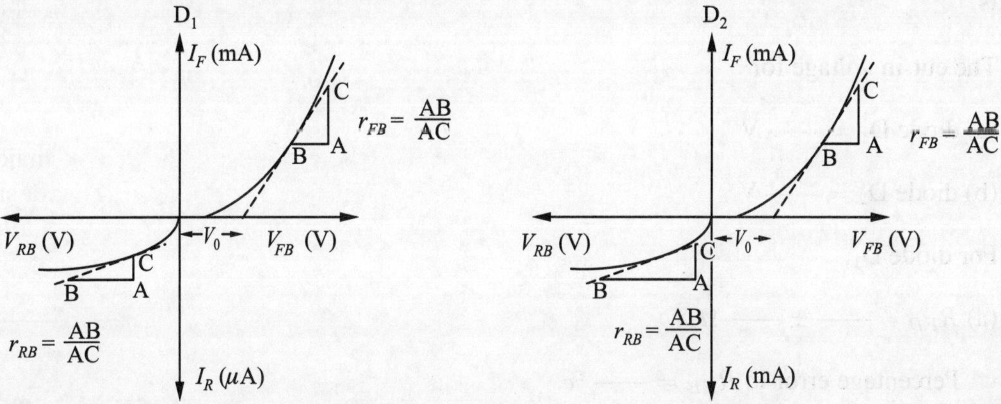

Figure 12.4 Static characteristics of diodes D_1 and D_2 in forward bias and reverse bias

Calculations

The resistances R_{FB} and R_{RB} are obtained from the graph as AB/AC (Fig. 12.4). The cut-in voltage V_0 is noted by drawing the tangent as shown in these figures.

Estimation of error

The error ΔR in the estimation of R can be determined by adopting the method outlined in Section 1.4.2. Thus,

$$\Delta R = 2\left[\frac{\Delta V}{(I_2 - I_1)} + \frac{(V_2 - V_1)\Delta I}{(I_1 - I_2)^2}\right] \tag{12.22}$$

where ΔV is the error in the reading of the voltmeter and ΔI is the error in the reading of the ammeter.

For example, for the following measured data,

$V_1 = 0.50$ V, $I_1 = 4.5$ mA, $V_2 = 0.80$ V, $I_2 = 13.5$ mA, least count of the voltmeter = 0.05 V, least count of ammeter = 0.5 mA,

$\Delta V = 0.025$ V and $\Delta I = 0.25$ mA and thus,

$$\Delta R = 2\left[\frac{0.025\ \text{V}}{(13.5-4.5)\times 10^{-3}\ \text{A}} + \frac{(0.8-0.5)\ \text{V}\times 0.25\times 10^{-3}\ \text{A}}{(13.5-4.5)^2\times 10^{-6}\ \text{A}^2}\right]$$

$$= 5.6\,\Omega + 1.8\,\Omega = 7.4\,\Omega$$

$$R = \frac{(0.8-0.5)\ \text{V}}{(13.5-4.5)\ \text{mA}} = 33.3\,\Omega$$

Thus the relative error in $R = \Delta R/R = 7.4\,\Omega/33.3\,\Omega = 0.22$ and the percentage error is 22%.

Results

1. The cut-in voltage for

   (a) diode D_1 = —— V

   (b) diode D_2 = —— V

2. For diode D_1,

   (a) R_{FB} = —— ± —— Ω.

   Percentage error in R_{FB} = —— %.

   (b) R_{RB} = —— ± —— MΩ.

   Percentage error in R_{RB} = —— %.

3. For diode D_2,

   (a) R_{FB} = —— ± —— Ω.

   Percentage error in R_{FB} = —— %.

   (b) R_{RB} = —— ± —— MΩ.

   Percentage error in R_{RB} = —— %.

Discussion

In this experiment we studied the response of a semiconductor diode under forward bias and reverse bias. Figure 12.4 shows the observed variation of current with voltage for Ge and Si diodes under these biases. In Chart 12.1 we applied band theory to understand the working of a diode in unbiased and biased conditions. In the following table we compare the theory and the experiment.

Theory	Experiment
In the unbiased state the barrier potential V_0 is developed across the junction.	It is observed that the diode does not conduct in forward bias immediately on application of voltage but requires certain minimum voltage for conduction to start.
The application of forward bias, V, should reduce the barrier potential to $V_0 - V$, facilitating a larger current flow. The current should increase exponentially with voltage.	The observed current in forward bias is large and hence a milliammeter is used to measure it. The current increases rapidly as the voltage is increased in steps of 0.1 V. The observed increase is almost exponential.
The application of reverse bias, V should enhance the barrier potential to $V_0 + V$, mitigating the current flow. However the flow of thermally generated minority carriers should not be affected by the enhancement of barrier potential.	The observed current in reverse bias is small and hence a microammeter is used to measure it. The current is almost constant even though the voltage is increased in steps of 1 V.

It can be seen from this table that there is a good qualitative agreement between the theory and the experiment. The reverse current, which is found to be almost constant here, can be enhanced if minority carrier concentration is increased using heat energy or light energy. This forms the basis of a thermal diode or a photodiode.

A notable point is the way the voltmeter and current meter are connected in circuits for forward and reverse bias measurements. It is to be remembered that a forward biased diode has a very small resistance, smaller than even the internal resistance of the ammeter. On the other hand, a reverse biased diode has a very high resistance, higher than the internal resistance of the voltmeter. The connections made for the voltmeter and the ammeter are dictated by these considerations. Suppose the voltmeter in the reverse biased circuit is connected across the diode, rather than as in the circuit (Fig. 12.2b), that is, across the series combination of the diode and the ammeter, then most of the current measured in the ammeter will be carried through the voltmeter as it offers less resistance than the diode. This will mean that the measured current will be very different from the current through the diode, which is what we want to measure. Similarly, if the voltmeter in Fig. 12.2a is connected as in Fig. 12.2b, then it will measure the voltage across the series combination of the diode and the ammeter, much of which appears across the ammeter which has larger resistance than offered by the diode in forward bias.

12.3 TRANSISTOR

A transistor is a three terminal device in which either an *n*-region is sandwiched between two *p*-regions (*pnp*-transistor) or a *p*-region is sandwiched between two *n*-regions (*npn*-transistor). Just as a *p-n* junction diode is not constructed by physically joining a *p*-type semiconductor with an *n*-type semiconductor, a *pnp*- or *npn*-transistor cannot be obtained by connecting two *pn*-junction diodes back-to-back. However, just as in the case of a diode, considering a transistor to be made of diodes in this way simplifies this study to a great extent.

The three regions of a transistor, namely, emitter, base and collector differ in their doping and thickness. Table 12.1 shows these differences and the reasons for them.

Table 12.1 Thickness and doping of emitter, base and collector

Region	Thickness	Doping	Reason
Emitter	Intermediate	Large	To supply the charge carriers
Base	Small	Small	To minimise the recombinations with the charge carriers coming from the emitter
Collector	Large	Intermediate	To achieve heat dissipation

Figure 12.5 shows the symbols for the *npn*- and *pnp*-transistors. In these symbols the arrow indicates the direction of the emitter current. Why the emitter current flows into the base in the *pnp*-transistor and away from the base in the *npn*-transistor can be understood after knowing how the transistor is biased. As there are two diodes and two ways of connecting a diode, there are four possibilities.

1. Both junctions can be forward biased,
2. Both junctions can be reverse biased,
3. Emitter base junction can be reverse biased, and base collector junction, forward biased,
4. Emitter base junction can be forward biased and base collector junction, reverse biased.

The transistor is normally biased with the last option (see Fig. 12.6) to achieve the desired transistor action.

The desired transistor action can be understood by knowing the meaning of the word, transistor. Transistor is a contraction of the name transfer resistor:

| TRANSFER | + | RESISITOR | = | TRANSISTOR |

As the emitter base junction is forward biased, its resistance is low, and hence a large emitter current flows. In an *npn*-transistor, electrons flow into the base, whereas in *pnp*-transistors the holes flow into the base. This explains the directions of the arrow in Fig. 12.5.

As the base is thin and lightly doped, very few charges are lost in the base region and most of the charges are swept across to the collector region. Now, the collector base junction is reverse biased and its resistance is high. But an interesting fact is that the majority of the carriers coming from the emitter become minority carriers in the base. Thus the current flowing through the reverse biased base collector junction is due to minority carriers, and we know that the reverse bias supports the flow of minority carriers! Thus the emitter current that flows through the low resistance of the emitter base junction flows as a collector current through the high resistance of the base collector junction. As the collector current is almost equal to the emitter current (due to the negligible base current), a very interesting situation arises. Almost the same current flows from a region of low resistance to a region of high resistance. This is the desired transfer resistor action or the transistor action.

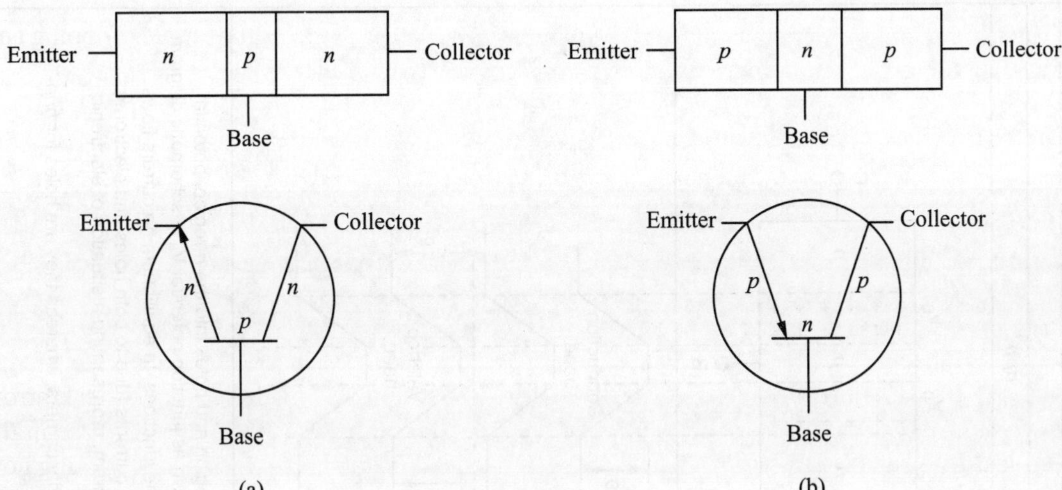

Figure 12.5 Symbols of (a) *npn*-transistor and (b) *pnp*-transistor

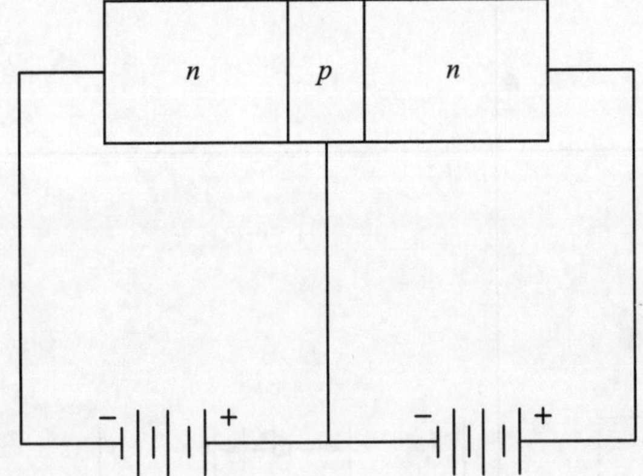

Figure 12.6 Normal biasing of a transistor. The emitter base junction is forward biased and the collector base junction is reverse biased.

As current always diminishes when flowing through a high resistance, it is impossible to construct an electrical circuit in which a high current can be made to flow through a high resistance. Thus it can be said that *what is electrically impossible is electronically possible in a transistor!* It is achieved in a transistor by three crucial factors:

1. the fact that current is contributed by two types of charge carriers—electrons and holes,
2. the construction of the base (very thin and lightly doped),
3. the mode of biasing.

The physics of transistors can be understood through crystal and band representations just as was shown for diodes in Chart 12.1. In Chart 12.2 we include similar illustrations for a transistor. The conclusions drawn out of the discussion in this chart can be verified experimentally. The experimental study of transistors is covered in Expt. 12.2.

Chart 12.2 Study of Transistor

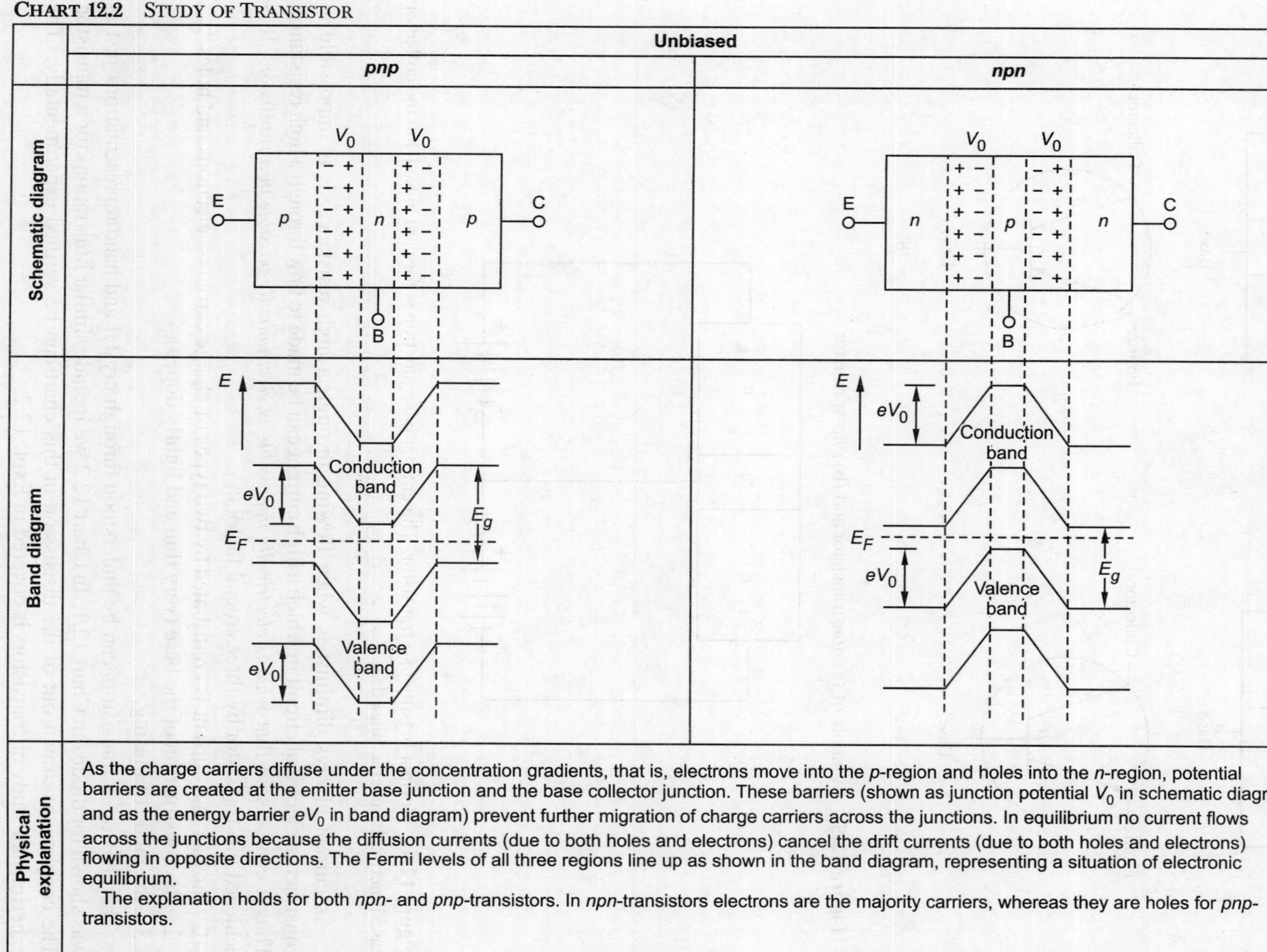

As the charge carriers diffuse under the concentration gradients, that is, electrons move into the *p*-region and holes into the *n*-region, potential barriers are created at the emitter base junction and the base collector junction. These barriers (shown as junction potential V_0 in schematic diagram and as the energy barrier eV_0 in band diagram) prevent further migration of charge carriers across the junctions. In equilibrium no current flows across the junctions because the diffusion currents (due to both holes and electrons) cancel the drift currents (due to both holes and electrons) flowing in opposite directions. The Fermi levels of all three regions line up as shown in the band diagram, representing a situation of electronic equilibrium.

The explanation holds for both *npn*- and *pnp*-transistors. In *npn*-transistors electrons are the majority carriers, whereas they are holes for *pnp*-transistors.

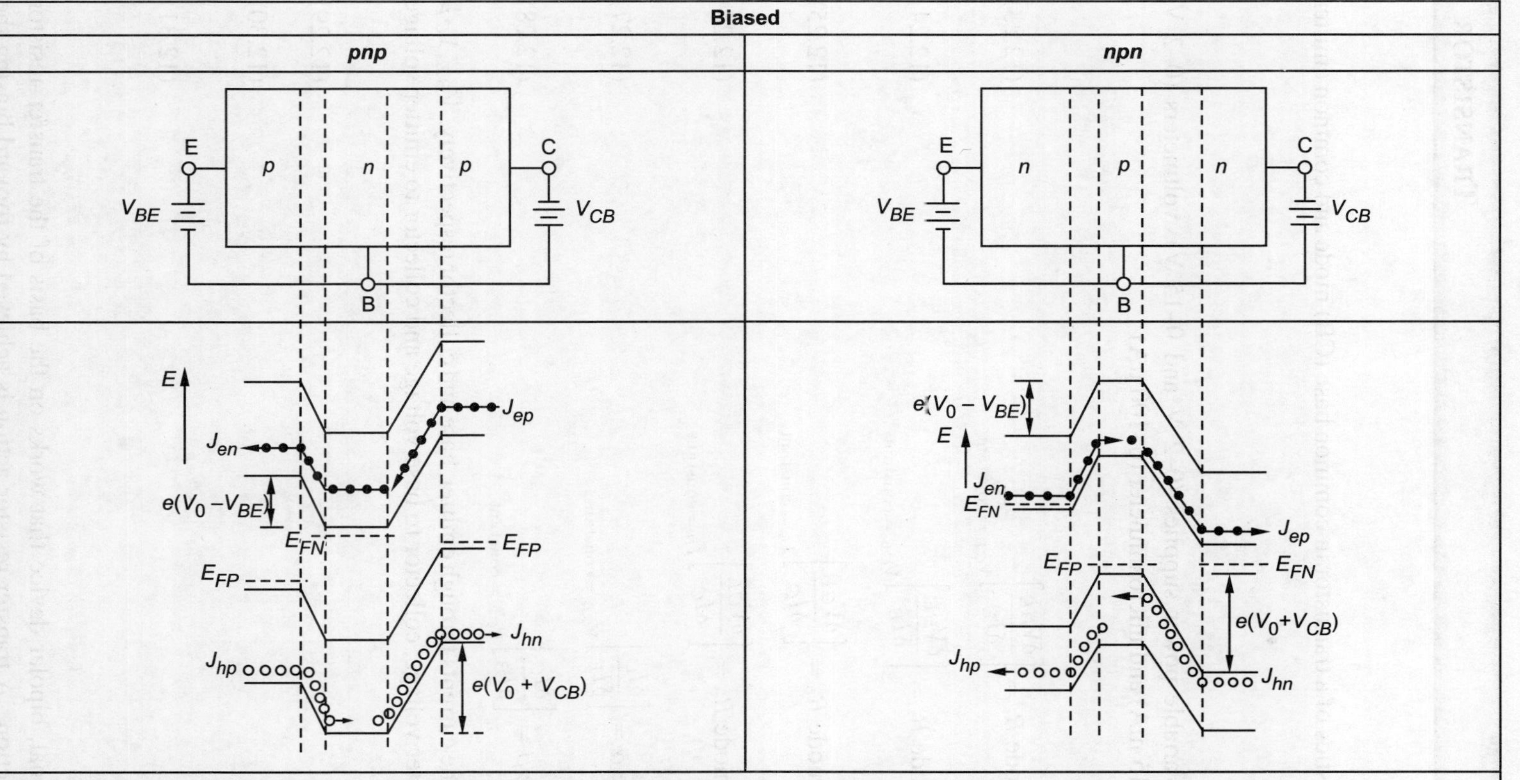

The effect of biasing is the same in both the cases. The basic operation is explained here for *npn*-transistor.

As the emitter base junction is forward - biased, the potential barrier is reduced to $e(V_0 - V_{BE})$ and a large number of electrons enter from emitter to base region.

A very few of these die in the base (*p*-region) due to recombination with holes. Most of the electrons diffuse across the thin base and are attracted by the positively biased collector. The electrons coming from the emitter are minority carriers in base and flow unhindered across the reverse biased base collector junction. The increased barrier height $e(V_0 + V_{CB})$ promotes this flow due to minority carriers (reverse saturation current in case of reverse biased diode, see Chart 12.1).

As the collector current is the reverse current it is independent of the reverse bias voltage applied, that is, V_{CB}. However, as the number of electrons emitted by the emitter and hence flowing through the collector is decided by the forward bias applied, that is, V_{BE} ; it is V_{BE}, rather than V_{CB} that can control I_C. This is the control valve action of a transistor.

The electron current density J_{en} >> the hole current density J_{hp}. Similarly $J_{ep} >> J_{hn}$. The total electron current density is $J_{en} + J_{ep}$ and the total hole current density is $J_{hp} + J_{hn}$.

EXPERIMENT 12.2 TRANSISTOR

Aim

To study the static characteristics of a transistor in common base (CB) mode and common emitter (CE) mode

Apparatus

An *npn*- or *pnp*-transistor, variable power supplies (0–2 V and 0–15 V), voltmeters (0–2 V, 0–15 V), milliammeters (0–15 mA) and microammeter (0–100 μA)

Formulae

Input resistance

$$\text{In CB mode } R_i = \left[\frac{\delta V_{BE}}{\delta I_E}\right]_{V_{CB}=\text{constant}} \tag{12.23}$$

$$\text{In CE mode } R_i = \left[\frac{\delta V_{BE}}{\delta I_B}\right]_{V_{CE}=\text{constant}} \tag{12.24}$$

Output resistance

$$\text{In CB mode } R_o = \left[\frac{\delta V_{CB}}{\delta I_C}\right]_{I_E=\text{constant}} \tag{12.25}$$

$$\text{In CE mode } R_o = \left[\frac{\delta V_{CE}}{\delta I_C}\right]_{I_B=\text{constant}} \tag{12.26}$$

Current gain

$$\text{In CB mode } \alpha = \left[\frac{\delta I_C}{\delta I_E}\right]_{V_{CB}=\text{constant}} \tag{12.27}$$

$$\text{In CE mode } \beta = \left[\frac{\delta I_C}{\delta I_B}\right]_{V_{CE}=\text{constant}} \tag{12.28}$$

where, I_E, I_B and I_C denote the currents through emitter, base and collector respectively. V_{BE}, V_{CB} and V_{CE} denote base to emitter voltage, collector to base voltage and collector to emitter voltage respectively.

$$I_E = I_C + I_B \tag{12.29}$$

$$\beta = \frac{\alpha}{(1-\alpha)} \tag{12.30}$$

$$\alpha = \frac{\beta}{(1+\beta)} \tag{12.31}$$

Theory

A transistor is a three terminal, bipolar device that works on the basis of the transfer–resistor action or the control valve action. A transfer resistor action is achieved by forward biasing the emitter base junction and reverse biasing the base collector junction. Ninety nine per cent of the input emitter current flowing through the low resistance of forward bias flows through the high resistance of reverse bias as the output collector current. A control valve action means that the output current, I_C, is independent of the output voltage (V_{CB} or V_{CE}), but depends on the input voltage (V_{EB}). The voltage range in which the output current is practically constant is known as the *active region*. The range corresponding to varying current is known as the *saturation region*.

A transistor can be connected either in CB mode or in CE mode, or in CC (common collector) mode. The term common here means either the base or the emitter or the collector is common to both the input and the output terminals. In CB mode the input variables are (V_{BE}, I_E) and the output variables are (V_{CB}, I_C). In CE mode they are (V_{BE}, I_B) and (V_{CE}, I_C). Input characteristics mean the graphs between the input variables with the output voltage held constant, whereas output characteristics mean the graphs between the output variables with the input current held constant.

The ratio of change in input voltage to the change in input current gives the input resistance R_i, and a similar ratio for output gives the output resistance R_o. Thus the values of R_i and R_o can be obtained from the input and the output characteristics respectively. The relation between current gains α and β in CB and CE modes respectively can be obtained as follows.

We know that the three currents I_E, I_B and I_C are related as follows:

$$I_E = I_C + I_B$$

Hence $\quad \delta I_E = \delta I_C + \delta I_B$

or $\quad \dfrac{\delta I_E}{\delta I_C} = \dfrac{\delta I_B}{\delta I_C} + 1$

that is, $\quad \dfrac{1}{\alpha} = \dfrac{1}{\beta} + 1 \text{ or } \alpha = \dfrac{\beta}{1+\beta} \text{ or } \beta = \dfrac{\alpha}{1-\alpha}$

Procedure

A. CB Mode

1. Connections are made as shown in Fig. 12.7a.
2. For input characteristics, V_{CB} is fixed at 0 V, and I_E is measured for different values of V_{BE}. V_{BE} is increased in steps of 0.05 V from 0 V to 0.1 V, and in steps of 0.005 V from 0.1 V to 0.15 V.
3. Step 2 is repeated by fixing the value of V_{CB} at 5 V.
4. For output characteristics, the emitter current is fixed at 2 mA and I_C is measured by varying the value of V_{CB} from 0 to 10 V in steps of 1V. It should be ensured that the value of emitter current is constant throughout this set of readings.
5. Step 4 is repeated by fixing the value of I_E at 4 mA and 6 mA.
6. The observations for input and output characteristics are tabulated in Tables 12.2 and 12.3 respectively.

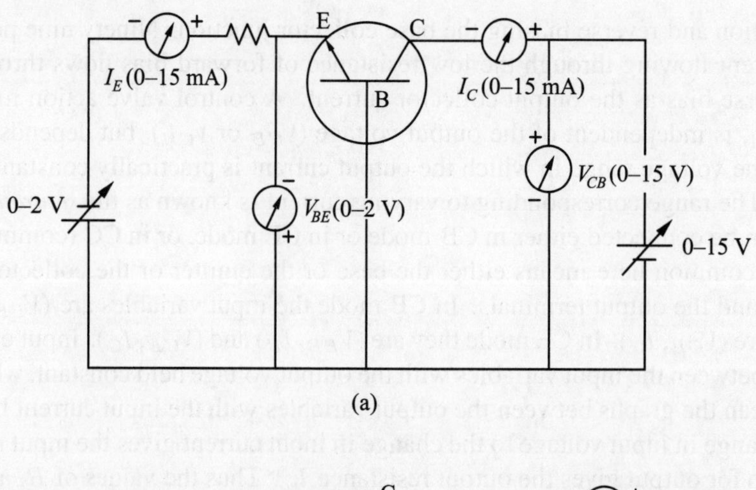

(a)

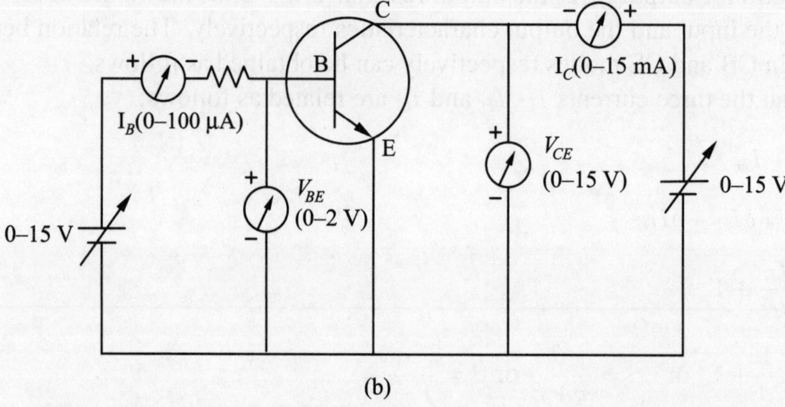

(b)

Figure 12.7 Circuit diagrams for study of an *npn*-transistor in (a) CB mode and (b) CE mode

B. CE Mode

1. Connections are made as shown in Fig. 12.7b
2. For input characteristics, V_{CE} is fixed at 0 V and I_B is adjusted to be 4 μA and the corresponding value of V_{BE} is measured. I_B is increased in steps of 4 μA and the corresponding values of V_{BE} are measured till I_B reaches 40 μA.
3. Step 2 is repeated by fixing the value of V_{CE} at 0.5 V.
4. For output characteristics, the base current is fixed at 10 μA and I_C is measured by varying the value of V_{CB} from 0 to 10 V in steps of 1V. It should be ensured that the value of base current is constant throughout this set of readings.
5. Step 4 is repeated by fixing the value of I_B at 20 μA and 30 μA.
6. The observations for input and output characteristics are tabulated in Tables 12.4 and 12.5 respectively.

Observations

A. CB Mode

Table 12.2 Input characteristics

S. No.		$V_{CB} = 0$ V	$V_{CB} = 5$ V
	V_{BE} (volts)	I_E (mA)	I_E (mA)

Table 12.3 Output characteristics

S.No.		$I_E = 2$ mA	$I_E = 4$ mA	$I_E = 6$ mA
	V_{CB} (volts)	I_C (mA)	I_C (mA)	I_C (mA)

B. CE Mode

Table 12.4 Input characteristics

S. No.		$V_{CE} = 0$ V	$V_{CE} = 0.5$ V
	V_{BE} (volts)	I_B (mA)	I_B (mA)

Table 12.5 Output characteristics

S.No.		$I_B = 10$ µA	$I_B = 20$ µA	$I_B = 30$ µA
	V_{CE} (volts)	I_C (mA)	I_C (mA)	I_C (mA)

Graph

Graphs are drawn as in Figs. 12.8 and 12.9.

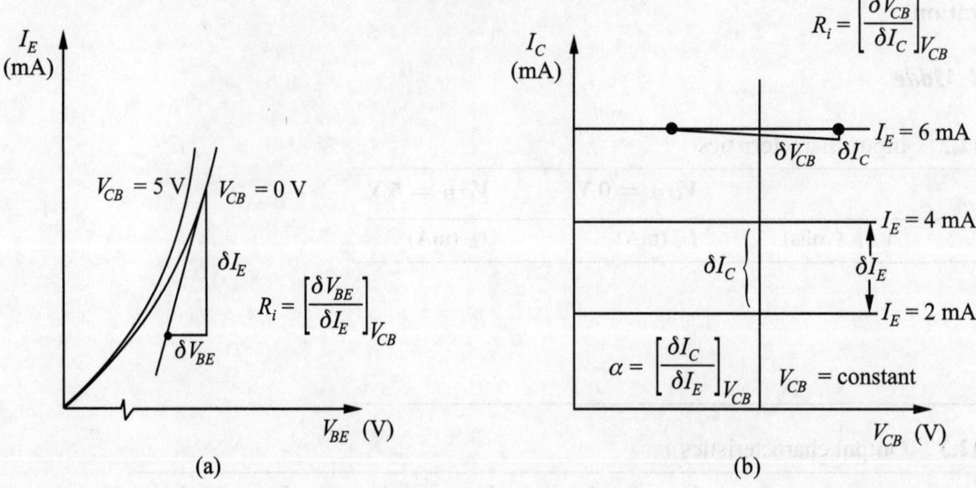

Figure 12.8 Transistor characteristics in CB mode: (a) Input characteristics and (b) Output characteristics

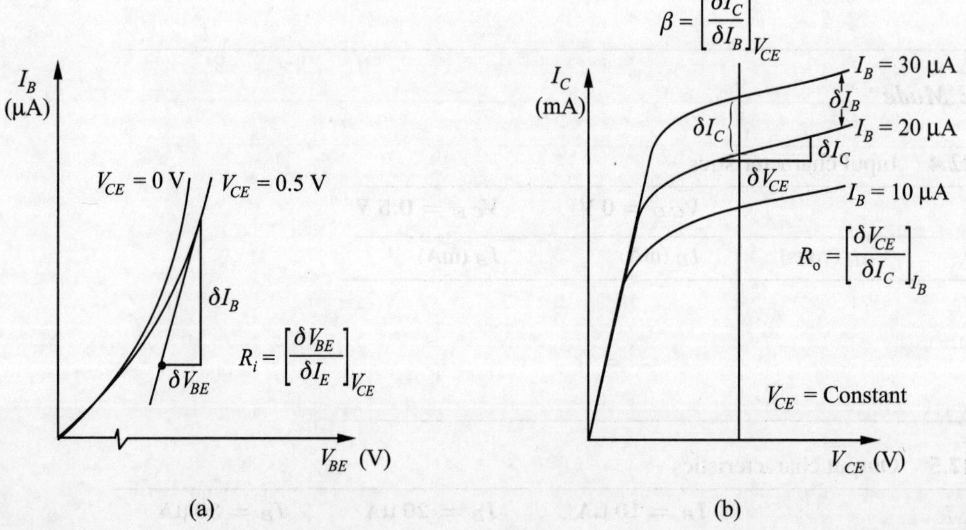

Figure 12.9 Transistor characteristics in CE mode: (a) Input characteristics and (b) Output characteristics

Calculations

From the graphs shown in Fig. 12.8 for CB mode we have,

$$R_i = (\delta V_{BE}/\delta I_E)_{V_{CB=\text{constant}}} = \text{——}$$

$R_o = (\delta V_{CB}/\delta I_C)_{I_E = \text{constant}} = \text{———}$

and $\quad \beta = (\delta I_C/\delta I_E)_{V_{CB} = \text{constant}} = \text{———}$

From the graphs shown in Fig. 12.9 for CE mode we have,

$R_i = (\delta V_{BE}/\delta I_B)_{V_{CE} = \text{constant}} = \text{———}$

$R_o = (\delta V_{CE}/\delta I_C)_{I_B = \text{constant}} = \text{———}$

and $\quad \beta = (\delta I_C/\delta I_B)_{V_{CE} = \text{constant}} = \text{———}$

Thus, $\quad \alpha/(1-\alpha) = \text{—————————}$ and $\beta/(1+\beta) = \text{—————————}$.

Estimation of Errors

The errors in the estimation of R_i, R_o, α and β can be determined by adopting the method outlined in Section 1.4.2. Thus,

In CB mode,

$$\Delta R_i = 2\left[\frac{\Delta V_{BE}}{I_{E1} - I_{E2}} + \frac{(V_{BE1} - V_{BE2})\Delta I_E}{(I_{E1} - I_{E2})^2}\right] \tag{12.32}$$

$$\Delta R_o = 2\left[\frac{\Delta V_{CB}}{I_{C1} - I_{C2}} + \frac{(V_{CB1} - V_{CB2})\Delta I_C}{(I_{C1} - I_{C2})^2}\right] \tag{12.33}$$

$$\Delta \alpha = 2\left[\frac{\Delta I_C}{I_{E1} - I_{E2}} + \frac{(I_{C1} - I_{C2})\Delta I_{\mathrm{E}}}{(I_{E1} - I_{E2})^2}\right] \tag{12.34}$$

In CE mode,

$$\Delta R_i = 2\left[\frac{\Delta V_{BE}}{I_{B1} - I_{B2}} + \frac{(V_{BE1} - V_{BE2})\Delta I_B}{(I_{B1} - I_{B2})^2}\right] \tag{12.35}$$

$$\Delta R_o = 2\left[\frac{\Delta V_{CE}}{I_{C1} - I_{C2}} + \frac{(V_{CE1} - V_{CE2})\Delta I_C}{(I_{C1} - I_{C2})^2}\right] \tag{12.36}$$

$$\Delta \beta = 2\left[\frac{\Delta I_{\mathrm{C}}}{(I_{B1} - I_{B2})} + \frac{(I_{C1} - I_{C2})\Delta I_B}{(I_{B1} - I_{B2})^2}\right] \tag{12.37}$$

where, $(V_{BE1}, I_{E1}), (V_{BE2}, I_{E2})$; $(V_{CB1}, I_{C1}), (V_{CB2}, I_{C2})$; (I_{E1}, I_{C1}), (I_{E2}, I_{C2}); (V_{BE1}, I_{B1}), (V_{BE2}, I_{B2}); $(V_{CE1}, I_{C1}), (V_{CE2}, I_{C2})$ and (I_{B1}, I_{C1}), (I_{B2}, I_{C2}) are the pairs of points shown in Figs. 12.8 and 12.9 and $\Delta V_{BE}, \Delta V_{CB}, \Delta V_{CE}, \Delta I_E, \Delta I_B$ and ΔI_C are, respectively, the errors in the readings of $V_{BE}, V_{CB}, V_{CE}, I_E, I_B$ and I_C.

Table 12.6 Estimation of errors in R_i, R_o, α and β

Least counts of the meters measuring V_{BE} in CB mode, V_{BE} in CE mode, V_{CB}, V_{CE}, I_E, I_B and I_C are, respectively, 0.005 V, 0.01 V, 0.5 V, 0.1 V, 0.01 mA, 1 μA and 0.01 mA. The corresponding errors are half of the least count.

	Measured data	Value	Absolute error	Relavtive error	Per cent error
CB Mode	$V_{BE1} = 0.145$ V $V_{BE2} = 0.170$ V $I_{E1} = 3.71$ mA $I_{E2} = 8.64$ mA	$R_i = 5.071\,\Omega$	$\Delta R_i = 1.024\,\Omega$	$\Delta R_i/R_i = 0.202$	20.2%
	$V_{CB1} = 1.0$ V $V_{CB2} = 9.0$ V $I_{C1} = 3.83$ mA $I_{C2} = 3.84$ mA	$R_O = 800$ KΩ	$\Delta R_O = 850$ KΩ	$\Delta R_O/R_O = 1.06$	106%
	$I_{C1} = 5.75$ mA $I_{C2} = 3.83$ mA $I_{E1} = 6.00$ mA $I_{E2} = 4.00$ mA	$\alpha = 0.96$	$\Delta\alpha = 0.0098$	$\Delta\alpha/\alpha = 0.01$	1%
CE Mode	$V_{BE1} = 0.57$ V $V_{BE2} = 0.58$ V $I_{B1} = 28$ μA $I_{B2} = 40$ μA	$R_i = 833.3\,\Omega$	$\Delta R_i = 902.8\,\Omega$	$\Delta R_i/R_i = 1.08$	108%
	$V_{CE1} = 4.0$ V $V_{CE2} = 10.0$ V $I_{C1} = 1.39$ mA $I_{C2} = 1.43$ mA	$R_o = 150$ KΩ	$\Delta R_o = 40$ KΩ	$\Delta R_o/R_o = 0.267$	26.7%
	$I_{C1} = 1.40$ mA $I_{C2} = 2.41$ mA $I_{B1} = 20$ μA $I_{B2} = 30$ μA	$\beta = 101$	$\Delta\beta = 11.1$	$\Delta\beta/\beta = 0.11$	11%

Results

In CB mode

1. Input resistance $R_i = — \pm — \,\Omega$
   Error in the determination of $R_i = —$ %
2. Output resistance $R_o = — \pm — \,\Omega$
   Error in the determination of $R_o = —$ %
3. Current gain $\alpha = — \pm —$
   Error in the determination of $\alpha = —$ %

In CE mode

1. Input resistance $R_i = — \pm — \,\Omega$
   Error in the determination of $R_i = —$ %
2. Output resistance $R_o = — \pm — \,\Omega$
   Error in the determination of $R_o = —$ %
3. Current gain $\beta = — \pm —$
   Error in the determination of $\beta = —$ %

The relations $\beta = \alpha/(1 - \alpha)$ and $\alpha = \beta/(1 + \beta)$ are verified.

Discussion

In this experiment we studied how a transistor behaves in a biased condition. A closer look at the experimental results and their theoretical understanding reveals why a transistor is considered a scientific marvel that brought about a revolution in technology.

As diodes were already in use before the discovery of transistors, we may naturally think that the transistor was discovered as an extension of the diode. We may also consider the transistor to be made up of two diodes connected back-to-back. However, we have already noted that a transistor cannot be manufactured in that way, and such an arrangement is used only for enhancing our understanding of the working of a transistor.

The most crucial aspect in the discovery of a transistor is the design of the base. The peculiar construction of the base region and a particular way of biasing makes a transistor behave quite differently from what could be made by merely connecting two diodes back-to-back. It can be said that *how well a transistor behaves is decided by how thin and lightly doped its base region is!*

In the following table we compare the experiment and the theory.

Theory	**Experiment**
As the emitter base junction is forward biased, we expect the input characteristics to be similar to the forward characteristics of a diode for both CB and CE modes.	Figures 12.8a and 12.9a show that the observed input characteristics are similar to the forward characteristics of the emitter base diode.
As the base collector junction is reverse biased, we expect the output characteristics to be similar to the reverse characteristics of a diode for both CB and CE modes.	Figures 12.8b and 12.9b show that the observed output characteristics are similar to the reverse characteristics of the base collector diode.

It can be seen from this table that there is a qualitative agreement between the theory and the experiment. However, it should be noted here that the input characteristics also depend on the output voltage and the output characteristics also depend on the input voltage. Thus the two-diode analogy cannot be stretched too far because the input interacts with the output. This interaction makes the analysis of a transistor difficult but interesting.

From the input characteristics we see that at constant V_{BE}, increase in V_{CB} increases the emitter current in CB mode, whereas increase in V_{CE} decreases the base current in CE mode. This can be explained on the basis of what is known as the *early effect*. According to this effect, as the reverse bias voltage across the base collector junction increases, the width of the depletion region across this junction increases, thus reducing the effective width of the base region. This reduces the chances of recombination of electrons and holes in the base region, resulting in increase in I_C leading to an increase in I_E in CB mode and decrease in I_B in CE mode (see Fig. 12.10).

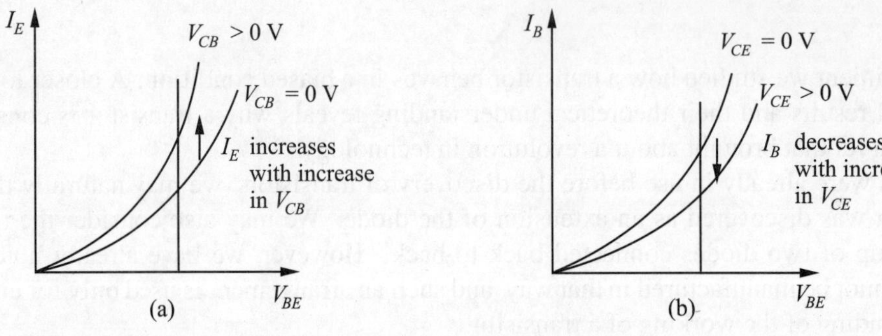

Figure 12.10 Input characteristics in (a) CB mode and (b) CE mode

The output characteristics in both the modes can be divided into three regions: active region, saturation region and *cut-off* region (Fig. 12.11).

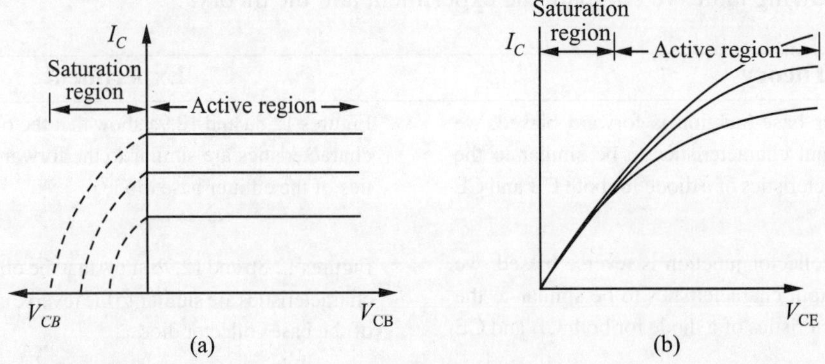

Figure 12.11 Output characteristics in (a) CB mode and (b) CE mode

Active region: In the active region, the collector current is essentially independent of the output voltage. This region corresponds to a *forward biased emitter base junction and a reverse biased base collector junction*. As the collector current flows through the reverse biased base collector junction, it is independent of the bias voltage. However, as this current is contributed mainly by the charge carriers supplied by the emitter, the emitter base voltage can control it. The slight increase in I_C in this region is due to early effect.

Saturation region: The saturation region corresponds to *both the emitter base junction and the base collector junction forward-biased*. In CB mode we find that active region is present up to zero V_{CB}, that is, I_C has a non-zero value even if V_{CB} is made zero. This is because the barrier field across the collector base junction helps the carriers coming from the emitter to be swept across the collector. Thus, to get the saturation region in CB mode we have to *actually* bias the base collector junction in the forward direction. As the forward bias across the base collector junction is increased, we can reduce I_C to zero as shown in Fig. 12.11a.

In CE mode, however, the saturation region does not correspond to the actual forward biasing of the base collector junction. This difference arises because in CB mode the entire applied voltage appears across the base collector junction, whereas in CE mode we have $V_{CE} = V_{CB} + V_{BE}$ (Fig. 12.12). The forward biasing of the base collector junction here means reducing V_{CB} below V_{BE} so that collector becomes less positive or more negative with respect to base, that is, forward biased.

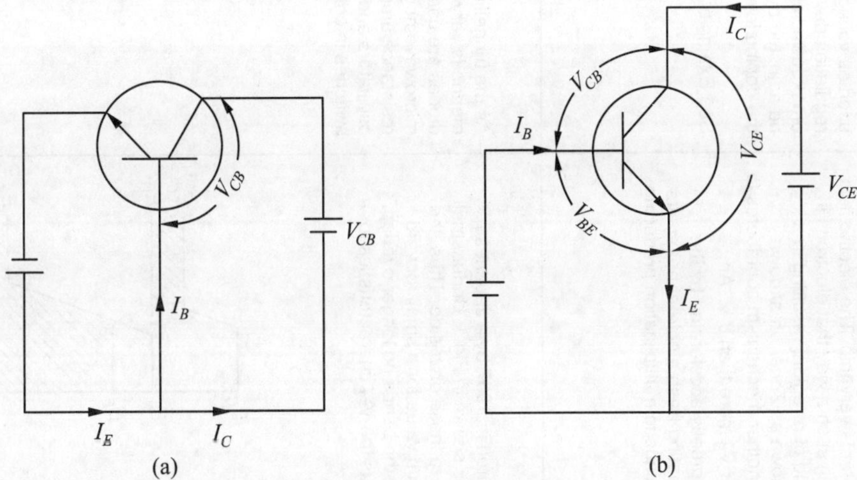

Figure 12.12 Voltage drops in (a) CB mode and (b) CE mode

Cut-off region: The cut-off region corresponds to *both emitter base and base collector junctions being reverse biased*. Even if the input current is zero, the collector current show some value. This current—usually in nanoamperes or microamperes—is known as the *reverse leakage current* and arises due to the flow of minority carriers across the base collector junction. In CB mode this current is referred to as I_{CBO}, the current from collector to base with emitter open, and in CE mode it is referred to as I_{CEO}, the current from collector to emitter with base open.

The errors in the determination of various parameters were found in Table 12.6. The error in the determination of output resistance in CB mode and input resistance in CE mode exceeds 100%. The reason for these large errors is that the changes in current/voltage are comparable to the least counts of the meters used for the measurement of these quantities.

12.4 APPLICATIONS

In this chapter we studied the physics of two important semiconductor devices—diodes and transistors. In Section 12.2 we studied the *p-n* junction diode. These semiconductor diodes replaced the existing vacuum tube diodes due to advantages in cost, size, reliability and efficiency. Several diodes that serve a wide range of applications have been developed. In Chart 12.3 the working and applications of the photodiode, tunnel diode and Zener diode are discussed. The applications of diodes for purposes such as rectification and voltage regulation can also be studied in a laboratory. Experiment 12.3 describes these applications.

CHART 12.3 JUNCTION DEVICES AND APPLICATIONS

Device	Description	Applications
Zener diode/ Avalanche diode	A Zener diode is based on the mechanism of Zener breakdown, explained for the first time by C Zener in 1934. If the *p-n* junction diode is heavily doped, then the depletion layer is very thin. The electric field across the junction becomes very high and the band diagram modifies in such a way that electrons in the valance band of *p*-region can directly tunnel into the conduction band of *n*-region, resulting in a sharp increase in current. The voltage at which breakdown occurs is known as *Zener breakdown voltage* V_B and is of the order of 4–6 volts. If the junction is thick then another mechanism contributes to breakdown, called the *avalanche breakdown* or secondary emission, for V_B more than 8 V. An avalanche arises due to the fact that electrons near the junction are so greatly accelerated by the electric field that they cause ionisation of the atoms they collide with creating fresh electron–hole pairs. This chain continues and the process of impact ionisation results in avalanche multiplication producing a very large current at breakdown.	Zener diodes are widely used as voltage regulation devices. The breakdown voltage can be adjusted by changing the doping levels. See Example 12.4.
Tunnel diode Fig. 4	A tunnel diode is based on the tunnelling phenomenon and was discovered in 1957 by a Japanese scientist, Esaki. Under no bias electrons can tunnel in both directions as shown in Fig.1. Under small forward bias (Fig. 2) electron energy levels on *n*-side are raised relative to those on *p*-side. Thus the tunnelling is from *n* to *p* only. The tunnelling current reaches its maximum value for a small forward bias. As the forward bias is increased energy levels shift further up and the bands no longer overlap. The diode now behaves as an ordinary one as no tunnelling occurs. Thus the *V–I* characteristics for a tunnel diode display a negative resistance characteristics shown in Fig. 4. Fig. 1 Fig.2 Fig.3	As the tunnelling phenomenon is very fast, tunnel diodes are used as microwave negative resistance amplifiers and oscillators and as fast switches in computers.

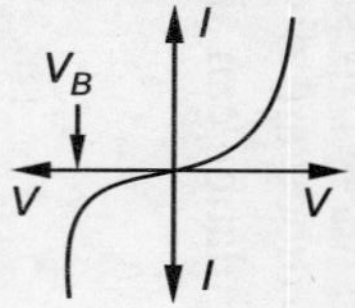

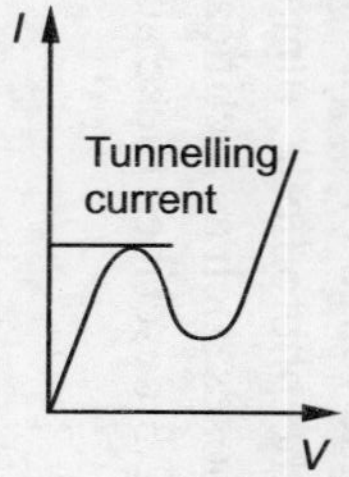

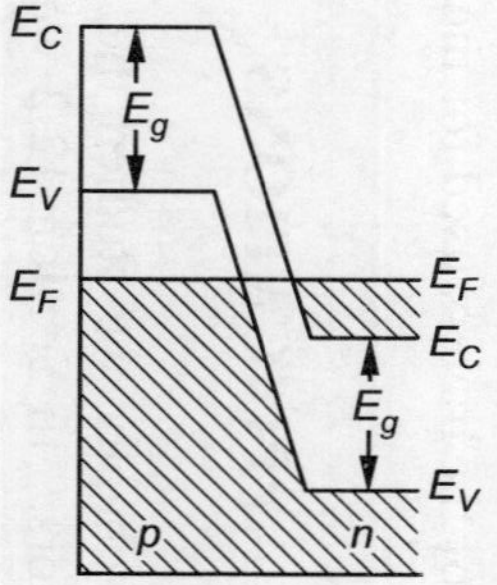

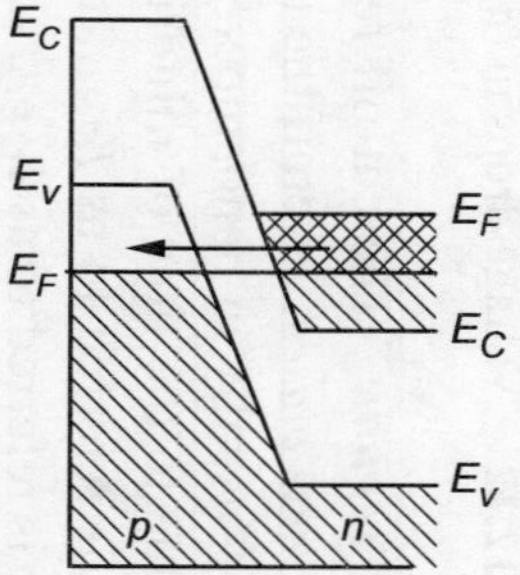

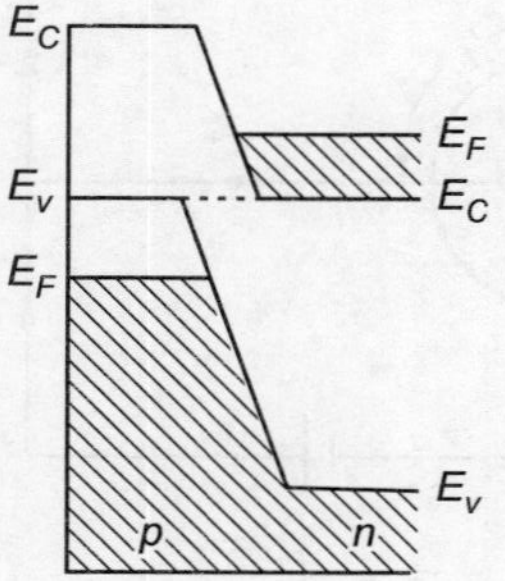

Device	Description	Applications
Varactor diode	In a *p-n* junction diode there are immobile positive and negative ions on both sides of a junction. Thus a *p-n* junction possesses junction capacitance, the depletion region serving as a dielectric medium. In reverse bias, the depletion region widens, thus decreasing the junction capacitance. A varactor diode has the impurity concentration profile such that the junction capacitance varies with the applied reverse bias voltage. Such diodes are also known as *varicaps* or *voltacaps*.	Varactor diodes find wide applications as low noise parametric amplifiers, har-monic frequency generators and frequency converters or mixers.
Photodiode	It is a type of photodetector (see Chart 11.6) in which electron–hole pairs are created in or near the depletion region under reverse biased condition. The electrons and holes separate quickly due to the high electric field in the depletion region and thus most of them contribute to the photocurrent. An avalanche photodiode is one operated at high reverse bias voltage so that a large photocurrent is obtained due to avalanche multiplication.	Photodiodes are used in sound films, computer cards and light driven switches.
Solar cell	If an unbiased photodiode is illuminated by radiation of suitable frequency, charge separation results and a potential difference is produced across the junction that forward biases the diode. This is known as *photovoltaic effect* and the diode is known as *photovoltaic cell* or *solar cell*.	Solar cells are widely used in space applications for con-verting solar energy into electrical power.
LED	When an electron recombines with a hole, a photon of energy equal to the band gap of the semiconductor is emitted. In Si and Ge the recombination energy is in the IR range, but in GaAsP and GaP it is in the visible range. LEDs have many advantages over ordinary incandescent sources like low power consumption, fast response, small and light weight, long life, emission of monochromatic light and stability against mechanical vibrations.	LEDs are used as light emi-tters in visual display units and in production of optically coupled circuits.
Semi-conductor laser	See Chart 14.2	

EXAMPLE 12.4
Voltage regulation using Zener diode

For the circuit shown in Fig. 12.13 find the current through the Zener diode. What will happen if (a) the input voltage increases to 41 V and (b) the load resistance increases to 4 KΩ?

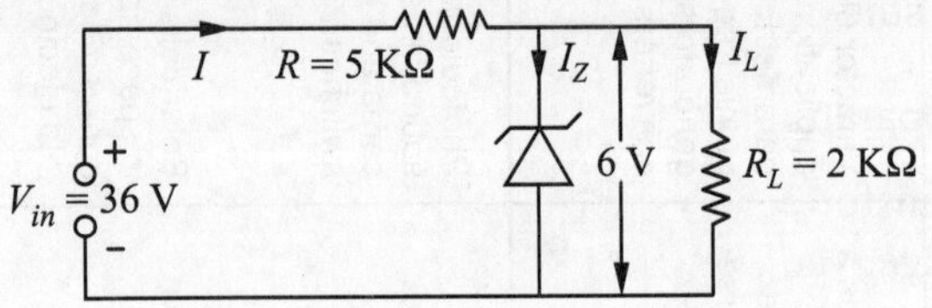

Figure 12.13

Solution

The input voltage, $V_{\text{in}} = 36$ V and the Zener breakdown voltage, $V_B = 6$ V.

Thus the voltage drop across the resistor R,

$$V_R = 36 - 6 = 30 \text{ V}$$

The current through R,

$$I = \frac{30 \text{ V}}{5 \times 10^3 \ \Omega} = 6 \text{ mA}$$

Now the voltage across R_L is the same as V_B, that is, 6 V. Hence the current through R_L,

$$I_L = \frac{6 \text{ V}}{2 \times 10^3 \ \Omega} = 3 \text{ mA}$$

The current through Zener diode,

$$I_Z = I - I_L = 6 - 3 = 3 \text{ mA}$$

(a) Even if the input voltage changes, the voltage across the Zener diode remains constant. Thus $V_L = 6$ V and hence

$$I = \frac{(41 - 6) \text{ V}}{5 \times 10^3 \ \Omega} = 7 \text{ mA}$$

Thus, $I_Z = 7 - 3 = 4$ mA

(b) For $R_L = 4$ KΩ

$$I_L = \frac{6 \text{ V}}{4 \times 10^3 \ \Omega} = 1.5 \text{ mA}$$

Thus, $I_Z = 6 - 1.5 = 4.5$ mA

Thus in both cases the *Zener diode maintains a constant voltage across it by absorbing the current changes through it*. This is possible because the resistance of Zener diode changes accordingly.

EXPERIMENT 12.3 APPLICATIONS OF DIODES

Aim

To study (a) rectification and (b) voltage regulation using diodes

Apparatus

Transformer of range 12–0–12 V, diodes (both ordinary and zener diodes), resistances of suitable range, variable dc power supply (0–20 V), milliammeter (0–25 mA), voltmeter (0–10 V), CRO

Formulae

$$\text{Ripple factor, } r = \sqrt{\frac{V_{rms}^2}{V_{dc}^2} - 1} \tag{12.38}$$

$$\text{Rectification efficiency, } \eta = \frac{V_{dc}^2}{V_{rms}^2} \tag{12.39}$$

where V_{rms} is the rms value of the ripple in the output voltage and V_{dc} is the dc value of the output voltage.

$$V_{\text{dc}} = \left(\frac{1}{T}\right) \int_0^T V_o dt \tag{12.40}$$

$$V_{\text{rms}} = \left[\left(\frac{1}{T}\right) \int_0^T V_o^2 dt\right]^{1/2} \tag{12.41}$$

where $\int V_o dt$ is the area under the graph of output voltage V_o versus t, traced from the CRO screen, and $\int V_o^2 dt$ is the area under the graph of V_o^2 versus t. These areas are calculated using the expression

$$\text{Area} = \left(n_1 + \frac{n_2}{2}\right) \times P \times Q \times \frac{1}{100} \tag{12.42}$$

where n_1 = number of 1 mm² area square cells fully under the graph and n_2 = number of 1 mm² area square cells partially under the graph, P is the reading of volts/div knob and Q is the reading of time/div knob of CRO.

The amplitude of the output voltage is given by

$$V_{\text{m}} = D_1 \times P \tag{12.43}$$

where D_1 is the distance on the voltage axis corresponding to the peak height of the signal.

The time period of the output voltage is given by

$$T = D_2 \times Q \tag{12.44}$$

where D_2 is the distance on the time axis between the starting points of two consecutive cycles.

Theory

Rectification: The electric supply available to us is in the form of ac but most of the domestic and other gadgets require dc for their operation. *The process of converting ac into dc is known as rectification* and the circuits that enable us to achieve rectification are called *rectifier circuits*. The common requirements of all rectifier circuits are transformer, diode (or diodes) and load resistance. The transformer is used for stepping down the 230 ac mains voltage to the level required.

Figure 12.14a shows a half wave rectifier circuit. The diode conducts heavily when it is forward biased and poorly when it is reverse biased. Thus, it conducts during the positive half cycle of the input voltage (Fig. 12.15a) and does not conduct in the negative half cycle. The output voltage is a unidirectional pulsating voltage as shown in Fig. 12.15b. This voltage can be expressed as,

$$V_o(t) = \begin{cases} V_m \sin \omega t & 0 \le t \le \dfrac{\pi}{\omega} \\ 0 & \dfrac{\pi}{\omega} \le t \le \dfrac{2\pi}{\omega} \end{cases} \tag{12.45}$$

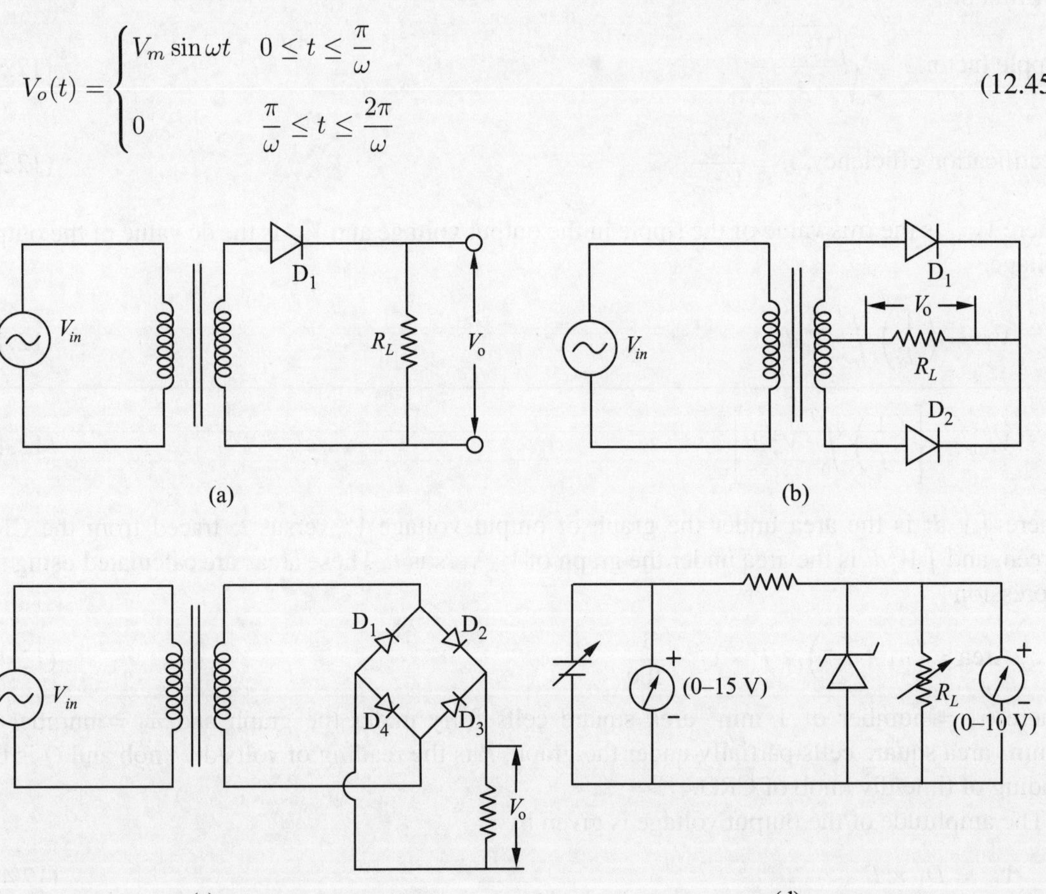

Figure 12.14 Experimental circuit diagrams: (a) Half wave rectifier, (b) Full wave rectifier, (c) Bridge rectifier and (d) Voltage regulator

Substituting Eq. 12.45 in Eqs. 12.40 and 12.41, solving and then using Eqs. 12.38 and 12.39 we get,

$$V_{\mathrm{dc}} = \frac{V_m}{\pi}, V_{rms} = \frac{V_m}{2}, r = \sqrt{\frac{\pi^2}{4} - 1} \text{ and } \eta = \frac{4}{\pi^2} \tag{12.46}$$

for a half wave rectifier.

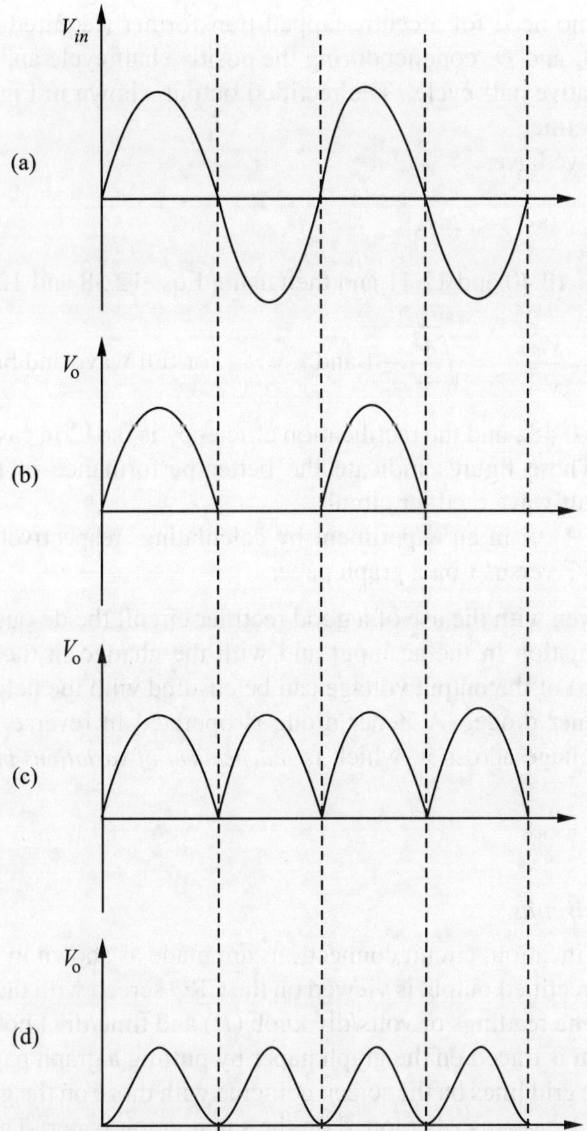

Figure 12.15 Voltage waveforms of (a) input voltage; output voltage in the case of (b) half wave rectifier, (c) full wave rectifier and (d) bridge rectifier

Thus the ripple factor, $r = 1.21 > 1$, and the ripple voltage exceeds the dc voltage indicating the poor performance of a half wave rectifier. Its rectification efficiency is 0.406, that is, only 40.6%

of the ac input power is converted into dc power.

In case of a full wave rectifier (Fig. 12.14b), two diodes are used so that in the positive half cycle diode D_1 gets forward biased (and D_2 gets reverse biased), and in the negative half cycle, diode D_2 gets forward biased. The rectified output is as shown in Fig. 12.15c. For a bridge rectifier (Fig. 12.14c), there is no need for a centre-tapped transformer (required in case of a full wave rectifier). The diodes D_2 and D_4 conduct during the positive half cycle and the diodes D_1 and D_3 conduct during the negative half cycle. The rectified output, shown in Fig. 12.15d, is similar to that for the full wave rectifier.

For both these cases we have,

$$V_o(t) = |V_m \sin \omega t| \quad 0 < t < 2\pi/\omega \tag{12.47}$$

Using Eq. 12.47 in Eqs. 12.40 and 12.41 and then using Eqs. 12.38 and 12.39 we get,

$$V_{dc} = \frac{2V_m}{\pi}, V_{rms} = \frac{V_m}{\sqrt{2}}, r = \sqrt{\frac{\pi^2}{8} - 1} \text{ and } \eta = \frac{8}{\pi^2} \text{ for full wave and bridge rectifiers} \tag{12.48}$$

Thus the ripple factor is 0.482 and the rectification efficiency is 0.812 in case of full wave rectifier and bridge rectifier. These figures indicate the better performance of these rectifier circuits compared to that of a half wave rectifier circuit.

We can find V_{dc} and V_{rms} in an experiment by calculating, respectively, the areas under the curves V_o versus t and V_o^2 versus t on a graph paper.

Voltage regulation: Even with the use of a good rectifier circuit the dc output voltage may show variation with the fluctuation in the ac input and with the change in the load resistance. The constancy (or regulation) of the output voltage can be ensured with the help of voltage regulation circuits made using Zener diode. A Zener diode is operated in reverse breakdown mode and maintains a constant voltage across it, which is *independent of variations in either input voltage or load resistance*.

Procedure

A. Study of rectifier circuits

1. For half wave rectification, circuit connections are made as shown in Fig. 12.14a.
2. The waveform of rectified output is viewed on the CRO screen with the AC-GND-DC switch on DC position. The readings of volts/div knob (P) and time/div knob (Q) are noted.
3. Now the waveform is traced on the graph paper by putting a graph paper on the screen such that the centimetre grid lines on the screen coincide with those on the graph paper. Using this trace the graph of V_o^2 versus t is plotted on the same graph paper. The values of D_1, D_2, n_1 and n_2 are measured from the graph.
4. The observations are tabulated.
5. Steps 1 to 4 are repeated using the full wave rectifier circuit (Fig. 12.14b) and the bridge rectifier circuit (Fig. 12.14c).

B. Study of voltage regulation circuit

1. The circuit is connected as shown in Fig. 12.14d.

2. Keeping the input voltage fixed at say 12 V (V_{in} should be more than the breakdown voltage), the load resistance is varied in suitable steps and the output voltage is noted.
3. The observations are tabulated and a graph is plotted between the output voltage along the y-axis and the load resistance along the x-axis. This graph is known as the *load regulation characteristics*. The value of load resistance above which V_o becomes constant is noted down.
4. Now this value of R_L is connected in circuit as shown in Fig. 12.14d.
5. The input voltage is increased from 0 to 15 V with a suitable interval and the corresponding output voltage is measured.
6. The observations are tabulated and a graph is plotted between V_o along the y-axis and V_{in} along the x-axis. This graph is known as the *line regulation characteristics*. The value of V_{in} above which V_o becomes constant is noted down.

Observations

Table 12.7 Rectifier circuits

Circuit	**P** (volts/div)	**Q** (s/div)	D_1 (div)	D_2 (div)	n_1	n_2	n_1	n_2
					For calculation of V_{dc}		For calculation of V_{rms}	
Half wave rectifier								
Full wave rectifier								
Bridge rectifier								

Table 12.8 Voltage regulation circuit

Load regulation		**Line regulation**	
$V_{\text{in}} = 12\ V$		$R_L = —\ \Omega$	
$R_L\,(\Omega)$	V_0 (volts)	V_{in} (volts)	V_0 (volts)

Calculations

Table 12.9 Calculations of V_{dc}, V_{rms}, r and η

Circuit	**Area under the curve =** $(n_1 + \frac{1}{2}n_2) \times P \times Q \times 1/100$		$V_m(V) = D_1 \times P$	$T(s) = D_2 \times Q$	V_{dc} (V) = $(1/T) \times \int V_0 dt$	V_{rms} (V) = $[(1/T) \times \int V_0^2 dt]^{1/2}$	$r = \sqrt{\frac{V_{rms}^2}{V_{dc}^2} - 1/T}$	$\eta = \frac{V_{dc}^2}{V_{rms}^2}$
	$\int V_0 dt$	$\int V_0^2 dt$						
Half wave rectifier								
Full wave rectifier								
Bridge rectifier								

Graph

Graphs are drawn as in Figs. 12.15 and 12.16

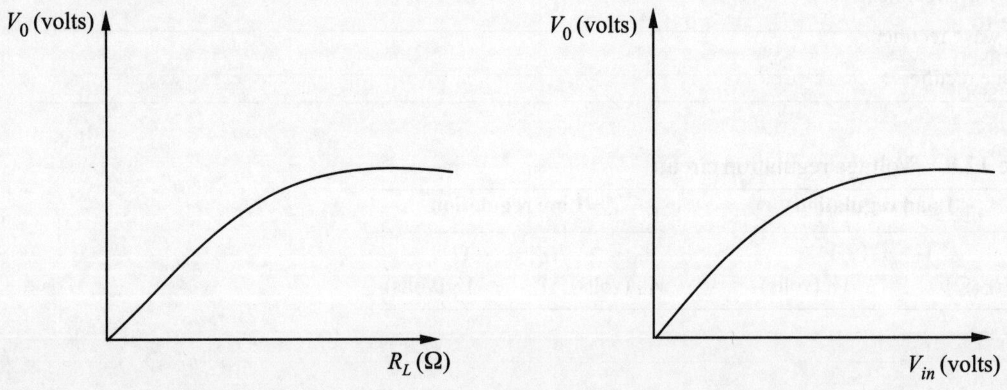

Figure 12.16 Variation of output voltage with input voltage and load resistance

Results

1. The experimental values of the ripple factor, r, and rectification efficiency, η, are determined as under.

   $r =$ ——, $\eta =$ —— for half wave rectifier

   $r =$ ——, $\eta =$ —— for full wave rectifier

   $r =$ ——, $\eta =$ —— for bridge rectifier

2. The output voltage is found to be fairly constant in a Zener regulated circuit for input voltage ranging from —— *V* to —— *V* and for load resistance ranging from —— Ω to —— Ω.

Discussion

In this experiment we studied the rectification and regulation properties of diodes. These properties arise because of the *bias dependent resistance of diodes*.

In the following table the theory and the experimental observations are compared

Circuit	**Theory**	**Experiment**
Half wave rectifier	$r = 1.21$	$\eta = 40.6\%$
Full wave rectifier	$r = 0.482$	$\eta = 81.2\%$
Bridge rectifier	$r = 0.482$	$\eta = 81.2\%$

The theoretical values of r and η are generally in agreement with the respective experimental values within the experimental limits for all the three rectifier circuits. The possible reasons for discrepancies are:

1. Voltage from ac mains is not truly sinusoidal and
2. The step down transformer introduces distortions in the input voltage to be fed to the rectifier circuit.

It is noted that the rectification efficiency increases as we refine the rectifier circuit from a half wave to a full wave/bridge rectifier. The rectification efficiency of rectifier circuits can be further increased by feeding the output of these circuits to what are known as the *filter circuits*. They consist of either a series inductance, a shunt capacitor or a combination of the two. Fig. 12.17 shows a π filter which consists of two shunt capacitors and an inductance.

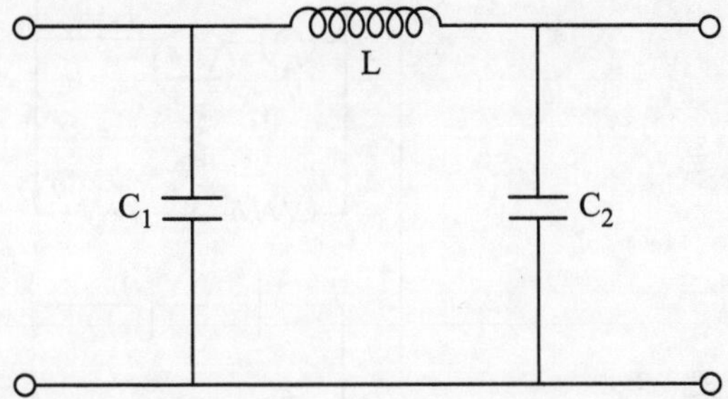

Figure 12.17 A π filter

The experimental curves shown in Fig. 12.16 show that the voltage across Zener diode is almost constant irrespective of the input voltage and the load resistance. This behaviour can be understood on the basis of the mechanism of Zener breakdown discussed in Chart 12.3 and Ex. 12.4.

CHART 12.4 APPLICATIONS OF TRANSISTOR

Application	Circuit diagram	Working	Uses
Amplifier	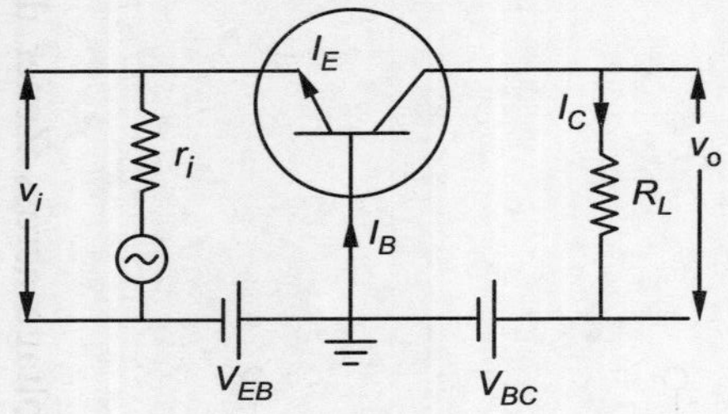	An amplifier boosts the voltage or power level of an electrical signal. The ratio of output voltage to the input voltage is called the gain of amplifier $A = v_0/v_i$. Now the following relations can be written by looking at the circuit diagram: $v_i = r_i \Delta I_E$, $v_0 = \Delta I_C R_L$. Using these relations and the expression for α ($\alpha = \Delta I_C/\Delta I_E$), we get, $A = \alpha (R_L/r_i)$. (1) Thus a large gain results because $R_L >> r_i$ and $\alpha \sim 1$. Thus larger gain result with the use of a transistor due to its property of $\alpha \sim 1$, that is, the collector current and emitter current are approximately the same, a result of the thin and lightly doped base. See Example 12.5.	Useful in the amplification of voltage or current or power. Used for the amplification of audio, radio or video signals required in radio or television sets. An amplifier can also be made to work as an oscillator using proper feedback circuit.
Switch	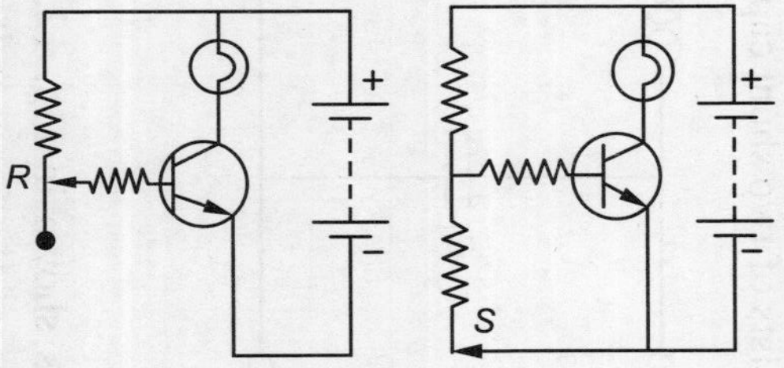 (a) (b)	A transistor can act as a switch, that is, it can have two states—ON and OFF. The figure shows two switching circuits. In (a) the ON state is obtained by reducing the resistance R until the base current is large enough to make the collector light the lamp. The OFF state is when the lamp does not glow. In (b) the ON state is obtained by adjusting the value of S such that the potential difference across S is enough to forward bias the emitter base junction.	The switching property of transistors is widely used in digital circuits. Multivibrator circuits use two transistors as switches to produce simultaneous ON–OFF states.

Application	Circuit diagram	Working	Uses
Light operated circuit	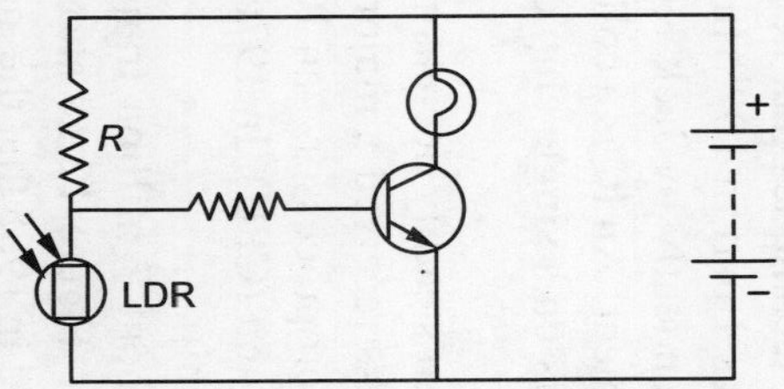 R, LDR, +, –	In the light operated circuit the LDR (light dependent resistor) is a part of the potential divider arrangement. If the resistance of LDR is more so that the potential difference across it is more than 0.6 V (or the forward biasing voltage of the emitter base junction), then the circuit operates and the lamp in the collector circuit glows. Thus the lamp glows if LDR is shielded from the light so that its resistance is more.	Light operated circuits can be used as alarm circuits that can detect small changes in light intensity. If LDR and R are inter-changed, then the lamp will glow when LDR detects light. Thus the circuit can be used as a light operated intruder alarm.
Temperature operated circuit	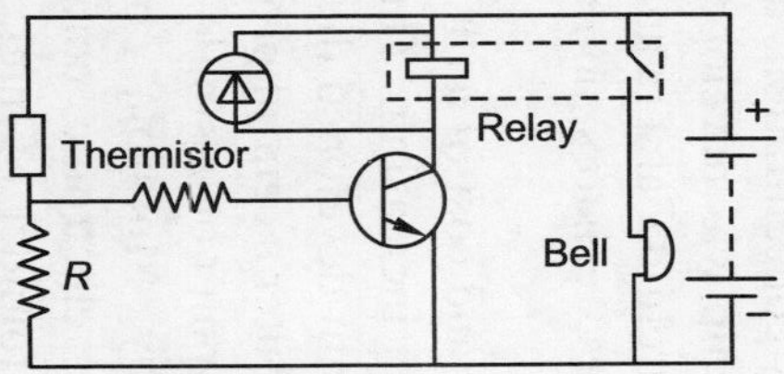 Thermistor, R, Relay, Bell, +, –	In the temperature operated circuit the thermistor is a part of the potential divider arrangement. When the temperature of the thermistor increases, its resistance decreases, and a larger fraction of the 6 V supply is dropped across R. When the potential difference across R is enough to forward bias the emitter base junction, the transistor is ON. The collector current is usually small and hence cannot ring a bell. Thus a relay circuit is used that is activated by the collector current and draws the larger current from the 6 V supply when activated.	Temperature operated circuits can be used to detect small changes in temperature. By inter-changing the thermistor and R, the circuit can be used as a frost-warning alarm.
Sound operated circuit	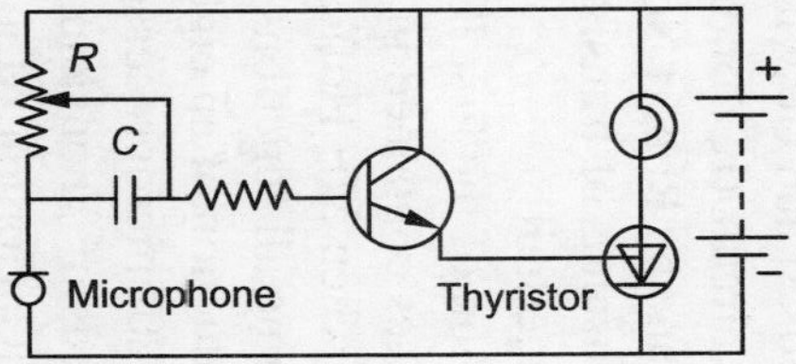 R, C, Microphone, Thyristor, +, –	In a sound operated circuit the microphone is a part of the potential divider arrangement. The resistance R is adjusted such that when someone speaks into the microphone, the emitter base junction gets forward biased and the transistor is switched ON. The emitter current then makes the lamp glow. The capacitor blocks the dc from the battery passing via R through the microphone but it allows the ac produced in the microphone (due to sound) to pass to the base.	Sound operated circuits can be used to detect changes in sound intensity. If a relay and bell are used instead of a lamp, then the circuit can be used as a sound operated intruder alarm.

EXAMPLE 12.5
Voltage gain in a transistor amplifier

Find the voltage gain of a transistor in common base mode if the input resistance is 200 Ω, load resistance is 200 KΩ and the change in base current ΔI_B, corresponding to the change in emitter current $\Delta I_E = 2$ mA, is 5 μA.

Solution

In a transistor, $I_E = I_B + I_C$

Thus, $\Delta I_C = \Delta I_E - \Delta I_B = 2 \text{ mA} - 5 \times 10^{-3} \text{ mA} = 1.995 \text{ mA}$

Therefore, $\alpha = \dfrac{\Delta I_C}{\Delta I_E} = \dfrac{1.995}{2} = 0.9975$

The voltage gain of an amplifier,

$$A = \alpha \times \frac{R_L}{r_i} = 0.9975 \times \frac{200 \text{ K}\Omega}{200\ \Omega} = 997.5$$

In Section 12.3 it was seen how the discovery of transistors revolutionised the world of electronics. This had a profound impact on society in the second half of the last century. By 1960 transistors replaced the vacuum tubes in a majority of the electronic applications. In Chart 12.4 five typical applications of a transistor—out of the myriad number possible—are discussed.

The advent of transistors initiated a progressive miniaturisation of electronic components and circuits. This progress was accelerated by the development of refining techniques that helped in preparing samples of Si and Ge with impurity concentrations as small as 1 part in 10^{10}. In 1958, the first integrated circuit (IC) on a silicon chip was developed independently by Jack Kilby at Texas Instruments and by Robert Noyce at Fairchild Camera and Instrument. An IC is a collection of hundreds of thousands of transistors, diodes, resistors, and capacitors on a single chip of area no larger than a fingernail!

Since the invention of the first IC, the size and cost of an IC has been steadily decreasing and the density of devices contained in it has been increasing. This progress received a major boost in the early 1970s when ME Hoff and his team developed the first microprocessor—an IC chip capable of performing all functions of a computer's central processing unit (CPU). In 1976, Intel announced the production of an eight-bit computer on a single silicon chip.

A power PC 620 microprocessor built by Motorola contains about 7 million transistors along with hundreds of thousands of other electronic components. Figure 12.18 shows the photograph of this IC. Gordon Moore, the co-founder of Intel, predicted in 1965 that the number of transistors on a chip would double every year. This is known as *Moore's law*. Recently, Intel engineers have designed and manufactured transistors of size 20 nm—the smallest and the fastest ever.

Figure 12.18 Motorola PowerPC 604e 233MHz
(Source: http://en.wikipedia.org/wiki/File:Motorola_PowerPC_604e_233MHz_2.jpg)

The microprocessors offered a great leap towards circuit miniaturisation. This brought about major advances in aerospace and defence applications where size and weight are of critical importance. The other fields in which microprocessors have found wide applications are computer technology (CPU and RAM chips: Random Access Memory chips), automobile engineering (chips that improve fuel economy and reduce pollution), domestic appliances (chips for controlling household appliances, games and entertainment), industry (chips for instrumentation and automation) and medical technology (device chips and chips for implants). Microprocessors have also brought about computerisation of many consumer products such as self-tuning TVs, programmable VCRs and security systems. Today mciroprocessors have pervaded virtually all fields of human endeavour.

Though the development in technology of semiconductors and semiconductor devices is vast, the potential of motion of charge carriers in various semiconducting materials has not been fully exploited as yet. Research in this area can provide a breakthrough in the future.

EXERCISES

1. Explain the mechanism of formation of depletion layer in *p-n* junction diode. How does the potential barrier arise in the process?
2. Draw the energy band diagram of a *p-n* junction diode in equilibrium. Hence obtain an expression for the height of the potential energy barrier.
3. Draw energy band diagrams of a *p-n* junction diode in the following modes: (i) unbiased, (ii) forward biased and (iii) reverse biased.
4. Derive the rectifier equation for the *p-n* junction diode.
5. Explain diffusion current and drift current.
6. With a neatly labelled diagram explain transistor action when it is biased to operate in the active region. Draw the energy band diagram of a ***npn***-transistor in common base mode.
7. Draw a neat energy band diagram for ***npn***-transistor when (i) unbiased or (ii) biased in common base mode showing various currents.
8. Give reasons for the following: (i) The emitter base junction is forward biased and collector base junction is reverse biased. (ii) In common base mode of a transistor the current gain is close to unity.

PROBLEMS

1. The resistivities of the two sides of an abrupt Ge diode are 1.5 Ω cm on *p*-side and 1 Ω cm on *n*-side. Calculate the height of the potential barrier. Assume: $n_i = 2.25 \times$ 1019/cm^3, $\mu_e = 0.38$ m^2/Vs and $\mu_h = 0.18$ m^2/Vs.
2. The current flowing in a certain reverse-biased junction at room temperature is 2×10^{-7} A. What will be the current flowing through the junction if forward bias of 0.1 V is applied?
3. The saturation current density of a *p–n* junction diode is 250 mA/m^2 at 300 K. Find the voltage that should be applied across the junction to cause a forward current density of 105 A/m^2 to flow.
4. In an ***npn***-transistor connected in common base configuration, emitter current is 2 mA and base current is 20 μA, what are the values of collector current and current gain?

13

Charged Particles in Electric and Magnetic Fields

13.1 THE STORY OF CHARGED PARTICLES IN MOTION

In the last two chapters we saw how our knowledge of physics of semiconductors has contributed to the amazing progress in engineering and technology. Curiously the roots of this progress can also be traced in our quest for understanding how charges move in fields that affect them. Here we capture the story of this quest and seek to learn about the motion of charged particles in electric and magnetic fields.

What is a charge? The fascination with this word—charge—can be traced to the Greek era. The Greek philosophers knew that amber, when rubbed, can attract bits of straw. The word 'electron' was derived from the Greek word meaning amber. It was long suspected that electrical charge had something to do with atomic structure but the connection remained hazy until the end of the nineteenth century. However, the science of electricity—based on the concept of charge—had by then advanced greatly through the pioneering work of Charles Coulomb (1736–1806), who proposed the law of electric force between charges in 1785, Andre Ampere (1775–1836), Carl Gauss (1777–1855) and others.

Charge was earlier thought to be some sort of a fluid. This notion arose from the fact that a conducting body carrying a charge can transfer part of the charge to another conducting body upon contact with it. In 1897 JJ Thomson discovered the electron, which carries a charge of 1.6×10^{-19} C. This ended the mystery. It was recognised that the electron is an elementary particle. Proton, the only other elementary particle known at that time carried the same charge as the electron, but of opposite sign. It is only in the last quarter of the twentieth century that particles carrying fractional charges $e/3, 2e/3$ and so on were discovered.

The science of magnetism had also been studied for centuries. In fact, magnetism is one of the earliest known physical phenomena of solid materials. The Greeks used the name 'magnet' for a stone (loadstone or magnetite), which was capable of attracting other pieces of the same material and iron as well. Though magnetite was discovered before 600 BC, a proper understanding of the physics of magnetic phenomena was gained only a few centuries back. William Gilbert (1540–1603) was the first to explore magnetic phenomena using scientific methods. He discovered the earth's magnetism. Quantitative studies of magnetism began in the early eighteenth century when Coulomb established the inverse square law of force between magnetic poles. SD Poisson (1781–1840) pioneered the development of magnetostatics.

Electricity and magnetism were studied in the beginning as independent disciplines. They were brought together by Hans Oersted in 1820. Interestingly, the connection was established by Oersted almost by accident. While preparing for a lecture demonstration, Oersted observed that a compass needle placed nearby was deflected when current passed through one of the wires. The field of electromagnetism was developed further through the pioneering unifying contributions of Michel Faraday and James Maxwell.

Alongside these developments in electromagnetism, Newton's laws had been applied to various problems in mechanics and had almost been accepted universally by that time. It was known that with the knowledge of the forces acting on a body the details of the motion of that body could be worked out. With the discovery of an electron and identification of the elementary charge associated with it, physicists became curious about the motion of electrons and other charged particles due to the application of electric/magnetic field.

Simple cases of motion of charged particles in constant electric and magnetic fields were first studied. This was followed by the more complicated but interesting cases of motion in non-uniform fields. An interesting outcome of this research was that electric/magnetic fields were found to affect the electron beams (as well as beams of other charged particles) in a manner analogous to the way optical lenses affect light beams. The study of motion of charged particles in non-uniform fields thus gained a lot from parallel concepts in optics. This resulted in the emergence of 'electron optics' as a new discipline of physics. Hamilton (1827), H Busch (1927), and CJ Davisson and CJ Calbick (1931) made pioneering contributions to this discipline.

In this chapter we study the trajectories of charged particles under the influence of uniform electric/magnetic fields (Sections 13.2–13.4). This is followed by a discussion of motion in non-uniform fields (Section 13.5). The application of this knowledge to engineering practices is discussed in the last section.

13.2 MOTION UNDER A FORCE

The study of motion of a body under the influence of a certain force generally follows the route

$$\text{force} \longrightarrow \text{acceleration} \longrightarrow \text{velocity} \longrightarrow \text{displacement}$$

The main aim of this study is to predict the position of a body as a function of time, called the path (or trajectory) of a body, under the action of a given force. The change in energy (kinetic or potential) of a body during its motion is an important aspect of this study. In analysing the motion of a body one has to be careful regarding

1. The number of dimensions in which motion takes palce
2. The vector nature of force and related quantities,
3. The scalar nature of energy.

A review of the various parameters of interest is made in Table 13.1.

It can be seen from this table that for a constant force, the acceleration is constant and thus can be taken out of the integral sign in Eqs. 13.5 and 13.6. These equations thus reduce to the following familiar Newton's equations of motion.

$$\mathbf{v} = \mathbf{v}_o + \mathbf{a}t \tag{13.14}$$

$$\mathbf{r} = \mathbf{v}_o t + \frac{1}{2}\mathbf{a}t^2 \tag{13.15}$$

It should be kept in mind that Eqs. 13.14 and 13.15 hold good only for constant forces, and if the force is not constant, then we have to use the more general Eqs. 13.5 and 13.6 for determination of velocity and displacement. Example 13.1 makes this point clear.

Apart from those parameters considered in Table 13.1, another important parameter is *field.* Field is the *cause* of the force. In Table 13.2 the main features of electric field and magnetic field are discussed. Gravitational field is also included in this table to underline the similarity of treatment. It should be noted that all these are vector fields (distribution of vectors), whereas distribution of temperature or pressure are examples of scalar fields.

Table 13.2 brings out an interesting and unifying aspect. *A gravitational field arises from a mass and exerts a force on any mass. An electric field arises from a charge and exerts a force on any charge. A magnetic field arises from a current, that is, a moving charge, and exerts a force on any current, that is, any charge in motion.*

EXAMPLE 13.1
Electron in an electric field

Consider a parallel plate capacitor whose plates are 5 cm apart as shown in Fig.13.1. An electron starts at rest on one plate of the capacitor. The applied voltage is zero at the instant the electron is released, and it increases linearly from 0 to 15 V in 0.15 μs. (a) If the opposite plate is positive, what speed will the electron attain in 50 ns? (b) What distance will it cover at the end of this time? (c) With what speed will the electron strike the positive plate?

Table 13.1 Various parameters of interest

Parameter	Definition	Expression
Force	An interaction that brings about acceleration of a body. For example, the earth's gravity brings about an acceleration g in a body. Force is also defined as the negative rate of change of potential energy, U.	$F = mg$ (13.1) m is the mass of the body $F = -\nabla U\ (= -dU/dx$ for 1D motion$)$ (13.2)
Accele-ration	Rate of change of velocity with time	$\mathbf{a} = d\mathbf{v}/dt = \mathbf{F}/m$ (13.3)
Velocity	Rate of change of position with time	$\mathbf{v} = d\mathbf{r}/dt$ (13.4) Using Eq. 13.3, $\mathbf{v} = \mathbf{v}_0 + \int \mathbf{a}dt$ where $\mathbf{v}_0 = \mathbf{v}(0)$, the initial velocity (13.5)
Dis-placement	Change from one position to another position	Using Eq. 13.4, $\mathbf{r} = \mathbf{r}_0 + \int \mathbf{v}\, dt$ (13.6) where $\mathbf{r}_0 = \mathbf{r}(0)$, the initial position
Energy	**Potential energy**: Energy due to position, that is, positional or configurational energy means hidden/stored energy **Kinetic energy**: Energy due to motion, that is motional energy means manifestable/usable/expendible energy Change in both energies is related to work done. **Potential energy**: When the work is done by the body against the force, its potential energy increases and kinetic energy decreases. It results in negative work. **Kinetic energy**: When the work is done by the force on the body, its kinetic energy increases and potential energy decreases. It results in positive work.	The work done when a force $\mathbf{F}$ displaces a body through an infinatesimal displacement $d\mathbf{r}$ is $dW = \mathbf{F} \cdot d\mathbf{r}$ (13.7) For $\mathbf{F} = F_x\mathbf{i} + F_y\mathbf{j} + F_z\mathbf{k}$ and $dr = dx\mathbf{i} + dy\mathbf{j} + dz\mathbf{k}$, $dW = F_x\, dx + F_y\, dy + F_z\, dz.$ (13.8) For motion along one dimension the work done in displacing a body from point 1 to point 2 is $W = \int_1^2 F_x\, dx$ (13.9) Making two different substitutions for F_x from Eqs. 13.2 and 13.3 $W = \int_1^2 -(dU/dx)dx = -\int_1^2 dU$ $= U1 - U2 = -\Delta U$ (13.10) $W = \int_1^2 m\,(dv/dt)dx = -m \int_1^2 (dx/dt)dv = m \int_1^2 v\, dv$ $= \frac{1}{2} mv_2^2 - \frac{1}{2} mv_1^2 = K_2 - K_1 = \Delta K$ (13.11) From eqs. (13.10) and (13.11), we get, $\Delta K = -\Delta U$ (13.12) or $K_1 + U_1 = K_2 + U_2$ (13.13)
	Equation 13.12 suggests that an increase in kinetic energy means decrease in potential energy and vice versa. Equation 13.13 is a statement of conservation of energy. It suggests that the total energy of a body always remains constant. However, it may alternate between its two forms, that is, kinetic and potential.	

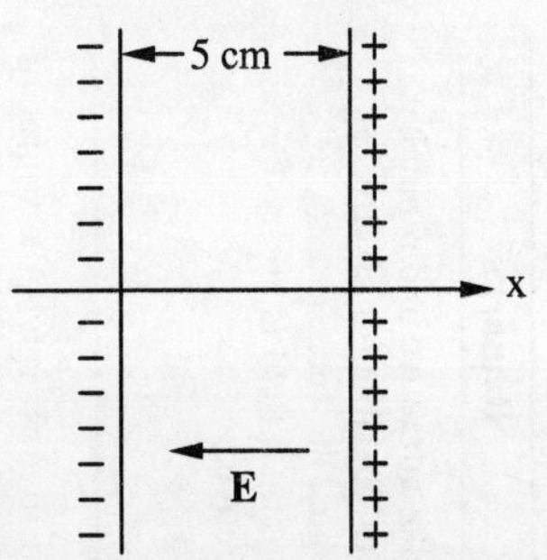

Figure 13.1

Solution

Here the electric field is not constant but is dependent on time.

At $t = 0$, V = 0 or $E = 0$ V/m

and at $t = 0.15$ μs, V = 15 V or $E = \dfrac{15\text{ V}}{(5 \times 10^{-2}\text{ m})} = 300$ V/m.

Thus the linear equation between E and t can be written as

$$(E - 0) = \frac{(300\text{ V/m}) - (0\text{ V/m})}{0.15\ \mu\text{s} - 0\ \mu\text{s}} \times (t - 0)$$

or $\quad E = 2 \times 10^9 \times t$ V/m

$$\text{Thus, } a_x = \frac{eE}{m}$$

$$= (1.76 \times 10^{11}\text{ C/kg})(2 \times 10^9 \times t\text{ V/m})$$

$$= 3.52 \times 10^{20} \times t\text{ m/s}^2$$

(a) $\quad v_x = \displaystyle\int a_x\, dt$

$$= 3.52 \times 10^{20} \int t\, dt$$

$$= 1.76 \times 10^{20} \times t^2\text{m/s}$$

Thus at $t = 50$ ns, the electron will attain the speed

$$v_x = (1.76 \times 10^{20})(50 \times 10^{-9})^2 = 4.4 \times 10^5\text{ m/s}$$

(b) Distance covered at the end of time t, $x = \displaystyle\int v_x dt = \int 1.76 \times 10^{20} \times t^2 dt$

$$= 5.87 \times 10^{19} \times t^3\text{m}$$

At $t = 50$ ns, $x = (5.87 \times 10^{19})(50 \times 10^{-9})^3$ m

$$= 7.34 \times 10^{-3}\text{ m} = 7.34\text{ mm}$$

(c) As $x = 5.87 \times 10^{19} \times t^3$, the time required to reach the plate is

$$t = \left[\frac{x}{5.87 \times 10^{19}}\right]^{1/3} = \left[\frac{5 \times 10^{-2}}{5.87 \times 10^{19}}\right]^{1/3}$$

$$= 0.95 \times 10^{-7}\text{s}$$

Table 13.2 Fields: Gravitational, electric and magnetic

Field	Cause	Effect	Strength of the field	Remarks
Gravitational field	Mass	The field caused by a mass $m_1 = -Gm_1/r^2$ exerts a force $F = m_2\,(-Gm_1/r^2) = -Gm_1m_2/r^2$ (1) on mass m_2. G is the gravitional constant and r is the distance between two masses.	The force experienced by a unit mass placed at the point at which the field is to be evaluated.	Unlike attractive and repulsive forces possible for electric charges, the force due to gravitional field is always attractive.
Electric field	Electric charge	The field caused by a charge $q_1 = (1/4\pi\varepsilon_0)q_1/r^2$ exerts a force $F = q_2\,(1/4\,\pi\varepsilon_0)\,q_1/r^2 = (1/4\pi\varepsilon_0)q_1q_2/r^2$ (2) on charge q_2 in vacuum. ε_0 is the permittivity of free space and r is the distance between two charges.	The force experienced by a unit positive charge placed at the point at which the field is to be evaluated. Electric field is also defined as the maximum rate of decrease of potential, $\mathbf{E} = -dV/d\mathbf{r}$ or the voltage applied per unit length (see example 13.2).	A positive charge is the source of an electric field (that is, field lines emanate from it) and a negative charge is the sink (that is, field lines terminate at it). Field lines are crowded for stronger fields and are rarer for weaker fields.
Magnetic field	Moving charge (current)	The field caused by current i_1 flowing through conductor 1 $= \mu_0 i_1/2\pi d$, exerts a force $F = i_2(\mu_0 i_1/2\pi d) = (\mu_0 i_1 i_2/2\pi d)$ (3) on another conductor 2 placed parallel to conductor 1 and through which the current i_2 is flowing. μ_0 is the permeability of free space and d is the distance between two conductors.	It could be defined as the force experienced by the unit north pole, but since magnetic monopoles do not exist, it is defined alternatively as a vector along the zero force axis whose magnitude is given by F/qv (see Eq. 13.17). The north pole of a bar magnet is the source of magnetic field lines and the south pole is the sink of field lines. Field lines travel from north to south outside the magnet and from south to north inside the	

* This expression is derived for an infinite wire using a more basic expression which follows from Biot–Savart law. According to it the force experienced by a current element $i_2\,\mathbf{dl}_2$ at $\mathbf{r}_2$ due to the current element $i_1\,\mathbf{dl}_1$ at $\mathbf{r}_1$ is $i_2\,dl_2 \times [(\mu_0/4\pi)\,i_1\,\mathbf{dl}_1 \times (\mathbf{r}_2 - \mathbf{r}_1)/|\mathbf{r}_2 - \mathbf{r}_1|^3]$

Thus, the speed with which the electron will strike the positive plate,

$$v_x = 1.76 \times 10^{20} \times t^2$$

$$= (1.76 \times 10^{20})(0.95 \times 10^{-7})^2$$

$$= 1.59 \times 10^6 \text{ m/s}$$

EXAMPLE 13.2

For constant $E, E = V/d$

Show that the electric field between two plates of a capacitor separated by distance d and having a potential difference of V volts is given by $E = V/d$.

Solution

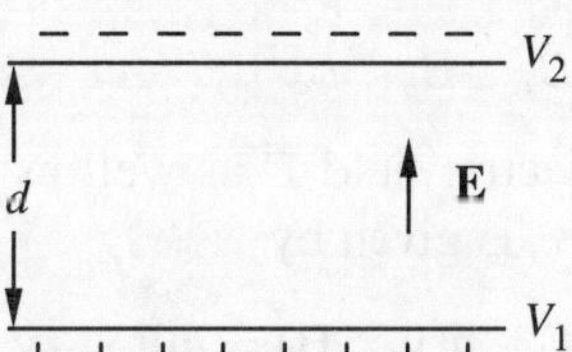

Figure 13.2 **E** and **dr** are along the same direction

Change in potential is negative of the work done by **E** on the unit positive charge.

Thus, $dV = -\mathbf{E} \cdot \mathbf{dr}$

and $\int dV = \int -\mathbf{E} \cdot \mathbf{dr}$

Choosing the path of integration along the electric field lines from positive plate to the negative plate (Fig.13.2), **E** and **dr** are along the same direction.

Hence $\mathbf{E} \cdot \mathbf{dr} = E \times dr$ and $V_2 - V_1 = -V$,

V being the absolute value of potential difference.

Thus, $-V = -E \int dr$ (E is taken out of the integral sign as it is constant.)

Hence, $E = V/d$.

13.3 MOTION OF CHARGED PARTICLES IN ELECTRIC AND MAGNETIC FIELDS

In the last section a general procedure for working out the motion of a body under a given force was studied. Now we apply this procedure to study the motion of charged particles in electric and magnetic field. The first question to be answered is, 'what is the force?'

The force experienced by a particle of charge q in an electric field of strength **E** is given by

$$\mathbf{F} = q\mathbf{E} \tag{13.16}$$

For a unit positive charge, $\mathbf{F} = \mathbf{E}$. Thus Eq. 13.16 is actually a restatement of the definition of electric field strength given in Table 13.2. In case of a magnetic field the charge q experiences a force given by

$$\mathbf{F} = q(\mathbf{v} \times \mathbf{B}) \tag{13.17}$$

where $\mathbf{v}$ is the velocity of the charged particle and $\mathbf{B}$ is the strength of the magnetic field, also called as magnetic flux density, magnetic induction or simply magnetic field. It is obvious from Eq. 13.17 that the magnetic field does not affect static charges. Using the definition of cross product, Eq. 13.17 can be expressed as

$$\mathbf{F} = qvB\sin[\angle(\mathbf{v}, \mathbf{B})]\mathbf{e} \tag{13.18}$$

$\mathbf{e}$ being the unit vector in the direction of force.

$$\mathbf{F} = \begin{vmatrix} \mathbf{i} & \mathbf{j} & \mathbf{k} \\ v_{\mathrm{x}} & v_{\mathrm{y}} & v_{\mathrm{z}} \\ B_x & B_y & B_z \end{vmatrix} \tag{13.19}$$

When an electric field $\mathbf{E}$ as well as magnetic field $\mathbf{B}$ are both present, the net force, called the *Lorentz force*, is given by

$$\mathbf{F} = q\mathbf{E} + q(\mathbf{v} \times \mathbf{B}) = q[\mathbf{E} + (\mathbf{v} \times \mathbf{B})] \tag{13.20}$$

Now we can proceed to find the acceleration, velocity, displacement and finally the trajectory of motion. We do this in a comparative manner in Chart 13.1A. This chart also includes the motion of a body (projectile) under a gravitational field. This case is included to facilitate a wider comparison and also because the motion of a projectile is covered in preliminary classes. In Chart 13.1B, we make a comparative study of energy in various fields.

In these charts it is assumed that the fields are uniform. This makes the analysis simple for electric and gravitational fields. However, for a magnetic field, such simplification is not possible because the force is velocity dependent and even with uniform $\mathbf{B}$, the force changes due to change in velocity. For non-uniform fields, the motion is complex and can be worked out approximately by incorporating the details of non-uniformity.

From Chart 13.1A we see that *a charged particle generally follows a parabolic path in an electric field and a helical path in a magnetic field.* The trajectory specialises to a straight line in an electric field and to a point circle or a straight line in a magnetic field, depending on the angle of projection.

It is interesting to note that trajectories of a charged particle in a uniform electric field are exactly like those of a mass in uniform gravitational field. The reason is, of course, that in both cases the force on the moving object is constant. The only difference is that acceleration in the electric field depends on the mass, whereas in the gravitational case, due to equality of inertial and gravitational masses it depends only on the gravitational field. So, in a given gravitational field, all masses have the same acceleration, whereas in a given electric field all charges do not have the same acceleration, as the acceleration depends on the masses of particles carrying those charges. Thus, an electron and a proton both experience forces of the same magnitude—$e\mathbf{E}$—in an electric field $\mathbf{E}$, but their accelerations are vastly different as the proton is about 2000 times heavier than an electron.

CHART 13.1A MOTION UNDER VARIOUS FIELDS

Field	Gravitational field	Electric field	Magnetic field
Initial considerations	A body of mass m is projected with initial velocity **u** making an angle α with the y-axis. Thus components of initial velocity are $u_y = u\cos\alpha$ and $u_z = u\sin\alpha$. Components of initial displacement are (x_0, y_0, z_0). The gravitational field is along the negative z-direction.	A particle of charge q is projected with initial velocity **u** making an angle α with the y-axis. Thus components of initial velocity are $u_y = u\cos\alpha$ and $u_z = u\sin\alpha$. Components of initial displacement are (x_0, y_0, z_0). The electric field is along the negative z-direction. Here we consider only the magnitude of the electric field.	A particle of charge q is projected with initial velocity **u** making an angle α with the z-axis. Thus components of initial velocity are $u_y = u\sin\alpha$ and $u_z = u\cos\alpha$. Components of initial displacement are (x_0, y_0, z_0). The magnetic field is along the negative z-direction.
Force	$F_x = 0, F_y = 0, F_z = -mg$ (1)	$F_x = 0$ $F_y = 0$ $F_z = qE$ (9)	Using Eq. 13.19, $F_x = q(\mathbf{v} \times \mathbf{B})_x = -qv_yB$ $F_y = q(\mathbf{v} \times \mathbf{B})_y = qv_xB$ $F_z = q(\mathbf{v} \times \mathbf{B})_z = 0$ (17)
Newton's equation	$mdv_x/dt = 0, mdv_y/dt = 0,$ $mdv_z/dt = -mg$ (2)	$mdv_x/dt = 0$ $mdv_y/dt = 0$ $mdv_z/dt = qE$ (10)	$mdv_x/dt = -qv_yB$ $mdv_y/dt = qv_xB$ $mdv_z/dt = 0$ (18)
Acceleration	$a_x = 0, a_y = 0, a_z = -g$ (3)	$a_x = 0, a_y = 0, a_z = qE/m$ (11)	$a_x = -qv_yB/m, a_y = qv_xB/m, a_z = 0$ (19)
Velocity	$v_x = 0, v_y = u_y, v_z = u_z - gt$ (4) (Velocities along the x- and y-directions are unchanged as there is no force along these directions.)	$v_x = 0, v_y = u_y, v_z = u_z + (qE/m)t$ (12) (Velocities along the x- and y-directions are unchanged as there is no force along these directions.)	$v_x = -u_y \sin\omega t, v_y = u_y \cos\omega t, v_z = u_z,$ where $\omega = qB/m$ (20) (Velocity along the z-direction is unchanged as the force along this direction is zero.) See Example 13.3.
Displacement	$x = x_0, y = y_0 + u_y t$ $z = z_0 + u_z t - (1/2)gt^2$ or $x = x_0, y = y_0 + (u\cos\alpha)t$ $z = z_0 + (u\sin\alpha)t - (1/2)gt^2$ (5)	$x = x_0, y = y_0 + u_y t, z = z_0 + u_z t + (1/2)(qE/m)t^2$ or $x = x_0, y = y_0 + (u\cos\alpha)t,$ $z = z_0 + (u\sin\alpha)t + ½(qE/m)t^2$ (13)	$x = x_0 + (u_y/\omega)(\cos\omega t - 1), y = y_0 + (u_y/\omega)$ $\sin\omega t, z = z_0 + u_z t$ or $x = x_0 + (u\sin\alpha/\omega)$ $(\cos\omega t - 1), y = y_0 + (u\sin\alpha/\omega)\sin\omega t,$ $z = z_0 + u\cos\alpha t$ (21)

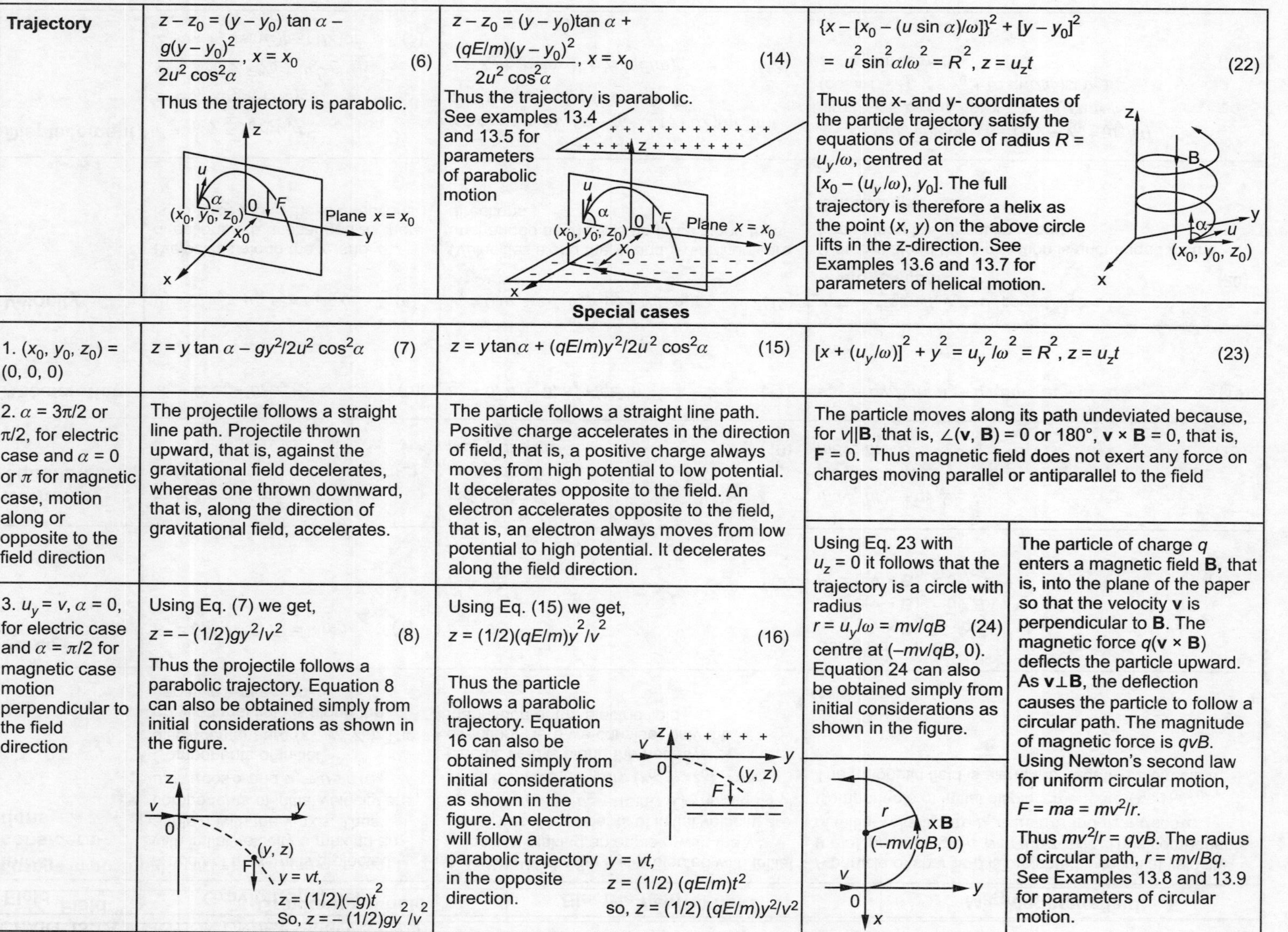

Trajectory	$z - z_0 = (y - y_0)\tan\alpha - \dfrac{g(y-y_0)^2}{2u^2\cos^2\alpha}$, $x = x_0$ (6) Thus the trajectory is parabolic.	$z - z_0 = (y - y_0)\tan\alpha + \dfrac{(qE/m)(y-y_0)^2}{2u^2\cos^2\alpha}$, $x = x_0$ (14) Thus the trajectory is parabolic. See examples 13.4 and 13.5 for parameters of parabolic motion	$\{x - [x_0 - (u\sin\alpha)/\omega]\}^2 + [y - y_0]^2 = u^2\sin^2\alpha/\omega^2 = R^2$, $z = u_z t$ (22) Thus the x- and y- coordinates of the particle trajectory satisfy the equations of a circle of radius $R = u_y/\omega$, centred at $[x_0 - (u_y/\omega), y_0]$. The full trajectory is therefore a helix as the point (x, y) on the above circle lifts in the z-direction. See Examples 13.6 and 13.7 for parameters of helical motion.
		Special cases	
1. $(x_0, y_0, z_0) = (0, 0, 0)$	$z = y\tan\alpha - gy^2/2u^2\cos^2\alpha$ (7)	$z = y\tan\alpha + (qE/m)y^2/2u^2\cos^2\alpha$ (15)	$[x + (u_y/\omega)]^2 + y^2 = u_y^2/\omega^2 = R^2$, $z = u_z t$ (23)
2. $\alpha = 3\pi/2$ or $\pi/2$, for electric case and $\alpha = 0$ or π for magnetic case, motion along or opposite to the field direction	The projectile follows a straight line path. Projectile thrown upward, that is, against the gravitational field decelerates, whereas one thrown downward, that is, along the direction of gravitational field, accelerates.	The particle follows a straight line path. Positive charge accelerates in the direction of field, that is, a positive charge always moves from high potential to low potential. It decelerates opposite to the field. An electron accelerates opposite to the field, that is, an electron always moves from low potential to high potential. It decelerates along the field direction.	The particle moves along its path undeviated because, for $v \parallel$ **B**, that is, $\angle$(**v**, **B**) = 0 or 180°, **v** × **B** = 0, that is, **F** = 0. Thus magnetic field does not exert any force on charges moving parallel or antiparallel to the field
3. $u_y = v$, $\alpha = 0$, for electric case and $\alpha = \pi/2$ for magnetic case motion perpendicular to the field direction	Using Eq. (7) we get, $z = -(1/2)gy^2/v^2$ (8) Thus the projectile follows a parabolic trajectory. Equation 8 can also be obtained simply from initial considerations as shown in the figure. $y = vt$, $z = (1/2)(-g)t^2$ So, $z = -(1/2)gy^2/v^2$	Using Eq. (15) we get, $z = (1/2)(qE/m)y^2/v^2$ (16) Thus the particle follows a parabolic trajectory. Equation 16 can also be obtained simply from initial considerations as shown in the figure. An electron will follow a parabolic trajectory in the opposite direction. $y = vt$, $z = (1/2)(qE/m)t^2$ so, $z = (1/2)(qE/m)y^2/v^2$	Using Eq. 23 with $u_z = 0$ it follows that the trajectory is a circle with radius $r = u_y/\omega = mv/qB$ (24) centre at $(-mv/qB, 0)$. Equation 24 can also be obtained simply from initial considerations as shown in the figure. The particle of charge q enters a magnetic field **B**, that is, into the plane of the paper so that the velocity **v** is perpendicular to **B**. The magnetic force q(**v** × **B**) deflects the particle upward. As **v**⊥**B**, the deflection causes the particle to follow a circular path. The magnitude of magnetic force is qvB. Using Newton's second law for uniform circular motion, $F = ma = mv^2/r$. Thus $mv^2/r = qvB$. The radius of circular path, $r = mv/Bq$. See Examples 13.8 and 13.9 for parameters of circular motion.

CHART 13.1B ENERGY IN VARIOUS FIELDS

Field	Gravitational field	Electric field	Magnetic field
Change in potential energy $dU = -\int_1^2 F_z\, dz$	$dU = -\int_1^2 (-mg)dz = mg \int_1^2 dz$ or $dU = mg\,(h_2 - h_1)$ (1) where h_1 and h_2 correspond to the values of z (that is, height) at points 1 and 2 respectively.	$dU = -\int_1^2 F_z\, dz = -\int_1^2 qE\, dz$ $= -\int_1^2 q(-dV/dz)dz = q\int_1^2 dV$ or $dU = q(V_2 - V_1)$ (4) where V_2 is the potential and qV_2 is the potential energy at point 2. V_1 and qV_1 are the respective quantities at point 1.	Using Eq.13.7 we have, $dW = \mathbf{F}.d\mathbf{r} = q(\mathbf{v} \times \mathbf{B}).d\mathbf{r} = q(\mathbf{v} \times \mathbf{B}).\mathbf{v}dt.$ (11) As the cross product **v** × **B** is perpendicular to both **v** and **B**, its scalar product with **v**dt is zero. Hence $dW = 0$. Thus magnetic field does no work on moving charges.
Energy conservation equation $\Delta K = -\Delta U$ or $(1/2)mv_1^2 + U_1$ $= (1/2)mv_2^2 + U_2$	$(1/2)mv_2^2 - (1/2)mv_1^2 = -mg\,(h_2 - h_1)$ or $(1/2)mv_1^2 + mgh_1 = (1/2)mv_2^2 + mgh_2$ (2)	$(1/2)mv_2^2 - (1/2)mv_1^2 = -q(V_2 - V_1)$ or $(1/2)mv_1^2 + qV1 = (1/2)mv_2^2 + qV_2$ (5) For an electron, $(1/2)mv_1^2 - eV_1 = (1/2)mv_2^2 - eV_2$ (6)	$(1/2)mv_2^2 - (1/2)mv_1^2 = 0$ (12) Thus, the kinetic energy of a charged particle remains constant when it moves through a magnetic field. The magnetic field hence changes the velocity of a charged particle without changing its speed. The magnetic field only alters the direction of a charged particle while making it move in a circular orbit. Thus circular motion is an accelerated motion performed with a constant speed.
Special cases	for $h_1 = h$, $v_1 = 0$, $h_2 = 0$ and $v_2 = v$, we get $(1/2)mv^2 = mgh$ (3) So $v = \sqrt{2gh}$. Thus the projectile dropped from height h gains the speed $\sqrt{2gh}$.	For $V_1 = 0$, $v_1 = v$, $V_2 = V$ and $v_2 = 0$ in Eq. 5, we get $(1/2)mv^2 = qV$ (7) Thus $v = \sqrt{2qV/m}$ (8) For an electron, substituting $V_1 = 0$, $v_1 = 0$, $V_2 = V$ and $v_2 = v$ in Eq. 6, we get $(1/2)mv^2 = eV$ (9) Substituting $e/m = 1.76 \times 10^{11}$ C/kg, we get, $v = 5.94 \times 10^5\sqrt{V}$ (10) Thus an electron accelerated through a potential difference of V volts gains a speed equal to $5.94 \times 10^5\sqrt{V}$ m/s.	

In Chart 13.1B also, we note similar features regarding the energy in a gravitational field and an electric field. In both cases an object experiences a constant force and hence the work done or the change in kinetic energy is directly proportional to the displacement of the particle in the direction of the field. However, in a magnetic field the situation is very different. Magnetic field does no work on the particle as magnetic force $qV \times B$ is always perpendicular to velocity and instantaneous displacement of the particle.

EXAMPLE 13.3

Velocity and displacement in a magnetic field

Determine the expressions for velocity and displacement if the components of force due to magnetic field are $F_x = -qv_yB$ and $F_y = qv_xB$.

Solution

$$F_x = \frac{mdv_x}{dt} = -qv_yB \tag{13.21}$$

$$F_y = \frac{mdv_y}{dt} = qv_xB \tag{13.22}$$

Equations 13.21 and 13.22 are coupled equations. We decouple them by differentiating Eq. 13.22 with time and using Eq. 13.21. Thus,

$$\frac{d^2v_y}{dt^2} = \frac{qB}{m} \times \frac{dv_x}{dt} = \frac{qB}{m}\left(\frac{-qv_yB}{m}\right) = -\frac{q^2B^2v_y}{m^2}$$

or $$\frac{d^2v_y}{dt^2} = -\omega^2 v_y, \text{ where } \omega = \frac{qB}{m} \tag{13.23}$$

The solution of Eq. 13.23 is $v_y = A\sin\omega t$ or $v_y = A\cos\omega t$. At $t = 0, v_y = u_y$, the y component of initial velocity. Thus the first solution is not possible. Using the initial condition in the second solution we get $A = u_y$.

Thus, $v_y = u_y \cos\omega t$ (13.24)

Using Eq.(13.21), $\frac{dv_x}{dt} = -\omega u_y \cos\omega t$

Integrating we get,

$$v_x = -u_y \sin\omega t \tag{13.25}$$

Equations 13.24 and 13.25 are the expressions for velocity components. To determine the expressions for components of displacement, we integrate these equations. Thus we get,

$$x = x_0 + (u_y/\omega)(\cos\omega t - 1) \quad \text{and} \quad y = y_0 + (u_y/\omega)\sin\omega t,$$

where x_0 and y_0 are the components of initial displacement.

EXAMPLE 13.4
Parameters of parabolic motion

Find the coordinates of the apex and the focus of a parabolic trajectory followed by a charged particle in an electric field.

Solution

The equation of the trajectory of a charged particle in an electric field is (Eq. 14 of Chart 13.1A)

$$z - z_0 = (y - y_0)\tan\alpha + \left(\frac{qE}{m}\right)\frac{(y - y_0)^2}{2u^2\cos^2\alpha},\ x = x_0$$

To find the coordinates of the apex and the focus we put the above equation in the standard form. Thus,

$$(y - y_0)^2 + \frac{2mu^2\cos^2\alpha}{qE}(y - y_0)\tan\alpha = \frac{2mu^2\cos^2\alpha}{qE}(z - z_0)$$

or $$(y - y_0)^2 + \frac{mu^2\sin 2\alpha}{qE}(y - y_0) = \frac{2mu^2\cos^2\alpha}{qE}(z - z_0)$$

or $$\left[(y - y_0) + \frac{mu^2\sin 2\alpha}{2qE}\right]^2 = \frac{2mu^2\cos^2\alpha}{qE}(z - z_0) + \frac{m^2u^4\sin^2 2\alpha}{4q^2E^2}$$

or $$\left[(y - y_0) + \frac{mu^2\sin 2\alpha}{2qE}\right]^2 = \frac{2mu^2\cos^2\alpha}{qE}\left[(z - z_0) + \frac{mu^2\sin^2\alpha}{2qE}\right]$$

Comparing this equation with the standard equation,

$$(y - y')^2 = 4A(z - z')$$

Coordinates of the apex are

$$(x_0, y', z') = \left[x_0, y_0 - \frac{mu^2\sin 2\alpha}{2qE}, z_0 - \frac{mu^2\sin^2\alpha}{2qE}\right]$$

Coordinates of focus are (Fig.13.3.)

$$(x_0, y', z' + A) = \left[x_0, y_0 - \frac{mu^2\sin 2\alpha}{2qE}, z_0 - \frac{mu^2\sin^2\alpha}{2qE} + \frac{mu^2\cos^2\alpha}{2qE}\right]$$

$$= \left[x_0, y_0 - \frac{mu^2\sin 2\alpha}{2qE}, z_0 + \frac{mu^2\cos 2\alpha}{2qE}\right]$$

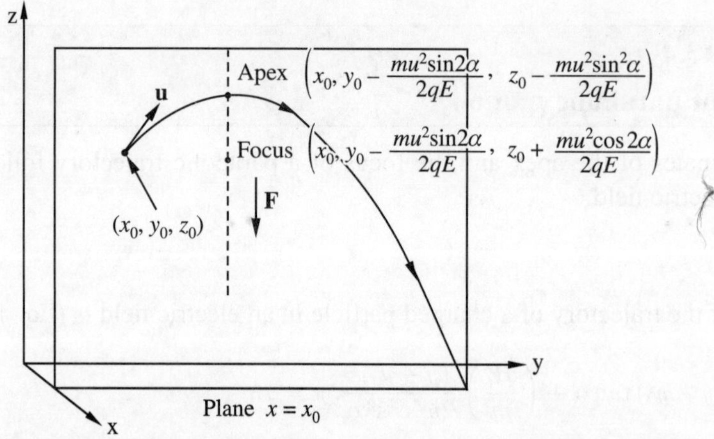

Figure 13.3 Parameters of parabolic trajectory in an electric field

EXAMPLE 13.5
Maximum height, time of flight and horizontal range

1. For a particle of charge q entering a region of uniform electric field E, at an angle α with the direction perpendicular to the field, find expressions for (a) maximum penetration against the direction of force, (b) the time of flight, that is, the time required to hit the direction perpendicular to the field and (c) the distance covered along this direction in the time of flight, called the horizontal range.
2. Find these parameters for an electron projected at an angle of 35° to the horizontal at an initial speed of 5×10^5 m/s in a region of uniform electric field of intensity 200 V/m oriented vertically upward.

Solution

Figure 13.4 shows the parameters maximum penetration, $z_{\max}$, and horizontal range, H, for trajectories of a positive charge ($q > 0$) and a negative charge ($q < 0$).

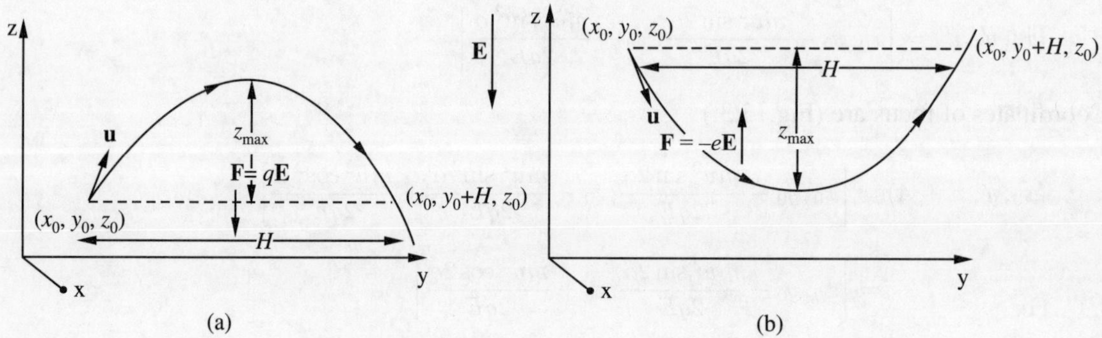

Figure 13.4 Trajectory of a particle of charge (a) $q > 0$ and (b) $q < 0$

(a) At the point of maximum penetration against the force direction, the velocity $v_z = 0$. (see Eq. 12, Chart 13.1A)

$$u\sin\alpha + \frac{qEt}{m} = 0$$

$$t = -\frac{mu\sin\alpha}{qE} \tag{13.26}$$

This is the time required to reach the point of maximum penetration against the force direction.

Now the expression for the distance covered z (Eq. 13, Chart 13.1A) is

$$z = z_0 + (u\sin\alpha)t + \frac{qEt^2}{2m} \tag{13.27}$$

Substituting for t from Eq. 13.26 we get,

$$z_{\max} = z_0 + u\sin\alpha\left(\frac{-mu\sin\alpha}{qE}\right) + \left(\frac{qE}{2m}\right)\left(\frac{-mu\sin\alpha}{qE}\right)^2$$

$$= z_0 - \frac{1}{2}\frac{m}{qE}u^2\sin^2\alpha$$

Thus the maximum penetration attained against the force direction

$$z_{\max} = z_0 - \frac{1}{2}\frac{u^2\sin^2\alpha}{a}, \text{ where } a = \frac{qE}{m} \tag{13.28}$$

This result can also be obtained by differentiating Eq. 13.27 with respect to t and setting it equal to zero. The resulting expression for t is then substituted in Eq. 13.27 to yield the expression for $z_{\max}$.

(b) For finding the time of flight, $z = z_0$. From Eq. 13.27 we get the time of flight

$$T = \frac{-2u\sin\alpha}{a}$$

Thus the time of flight is twice the time required to attain $z_{\max}$.

(c) Horizontal range,

$$H = u\cos\alpha \times T$$

$$= u\cos\alpha \times -\frac{2u\sin\alpha}{a} = \frac{-u^2\sin 2\alpha}{a}$$

For $u = 5 \times 10^5$ m/s, $\alpha = 35°$, and $E = 200$ V/m

Thus, $a = \dfrac{-eE}{m}$

$$= -1.76 \times 10^{11} \text{ C/kg} \times 200 \text{ V/m}$$

$$= -3.5 \times 10^{13} \text{ m/s}^2$$

(a) Maximum penetration against the force direction assuming $z_0 = 0$,

$$z_{\max} = -\frac{(5 \times 10^5 \text{ m/s})^2 \sin^2 35°}{2 \times -3.5 \times 10^{13} \text{ m/s}^2}$$

$$= 1.2 \text{ mm}$$

(b) Time of flight, $T = \dfrac{-2 \times 5 \times 10^5 \text{ m/s} \times \sin 35°}{-3.5 \times 10^{13} \text{ m/s}^2}$

$$= 1.64 \times 10^{-8} \text{s}$$

(c) Horizontal range, $H = \dfrac{-(5 \times 10^5 \text{m/s})^2 \sin 70°}{-3.5 \times 10^{13} \text{ m/s}^2}$

$$= 6.7 \times 10^{-3} \text{ m} = 6.7 \text{ mm}$$

EXAMPLE 13.6
Parameters of helical motion

Find the parameters of helical trajectory followed by a charged particle in a magnetic field, governed by Eq. 22 of Chart 13.1A.

Solution

Equation 22 of Chart 13.1A is

$$\left[x - \left(x_0 - \frac{u\sin\alpha}{\omega}\right)\right]^2 + [y - y_0]^2 = \frac{u^2 \sin^2 \alpha}{\omega^2}$$

$$= R^2, \; z = u_z t$$

and the corresponding helical trajectory is shown in Fig. 13.5. The parameters of the helix are its radius and pitch. The radius is given by

$$r = \frac{u\sin\alpha}{\omega} = \frac{mu\sin\alpha}{qB}$$

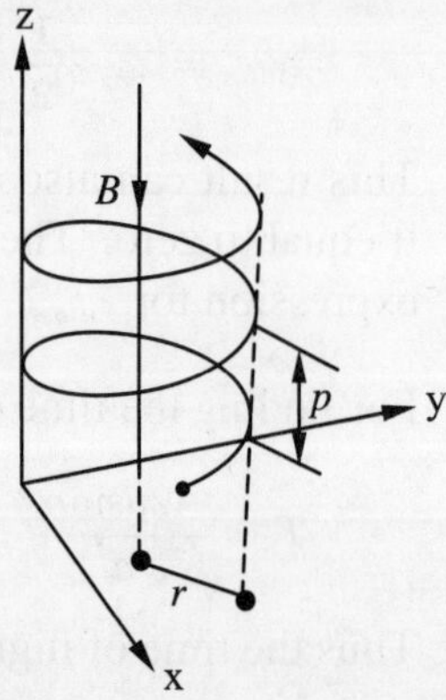

Figure 13.5 Parameters of helical trajectory in a magnetic field

The pitch is defined as the distance travelled along the direction of the magnetic field in one revolution. As the time period of revolution is $T = 2\pi m/Bq$ and the velocity along the direction of magnetic field is $u \cos \alpha$, the pitch is given by $p = (2\pi m/Bq)u\cos\alpha$.

EXAMPLE 13.7

Time period and pitch of the helical path of an electron

An electron with a speed of 5×10^7 m/s enters a uniform magnetic field at an angle of 30° with it. Calculate (a) the magnetic flux density required to make the electron follow a helical path of 0.5 m diameter, (b) the time taken by the electron for completing one revolution and (c) the pitch of the helix.

Solution

(a) The radius of the helix is given by

$$r = \frac{musin\alpha}{Be}$$

$$\text{So, } B = \frac{musin\alpha}{er} = \frac{5 \times 10^7 \text{ m/s} \times \sin 30°}{1.76 \times 10^{11} \text{ C/kg} \times 0.25 \text{ m}}$$

$$\text{Thus, } B = 0.57 \times 10^{-3} \text{ Wb/m}^2$$

(b) The time to complete one revolution,

$$T = \frac{2\pi \text{m}}{Be} = \frac{2\pi}{0.57 \times 10^{-3} \text{ Wb/m}^2 \times 1.76 \times 10^{11} \text{ C/kg}}$$

$$= 6.26 \times 10^{-8} \text{ s} = 62.6 \text{ ns}$$

(c) The pitch of the helix is given by

$$p = Tucos\alpha = 62.6 \times 10^{-9} \text{ s} \times 5 \times 10^7 \text{ m/s} \times 0.866 = 2.71 \text{ m}$$

EXAMPLE 13.8

Parameters of circular orbit

Find the various parameters of circular motion of a charged particle in a magnetic field. Bring out the significance of the expressions for T, ω and v.

Solution

The radius of a circular path in a magnetic field is given by $r = \dfrac{mv}{Bq}$, so, $\dfrac{v}{r} = \dfrac{Bq}{m}$

But $v/r = \omega$, the angular frequency of the motion.

Hence $\omega = \dfrac{Bq}{m}$ or $\nu = \dfrac{Bq}{2\pi m}$ (because $\omega = 2\pi\nu$)

Hence the time period of motion is $T = \dfrac{1}{\nu} = \dfrac{2\pi m}{Bq}$.

Thus the quantities T, ν and ω do not depend on the speed of a particle. This means faster particles move in larger circles and the slower ones in smaller circles so that they take same time in completing their orbits. This may be contrasted with the orbital motion of planets. In planetary motion, if the object goes nearer a massive centre of gravitational attraction, the smaller radius orbits have larger orbit speeds.

EXAMPLE 13.9

Handy formula for time period of an electron's orbit in a magnetic field

Show that for an electron moving in a uniform magnetic field B, the time period of circular orbit is given by $T = 3.57 \times 10^{-11}/B$.

Solution

For an electron the time period of circular orbit in a magnetic field is given by

$$T = \frac{2\pi m}{Be} = \frac{2\pi}{B \times 1.76 \times 10^{11}\ \text{C/kg}} = \frac{3.57 \times 10^{-11}}{B}, \text{ where } B \text{ is in Wb/m}^2$$

13.4 MOTION OF CHARGED PARTICLES IN COMBINED ELECTRIC AND MAGNETIC FIELDS

A charge q subjected simultaneously to both the electric and magnetic field experiences the Lorentz force given by Eq. 13.20. The two fields can be adjusted in such a manner that no net force acts on the particle and the particle continues to move undeviated. Figure 13.6 shows such an arrangement for electrons. The fields **B** and **E** are crossed, that is, they are perpendicular to each other. This arrangement is called *velocity selector* or *velocity filter* as it allows only those electrons to come out of the slit S which have the same velocity. The expression for this velocity can be easily derived as follows.

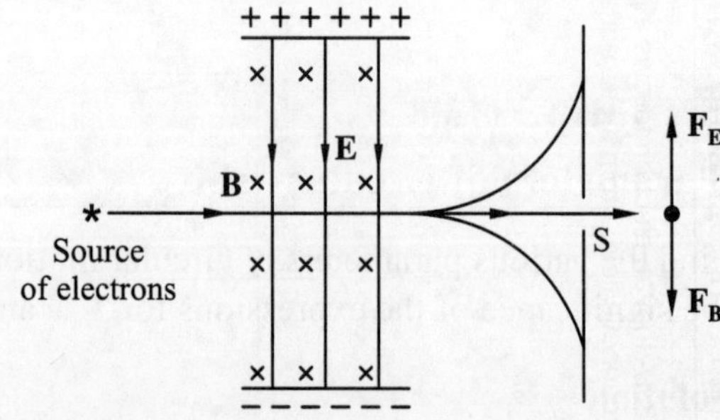

Figure 13.6 Velocity selector arrangement

Electrons coming from the source experience the force, $\mathbf{F_E} = -e\mathbf{E}$ acting upwards and the force $\mathbf{F_B} = -e(\mathbf{v} \times \mathbf{B})$ acting downwards. If the fields are so adjusted in their magnitude that

$$F_E = F_B$$

CHART 13.2 ELECTRIC AND MAGNETIC DEFLECTIONS SYSTEMS AND THOMSON'S EXPERIMENT

<table>
<tr><td>Principle of Thomson's method</td><td>We know from Chart 13.1A that an electron follows a parabolic path in electric field governed by the equation, $y = ½(eE/m)\ (l^2/v^2)$ and a circular path of radius $r = mv/Be$ in magnetic field. From these equations we get,
$e/m = 2yv^2/El^2$ (1)
in case of electric field or
$e/m = v/Br$ (2)
in case of magnetic field.
Thus determination of e/m looks so simple! But a careful examination of the right-hand sides of these equations reveals that the quantities y, r and v are not directly measurable in an experiment. These quantities should thus be related to some other experimentally measurable quantities. This is possible using a device known as the cathode ray tube (CRT).</td></tr>
<tr><td>CRT and Thomson's experiment</td><td>A CRT is an evacuated glass tube consisting of three main parts: (i) an electron gun for providing a collimated beam of electrons, (ii) an arrangement for producing electric and magnetic field with a facility that either electric or magnetic or both the fields can be applied, (iii) a fluorescent screen that produces a visible spot at the place where the electron beam strikes the screen. The position of the spot can thus be controlled by the magnitude and direction of E or B or both. In Thomson's method the velocity of electrons appearing in the above expression is found by simultaneous application of E and B so that electron beam strikes the screen undeviated, as explained in Section 13.4 (Fig. 13.6). The velocity is thus given by
$v = E/B$ (3)
Secondly either electric or magnetic field is switched on and the deflection Y of the spot on the screen is measured. This deflection can be related with y in case of electric field and with r in case of magnetic field and e/m can thus be determined experimentally using the above expressions. These relations and the resulting expression for e/m are derived in the following.
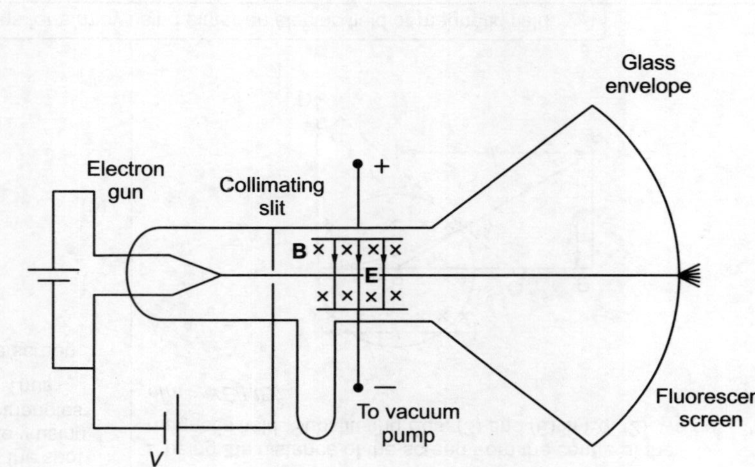
</td></tr>
</table>

	Using electric deflection system	Using magnetic deflection system
Derivation of expression for e/m	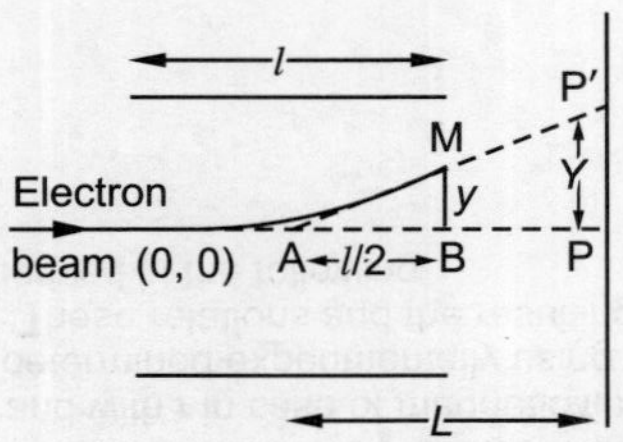 After emerging from the region of the electric field — extending up to the length l of the plates, — the electron will fly tangentially to the parabola at M and will strike the screen at point P′. The observed deflection, Y, of the spot can be related to y using the similarity of triangles ABM and APP′. Thus, y/Y = AB/AP. The distance AP = L is the distance of the screen from the centre of the plates. It can be shown that (see Example 13.10) AB = $l/2$. Thus, $y/Y = l/2L$ (4) Substituting Eqs. (3) and (4) in Eq.(1) we get, $e/m = YE/LlB^2$ (5)	After emerging from the region of magnetic field extending up to the length l, electrons will fly tangentially to the circle at M and will strike the screen at P′. The deflection Y of the spot can be related to r in the following way. The circular arc covered in the magnetic field has the radii AD and MD. AD ⊥ AP and BP′ ⊥ CD. Hence ∠ADC = ∠PBP′ = α. Further, ∠CAD = ∠BPP′ = 90°. Hence triangles ADC and BPP′ are similar. Thus, $l/r = Y/L$ (6) L being the distance of the screen from the centre of the magnetic field. Substituting Eqs. (3) and (6) in Eq. (2), we get $e/m = YE/LlB^2$ (7) 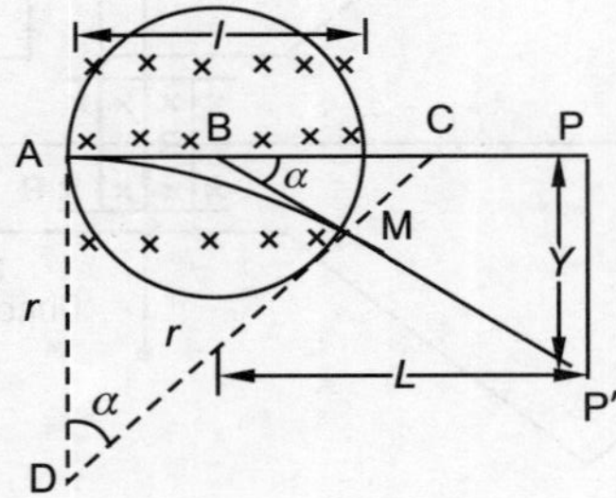
	Thus we get symmetrical relations for e/m by using either an electric field or magnetic field.	
1. Angular displacement, θ	$\theta_E = \tan^{-1}(Y/L) = \tan^{-1}(2y/l) = \tan^{-1}(lV/2dV_A)$ (8)	$\theta_B = \alpha = \tan^{-1}(Y/L) = \tan^{-1}(l/r) = \tan^{-1}(lB\sqrt{e/2mV_A})$ (9)
	Here we have used Eqs. 1, 2, 4, 6 and the relations $E = V/d$ (10) and ½ $mv^2 = eV_A$ (11), where V_A is the anode voltage.	
2. Deflection sensitivity, S (defined as the amount of deflection per unit applied field.)	$S_E = Y/V$. Using Eqs. (1), (4), (10) and (11) we get, $S_E = Ll/2dV_A$ (12) Thus the sensitivity varies inversely with the anode voltage and is independent of the electric field E, that is, voltage V across the plates. It is also independent of the ratio e/m.	$S_B = Y/B$. Using Eqs. (2), (6) and (11) we get, $S_B = Ll\sqrt{e/2mV_A}$ (13) Thus the sensitivity varies inversely with the square root of the anode voltage and is independent of the magnetic field B. It is, however, dependent on the ratio e/m.
	In both cases sensitivity can be increased by increasing L, the distance between the centre of plates/field width and the	

that is, $eE = evB$ (13.29)

then the electrons will move undeviated. Thus the electrons coming out of the slit S will have the same velocity given by Eq. 13.29, that is,

$$v = \frac{E}{B} \tag{13.30}$$

Electrons which do not satisfy Eq. 13.30 experience an unbalanced force (that is either $\mathbf{F_E} > \mathbf{F_B}$ or $\mathbf{F_B} > \mathbf{F_E}$). Hence they are deflected sideways and are blocked by the slit S.

The crossed-field arrangement was used by Thomson in his experimental arrangement for the determination of e/m of an electron. Thomson's experiment is described in Chart 13.2. The expression used for the determination of e/m can be derived either using the magnetic deflection system or the electric deflection system and the combination of the two. These deflection systems and derivations are similar in many respscts and are also included in Chart 13.2.

EXAMPLE 13.10

To show $AB = l/2$ for electric deflection system in Chart 13.2

Show that the tangent to the parabola described by an electron in a CRT, drawn at the end of the deflection plates (see Chart 13.2) passes through the point halfway between the plates along the undeviated trajectory.

Solution

The equation of the parabola described by an electron is

$$y = \frac{1}{2}\left(\frac{eE}{m} \times \frac{x^2}{v^2}\right)$$

Thus the slope of the line AP′ = slope of the parabola at point $M = \left.\dfrac{dy}{dx}\right|_{x=l}$

$$= \frac{eE}{m}\frac{l}{v^2}$$

Now the point M has coordinates $\left(l, \dfrac{eEl^2}{2mv^2}\right)$

Hence the equation of line AP′ is [equation of line with slope m and point (x_0, y_0) on it is $y - y_0 = m(x - x_0)$]

$$y - \frac{1}{2}\left(\frac{eE}{m}\right)\left(\frac{l^2}{v^2}\right) = \left(\frac{eE}{m}\right)\left(\frac{l}{v^2}\right)(x - l)$$

If the line MP′ is extended it touches the x-axis at point A for which the y coordinate is zero. Substituting $y = 0$ in the above equation and simplifying, we get $x = l/2$.

Thus, AB $= l/2$, that is, the tangent meets the point half way between the plates along the undeviated trajectory.

13.5 ELECTRON OPTICS

What does a lens do to a light beam? It deviates the beam. What does an electric field or a magnetic field do to an electron beam? They also deflect the electron beam. Thus parallels can be drawn between propagation of light beams and electron beams. The results of optics can be extended to the study of motion of charged particles in electric and magnetic fields. These aspects comprise the subject of *electron optics*. In electron optics we deal mainly with non-uniform fields. But the results obtained in the last section for uniform fields can be extended easily for non-uniform fields as explained below.

13.5.1 Electrostatic Lens

The principle of electron refraction governs the working of an electrostatic lens. Consider an electron beam entering from region I of potential V_1 with velocity v_1 into region II of higher potential V_2 with velocity v_2 as shown in Fig. 13.7.

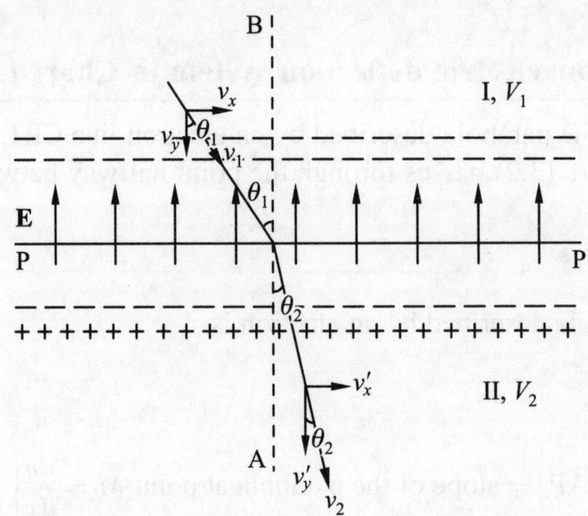

Figure 13.7 Refraction of electron beam across an equipotential surface

The surface PP′ is an equipotential surface, that is, it is a surface on which all points have the same potential. An equipotential surface is always normal to the direction of the electric field because if it were not so then there would be a non-zero component of the electric field on this surface. As an electric field is the rate of fall of potential, this surface would not be an equipotential surface.

In the present case the equipotential surface is a plane one because the electric field is produced by two parallel plates. An equipotential surface can also be spherical or cylindrical depending on whether the field is created by charged concentric spheres or coaxial cylinders. It can be of any other shape also depending on the geometry of the sources producing the electric field. The concept of an equipotential surface is analogous to that of a wavefront in physical optics. (A wavefront is a surface on which all points have the same phase and is always normal to the propagation of light rays.)

As the electric field is present along AB, the y-component of velocity undergoes a change. The x-component of velocity remains unaltered.

Thus $v_x = v'_x$ or $v_1 \sin\theta_1 = v_2 \sin\theta_2$ (13.31)

where θ_1 and θ_2 are the angles made by the electron beam with the electric field in region I and region II respectively.

Now, suppose that the electron beam in region I is made of electrons accelerated from rest in region of zero potential. Then, the total energy of the electrons is zero, and using Eq. 6 of Chart 13.1B we have,

$$\frac{1}{2}mv_1^2 = eV_1 \tag{13.32}$$

and $$\frac{1}{2}mv_2^2 = eV_2 \tag{13.33}$$

Thus $$\frac{v_1}{v_2} = \sqrt{\frac{V_1}{V_2}} \tag{13.34}$$

Now from Eq. 13.31, $$\frac{\sin\theta_1}{\sin\theta_2} = \frac{v_2}{v_1} \tag{13.35}$$

Combining Eqs. 13.34 and 13.35, we get

$$\frac{\sin\theta_1}{\sin\theta_2} = \frac{v_2}{v_1} = \sqrt{\frac{V_2}{V_1}} \tag{13.36}$$

Equation 13.36 is the 'law' of refraction of electrons in an electric field. It states that *an electron beam moving from a region of low potential to a region of high potential gets accelerated and bends towards the direction of the electric field.* This law is a counterpart of Snell's law in optics. It must be clearly undustood that Eq. 13.36 holds only under the condition of zero total energy stated earlier. Thus V_1 and V_2 are potentials relative to the point where the electrons are at rest. Table 13.3 compares these two laws. Example 13.11 illustrates an application of Eq. 13.36.

Table 13.3 Snell's law and law of electron refraction

	Snell's law	**Law of electron refraction**
Expression	$\sin i/\sin r = v_1/v_2 = \mu_2/\mu_1$, μ_1 and μ_2 are refractive indices of regions I and II respectively.	$\sin\theta_1/\sin\theta_2 = v_2/v_1 = \sqrt{V_2/V_1}$ V_1 and V_2 are potentials of regions I and II respectively, relative to regions where the electrons are at rest.
	For $\mu_2 > \mu_1, v_2 < v_1, r < i$	For $V_2 > V_1, v_2 > v_1$ and $\theta_2 < \theta_1$
Bending	A light beam entering from a rarer medium into a denser medium bends towards the normal.	An electron beam entering from a region of low potential to a region of high potential bends towards the normal (that is, towards the direction of electric field).
Velocities	A light beam entering from a rarer medium into a denser medium decreases in velocity.	An electron beam entering from a region of low potential to a region of high potential increases in velocity.

EXAMPLE 13.11
Angle of refraction for an electron beam

An electron enters from a region of potential 250 V into a region of potential 500 V. Find the angle through which the beam is refracted if the angle of incidence is 45°.

Solution

$V_1 = 250$ V; $V_2 = 500$ V; $\theta_1 = 45°$

Using the law of electron refraction, $\dfrac{\sin \theta_1}{\sin \theta_2} = \sqrt{\dfrac{V_2}{V_1}}$

We get, $\sin \theta_2 = \left[\sqrt{\dfrac{250\ V}{500\ V}}\right] \sin 45° = \dfrac{1}{2}$

Thus $\theta_2 = 30°$

Thus the electron beam is refracted through 30°.

A typical electrostatic lens is shown in Fig. 13.8. The regions of different potentials are produced by two coaxial cylinders having some separation. The figure also depicts the electric field lines and equipotential surfaces resulting due to this geometry.

The electrons moving along the axis of the cylinders, labelled as A in the figure, experience no bending as the electric field lines are along the same direction. The axial electrons are thus only accelerated in the forward direction. The off-axis electrons, labelled O in the figure, suffer

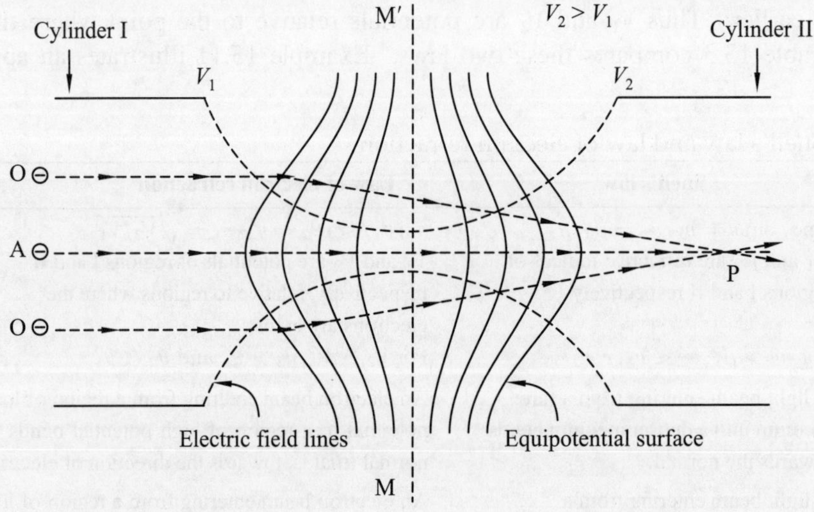

Figure 13.8 Working of an electrostatic lens

bending at the equipotential contours towards the direction of field lines and also get accelerated in the forward direction. Thus the off-axis electrons experience a converging action.

However, after crossing the mid-plane MM′, the electrons tend to diverge due to the opposite shape of equipotential surfaces. But the diverging (defocussing) action is weaker than the converging (focussing) action, because the electrons spend less time in the second half due to their increased speeds and the net result is the focussing of electrons at point P as shown in the figure.

13.5.2 Magnetostatic Lens

A magnetostatic lens is based on the helical motion of electrons, explained in Chart 13.1. The pitch of the helix is given by (see Ex 13.7)

$$p = \left(\frac{2\pi m}{Be}\right) v \cos\alpha \tag{13.37}$$

For small values of α, $\cos\alpha \sim 1$ and therefore,

$$p = \frac{2\pi mv}{Be} \tag{13.38}$$

Thus for electrons which make small angles with the direction of **B**, the pitch is independent of the angle. The divergent electrons can therefore be brought to focus at the same point, provided the divergence is small. This is the principle of magnetostatic focussing. For zero magnetic field, the electrons are defocussed and form a smudge at point P on the screen (see Fig. 13.9). As the field is applied and its magnitude is so adjusted that the pitch is equal to the anode-to-screen distance, then all electrons are brought to focus at the same point and an image of the anode hole will be observed on the screen. As **B** is increased further, the spot gets defocussed because electrons travel through more than one complete revolution. However, **B** can be increased some more to get a focussed spot again by making the electrons travel two complete revolutions before coming to focus at P. In this way **B** can be adjusted to several values to obtain a focussed spot by making the anode-to-screen distance, an integral multiple of the pitch. Figure 13.10 shows a solenoidal magnetic lens. It is best suited for confinement and focussing of cylindrical paraxial particle beams. The magnetic field profile of the lens is such that there is a strong axial component (z-component in Fig. 13.10) and a small radial component B_ρ, that is, B_x, that increases away from the axis.

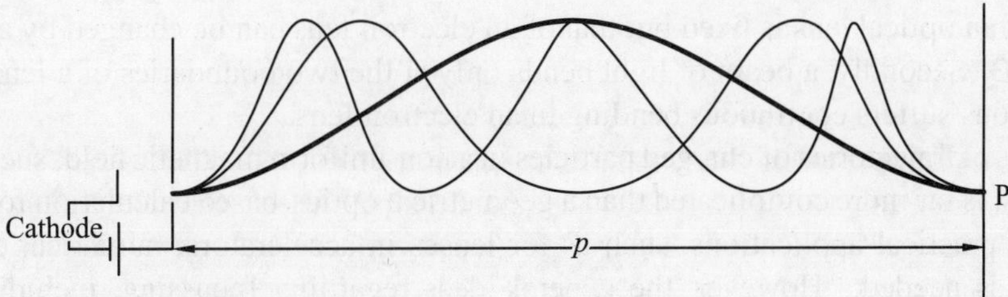

Figure 13.9 Magnetostatic focussing

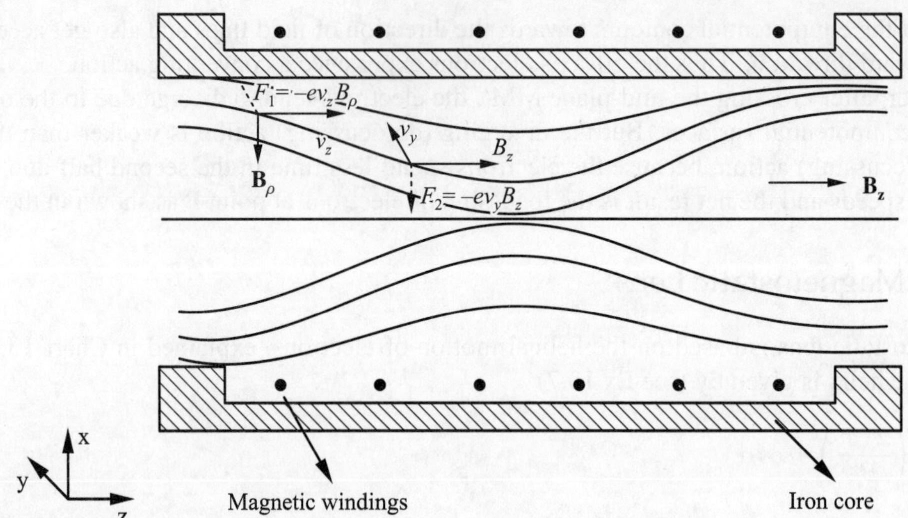

Figure13.10 Lens action in a solenoidal magnetic lens: $\mathbf{F}_1$ results in $\mathbf{v}_y$, $\mathbf{v}_y$ interacts with $\mathbf{B}_z$ and the resulting force $\mathbf{F}_2$ pulls the electron towards the axis.

If a charged particle, say an electron, travels away from the axis, that is, develops a radial component of velocity, v_ρ or v_x, then the z-component of its velocity interacts with B_ρ and exerts a force, $\mathbf{F}_1 = -e(\mathbf{v_z} \times \mathbf{B}_\rho)$ (of magnitude $ev_z B_\rho$), on the electron. This force develops an azimuthal component of velocity that pulls the electron along the y-axis and makes it spiral around the z-axis. As B_ρ is larger, the azimuthal velocity developed by electrons farther away from the axis is larger. Now the axial component of the field, that is, B_z, acts on v_y and exerts a force, $\mathbf{F}_2 = -e(\mathbf{v_y} \times \mathbf{B_z})$ (of magnitude $ev_y B_z$), towards the negative x-direction, that is, it leads to an inward radial acceleration. Again, inward pull on electrons farther from the axis is larger.

In this way the electrons drifting away from the axis are brought back towards the axis, thus achieving the confinement and focussing of the electron beam towards the field axis. Since the magnetic field is static, there is no change in energy of the particle passing through the lens.

13.5.3 Comparison with Optical Lens

The electron lenses discussed so far differ from the optical lenses mainly in two respects. The focal length of an optical lens is fixed but that of an electron lens can be changed by adjusting the value of **E** or **B**. Secondly, a beam of light bends only at the two boundaries of a lens, whereas a beam of electrons suffers continuous bending in an electron lens.

Calculations of trajectories of charged particles in a non-uniform magnetic field, such as that in a solenoidal lens, is far more complicated than a geometrical optics-based calculation for the optical case. In most practical applications, such as for lenses in accelerators, numerical computation of trajectories is needed. However, the general ideas regarding focussing, including those of abberations like astigmatism, apply to electron lenses also.

13.6 APPLICATIONS

In this chapter we studied the motion of charged particles in electric and magnetic fields. In Sections 13.2–13.4, the various aspects considering uniform fields were analysed. These aspects can be summarised into a single principle that embodies this study: *charges produce fields and fields control charges*. Scientists, technologists and engineers employed this principle to develop many novel and useful applications. Most of the progress in the disciplines of electrical and electronic engineering can be attributed to the clever exploitation of this principle. In fact the word electronics has been coined as a contraction of electron mechanics.

As the subject of applications of this technology is extremely vast, we restrict ourselves here to a discussion of only a few applications based on this knowledge. Figure 13.11 shows the use of an electric field as an electrostatic precipitator to prevent air pollution due to industrial effluents. An electric field exerts a force on charged ash as it ascends the stack, so that much of it gets collected in the stack and thus does not pollute the atmosphere. Figure 13.12 shows an inkjet printer in which an electric field is used to deflect the negatively charged drop onto the paper. The field applied is uniform but the drops have different charges. Hence they each experience a different amount of force and get deflected to different positions on the paper. The charging unit charges the drops in accordance with the input signal received from a computer that encodes the material to be printed.

Figure 13.11 Photo of the Hazelwood Power Station in the Latrobe Valley, Victoria Australia. One of the electrostatic precipitators has been taken offline, resulting in the brown smoke (Source: http://en.wikipedia.org/wiki/File:Hazelwood_Power_Station_ESP.jpg)

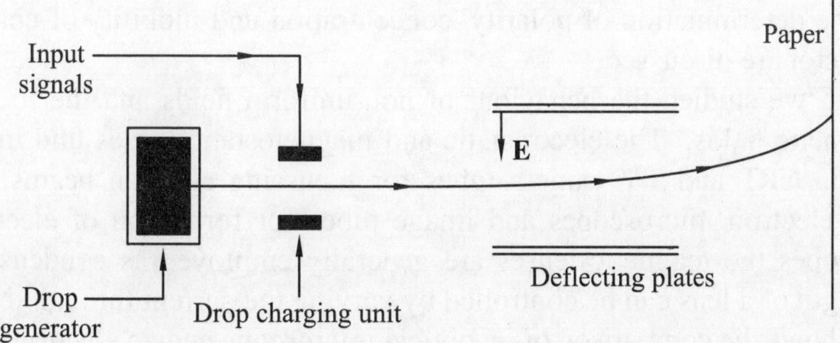

Figure 13.12 The essential features of an inkjet printer. An input signal from a computer controls the charge given to each drop and thus the position on the paper at which the drop lands.

Magnetic fields are used in industry for generation of power, in magneto-hydrodynamic (MHD) devices for confinement of plasma and in accelerators for confinement of charged particles to circular orbits. A modern television tube has two sets of magnetic deflection coils similar to the two sets of plates in the CRT (see Chart 13.2). The currents flowing through the coils are adjusted such that uniform intensity is imparted to the screen. A video signal, applied to the electron gun, modulates the intensity of the electron beam and thus forms the television picture.

Three important inventions based on simultaneous applications of electric and magnetic field are discussed here. They are the Bainbridge mass spectrograph, cyclotron and cathode ray oscilloscope (CRO). As the working of these devices is based on the same fundamental principle they are discussed in a comparative way in Chart 13.3. Among these, the CRO is the most indispensable in the modern electronic industry. It is often described as the 'eye' of an electronic engineer as it enables him/her to know the electrical details of the minute components of complicated circuitaries which cannot be comprehended visually by the eye. In Chart 13.4 we study the use of the CRO in the analysis of Lissajous patterns and waveforms.

The CRO has become an indispensable instrument in any physics (or physics-based) and electronics (or electronics-based) laboratory because of its ability to provide the visual representation of voltage signals under study or testing. Secondly, as the electronic mass is negligible, the CRO is also useful in the study of high frequency signals. In Expt. 13.1 the applications of CRO for the determination of amplitude, frequency and phase of an electrical voltage are studied.

The technology of the cathode ray tube (CRT) played an important role in the development of television. Early experimenters realised the potential usefulness of the CRT as a display devise for television. In 1920, VK Zworykin demonstrated a CRT, called the kinescope that was suitable for display of television images. It was modified for use as a monochrome television picture tube in the late 1940s and 1950s. In 1953, the first colour picture tube was introduced. By 1960 the CRT also found application as a data display tube in computers. Apart from these, the CRT found wide application in radar technology. Photorecording applications of CRT include electronic phototypesetting and the storage of computer output information on microfilm.

In 1879, Edwin H Hall, then a 24-year old graduate student at the Johns Hopkins University, extended the analysis of motion of charged particles in vacuum to that in solids. A phenomenon known by his name, *Hall effect*, is particularly important in semiconductors and enables one to know whether a semiconductor is p-type or n-type. In Expt. 13.2, the Hall effect and its applications in the determination of polarity, concentration and mobility of charge carriers in a given semiconductor are discussed.

In Section 13.5 we studied the behaviour of non-uniform fields and the focussing action of electric and magnetic fields. The electrostatic and magnetostatic lenses find many applications including those in CRT and TV camera tubes for focussing electron beams, and in infrared converter tubes, electron microscopes and image tubes for formation of electron images. In electron microscopes the magnetic lenses are generally employed as condenser and projector lenses. The strength of a lens can be controlled by varying the current through the electromagnet.

Figure 13.13 shows the comparison of an optical microscope and an electron microscope. The use of magnetostatic lenses in the electron microscope has helped to attain the highest resolution and magnification.

CHART 13.3 DEVICES

Devices	Mass spectrograph
Invention	KT Bainbridge (1935)
Diagram	Source of positive ions; Collimating slits; S_1; S_2; **E**; +; –; **B**; Photographic plate; S_3; *r*; **B′**; Vacuum chamber; **B** and **B′** out of the plane of paper
Construction	The Bainbridge mass spectrograph consists of a source of positive ions collimated into a thin pencil by slits S_1 and S_2. The ions enter the velocity selector where both electric and magnetic fields are present (Section 13.4) . The monovelociy ions coming out of the slit S_3 follow different circular paths due to the magnetic field B' in a vacuum chamber. A photographic plate registers their arrival.
Working	The source of positive ions contains ions with different velocities. When they pass through the velocity selector, both **E** and **B** act upon them and if their magnitudes are adjusted such that $qE = qvB$, then the ions coming out of the slit S_3 have the same velocity given by E/B. As the ions come out of the slit S_3 only the magnetic field B' exerts a force upon them and the ions follow circular orbits of radii $r = mv/qB'$. Thus the ions with same m/q ratio come to focus at a particular point on the photographic plate. If the charges of ions are the same, then the radii of circular orbits are proportional to the masses of ions. From a knowledge of radii the masses of different species of ions can be determined.
Derivation of expressions	The force experienced by the ion in the vacuum chamber $B'qv = mv^2/r$. In the velocity selector, $Bqv = qE$. Hence $v = E/B$. Thus $r = mE/BB'q$ or $m = BB'qr/E$ (1) If the distance of the line on the photographic plate from the centre of the slit S_3 is s, then $s = 2r$. Thus, $m = BB'qs/2E$ (2) (see Example 13.12)
Applications	Used for measurement of atomic masses and for separation of isotopes.
Limitations	

Cyclotron
EO Lawrence and MS Livingston (1932)
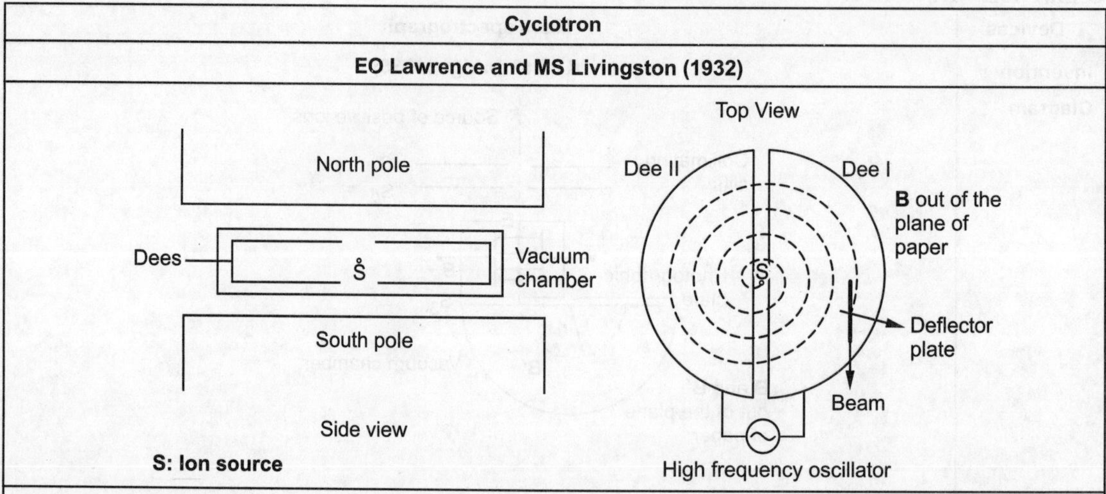 **S: Ion source**
The cyclotron consists of two hollow 'dee' shaped metallic objects placed in a vacuum chamber placed in a strong magnetic field perpendicular to the semicircular faces of dees (out of the plane of the page in the present case) and a high frequency ac potential applied to the dees. A source of positive ions is kept at the centre of the dees.
If dee I is at negative potential at a particular instant, the positive ion will accelerate towards it. Once it enters dee I it is shielded by the metallic walls of the dee and the electric field does not affect it. However, the magnetic field exerts a force upon it that makes it follow a semicircular path inside the dee I. After completion of this path, when the ion is in the gap between the two dees, if the polarity of the voltage reverses so that dee II now becomes negative, then the ion will again get accelerated and will follow a semicircular path of larger radius inside dee II. This condition is the key to the operation of the cyclotron and is called *resonance* condition or *synchronisation* condition. It is achieved by making the frequency of the oscillator equal to the frequency of the revolution of the ion in the magnetic field. Thus the resonance condition ensures that each time the ion comes in the gap between the two dees the ac voltage reverses its polarity and the ions follow a circular orbit of larger and larger radius, each time gaining energy qV from the field. The ions thus spiral and gain a tremendous energy before coming out of the cyclotron, due to a deflector plate.
The resonance condition is: the frequency of ac voltage $f = Bq/2\pi m$ (3) If the ion makes N revolutions before coming out of the cyclotron, then the final energy of the ion = $2NqV$ (4) (because an ion comes twice in the gap between the two dees in one revolution and each time it gains energy qV from the field). The final energy is also given by $E = ½\, mv^2$ where v is the final velocity of the ion which is given by $mv^2/R = Bqv$ where R is the radius of the dee. Thus, $E = B^2R^2q^2/2m$ (5) (see Example 13.13).
Used for accelerating positive ions to extremely high energies. These high energy ions are used in nuclear science and engineering for bringing about the desired nuclear reactions. For this reason a cyclotron is also called an *atom smasher.*
A cyclotron cannot accelerate electrons because electrons—due to their very small mass—gain speeds of the order of speed of light. Hence their mass $m = m_0/\sqrt{[1 - (v^2/c^2)]}$ increases. The resonance condition is thus spoiled and electrons fall out of step with the applied field, thus decelerating instead of getting accelerated.

CHART 13.3 (Continued)

Devices	Cathode ray oscilloscope
Invention	Product of several experiments in the late nineteenth century that used the cathode ray tube. Notable efforts arc by Roentgen (1895), Thomson (1897) and Ferdinand Braun (1897).
Diagram	Y input; Vertical amplifier; Power supply; Electron gun; Y; X; Time base generator; INT; EXT; Horizontal amplifier; X input
Con-struction	The CRO consists of a cathode ray tube (CRT), vertical and horizontal amplifiers connected to the Y and X-plates of the CRT respectively, time base generator and the power supply. The CRT consists of an electron gun, pairs of X- and Y-deflection plates and a fluorescent screen. The time base generator is often provided with a trigger circuit that triggers the generator. The high voltage power supply is connected to the CRT; whereas other components are connected to a low voltage supply. The screen is provided with calibrated x- and y-axes.
Working	The electron beam issued by an electron gun deflects in response to the voltages applied to the X- and Y- plates of the CRT. When no voltage is applied to both the plates, the beam does not experience any force, goes undeviated and strikes the centre of the screen (origin of xy-axes). If dc voltage is applied only to the Y-plates, the electron beam moves either up or down depending on whether the upper plate or the lower plate is positive. Similarly, the spot moves either to the left or to the right along the x-axis if dc voltage is fed to the X-plates only. If ac voltage is fed to the Y-plates only, then the electron beam will move up and down, and if the corresponding movement of spot up and down is fast enough, then a vertical straight line is seen due to persistence of vision and phosphorescence of the screen material. Similarly, the spot will trace a horizontal straight line if ac voltage is applied to the X-plates. If ac voltages are applied to both the X- and the Y-plates, then the resultant pattern on the screen is decided by the relative amplitudes, frequencies and phases of the two voltages. If the internal time base voltage is applied to the X-plates, then the actual waveform of the voltage fed to the Y-plates is traced on the screen. These aspects are illustrated further in Chart 13.4.
Derivation of expre-ssions	See Chart 13.4 and Expt. 13.1
Appli-cations	Used for measuring voltages, currents, frequency and phase of ac voltages; for observing the waveforms of voltages and finding out distortions; for measuring different physical quantities like temperature and pressure using appropriate transducers.
Limita-tions	Every CRO has limitations depending on the circuit design and the components used. The specification such as 10 kHz means that the performance of the CRO is not reliable above these frequencies.

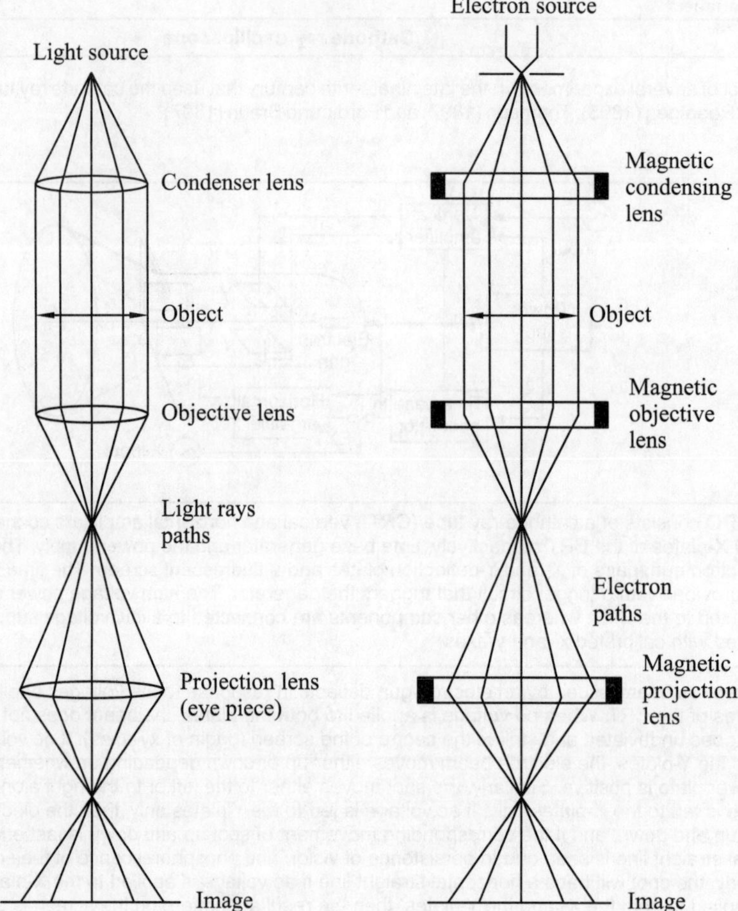

Figure 13.13 Comparison between (a) light microscope and (b) electron microscope

EXAMPLE 13.12

Linear separation of lines formed by isotopes in Bainbridge mass spectrograph

In a Bainbridge mass spectrograph, the electric field and the magnetic field in the velocity selector are 7×10^4 V/m and 0.50 Wb/m^2 respectively. The magnetic field that the ions face after coming out of the velocity selector is 0.75 Wb/m^2. Find the linear separation of the lines formed on the photographic plate if the ion source consists of singly ionised neon isotopes of atomic masses 20 amu and 22 amu.

Solution

The linear separation of the lines formed on the photographic plate is given by (see Chart 13.3)

CHART 13.4 STUDY OF LISSAJOUS PATTERNS AND WAVEFORMS

In a CRO if a sawtooth voltage is connected to the X-plates and a voltage of any other waveform is fed to the Y-plates, then the actual waveform of the voltage applied to the Y-plates is displayed on the screen. If sinusoidal voltages are applied to both X- and Y- plates, then the resulting pattern on the screen—called the *Lissajous pattern*—is decided by the relative amplitudes, frequencies and phases of the two voltages. The resultant can be obtained by two methods: mathematical method and graphical method. In the mathematical method we represent the voltages applied to the X-plates and the Y-plates, respectively as $v_x = V_1 \sin \omega t$ and $v_y = V_2 \sin(\omega t + \alpha)$. Here we have assumed the voltages to have the same frequency, ω. V_1 and V_2 are the respective amplitudes, and α is the phase difference.

Now $v_y = V_2 [\sin\omega t \cos\alpha + \cos\omega t \sin\alpha]$.

Substituing $\sin\omega t = v_x/V_1$ and $\cos\omega t = \sqrt{1-(v_x^2/V_1^2)}$, we get

$$v_y/V_2 = v_x \cos\alpha/V_1 + \sqrt{1-(v_x^2/V_1^2)} \sin\alpha \quad \text{or} \quad (v_y/V_2 - v_x \cos\alpha/V_1)^2 = (1 - v_x^2/V_1^2) \sin^2\alpha.$$

Thus, $$v_x^2/V_1^2 + v_y^2/V_2^2 - 2 v_x v_y \cos\alpha/V_1 V_2 = \sin^2\alpha \quad (1)$$

This is the equation of the resultant pattern which is an inclined ellipse. The pattern for a particular α can be predicted by substituting that value of α in the above equation.

In the graphical method the horizontal input wave and the vertical input wave are drawn on the two axes of a graph. The resultant is then traced by joining the points obtained corresponding to the same time instant on the horizontal and vertical waves. In the following the resultant Lissajous figures are obtained using both these methods for different special cases.

Case	$\alpha = 0°$ Eq. 1 becomes $v_y = (V_2/V_1)v_x$ (2)	$\alpha = 90°$ Eq. 1 becomes $v_x^2/V_1 + v_y^2/V_2^2 = 1$ (3)	$\alpha = 90°$ and $V_1 = V_2 = V$ Eq. 1 becomes $v_x^2 + v_y^2 = V^2$ (4)	$\alpha = 0°$ and $\omega_2 = 2\omega_1$
Resultant Lissajous figure	Straight line of slope V_2/V_1. If amplitudes are equal then slope = 1	Ellipse having V_1 and V_2 as the semi-major and semi-minor axes, respectively	Circle of radius V	Figure of eight

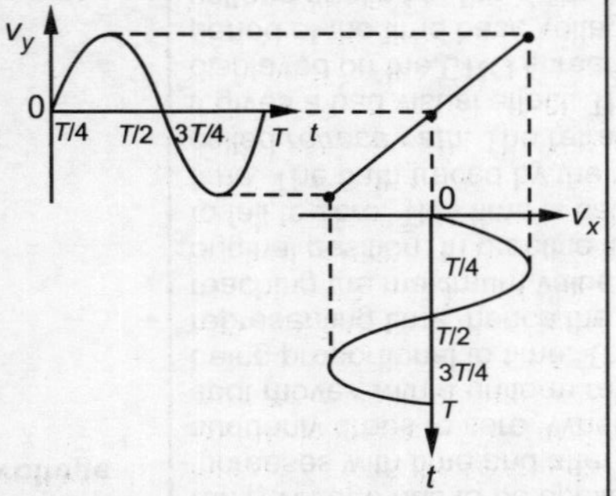

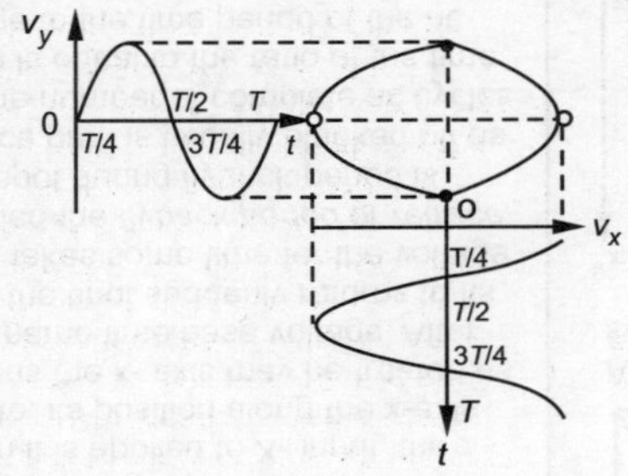

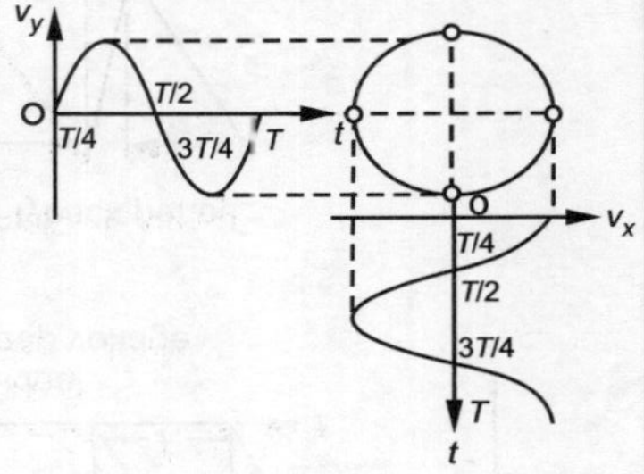

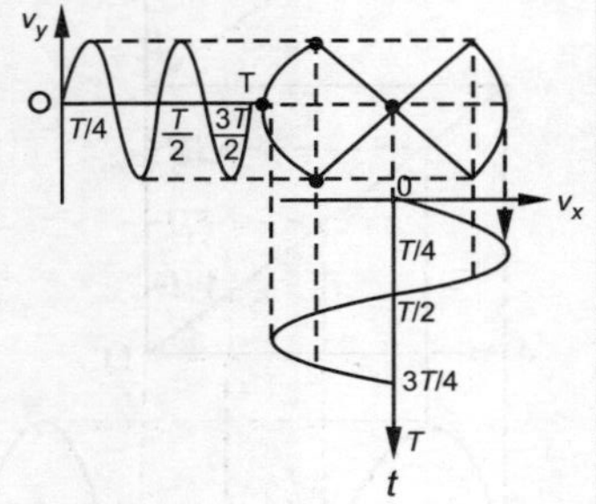

Display of actual waveform using time base voltage	If a sinusoidal voltage is applied to the Y-input with no voltage to the X- input, only a vertical straight line is seen. To see the actual waveform of the voltage, internal *time base voltage*, also called as the *sweep voltage* or the *saw tooth voltage* or *ramp voltage* has to be applied. The time base voltage increases with time and after reaching a maximum value V suddenly drops to zero. When it is applied to X- input, the spot moves with a uniform rate, its position along the x-axis being proportional to time. Thus the x- axis may be treated as representing time, hence the name time base voltage. After reaching the maximum value the spot suddenly returns to its original position. In practice it takes some time for the voltage to fall to zero. This time is called the *flyback period* or *retrace time*. The path traced by the spot during flyback period is called *retrace path*. The retrace path is usually blanked off as it gives a bad visual effect. The number of complete ac cycles displayed on the CRO screen is equal to the ratio of the time period of the time base voltage to the time period of the ac voltage applied to the Y- input.	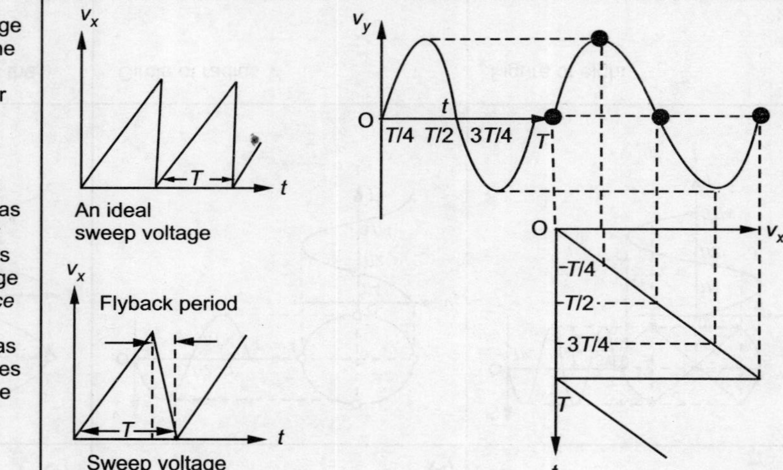 Mathematically $v_x = kt$, where k is a constant and $v_y = V \sin(\omega t + \alpha) = V \sin(\omega v_x/k + \alpha)$. Thus v_y varies sinusoidally with v_x.

$$S_2 - S_1 = \frac{2E(M_2 - M_1)}{BB'e}$$

$$= 2 \times 7 \times 10^4 \text{ V/m} \times \frac{(22 - 20) \times 1.66 \times 10^{-27} \text{ kg}}{0.5 \text{ Wb/m}^2 \times 0.75 \text{ Wb/m}^2 \times 1.6 \times 10^{-19} \text{ C}}$$

$$= 77.5 \times 10^{-4} \text{ m} = 8 \text{ mm}$$

EXAMPLE 13.13

Various parameters for deuteron motion in a cyclotron

A certain cyclotron produces 5 MeV deuterons. The oscillator that supplies the power to the dees imparts 50 keV to the deuterons, each time when they come in the gap between the two dees. Calculate (a) the magnetic field intensity, (b) the frequency of the oscillator, and (c) the time it will take for an ion introduced at the centre of the chamber to emerge at the rim of the dee. Assume the radius of the last semicircle to be 15 inches.

Solution

The mass of the deuteron is, $m = 2.01 \times 1.66 \times 10^{-27} = 3.34 \times 10^{-27}$ kg.
The charge on the deuteron is, $q = 1.6 \times 10^{-19}$ C.
The velocity of 5-MeV deuteron is given by the equation $\frac{1}{2}mv^2 = qV$.

$$\text{Hence, } v = \left[\frac{2 \times 5 \times 10^6 \times 1.6 \times 10^{-19} \text{J}}{3.34 \times 10^{-27} \text{ kg}}\right]^{1/2}$$

$$= 2.2 \times 10^7 \text{ m/s}$$

(a) The magnetic field intensity needed is given by

$$B = \frac{mv}{qR} = \frac{(3.34 \times 10^{-27} \text{ kg})(2.2 \times 10^7 \text{ m/s})}{(1.6 \times 10^{-19} \text{ C})(15 \times 2.54 \times 10^{-2} \text{ m})}$$

$$= 1.2 \text{ Wb/m}^2$$

(b) The frequency of the oscillator must be equal to the frequency of revolution of the ion,

$$f = \frac{qB}{2\pi m} = \frac{1.6 \times 10^{-19} \text{ C} \times 1.2 \text{ Wb/m}^2}{2\pi \times 3.34 \times 10^{-27} \text{ kg}}$$

$$= 9.15 \times 10^6 \text{ Hz} = 9.15 \text{ MHz.}$$

(c) The final energy of deuterons is 5 MeV. The energy received from the oscillator in each step is 50 keV. Thus the deuterons must pass across the accelerating gap $(5 \times 10^6 \text{ eV})/(50 \times 10^3 \text{ eV})$ times, which is 100 times. As the ions pass twice in the accelerating gap in one revolution, the number of complete revolutions necessary to gain the full energy is 50. Thus the time required to emerge at the rim of the dee is

$$t = 50T = \frac{50}{f} = \frac{50}{9.15 \times 10^6 \text{ Hz}} = 5.46 \text{ μs}$$

EXPERIMENT 13.1 — CATHODE RAY OSCILLOSCOPE

Aim

To determine the amplitude and frequency of a sinusoidal voltage and phase difference between two sinusoidal voltages using a CRO

Apparatus

CRO, step down transformer, signal generator, AC voltmeter and a C–R circuit for determination of phase difference

Formulae

Amplitude, $A = \dfrac{np}{2}$ (13.39a)

or $A = \sqrt{2}V_{\text{rms}}$ (13.39b)

where n is the vertical length of the display, p is the reading corresponding to volts per division of the knob and V_{rms} is the rms voltage measured by an AC voltmeter.

Frequency, $F_y = \dfrac{1}{nq}$ (13.40a)

or $F_y = \dfrac{T_x}{T_y} \times F_x$ (13.40b)

where n is number of divisions on the horizontal axis, q is the reading corresponding to time/div knob, T_x and T_y denote the number of tangency points along the x- and y-axis, respectively, and F_x and F_y are frequencies of the respective sinusoidal voltages.

Phase difference, $\alpha_P = \dfrac{s}{T} \times 360°$ (13.41a)

or $\alpha_P = \sin^{-1}\left(\dfrac{y_1}{y_2}\right)$ (13.41b)

and $\alpha_T = 90° - \tan^{-1}(1/2\pi fCR) = \tan^{-1}(2\pi fCR)$ (13.41c)

where α_P and α_T denote, respectively, the practical and the theoretical value of phase difference, s is the time difference between the two waves, T is the time period of each wave, f is the frequency of signal generator and C and R denote the values of capacitance and resistance used.

Theory

The CRO (cathode ray oscilloscope) is the most versatile, all-in-one instrument capable of measuring amplitude, frequency and phase of currents and voltages in electrical circuits. It can also produce a visual presentation of varying parameters such as AC current and voltage. Thus it is expected to have adequate familiarity with its basic operations. In the following page we discuss the functions of various knobs on the front control panel of a CRO to gain familiarity. The front panel view of a *single trace CRO* and a typical list describing functions of various controls provided in a manual are given below. A *dual trace CRO*, as the name suggests, enables viewing of two traces simultaneously on the screen. This is facilitated by additional controls for operation of its two channels. A dual trace CRO thus operates similar to a single trace CRO if only one of its channels is used.

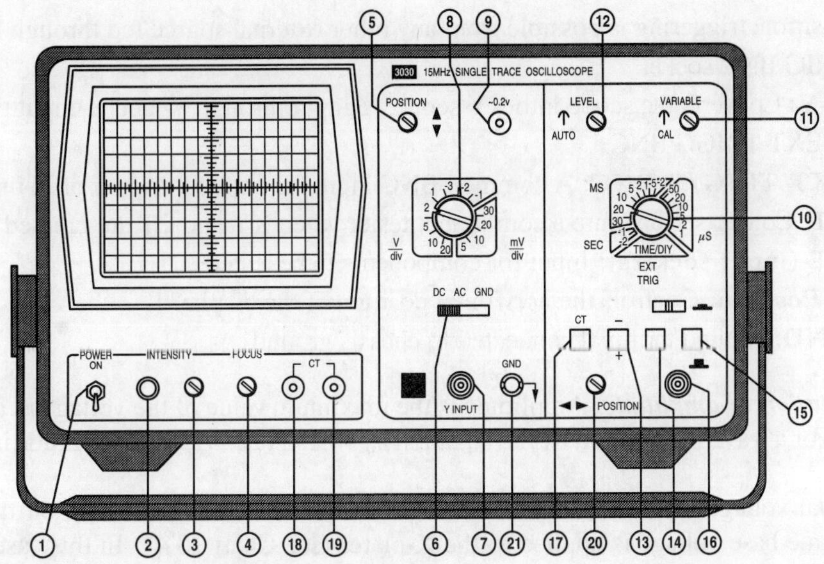

1. **Power**: Turns the power on
2. **LED**: Glows when power is turned on
3. **Intensity**: Controls the brightness of the display
4. **Focus**: Controls the sharpness of the display
5. **⬍ Position**: Controls the vertical position of the display
6. **DC–AC–GND**: Selects the input coupling mode. AC: couples the signal through the capacitor DC: couples the signal directly to the input of the attenuator. GND: grounds the input to the vertical amplifier
7. **Y input**: Feeds input signal to vertical amplifier through DC–AC–GND coupling switch
8. **Attenuator**: 12-step compensated attenuator from 5 mV/div to 20 V/div in 1, 2, 5 sequence
9. **⊓⊔ 0.2 V**: Provides a square wave of 200 mV_{P-P} (peak-to-peak voltage), 1 kHz for checking the calibration of the Y amplifier
10. **Time/div**: Selects the speed of sweep voltage from 0.5 μs/div to 0.2 s/div in 18 calibrated steps
11. **Variable**: In CAL position (fully anticlockwise), the selected sweep speed holds indicated calibration. Clockwise rotation of the control increases the sweep speed. In fully clockwise position the speed increases by a factor of 2.5.
12. **AUTO/LEVEL**: Selects the mode of triggering. In AUTO mode signals above

50 Hz, greater than 0.5 div in amplitude will trigger the sweep with input signal. Automatic base line is produced with no signal input. When the knob is switched out of AUTO position, it selects the trigger level on a displayed waveform.

13. +/−: Selects whether the waveform is to be triggered on positive or negative steps.
14. **EXT TRIG**: Normally (switch released - INT TRIG) display triggers from signals internally derived in coincidence with the displayed waveform. In EXT TRIG position, triggering is possible from any other external source fed through EXT TRIG BNC socket.
15. X-Y: Converts the scope into XY scope. Y-input through Y- and X-input through X/EXT TRIGG BNC.
16. **EXT TRIG/X BNC**: A common BNC input for external trigger or X-input signal
17. CT: Converts scope into a component tester when XY and CT are pressed

18, 19. **CT (input sockets)**: Input for components to be tested

20. ◂▸ **Position**: Controls the horizontal position of the display
21. GND: Ground terminal connected to chassis ground

A. Determination of amplitude: Amplitude is the maximum value of the voltage in a sinusoidal AC voltage. For a wave represented by the equation $v_y = A\sin(\omega t + \alpha)$, the amplitude is A because $(v_y)_{\max} = A$.

A sinusoidal voltage, applied to the Y-plates produces the actual waveform on the screen if the internal time base voltage is applied to the X-plates (see Chart 13.4). In this case the x-axis automatically gets calibrated in terms of time. If the internal voltage is switched off then a vertical line is seen, the length (n) of which is equal to the peak-to-peak distance of a sine wave (Fig 13.14). The amplitude is then determined easily by multiplying $n/2$ by the reading p of the deflection sensitivity (volts/div) knob. Thus, $A = np/2$. Amplitude can also be measured using an AC voltmeter by multiplying the rms voltage, V_{rms} by $\sqrt{2}$.

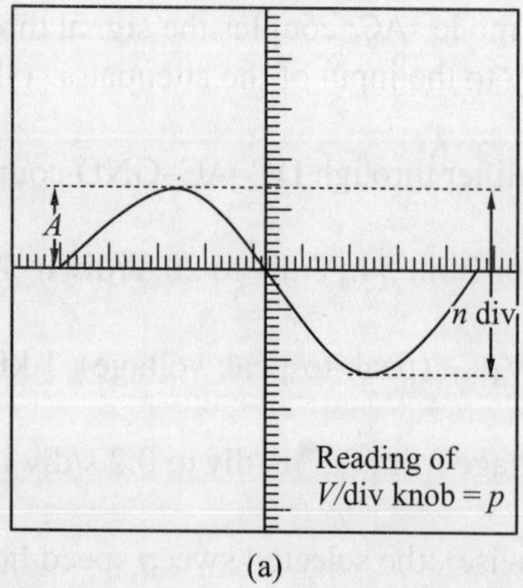

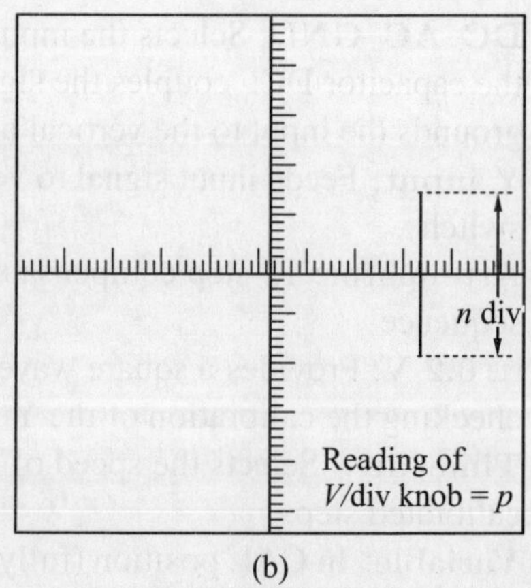

Figure 13.14 Determination of amplitude: (a) Display of the actual waveform when the time base circuit is on; (b) A vertical straight line is displayed if the time base circuit is off. n is the same in both the cases.

B. Determination of frequency: The frequency of a sinusoidal voltage is the number of AC cycles per second. In the expression $v_y = A\sin(\omega t + \alpha), \omega = 2\pi f$ is the angular frequency and f is the natural frequency. The reciprocal of the frequency is the time period, that is, the time required to complete one cycle of an AC wave. The frequency can easily be found in the following way.

An AC voltage of unknown frequency is fed to the Y-input and the frequency of the time base voltage is adjusted in such a way that a single wave fits exactly one division on the x-axis, as shown in the Fig. 13.15a. The reading of time base sensitivity (time/div) knob in this setting is

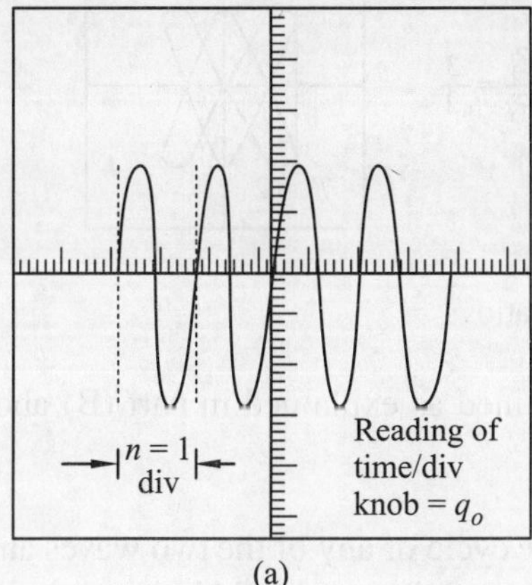

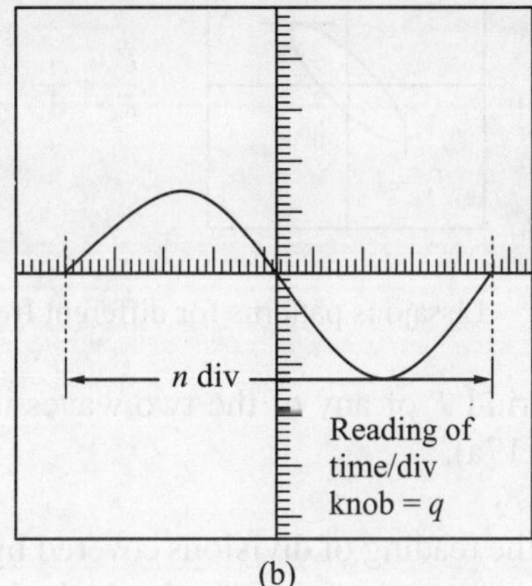

Figure 13.15 Determination of frequency: In (a) $T = q_0$ and in (b) $T = nq$

the time period and its reciprocal is the frequency of the voltage. If the single cycle covers n divisions (Fig. 13.15b), then the time period is n multiplied by the reading of time/div knob.

Another method for the determination of unknown frequency uses the Lissajous patterns. The AC voltage whose frequency, F_y, is to be determined, is fed to the Y-input and another AC voltage of known frequency, F_x, is fed to the X-input. As the two voltages are taken from two independent sources, the phase difference keeps changing with time and a stable Lissajous pattern cannot be seen on the CRO screen. The Lissajous patterns can, however, be made nearly stable if the ratio of the two frequencies is an integer or a ratio of integers. The unknown frequency is given by $F_y = (T_x/T_y)F_x$ where T_x and T_y are the number of tangency points along the x- and y-axes, respectively. Figure 13.16 shows some Lissajous patterns which satisfy the above relation.

C. Determination of phase difference: The phase difference between two sinusoidal voltages can be obtained directly using a dual trace CRO. The two voltages are fed to the two channels of a dual trace CRO. If the two waves differ by n_1 divisions on the horizontal axis and r_1 is the reading of time/div knob, then $s = n_1r_1$ represents the time difference between two waves. Thus the corresponding phase difference is

$$\alpha_P = \omega n_1 r_1 = \left(\frac{2\pi}{T}\right) n_1 r_1 = \frac{n_1 r_1}{T} \times 360^\circ$$

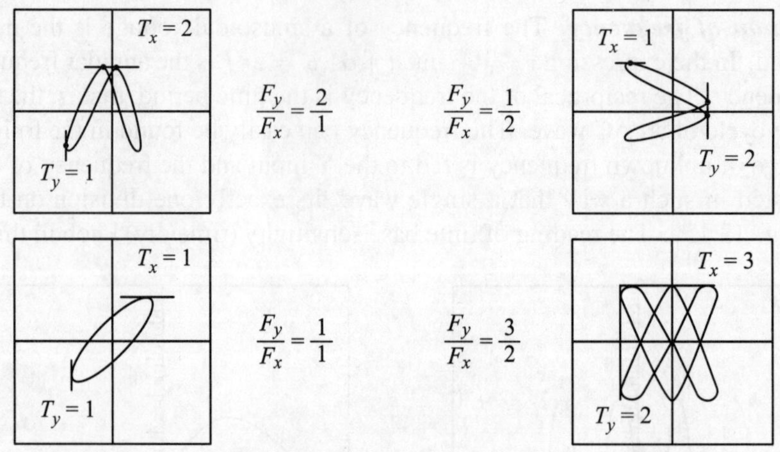

Figure 13.16 Lissajous patterns for different frequency ratios

The time period T of any of the two waves is determined as explained in part (B) above. Thus (see Fig. 13.17a),

$$T = n_2 r_2$$

where n_2 is the reading of divisions covered by a single cycle of any of the two waves and r_2 is the corresponding reading of time/div knob. It should be noted that r_1 should be set appropriately so that the distance s is larger (Fig. 13.17b) and is thus measured with better accuracy (see discussion).

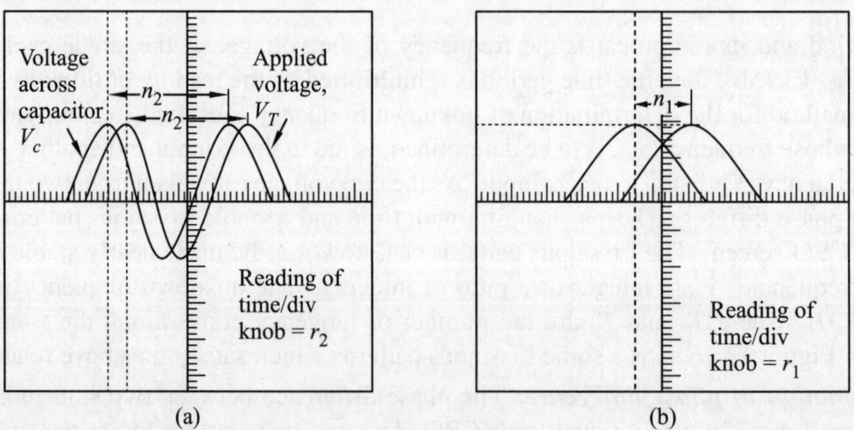

Figure 13.17 Determination of frequency using dual trace CRO: (a) Time period of each wave, $T = n_2 r_2$; (b) Time difference between two waves, $s = n_1 r_1$

Another method for determination of phase difference is using the Lissajous pattern. If two sinusoidal voltages with phase difference α are fed as the two inputs of a CRO, then different stable Lissajous figures can be observed on the CRO screen for different values of α, as explained in Chart 13.4. For $0 < \alpha < 90°$, the resultant figure is an inclined ellipse (Fig. 13.18), governed by the equation,

$$\frac{v_x^2}{V_1^2} + \frac{v_y^2}{V_2^2} - 2\left(\frac{v_x v_y}{V_1 V_2}\right)\cos\alpha = \sin^2\alpha$$

Now let y_1 be the intercept on the y-axis so that the coordinates of the point P on the ellipse are $(0, y_1)$. Substituting these coordinates in the above equation, we get

$$p^2 y_1^2 = V_2^2 \sin^2\alpha \quad \text{or} \quad py_1 = \pm V_2 \sin\alpha, \; p \text{ being the reading of volts/div knob.}$$

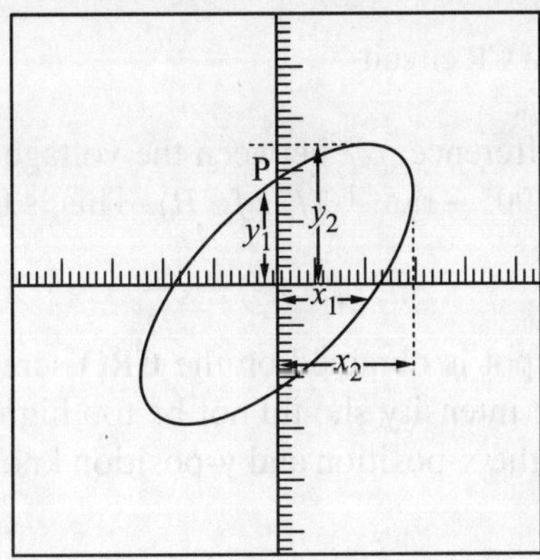

Figure 13.18 Inclined ellipse observed on the CRO screen in phase determination

As the positive and negative intercepts are equal, the ellipse is symmetrical. Now $V_2 = py_2$, y_2, being the maximum vertical deflection, as shown in the figure.

$$\text{Thus, } y_1 = y_2 \sin\alpha \quad \text{or} \quad \alpha = \alpha_P = \sin^{-1}(y_1/y_2).$$

A similar result, $\alpha_P = \sin^{-1}(x_1/x_2)$, can be obtained by considering the point $(x_1, 0)$ on the ellipse, x_2 being the maximum horizontal deflection. Thus the practical value of phase difference can be determined by measuring y_1 and y_2 (or x_1 and x_2) on the CRO screen.

The theoretical formula for phase difference can be derived as follows. We know that the current through the capacitor leads the applied voltage by 90°, current through the inductor lags behind the applied voltage by 90° and the current through the resistance is in phase with the applied voltage. Thus for a CR or LR circuit, the phase difference between applied voltage and current is between 0° and 90°.

Using the phasor diagram shown in Fig. 13.19, $\theta = \tan^{-1}\left(\frac{X_C}{R}\right)$

As X_C, the capacitive reactance, is given by $X_C = \frac{1}{2\pi fCR}$, we get,

$$\theta = \tan^{-1}\left(\frac{1}{2\pi fCR}\right)$$

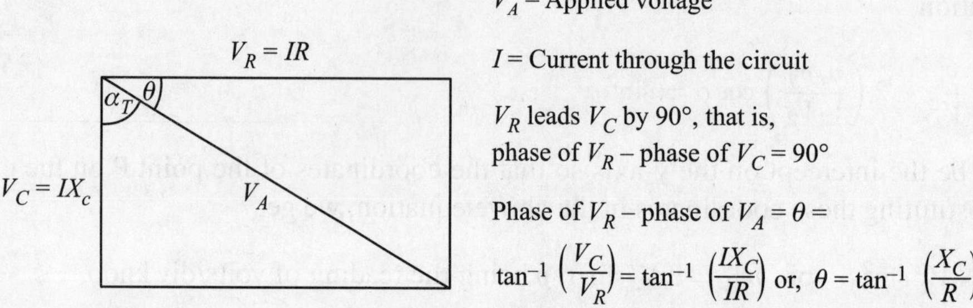

Figure 13.19 Phasor diagram for a CR circuit

Hence the theoretical phase difference, α_T between the voltage across the capacitance and the applied voltage is given by $\alpha_T = 90° - \tan^{-1}(1/2\pi fCR)$. This is the same as $\tan^{-1}(2\pi fCR)$.

Procedure

The CRO is switched on and a spot is obtained on the CRO screen. The focus control knob is adjusted to get a sharp spot. The intensity should not be too high. The spot is positioned at the centre of the screen by adjusting the x-position and y-position knobs.

A. *Measurement of amplitude*

1. Connections are made as shown in Fig. 13.20a.
2. A sinusoidal waveform is obtained on the screen by suitably adjusting the volts/div and the time/div knobs. Now the X-input switch is set to EXT position so that the internal time base voltage is disconnected and a vertical line is observed on the CRO screen. The length of this line is measured on the y-axis scale.
3. The reading corresponding to volts/div knob is noted.
4. The AC voltage, t, is also measured using an AC voltmeter.
5. Steps 2 to 4 are repeated for three more AC voltages tapped from the secondary of the transformer.
6. The observations are tabulated in Table 13.4.

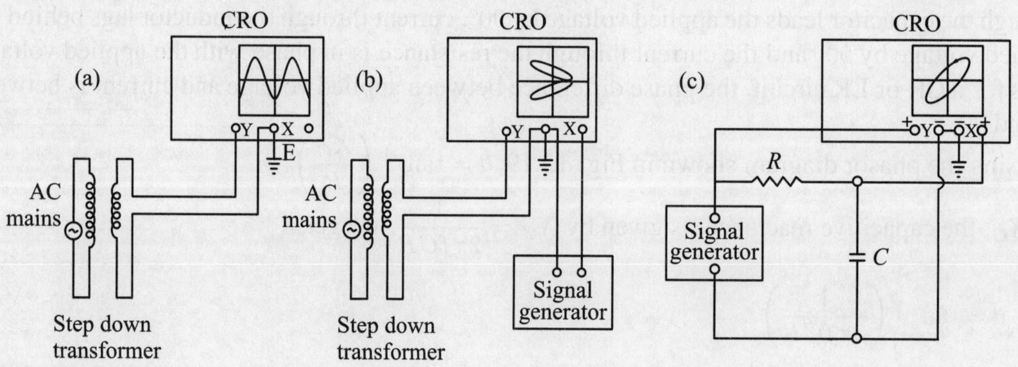

Figure 13.20 Circuit diagram for determination of (a) amplitude, (b) frequency and (c) phase difference

B. Measurement of frequency

1. Connections are made as shown in Fig. 13.20b.
2. The frequency of the signal generator is varied in a gradual manner from 10 Hz onwards to obtain a nearly stable Lissajous pattern.
3. The values of F_x and those of T_x and T_y as explained in part (B) of theory are noted.
4. Steps 2 and 3 are repeated for different Lissajous patterns. At least eight such sets of readings are taken.
5. The observations are tabulated in Table 13.5.
6. The frequency is also determined directly as explained in part B of the theory.

C. Measurement of phase difference

1. Connections are made as shown in Fig. 13.20c
2. The volts/div knob and the voltage knob of signal generator are adjusted to obtain an inclined ellipse of suitable maximum size.
3. The ellipse is centred around the origin. This is easily done by disconnecting the X- and Y-inputs and bringing the spot to the centre of the screen.
4. The values of y_1 and y_2 are noted as explained in part C of the theory.
5. The procedure is repeated for another value of R.
6. Phase difference is also determined directly as explained in part C of the theory.
7. The observations are tabulated in Tables 13.6A and 13.6B.

Observations and calculations

A. Measurement of amplitude

Table 13.4

S.No.	n **(div)**	p **(volts/div)**	t **(volts)**	A_{CRO} **(volts)** $= np/2$	$A_{\mathrm{voltmeter}}$ **(volts)** $= \sqrt{2}t$
1.					
2.					
3.					
4.					
5.					
6.					
7.					

B. Measurement of frequency

Number of divisions covered by one cycle = n = —— div

Reading of time/div knob $= q =$ —— s/div

Time period $= T = nq =$ —— s

Frequency $= F_y = 1/T =$ —— Hz

Table 13.5

S.No.	Frequency of signal generator F_x (Hz)	T_x	T_y	$F_y = (T_x/T_y)F_x$	Mean F_y
1.					
2.					
3.					
4.					
5.					
6.					
7.					
8.					

C. Measurement of phase difference

Table 13.6A

S.No.	$R(\Omega)$	n_1 (div)	r_1 (s/div)	$T(\text{s}) = n_2r_2$	$\alpha_P = (n_1r_1/T) \times 360°$
1.					
2.					
3.					
4.					
5.					
6.					
7.					
8.					

Table 13.6B $C =$ —— µF, $f =$ —— Hz

S.No.	$R(\Omega)$	y_1 (div)	y_2 (div)	$\alpha_P = \sin^{-1}(y_1/y_2)$	$\alpha_T = 90° - \tan^{-1}(1/2\pi fCR)$
1.					
2.					
3.					
4.					
5.					
6.					
7.					
8.					

Estimation of error

The errors ΔA, ΔF_y, and $\Delta\alpha_P$ in the estimation of amplitude, frequency and phase difference can be determined by adopting the method outlined in Section 1.4.2. Thus,

$$\Delta A = \frac{p\Delta n}{2} \tag{13.42a}$$

$$\Delta A = \sqrt{2}\Delta V_{\text{rms}} \tag{13.42b}$$

$$\Delta F_y = \frac{\Delta n}{qn^2} \tag{13.43a}$$

$$\Delta F_y = \left(\frac{T_x}{T_y}\right)\Delta F_x \tag{13.43b}$$

$$\Delta\alpha_P = 360^\circ\left(\frac{r_1}{r_2}\right)\left[\frac{n_1+n_2}{n_2^2}\right]\Delta n \tag{13.44a}$$

$$\Delta\alpha_P = \left[1+\frac{y_1}{y_2}\right]\left[\frac{\Delta n}{\sqrt{y_2^2-y_1^2}}\right] \tag{13.44b}$$

where Δn is the error in the reading of vertical or horizontal scale on the CRO screen, and ΔF_x is the error in the measurement of F_x. Table 13.7 illustrates the estimation of errors using the above expressions.

The measurement of distances using the vertical and horizontal scales of CRO involves reading of the end points and then taking their difference. Thus errors in both these readings should be considered in the estimation of total error. But in Eqs. 13.42a–13.44b we have considered the error in only one of these readings. Thus precaution should be taken to coincide the display with the gridline on one side such that the reading on the other side is only in error.

Table 13.7 Estimation of errors in A, F_y and α_P

Least count of the scale on CRO screen = 0.2 div
Least count of the signal generator = 1 Hz
Least count of AC voltmeter = 0.1 V
Thus, $\Delta n = 0.1$ div, $\Delta F_x = 0.5$ Hz and $\Delta V_{\text{rms}} = 0.05$ V

Parameter	Measured data	Absolute error	Relative error	Error (%)
Amplitude	$p = 2$ V/div, $n = 6$ div	$\Delta A = p\Delta n/2 = 0.1$ V	$\Delta A/A = \Delta n/n = 0.017$	1.7
	$V_{\text{rms}} = 4.4$ V	$\Delta A = \sqrt{2}\Delta V_{\text{rms}} = 0.07$ V	$\Delta A/A = \Delta V_{\text{rms}}/V_{\text{rms}} = 0.011$	1.1
Frequency	$q = 5$ ms/div, $n = 4$ div	$\Delta F_y = \Delta n/qn^2 = 1.25$ Hz	$\Delta n/n = 0.025$	2.5
	$T_x = 1$; $T_y = 1$ $F_x = 50$ Hz	$\Delta F_y = (T_x/T_y)\Delta F_x =$ 0.5 Hz	$\Delta F_x/F_x = 0.01$	1
Phase difference	$r_1 = 1$ ms/div, $n_1 = 1.8$ $r_2 = 5$ ms/div $n_2 = 4$ div	$\Delta\alpha_p = 360^\circ \times (r_1/r_2) \times [(n_1+n_2)/n_2^2] \times \Delta n = 2.6^\circ$	$\Delta\alpha_p/\alpha_p = [(n_1+n_2)/n_1n_2] \times \Delta n = 0.08$	8.0
	$y_1 = 0.8$ div $y_2 = 1.5$ div	$\Delta\alpha_p = [1+(y_1/y_2)] \times [\Delta n/\sqrt{y_2^2-y_1^2}] = 0.121$ rad $= 6.9^\circ$	$\Delta\alpha_p/\alpha_p = \{[1+(y_1/y_2)] \times [\Delta n/\sqrt{y_2^2-y_1^2}]\}/[\sin^{-1}(y_1/y_2)] = 6.9^\circ/32.2^\circ = 0.214$	21.4

Results

1. The values of amplitude determined using the CRO and those measured using a voltmeter are in close agreement. The error in the determination of amplitude is —— % using CRO, and —— % using voltmeter.
2. The frequency of AC mains is found to be —— ± —— Hz. The error in the determination of frequency is —— % by direct method, and —— % by Lissajous method.
3. The phase difference for a given combination of R and C is found as given below.

C (μF)	R (Ω)	α_p	Error in α_T	α_T	Error in α_T
		—— ± —— °	—— %	—— ± —— °	—— %
		—— ± —— °	—— %	—— ± —— °	—— %

Discussion

This experiment is convincing proof of the versatility of the CRO as a measuring instrument. We studied in Chart 13.3 that in the CRO the electron beam shot out by the electron gun experiences a force, which is a resultant of the forces due to the voltages applied to the X- and Y-plates. Thus the electron beam moves along some resultant path. The fluorescent screen converts this path into a visual display. In other words the display on the screen tells us about the voltage(s) present across the X- and Y-plates.

For example, if a circle is displayed on the screen, then the voltages present at the X- and Y-plates have equal amplitudes and differ in phase by 90°. If the display is a vertical straight line, then it means that there is an AC voltage across the Y-plates and there is no voltage across the X-plates. In this way we can analyse the various details of the unknown voltages fed to the inputs on the basis of what we see on the screen. The CRO is thus aptly described as the *eye of an electronic engineer*. The versatility of the CRO can be appreciated even more if one takes into account the fact that many physical quantities such as temperature and pressure are convertible to electrical form with suitable transducers and hence can be studied using the CRO.

The theory of the CRO is basically the theory of the motion of electrons in an electric field, which we have discussed in the preceding sections. The recipe that we learnt at the outset was to start with the knowledge of force and to finally draw a conclusion about the resultant path of the particle. In using the CRO we follow the reverse route: from the resultant path we decipher what the forces acting on the particle are.

For example, if the spot deflects upwards, we conclude that the force is acting in the upward direction [see Chart 13.2, $y/Y = l/2L, y = \frac{1}{2}(F/m)(l^2/v^2)$]. This means an electric field points in the downward direction ($\mathbf{F} = -e\mathbf{E}$), or, the upper plate is at a higher potential ($\mathbf{E} = -\boldsymbol{\Delta} V$). Thus a larger vertical deflection of the spot means a larger force along the y-axis, that is, larger electric field or larger voltage between the vertical deflection plates. A straight line making an angle of 45° with the x-axis means that the electrons are subjected to forces of equal magnitude along the x- and the y-axis, that is, AC voltages of equal amplitude and no phase difference are present at the two inputs.

The following simple illustration helps us understand the phenomenon of persistence of vision on the basis of interpretation of the display. The internal time base circuit is switched on and the time period of the saw tooth voltage is gradually increased. The spot starts moving to and fro along the x-axis getting faster and faster. As the spot begins to move rapidly, the actual movement of the spot is not seen, but only a single continuous horizontal straight line is observed. This happens due to persistence of vision. The movement of the spot is perceived by the eye if the time period of the saw tooth voltage is more than the time of persistence of vision, that is 0.1 s. If the time period is less than 0.1 s, then the eye fails to perceive the actual movement of the spot and instead a continuous line is seen.

There is an interesting analogy between Lissajous patterns in CRO and interference patterns in optics. In part B, that is, the measurement of frequency, voltages are fed from two independent sources (transformer and signal generator) as inputs, and unstable Lissajous patterns are observed on the screen. The patterns come closer to stability if the frequency ratio of the two voltages is an integer, but still they are not perfectly stable. The corresponding situation in optics is that two independent light sources cannot produce a stable interference pattern. The reason for instability in both the cases is that the frequencies of two independent sources can never match exactly and thus the phase difference between them keeps changing with time.

There is a slight residual difference between frequencies, which cannot be made to vanish as the sources are independent and each has a slightly fluctuating frequency. In fact the Lissajous figure is seen to tumble in one direction, become momentarily stationary and then tumble in the opposite direction as the oscillator frequency is fine tuned. The reason this happens is that because of the slight difference between the frequencies of the two signals, the Lissajous figure is not quite a closed figure. As the phosphor has a finite glow time, the figure seen at a time is the almost closed part of the full Lissajous figure at that time. We thus see the tumbling of the figure. As the oscillator frequency increases or decreases through the mains frequency value, the tumbling reverses direction.

There is another source of instability of the Lissajous pattern. This is the possibility of a random phase difference between the signals causing the pattern. The signals are from independent sources. Thus their phases are not correlated. If this random part of the phase difference were a considerable fraction of 2π, then the pattern would be seen as 'jumping' randomly. However, we do *not* see this. This is because the phase-noise in both the sources—mains and oscillator—is small, so that the random part of the phase difference between the signals is never so large as to manifest in a jumping pattern.

On the other hand we observe a stable ellipse in part (C), that is, the phase part. In this part the two voltages are not independent: they are voltage drops across two components in a CR circuit and are thus derived from the same voltage source. This situation is analogous to two sources of light derived from a single source leading to a stable interference pattern. The reason for stability in both the cases is that the two sources have the same frequency and thus have a constant phase difference.

In Chart 13.4 we studied various Lissajous patterns due to different values of phase difference between the two input voltages. These are theoretical patterns obtained using the graphical or mathematical method. We can verify these patterns using various combinations of voltages as inputs in part (C) of the experiment. This also helps us to study the various phase relationships in electrical circuits. In Table 13.8 these combinations and the resultant Lissajous patterns are displayed.

Table 13.8 Lissajous patterns due to different phase differences in electrical circuits

X- and Y-inputs	Phase relationship	Lissajous figure
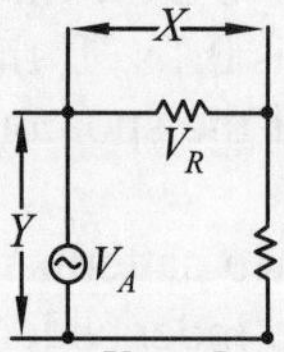 X, Y, V_R, V_A; $V_R \propto I$	The voltage across resistance is, according to Ohm's law, simply proportional to current through it and therefore in phase with it. Thus the applied voltage and voltage across the resistance are in phase.	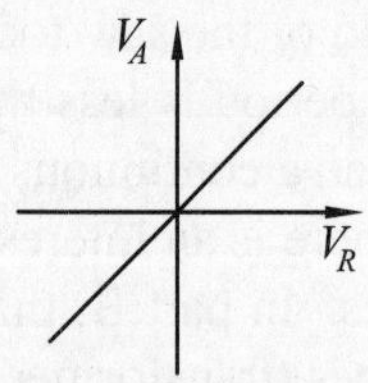 V_A, V_R
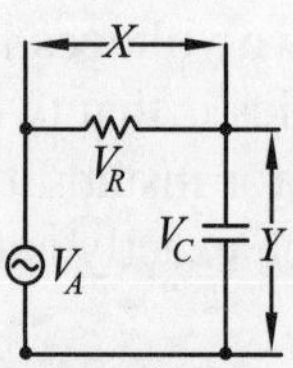 X, Y, V_R, V_C, V_A; $V_C \propto \int I\,dt$	The current through the capacitor charges the capacitor. As the charge builds up it opposes the current flow and the current reduces. Thus the charge on the capacitor and hence the voltage across it lags behind the current by 90°. Thus V_A and hence V_R leads V_C by 90°.	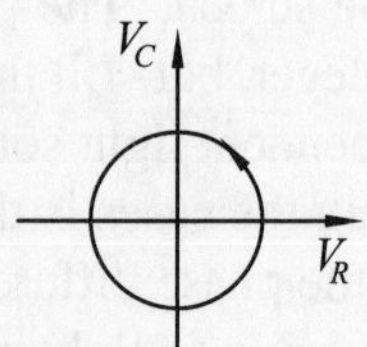 V_C, V_R
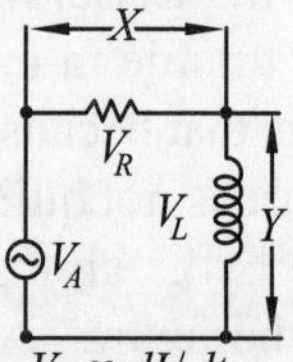 X, Y, V_R, V_L, V_A; $V_L \propto dI/dt$	The current through the inductor induces voltage in it that opposes the change in current. The induced voltage is proportional to the rate of change of current. The current is changing fastest when it is zero. Thus the voltage across inductor leads the current by 90°. Thus V_A, and hence V_R, lags behind V_L by 90°.	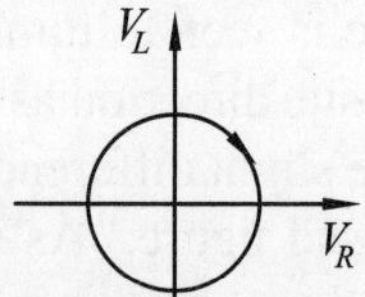 V_L, V_R
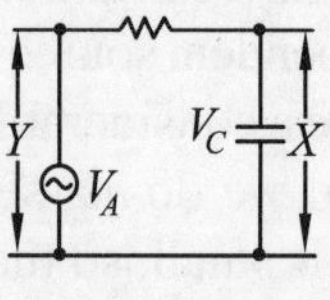 Y, X, V_C, V_A	V_A leads V_C by an angle smaller than 90°.	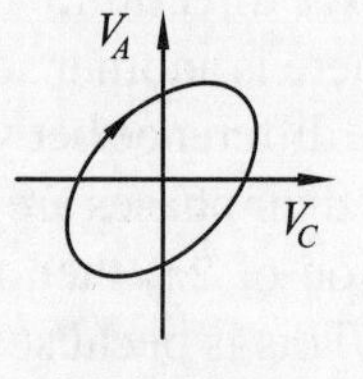 V_A, V_C
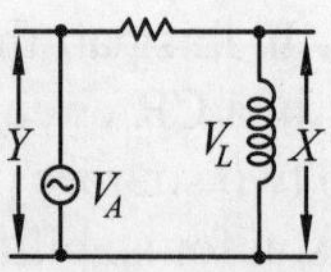 Y, X, V_L, V_A	V_A lags behind V_L by an angle smaller than 90°.	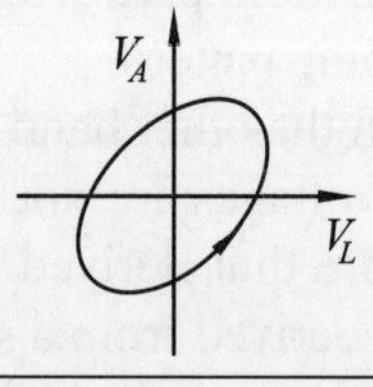 V_A, V_L

For measurements on the CRO screen care should be taken that the pattern on the screen occupies as much of the screen as possible. For example, in Table 13.7, if we use an enlarged ellipse such that $y_1 = 4 \times 0.8 = 3.2$ div and $y_2 = 4 \times 1.5 = 6$ div, then $\Delta\alpha_P = 1/4 \times 6.9°$ and $\Delta\alpha_P/\alpha_P = 1/4 \times 0.214$, that is, the error reduces to 1/4 th of the value obtained using a smaller ellipse.

Another precaution in the measurement of phase using dual trace CRO is that appropriate values of r_1 and r_2 should be selected for measurement of T in phase determination using a dual trace

CRO. For example, in Table 13.7, if we use $r_1 = r_2 = 5$ ms/div so that $n_1 = 0.36 \sim 0.4$ and $n_2 = 4$ then, $\Delta\alpha_P = 360° \times (4.4/16) \times 0.1 = 9.9°$ and $\Delta\alpha_P/\alpha_P = [4.4/(4 \times 0.4)] \times 0.1 = 27.5\%$, that is, the error increases to about 3.4 times the value obtained by considering the more appropriate values, $r_1 = 1$ ms/div and $r_2 = 5$ ms/div.

We can also apply the method of Section 1.4.2 to estimate the error in α_T. Thus we have,

$$\Delta\alpha_T = \frac{2\pi fCR}{1 + (2\pi fCR)^2}\left[\frac{\Delta R}{R} + \frac{\Delta C}{C} + \frac{\Delta f}{f}\right]$$

For example, if the component values, $R = 10^3\ \Omega, C = 0.022\ \mu\text{F}$ are known to 1% accuracy, that is, $\Delta R = 10\ \Omega, \Delta C = 0.00022\ \mu\text{F}$ and the frequency $f = 10^3$ Hz is measured with the scale of least count 100 Hz, that is, $\Delta f = 100/2 = 50$ Hz, then,

$$\Delta\alpha_T = \left[2\pi \times \frac{0.022}{1 + (2\pi \times 0.022)^2}\right]\left[\frac{10}{10^3} + \frac{0.022 \times 10^{-2}}{0.022} + \frac{50}{10^3}\right]$$

$$= \left[\frac{0.14 \times 10^{-2}}{1 + 0.02}\right][1 + 1 + 5] = 0.0096 \text{ rad} = 0.55°$$

EXPERIMENT 13.2 HALL EFFECT

Aim

To study the variation of Hall voltage with current and magnetic field and to determine the Hall coefficient and concentration, polarity and mobility of charge carriers in a given semiconductor

Apparatus

Constant current source, Hall probe, electromagnet with a calibrated gaussmeter, digital microvoltmeter

Formulae

Hall coefficient, $$R_H = \text{slope} \times \frac{t}{I}, \ \text{slope} = \frac{\delta V_\text{H}}{\delta B_\text{Z}} \tag{13.45}$$

where V_H is the Hall voltage, t is the thickness of the sample, B_Z is the applied magnetic field and I is the constant current through the sample.

Density of charge carriers, $$\rho = \frac{1}{R_H e} \tag{13.46}$$

where e is the electronic charge.

Mobility of charge carriers, $$\mu = R_\text{H}\sigma \tag{13.47}$$

where σ is the conductivity of the sample.

Theory

If a current is passed through a semiconductor placed in a magnetic field applied perpendicular to the direction of current flow, then an electric field is induced in a direction that is mutually perpendicular to the directions of both the current and the magnetic field. This phenomenon is known as the Hall effect. The induced field and the corresponding voltage is known, respectively, as the *Hall field* and the *Hall voltage*.

Consider a bar of p-type semiconductor in which the holes are majority carriers to be placed in a magnetic field $\mathbf{B_z}$, applied in the z-direction as shown in Fig. 13.21. Application of an electric field $\mathbf{E_x}$ produces current density J_x in the material. This means that the majority holes move in the x-direction with the drift velocity $\mathbf{v_d}$. These holes experience a Lorentz force due to the magnetic field $\mathbf{B_z}$ given by,

$$\mathbf{F_l} = e(\mathbf{v_d} \times \mathbf{B_z}) \tag{13.48}$$

Thus $|\mathbf{F_l}| = ev_d B_z$ and this force tends to drive holes towards the front face of the block. This results in an excess of holes at the front face and a deficiency of holes at the back face, leading to the creation of Hall field $\mathbf{E_y}$. In equilibrium, the force, $e\mathbf{E_y}$ counterbalances the Lorentz force such that there is no transverse current flow. Thus,

$$eE_y = F_l = ev_d B_z \tag{13.49}$$

or $$E_y = v_d B_z \tag{13.50}$$

Now, the current density J_x is given by (see Ex. 13.1)

$$J_x = pev_d \tag{13.51}$$

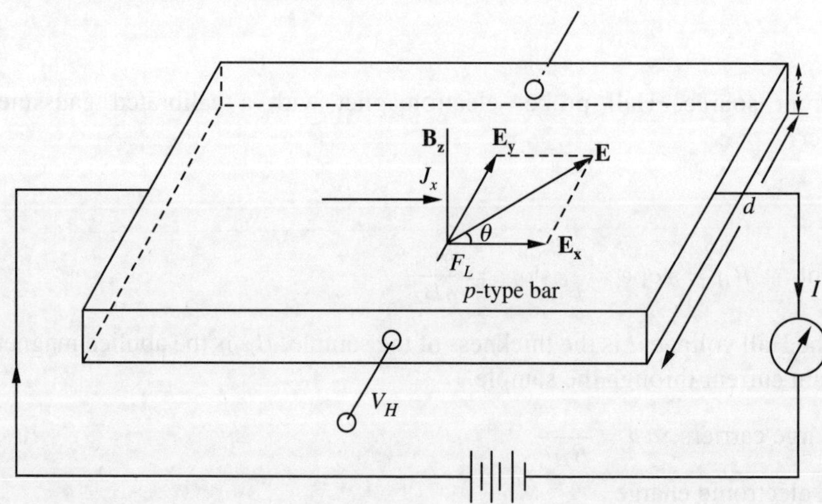

Figure 13.21 Hall effect in a p-type semiconductor: E_y is the Hall field, V_H is the Hall voltage and θ is the Hall angle

where p is the concentration of holes. Substituting for v_d from Eq. 13.51 in Eq. 13.50 we get,

$$E_y = \frac{J_x B_z}{pe} \tag{13.52}$$

Substituting $E_y = V_H/d$, where V_H is the Hall voltage and d is the breadth of the bar, we get

$$V_H = \frac{J_x B_z d}{pe} \tag{13.53}$$

But $J_x = I/td$, I being the current through the bar and t, the bar thickness. Hence,

$$V_H = \frac{IB_z}{pet} \tag{13.54}$$

The *Hall coefficient* is defined as $R_H = \dfrac{E_y}{J_x B_z} = \dfrac{1}{pe}$ (13.55)

$$\text{or} \quad R_H = \frac{V_H t}{IB_z} \tag{13.56}$$

In case of an n-type bar the Hall coefficient is

$$R_H = \frac{-1}{ne} \tag{13.57}$$

n being the concentration of electrons.

The *Hall angle*, θ as shown in Fig. 13.21 is given by

$$\tan\theta = \frac{E_y}{E_x} = \frac{B_z \sigma}{pe} \tag{13.58}$$

where we have used Eq. 13.52 and substituted

$$J_x = \sigma E_x \tag{13.59}$$

σ being the conductivity of the sample. The mobility is defined as

$$\mu = \frac{v_d}{E_x} \tag{13.60}$$

Using Eqs. 13.51, 13.55, 13.59 and 13.60 we have

$$\sigma = pe\mu \quad \text{or} \quad \mu = R_H \sigma \tag{13.61}$$

In the experimental arrangement (Fig. 13.22) current is passed through the sample, called the *Hall probe*, through the faces 1 and 2 using a constant current source. The resulting Hall voltage across the faces 3 and 4 is measured using a digital microvoltmeter.

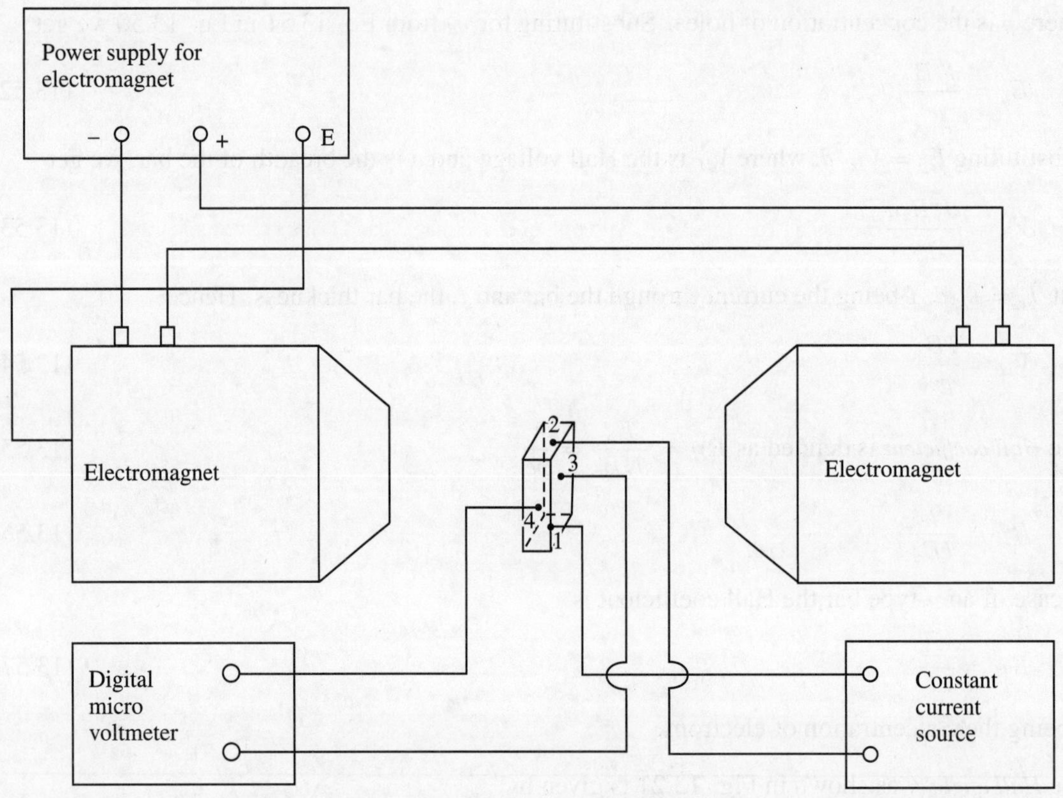

Figure 13.22 Experimental arrangement for the study of Hall effect

Procedure

1. The connections are made as shown in Fig. 13.22 and the Hall probe is placed so that the sample is parallel to the pole pieces of the electromagnet, that is, normal to B_z.
2. The current through the probe is adjusted to a value of 0.4 mA with the help of a knob provided on the front panel of the constant current source.
3. The magnetic field is varied from 0 to 5 kG in steps of 0.5 kG and the corresponding Hall voltage is measured by the digital microvoltmeter.
4. Steps 2 and 3 are repeated for the currents 0.8 mA and 1.2 mA through the probe.
5. The values of Hall voltage are noted for the currents 0.4 mA, 0.8 mA and 1.2 mA by switching off the power supply to the electromagnet. These spurious values should be subtracted from the observed values to obtain the actual values of voltage.
6. The observations are tabulated and the graphs are plotted between magnetic field (along the x-axis) and Hall voltage (along the y-axis) for different values of current.
7. The polarity of charge carriers in a given semiconductor is determined using steps explained in Fig. 13.23.

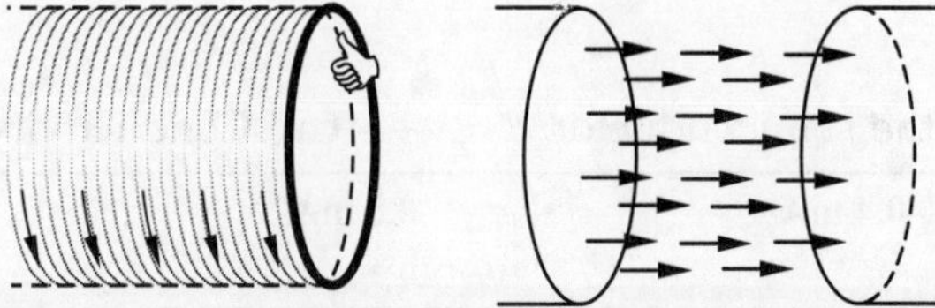

Step 1: Use the right-hand rule to determine the direction of **B**, the magnetic field. Here it is to the right.

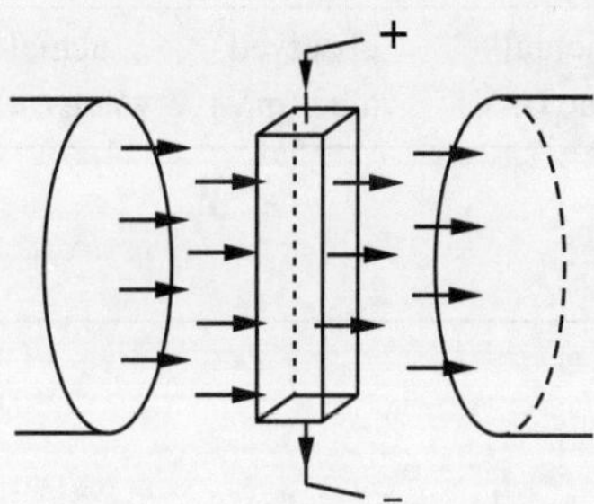

Step 2: Determine the direction of I, the current from the polarity of power supply providing the current in the sample.

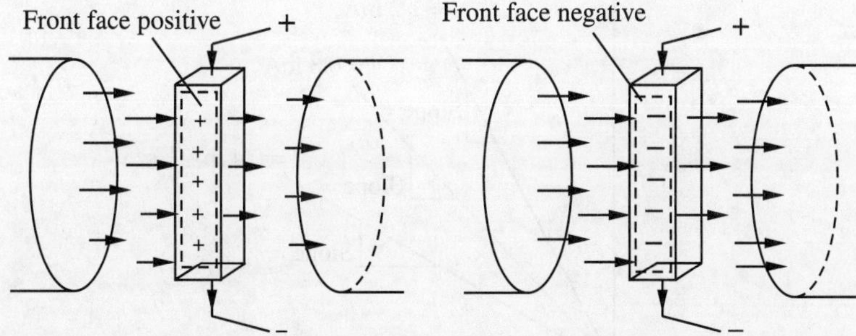

Step 3: Use the Hall voltage mea-surement to determine the polarity of front face relative to the back face of the sample. If the front face is positive relative to the back face, the sample is *p*-type; if it is negative, the sample is *n*-type. (See the following figures.)

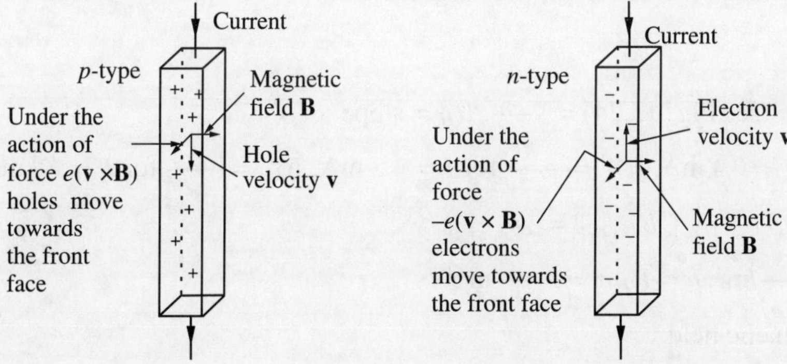

Figure 13.23 Determination of type of semiconductor from Hall voltage polarity

Observations

Table 13.9 Thickness of the semiconductor, t = —— cm, Conductivity, σ = —— $\text{ohm}^{-1}\text{cm}^{-1}$

Current I through the sample ⟶	**0.4 mA**		**0.8 mA**		**1.2 mA**	
Magnetic field, B_z (kG) ↓	Hall voltage, V_H					
	observed value (mV)	actual value (mV)	observed value (mV)	actual value (mV)	observed value (mV)	actual value (mV)

Graph

See Fig. 13.24.

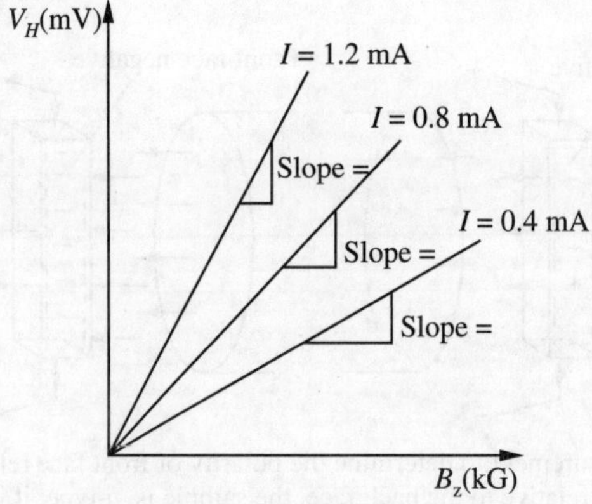

Figure 13.24 Variation of Hall voltage with magnetic field

Calculations

Slope = $\delta V_H/\delta B_z$ (see Fig. 13.24) = ——, R_H = slope × t/I

R_{H_1} = —— for $I = 0.4$ mA, R_{H_2} = —— for $I = 0.8$ mA, R_{H_3} = —— for $I = 1.2$ mA

Mean $R_H = (R_{H_1} + R_{H_2} + R_{H_3})/3$ = ——

$\rho = 1/R_H e$ = —— /m^3, $\mu = R_H\sigma$ = —— m^2/Vs

Direction of magnetic field:
Direction of current:
Sign of Hall voltage:
Direction of Lorentz force:
Thus, the polarity of charge carriers:

Estimation of error

The error ΔR_H in the estimation of R_H can be determined by adopting the method outlined in Section 1.4.2. Thus,

$$\Delta R_H = \left(\frac{t}{I}\right) \text{slope} \left[\frac{\Delta I}{I} + \frac{2\Delta V_H}{(V_{H_2} - V_{H_1})} + \frac{2\Delta B_Z}{(B_{Z_2} - B_{Z_1})}\right] \tag{13.62}$$

where slope $= (V_{H_2} - V_{H_1})/(B_{Z_2} - B_{Z_1})$ and $\Delta I, \Delta V_H, \Delta B_Z$ are the errors in the readings of current, Hall voltage and magnetic field respectively. Table 13.10 illustrates the estimation of error using the above expression.

Table 13.10 Estimation of error in Hall coefficient

Least count of the constant current source = 0.1 mA
Least count of microvoltmeter = 0.1 mV
Least count of gaussmeter = 0.1 kG
Thus, $\Delta I = 0.05$ mA, $\Delta V_H = 0.05$ mV and $\Delta B_Z = 0.05$ kG

Measured data	Absolute error	Relative error	% error
$t = 0.4$ mm $I = 0.8$ mA $V_{H1} = 12.1$ mV $V_{H2} = 15.9$ mV $B_{Z1} = 3.0$ kG $B_{Z2} = 4.0$ kG	$\Delta R_H = (0.4 \times 10^{-3}\ \text{m}/0.8 \times 10^{-3}\ \text{mA}) \times 3.8 \times 10^{-2}\ \text{V/Wbm}^{-2} \times [0.05\ \text{mA}/0.8\ \text{mA} + 2 \times 0.05\ \text{mV}/(15.9-12.1)\ \text{mV} + 2 \times 0.05\ \text{kG}/(4-3)\ \text{kG}] = 1.9 \times 10^{-2}\ \text{C}^{-1}\ \text{m}^3\ (0.062 + 0.026 + 0.1) = 0.357 \times 10^{-2}\ \text{C}^{-1}\ \text{m}^3$	$\Delta R_H/R_H = 0.188$	18.8%

Results

1. The Hall voltage varies linearly with both the current and the magnetic field.
2. The value of Hall coefficient is found to be —— ± —— m^3/C . The error in the determination of Hall coefficient = —— %.
3. The concentration of majority carriers is found to be —— ± —— /m^3. The error in the determination of concentration = —— %.
4. The mobility of charge carriers is found to be —— ± —— m^2/Vs. The error in the determination of mobility = —— %.
5. The given semiconductor is of —— type because the majority charge carriers have —— polarity.

Discussion

This experiment is a novel application of our knowledge of motion of charged particles in electric and magnetic fields to the determination of solid state properties such as polarity, mobility and density of charge carriers in a solid under study. If these properties are known accurately for a particular sample then that sample can be used as a probe for measurement of magnetic field.

The crux of the theory of Hall effect is the Lorentz force. Thus the various observations in this experiment can be explained in terms of this force. In the following table we compare the experiment and the theory.

Experiment	Theory
Voltage is observed across faces 3 and 4 using the experimental arrangement shown in Fig. 13.22.	The charge carriers in the Hall probe (sample) experience the Lorentz force $\mathbf{F} = q(\mathbf{v_d} \times \mathbf{B})$ due to the magnetic field and hence accumulate on face 3 or 4 (depending on whether the carriers are holes or electrons). This gives rise to Hall voltage.
Hall voltage is found to increase with both the current and the magnetic field (Fig. 13.24).	Lorentz force increases with both the current and the magnetic field thus causing more and more charge carriers to be accumulated, that is, increasing the Hall voltage.

It can be seen from this table that there is a qualitative agreement between the experiment and the theory. If $B_z = 0$ then $F = 0$ and we should not observe any Hall voltage. But we do get some voltage that is not due to the Hall effect and is thus called the *spurious voltage*.

From Table 13.10 we see that error contributions due to measurement of magnetic field, current and Hall voltage to the error ΔR_H are of the same order. If the slope $\Delta V_H/\Delta B_Z$ is calculated using the method of least squares (see Section 1.4.3), then the best possible value of the Hall coefficient can be estimated from the given set of observations.

EXERCISES

1. What is Lorentz force? Discuss the path travelled by an electron in a uniform transverse electric field.
2. State the shape of trajectories of an electron of velocity v moving in a uniform electric field E when (i) v is perpendicular to B and (ii) v makes an acute angle with E. What are these shapes if E is replaced by a magnetic field B?
3. Discuss the motion of an electron projected into a uniform electric field at an acute angle with the field direction, and hence obtain the formula for range, time of flight and maximum height attained by a particle.
4. Show that an electron moves along a circular path when it enters a transverse uniform magnetic field.
5. In a transverse uniform magnetic field, show that the radius of curvature of the path of the charged particle is proportional to its momentum.
6. Obtain the expressions for the radius and pitch of a helical path described by the charged particle when it enters the uniform magnetic field, making an acute angle 'θ' with the direction of the magnetic field.
7. Show that the velocity acquired by an electron in a uniform electrostatic field varies as the square root of potential difference through which it is accelerated. (Given $v||E$)
8. Explain why slower particles and faster particles require the same time for completing one rotation in a magnetic field.
9. Show that the kinetic energy of a charged particle remains constant when it moves in a magnetic field.
10. How is it possible for a charged particle to pass through a combination of electric and magnetic field without any deviation?
11. Explain Thomson's method for determination of *e/m* of an electron.
12. State the law that governs the refraction of electrons. How does it resemble Snell's law and how does it differ from it?
13. Explain with a labelled diagram, the construction and working of a cyclotron. Obtain the expressions for (i) resonance condition and (ii) maximum kinetic energy gain for a charged particle. Why can electrons not be accelerated to high energies in a cyclotron?
14. Explain in brief the principle, construction and working of a Bainbridge mass spectrograph.
15. Draw a neat and well-labelled schematic of the cathode ray tube. How can the intensity and focussing of the trace on the screen be controlled?
16. Draw a schematic of an electrostatic CRT. Describe the role of (i) an electron gun, (ii) a deflection system, (iii) a fluorescent screen and (iv) an aquadag coating.
17. Which voltage controls the intensity of the electron beam on the screen in CRT? What is the function of the voltage given to the anode cylinders in it? What is the use of aquadag coating in CRT?
18. Draw a block diagram of CRO. State the function of different blocks. Is it possible to display the waveform on the CRO screen without a time base generator? Explain.

19. If the two voltages V_x and V_y with amplitudes A_1 and A_2, same frequency and a phase difference θ are fed to the X and Y plates of CRO, respectively, state the expression for the resultant figure displayed on the CRO screen . Hence, which figure will be observed on the CRO screen if $\theta = 0°$ and $\theta = 90°$ for $A_1 = A_2$.
20. What is synchronisation? How is it achieved in the CRO? Explain how the sharpness of the trace is controlled in the CRO.

PROBLEMS

1. An electron is liberated at rest from one of two large parallel plates separated by 2 cm. The other plate has a relative potential of 2.4 kV. How long does the electron take to reach it?
2. A proton accelerates from rest in a uniform electric field of 500 V/m. At some later time, its speed is 2.5×10^6 m/s. Find (i) the acceleration, (ii) the time taken by the proton to reach this speed and (iii) the distance travelled by the proton during this time.
3. A proton travelling at 23° with respect to the magnetic field strength 2.63 mT experiences a magnetic force of 6.48×10^{-17} N. Calculate (i) the speed and (ii) the kinetic energy in eV of the proton.
4. The electrons are circulating in a uniform magnetic field of 4.55×10^{-4} Wb/m^2 with kinetic energy 22.5 eV. What is the radius and the period of revolution of the electron path?
5. An electron describes a helix of radius 6 cm and pitch 2 cm in a magnetic field of intensity 30 gauss. Obtain the component of its velocity along the perpendicular to the magnetic field.
6. An electron moving with a velocity of 10^6 m/s enters a magnetic field at an angle of 20° to the direction of the field. Calculate (i) the value of the magnetic induction for the radius of the helical path to be 2 cm, (ii) the time required to execute one revolution, and (iii) the pitch of the spiral.
7. A positive ion beam moving along the x-axis enters a region of uniform electric field of 3 kV/m along the y-axis and magnetic field of 1 kG along the z-axis acting simultaneously. Calculate the speed of those ions which pass undeviated. What will happen if ions are moving slower or faster than this velocity?
8. In an experiment to determine the e/m of electrons by Thomson's method, the electric field between plates of 6 cm length and 2 cm separation at a potential difference of 480 V produces a deflection of 10.9 cm on the screen 40 cm away. A magnetic field of 8×10^{-4} T restores the beam to its original position. Calculate e/m.
9. In a Bainbridge mass spectrograph, the magnetic field in the velocity selector is 1 T and the ions with a speed of 4×10^6 m/s pass through it undeflected. (i) What should be the electric field between the plates? (ii) If the separation of the plates is 0.5 cm, what is the potential difference between the plates?
10. A particle cyclotron is designed with dees of radius 75 cm and magnets that can provide a field of 1.5 T. (i) To what frequency should the oscillator be set if deuterons are to be accelerated? (ii) What is the maximum energy of deuterons that can be obtained?

11. An n-type Ge sample has a donor density of $10^{21}/\text{m}^3$. It is arranged in a Hall effect experiment with magnetic field of 0.5 T and current density 500 A/m^2. Find the Hall voltage if the sample is 3 mm wide.
12. A specimen of a semiconductor is 1 mm thick and 1 cm wide. A magnetic flux density of 0.5 T is applied parallel to the 1 mm edge and Hall voltage contacts are attached to measure the voltage across the width of the sample. The current flowing lengthwise through the sample is 10 mA. If the Hall coefficient of the material is 3.66×10^{-4} m^3/C, compute the voltage between the Hall contacts. If the resistivity of the material is 8.93×10^{-3} Ωm, find the Hall angle.

14

Lasers

14.1 THE STORY OF LASERS

In Chapter 6 we studied about the quantum theory—a new theory that endowed physicists a fresh perspective of light. Light can also be looked upon as a stream of photons. This helped physicists to explain many of the observed properties of light. Research in quantum optics led to the crowning success of modern physics—the discovery of lasers.

The term LASER conjures up images of incredible sources of light energy, and magic light machines that can destroy satellites, cut and drill metals, perform complicated surgeries and provide sophistication to almost all walks of life. The immense publicity of lasers in the last few decades sometimes make people think of it as an extraordinary invention based on magical concepts. However, it is not so and the discovery of lasers (like all other discoveries) can be attributed to systematic research endeavours of scientists and engineers for the last several years.

The story of the development of lasers is the story of physicists' quest to 'purify' light. Understanding the nature of light, which we have covered in Chapters 2–6, has been a part of this story. Natural light is unpolorised, incoherent (disordered), divergent and weak. Thus research had been aimed at obtaining light that is polarised, coherent, convergent and intense. The breakthrough in achieving all these properties did not come at once.

The earliest sources of light were based on the thermionic heating of the filament in evacuated glass bulbs. After the advent of the quantum theory, fluorescent lamps were developed on the basis of discharge of electricity through gases. These lamps gave 'cooler' light compared to that from an electric bulb, because it consisted of purely visible radiation with an extremely small infrared component. The intensity of these sources was limited by the Wattage of the filament and the quality of the vacuum used. Moreover, the light was incoherent, divergent and unpolarised. Thus no real progress was made and a further breakthrough required a 'peep into the deep', that is, a new insight into the mechanism of emission of light. It was once again the genius of Einstein that provided this breakthrough.

In 1917, Einstein made a theoretical speculation that much intense and coherent light can be obtained if the emission process can be stimulated instead of allowing it to take place spontaneously as in the case of ordinary light. However, due to the enormous practical difficulties involved in the fabrication of a real laser, it took 43 years before the first operational laser source could see the light of day in 1960.

The history of these 43 years is, therefore, a story in technology. It is replete with intermediate achievements of several physicists and engineers (mainly from the USA and the USSR), that led to the invention of the first practical laser source. In 1939, V Fabrikant of the USSR suggested that amplification of electromagnetic radiation could be achieved through stimulated emission. The Soviet physicists N. Basov and A. Prokhorov, and the American physicists, Charles Townes and Arthur Schawlow, suggested an amplifier in the microwave range which later came to be known as Maser (Microwave Amplification by Stimulated Emission of Radiation). The theory and development of maser is credited to these physicists and was known by 1954. In 1958, these physicists extended these concepts to the optical range and suggested the possibility of a laser as an optical maser.

The culmination of these efforts happened in 1960 when Theodore Maiman, an American physicist, demonstrated the first operational laser, known as Ruby Laser, at the Hughes Research Laboratory. Since 1960, several lasers have been developed so far and are continuing to be developed even today. Though the laser design technology has aided this development, the physics behind the lasing process continues to be the same. In this chapter we will study the basic principles required for understanding the laser action (Section 14.2), a few important laser sources (Section 14.3), the characteristics of lasers (Section 14.4), semiconductor photonic devices (Section 14.5) and some major applications of lasers (Section 14.6).

14.2 INTRODUCTION

Laser is an acronym for Light Amplification by Stimulated Emission of Radiation. As the name suggests the amplification is achieved through the process of stimulated emission. The process of stimulated emission can be understood as one of three possibilities arising out of the interaction of matter with radiation. Figure 14.1 depicts these three possibilities.

If a photon of energy $h\nu = E_2 - E_1$ is incident on an atom present in the lower state E_1, then the photon is absorbed by an atom, raising it to a higher state E_2 (see Fig. 14.1a). This is the familiar *absorption* process.

Emission is due to the transition of an atom from level E_2 to level E_1. If the emission takes place of its own accord, that is, an atom comes to state E_1 from E_2 of its own accord (Fig. 14.1b), it leads to another familiar process—*spontaneous emission.* This is a random, uncontrolled emission resulting in incoherent light. Spontaneous emission dominates the mechanism producing light from all ordinary sources of light.

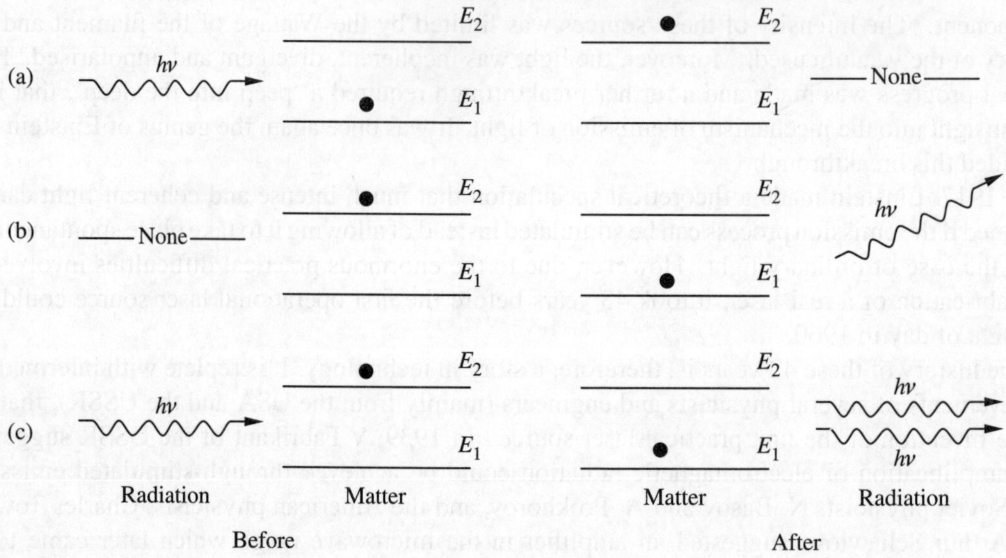

Figure 14.1 Interaction of matter with radiation resulting in (a) absorption (b) spontaneous emission and (c) stimulated emission

In a process of *stimulated emission* (Fig. 14.1c) an incident photon forces an atom in the state E_2 to move to state E_1. The photon emitted in this process has the same frequency, same polarisation and the same direction of propagation as that of the incident photon. Stimulated emission is thus an ordered, controlled process leading to coherent light.

For laser action *stimulated emission should dominate over both absorption and spontaneous emission.* Einstein applied PlanckŠs quantum hypothesis to show that the probability of stimulated emission is the same as the probability of absorption in a system of atomic oscillators in equilibrium with a large number of photons. Example 14.1 illustrates how this result can be obtained.

EXAMPLE 14.1

Relations between Einstein's coefficients

According to Planck's radiation law, the energy density $u(\nu)$ is given as

$$u(\nu) = \frac{8\pi h\nu^3}{c^3}\left[\frac{1}{\exp(h\nu/kT) - 1}\right]$$

Consider a system having N_1 atoms in the energy state E_1 and N_2 atoms in the state E_2, in equilibrium with $u(\nu)$. Obtain the following Einstein's relations for this case

$$B_{12} = B_{21}$$

$$\frac{8\pi h\nu^3}{c^3}B_{21} = A_{21}$$

where, A_{21}, B_{12}, B_{21} are Einstein coefficients, defined as,

A_{21} = probability per unit time for spontaneous transition from E_2 to E_1,

$B_{12}u(\nu)$ = probability per unit time for induced transition from E_1 to E_2 with the absorption of a photon

$B_{21}u(\nu)$ = probability per unit time for induced transition from E_2 to E_1 with the emission of a photon.

Solution

The transition rate for photon emission = Probability of emission $\times N_2$

$$= A_{21}N_2 + B_{21}u(\nu)N_2 = [A_{21} + B_{21}u(\nu)]N_2$$

The transition rate for photon absorption = Probability of absorption $\times N_1$

$$= B_{12}u(\nu)N_1$$

At thermal equilibrium, the rate of the emission process equals that of the absorption process.

$$\text{Hence } [A_{21} + B_{21}u(\nu)]N_2 = B_{12}u(\nu)N_1 \tag{14.1}$$

$$\text{or} \quad u(\nu) = \frac{A_{21}}{[(N_1B_{12}/N_2) - B_{21})]} \tag{14.2}$$

Now the ratio,

$$N_2/N_1 = \exp\left[\frac{-(E_2 - E_1)}{kT}\right] = \exp\left(\frac{-h\nu}{kT}\right) \quad \text{(see Eq. 14.5)}$$

Substituting in Eq. 14.2 we get,

$$u(\nu) = A_{21}/[\exp\left(\frac{h\nu}{kT}\right)B_{12} - B_{21}] \tag{14.3}$$

Comparing Eq. 14.3 with Planck's formula,

$$u(\nu) = \left(\frac{8\pi h\nu^3}{c^3}\right)\left\{1/\left[\exp\left(\frac{h\nu}{kT}\right) - 1\right]\right\}, \text{ we get,}$$

$$\left(\frac{8\pi h\nu^3}{c^3}\right)B_{21}\left\{1/\left[\exp\left(\frac{h\nu}{kT}\right) - 1\right]\right\} = A_{21}/\left[\exp\left(\frac{h\nu}{kT}\right)\left(\frac{B_{12}}{B_{21}}\right) - 1\right]$$

$$\text{Thus, } B_{12} = B_{21} \text{ and } \left(\frac{8\pi h\nu^3}{c^3}\right)B_{21} = A_{21} \tag{14.4}$$

These are the relations for Einstein's coefficients. The relation $B_{12} = B_{21}$ signifies that the probability of stimulated emission equals that of stimulated absorption.

14.2.1 Population Inversion

Stimulated emission leads to coherent light and is thus the key to laser operation. Figure 14.1c shows two coherent photons as a result of stimulated emission. For the amplified output we need several such coherent photons. This is possible only if there are a large number of atoms in an excited state. However, this is not the equilibrium situation. In thermal equilibrium at temperature T, the ratio of number of atoms in level E_2 to the number in level E_1 is given by the *Boltzman factor* as

$$\frac{N_2}{N_1} = \exp\left[-\frac{(E_2 - E_1)}{kT}\right] \tag{14.5}$$

Thus, for $E_2 > E_1, N_2 < N_1$, that is, the number of atoms in the higher energy level is always smaller than the number in the lower level (Fig. 14.2a). If this situation is reversed by some means and the population of the higher level is more than the population of the lower level (Fig. 14.2b), then *population inversion* is achieved. This condition is also sometimes called as *negative temperature condition*, because if T is made negative in Eq. 14.5, then $N_2 > N_1$, for $E_2 > E_1$. However, this is a misnomer as T(K) cannot be made negative. Here T should be understood as a non-physical temperature referring to a situation that is not a thermal equilibrium.

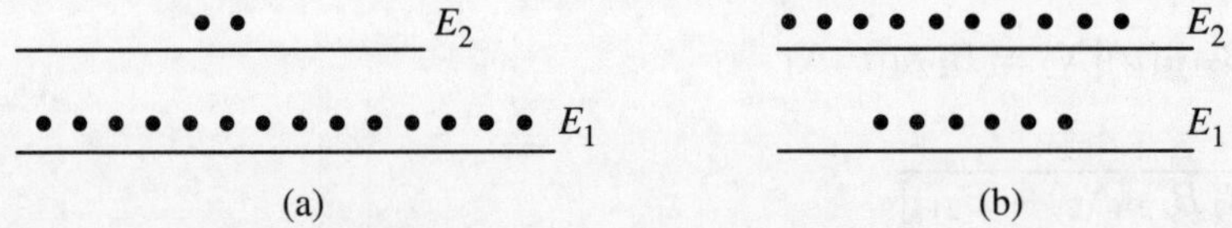

Figure 14.2 (a) The normal equilibrium distribution of atoms in thermal equilibrium, $N_1 > N_2$ and (b) The distribution after population inversion, $N_2 > N_1$

14.2.2 Metastable State

To achieve population inversion we must raise atoms to higher energy states. But atoms cannot stay in an excited state for a long time. The lifetime of an excited state is of the order of 10^{-8} s. Thus atoms return to the ground state in times less than 10^{-8} s and populations cannot be inverted normally. To achieve population inversion one has to discover states which are relatively long lived. These levels are called *metastable states* meaning that they have intermediate stability. They have life times of the order of 10^{-3}s (Fig. 14.3).

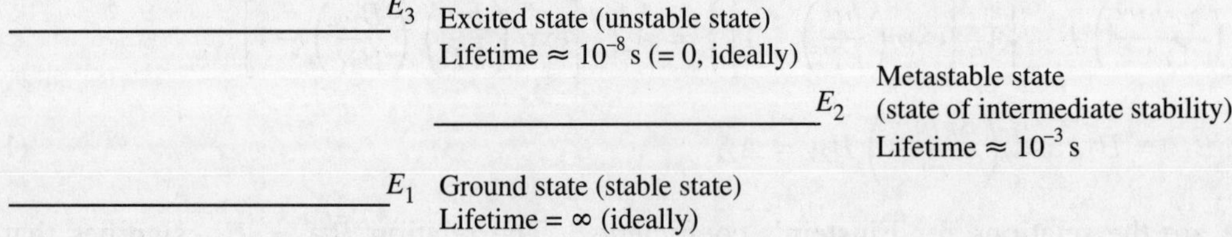

Figure 14.3 Metastable states have intermediate lifetime and stability compared to the ground state and the excited state

14.2.3 Pumping

Population inversion involves raising atoms to higher energy levels. This is achieved by supplying energy from outside the system and is usually referred to as 'pumping'. The atoms are pumped to the higher level by supplying the adequate pump energy. Example 14.2 illustrates why 'thermal pumping' is not possible. The various pumping mechanisms employed in different lasers are mentioned in Chart 14.2.

EXAMPLE 14.2

Why 'thermal pumping' is not possible

A beam of light emitted by a certain laser source has a wavelength of 6000 Å. (a) What will be the equilibrium ratio of the population of the upper level E_2 to that of the lower level E_1 at room temperature? (b) If 'thermal pumping' is assumed, that is, if one wishes to increase the population of the upper level by raising the temperature, at what temperature will the ratio be ‰?

Solution

(a) The energy of the emitted photon which corresponds to the difference $E_2 - E_1$ is given by hc/λ,

$$12422/6000\ \text{Å} = 2.07\ \text{eV}$$

Now kT at room temperature = $(8.62 \times 10^{-5}\ \text{eV/K})(300\ \text{K}) = 0.026\ \text{eV}$

Using Eq. 14.5,

$$\frac{N_2}{N_1} = \exp\left[\frac{-(E_2 - E_1)}{kT}\right] = \exp\left(\frac{-2.07\ \text{eV}}{0.026\ \text{eV}}\right) = 2.65 \times 10^{-35}$$

Thus at room temperature there is a negligible fraction of atoms present in level E_2. (For 10^{35} atoms in state E_1 there is one atom in state E_2!). This is to be expected, however, because an atom with mean thermal energy equal to 0.026 eV at room temperature has a negligible chance of raising another atom to an energy level 2.07 eV higher through a collision.

(b) $$\frac{N_2}{N_1} = \exp\left[\frac{-(E_2 - E_1)}{kT}\right]$$

or $$\ln\frac{N_2}{N_1} = \frac{-(E_2 - E_1)}{kT}$$

For $\frac{N_2}{N_1} = \frac{1}{2}$ we get,

$$T = \frac{E_2 - E_1}{k \ln 2} = \frac{2.07\ \text{eV}}{(8.62 \times 10^{-5}\ \text{eV/K}) \ln 2} = 34{,}645\ \text{K}$$

Such an enormous temperature (greater than the temperature of the sun's surface) is required to raise only half of the atoms in level E_1 to level E_2. This example thus demonstrates why *thermal pumping is not a possible mechanism of pumping*.

14.2.4 Basic Laser Action

Figure 14.4 illustrates the basic laser action of a typical laser. Atoms in the ground state E_1 are pumped to an excited state E_3 using a suitable pumping source. These atoms do not stay in state E_3 for longer than 10^{-8} s and make *non-radiative* transitions to the metastable state E_2. A non-radiative transition is one in which the atom shifts to another energy level not by emitting a photon but via some other mechanism such as, for example, atomic collision. A radiative transition is one in which a photon is emitted. In time the state E_2 gets more and more populated till its population far exceeds that of E_1. A population inversion is thus achieved between levels E_1 and E_2. A spontaneous emission from E_2 to E_1 then produces a photon of the correct energy to stimulate further emission and produce a second identical photon. This photon stimulates further transition. An avalanche of coherent photons is thus produced. The state E_1 where the laser action ends is called the *laser terminal state*. Figure 14.4 illustrates the three-level scheme of laser operation. There can be a four-level laser also in which the laser terminal state is not the ground state but a state above the ground state. Chart 14.1 illustrates the difference between a three-level and a four-level laser. It also includes an explanation of why a two-level laser is not possible.

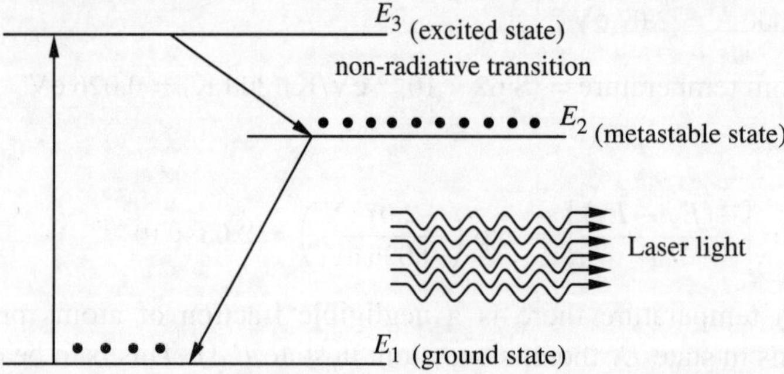

Figure 14.4 Basic laser action

14.2.5 Resonator

The discussion so far explains how light can be amplified by stimulated emission. In order to obtain a laser source, which could supply light energy, a part of the output must be fed back as an input so that the mechanism is self-sustaining. This is similar to the conversion of an electronic amplifier into an electronic oscillator using positive feedback. This can be realised using a system of two opposite parallel mirrors, one of which is completely reflecting and the other is partially reflecting. Such a system is called a resonant cavity or a resonator. Figure 14.5 illustrates how a resonator helps achieve oscillator action for a laser amplifier.

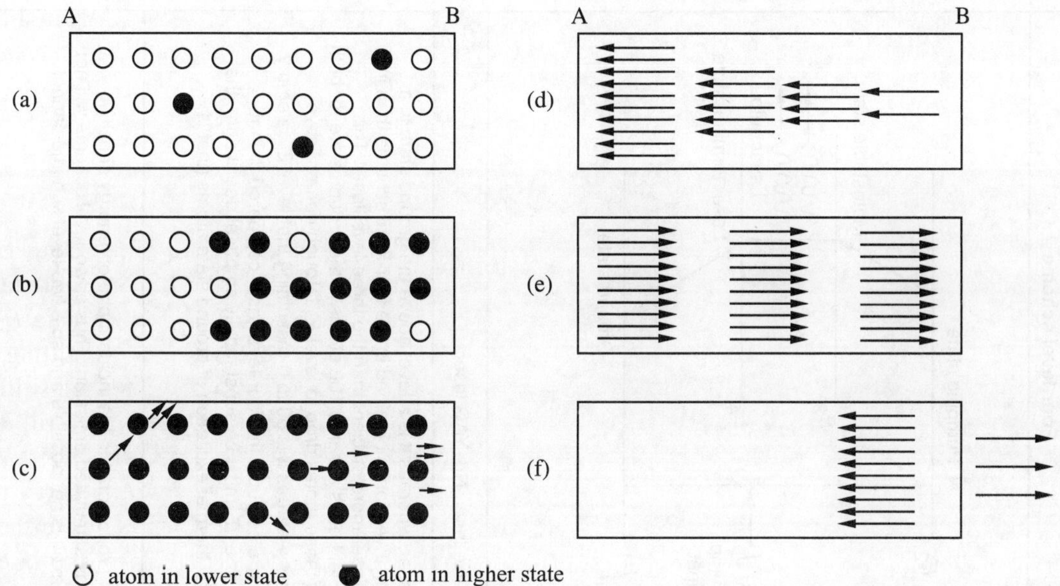

Figure 14.5 Resonator showing the oscillator action for a laser amplifier. Mirror A is perfectly reflecting and B is partially reflecting.

The normal equilibrium distribution of atoms (Fig. 14.5a) in which most of the atoms are in a lower state is taken to an inverted population state (Fig. 14.5b) by means of some pumping mechanism. Initially photons are emitted randomly and some of them are lost through the sides. But one of the photons moving perpendicular to the mirrors initiates stimulated emission (Fig. 14.5c). The photons reflected by mirror B continue to augment the number of coherent photons through further stimulated emission (Fig. 14.5d). This process continues further after reflection from mirror A. Thus the photons directed along the axis of the resonator bounce back and forth and build strong oscillations (Fig. 14.5e). Finally, the laser output leaks through the partially reflecting mirror B (Fig. 14.5f). Multiple reflection between mirrors results in the formation of standing waves and the cavity supports those modes of oscillation for which

$$2L = n\lambda, n = 1, 2, 3, \ldots \tag{14.6}$$

where L is the cavity length and λ is the wavelength of the radiation in the medium enclosed by the cavity. Substituting $\lambda = c/\nu$ in Eq. 14.6 we get the resonant frequencies of the cavity as

$$\nu = \frac{nc}{2L} \tag{14.7}$$

These frequencies are also called the *longitudinal modes* of the cavity, n being the *mode number*; (See Ex. 14.3.)

Part of the amplified light escapes through a mirror B, giving a highly intense, directional and coherent laser output. Mirror B is also responsible for feeding a part of the radiation back into the cavity so that the mechanism of build-up of radiation sustains itself.

Chart 14.1 Different Level Schemes in Lasers

<table>
<tr><th>Scheme</th><th>Three-level scheme</th><th>Four-level scheme</th></tr>
<tr><td>Diagram</td><td> 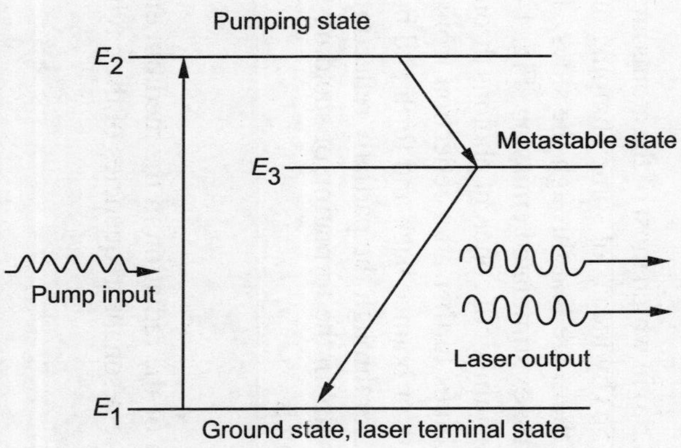Pumping state
E_2
Metastable state
E_3
Pump input
Laser output
E_1
Ground state, laser terminal state</td><td> 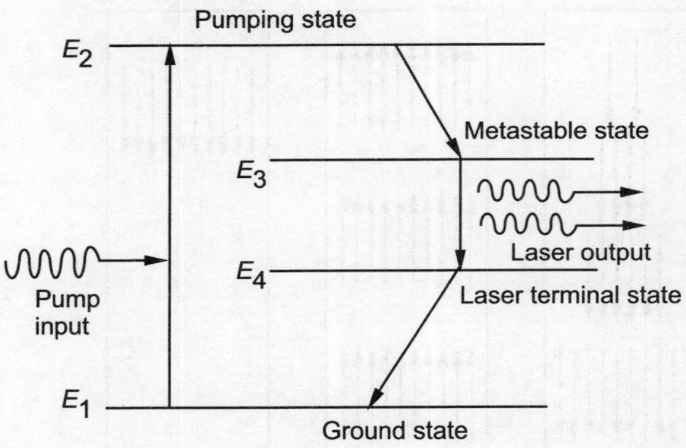Pumping state
E_2
Metastable state
E_3
Laser output
Pump input
E_4
Laser terminal state
E_1
Ground state</td></tr>
<tr><td>Example</td><td>Ruby laser</td><td>He–Ne laser, Nd–YAG laser</td></tr>
<tr><td>Why is a four-level laser more efficient than a three-level laser?</td><td colspan="2">The reason is that the laser terminal state is a ground state in a three-level laser, whereas it is a level above the ground state in a four-level laser. As the laser terminal state is relatively empty, more population inversion is obtained in the four-level scheme than in the three-level scheme with the same pump input. This makes the four-level laser more efficient than the three-level laser. This situation can be understood with the help of the following hypothetical example. Let us assume 100 atoms in the ground state for both the schemes. Let the pump input raise 75 atoms in the state E_2 and let 50 of these atoms drop to level E_3 through non-radiative transitions. The net population inversion is thus 50 to 25 in the three-level laser whereas it is 50 to 0 (assuming that level E_4 is empty at room temperature) in the four-level laser. Thus the same input produces more output in the four-level laser compared to that produced in the three-level laser. For maximum efficiency with four-level scheme, E_4 must be substantially empty at equilibrium, that is, it must be at the value of energy that will prevent it from being populated by thermal excitation of ground state atoms (that is, $E_4 - E_1 >> kT$).</td></tr>
<tr><td>Why is a two-level scheme not possible?</td><td colspan="2">We know that the probability of absorption and stimulated emission are both proportional to the incident photon energy density. An increase in the incident photon flux will increase both absorption as well as stimulated emission. Thus a two-level scheme will not have more than half the atoms in the metastable state. Thus population inversion is not possible and so a two-level scheme cannot result in laser light.</td></tr>
</table>

EXAMPLE 14.3

Wavelength difference between adjacent modes in a resonant cavity

The mirrors in a resonant cavity of a laser are separated by 8 cm. In the vicinity of $\lambda = 5330$ Å how far apart in wavelength are the adjacent allowed operating modes?

Solution

The allowed modes in a resonant cavity have wavelengths given by

$$\lambda = \frac{2L}{n}$$

Thus, $n = \dfrac{2L}{\lambda} = \dfrac{2 \times 8 \text{ cm}}{5330 \times 10^{-8} \text{ cm}}$

$$= 30.0188 \times 10^4 = 300188$$

For the adjacent mode, $n = 300188 + 1 = 300189$

The corresponding wavelength is

$$\lambda = \frac{2L}{n} = \frac{2 \times 8 \text{ cm}}{300189} = 5329.975 \text{ Å}$$

Thus the difference between two wavelengths is (5330 – 5329.975) Å = 0.025 Å or 2.5 pm.

14.3 DIFFERENT TYPES OF LASERS

Since the discovery of the first practical laser by Maimon in 1960, a large number of laser sources have been developed. However, the various features common to different types of lasers make their comparative study easier. In Chart 14.2 three important lasers—Ruby laser, He–Ne laser, and semiconductor laser—are described. These laser sources differ mainly in respect of the choice of pumping mechanism and the choice of amplifying medium. An amplifying medium—also called the active medium—is a material containing particles (atoms/molecules) that have metastable energy states. These particles are called the active centres responsible for the stimulated emission. The resonant cavity described in the last section is a common requirement of all the practical laser sources. There are numerous other possibilities for designing a laser source. Chart 14.3 gives a brief idea of some other important laser sources.

14.4 CHARACTERISTICS OF LASER LIGHT

Light from a laser source differs from that from ordinary sources. It has high coherence, monochromaticity, directionality and intensity. The most striking characteristic of a laser beam

CHART 14.2 TYPES OF LASERS

Type	Ruby laser
Invention	**Maimon, 1960**
Diagram **Schematic**	Glass tube Xenon flash lamp Light shield Laser beam Ruby rod Tap key Capacitor DC power supply
Energy level	E_2 E_1 Non-radiative transitions $\lambda \sim$ 6600 Å 4000 Å E_3 E_0 λ = 6943 Å
Active medium **Active centres**	Ruby: Al_2O_3 crystal with some (~0.05%) Al^{3+} ions replaced by Cr^{3+} ions Cr^{3+} ions
Pumping method	Optical pumping
Resonator design	The ends of the ruby rod are polished, silvered and made optically parallel. One end is partially silvered and the other end is completely silvered.
Construction and working	A xenon flash lamp connected to the dc power supply with a capacitor connected in parallel provides the pump input for a cylindrical ruby rod enclosed in a glass tube. The chromium ions absorb photons of wavelengths 6600 Å and 4000 Å, and make transitions to the excited energy bands E_1 and E_2, respectively. They make non-radiative transitions to the metastable state, E_3. Population inversion thus takes place between E_3 and E_0. One of the photons emitted spontaneously due to transition from E_3 to E_0 triggers laser action and a red light of wavelength 6943Å is obtained in the output.
Properties	Output is pulsed (pulse duration $\approx$ 10 nanoseconds), wavelength = 6943 Å, peak power = 100 mW Efficiency is poor as only two wavelengths in the pump input are utilised; the rest are unutilised.

Helium neon laser
Ali Javan, 1961
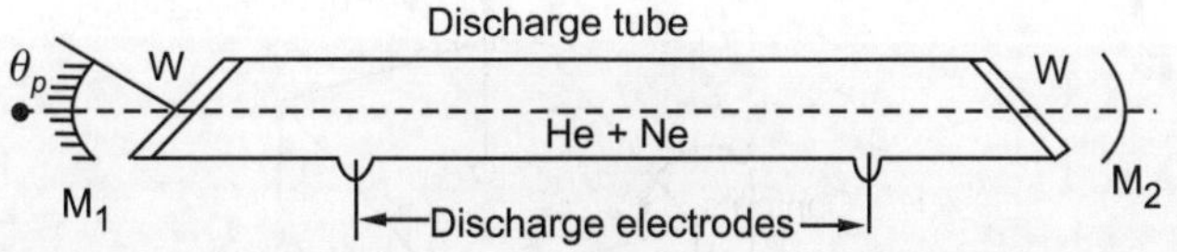 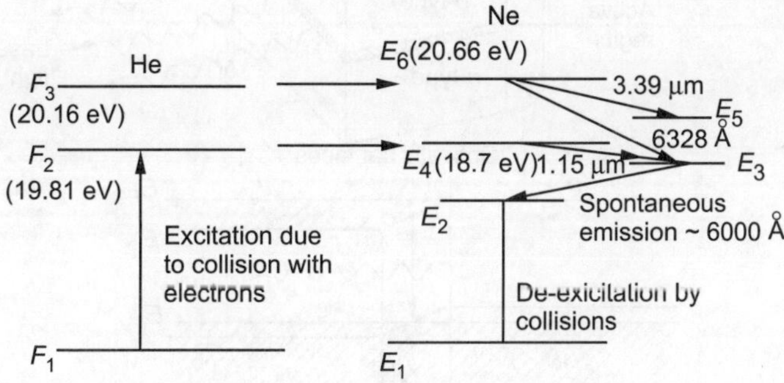
Mixture of helium and neon in the ratio 10:1 Neon atoms
Electron impact pumping or pumping by electric discharge
Mirrors M_1 and M_2 are concave and their focal points nearly coincide at the centre of the tube. M_1 is perfectly reflecting and M_2 is partially reflecting. The windows W of the discharge tube are inclined at Brewster's angle, θ_p, to minimise reflections (Section 5.4B).
It consists of a long narrow discharge tube containing He and Ne in the ratio 10:1 at the pressure of about 1 mm Hg. The discharge produces high energy electrons which excite the He atoms to the metastable levels F_2 and F_3 (He atoms being light are more excitable than Ne atoms). These He atoms collide with the unexcited Ne atoms raising them to levels E_4 and E_6 that are comparable to F_2 and F_3. A state of population inversion is thus achieved between states E_4, E_6 and E_3, E_5 and lasing action takes place resulting in three wavelengths— 3.39 μm, 6328 Å and 1.15 μm—out of which the wavelength 6328 Å lies in the visible range and corresponds to red colour. The Ne atoms then drop down to level E_2 through spontaneous emission. These atoms then come down to the ground state by deexcitation with walls. To enhance the process of depopulation of state E_2, the tube is made narrower so that the probability of collisions with walls increases. The ratio 10:1 of He to Ne reduces the chance of reverse energy transfer from Ne to He.
Output is continuous. There is no need for cooling. Due to the absence of effects such as thermal distortions and crystalline imperfections in solid state lasers, He–Ne (and other gas lasers too) emit light that is more directional and more monochromatic.

CHART 14.2 (Continued)

Type	Semiconductor laser (see Chapter 12)
Invention	**Hall, Nathen, 1962**
Diagram	
Schematic	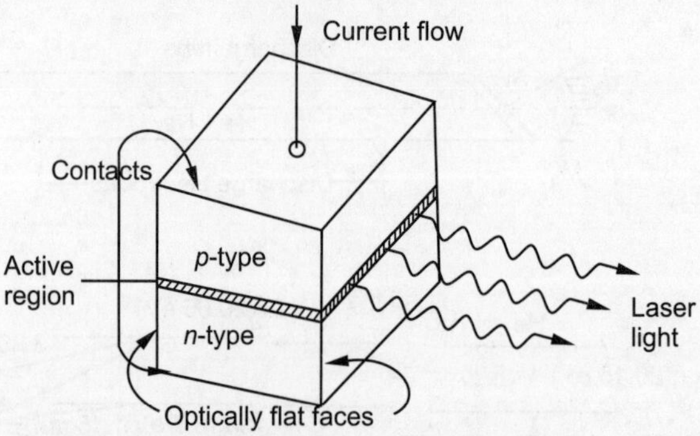
Energy level	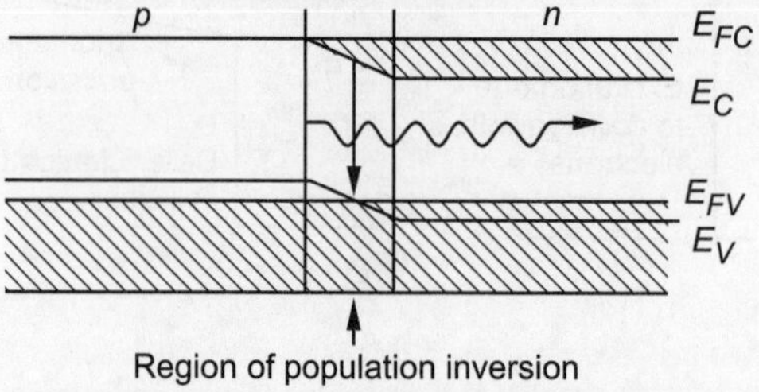
Active medium **Active centres**	Depletion region
Pumping method	By forward bias current and heavy doping of *p*- and *n*- regions.
Resonator design	The end faces of the semiconductor crystal are cleaved and polished to make them optically parallel.
Construction and working	In a *p–n* junction both *p*- and *n*-regions are heavily doped. The faces perpendicular to the plane of the junction are polished optically flat and the other sides are roughened to eliminate laser emission in those directions. Due to the heavy doping the Fermi levels of *n*-region and *p*-region lie respectively in the conduction band and valance band. Forward biasing results in injection of large number of minority carriers in the depletion region. This results in significant population inversion. (Population inversion is not as well defined in bands as for discrete levels). Initial emission of light (at low forward currents) is due to spontaneous recombination. When the forward bias is large enough population inversion and lasing action results. The coherent laser light comes out in the form of a narrow sheet along the plane of the junction.
Properties	Small size, high efficiency, ease of modulation, high durability and low cost. Output is continuous. For GaAs the wavelength is 8400 Å (IR range), for GaAs-GaAlAs, the wavelength is 7100 Å giving red light.

CHART 14.3 BRIEF IDEA OF SOME OTHER IMPORTANT LASERS

Laser	Description
CO_2 laser	It is a four-level gas laser which employs the molecular energy levels of CO_2 for laser transitions. Its high efficiency is due to the fact that vibrational and rotational states of CO_2 molecule require little energy for excitation. Compared to 20 V of energy required to excite the He atom in He–Ne laser, the energy required to excite a CO_2 molecule is just 1/3 V. The discharge tube contains CO_2, N_2 and He gases in the ratio 1:2:3. The N_2 molecules raise the CO_2 molecules to higher energy levels thus inverting the latter's population. Laser transitions result in IR wavelengths of 9.6 μm and 10.6 μm. He atoms help in depopulating the laser terminal state by taking energy from CO_2 molecules through inelastic collisions.
Nd-YAG laser	It is a four-level solid state laser operable in both pulsed and continuous wave (CW) mode. As in the case of the Ruby crystal, Nd^{3+} ions here replace the Y^{3+} ions in the host crystal of YAG. The krypton flash lamp provides the pump input. The use of the ellipsoidal reflector results in the proper focussing of the pump radiation on the Nd-YAG rod. Laser transitions occur between energy levels of Nd^{3+} and laser output of the wavelength 1.06 μm is obtained in the IR range.
Dye lasers	Dye lasers are liquid lasers that use organic dye solutions as active media. Flash lamps or other lasers are used as pump sources. As the dye molecules have very closely spaced energy levels, lasing can be achieved over a continuous range of wavelength. Such lasers are called as *tunable lasers*. The output wavelength can be finely adjusted by using a diffraction grating in place of one of the laser mirrors. These lasers can also be mode locked to generate ultra-short pulses.
Chemical lasers	Chemical lasers are based on energy released in some fast chemical reactions. Atoms or molecules are in metastable excited states during these reactions. The reacting gases are mixed and ignited by UV light or fast electrons. The output can be pulsed or continuous wave. The reaction of hydrogen and fluorine to produce HF yields IR laser light of over 2 MW.
Nuclear lasers	The fast moving ions produced in nuclear fusion can cause laser action in several gases. The gas atoms are excited and ionised by the fast moving ions and direct conversion of nuclear energy into optical energy is possible.
Gas dynamic lasers	They are based on sudden expansion of a hot molecular gas through a nozzle. The different excited states of molecules lose energy at different rates and the state of population inversion can result. Gas dynamic lasers employing CO_2 gas can yield very high CW output.
Photo-dissociation lasers	Using high intensity UV light the molecules can be dissociated in such a manner that one component is left in an excited state and is capable of sustaining laser action. Iodine molecules have been used to yield laser output of more than 10^9 W in the IR range with 1.3 μm wavelength.

is its high degree of coherence. Coherence was discussed in Sections 2.3 and 3.3, and wavepackets were discussed in Sections 6.6 and 6.7. Here we weave these concepts to appreciate the underlying unification.

14.4.1 Coherence and Monochromaticity

The general idea espoused by the term 'coherenceŠ is that of harmony over extended space and time. For an ideally coherent wave the phases at different locations at any given time are completely correlated or in complete concord (Fig. 14.6a). It means that if the phase at a point is known at any time, the phase at any other point is also known at that time. Similarly, if the phase at a time is known at any point, the phase at any other time is also known at that point. A completely coherent wave is thus an infinite wavetrain existing for all time. It follows that such a wave is also a truly monochromatic wave, that is, a wave, which has an exactly defined wavelength.

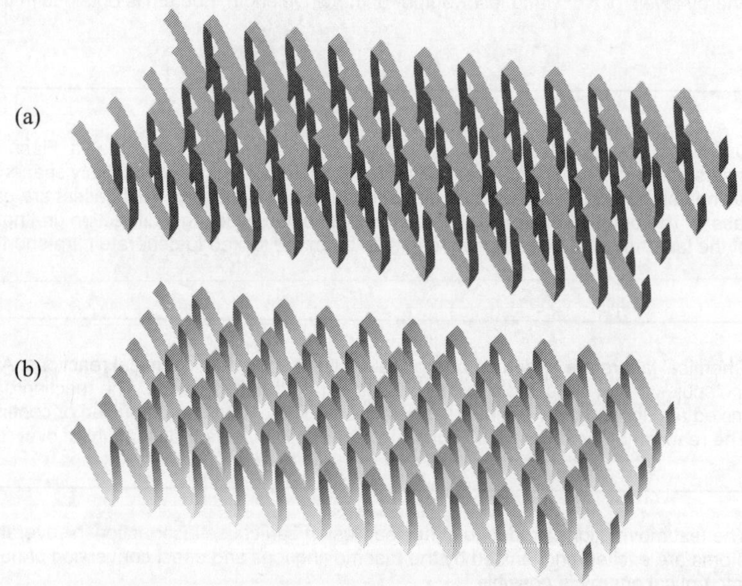

Figure 14.6 (a) Coherent wavetrain showing phase concord; (b) Incoherent wavetrain showing phase discord (Note the increase in mutual discord between waves as the wavetrain propagates.)

A real light wave from an ordinary source, on the other hand, is a finite wavetrain, existing for a short time and having a smaller degree of coherence and monochromaticity. Thus the phase correlation exists over limited time and space. The overall wave is thus a jumble of waves having chaotic or discordant phases (Fig. 14.6b). It follows that such a wave does not have an exactly defined wavelength λ but has a range, $\Delta\lambda$, of wavelengths. That is, a finite wavetrain is a superposition of many wavelengths, it is a wavepacket of the type we studied in Sections 6.6 and 6.7, in connection with de Broglie theory and Heisenberg uncertainty principle.

In Section 6.7 we understood that the width of the wavepacket, Δx, is inversely proportional to $\Delta\lambda$ (because, from Eq. 6.22, $\Delta x \propto 1/\Delta k$ or $\Delta x \propto 1/\Delta(2\pi/\lambda)$ or $\Delta x \propto 1/\Delta\lambda$). Thus an exactly defined wavelength λ for which $\Delta\lambda = 0$ implies an infinitely wide wavepacket. We get the same result using the de Broglie equation and the Uncertainty relation, because

$$\Delta\lambda = \Delta\left(\frac{h}{p}\right) = \frac{h}{p^2}\Delta p$$

$$= \frac{h}{p^2} \times \frac{h}{\Delta x} \text{ (as } \Delta p \Delta x = h\text{)}$$

or $$\Delta\lambda = \frac{\lambda^2}{\Delta x} \tag{14.8}$$

Comparing Eq. 14.8 with Eq. 3.17 we understand that the terms *coherence length* (l_c) *width of wave packet* (Δx) *and the uncertainty in the position of a particle* (Δx) *have equivalent meaning*. Equation 14.8 can thus be written as

$$\Delta\lambda = \frac{\lambda^2}{l_c} = \frac{\lambda^2}{c\tau_c} \tag{14.9}$$

We also understand that the terms, *coherence time* (τ_c), *lifetime of the energy level from which emission takes place (lasing level in case of laser emission)* (τ) and *the uncertainty in time* (Δt) *have equivalent meaning*. Using Eq. 14.8 and the uncertainty relation between energy and time it can be shown that (see Ex. 14.4)

$$\Delta x = c\Delta t \tag{14.10}$$

which is similar to the relation $l_c = c\tau_c$. The equivalence of these various terms is illustrated in Fig. 14.7.

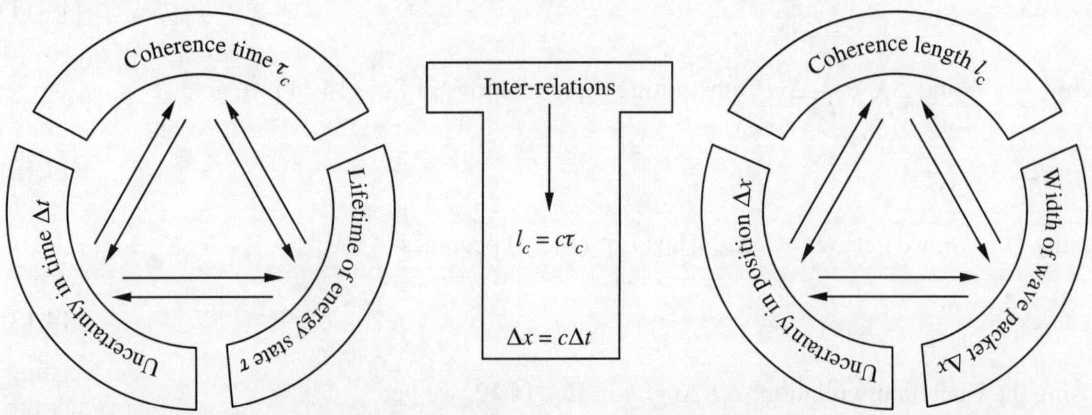

Figure 14.7 Equivalence of τ_c, τ, and Δt, and l_c and Δx and their inter-relations. We understand τ_c as the time for which a single photon wave is emitted, τ as the time over which decay takes place from higher level to lower level, Δt as the time for which energy level lasts, that is, time available for measurement of energy of the level $h/\Delta E$ by uncertainty principle, l_c as the distance covered by a single photon wave during time τ_c, Δx as the width of the single photon wave packet, and also as the uncertainty in the position of the photon.

It is important to understand here that the degree of monochromaticity of laser beam arises from the long life of lasing levels and is strictly not a consequence of stimulated emission. Stimulated emission produces a bunch of identical photons but each photon in this bunch is a wavepacket whose width depends on the lifetime of lasing levels. Stimulated emission thus strengthens the laser beam but does not make it more monochromatic. However, it is interesting to note that population inversion and hence the *effective* use of stimulated emission in laser action depends critically on the long lifetime of lasing levels. That is, both monochromaticity and stimulated emission depend on lifetime but monochromaticity does not depend just on stimulated emission!

For ordinary sources of light, such as a tungsten filament, the degree of monochromaticity is quite small as all wavelengths in the visible range are emitted. The light emitted by gas discharge lamps is comparatively more monochromatic as the wavelengths radiated are restricted by the allowed quantum energy levels in atoms. Example 14.5 illustrates how longer lifetime of lasing levels results in higher degree of monochromaticity of laser sources.

EXAMPLE 14.4

To show $\Delta x = c\Delta t$

Obtain a relation between the uncertainty in time and uncertainty in position using Eq. 14.8.

Solution

Using Eq. 14.8,

$$\Delta x = \frac{\lambda^2}{\Delta\lambda} \tag{14.11}$$

Now, $\lambda = \frac{c}{\nu}$ and $\Delta\lambda = \frac{c}{\nu^2}\Delta\nu$. Substituting these relations in Eq. (14.11) we get,

$$\Delta x = \frac{c}{\Delta\nu} \tag{14.12}$$

Using $E = h\nu$, we get $\Delta E = h\Delta\nu$. Thus Eq. (14.12) becomes

$$\Delta x = \frac{hc}{\Delta E} \tag{14.13}$$

Using the Uncertainty relation, $\Delta E\Delta t \approx h$ in Eq. 14.13, we get

$$\Delta x = c\Delta t$$

which is the desired relation.

EXAMPLE 14.5

Comparison of $\Delta\lambda/\lambda$ for an ordinary source and a laser source

Find $\Delta\lambda/\lambda$ for a laser source having lifetime of lasing energy level τ_c equal to 10^{-5} s. Assume $\lambda = 5000$ Å. Find the corresponding value for an ordinary source for which $\tau_c \approx 10^{-8}$ s.

Solution

Using Eq. 14.9, $\dfrac{\Delta\lambda}{\lambda} = \dfrac{\lambda}{c\tau_c}$

Thus,

(a) For a laser source,

$$\frac{\Delta\lambda}{\lambda} = \frac{5000 \times 10^{-10}\ \text{m}}{(3 \times 10^{8}\ \text{m/s})(10^{-5}\ \text{s})} \approx 10^{-10}$$

(b) For an ordinary source,

$$\frac{\Delta\lambda}{\lambda} = \frac{5000 \times 10^{-10}\ \text{m}}{(3 \times 10^{8}\ \text{m/s})(10^{-8}\ \text{s})} \approx 10^{-7}$$

Thus the *wavelength is well defined upto one part in 10 billion for a laser source*, whereas it is defined upto one part in 10 million for an ordinary source.

We know that the formation of a stationary interference pattern is mainly decided by the coherence of the interfering waves. Thus due to the high degree of coherence of laser beams stable fringe patterns can be easily obtained using two laser beams whose paths may differ by distances of the order of kilometres! This is the reason why a laser lamp is not recommended for reading. For reading average illumination is required; laser light will form dark and bright interference fringes thus blurring the vision.

14.4.2 Directionality

Light from an ordinary source diverges in all possible directions. In laser light directionality is achieved by a process of stimulated emission. A small beam divergence of laser light arises due to diffraction effects (depending on the wavelength of light) and the diameter of the exit aperture. The output of a laser source is very close to an ideal plane wave. Thus the phase at all points on a plane wavefront is very nearly the same at all times. Laser light can be sharply focussed due to its high directionality.

14.4.3 Intensity

The high directionality of a laser output means energy is not spread over all directions (as in ordinary light) but is concentrated only in a small region of space. This results in extremely high energy density of a laser beam compared to that of an ordinary light beam (intensity is energy per unit area, so the same energy gives larger intensity in a laser beam due to its smaller area (Ex. 14.6). Moreover the intensity of laser remains undiminished over very large distances whereas the intensity falls off rapidly in ordinary light sources due to their divergent nature.

EXAMPLE 14.6

Intensity of a laser beam

Find the intensity of a 10 W laser focussed to an area $\sim \lambda^2$ assuming the wavelength to be 5000 Å. Compare this intensity with the intensity of the sun's radiation, which is 7 kW/cm².

Solution

The intensity of the laser beam is

$$I = \frac{10\ \text{W}}{(5 \times 10^{-7}\ \text{m})^2}$$

$$= 4 \times 10^{13}\ \text{W/m}^2$$

$$= 4 \times 10^{6}\ \text{kW/cm}^2$$

Thus *the intensity of the given laser source is* $\sim 10^6$ *times that of the sun's radiation.*

In summary, it can be said that the coherence and monochromaticity of the laser is primarily because of the lasing medium—its energy level structure and level lifetimes—whereas its intensity is a consequence of the technology of pumping and that of the resonant cavity.

A large number of experiments can be performed using laser light. All optics experiments discussed in earlier chapters can be performed with precision and better results can be obtained using laser light.

14.5 SEMICONDUCTOR PHOTONIC DEVICES

As we know (see Chapter 11) a semiconductor crystal provides a system of energy bands to its electrons. Two of these bands—the valence and conduction bands—are of particular importance in useful functioning of a semiconductor device. Such a device works by electrically injecting electrons from an n-type region to a p-type one. Upon such injection, these electrons find themselves in energy levels that are higher than the majority of other occupied energy levels on the p-side. These electrons can fall down to lower energy levels, radiating a photon in the process.

Two devices working on this mechanism find wide applications in various technologies. These are the light emitting diode (LED) and a laser diode (LD). The LED functions by spontaneous emission of photons when an electron in the conduction band falls into a valence level. In the LD, such spontaneous emission is harnessed by specific choice of semiconductor material and by geometrical design and above-threshold electrical pumping to produce a laser beam (see Section 14.5.2 and Fig. 14.11).

14.5.1 LED

Figure 14.8 shows the schematic of a typical LED. The Te-doped donor GaAsP layer sits on an n-type GaAs substrate. The Zn-doped GaAsP acceptor on top of the donor is created by Zn diffusion through the mask. The active region is the boundary between the Zn-doped acceptor and the Te-doped donor. Such a device is driven under a forward bias. The output radiation is proportional to the forward current. The central wavelength of the emission spectrum is determined by the energy gap $E_G : \lambda = ch/E_G$. The emissive efficiency of such a device is determined by the probabilities of radiative and non-radiative transitions. It is limited by external factors like total internal reflection and re-absorption of the photon within the diode. Total internal reflection is considerable because of the high refractive index (3.5) of GaAs. Re-absorption can reduce the photon number to 10%.

Table 14.1 lists typical materials used in LEDs along with the band gap, central wavelength in the emission spectrum and the colour of emitted radiation.

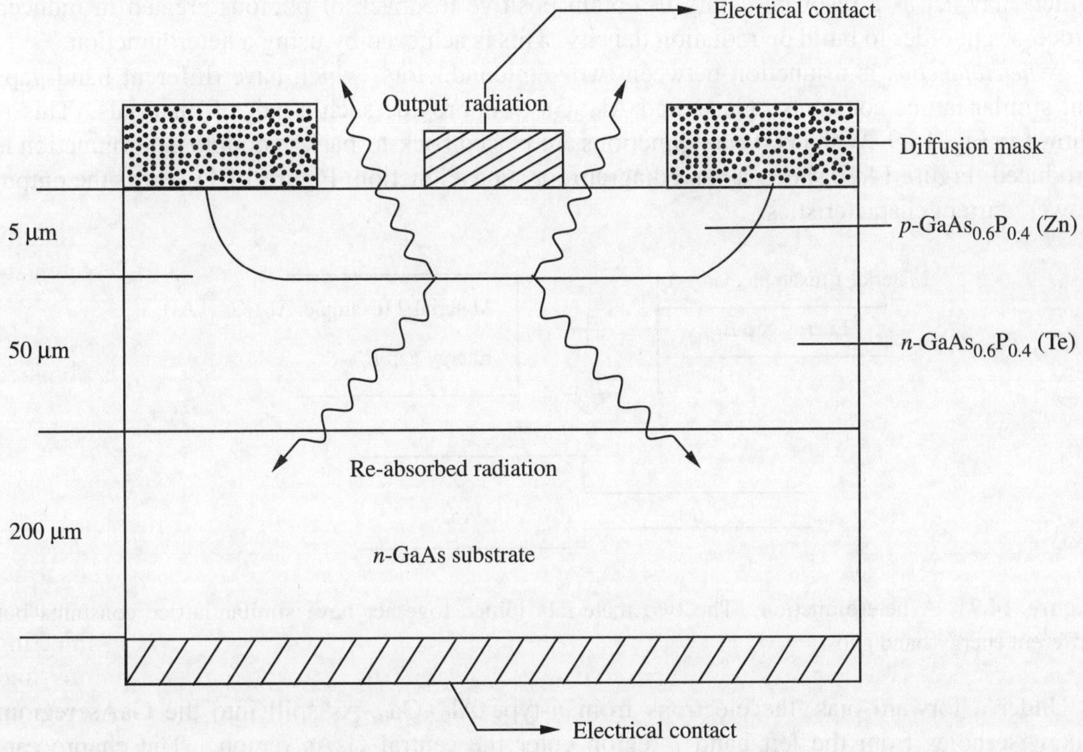

Figure 14.8 Schematic of a typical LED

Table 14.1 Typical materials used in LEDs

Material	Band gap (eV)	Central wavelength (nm)	Colour
GaAs	1.43	870	–
InGaAs	0.95–1.3	1000–1300	IR
InGaAsP	0.75–1.35	900–1700	IR
AlGaAs	1.4–1.7	700–900	Red/IR
InGaP	1.82–1.94	640–880	Red
GaAsP	1.9–2.05	600–650	Orange/Red
GaP	2.26	549	Yellow
GaN/AlGaN	2.5–2.75	450–500	Blue/Green

14.5.2 Laser Diode

A laser diode is a refashioned LED in which the injection of charge carriers exceeds a certain threshold after which the induced radiation processes are so increased as to take care of losses and sustain a stable monochromatic output beam. To achieve this the device must contain a region, which may act as a resonant cavity to obtain positive feedback of photons created in induced processes in order to build up radiation density. This is achieved by using a heterojunction.

A *heterojunction* is a junction between two semiconductors, which have different band gaps but similar lattice constants. GaAs and $Al_{0.3}Ga_{0.7}As$ provide such a pair of materials. This is shown in Fig. 14.9. When two such junctions are placed back-to-back, a double heterojunction is produced. Figure 14.10 shows a band diagram for such a junction. Figure 14.11 shows the output power–current characteristics.

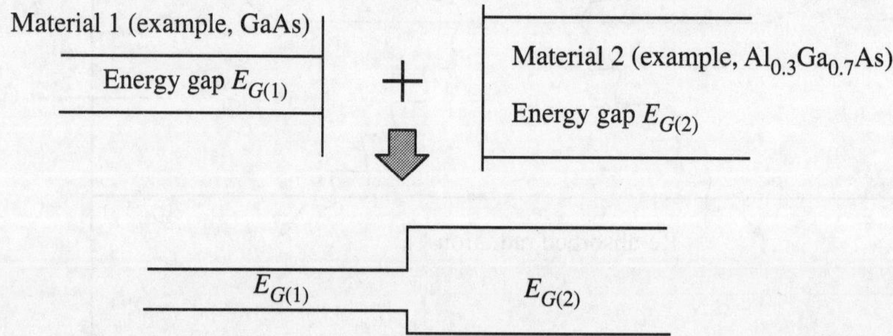

Figure 14.9 A heterojunction: The two materials joined together have similar lattice constants but different energy band gaps.

Under a forward bias, the electrons from n-type $Al_{0.3}Ga_{0.7}As$ spill into the GaAs region. Likewise holes from the left hand p-region enter the central GaAs region. The charge carriers in GaAs now face a barrier on both sides and are trapped. This is the basis of population inversion. The trapped electrons and holes combine efficiently to produce photons of energy $\hbar\omega = E_{G(1)}$. The central region then serves as a cavity to feed these photons back into itself, thus stimulating more electron–hole combinations. The radiation intensity thus builds up.

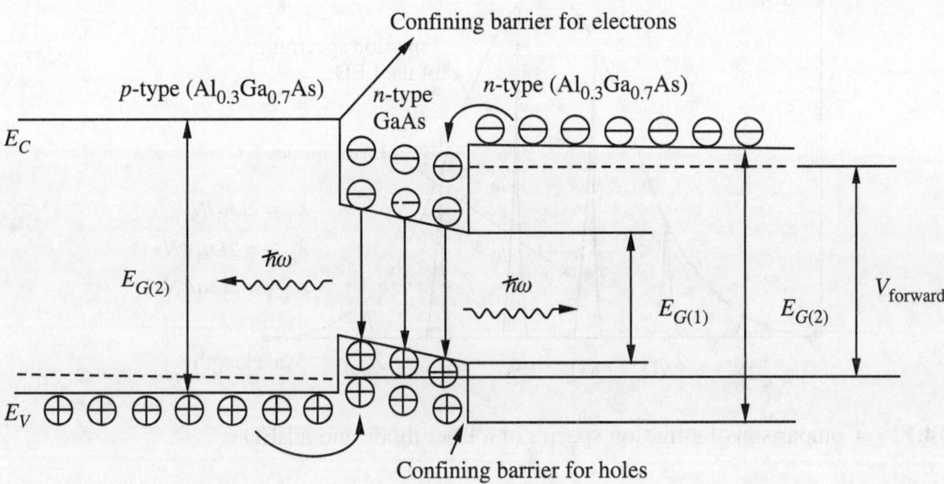

Figure 14.10 A double heterojunction: The central region has a lower band gap compared to the side regions. This helps trapping of electrons injected from the right and holes injected from the left in this region for longer times. Population inversion is thus achieved more readily.

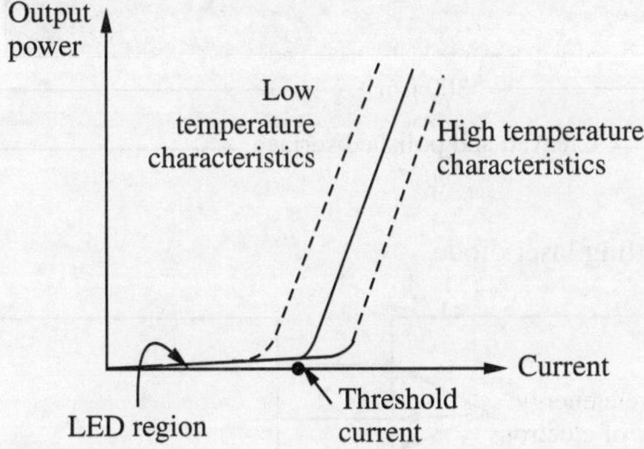

Figure 14.11 Power–current characteristics of a laser diode

From the LED spectrum with central wavelength corresponding to $E_{G(1)}$, the cavity resonator selects only a discrete set of frequencies $\lambda_N = 2Ln_1/N$, n_1 being the refractive index of the central spectrum region and N, some integer. Figure 14.12 shows the LED spectrum and overlapped laser diode spectra. In an *edge-emitting laser diode* (see Fig. 14.13), the central region is a thin guide. The end surfaces are polished to make the region into a resonant cavity.

The monochromaticity of a heterojunction laser diode can be increased by a *quantum well*. In a quantum well the central region is made very thin—about 2–20 nm (see Fig. 14.14). In a cavity of this size, the electronic energy levels become discrete. The energies are determined by the size of the central region, just as for a particle confined to a box (see Chapter 6). The transitions between such levels are strongly monochromatic.

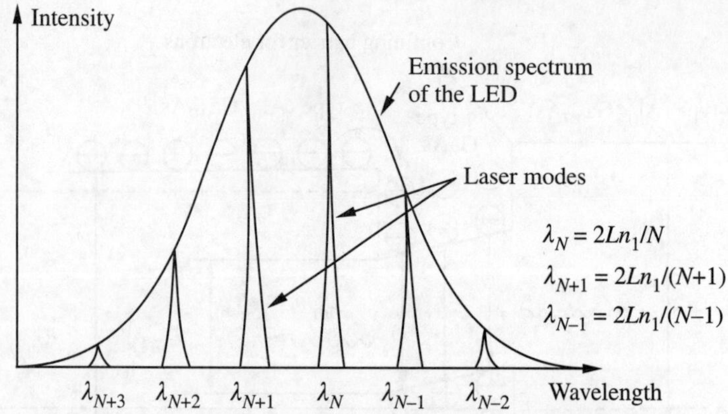

Figure 14.12 Comparison of emission spectra of a laser diode and a LED

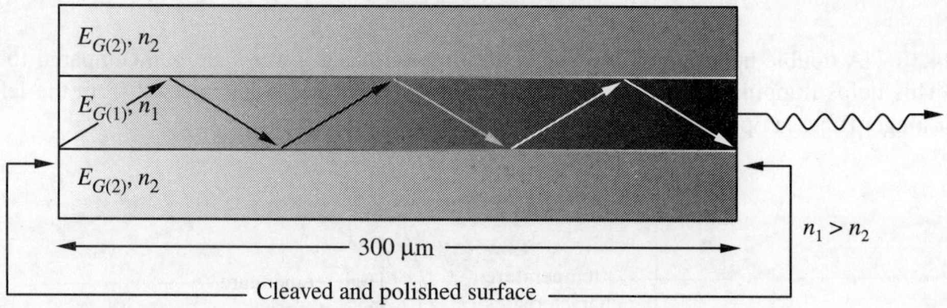

Figure 14.13 An edge-emitting laser diode

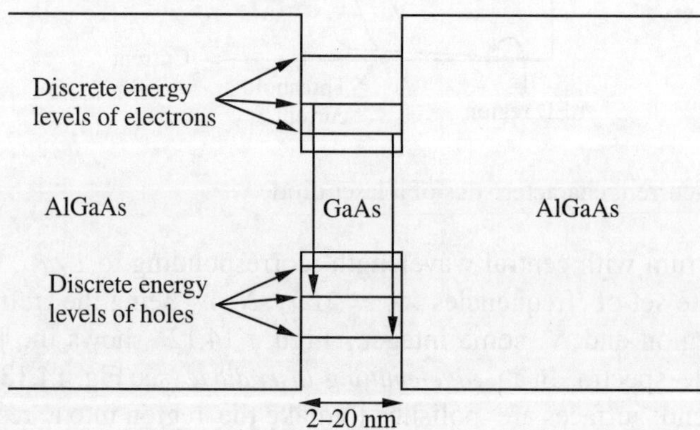

Figure 14.14 A quantum well

Such a quantum well laser diode has many positive features. The tiny active region gives very efficient electron–hole recombination. It is easier to create a large electron–hole density.

Non-radiative transitions become less probable. A much smaller threshold current—about a tenth of what is needed ordinarily—suffices for population inversion and lasing. Also, the threshold current becomes less temperature sensitive leading to more stable performance. Most importantly, the quantum well can be tailored to the desired emission wavelength.

14.6 APPLICATIONS

In this chapter we discussed the basic principles, characteristics, types of laser sources and semiconductor photonic devices. The discovery of the laser is considered as one of the two most revolutionary discoveries of twentieth-century physics. The discovery of the transistor in 1947 gave birth to a revolution called 'Electronics'. The discovery of the laser is at the base of another fledgling revolution referred to as 'Photonics'. Interestingly both these discoveries are based on the quantum theory. Currently lasers have found application in almost all walks of life and it is impossible to discuss all these applications. Here we discuss a few relevant applications.

14.6.1 Applications in Measurement

Lasers are used for precision long distance measurement due to their high degree of coherence. *The distance between the moon and the earth was measured using a laser beam to a probable error of less than 10 cm!* A specially constructed mirror was placed on the moon's surface by the astronauts of the Apollo mission. A laser beam was shot at this mirror through a telescope from Texas and the reflected beam was received by the same telescope. Using the known speed of light and the measured round trip travel time of the beam, the distance could be determined with great precision. Such precise measurements are useful for astronomical and geophysical purposes. These data, for example, can be used to make more reliable earthquake predictions, to measure drift of continents and to determine the depths of points on the earth's surface, which are otherwise inaccessible.

Lidars (light detection and ranging) are devices, which measure the locations using optical means. They are optical counterparts of radar and are also called *optical radar* or *laser radar* for this reason. They have many applications that include airborne bathymeters to measure the depth of water or the location of any submerged object, monitoring of volcanic aerosols that may affect global climate, monitoring of atmospheric pollution and detection and characterisation of fog layers.

Another interesting application is the creation of ultra-short laser pulses of the order of 10^{-14} s using dye lasers. These are much smaller than the electronic response times, which are of the order of 10^{-9} s. These femtosecond pulses are used for gathering new knowledge in the femtosecond domain. They are of particular interest in spectroscopy and optoelectronics. Applications in measurements also include the use of laser beams in optical fibres to sense changes in various physical quantities. Many fibreoptic sensors have been developed for various engineering and medical applications (see Chapter 15).

14.6.2 Applications in Information Processing

Laser printers and compact discs (CDs) are two of the most common applications of lasers in computer technology. Compact discs have enormously reduced the volume required for information storage and one CD can store as much information as a complete set of the *Encyclopaedia Britannica.* In a CD the digitised information is coded in tracks of microscopic indentations on a metal foil encased in a plastic cover. As the disc spins, a laser beam which can be focussed to the accuracy of wavelength of light decodes this information using a photocell and digital-to-analogue circuits. The information is reproduced with extreme fidelity due to the accurate focussing and high monochromaticity possible with a laser beam.

In telecommunication optical fibres are used for carrying laser light, which represents information such as telephone conversation, audio and video signals or computer data. (see Ex. 14.7 to get a 'feel' of the tremendous information capacity of laser light.) More efficient methods of processing information are being probed by using optical means instead of the existing electronic techniques. This is a frontier research area in photonics.

EXAMPLE 14.7

Information capacity of a laser beam

Assume that lasers whose wavelengths can be precisely 'tuned' to anywhere in the visible range are available. If a telephone conversation requires a frequency band of 1 kHz, and a TV video signal needs a bandwidth of 10 MHz, find the telephone conversations or television programmes that could be transmitted simultaneously on a single laser beam.

Solution

The frequency range corresponding to the visible wavelength range of 4000 Å to 7000 Å is

$$\frac{3 \times 10^8 \text{ m/s}}{4000 \times 10^{-10} \text{ m}} \text{ to } \frac{3 \times 10^8 \text{ m/s}}{7000 \times 10^{-10} \text{ m}} \text{ Hz},$$

that is, 7.5×10^{14} Hz to 4.3×10^{14} Hz, or $\Delta\nu = 3.2 \times 10^{14}$ Hz

Now, $\dfrac{3.2 \times 10^{14} \text{ Hz}}{10^3 \text{ Hz}} \sim 10^{11}$ and $\dfrac{3.2 \times 10^{14} \text{ Hz}}{10^7 \text{ Hz}} \sim 10^7$

Thus a *single laser beam can be used to transmit* 10^{11} *(hundred billion) telephone conversations or* 10^7 *(ten million) television programmes.*

Lasers are used widely for deciphering the bar code on pre-packaged items (Fig. 14.15). The laser beam from a He–Ne laser is focussed to a fine spot and scans the bar code. The variation of the reflected light is fed as an electrical signal to a computer, which in turn prints the decoded information of the product, such as cost, manufacturer and the type.

An application area, which is still in the development stage, envisages the use of semiconductor lasers for on-board communication. This development promises replacement of conducting tracks in circuit boards by lasers. Such technology will also allow larger chip packing as laser can travel in 3D from chip to chip without needing a track on the board material.

Figure 14.15 Universal product code (UPC) or bar code.

14.6.3 Applications in Industry

The high intensity and sharp focussing of laser beams make them useful for various industrial applications like light- to heavy-duty industrial drilling, cutting, welding, marking and alloying. They are also used in industrial pollution monitoring, wireless initiation of thermal batteries, explosives and propellants, industrial wire stripping, precision micro-machining and polishing. Laser diodes find application in fibreoptic communication, CD drives, laser printers and several optoelectronic devices.

The focussed beam from a high power laser like a CO_2 laser can cut diamonds or thick metal plates in just a few seconds. The intense heat of the laser (for example, a CO_2 laser) facilitates the welding of materials with a large difference in thermal conductivity. For example, the welding of stainless steel handles onto the cooking pots can be accomplished with lasers.

Lasers are used in the garment industry for automated cloth cutting. They are used in automobile industry for hardening parts of automobile cam shafts using localised heating. Lasers also find wide use in the semiconductor industry for various fabrication jobs. High power laser beams have also been used for production of very high temperatures in fusion reactors.

14.6.4 Applications in Medicine

Lasers have been used as a powerful surgical and diagnostic tool due to their high intensity and very sharp focussing, and also for the reason that light cannot infect wounds. They are used for cutting and sealing nerves in neurosurgery, for raising the temperature of localised areas of tissue in physiotherapy and for removing lesions, tongue tumours, different types of skin cancers, tattoos, birthmarks and benign pigmented marks. Physicians use laser beams to cauterise bleeding ulcers by using fibreoptic endoscopes that can be inserted into the gastrointestinal tract through the mouth. The advantage of the laser is that it cuts and coagulates at the same time thus leading to minimum blood loss.

Lasers have been widely used in eye surgery, spinal surgery, brain surgery, plastic surgery and angioplasty. Laser surgery has proved to be very useful in curing glaucoma and blindness induced by diabetes. In glaucoma, a laser beam burns open a tiny hole in the clogged membrane thus reducing the excess pressure. In case of diabetic blindness, a laser beam can seal the leaking vessels by photocoagulation. For diagnostic purposes, the laser-induced fluorescence method is normally employed, in which the normal and diseased tissue can be distinguished by correlating their spectral features with other pathological data. This technique has been used to

CHART 14.4 PRINCIPLE AND APPLICATIONS OF HOLOGRAPHY AND NONLINEAR OPTICS

	Holography	Nonlinear Optics
Principle	A photograph is a two-dimensional record of intensity distribution of the light scattered from a three-dimensional source object. The details of phase distribution are lost in photography. This causes the different parts of the object that are at different depths to appear in the same plane. For example, the eyes and the nose are in the same plane in a human photograph whereas the phases of light coming from these parts are different. Thus phase details which convey 'depth' are lost in a photograph. Holography—also called as *lensless photography*—was discovered by Gabor in 1947. It uses the whole (*holos* = whole) information (phase as well as amplitude) of the wave from the object in creating its record. A *hologram* is a record of the interference pattern formed by the superposition of two coherent waves; one scattered by the object called as *object wave* or *signal wave* and the other called as an *auxiliary* or a *reference wave* (see Fig. 14.16a) This interference pattern acts like a diffraction grating. Upon illumination by a coherent wave identical to the reference wave—called as *readout wave* or *reconstruction* wave—a virtual and real image of the object can be seen (Fig. 14.16b). A holographic image presents a remarkable three-dimensional appearance. The images stand in space and may be viewed from different angles and distances.	For a dielectric placed in an electric field E, the induced polarisation is given by $P = \chi_0 E$, where χ_0 is the dielectric susceptibility. The refractive index of a material is given by $n = \varepsilon^2 = (1 + 4\pi\chi_0)^2$. The linear dependence of P on E belongs to the domain of what is known as linear optics. Linear optics is pre-laser optics. With the advent of lasers it was discovered that the optical properties of a medium can change with the intensity of the laser. The reason is that atomic field strengths are of the order of $10^7 - 10^{10}$ V/m; ordinary light sources generate much smaller fields not exceeding 10^3 V/m and thus can't influence the atomic fields. Lasers can generate fields from 10^5 to 10^9 V/m and so the optical properties of the material change with the intensity of laser light. This feature takes us from the realm of linear optics into the realm of nonlinear optics. In nonlinear optics the dielectric susceptibility is not constant but depends on the field strength as $\chi(E) = \chi_0 + \chi_1 E + \chi_2 E^2 + —$ (For small fields, $\chi(E) = \chi_0$ which is a case of linear optics).
Applications	The holographic technique finds wide application due to its ability to record information about the depth. It is used to study transient microscopic events. The distribution of strain in a stressed object can be studied using a technique called *double exposure holographic interferometry*. This technique can also be used for applications such as determination of Young's modulus and Poisson's ratio of material. A hologram can be used to store a huge amount of data. The reliability of data storage is high due to the fact that each part of a hologram contains information of the whole object. Thus, destruction of a part of a hologram does not erase a specific portion of image on reconstruction. The hologram can also be generated using computers by feeding the necessary data. These simulations eliminate the need for making an actual model in the design stage. This has given rise to the field of digital holography. Holography is also used in the fabrication of microcircuits. Thus holography can be a major contributor to computer technology through its use in microcircuit fabrication and data storage, and promises a major role in the development of future optoelectronic computer systems.	Nonlinear optics enhanced our understanding of optical phenomena and paved the way for new applications that could not be thought of in pre-laser optics. With extremely high intensities possible using a laser beam, an electron could absorb two (or more) photons. Thus a multiquantum photoelectric effect is observable using high-intensity laser light. Thus laser can induce transparency or opacity in materials. This effect is used in the fabrication of phototropic gates or optical shutters. Nonlinear optical phenomena are used in the construction of several new devices such as second harmonic generators, parametric light oscillators, and oscillators for various operations of frequency mixing such as summation and subtraction.

distinguish between normal, benign and malignant human breast tissues and for the detection of lung cancer.

14.6.5 Applications in Defence

Lasers are used in defence for target range finding, target designation and tracking and guidance. Battlefield tanks and armoured combat vehicles are mostly equipped with a laser range finder. Lasers are widely used as directed energy weapons (DEWs) or laser weapons. They are basically high power lasers with suitable control mechanisms that can disable the guidance system of warheads, or trigger an explosion of the fuel or warhead, or cause temporary or permanent damage to the target. Laser systems that serve as electrooptic counter measure (EOCMs) devices have been developed for use in the battlefield. These systems are used to disable, temporarily or permanently, similar systems used by the enemy, ensuring at the same time reliable and uninterrupted functioning of the system.

Apart from the above applications, lasers have guided the developments in two important fields of recent origin: *holography* and *non-linear optics*. A brief discussion of these fields is included in Chart 14.4.

The applications of lasers are vast, diverse and growing. The laser has aptly been described as *the solution looking for a problem*, when it was discovered almost five decades ago. The range of applications of laser technology is increasing everyday. Newer insights into the physics involved are expected to accelerate this growth.

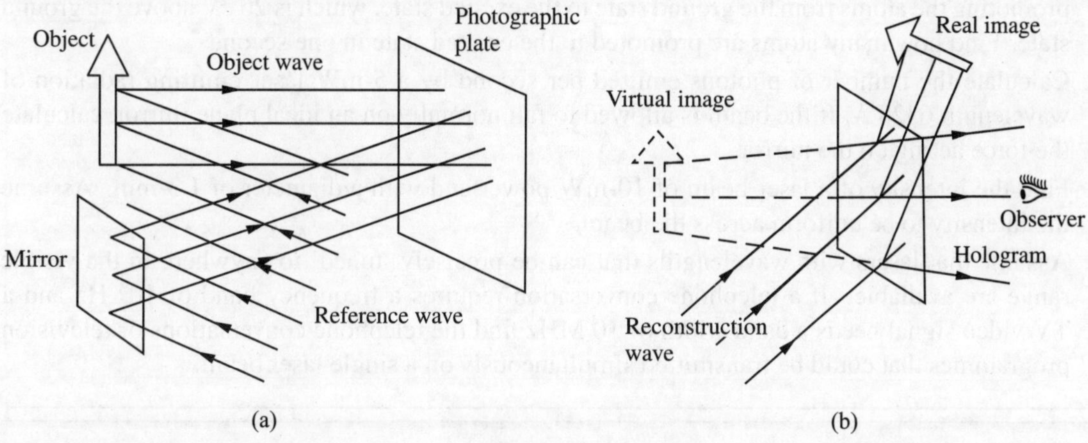

Figure 14.16 (a) Recording a hologram and (b) Reconstruction of an image using a hologram

EXERCISES

1. Explain the terms: (i) Spontaneous emission, (ii) stimulated emission, (iii) metastable state, (iv) optical pumping, (v) population inversion and (vi) temporal coherence.
2. Distinguish between (i) spontaneous and stimulated emission, and (ii) three-level and four-level lasers.
3. With the help of neat sketches explain the three quantum processes that may occur when light radiation interacts with matter.
4. What is population inversion? How is it achieved in the He–Ne laser?
5. What is optical resonator cavity? With a neat diagram explain its importance in producing a laser beam.
6. Explain the role of end mirrors in lasers.
7. Describe the construction and working of a solid state ruby laser.
8. Explain with an energy level diagram, the construction and working of the He–Ne laser.
9. How is laser light different from ordinary light?
10. State the important characteristics of lasers.

PROBLEMS

1. A 10-mW He–Ne laser has an efficiency of 1%. Assume that all input energy is utilised in promoting the atoms from the ground state to the excited state, which is 20 eV above the ground state. Find how many atoms are promoted to the excited state in one second.
2. Calculate the number of photons emitted per second by a 5-mW laser emitting radiation of wavelength 6328 Å. If the beam is allowed to fall normally on an ideal plane mirror, calculate the force acting on the mirror.
3. Find the intensity of a laser beam of 10-mW power and with a diameter of 1.3 mm. Assume the intensity to be uniform across the beam.
4. Assume that lasers with wavelengths that can be precisely 'tuned' to anywhere in the visible range are available. If a telephone conversation requires a frequency band of 10^3 Hz and a TV video signal needs a band width of 10 MHz find the telephone conversations or television programmes that could be transmitted simultaneously on a single laser beam.

15

Fibre Optics

15.1 THE STORY OF FIBRE OPTICS

Fibre optics is a branch of optics that deals with the study of propagation of light through transparent dielectric waveguides, for example, optical fibres. Although the beginning of the modern fibre optic technology can be traced to the mid-twentieth century, light had been used for communication since antiquity. Communication through hand signals, smoke signals and beacons are the simplest examples of optical communication. These forms of communication could carry information only when within sight and when there was no obstruction in the path of light, as they used the property of rectilinear propagation of light in air.

There had been curious attempts to find out whether light could be piped and transported from one place to another like electricity, water and domestic gas. In 1880, William Wheeler, an American engineer, when still in his twenties, received a patent for piping light to illuminate distant rooms using pipes with reflective linings (Fig. 15.1). Wheeler's suggestion did not receive much attention, because the use of electric wires and incandescent bulbs for illumination was far more practical than piped light. However, Wheeler's idea was not totally forgotten and kept reappearing in one form or the other, and finally culminated in the discovery and development of today's optical fibres.

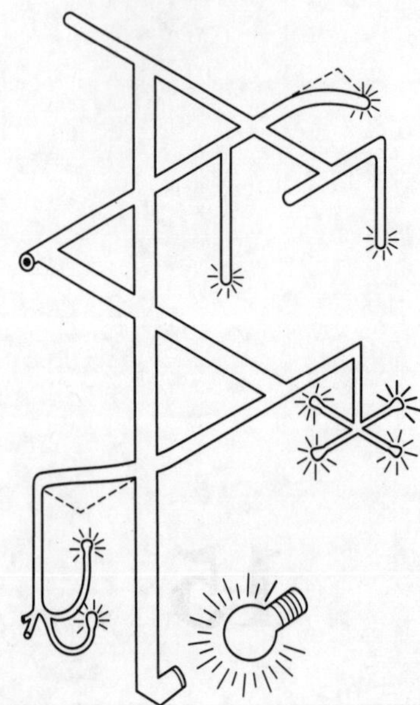

Figure 15.1 Wheeler's idea of piping light

Interestingly, the fundamental concept underlying the propagation of light through optical fibres was well known before Wheeler's time. The underlying concept is the total internal reflection of light by means of which light can be confined in any transparent medium denser than air. It could be understood using the theory of refraction proposed by Willebrord Snell, a Dutch mathematician, in 1621. In 1854, John Tyndall demonstrated the confinement of light to a stream of water by total internal reflection (Fig. 15.2). Perhaps this concept did not appeal to the engineer's mind, or Wheeler may have ignored it as being practically nonviable.

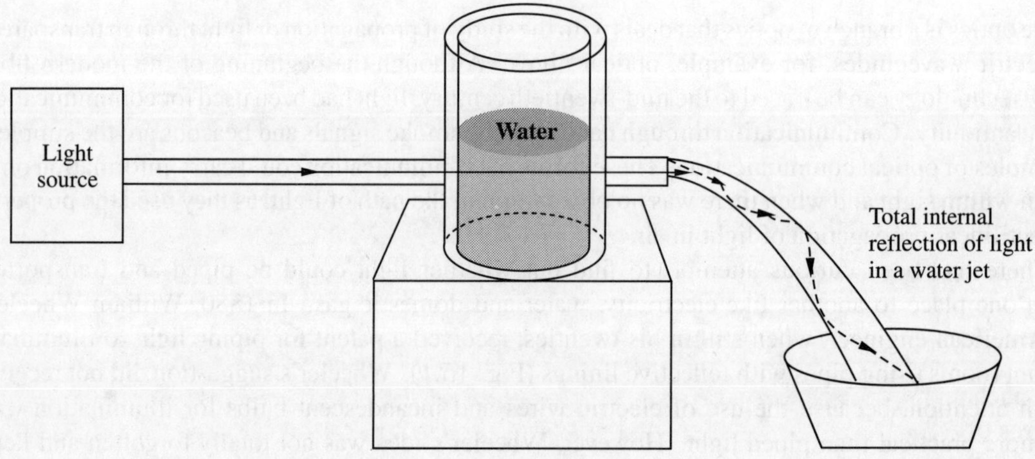

Figure 15.2 Tyndall's demonstration of guiding light through a stream of water

Another important development took place in 1880: Alexander Graham Bell invented the photophone. Photophone was the first device to use light waves to encode and decode information—the first wireless voice communication! It reproduced voice signals by detecting variations in the amount of light reaching a receiver. The photophone again proved to be impractical when compared to other devices such as the telegraph and Bell's earlier invention: the telephone. This was because atmosphere did not transmit light as reliably as wires carried electricity.

It was realised by the end of the century that bent quartz rods could carry light and many inventors patented them as dental illuminators. In 1930, Heinrich Lamn, a medical student in Munich, for the first time reported transmission of the image of a light bulb through a bundle of optical fibres. However, the transmitted images were poor, and due to World War II Lamn could not pursue further research to improve the images. In 1954, van Heel of the Technical University of Delft, Holland, and Harold Hopkins and Narinder Kapany of Imperial College, London, separately reported imaging bundles in the prestigious British journal, *Nature*. It was Kapany who coined the term 'fibre optics' in 1956.

Van Heel was motivated by a conversation with the American optical physicist Brian O'Brien and made the crucial innovation of covering the glass/plastic fibre with a transparent cladding of lower refractive index. This cover protected the total reflection surface from contamination and reduced crosstalk between fibres. Later, Lawrence Curtiss, an undergraduate student at the University of Michigan, developed glass-clad fibres while working part-time on a project to develop an endoscope to 'see' inside the stomach. By 1960, glsss-clad fibres suitable for medical imaging were developed, but they were unsuitable for communication purposes due to high attenuation.

Developments in the field of fibre optics took a huge stride when T Maiman developed the first operational laser in 1960, followed by the development of the semiconductor laser in 1962, and the laser diode eight years later by researchers of various laboratories. Lasers provided controlled, intense and directed light that could be sent over glass fibres as encoded information. This development motivated scientists to create better fibres that could transmit optical information over long distances without significant losses.

The task of developing better fibres requires deeper insights into the characteristics of both the light as well as the fibre. In this chapter we first review the phenomenon of total internal reflection, the basic principle that governs propagation of light in fibres (Section 15.2) In Section 15.3 we understand the basic structure of optical fibres. Section 15.4 deals with the ray optics treatment of light propagation in fibres and Section 15.5 explains the wave optics treatment. Various power losses and distortions that light undergoes during propagation are discussed in Sections 15.6 and 15.7. In Section 15.8 the basic aspects of fibre optic communication systems are discussed, and in the last section, the various applications of fibre optics are discussed.

15.2 TOTAL INTERNAL REFLECTION

The phenomenon of refraction of light at the interface between two transparent media of uniform indices of refraction is governed by Snell's law. Consider a ray of light passing from a medium of refractive index n_1 into a medium of refractive index n_2 (Fig. 15.3a). Assume that $n_1 > n_2$ and that the angles of incidence and refraction with respect to the normal to the interface are, respectively, ϕ_1 and ϕ_2. Then, according to Snell's law

$$n_1 \sin \phi_1 = n_2 \sin \phi_2 \tag{15.1}$$

As $n_1 > n_2$, if we keep increasing the angle of incidence ϕ_1, the angle of refraction ϕ_2 will also go on increasing until a critical situation is reached, when for a certain value of $\phi_1 = \phi_c, \phi_2$ becomes $\pi/2$, and the refracted ray passes along the interface (Fig. 15.3b). This angle, ϕ_c, is called the *critical angle*. If we substitute $\phi_1 = \phi_c$ and $\phi_2 = \pi/2$ in Eq. 15.1 we get

$$\sin \phi_c = \frac{n_2}{n_1} \tag{15.2}$$

If the angle of incidence is increased beyond ϕ_c, the ray is no longer refracted but is reflected back into the same medium (Fig. 15.3c). This is called *total internal reflection*. It is this phenomenon that is responsible for the propagation of light through optical fibres.

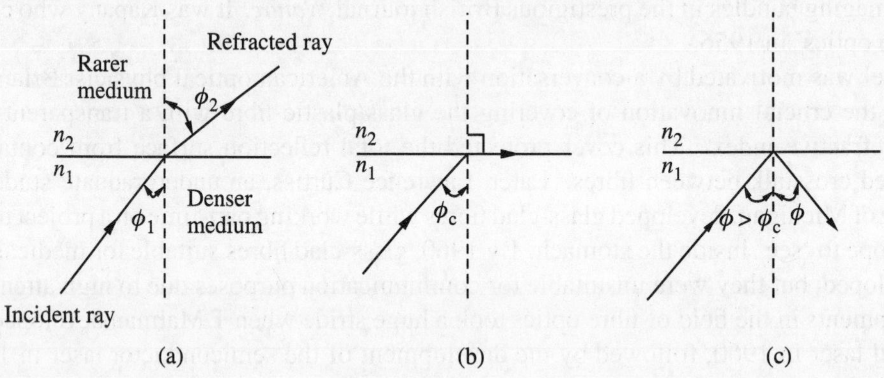

Figure 15.3 (a) Refraction of a ray of light; (b) Refraction along the interface at the critical angle of incidence; (c) Total internal reflection

15.3 STRUCTURE OF AN OPTICAL FIBRE

The optical fibre, which is at the heart of any fibre optic system, works on the basis of internal reflection at the boundary between two optical media. In essence, the fibre is a waveguide, with a tubular structure providing the boundary. A typical fibre is a flexible wire of cylindrical shape drawn from glass. It has a coaxial cylindrical structure (Fig. 15.4). The *core*—made of an optically denser material—has a diameter ranging over 50–100 μm. The optically rarer *cladding* encapsulating the core has a diameter which ranges over 120–150 μm. The cylindrical interface between the core and cladding reflects light travelling in the core back into the core. The cladding serves a twofold purpose. It prevents damage to the reflecting interface. More importantly it determines the features of propagation in the fibre (Section 15.4).

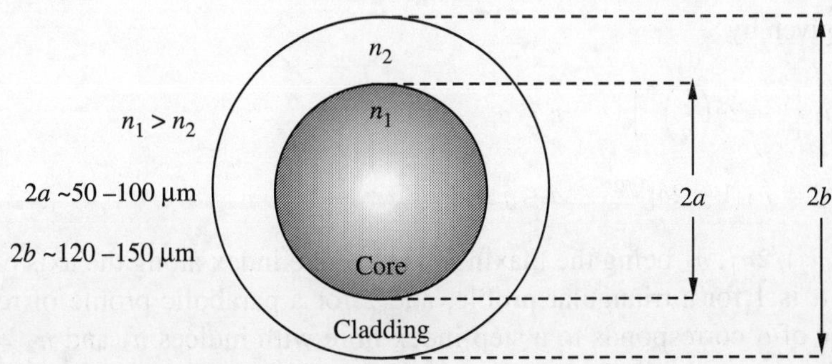

Figure 15.4 Core and cladding of an optical fibre

Modern cables have additional coaxial layers on top of the cladding. Figure 15.5 shows the cross section of a modern optical fibre. A buffer layer provides insulation. A strengthening material provides support to the delicate structure inside and finally a jacket layer separates the structure from other fibres. With the jacket each fibre is thus housed in a cylindrical region of a diameter of about 1 mm. Just about 0.25%–1% volume of this space around its axis actually carries the optical signal. Another 1%–2% is occupied by the cladding, and an overwhelming 97%–99% of the space is used for protection of the channel carrying the signal. Such jacketed fibres are deployed in three types of packaging: distribution, break-out style and loose buffer cables. Of these, the loose-buffered packaging alone allows individual fibres to move, thus reducing tension due to bending. This packaging is used in outdoor applications.

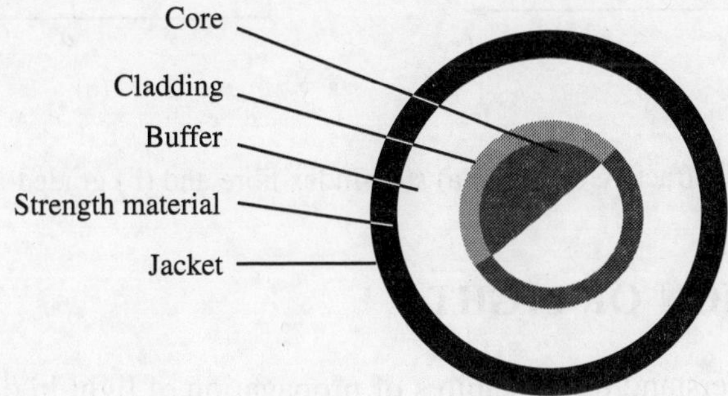

Figure 15.5 Cross section of a modern optical fibre

The characteristics of signal propagation are fixed by the properties of the small region occupied by the core and cladding. The core is generally made from doped silica glass. It has a refractive index higher than pure silica glass, which normally provides the cladding material. The level of doping controls the refractive index of the core. The variation of refractive index from the fibre axis outward to the cladding boundary is shown in Fig. 15.6a. Such a fibre is called a *step-index* fibre. In *graded-index* fibres the refractive index in the fibre varies continuously between $r = 0$ and $r = a$, that is, the axis and cladding-core boundary as shown in Fig. 15.6b.

The variation is given by

$$n(r) = \begin{cases} n_1 \left[1 - 2\delta\left(\dfrac{r}{a}\right)^{\alpha}\right]^{1/2} & r \leq a \\ n_2 = n_1[1 - 2\delta]^{1/2} & a \leq r \leq b \end{cases} \tag{15.3}$$

where $\delta = (n_1^2 - n_2^2)/2n_1^2$, n_1 being the maximum refractive index along the axis of the core. The *profile parameter* α is 1 for a triangular profile, and 2 for a parabolic profile of refractive index. The infinite value of α corresponds to a step-index fibre with indices n_1 and $n_2 = n_1[1 - 2\delta]^{1/2}$. If $\delta << 1$ then $\delta = (n_1 - n_2)/n_1$ is the normalised refractive index difference between the core and the cladding, which we represent by Δ later. The variation of refractive index in graded-index fibres reduces distortion in multimode propagation (see Section 15.7).

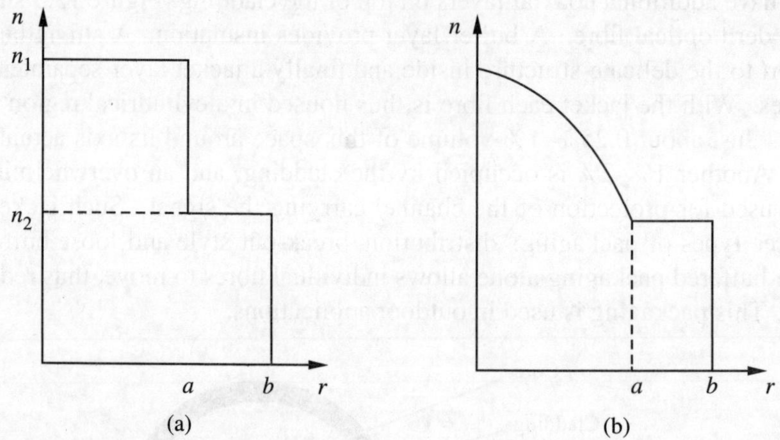

Figure 15.6 Variation of refractive index in (a) step-index fibre and (b) graded-index fibre

15.4 PROPAGATION OF LIGHT

The simplest way to understand basic features of propagation of light in optical fibres is through the use of geometrical optics or ray optics. A more complete treatment requires us to look at light signals as electromagnetic waves and treat the fibre as a waveguide which can carry different modes of these waves as determined by the geometry of the fibre. Here, we focus mostly on ray-optics treatment of propagation. The wave-optics picture of propagation is discussed in the next section.

From a ray-optics viewpoint the fibre is essentially a channel, the boundary of which repeatedly reflects back into the channel any light rays falling on it. The rays that are not so reflected are lost into the cladding and do not contribute to the light travelling down the channel. In Chart 15.1 the propagation of light in a step-index fibre is discussed. This includes the discussion of acceptance angle, numerical aperture and skip distance. The important results of this discussion are mentioned on p. 524 (symbols as in figure in Chart 15.1).

CHART 15.1 PROPAGATION OF LIGHT IN A STEP-INDEX FIBRE

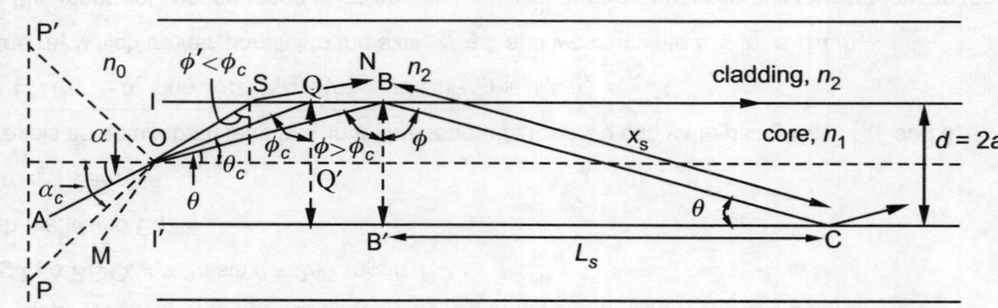

The figure shows a side-view of the core and cladding with paths of a few typical rays from a source adjacent to the core cross section at the opening end of the fibre.

The ray PO making an angle α_c with the fibre axis refracts at the interface II′ of outer medium with refractive index n_0 and the core of index n_1. The angle of refraction is θ_c resulting in angle OQQ′ $= \phi_c$, which is the critical angle of incidence at the core–cladding interface: $\phi_c = \sin^{-1}(n_2/n_1)$. This ray grazes the interface (ray POQB).

Any ray, like AO, making an angle of incidence $\alpha < \alpha_c$ at the interface of the outside medium—usually air—and the core cross section, falls on the core–cladding interface at an incidence angle $\phi > \phi_c$. Such a ray is totally internally reflected back into the core (ray AOBC).

On the other hand, any ray like MO that falls on the core–cladding interface at an incidence angle $\phi' < \phi_c$ and travels into the cladding and is lost (ray MOSN). Such rays do not contribute to signals propagating down the fibre.

Angle relations	The angle relations for the fibre propagation can be easily worked out from the core and cladding indices and fibre geometry. For the ray POQB, $n_0 \sin \alpha_c = n_1 \sin \theta_c$ (1) From ΔOQQ′, $\theta_c = \pi/2 - \phi_c$, so that $\sin \phi_c = \cos \theta_c = n_2/n_1$ (2) Similar relations hold for a general ray like AOBC within the acceptance cone, except for the last equality in Eq. (2). Thus, we have, $n_0 \sin \alpha = n_1 \sin \theta$ and $\sin \phi = \cos \theta$ (3)
Acceptance angle	All the rays incident at an interface II′ and lying within a cone like POP′ at the point of incidence O are accepted by the fibre for propagation. Those lying outside are rejected. The semivertical angle of this cone, α_c, is called the fibre acceptance angle.

Numerical aperature	The quantity $n_0 \sin \alpha_c$ is called the numerical aperture, NA, of the fibre. Using Eqs. 1 and 2, we can write, $\text{NA} = n_0 \sin \alpha_c = n_1 \sin \theta_c = n_1 \cos \phi_c = \sqrt{(n_1^2 - n_2^2)}$ (4) and $\alpha_c = \sin^{-1} (\text{NA}/n_0)$ (5) Normally, the requirement of fibre usage is to prevent stray radiation from being accepted by the fibre at the feeding end of the fibre as well as at joins—called *splices*—of two fibres. This means that the acceptance angle and the numerical aperature should be small so that the acceptance cone POP′ is narrow. This is achieved by keeping the difference between indices of the core and the cladding small. Thus, usually, $n_1 - n_2 << n_1, n_2$. We can then approximate $\sqrt{n_1^2 - n_2^2}$ as $\sqrt{(n_1 + n_2)(n_1 - n_2)}$ as $\sqrt{2n_1^2 (n_1 - n_2)/n_1} = n_1\sqrt{2\Delta}$, where, Δ is the normalised difference between core and cladding indices. We can then write $\text{NA} = n_1\sqrt{2\Delta}$, $\Delta << 1$ (6)
Skip distance	An important parameter in fibre propagation is the distance travelled by a ray along the fibre between two internal reflections at the core-cladding interface. This distance is called the skip distance. It is obvious from the ray optics of propagation that the minimum skip distance would occur for a ray, which makes just more than the critical angle at the core-cladding interface. The maximum skip distance will, of course, be ∞ and will occur for the ray travelling parallel to the fibre axis. From ΔBB′C in the above figure, we can write the skip distance L_s as $L_s = 2a \cot \theta$. Using Eq. 3 we get $L_s = (2n_1 a/n_0 \sin \alpha)\sqrt{1 - [(n_0^2 \sin^2\alpha)/n_1^2]} = 2a\sqrt{(n_1^2 - n_0^2 \sin^2\alpha)}/n_0 \sin\alpha$ As $\alpha \le \alpha_c$ we have $L_s \ge 2a\sqrt{(n_1^2 - n_0^2 \sin^2\alpha_c)}/n_0 \sin\alpha_c = 2an_2/\sqrt{(n_1^2 - n_2^2)} \approx a\sqrt{2/\Delta}$ (using Eqs. 4 and 6) The minimum value of skip distance is important as it determines the extent of total power loss due to less than total reflection at the interface. This minimum value is itself is generally referred to as the skip distance. Thus, $L_s = 2a\sqrt{(n_1^2 - n_0^2 \sin^2\alpha_c)}/n_0 \sin \alpha_c = a\sqrt{2/\Delta}$ (7) A related concept is that of path x_s travelled by a ray in the fibre between two successive internal reflections. x_s obviously depends on angle ϕ and consequently on θ and α. x_s gains importance because the phase of the ray—considered as an electromagnetic wave (Section 15.5)—depends on it. This has consequences regarding signal distortions in the fibre. Again, referring to Δ BB′C, $x_s = 2a/\sin \theta = 2an_1/n_0 \sin \alpha$. As $\alpha \le \alpha_c$ we have, using Eq. 4, $x_s \ge 2an_1/n_0 \sin \alpha_c = 2an_1/\text{NA}$. (8) Also, if a ray travels an actual path length x_L in a fibre length L, then we can write using Eqs. 4, 7 and 8, $x_L = x_s L/L_s = n_1 L/\sqrt{(n_1^2 - n_0^2 \sin^2\alpha)} \le n_1 L/\sqrt{(n_1^2 - n_0^2 \sin^2\alpha_c)} = n_0 L/n_2$ (9) Obviously for the ray which travels parallel to the axis, $x_L = L$ and we can write $L \le x_L \le n_1 L/n_2$ This result has the important consequence of determining maximum delay between two rays traversing the fibre at different angles with the axis, a concept which has repercussions on the bandwidth usable with a given fibre length (Section 15.7.1).

Basic relations	$n_0 \sin\alpha_c = n_1 \sin\theta_c$ $\sin\phi_c = \cos\theta_c = n_2/n_1$	(15.4) (15.5)

Parameter	**Meaning**	**Expression**
Acceptance angle	The semi-vertical angle of a cone of light rays that are accepted by the fibre for propagation	$\alpha = \sin^{-1}\sqrt{(n_1^2 - n_2^2)}/n_0$ (15.6)
Numerical aperture	The quantity $n_0 \sin\alpha_c$, that is, the sine of the acceptance angle if the outside medium is air, determines the light gathering capacity of the fibre.	$\text{NA} = \sqrt{(n_1^2 - n_2^2)} = n_1\sqrt{2\Delta}$ for $\Delta \ll 1$ (15.7) where $\Delta = (n_1 - n_2)/n_1$ (15.8)
Skip distance	Distance travelled by the ray along the fibre between two internal reflections at the core cladding interface	$L_s = 2a\sqrt{(n_1^2 - n_0^2 \sin^2\alpha_c)}/$ $n_0 \sin\alpha_c \approx a\sqrt{(2/\Delta)}$ (15.9)

Examples 15.1 and 15.2 provide the numerical appreciation of the above expressions.

EXAMPLE 15.1

Crucial importance of cladding material

A step-index optical fibre has core and cladding refractive indices given by 1.5025 and 1.4975, respectively. Find the numerical aperture and acceptance angle if the fibre is used in air. What would these be if the cladding was absent?

Solution

$$\text{NA} = \sqrt{n_1^2 - n_2^2} = n_1\sqrt{2\Delta} \text{ as } \Delta = \frac{1.5025 - 1.4975}{1.5025} = 0.0033 << 1$$

Thus, $\quad \text{NA} = 1.5025\sqrt{2 \times 0.0033} = 0.1221$

$$\alpha_c = \sin^{-1}\left(\frac{\text{NA}}{n_0}\right) = \sin^{-1}\left(\frac{0.1221}{1}\right) = 7.013^\circ$$

In the absence of the cladding $\text{NA} = \sqrt{(1.5025^2 - 1)} = 1.1214 > 1$.

Therefore $\alpha_c = \sin^{-1}(\text{NA}/n_0)$ has no solution between 0 and $\pi/2$. This means that a ray incident at any angle of incidence on the air–core boundary will refract to encounter the core-cladding boundary at greater than the critical angle of incidence, $\sin^{-1}(n_2/n_1)$. Thus, any such ray will undergo total internal reflection back into the core, that is, any ray will be accepted by the fibre. Cladding thus plays a crucial role in restricting the acceptance cone.

EXAMPLE 15.2
Number of reflections per metre of fibre

Using data of the previous problem find the skip distance and number of internal reflections per metre length of fibre. Assume radius of the core to be 50 μm.

Solution

$$L_s = 2a\frac{\sqrt{(n_1^2 - \text{NA}^2)}}{\text{NA}} \cong a\sqrt{\frac{2}{\Delta}} = 50\ \mu\text{m}\sqrt{\frac{2}{0.0033}} = 1230.9\ \mu\text{m}$$

$$\text{Number of reflections per metre} = 1\,\text{m}/L_s = \frac{1\ \text{m}}{1230.9 \times 10^{-6}\ \text{m}} = 812$$

In the case of a graded-index fibre the details of propagation of light depend on the refractive index profile. However, qualitatively it is obvious that there is a continuous bending of an oblique ray towards the axis, just as in the phenomenon of mirage. Thus longer paths are through lower index regions, compared to shorter paths which are through higher index regions. (Fig. 15.7).

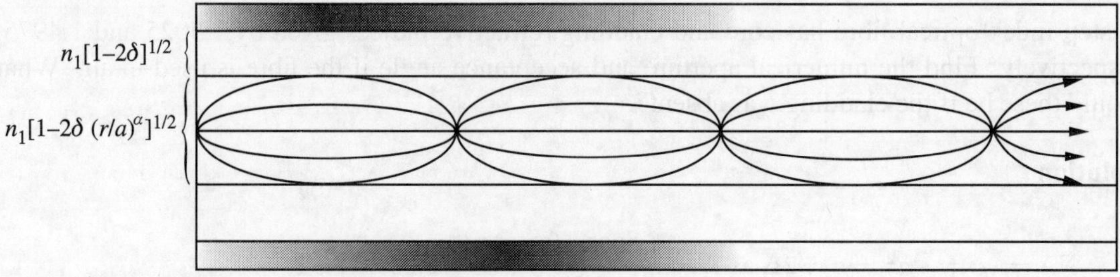

Figure 15.7 Propagation of light in a graded-index fibre

15.5 WAVE OPTICS: MODES

Ray optics does not give a complete picture of propagation in the fibre. Since light is an electromagnetic wave, the picture must be supplemented by waveguide theory of propagation. Here we will restrict ourselves to an outline to understand the concept of modes in the guide.

The simplest kind of electromagnetic wave in boundless media is a uniform plane wave, also called a transverse electromagnetic wave or a TEM. This wave has electric and magnetic fields pointing in mutually perpendicular directions at each point in space (see Fig. 2.9). The plane of these fields is normal to the direction in which the wave travels. In general, such a wave is not sustainable in a bounded region such as the core of an optical fibre. This is because in bounded regions the electric and magnetic fields must satisfy certain continuity conditions, called the *boundary conditions*, on boundaries between media. This is the case with a boundary such as the one between the core and the cladding in an optical fibre. The general solution of the problem of finding fields in the fibre is beyond the scope of this book. Here we confine ourselves to a narration of the main results of such calculation.

The upshot of application of electromagnetism to fibres is the *mode theory*. The waves sustained by the fibre can be classified into modes. A general wave in the guide is a mixture of modes. Each mode has a cut-off frequency below which it is not sustained in the fibre core, and the entire power leaks out into the cladding. One of the parameters for the wave corresponding to a frequency f, and free-space wavelength $\lambda = c/f$ is the *normalised frequency*—also called the V *number*—given by

$$V = \left(\frac{2\pi a}{\lambda}\right)\sqrt{n_1^2 - n_2^2} = \left(\frac{2\pi a}{\lambda}\right)\text{NA} \tag{15.10}$$

for a step-index fibre. Each mode has a cut-off value V_c determined by the mode parameters. For $V < 2.405$, only a single mode propagates in the fibre. This serves as a guide to choosing a fibre geometry and composition if single mode propagation is desired at a given frequency. Also, when the value of V for a given fibre at the operating wavelength is large enough, one can estimate the number of modes supported by the fibre by using the following relation.

$$N_{\text{modes}} \cong \begin{cases} V^2/2 \text{ (step-index fibre)} \\ \alpha/(\alpha+2)[(V^2/2)] \text{ (graded-index fibre with profile parameter } \alpha) \end{cases} \tag{15.11}$$

For smaller values of V we have to look at low mode cut-offs to find the number of modes. For $2.405 < V < 3.832$, only two modes can propagate. This number becomes four for $3.832 < V < 5.136$ and five for $5.136 < V < 5.520$. From these considerations it is easy to see that for a given composition and wavelength, the core diameter determines how many modes the fibre will carry (see Ex. 15.3).

As a design consideration for single mode propagation in step-index fibre, V is considered to be 2. We can also define a *cut-off wavelength*, λ_c, defined by

$$\lambda_c = \left(\frac{2\pi a}{2.405}\right)\text{NA} \tag{15.12}$$

Only if $\lambda > \lambda_c$, does the fibre operate in single mode.

EXAMPLE 15.3
Limiting diameter for single mode propagation

Determine the limiting diameter such that a step-index fibre with smaller core diameter will squeeze out all modes but one. Given $\lambda = 1.25$ μm, $n_1 = 1.462$, $n_2 = 1.457$.

Solution

For single mode propagation

$$V = \frac{2\pi a}{\lambda}\sqrt{n_1^2 - n_2^2} < 2.405$$

$$\text{Thus, } a < \frac{2.405}{2\pi}\left[\frac{\lambda}{\sqrt{n_1^2 - n_2^2}}\right] < \frac{2.405}{2\pi}\left[\frac{1.25\ \mu\text{m}}{\sqrt{1.462^2 - 1.457^2}}\right] < 3.96\ \mu\text{m}.$$

The diameter therefore has to be less than 7.92 μm to squeeze out all modes but one.

15.6 ATTENUATION

As light propagates in a fibre, several mechanisms lead to a loss of power. Consideration of such losses is critical in any application of fibre technology. The overall effect of these losses can be characterised in a straightforward manner. Wave theory tells us that the electric and magnetic fields lose amplitude as the light wave progresses in the fibre. Moreover, this amplitude loss is exponential in the distance covered in the fibre. Since electromagnetic power flow across unit area in the wave is proportional to the square of field amplitudes, wave attenuation is exponential. Thus, a meaningful measure of attenuation is logarithmic. Power lost per unit length of the fibre is usually measured in decibels per kilometre. It is defined as

$$A = -10\log_{10}\left(\frac{P_{z+1}}{P_z}\right) = 10\log_{10}\left(\frac{P_z}{P_{z+1}}\right) \tag{15.13}$$

where P_z is power flow at z km and P_{z+1} is power flow at $z + 1$ km along the fibre (see Ex. 15.4).

EXAMPLE 15.4
Calculation of attenuation

A step-index fibre with a core diameter of 30 μm and $n_1 = 1.525$, $n_2 = 1.500$ shows absorption of 0.00001% of the incident power at each reflection at the core boundary. Find the attenuation in dB/km for such a fibre for a ray entering just below the acceptance angle. Assume that there are no other losses.

Solution

$$\Delta = \frac{1.525 - 1.500}{1.525} = 0.01639$$

$$L_s = a\sqrt{\frac{2}{\Delta}} = 15\ \mu\text{m}\sqrt{\frac{2}{0.01639}} = 165.7\ \mu\text{m}$$

$$\text{Number of reflections per km} = \frac{1000\text{ m}}{165.7\ \mu\text{m}} \approx 6 \times 10^6.$$

At reach reflection the power reduces by the factor $1 - 0.0000001 = 0.9999999$.

So we can write

$$\text{Power } P_{1\text{km}} = P_0 \times 0.9999999^{6\times10^6}$$

$$\text{Hence attenuation, } A = 10 \log \left[\frac{P_0}{P_{1\text{km}}}\right]$$

$$= 10 \log \left[\frac{P_0}{0.9999999^{6\times10^6} P_0}\right] = 2.6\text{ dB/km}$$

15.6.1 Loss Mechanisms

There are many ways in which power can be lost as light travels in an optical fibre. The core material can absorb power. Light waves scatter from impurities and defects. In addition, in a fibre network, fibre joins and connectors dissipate some power. Bends in fibre lines can also lead to power losses. In general, material and scattering losses grow exponentially with line length. On the other hand, losses due to connectors and bends are point losses and depend on the total number of such discontinuities, rather than the fibre length.

Material absorption is of two kinds: intrinsic and extrinsic. Intrinsic absorption is caused by one or more of the main components of the fibre material. Pure silica exhibits vibrational resonances in the UV region and occasionally in the IR region of the electromagnetic spectrum.

Extrinsic material absorption is caused by Rayleigh scattering at impurities. Such scattering is proportional to the fourth power of frequency and generally limits the operating wavelength from below. Wavelengths smaller than about 0.8 μm are consequently ruled out for fibre use. Certain transition metal ion impurities such as Cr^{3+}, Cu^{2+}, Fe^{2+}, Ni^{2+} and Mn^{2+} show strong absorption in the 0.6-1.6 micron wavelength range. If attenuation below 1 dB/km is desired, the extent of such impurities has to be reduced to below 1 part per billion (ppb). One of the most important causes of attenuation is the hydroxyl ion. Absorption by OH^- peaks at 2.73 μm. Overtones at 1.38 μm, 0.95 μm, and 0.72 μm also show strong hydroxyl ion absorption. At the operating wavelength of 1.38 μm requirement of less than 10 dB/km losses necessitates hydroxyl ion presence to be below 10 ppb.

Figure 15.8 shows a typical loss curve in a silica-based fibre against wavelength in the range 0.6 μm to 1.6 μm showing hydroxyl ion peaks. Also shown are wavelength windows, which are generally used for fibre communication. The other curves in the figure show background UV, IR and waveguide losses. Figure 15.9 shows the loss curve due only to hydroxyl ion absorption including overtone peaks mentioned above.

Waveguide imperfections such as core-cladding interface irregularities and variations in core diameter lead to scattering, called *Mie scattering*, and attendant power loss. At a join between

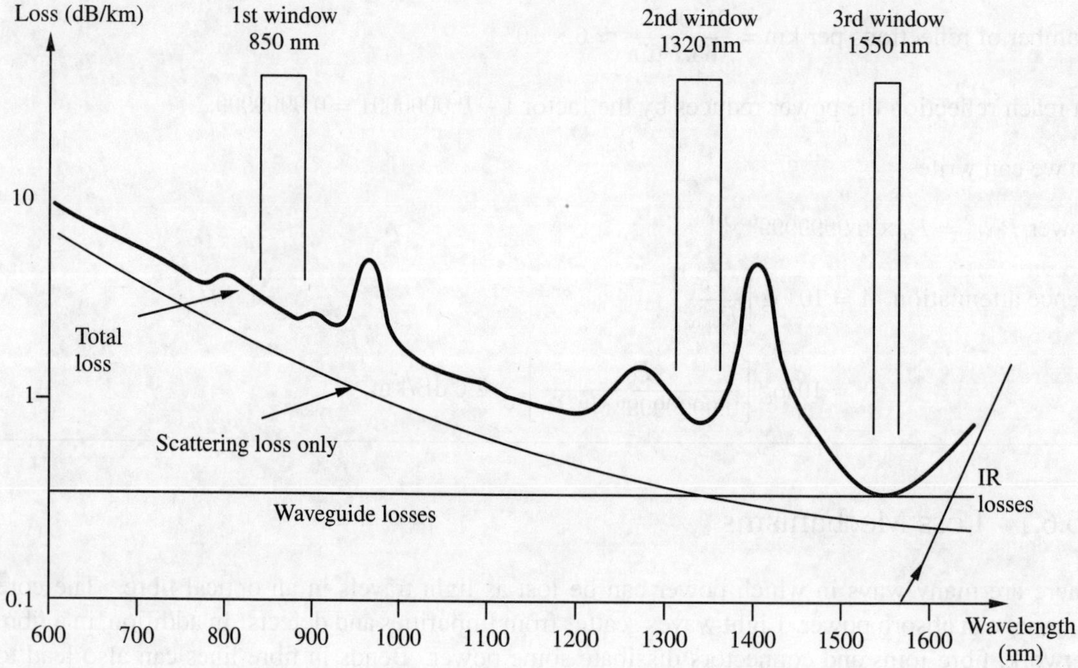

Figure 15.8 Loss curve in a silica-based fibre

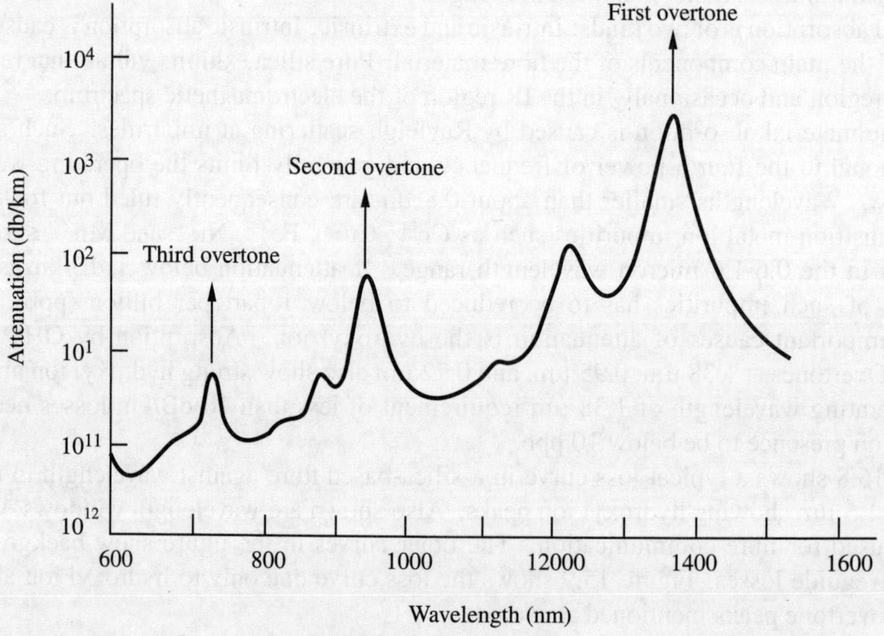

Figure 15.9 Loss curve due to hydroxyl ion absorption

two fibre sections, that is, a splice, power is lost due to impedance mismatch. At a physical splice in which the fibre ends are aligned visually under a microscope, the extent of loss is around 0.2 dB. At a fusion splice, in which fibre ends are fused together in an arc, the loss could be as low as 0.02 dB. At a bend in a fibre length, the angle of incidence changes abruptly and may become less than the critical angle. This causes the phenomenon called *frustrated total internal reflection* giving rise to an evanescent field in the cladding and attendant power loss. Table 15.1 shows typical losses in modern fibres and Table 15.2 shows attenuation in (dB/km) caused by 1 ppb of various common metallic inpurity ions.

Table 15.1 Losses in modern fibres

Fibre type	**Dimensions (core/cladding dia in μm)**	**Attenuation (dB/km)**	
		At 850 nm	At 1550 nm
Single mode fibre	8/125	—	0.25
Multimode fibre	50/125	2.4	—
Multimode fibre	62.5/125	2.8	

Table 15.2 Attenuation caused by metallic impurities

	Peak wavelength (nm)	**Attenuation caused by 1 ppb (dB/km)**
Cr^{3+}	625	1.6
Cu^{2+}	850	1.1
Fe^{3+}	400	0.15
Ni^{2+}	650	0.1
Mn^{3+}	460	0.2
V^{4+}	725	2.7

15.7 SIGNAL DISTORTION

In very general terms the term ‘distortion’ refers to unwanted modification in the content of any communication. Communication over a fibre is in the form of light signals or waves. Thus, in the present context, distortion refers to undesired changes in the waveform. The profile of the wave travelling down a fibre undergoes changes due to the phenomenon referred to as *dispersion*. As information travelling in the fibre is coded in the waveform, changes in it imply modification, and therefore loss of information.

In a fibre optic communication system, such waveform distortion specifically presents itself as loss of digital information. This can be understood as follows. Information communicated through a fibre is usually digital. This implies that it is coded into a sequence of light pulses sent along a fibre. Dispersion in the fibre leads most generally to broadening of the pulses (see Fig. 15.10). Such broadening can kill the coded information because the neighbouring pulses can overlap at the output end of the fibre. The overlap destroys the fidelity of the decoder operation as a sequence of two pulses might be misunderstood as a single pulse. These considerations imply that dispersion limits the density with which pulses can be packed sequentially. A decrease in packing density reduces the amount of information, which may be communicated per unit time.

15.10 Dispersion in the fibre leads to broadening of the pulse

15.7.1 Mechanisms of Dispersion

Broadly, there are two dominant ways in which optical signals in a fibre distort. They are referred to as *chromatic dispersion* and *modal dispersion*. Both the mechanisms emerge from the fact that pulsed digital signals contain a large number of wavelengths. This situation is easily understood if we recall that a pulse is a wave packet, which is a superposition of a large number of harmonic waves, each of a different wavelength.

Chromatic dispersion

The dielectric constant of the fibre material varies with wavelength. Consequently, each wavelength component of the pulsed signal in the fibre sees a different refractive index and travels with a different velocity. This has the result that different wavelength components take different times to travel any particular fibre length. The phase relationship between any two such components is then not the same at the output end of the fibre as it was at the input end. Therefore, at the output end these components superpose to give a broadened pulse. This mechanism of dispersion is called chromatic dispersion.

The importance of chromatic dispersion thus depends on the nature of dependence of the refractive index n on the wavelength λ, that is, on the function $n(\lambda)$. Specifically, the basic measure of chromatic dispersion is the size of the second derivative $d^2n/d\lambda^2$ of $n(\lambda)$. Chromatic dispersion is then smallest at wavelengths near the inflexion points in $n(\lambda)$ versus λ graphs where $d^2n/d\lambda^2$ vanishes. For fused silica fibres this occurs at $\lambda = 1.3$ μm, which is in the second window in Fig. 15.8. For a wavelength range 1nm, around λ nm the chromatic dispersion in the fibre is given by

$$D_c = 10^9 \left(\frac{\lambda}{c}\right) \frac{d^2n(\lambda)}{d\lambda^2} \text{ns/km–nm} \tag{15.14}$$

This means that the delay introduced by 1 km length of fibre between the components with wavelengths between $(\lambda - \frac{1}{2})$ nm and $(\lambda + \frac{1}{2})$ nm of the range is D_c in nanoseconds.

Modal dispersion

We noted earlier that an accurate description of light propagation requires us to look at propagation as wave propagation. This theory classifies the waves in the guide in terms of modes. Each mode has its own propagation constant and wave number. In ray approximation these modes can be conceived as rays propagating at definite angles to the fibre axis and undergoing multiple internal reflections at the core-cladding boundary. In this simple picture, dispersion arises because rays travelling at different angles travel different path lengths in the fibre and undergo different phase lags before they recombine at the receiving end of the fibre.

This phenomenon of differential phase lag for different modes is responsible for signal dispersion. If we examine a digital signal, we find that it is a sequence of pulses. Such pulses contain fourier components of different wavelengths. Depending on fibre geometry, these components will excite different modes. These modes undergo different phase lags. Recombination of these components gives a waveform of spread out pulses. If the pulse spread is large enough for neighbouring pulses to overlap, signal fidelity is affected and loss of transmitted information follows. Such considerations mean that dispersion in fibre puts a limit on the number, f, of pulses that can be packed per unit time—the so called *band width* supported by the fibre. The larger f is, the more the information which can be transmitted per unit time. Dispersion degrades the band width.

We can estimate delays by a simple calculation using ray optics. If the acceptance angle of the fibre is just less than α_c, then the path length travelled by this ray in fibre length L is $x_{cL} = L/cos\theta_c$, where θ_c is the refraction angle in the fibre corresponding to incidence at α_c at the transmission end of the fibre. $x_{cL} = x_L$ for $\alpha = \alpha_c$ (see Chart 15.1). Using Eq. 15.5 we get $x_{cL} = n_1 L/n_2$. The ray that is parallel to the axis travels distance L in the fibre. So the delay of the critical ray is due to the path difference $x_{cL} - L$ between the two rays. As all the rays in the fibre core travel at speed c/n_1, the maximum delay is

$$\Delta t = \frac{x_{cL} - L}{c/n_1} = \frac{n_1 L}{c}\left[\frac{n_1}{n_2} - 1\right] \tag{15.15}$$

If $n_1 - n_2 << n_1$, we can approximate the maximum delay as

$$\Delta t = \frac{L n_1 \Delta}{c} \tag{15.16}$$

It is clear from this expression that the delay is proportional to the fibre length L. Now, a simple estimate of band width f allowed in such a fibre is obtained by requiring that this delay be not more than half the time gap $1/f$ between pulses, that is, $\Delta t \leq 1/2f$. This gives the restriction on band width

$$f \leq \frac{1}{2\Delta t} = \frac{c}{2Ln_1\Delta} \tag{15.17}$$

A useful parameter for the fibre can be distilled out from this by multiplying the band width by distance L, since the product depends only on fibre parameters. This product is called the *band width–distance product*, or BDP.

So, we can write

$$\text{BDP} = fL = \frac{c}{2n_1\Delta} \tag{15.18}$$

The BDP of a fibre restricts the band width or *transmission rate*—also called the *bit-rate* of pulses—which can be used for communication over a given fibre length. On the other hand, for a given transmission rate for information, it limits the distance over which communication can be effectively done (see Ex. 15.5).

Multimode step-index fibres generally have BDP ranging over 6–25 MHz-km, which is suitable for local area networks. Single mode fibres have BDP ranging over 500–1500 MHz-km, which makes them appropriate for telecom applications.

EXAMPLE 15.5

Calculation of maximum delay

A step-index fibre has a core of refractive index 1.5025 and cladding of index 1.4975. Find the maximum delay over a fibre of length 1 m. How far can communication using a band width of 2 MHz be reliably sustained over such a fibre?

Solution

We have $n_1 = 1.5025$ and $n_2 = 1.4975$

Maximum delay over length L is $\Delta t = (n_1 L/c)[(n_1/n_2) - 1]$.

So, $\Delta t = [(1.5025 \times 1\ \text{m})/(3 \times 10^8\ \text{m/s})] \times [(1.5025 - 1.4975)/1.4975] \approx 1.67 \times 10^{-11}\ \text{s} = 16.7\ \text{ps}$

So, band width for propagation over 1 m length is $f = \dfrac{1}{2\Delta t} = \dfrac{1}{2 \times 16.7 \times 10^{-12}\ \text{s}}$

This is the BDP in m-Hz. Therefore, the distance L of propagation at 2 MHz band width is, $L \leq \dfrac{\text{BDP}}{2 \times 10^6}$, or $L \leq [1/(2 \times 2 \times 10^6 \times 16.7 \times 10^{-12})]$ m.

Thus $L \leq 10^6$ m/66.8 = 15 km.

Thus, with a band width of 2 MHz, that is, 2×10^6 pulses per second, we can communicate reliably over a distance of 15 km.

15.7.2 Measure of Dispersion

From the above discussion it should be clear that in a single mode fibre, generally chromatic dispersion is the main cause of dispersion. In such fibres dispersion is measured in picoseconds per kilometre per nanometre of spectral wavelength—width of light input into the fibre. The empirical relation between dispersion D_c (ps/km-nm) in a single mode fibre and the band width BW which may be used with length L of this fibre is

$$BW = \frac{0.187}{D_c} \times \Delta\lambda \times L \tag{15.19}$$

where $\Delta\lambda$ is the wavelength width of the source.

In a multimode fibre, on the other hand, manufacturers normally specify a BDP in MHz-km, rather than delays. Thus a 400 MHz-km BDP fibre will have a band width of 400 MHz when only 1 km length is used, but a BW of only 200 MHz over a 2 km length. The general rule is BW $\times$ L should be less than the specified BDP for a multimode fibre.

15.8 FIBRE OPTIC COMMUNICATION SYSTEMS

A general optical fibre communication system can be visualised in terms of the block diagram shown in Fig. 15.11. The fibre optic part lies between the transmitter and the receiver. The information source and destination could be remote personal computers generating and receiving signals at their output and input ports.

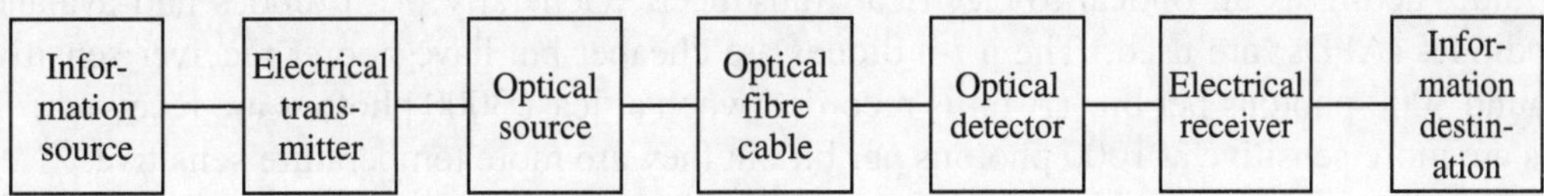

Figure 15.11 A general optical fibre communication system

In the early days of fibre optic communications the optical sources operated in three main narrow bands at wavelengths of 850 nm, 1300 nm and 1500 nm. The later generation systems operated by exploiting the entire band of wavelengths. Table 15.3 shows the bit rates and transmission ranges of typical sources, which are light emitting diodes (LEDs) and laser diodes (LDs). Table 15.4 shows the wavelength windows in use.

Table 15.3 Optical sources

Device	Material	Bit rate (Mbps)	Transmission range (km)
LED	AlGaAs	0–40	0–5
LED	InGaAsP	0–300	5–10
LD	InGaAsP	30–800	10–50
LD	InGaAsP	100–800	50–100

Table 15.4 Wavelength windows in use

Band	λ range (nm)
850	810–890
O	1270–1350
E	1350–1470
S	1470–1530
C	1530–1565
L	1565–1625

These optical sources couple to electrical transmitters thus acting as electrical to optical transducers. Laser diodes are more directive and provide efficient coupling to fibre cable. LEDs are cheaper, have simpler driving circuits and can operate under a wide range of environments. They have a much wider wavelength spectrum—50 to 100 nm—compared to laser diodes—0.1 to 1 nm (see Fig. 14.12). The wider spectrum is not necessarily a disadvantage, particularly when multimode fibre is to be used.

The optical detector couples to the fibre on one side and to the electrical receiver on the other side, acting as an optical-to-electrical transducer. Generally p-i-n diodes and avalanche photodiodes (APDs) are used. The p-i-n diodes are cheaper but have poorer receiver sensitivity at around 3000 photons per bit. (A bit is recorded when at least 3000 photons are received.) The APDs are more sensitive at 1000 photons per bit but they are more temperature sensitive.

There are two main considerations in choosing the components of the fibre optic system: (i) power budget and (ii) band width budget. In any specific design, the cost budget is of course, the third major, but nontechnical consideration. The power budget must consider coupling losses at transducer couplings and at connectors and splices of fibres. LD coupling loss is 0 dB, whereas an LED coupling loss may be 15–20 dB. The splice and connector losses are 0.02–0.2 dB a piece. So this loss must be calculated with the total number of such connectors and splices in the whole system. The loss in fibre depends on the dB/km loss rating of the fibre. To find this loss the total length of cable in use must be taken into account.

The band width budget concerns calculation of the total delays in the system. This includes the response times of the transmitter τ_{tx}, and the receiver τ_{rx} as well as the fibre delays τ_f. The total delay is computed as

$$\tau = \sqrt{\tau_{tx^2} + \tau_f^2 + \tau_{rx}^2} \tag{15.20}$$

This is because each of these delays involves physically independent mechanisms. The delay τ_f consists of delays due chiefly to chromatic (τ_c) and modal dispersion (τ_m). We can estimate these by

$$\tau_c = D_c \Delta\lambda L \text{ and } \tau_m = D_m L \tag{15.21}$$

where D_m = 1/BDP, and D_c is as defined earlier in Eq.15.14. Again $\tau_f = \sqrt{\tau_c^2 + \tau_m^2}$.

Once the total delay time τ is available, the usable band width is determined by BW $< 1/2\tau$. The band width is an important design decision as it determines the rate at which information can be transmitted with fidelity.

15.9 APPLICATIONS

In this chapter we discussed the basic aspects of the physics of fibre optics. The development of better fibres over the last few decades led to the evolution of fibre optic technology—a key technology that produced amazing advances in engineering and medical fields. The application potential of fibre optics became evident soon after the development of an image-transmitting device called the *flexible fibrescope* by Kapany and his colleagues at Imperial College of Science and Technology, London, in 1956. The device found application in inspecting inaccessible points inside reactor vessels and jet aircraft engines, and a similar device—the *flexible endoscope*—found application in medical diagnostics and surgery. As the applications of fibre optic systems are too many, we restrict ourselves to a brief discussion of a few of the major applications.

15.9.1 Advantages of the Fibre Optic Systems

Fibre optic systems have several advantages over the traditional ones. They allow information to travel greater distances without the need for strengthening of the signal. Fibre optic cables are small, lightweight, chemically passive and require less maintenance than conventional copper wires. These advantages have made fibre optics ubiquitous in the modern world due to their wide-ranging applications.

As optical fibres are made of glass or plastic, both of which are insulators, the fibre optic cables have a greater immunity to electromagnetic noise. Thus the optical signals traversing through these cables are free from any radio frequency interference (RFI) and electromagnetic interference (EMI). The optical fibres do not pick-up or propagate electromagnetic pulses (EMPs). Hence they are suitable for reliable monitoring and telemetry in industrial environments where EMI and EMPs cause problems for metallic cables. The optical signal propagating through the fibre is well protected from interference and coupling with other communication channels giving a high degree of signal security. The fibre optic systems are hence suitable for military and banking applications and for computer networks.

Fibre optic systems have been found to be suitable for large-scale exploitation, because silica, the starting raw material for the production of glass fibres, is easily available and the costs of transportation and installation of fibre optic cables for long distance communication are estimated to be lower than that for metallic cables.

15.9.2 Optical Fibre Sensors

Fibre optic sensors are devices, which use light guided within an optical fibre to detect any external physical, chemical, biomedical or any other parameter, such as, pollutant, blood pressure and position of earthquake faults. They have the ability to interface with a variety of systems. Distributed fibre optic sensors can be embedded in composites or building materials, for example, concrete, to create a ‘smart structure’ that can monitor the internal conditions of the component throughout its life. The fibre optic smart structures have the potential to revolutionise the field of future composite materials and intelligent structures.

15.9.3 Medical Applications

Optical fibres have also become indispensable in the medical field where they are used for imaging, diagnosis and surgery. They can be inserted through natural openings or by making a small incision, and threaded along the body's established pathways to peer into the lungs, airways, intestinal folds, heart chambers and other areas that were previously inaccessible. The surgical procedures that were performed earlier with a scalpel, leading to large incisions and great deal of soreness, are now possible using fibres that require very small incisions leading to speedy and aesthetically pleasing recovery. Optical fibres are thus widely used in photodynamic therapy for cancer, in the treatment of lung disorders and bleeding ulcers, and in arthroscopic surgery.

Arthroscopic surgery is used to repair damaged cartilage, ligaments and tendons in major joints such as knees and shoulders. Whereas traditional surgery might have a scar upto a couple of inches long, typical arthroscopic incisions are only about 1 cm long. Surgeons send a small fibre optic telescope into one of the incisions and the scope is connected to the monitor to enable the surgeon to see inside the patient's joint. Large parts of the joint can be seen without tearing open the patient's body.

15.9.4 Applications in Communications and Information Technology

The fibre optic revolution has had the most powerful impact in the field of telecommunication engineering and information technology. The radio and microwave frequencies were in use earlier for the telecommunication purposes. However, the increasing television and telephone traffic placed heavy demand on these frequencies and people began to invest in optical communications with the invention of the laser in 1960. Of late, the fibre optic systems are steadily replacing copper wires for telecommunication purposes. The users of these systems include telephone services, cable television services, university campuses, office buildings, industrial plants and electric utility companies.

The transmission band width, and hence the information carrier capacity of a fibre optic system, is much greater than the coaxial copper cable, wide-band radio, or microwave systems. The low transmission losses (typically around 0.2 dB/km) in optical fibres reduce the system cost as well as the need for numerous repeaters in long-haul telecommunication applications.

In fibre optic systems, the carrier frequencies, usually corresponding to the infrared part of the electromagnetic spectrum, are of the order of 10^{14} Hz, which is 10,000 times the microwave frequencies. The immense band width (more than 50 THz) of optical fibres opens up extraordinary possibilities for future fibre optic systems and their applications.

In 1970, Robert Maurer, Donald Keck and Peter Schultz of Corning Glass Corporation, USA, provided a breakthrough in fibre optics when they produced pure glass fibres that exhibited an attenuation of less than 20 dB/km. The development of fibre optics technology over the past three decades can be understood in terms of a series of generations, mainly identified by the operating wavelengths they employed. The first generation fibre optic systems employed a wavelength of 0.85 μm, corresponding to the 'first window' in a silica-based fibre. It exhibited a relatively high loss of the order of 3 dB/km. The second generation systems used a 'second window' at 1.3 μm with theoretically zero dispersion for silica fibres and attenuation around

0.5 dB/km. In 1977, Nippon Telegraph and Telephone succeeded in developing a 'third window' at 1.55 μm that offered a loss of about 0.2 dB/km but non-zero dispersion for silica fibres. The researches in dispersion management technologies have played a very important role in the explosive progress in optical fibre communications in the past decade.

In 1977, AT and T Corp. and GTE Corp. succeeded in the commercial deployment of fibre optic systems. In the 1980s, fibre optic networks first used single mode fibres operating at 1.31 μm. In 1990, Bell Laboratories demonstrated the transmission of 2.5 Gbits (2.5×10^9 bits) per second over 7500 km without regeneration. In 1998, Bell Laboratories achieved another milestone of transmitting 100 simultaneous optical signals, each at a data rate of 10 Gbits per second, for a distance of nearly 400 km.

The primary motivation behind fibre optic technology is increased band width for telecommunications. Increased band width facilitates more data transfer at a quicker rate without interference. For example, FM radio stations that broadcast at frequencies in the order of 100 MHz are permitted a band width of 200 kHz, whereas AM stations with frequency around 1000 kHz have a band width of only 5 kHz. Thus FM stations can send more information resulting in better stereo quality sound. Therefore, higher frequencies are imperative for higher band width and more efficient communication.

Just as FM is better than AM, higher frequencies would be even better than FM. Thus encoding information in light would be ideal because of relatively higher frequency. This is the reason for the use of fibre optics.

From wired society to fibred society

In today's information world the efficient transfer of information is highly relevant to our well-being. There is an ever growing demand for telephones, televisions and communication networks. Fibre optics offers efficient solutions to these demands, because the transmission band width, and hence the information carrier capacity of a fibre optic system, is much greater than the coaxial copper cable, wide-band radio, or microwave systems.

As the demand for more capacity, more speed and more clarity increases—driven mainly by the phenomenal growth of the Internet—the move towards optical networking is the focus of new technology. The next-generation fibre optic systems will revamp the existing network architectures to develop all optical networks. Fibre optics is spreading into homes, offices and stores and is destined to play an important role in the developments in almost every sphere of life. It is only apt to say that we are living in the stage of transition from a wired society into a fibred society.

EXERCISES

1. What is an optical fibre? What is the principle involved in its working?
2. Explain the phenomenon of total internal reflection of light. How is it used in fibre-optics communication?
3. Explain what is step-index, graded-index, monomode and multimode fibre. Draw relevant sketches.
4. Explain in detail how optical fibres are classified.
5. Briefly describe the classification of the optical fibres based on material.
6. Explain the following terms: (i) Critical angle, (ii) acceptance cone and (iii) numerical aperture.
7. Deduce an expression for the acceptance angle of an optical fibre.
8. Derive an expression for the numerical aperture of a step-index fibre in terms of Δ.
9. Describe various mechanisms of attenuation in optical fibres.
10. What is meant by attenuation in optical fibres? State the factors responsible for losses in fibres.
11. What is dispersion in optical fibres? State the different mechanisms that contribute to dispersion. How does it affect the transmission of data through optical fibres?
12. What are the advantages of optical fibres over conventional cables?
13. Explain any one application of the optical fibre used as a sensor.
14. What are optical fibres? State at least three of their applications.
15. Discuss the application of optical fibres as temperature sensors.

PROBLEMS

1. A glass clad fibre is made with core glass of refractive index 1.5 and the cladding is doped to give a fractional index 1.5 and a fractional index difference of 0.0005. Find (i) the cladding index and (ii) the acceptance angle.
2. A step-index fibre has NA of 0.16 in air, a core refractive index of 1.45 and a core diameter of 60 μm. Determine the normalised frequency for the fibre when light at a wavelength of 0.9 μm is transmitted.
3. Find the core radius necessary for single-mode operation at wavelength 850 nm in a step-index fibre with $n_1 = 1.480$ and $n_2 = 1.47$.
4. A certain optical fibre has an attenuation of 3.5 dB/km at 850 dB/km at 850 nm. If 0.5 mW of optical power is initially launched into the fibre, what is the power level in microwatts after 4 km?
5. A certain optical fibre has an attenuation of 1.5 dB/km at 1300 nm. If 0.5 mW of optical power is initially launched into the fibre, what is the power level in microwatts after 8 km and 16 km?

16

Acoustics

16.1 THE STORY OF ACOUSTICS

Acoustics happens to be one of the oldest branches of physics. It derives its name from the Greek word meaning 'hearing'. The great variety of sounds that we hear in nature makes us wonder about this beautiful gift of nature. But a physicist's 'peep' into sound is concerned more with its physical origin and aspects.

The fact that certain sounds are more pleasing to the ear than others must have attracted attention since antiquity and led to the development of music. That the pitch of a musical sound has something to do with its frequency seems to have been understood by the sixth-century Greek philosopher Pythagoras. Aristotle and the first-century BC Roman architectural engineer Marcus Vitruvius Pollio have contributed to the understanding of sound.

In 1635, Pierre Gassendi— a French philosopher and mathematician—made the first measurements of the velocity of sound in air. Gassendi assumed that sound travels because of the transfer of a stream of very small, invisible particles that stimulate the ear. The application of the wave theory of sound was made, probably for the first time, by Sir Issac Newton who derived an expression for the velocity of sound in a gaseous medium. However, Newton's expression did not agree with the experimentally measured velocity. This discrepancy was resolved by a French mathematician, Pierre-Simon Laplace, who contended that Newton's assumption that pressure and density changes during sound propagation are isothermal was wrong. He argued that these changes should be adiabatic, and derived a new expression for sound velocity that was in better agreement with the experiment than Newton's formula.

The period between the second half of the eighteenth century and the first quarter of the nineteenth century produced a lot of theoretical research on the study of mechanical waves. This was based largely on the mathematical expression for wave motion proposed by d'Alembert in 1747. The pioneering contributors to this research were two mathematicians: Leonhard Euler from Switzerland and Joseph-Louis Lagrange from France.

In 1830, a French physicist, Felix Savart, showed that the audibility range of humans is about 20 Hz to 20 kHz. In 1843, Ohm, a German physicist, theorised that the human ear is capable of performing the Fourier analysis (breaking of any complex signal into its component fundamental tones as suggested by the French mathematician, Jean Fourier) of any complex sound. Hermann von Helmholtz, a German physician serving as a surgeon in the Prussian army, took keen interest in understanding the physics behind biological systems and extended Ohm's work. His book *Sensations of Tone* is the cornerstone of physiological acoustics. He advanced theories to explain the role of bones of the middle ear and the cochlea in the transmission of sound waves through the ear. Helmholtz also made important contributions to optics by developing the opthalmoscope and reviving the theory of colour vision proposed earlier by Young. Though early work on sound in enclosures was reported around 1850, the quantitative foundation of the subject was laid down through the pioneering efforts of US physicist Wallace Clement Sabine who derived, in 1900, an expression for reverberation time.

The theoretical developments in acoustics are summarised in the volume, *The Theory of Sound*, by the English physicist, Lord Rayleigh (1877). Most of the ideas in this treatise have been exploited in the further development of acoustics in the nineteenth and twentieth centuries.

In Chapters 2–5 we studied about light waves. Light waves are electromagnetic in nature and do not need a material medium for their propagation. Another type of familiar waves, which need a material medium for their propagation, are mechanical waves. Water waves, sound waves and seismic waves are examples of such waves. Acoustics, generally known as the science of sound, is broadly defined as the *study of generation, transmission and reception of energy in the form of mechanical waves*. Whereas sound waves, also known as *sonic* waves, are limited only to a narrow spectrum of 20 Hz to 20 kHz, the discipline of acoustics also encompasses the study of waves with frequency below 20 Hz, called the *infrasonic* waves and those above 20 kHz, called the *ultrasonic* waves. In this chapter we study the fundamentals of vibration (Section 16.2), sound waves and their characteristics (Section 16.3), mechanisms of speech and hearing (Section 16.4), classical ray theory (Section 16.5) and ultrasonics (Section 16.6). The applications are discussed in Section 16.7.

16.2 FUNDAMENTALS OF VIBRATIONS

Fundamental to acoustics is the physics of oscillatory or vibrational motion. The simplest vibrating system is a simple harmonic oscillator. A mass spring system (Fig. 16.1) is a typical example of such an oscillator. The restoring force F produced in the spring due to displacement of the mass from its equilibrium position by an amount x is proportional to this displacement,

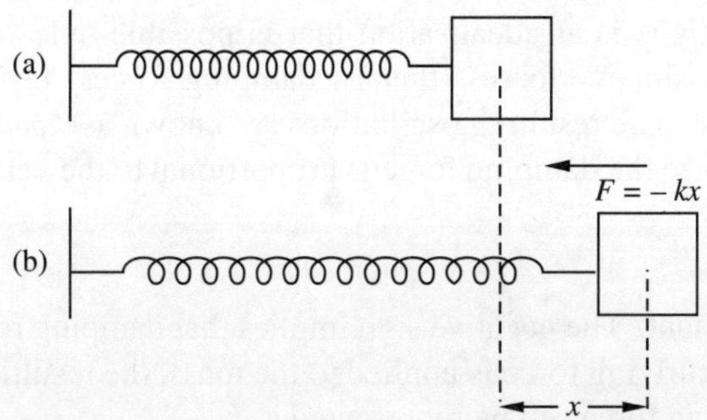

Figure 16.1 (a) The mass spring system in equilibrium; (b) The mass, displaced through x, produces the restoring force F, in the spring that pulls the mass towards the equilibrium position

that is, $F = -kx$ (16.1)

k being the force constant of the spring.

The ensuing motion of the mass is periodic and varies sinusoidally with time. The vibrations are *free* vibrations, meaning that the effects due to friction and viscosity are neglected and the amplitude of vibrations remains constant in time. The equation of motion in this case can be written as

$$-kx = \frac{md^2x}{dt^2} \tag{16.2}$$

Thus, $$\frac{d^2x}{dt^2} + \frac{k}{m}x = 0 \tag{16.3}$$

It can be readily seen that the equation,

$$x = A\cos(\omega_0 t + \phi) \tag{16.4}$$

is a solution of Eq. (16.3) if $\omega_0^2 = k/m$, that is, the angular frequency of the motion is identified as

$$\omega_0 = \sqrt{\frac{k}{m}} \tag{16.5}$$

The natural frequency ν_0 and the time period T_0 are thus given, respectively, by

$$\nu_0 = \frac{1}{2\pi}\sqrt{\frac{k}{m}} \tag{16.6}$$

and $$T_0 = 2\pi\sqrt{\frac{m}{k}} \tag{16.7}$$

The foregoing discussion can be applied to study free vibrations in a variety of vibrating systems by writing the equation of motion and solving it. This is illustrated in a *production line spirit* in Chart 16.1.

The case of free vibrations is an idealisation that is possible only in a perfect vacuum. In reality the presence of dissipative forces (that is, damping forces) results in the amplitude of motion decaying with time. The resulting oscillations are known as *damped* oscillations. In many practical cases of importance the damping force is proportional to the velocity of the body, that is,

$$f_d = -bv \tag{16.8}$$

where b is a damping constant. The negative sign implies that damping forces oppose the motion.

If an external harmonic driving force is applied to the mass, the resulting motion is known as a *forced* vibration. The driving force can be represented as

$$f = f_m \cos \omega t \tag{16.9}$$

The difference between the frequency ω of the driving force and the natural frequency of the vibrator decides the resultant amplitude of the vibrating system. When the two frequencies are equal, the amplitude is maximum and the vibrating system is said to be in *resonance.*

The analysis of damped and forced vibrations—though more complicated than that of the free vibrations—follows a classical approach, that is, write the equation of motion and solve it. Chart 16.2 is a comparative study of these cases.

16.3 SOUND WAVES AND THEIR CHARACTERISTICS

In the last section we studied the oscillations of a particle under a restoring force of type $-kx$. In Chart 16.1 various cases of oscillatory motion of this type are shown. In all these cases the oscillations are confined to the respective positions of equilibrium. Now, if such oscillations travel from point to point in a medium, then we get a *travelling* or a *progressive wave.* A sound wave is an example of a progressive wave. For sound waves to propagate, a medium must possess elasticity and inertia. The elasticity of a medium pertains to the restoring forces among particles of the medium, and inertia pertains to the masses of these particles.

A source of sound sets the molecules of the medium into oscillation. The elasticity of the medium tries to restore the molecules to their equilibrium positions, and the inertia of the molecules makes them overshoot their positions of equilibrium. When a molecule moves, as well as when it returns and overshoots the position of equilibrium, it *pushes* the next molecule away from its equilibrium position, which in turn goes into oscillation. Part of the energy of the first oscillating molecule is thus transferred to the next molecule. This process repeats for the second next molecule, third next molecule and so on. In this way the oscillation disturbance propagates in the medium, whereas each molecule oscillates about its mean position. This propagating disturbance is called a sound wave. Thus *a sound wave is the transport of oscillation disturbance of molecules in the medium without any net transport of the molecules themselves.* The sound wave thus appears as a series of pressure and density variations—compressions and rarefractions—as shown in Fig. 16.2. We studied the elementry aspects of such waves in Chapter 2. Here we study the propagation characteristics of sound waves.

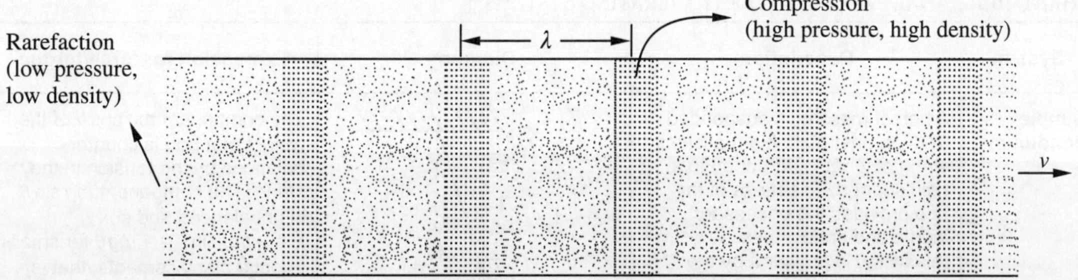

Figure 16.2 A sound wave consists of periodic compressions and rarefactions of the portions of the medium through which it is propagating

Sound waves always travel as longitudinal waves. Near the source of sound, assumed to be a point source, the waves are spherical waves as the wavefronts are spherical, and at far distances the wavefronts are plane so the waves are plane waves (Fig. 16.3).

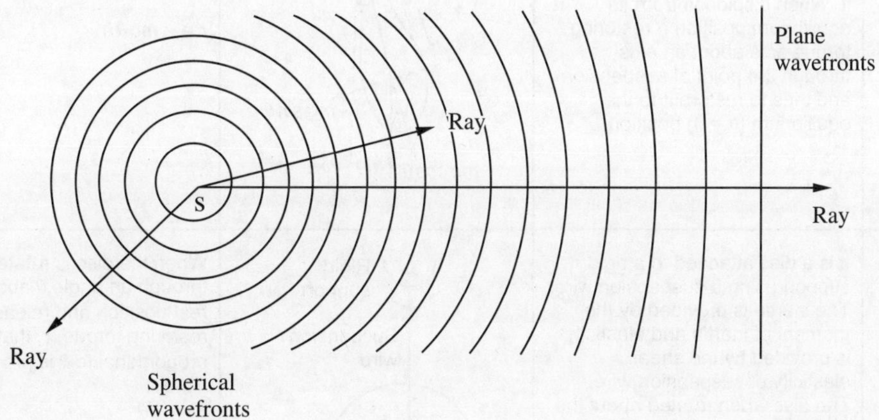

Figure 16.3 Sound waves travel as spherical waves near the source of sound, and as plane waves far from the source of sound

To study these plane waves, we basically use two physical principles. The first is the principle of conservation of fluid mass between two plane layers of fluid, separated along the direction of propagation of the wave, as these layers displace longitudinally due to the wave disturbance. The second is the thermodynamic principle which relates the pressure in the fluid at any point to the fluid density at that point. The pressure–density relation is determined by the type of fluid and by the type of thermodynamic process (example, adiabatic/isothermal) in the fluid as the wave propagates. We assume that this relation is contained in the function $P(\rho)$, which expresses the fluid pressure as a function of the fluid density. Thus the pressure at any point in the fluid at any time is determined by the density of the fluid at that point at that time. Now we consider the application of these two principles for deriving an equation satisfied by the instantaneous local disturbance $s(x, t)$ caused by the plane wave progressing in the x-direction.

CHART 16.1 ANALYSIS OF DIFFERENT VIBRATING SYSTEMS*

System	Description	Diagram	Expression for force/torque
Simple pendulum	A bob of mass m is suspended from an unstretchable massless string of length l, attached to a rigid support. The mass of the bob provides the necessary inertia and springiness (elasticity) is provided by the tension in the string. The bob when displaced from the equilibrium position and released, performs simple harmonic motion about the equilibrium position.	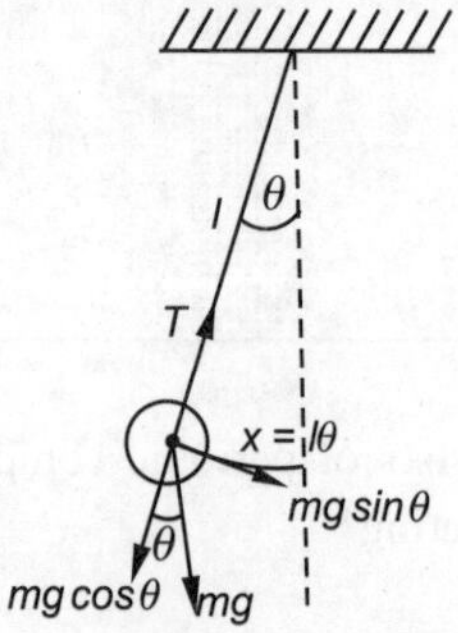	The component $mg\cos\theta$ of the weight of a bob is counter-balanced by the tension in the string. The component $mg\sin\theta$ is unbalanced and so $F = -mg\sin\theta = -mg\theta$ for small angular displacements, that is, $F = -mg\,x/l$, where x is the corresponding linear displacement. Thus, $F = -(mg/l)\,x$ (1a)
Physical pendulum	It is just a rigid body capable of oscillating freely about a horizontal axis passing through it. When displaced from its equilibrium position a restoring torque acts about an axis through the point of suspension and tries to restore it to the equilibrium ($\theta = 0$) position.	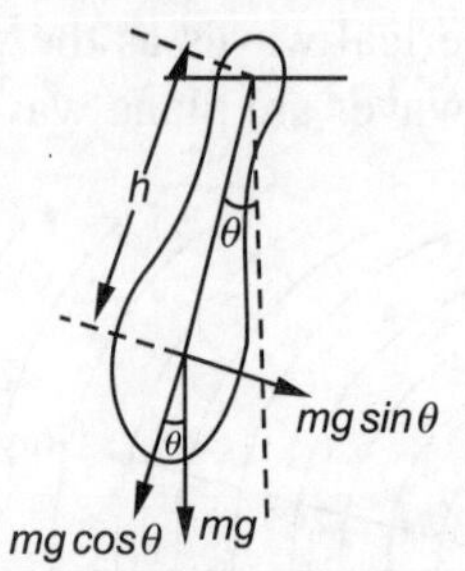	The restoring torque is given by $\tau = -mg\sin\theta\,h$. For small angular displacements, $\tau = -mg\theta h$ (1b)
Tortional pendulum	It is a disc attached to a rigid support using a suspension wire. The inertia is provided by the moment of inertia and elasticity is provided by the shear elasticity of suspension wire. The disc when rotated about the support (twisted) slightly and released, performs angular simple harmonic motion about the mean position.	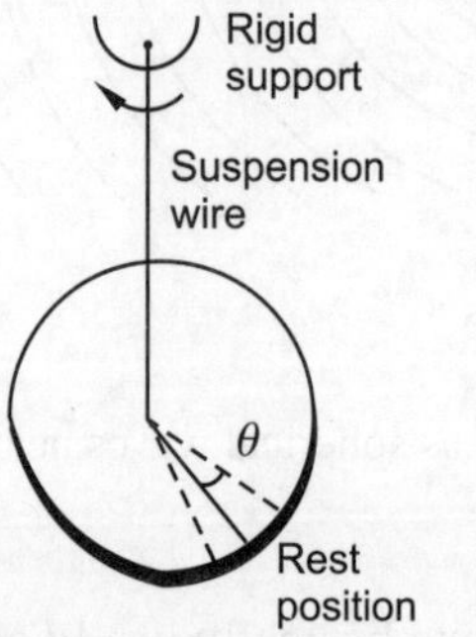	When the disc is rotated through an angle θ about its rest position and released, a restoring torque, τ, that is proportional to θ is produced $\tau = -\kappa\theta$ (1c) κ is called the tortion constant.
Acoustic vibrator (Helmholtz resonator)	It is a rigid enclosure of volume V having a narrow neck of length l and uniform cross section a. The volume $al << V$. Air in the neck provides the inertia and that in the enclosure provides the elasticity.	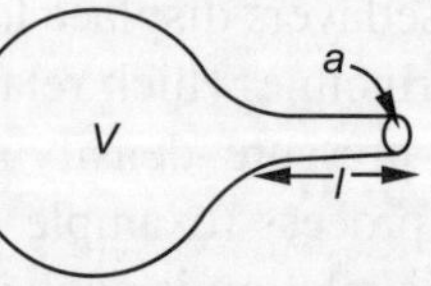	Let the displacement of air in the neck inward be x. If the pressure difference between the gas inside and outside the enclosure is p' then the net outward force on this air is ap'. Thus $F = ap'$ (1d)

* Though the examples considered here pertain to mechanical systems, the method applied here is useful in analysing many other cases of vibrations such as plasma vibrations, molecular vibrations and circuit oscillations.

Force/Torque equation	Solution of force/ torque equation	Expression for frequency and time period	Examples
$\frac{md^2x}{dt^2} = \frac{-mg}{l} x$ or $\frac{d^2x}{dt^2} + \frac{gx}{l}$ (2a)	$x = A\cos(\omega t + \phi)$ (3a)	$\omega = \sqrt{\frac{g}{l}}$ (4a) $T = 2\pi\sqrt{\frac{l}{g}}$ (5a)	A simple pendulum is an ideal concept. Any real pendulum is an example of a compound pendulum. A few examples are: rocking of a chemical balance about its pivot, roll of a ship and the pendulum of a clock.
$\frac{Id^2\theta}{dt^2} = -mg\theta h$ (2b) where I is the moment of inertia of the pendulum about an axis through the point of suspension and perpendicular to the plane of swing, and $d^2\theta/dt^2$ is the angular acceleration.	$\theta = A\cos(\omega t + \phi)$ (3b)	$\omega = \sqrt{mgh/I}$ (4b) $T = 2\pi\sqrt{I/mgh}$ (5b)	
$\frac{Id^2\theta}{dt^2} = -\kappa\theta$ or $\frac{d^2\theta}{dt^2} + \frac{\kappa\theta}{I} = 0$ (2c)	$\theta = A\cos(\omega t + \phi)$ (3c)	$\omega = \sqrt{\frac{\kappa}{I}}$ (4c) $T = 2\pi\sqrt{\frac{I}{\kappa}}$ (5c)	Spring driven clocks and watches, moving coil galvanometer
$ap' = (la\rho)\, d^2x/dt^2$ where $la\rho$ is the mass of the air, ρ being its density. Now the compressibility of air is $K = -(\partial V/V)/\partial p$. For the air in the enclosure, $\partial V = ax$ and $\partial p = p'$. Thus for small values of x and p' we have $K = -ax/Vp'$. Thus the above equation becomes $\frac{d^2x}{dt^2} + \frac{ax}{VKl\rho} = 0$ (2d)	$\theta = A\cos(\omega t + \phi)$ (3d)	$\omega = \sqrt{\frac{a}{lVK\rho}}$ (4d) $T = 2\pi\sqrt{\frac{lVK\rho}{a}}$ (4d)	These resonators are situated behind walls or in the ceiling of studios and auditoriums to provide satisfactory reverberation characteristics.

CHART 16.2 FREE, DAMPED AND FORCED VIBRATIONS

	Diagram showing the force	Equation of motion
Free vibrations	m, y, F $F = -ky$ (1)	$\frac{d^2y}{dt^2} + \frac{k}{m}y = 0$ (2)
Damped vibrations	F, m, Damping $F = -ky - bv = -ky - bdy/dt$ (4)	$\frac{d^2y}{dt^2} + \left(\frac{b}{m}\right)\frac{dy}{dt} + \left(\frac{k}{m}\right)y = 0$ (5)
Forced vibrations	Driving force *f*, F, m $F = -ky - bv + f = -ky - bdy/dt + f$ (7) where $f = f_m \cos \omega t$	$\frac{d^2y}{dt^2} + \left(\frac{b}{m}\right)\frac{dy}{dt} + \left(\frac{k}{m}\right)y = \left(\frac{f_m}{m}\right)\cos \omega t$ (8)

<table>
<tr><th>Solution</th><th>Solution plot</th></tr>
<tr><td>$y = A\cos(\omega_0 t + \phi)$ (3)
where A is the amplitude, ϕ is the phase angle and ω_0 is the angular frequency given by $\omega_0 = \sqrt{k/m}$. The natural frequency, $v_0 = (1/2\pi)\sqrt{k/m}$ and the time period, $T_0 = 2\pi\sqrt{m/k}$.
It can be seen that the amplitude remains constant (that is, undamped) in time.</td><td>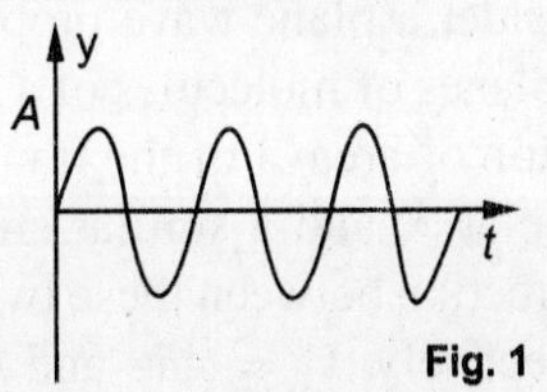
Fig. 1</td></tr>
<tr><td>Let the solution be $y = A\exp(\gamma t)$ where A and γ are to be determined. Substitution in the equation of motion yields
$\gamma^2 + b\gamma/m + k/m = 0$ or $\gamma = (-b/2m) \pm \sqrt{(b/2m)^2 - (k/m)} = -\alpha \pm \beta$
where $\alpha = b/2m$ and $\beta = \sqrt{(b/2m)^2 - (k/m)} = \sqrt{\alpha^2 - \omega_0^2}$
where $\omega_0 = \sqrt{k/m}$. Thus the solution is
$y = A_1 \exp[(-\alpha + \beta)t] + A_2 \exp[(-\alpha - \beta)t] =$
$\exp(-\alpha t)\,[A_1 \exp(\beta t) + A_2 \exp(-\beta t)]$.
In all cases of importance in acoustics, $\omega_0^2 > \alpha^2$,
so β is imaginary and γ is complex.
Let $\beta = i\sqrt{\omega_0^2 - \alpha^2} = i\omega_d$. Then $y = \exp(-\alpha t)\,[A_1 \exp(i\omega_d t) + A_2 \exp(-i\omega_d t)]$.
Taking the real part of this complex solution,
$y = A\exp(-\alpha t)\cos(\omega_d t + \phi)$ (6)
where A and ϕ are constants to be determined from the initial conditions.
It can be seen that the amplitude is no longer constant but decreases exponentially with time.</td><td>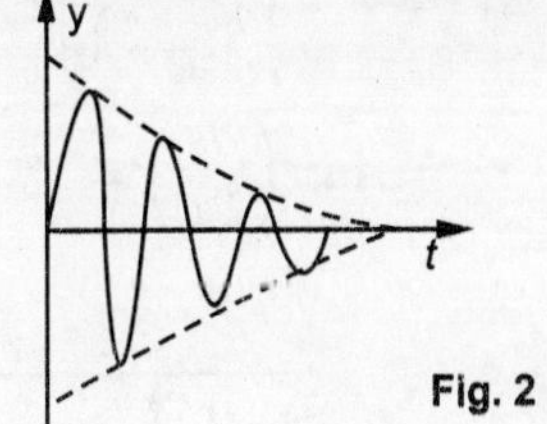
Fig. 2</td></tr>
<tr><td>Let the driving force be denoted as $f = f_m \exp(i\omega t)$ so that its real part is $f_m \cos\omega t$. Thus the equation of motion becomes $md^2y/dt^2 + bdy/dt + ky = f_m \exp(i\omega t)$. Assuming the solution to be $y = A\exp(i\omega t)$, where A is in general complex, substitution into above equation yields
$(-\omega^2 m + i\omega b + k)\,A\exp(i\omega t) = f_m \exp(i\omega t)$.
Thus, $A = f_m/[i\omega b + (k - \omega^2 m)]$, for the above equation to be true for all t.
Thus, $y = f_m \exp(i\omega t)/[i\omega b + (k - \omega^2 m)] = -if_m \exp(i\omega t)/\omega[b + i(\omega m - k/\omega)]$.
Substituting $Z_m = b + i(\omega m - k/\omega) = b + i\chi m$ where, $\chi m = (\omega m - k/\omega)$,
we get $y = -ifm \exp(i\omega t)/\omega Zm$. Zm is known as the *mechanical impedance* and consists of two parts, the resistive part b and the reactive part χm, χm is called the *mechanical reactance*. Substituting $b = zm\cos\phi$ and $\chi m = zm \sin\phi$, Zm can be written as $Zm = zm\exp(i\phi)$ where, $zm = \sqrt{b2 + \chi_m^2}$ and $\phi = \tan^{-1}(\chi m/b)$.
Thus, $y = -if_m \exp(i\omega t)/\omega z_m \exp(i\phi) = -if_m \exp[i(\omega t - \phi)]/\omega z_m$
Considering the real part of y, the actual displacement is given by
$y = f_m \sin(\omega t - \phi)/\omega z_m = f_m \sin(\omega t - \phi) / [\omega\sqrt{b^2 + (\omega m - k/\omega)^2}$ (9)
From this equation it can be seen that y is maximum when $\omega m = k/\omega$, that is, $\omega^2 = k/m$ or $\omega^2 = \omega_0^2$, or $\omega = \omega_0$, that is, when the frequency of the driving force is equal to the natural frequency of the vibrating system. Figure 3 shows the variation of amplitude with the factor ω/ω_0 for different values of b. The amplitude is maximum for least value of b, as expected.</td><td>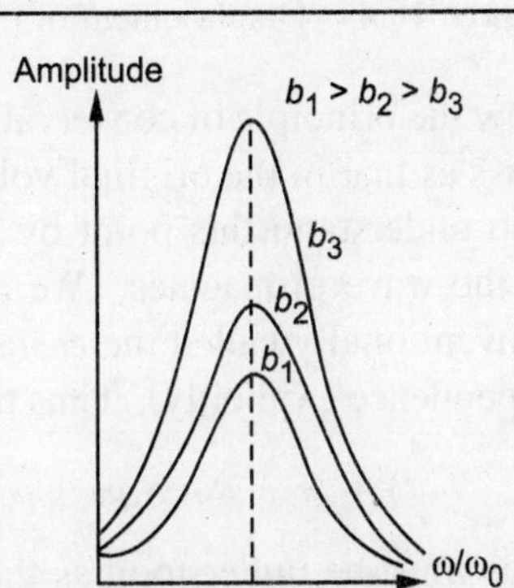
At resonance, $\omega = \omega_0$
Fig. 3</td></tr>
</table>

16.3.1 Wave Equation

Consider a plane wave propogating along the x-axis in a fluid (Fig. 16.4). As the wave moves the planes of molecules of a fluid are displaced from their equilibrium positions. Consider a cross section of area A of the transverse (that is, normal to x, the direction of wave propagation) fluid plane at x, and a similar cross section at $x + dx$. The volume and the mass of fluid along the x-direction between these two cross sections in equilibrium, that is, in the absence of any wave are, respectively, $V_0 = Adx$ and $m = \rho_0 Adx$, where, ρ_0 is the density of the fluid at x in equilibrium. As the wave propagates, the fluid plane at x is displaced to $x + s(x,t)$, and that at $x + dx$ to $x + dx + s(x + dx, t) = x + dx + s(x, t) + (\partial s/\partial x)dx$. Thus the new volume, V, of the fluid between the same two cross sections differs from the equilibrium volume by ΔV.

$$\Delta V = V - V_0 = A\left\{\left[x + dx + s(x,t) + \left(\frac{\partial s}{\partial x}\right)dx\right] - (x + s(x,t)\right\} - Adx$$

$$= Adx\left(\frac{\partial s}{\partial x}\right) = V_0\left(\frac{\partial s}{\partial x}\right) \tag{16.10}$$

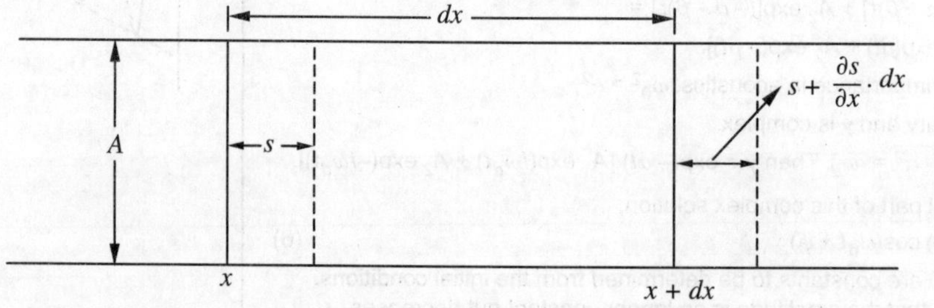

dx

$s + \frac{\partial s}{\partial x}dx$

A

s

x

x + dx

Figure 16.4 Displacement of planes in the propagation of a plane sound wave

Now the principle of conservation of fluid mass says that the new volume should contain the same mass as that in the original volume. For this to happen, the density of the fluid must change. We also understand this point by noting that the density changes because the layers of fluid move as the wave propagates. We denote the change in density at x by $\Delta\rho = \rho_0\eta(x)$, where $\eta(x)$ is conventionally called the *condensation*. (η is a function of both x and t but we do not show the t dependence explicitly). Thus the new density at x is

$$\rho(x) = \rho_0 + \Delta\rho = \rho_0 + \rho_0\eta(x) = \rho_0[1 + \eta(x)] \tag{16.11}$$

This must be understood as the density at $x + s(x)$, that is, at the displaced position of the fluid layer originally at x.

Therefore mass conservation can now be written as

$$m = \rho_0 V_0 = \rho(x)V \tag{16.12}$$

Using Eqs. 16.10 and16.11 we have,

$$\rho_0 V_0 = \rho_0[1 + \eta(x)]V_0\left[1 + \left(\frac{\partial s}{\partial x}\right)\right] \tag{16.13}$$

Simplifying, $\eta(x) = -\dfrac{\partial s}{\partial x}$ (16.14)

where we have neglected the product of $\eta(x)$ and $\partial s/\partial x$, both of which are very small in magnitude.

Now, we turn to the application of the thermodynamic principle. In order to write the equation of motion for the fluid mass between the oscillating layers, which, in equilibrium, are at x and $x + dx$, we need to find the force on the layer. This can be calculated from the pressure difference across this layer. To do this we first understand the concept of *acoustic pressure*. As the equilibrium density ρ_0 at x changes to the density $\rho_0[1 + \eta(x)]$, at the displaced position of the layer originally at x, the pressure changes from its equilibrium value, P_0 to

$$P(x) = P[\rho(x)] = P_0 + \Delta P = P_0 + \left(\frac{\partial P}{\partial \rho}\right)_{\rho=\rho_0} d\rho = P_0 + p(x) \tag{16.15}$$

$p(x)$ is called the acoustic pressure and denotes *the departure of local pressure from its equilibrium value due to the presence of an acoustic wave.* It is obvious that $p(x)$ is zero in the absence of the wave.

Now, the acoustic pressure on the layer at $x + dx$ is $p(x) + (\partial p/\partial x)dx$ and therefore the net force acting on the fluid layer of mass m is

$$dF_x = p(x)A - \left[p(x) + \left(\frac{\partial p}{\partial x}\right) dx\right] A = -\left(\frac{\partial p}{\partial x}\right) dxA \tag{16.16}$$

The fluid mass accelerates due to this force and the equation of motion can be written as

$$-\left(\frac{\partial p}{\partial x}\right) dxA = \rho_0 Adx \left(\frac{\partial^2 s}{\partial t^2}\right)$$

or $$-\frac{\partial p}{\partial x} = \rho_0 \frac{\partial^2 s}{\partial t^2} \tag{16.17}$$

Now the bulk modulus of the fluid is defined as

$$B = -\frac{p}{\Delta V/V_0} \tag{16.18}$$

Using Eq. 16.10, $B = \dfrac{-p}{(\partial s/\partial x)}$ or $p = -B\dfrac{\partial s}{\partial x}$ (16.19)

Hence $$\frac{\partial p}{\partial x} = -B\frac{\partial^2 s}{\partial x^2} \tag{16.20}$$

Substituting Eq. 16.20 in Eq. 16.17, we get

$$\frac{\partial^2 s}{\partial x^2} = \frac{\rho_0}{B}\frac{\partial^2 s}{\partial t^2} \tag{16.21}$$

Equation 16.21 represents the wave equation that governs the propagation of a sound wave in a medium of density ρ_0 and bulk modulus B.

16.3.2 Velocity

Equation 16.21 is similar to the general wave equation,

$$\frac{\partial^2 y}{\partial x^2} = \frac{1}{v^2}\frac{\partial^2 y}{\partial t^2} \tag{16.22}$$

where v is the wave velocity. Comparing Eqs 16.21 and 16.22 we get the expression for the velocity of sound waves as

$$v = \sqrt{\frac{B}{\rho_0}} \tag{16.23}$$

Thus we see that the elasticity (B) and the inertia (ρ_0) of the medium decide the velocity of sound in that medium, in conformity with our earlier qualitative remarks. The strength of the intermolecular forces decides the velocity of sound on a microscopic level. A larger strength implies more elastic stiffness and hence faster communication between molecules, and thus larger values of sound velocity. This is why sound travels fastest in solids and slowest in gases.

Table 16.1 displays the data of sound velocity in different media along with the values of B and ρ_0. It can be seen from this table that the measured velocities show some discrepancy with the values calculated using Eq. 16.23. The reason for this discrepancy is to be found in the simplistic model discussed above. The expression for velocity thus needs correction. Moreover, the velocity also depends on several parameters which are not considered in this model. In Chart 16.3 the corrections to Eq. 16.23 and the dependence of velocity on various parameters are discussed.

Table 16.1 Velocity of sound, bulk modulus and density of some media

Medium	**Density (ρ_0) (kg/m^3)**	**Bulk modulus (B) (N/m^2) $\times$ 10^9**	**Velocity of sound (v) (m/s)**	**Velocity calculated using Eq. 16.23 (m/s)**
Hydrogen (0°C)	0.09		1269.5	
Oxygen (°C)	1.43		317.2	
Air (20°C)	1.21		343	
Water	998	2.18	1481	1478
Glycerin	1260		1980	
Rubber (soft)	950	1	1050	1026
Ice	920		3200	
Concrete	2600		3100	
Steel	7700	170	6100	4699
Aluminium	2700	75	6300	5270
Glass (pyrex)	2300	39	5600	4118

CHART 16.3 DEPENDENCE OF SOUND VELOCITY ON VARIOUS FACTORS

The expression for velocity of sound,

$$v = \sqrt{\frac{B}{\rho_0}} \tag{1}$$

is not universal but needs to be corrected in a particular context taking several parameters into account. The derivations of v due to Newton and Laplace are discussed. It is followed by the discussion of dependence of v on various parameters.

Newton's formula	Laplace's formula
Newton assumed propagation of sound as a slow process in which the compressions and rarefactions do not lead to any thermal changes, that is, he considered propagation of sound as an isothermal process. Using Boyle's law, that is, PV = constant at constant temperature, $PdV + VdP = 0$ or $dV/V = -dP/P$ or $B = dP/(-dV/V) = P$ Substituting in Eq. 1, Newton's expression for v becomes $v = \sqrt{\dfrac{P}{\rho_0}}$ (2) The value of velocity obtained using this expression shows large discrepancy with the experimental value.	Laplace maintained that compressions and rarefactions take place so rapidly that no heat exchanges are possible and the process is adiabatic instead of isothermal. For this case PV^{γ} = constant where γ is the ratio of specific heats C_p and C_v, at constant pressure and constant volume respectively. Thus, we have $P\gamma V^{\gamma-1} dV + V^{\gamma} dP = 0$ or $dV/V = -dP/P\gamma$ or $B = dP/(-dV/V) = \gamma P$ Substituting in Eq. 1 Laplace's expression for v becomes $v = \sqrt{\dfrac{\gamma P}{\rho_0}}$ (3) The value of the velocity obtained using this expression is in close agreement with the experimental value.

See Example 1.4

Pressure, density and temperature	Pressure and density do not affect velocity at constant temperature because the change in pressure results in the change in density so that $P/\rho = P_0/\rho_0$. This can also be shown using the gas law as follows. The ideal gas law is $PV = (m/M)RT$ where, m is the mass of volume V, M is the molecular weight of the gas, R is the gas constant and T is the absolute temperature. Thus the pressure is given by $P = (m/V)(RT/M) = \rho_0 RT/M$ Substituting in Eq. 3, we get $v = \sqrt{\gamma RT/M}$ (4) Equation 4 tells us that the speed of sound is independent of pressure and density but depends only on the temperature of the gas and the values of γ and M for it. Example 16.1 shows that the velocity near the room temperature increases by 0.6 m/s for every 1°C rise in temperature.
Humidity	The presence of moisture in air decreases its density. This results in the increase in velocity of sound through moist (humid) air. The velocity in dry air, v_d, and that in moist air, v_m, are related as $v_d = v_m \sqrt{\rho_m/\rho_d}$; ρ_d and ρ_m being the corresponding densities.
Wind	The wind velocity adds to the sound velocity if the two are in the same direction, that is, the velocity of sound becomes $v + w$, where w is the wind velocity. If the wind blows in the opposite direction, then the resultant sound velocity is $v - w$. For the wind making an angle θ, the corresponding sound velocity will be either $v + w\cos\theta$ or $v - w\cos\theta$.

EXAMPLE 16.1
Rate of increase of sound velocity with temperature

Show that the speed of sound in air near room temperature increases by about 0.6 m/s for every 1°C rise in temperature. Assume that the speed at room temperature is 343 m/s.

Solution

The velocity of sound in air at absolute temperature T is

$$v = \sqrt{\frac{\gamma RT}{M}} = K\sqrt{T}, \text{ where, } K = \sqrt{\frac{\gamma R}{M}} \text{ is a constant}$$

Thus, $v = K\sqrt{(t + 273)}$, where, t is the temperature in degree Celsius.

Thus, $v_1 = K\sqrt{t_1 + 273}$ and $v_2 = K\sqrt{t_2 + 273}$

$$\text{or} \quad \frac{v_2}{v_1} = \sqrt{\frac{t_2 + 273}{t_1 + 273}}$$

$$= \sqrt{\frac{\Delta t + t_1 + 273}{t_1 + 273}}, \quad \Delta t = t_2 - t_1, \text{ being the increase in temperature.}$$

$$= \left[1 + \frac{\Delta t}{t_1 + 273}\right]^{1/2}$$

$$\approx \left[1 + \frac{\Delta t}{2(t_1 + 273)}\right]$$

For $t_1 = 27°\text{C}$ and $\Delta t = 1°\text{C}$,

$$\frac{v_2}{v_1} = 1 + \frac{1}{2(27 + 273)} = 1.0017$$

Now, $v_1 = 343$ m/s and hence

$$v_2 = 343 \times 1.0017 = 343.58 \text{ m/s.}$$

Therefore $\Delta v = v_2 - v_1 = 0.6$ m/s

Thus the speed of sound in air near room temperature increases by about 0.6 m/s for every 1°C rise in temperature.

16.3.3 Displacement Amplitude and Pressure Amplitude

Let the displacement s of the plane of fluid in Fig. 16.4 be represented by the equation,

$$s = s_m \cos(kx - \omega t) \tag{16.24}$$

which is a familiar representation of a wave (see Section 2.3), where s_m is the displacement ampltude. Now, using Eq. 16.19,

$$p = s_m Bk \sin(kx - \omega t) \tag{16.25}$$

This equation, representing the pressure variation can be expressed as

$$p = p_m \sin(kx - \omega t) \tag{16.26}$$

where $p_m = Bks_m$, is the pessure amplitude. Thus *the pressure variation is* $90°$ *out of phase with the displacement*. The pressure variation is zero for maximum displacement and vice-versa. The pressure amplitude can be written using Eq. 16.23 as

$$p_m = Bk_{sm} = (v^2 \rho_0 k) s_m$$

Using $v = \omega/k$, we get

$$p_m = (v\rho_0\omega) s_m \tag{16.27}$$

In Ex. 16.2 we use Eq. 16.27 to find the upper and lower limits of displacement amplitudes for a human ear.

EXAMPLE 16.2
Limits of displacement amplitudes for the human ear

Find the minimum and maximum displacement corresponding to the minimum detectable effective (root mean square) acoustic pressure equal to 0.0002 microbar and the maximum tolerable effective acoustic pressure equal to 20 pascal for a human ear. Assume the speed of sound as 343 m/s, frequency of sound = 1 kHz and density of air = 1.21 kg/m^3.

Solution

The root mean square pressure and the pressure amplitude are related as $p_{\text{rms}} = p_m/\sqrt{2}$

Hence pressure amplitude corresponding to the minimum detectable sound

$$(p_m)_{\text{min}} = \sqrt{2} \times 0.0002 \text{ microbar}$$

$$= \sqrt{2} \times 0.0002 \times 0.1 \text{ N/m}^2 = 2.8 \times 10^{-5} \text{ N/m}^2$$

(as 1 microbar = 10^{-6} bar = $10^{-6} \times 10^6$ dyne/cm^2 $= 1$ dyne/cm^2 $= 0.1$ N/m^2)

The pressure amplitude corresponding to the maximum tolerable sound

$$(p_m)_{\text{max}} = \sqrt{2} \times 20 \text{ pascal} = 28.3 \text{ N/m}^2 \text{ (as 1 pascal = 1 N/m}^2\text{)}$$

Displacement amplitude, $s_m = p_m / v\rho_0\omega$

$$\text{Thus, } (s_m)_{\min} = \frac{2.8 \times 10^{-5}\ \text{N/m}^2}{343\ \text{m/s} \times 1.21\ \text{kg/m}^3 \times 2\pi \times 1000\ \text{Hz}}$$

$$= 1.07 \times 10^{-11}\ \text{m} = 11\text{pm}$$

$$\text{and } (s_m)_{\max} = \frac{28.3\ \text{N/m}^2}{343\ \text{m/s} \times 1.21\ \text{kg/m}^3 \times 2\pi \times 1000\ \text{Hz}} = 1.08 \times 10^{-5}\ \text{m} = 11\ \mu\text{m}$$

Thus the *ear is a very sensitive detector of sound waves.*

16.3.4 Intensity

The intensity of a sound wave is defined as the average rate of flow of energy through unit area normal to the direction of wave propagation. If E is the average energy per unit volume in the wave, then $EAdx$ is the total energy in the volume element Adx. If this energy crosses the area A in time dt then the rate at which energy passes unit area is $EAdx/Adt = Ev$. Thus the intensity of a wave is the product of the average energy density in the wave and the wave velocity. To find the average energy density in the wave at any point x, we have to add the average kinetic and potential energy densities at x.

Now the instantaneous kinetic energy in the small volume V_0 at x is

$$dK = \frac{1}{2}(V_0\rho_0)v_s^2 \tag{16.28}$$

where v_s is the velociy of molecules in the plane consiedred, that is, for the wave of Eq. 16.24.

$$\text{Thus,} \quad v_s = \frac{\partial s}{\partial t} = -\omega s_m \sin(kx - \omega t) \tag{16.29}$$

$$\text{and so, } dK = \frac{1}{2}(V_0\rho_0)\omega^2 s_m^2 \sin^2(kx - \omega t) \tag{16.30}$$

Similarly the instantaneous potential energy for the volume element V_0 is

$$dU = -\int p dV \tag{16.31}$$

To calculate dU, dV should be expressed in terms of dp. To do this we use Eqs. 16.10 and 16.19 and write

$$V = V_0\left[1 - \frac{p}{B}\right]$$

$$\text{Differentiating,} \quad dV = -dp\frac{V_0}{B}$$

$$\text{Substituting in Eq. 16.31,} \quad dU = p^2V_0/2B \tag{16.32}$$

Using Eq. 16.25, $dU = \frac{1}{2} V_0 s_m^2 B k^2 \sin^2(kx - \omega t)$ (16.33)

From Eqs.16.30 and 16.33 it can be seen that dU = dK. Thus the instantaneous total energy, dE, in a small volume element V_0 at x is obtained by adding these two equations so that

$$dE = dK + dU = V_0 \rho_0 \omega^2 s_m^2 \sin^2(kx - \omega t)$$

Thus the time-averaged energy per unit volume at x is

$$E = \frac{< dE >}{V_0} = \rho_0 \omega^2 s_m^2 < \sin^2(kx - \omega t) >= \frac{1}{2} \rho_0 \omega^2 s_m^2 \tag{16.34}$$

because, $< \sin^2(kx - \omega t) > = \left(\frac{1}{T}\right) \int_0^T \sin^2(kx - \omega t) dt = \frac{1}{2}$

Thus the intensity is given by

$$I = Ev = \frac{1}{2} \rho_0 v \omega^2 s_m^2 \tag{16.35}$$

16.3.5 Sound Level

The range of sound intensities that can be heard by humans is enormous ($\sim 10^{12}$, see Ex. 16.3). This is possible because the human ear works like a *logarithmic device*: It perceives the relative loudness of two sounds by the ratio of their intensities. In engineering practices logarithmic scales known as *sound levels* are used instead of actual intensity scales. A more convenient quantity called the *intensity level* is used which is defined as

$$\beta = 10 \log \left(\frac{I}{I_0}\right) \text{ dB} \tag{16.36}$$

where dB, known as *decibel*, is the unit of intensity level and I_0 is a reference intensity which is taken as 10^{-12} W/m^2 (see Ex. 16.4). As the intensity is proportional to the square of the pressure, the *sound pressure level* (*SPL*) is defined as

$$SPL = 20 \log \left(\frac{p}{p_0}\right) \text{ dB} \tag{16.37}$$

EXAMPLE 16.3

I_{max}/I_{min} for the human ear

Find the ratio of the maximum tolerable and minimum detectable intensity of sound for the human ear using the data and results of Ex. 16.2.

Solution

The intensity of sound is given by

$$I = \frac{1}{2}\rho_0 v\,\omega^2 s_m^2$$

Thus, $I_{\max} = \frac{1}{2}\,\rho_0 v\,\omega^2\,[(s_m)_{\max}]^2$

$$= \frac{1}{2} \times 1.21\ \text{kg/m}^3 \times 343\ \text{m/s} \times (2\pi \times 1000\ \text{Hz})^2 \times (11 \times 10^{-6}\ \text{m})^2 = 1\ \text{W/m}^2$$

and $I_{\min} = \frac{1}{2} \times \rho_0 v\,\omega^2[(s_m)_{\min}]^2$

$$= \frac{1}{2} \times 1.21\ \text{kg/m}^3 \times 343\ \text{m/s} \times (2\pi \times 1000\ \text{Hz})^2 \times (11 \times 10^{-12}\ \text{m})^2 = 10^{-12}\ \text{W/m}^2$$

Hence $I_{\max}/I_{\min} = \dfrac{1\ \text{W/m}^2}{10^{-12}\ \text{W/m}^2} = 10^{12}$

EXAMPLE 16.4
Sound intensities corresponding to various sound levels

Find the sound intensities corresponding to the intensity levels for various environments as given below.

Hearing threshold	0 dB
Speech activity	60 dB
Pain threshold	120 dB

Solution

The intensity level is given by $\beta = 10\log(I/I_0)$, where $I_0 = 10^{-12}$ W/m^2

For hearing threshold, $\beta = 0$ or $10\log(I/I_0) = 0$

Thus, $I = I_0 = 10^{-12}\ \text{W/m}^2$

For speech activity, $\beta = 60$ dB or $10\log(I/I_0) = 60$

Thus, $I = 10^6 I_0 = 10^{-6}\ \text{W/m}^2$

For pain threshold, $\beta = 120$ dB or $10\log(I/I_0) = 120$

Thus, $I = 10^{12} I_0 = 1\ \text{W/m}^2$

Thus *it is possible for the human ear to perceive an enormous range of intensities because of its logarithmic operation.* The increase in intensity of sound from 10^{-12} W/m^2 to 1 W/m^2

corresponds to increase in intensity level from 0 dB to 120 dB. Doubling the intensity corresponds to an increase in intensity level of $10 \log 2 = 3$ dB. A hundred-fold change in intensity corresponds to an intensity level change of 20 dB; a thousand-fold change to 30 dB; and so on.

16.3.6 Loudness Level

Loudness is a subjective characteristic of sound that is determined by both the frequency and intensity of the sound waves. Figure 16.5 displays the equal loudness contours, the data for which are obtained by alternately sounding a reference tone of 1000 Hz and a second tone of some other frequency. The intensity level of the other tone is adjusted so that the two tones appear to be equally loud. From this figure it can be seen that a sound of frequency 200 Hz at the intensity level 40 dB appears to be equally loud as the sound of frequency 1000 Hz at the intensity level 20 dB.

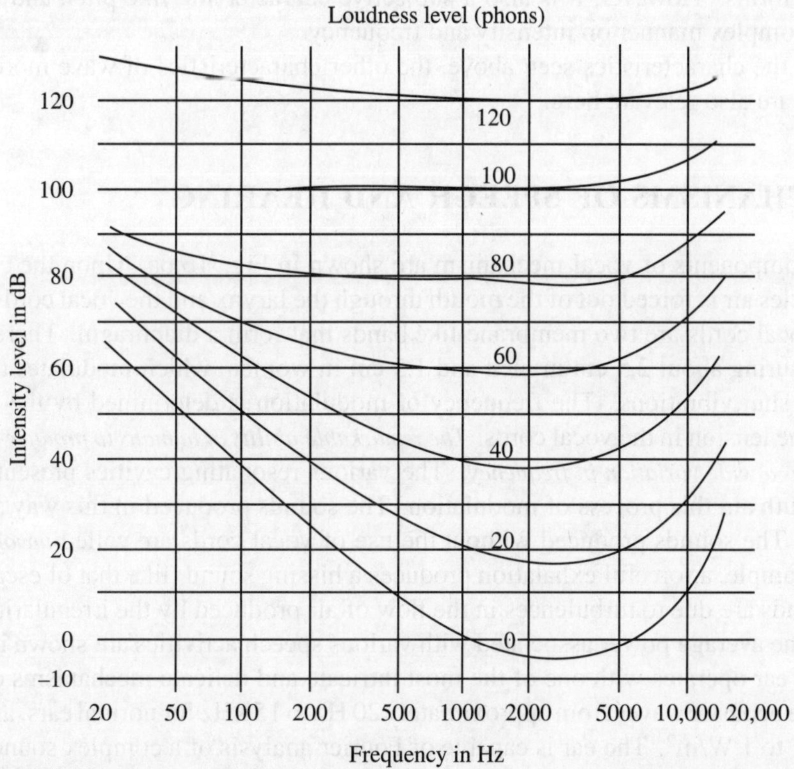

Figure 16.5 Equal loudness level contour
(Adapted from http://en.wikipedia.org/wiki/File:Lindos4.svg#file)

The loudness level of a sound is measured in units of *phon*, which is defined as *equal to the intensity level in dB of a pure 1000 Hz tone that is judged by the average observer to be equally loud.* The loudness level of the 200 Hz pure tone in the above example is thus 20 phons. Similarly a sound of frequency 500 Hz at the intensity level 45 dB has a loudness level equal to 40 phons. It can be seen that all equal loudness contours intersect the corresponding intensity level lines at a frequency of 1000 Hz.

16.3.7 Pitch

The pitch, like loudness, is a subjective characteristic complexly dependent on various physical quantities. However, it primarily depends on frequency and a higher frequency necessarily means higher pitch. The pitch classifies a shrill sound from a flat sound in musical acoustics.

16.3.8 Quality/Timbre

The quality or timbre is another subjective characteristic generally used in musical acoustics to distinguish between two tones having the same intensity level and fundamental frequency but different waveforms. However, it is also a subjective characteristic like pitch and loudness and depends in a complex manner on intensity and frequency.

Apart from the characteristics seen above, the other characteristics of wave motion discussed in Section 2.3 are also relevant here.

16.4 MECHANISMS OF SPEECH AND HEARING

The various components of vocal mechanism are shown in Fig. 16.6a. Upon the contraction of the chest muscles air is forced out of the mouth through the larynx and the vocal cords, resulting in sound. The vocal cords are two membrane-like bands that form a diaphragm. There is a slit-like opening, measuring about 2.5 cm in men and 1.5 cm in women, which modulates the air stream with its open–shut vibrations. The frequency of modulation is determined by the length of the opening and the tension in the vocal cords. *The remarkable ability of humans to produce many different sounds is due to a wide variation in frequency.* The various resonating cavities present in the nose, throat and mouth aid this process of modulation. The sounds produced in this way are called the *voiced sounds*. The sounds produced without the use of vocal cords are called *unvoiced* or *breath sounds*. For example, a forceful exhalation produces a hissing sound (like that of escaping steam). Unvoiced sounds are due to turbulences in the flow of air produced by the irregularities along the vocal track. The average power associated with various speech activities are shown in Table 16.2.

The human ear operates with one of the most intricate and delicate mechanisms of the human body to receive acoustic waves from approximately 20 Hz to 15 kHz for normal ears, and intensities between 10^{-12} to 1 W/m^2. The ear is capable of Fourier analysis of a complex sound heard by it. The accurate details of structure and working of the ear are too complicated, and are beyond the scope of this book. Only a heuristic account is presented here.

The three main divisions of the ear (Fig. 16.6b) are the outer, the middle and the inner ear. The outer ear consists of the pinna (external or visible ear), the auditory canal and the ear drum. The pinna collects the acoustic energy impinged upon it and leads it to the auditory canal.

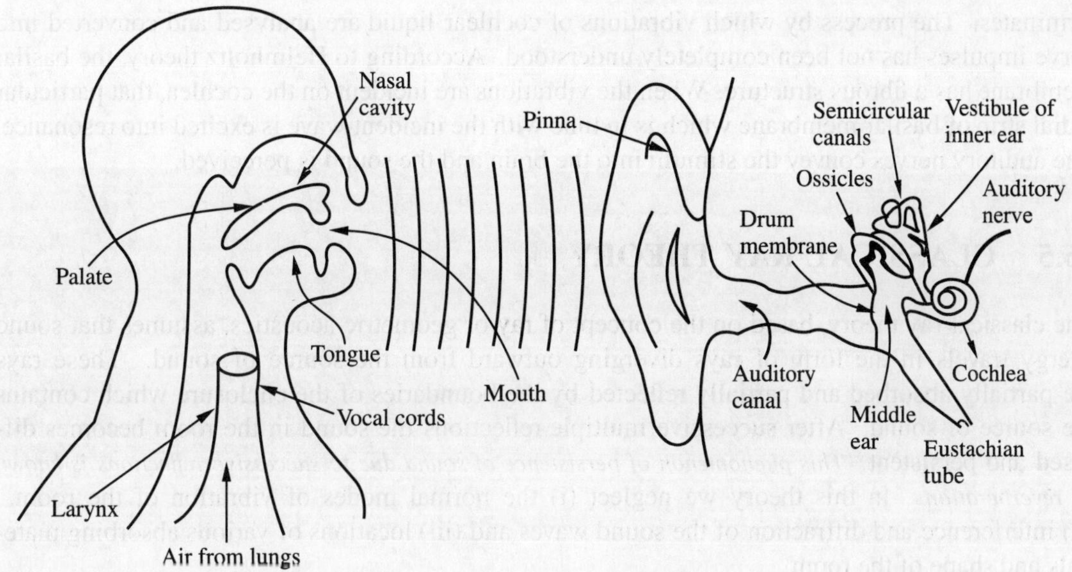

Figure 16.6 Sketch of (a) voice mechanism and (b) hearing mechanism

The canal terminates at the drum membrane, or the ear drum. The drum, also called as tympanic membrane, is positioned obliquely to the auditory canal and comprises of a middle ear that is an air-filled chamber which is bridged across by a train of three little bones called as ossicles. These are the malleus (hammer), the incus (anvil) and the stapes (stirrup). The vibrations cause corresponding vibrations in the tympanic membrane which are transmitted to the inner ear by the ossicles. The inner ear is a cavity filled with a liquid and is of complex shape; it lies within the bony structure of the skull. The footplate of the stirrup is attached to a membrane that closes an oval opening that connects the middle ear to the inner ear. It rocks back and forth due to a hinge-like support, thus transmitting the pressure waves to the liquid that fills the inner ear.

Table 16.2 Average power associated with speech activities

Activity	Average speech power in microwatts
Whisper	0.001
Normal conversation	10
Speaking loudly	200
Shout	1000

The inner ear, also called as labyrinth, consists of the vestibule, the semicircular canals and the cochlea. The vestibule is connected to the middle ear through the oval opening as mentioned earlier. The semicircular canals play no role in the hearing process but are responsible for providing a sense of balance. The cochlea, with the shape of a snail shell, has a complex partition consisting of a bony ledge and three membranes. The bony ledge projects into the liquid-filled tube and carries the auditory nerves. The nerve fibres enter the basilar membrane where the bony ledge terminates. The process by which vibrations of cochlear liquid are analysed and converted into nerve impulses has not been completely understood. According to Helmholtz theory, the basilar membrane has a fibrous structure. When the vibrations are incident on the cochlea, that particular radial strip of basilar membrane which is in tune with the incident wave is excited into resonance. The auditory nerves convey the stimulii into the brain and the sound is perceived.

16.5 CLASSICAL RAY THEORY

The classical ray theory, based on the concept of ray or geometric acoustics, assumes that sound energy travels in the form of rays diverging outward from the source of sound. These rays are partially absorbed and partially reflected by the boundaries of the enclosure which contains the source of sound. After successive multiple reflections the sound in the room becomes diffused and persistent. *This phenomenon of persistence of sound due to successive reflections is known as reverberation.* In this theory we neglect (i) the normal modes of vibration of the room, (ii) interference and diffraction of the sound waves and (iii) locations of various absorbing materials and shape of the room.

Live rooms are those in which the absorption is small and thus it takes a longer time for the sound to reach its ultimate intensity. The rooms for which the absorption is large and the reflected energy is thus small, require a short time for the sound to reach its ultimate intensity and are called *dead rooms.*

To understand the growth and decay of sound in live rooms, consider a source of sound kept in an enclosure of volume V. After a large number of reflections from the boundaries of the enclosure, the sound energy density, E, may be assumed to be the same throughout the room. The differential equation governing the growth of sound energy may be written as

$$\frac{VdE}{dt} + W = Q \tag{16.38}$$

where, $VdE/dt = d(VE)/dt$ is the rate at which the total energy in the room increases, W is the rate at which energy is absorbed by the boundaries and absorptive materials of the room, and Q is the rate at which sound energy is input into the room by the source.

To find the expression for W, consider a volume element dV that is a part of a hemispherical shell of radius r and thickness dr, in the medium, and an element of wall surface dS at the centre of this shell, such that r makes an angle θ with the normal to dS, as shown in Fig. 16.7. The energy present in dV is EdV. This energy is isotropically moving in all directions. The fraction of this energy that impinges directly upon dS is $(EdV/4\pi r^2)dS\cos\theta$ as the area offered by dS to this energy is $dS\cos\theta$ and $(EdV/4\pi r^2)$ is the energy falling per unit area. Thus the total energy that impinges upon dS from the entire shell is

$$\Delta E = \frac{EdS}{4\pi r^2}\int cos\theta dV$$

Now, considering a circular zone of radius $r\sin\theta$, for which θ is constant, $dV = 2\pi r\sin\theta rdrd\theta$.

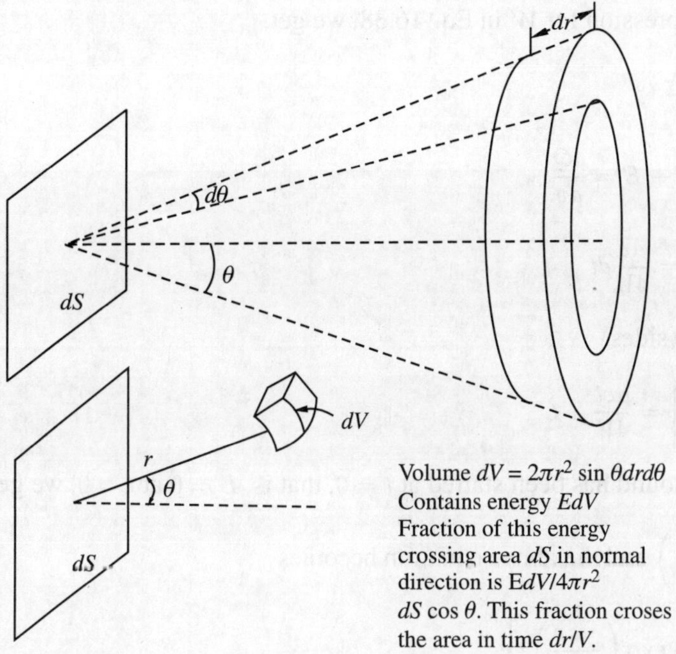

Figure 16.7 Volume element dV and surface element dS used in the derivation of E

Hence, $$\Delta E = \frac{EdSdr}{2}\int_0^{\pi/2} \sin\theta\cos\theta d\theta = \frac{EdSdr}{4} \tag{16.39}$$

is the total energy coming out from all directions to the area dS. This energy arrives at dS in time $\Delta t = dr/v$, the time for energy to cross the spherical shells of thickness dr at various distances from dS.

Thus, the rate at which energy falls on dS from all directions is $\dfrac{\Delta E}{\Delta t} = \dfrac{EvdS}{4}$

The rate at which energy falls per unit area on the walls, that is, the intensity is,

$$I = \frac{Ev}{4} \tag{16.40}$$

Therefore the rate at which energy is absorbed by the walls and other absorptive materials in the room is given by

$$\frac{Ev}{4}(\alpha_1 S_1 + \alpha_2 S_2 + \ldots + \alpha_n S_n) \tag{16.41}$$

where $\alpha_1, \alpha_2, \ldots, \alpha_n$ are the absorption coefficients of the surfaces of areas $S_1, S_2, \ldots, S_n$ respectively. Thus the rate of absorption is given by

$$W = \frac{Eva}{4} \tag{16.42}$$

where $$a = \sum_{i=1}^{N} \alpha_i S_i \tag{16.43}$$

Substituting this expression for W in Eq. 16.38, we get,

$$\frac{V\,dE}{dt} + \frac{Eva}{4} = Q \tag{16.44}$$

Thus, $\left(\frac{4V}{av}\right)\frac{dE}{dt} + E = \frac{4Q}{av}$

or $$\frac{dE}{(4Q/av) - E} = \frac{av}{4V}dt \tag{16.45}$$

Integrating both the sides,

$$-\ln\left[\frac{4Q}{av} - E\right] = \frac{avt}{4V} + C$$

Assuming that the sound has been started at $t = 0$, that is, $E = 0$ at $t = 0$, we get,

$C = -\ln\left(\frac{4Q}{av}\right)$ and the above equation becomes

$$E = \frac{4Qv}{a}\left[1 - \exp\left(\frac{-av}{4V}\right)t\right] \tag{16.46}$$

This equation governes the growth of sound in live rooms. The corresponding equation for intensity, using Eq. 16.40, is

$$I = \frac{Q}{a}\left[1 - \exp\left(\frac{-av}{4V}\right)t\right] \tag{16.47}$$

Similarly the decay of sound can be studied by assuming that the source of sound is shut off at time $t = 0$ and that at this instant the energy density and the intensity were E_0 and I_0 respectively. Thus the differential equation in this case is

$$\frac{V\,dE}{dt} + \frac{Eva}{4} = 0 \tag{16.48}$$

Thus, $$\frac{dE}{E} = -\frac{av}{4V}dt \tag{16.49}$$

Integrating both sides and using $E = E_0$ at $t = 0$, we get

$$E = E_0 \exp\left(\frac{-av}{4V}\right)t \tag{16.50}$$

and the corresponding equation for intensity is

$$I = I_0 \exp\left(\frac{-av}{4V}\right)t \tag{16.51}$$

Taking logarithm of both sides and multiplying by 10 we get,

$$10\log\left(\frac{I}{I_0}\right) = 10\log\left[\exp\left(\frac{-av}{4V}\right)t\right] = \left(\frac{10}{2.3}\right)\ln\left[\exp\left(\frac{-av}{4V}\right)t\right] = -1.087\frac{avt}{V}$$

Now, $|[10\log(I/I_0)]/t|$ represents the change in intensity level in time t, that is, the decay rate D in decibels per second.

Thus, $$D = 1.087\frac{av}{V} \tag{16.52}$$

Sabine defined the reverberation time as the time required for the sound level to decay by 60 dB. Thus, the reverbation time in seconds is given by

$$T = \frac{60}{D} = 55.2\frac{V}{av} \tag{16.53}$$

Substituting $v = 343$ m/s, we get

$$T = 0.161\frac{V}{a} \tag{16.54}$$

In engineering acoustics V is generally expressed in cubic feet and a in sabins, where a sabin represents a surface capable of absorbing sound at the same rate as does 1 ft^2 of a perfectly absorbing surface. In these units Eq. 16.54 becomes

$$T = 0.049\frac{V}{a} \tag{16.55}$$

Examples 16.5 and 16.6 illustrate the use of Eq. 16.55 for calculating the reverberation time and the unknown absorption coefficient of a material respectively.

Equation 16.55 is not applicable for dead rooms because for such rooms the reverberation time must be zero (absorption is maximum and reflected energy is zero), whereas Eq. 16.55 gives a finite value for T. For dead rooms, Eyring derived the following expression for reverberation time

$$T = \frac{0.049}{[-S\ln(1-\alpha)]} \tag{16.56}$$

where $\alpha = \sum \alpha_i S_i/S$ is the average absorption coefficient, S being the total surface area of the interior of the room. It can be seen from Eq. 16.56 that *for dead rooms, for which* $\alpha = 1$, *the reverberation time is zero.*

EXAMPLE 16.5
Determination of reverberation time

The dimensions of an auditorium are 200 ft × 50 ft × 30 ft and its interior surfaces have an average absorption coefficient of 0.25. Find the reverberation time for the auditorium. Find the reverberation time if 400 people are present in the auditorium, each equivalent to 4.5 sabins.

Solution

The reverberation time is given by Sabine's formula as $T = \dfrac{0.049V}{a}$

Now, $V = 200 \times 50 \times 30 = 300000\ \text{ft}^3$

and $a = \sum \alpha_i S_i = \alpha S$

where $\alpha = 0.25$ and $S = 2 \times [(200 \times 50) + (200 \times 30) + (50 \times 30)] = 35000\ \text{ft}^2$

Thus, $a = 0.25 \times 35000 = 8750$ sabins

Hence, $T = \dfrac{0.049 \times 300000\ \text{ft}^3}{8750\ \text{sabins}} = 1.68$ s

In the presence of 400 people the reverberation time is $T' = \dfrac{0.049V}{a'}$

where $a' = a + 400 \times 4.5 = 8750 + 1800 = 10550$ sabins

Thus, $T' = \dfrac{0.049 \times 300000\ \text{ft}^3}{10550\ \text{sabins}} = 1.39$ s

The presence of an audience thus alters the reverberation time and is an important acoustic consideration in the design of auditoriums and concert halls.

EXAMPLE 16.6
Determination of unknown absorption coefficient of a material

The reverberation time of a reverberation chamber (specially constructed chambers used for the measurement of absorption coefficients of materials) of size 9 ft × 10 ft × 11 ft is 4 s. Find the average absorption coefficient of the surfaces of the chamber. If the reverberation time reduces to 1.3 s by covering 50 ft^2 area of one of the walls by an acoustic material, find the effective absorption coefficient of the material.

Solution

$$\text{Reverberation time, } T = \frac{0.049V}{a} = \frac{0.049V}{\alpha S}\ \text{(s)} \tag{16.57}$$

where V and S are measured in ft^3 and ft^2, and α is a dimensionless number—the average absorption coefficient of the surfaces of the chamber.

$$\text{Thus, } \alpha = \frac{0.049V}{ST} = \frac{0.049 \times (9 \times 10 \times 11)}{2 \times [(9 \times 10) + (9 \times 11) + (10 \times 11)] \times 4}$$

$$= \frac{0.049 \times 990}{598 \times 4} = 0.02$$

If area S' is now covered with an acoustic material of effective absorption coefficient α_e then the new reverberation time is

$$T' = \frac{0.049V}{\alpha(S - S') + \alpha_e S'} \tag{16.58}$$

From Eqs. 16.57 and 16.58 we get,

$$\alpha_e = \left[\frac{0.049V}{S'}\right]\left[\frac{1}{T'} - \frac{1}{T}\right] + \alpha$$

$$= \left[\frac{0.049 \times 990 \text{ ft}^3}{50 \text{ ft}^2}\right]\left[\frac{1}{1.3 \text{ s}} - \frac{1}{4 \text{ s}}\right] + 0.02 = 0.52$$

Thus the effective absorption coefficient of the acoustic material is 0.52. Various acoustic materials are investigated in this way to tailor the acoustic performance of an enclosure in a suitable way.

16.6 ULTRASONICS

Ultrasonics is the study of high frequency acoustic waves beyond the limit of audibility. Though there is no standard limit it generally includes frequencies greater than 15,000 Hz. The name ultrasonics is given because sonic waves mean the audible sound waves. Bats and porpoises can sense ultrasonic waves (Ex. 16.7). Chart 16.4 explains the principles and methods used in the generation and detection of ultrasonic waves.

EXAMPLE 16.7
Use of ultrasound by bats

Bats use ultrasonic waves for 'seeing' objects. They send ultrasonic waves to the object to be 'seen' and decipher the echoes received from the object. If objects of the order of a centimetre are to be deciphered, what frequencies should bats generate? Assume the velocity of ultrasound to be 343 m/s.

Solution

The wavelength of the ultrasonic waves should be reasonably smaller than 1 cm. Thus, the corresponding frequency is reasonably larger than

$$f = \frac{v}{\lambda} = \frac{343 \text{ m/s}}{1 \times 10^{-2} \text{ m}}$$

$$= 34.3 \text{ kHz}$$

CHART 16.4 GENERATION AND DETECTION OF ULTRASONIC WAVES

Generation
Various methods are used for the generation of ultrasonic waves depending on the frequency and power required. Two most widely used methods are discussed here.

	Piezoelectric method
Discovery	Curie brothers (1880); generator designed by Langevin (1917) and Nikolson (1920).
Phenomenon	In piezoelectricity a piezoelectric material placed in an electric field undergoes a change in its dimensions. The change in dimensions depends on the direction of the electric field.
Materials	Piezoelectricity is observed in many crystals but it is strongest in Rochelle salt, quartz and ammonium dihydrogen phosphate (ADP). The material is taken in the form of a thin slice and the electric field is applied along its two parallel faces.
Principle of generation	If the applied electric field is alternating, the crystal alternately expands and contracts producing forced vibrations, thus generating ultrasonic waves. The amplitude is maximum at resonance when the frequency of the applied field coincides with the natural frequency of the crystal. Since the expansion and contraction depends on the direction of the applied electric field, the frequency of ultrasonic waves is the same as that of the applied field.
Generator	RF Oscillator; Ultrasonic waves; Piezoelectric crystal Alternating electrical voltage supplied to the crystal is obtained using an RF oscillator such as Hartely or Colpitts oscillator. Frequency of the oscillator can be varied by means of a variable capacitor used in the oscillator circuit. $\nu = (n/2t)\sqrt{\frac{E}{\rho}}$ (1) where t is the thickness of the crystal.
Fundamental frequency and higher harmonics	See Ex. 16.8 E is the modulus of elasticity in the particular direction within the material and ρ is the density of the material. The fundamental frequency corresponds to $n = 1$ for which t(or l) = $\lambda/2$. Higher harmonics correspond to n = 2, 3, ... for which t(or l) = λ, 3λ, 2λ ...

Magnetostriction method	Detection
Joule (1847); generator designed by Pierce (1925). In magnetostriction a ferromagnetic material placed in a magnetic field undergoes a change in its dimensions. The change in dimensions is independent of the direction of magnetic field. Magnetostriction is observed in ferromagnetic materials such as iron, cobalt and nickel. The material is taken in the form of a rod and a magnetic field is applied along the axis of the rod. If the applied magnetic field is alternating, the rod periodically shortens (or lengthens depending on the material) producing forced vibrations, thus generating ultrasonic waves. The amplitude is maximum at resonance when the frequency of the applied field coincides with the natural frequency of the rod. Since the shortening (or lengthening) is independent of the direction of the applied magnetic field, the frequency of ultrasonic waves is twice that of the applied field. 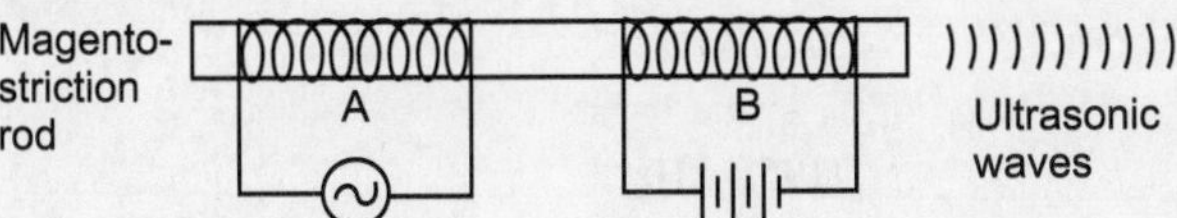 Alternating magnetic field supplied to the rod is obtained by passing an alternating current through a coil A wound around the rod. To get the frequency of ultrasound same as that of the applied field, a steady polarising magnetic field, greater than the amplitude of the alternating field, is produced by passing direct current through the second coil B. $\nu = (n/2l)\sqrt{E/\rho}$ (2) where, l is the length of the rod. See Ex. 16.8 Fundamental frequency is thus the first harmonic. The higher frequency modes are also called the *overtones*. The acoustic power of a generator is smaller for higher harmonics.	The most widely used method for detection of ultrasonic waves is based on the *direct piezoelectric effect*. According to this effect if ultrasonic waves are incident on a piezoelectric transducer, alternating emf is induced in it perpendicular to the direction of the stress applied due to ultrasound. (The converse of this which is the principle of generation of ultrasound is known as *the inverse piezoelectric effect*.) The induced voltage is usually very small and an RF amplifier is used to amplify this voltage. A transducer based on the magnetostriction effect can also be used because the stresses and strains developed in it due to impinging ultrasonic waves result in induction of emf in the surrounding coils which can be amplified. A probe of fine platimum wire can be used to detect ultrasonic waves. When it is exposed to ultrasonic waves its temperature and hence resistance undergoes alternating variations.

EXAMPLE 16.8

Frequencies generated by ultrasonic generators

Find the frequencies of the fundamental mode and the first two overtones, that is, the first three harmonics that can be generated with (a) piezoelectric generator and (b) magnetostriction generator using the following data.

	Quartz	**Nickel**
Modulus of elasticity (N/m^2)	8.55×10^{10}	21×10^{10}
Density (kg/m^3)	2650	8800
Dimension	thickness of crystal = 2 mm	length of rod = 50 mm

Solution

	Piezoelectric generator	**Magnetostriction generator**
Expression	$\nu = \left(\frac{n}{2t}\right)\sqrt{\frac{E}{\rho}}$	$\nu = \left(\frac{n}{2l}\right)\sqrt{\frac{E}{\rho}}$
Calculation	$\nu = \left[\frac{n}{(2 \times 10^{-3}\text{m})}\right] \times \sqrt{\frac{(8.55 \times 10^{10}\ \text{N/m}^2)}{(2650\ \text{kg/m}^3)}}$	$\nu = \left[\frac{n}{(50 \times 10^{-3}\text{m})}\right] \times \sqrt{\frac{(21 \times 10^{10}\ \text{N/m}^2)}{(8800\ \text{kg/m}^3)}}$
	$= n \times 2.84$ MHz	$= n \times 97.7$ kHz
Results: Frequency of Fundamental mode (first harmonic)	2.84 MHz	97.7 kHz
First overtone (second harmonic)	5.68 MHz	195.4 kHz
Second overtone (third harmonic)	8.52 MHz	293.1 kHz

Thus the piezoelectric generator generates ultrasonic frequencies 1–2 orders of magnitude higher than those generated by the magnetostriction generator.

16.7 APPLICATIONS

In this chapter we studied some preliminary aspects of the physics of acoustics. Though acoustics happens to be one of the oldest branches of physics, the physics of acoustics is young and its large-scale application to technology has a relatively recent origin.

In Sections 16.2 and 16.3 we studied the basic concepts of vibration and propagation of sound waves. These concepts found many applications in engineering and technology. This led to the emergence of branches like *engineering acoustics*, dealing with the technology of recording and reproduction of sound waves, and *communication acoustics*, dealing with the study of production and transmission of speech.

The early developments in the field of engineering acoustics were pioneered by Thomas A Edison of USA who invented the phonograph in 1877. There were many technical innovations in Europe and the United States in sound recording and reproduction, based mainly on the concept of Edison's phonograph. In the early 1920s developments in vacuum tube electronics made electrical recording possible and microphones, loudspeakers and other electroacoustic devices were developed. With the advent of microelectronics, engineering acoustics also encompassed the study of aspects related to sound in radio, television and motion pictures. Digital electronics paved the way for the development of *digital acoustics* in the recent past.

In Section 16.4 we studied the mechanisms of speech and hearing. In engineering acoustics *noise control* is one of the cardinal areas of study. Though noise has been studied earlier also as an unwanted sound, it has assumed alarming proportions of late and is considered as a serious environmental pollutant (see Ex. 16.9). For this reason, the study of noise control is also termed as *environmental acoustics*. It mainly deals with the major sources of noise, noise propagation, human response to noise and the physics of methods of noise control. The state of art in noise pollution control is a device which generates in real time a cancelling noise which is the same as the original noise but of opposite phase!

EXAMPLE 16.9

Noise pollution

Find the intensity of sound corresponding to the following noise environments.

Amplified rock music	110 dB
Jet plane at take off	150 dB
Rocket engine	180 dB

Solution

The intensity level is given by

$$\beta = 10 \log\left(\frac{I}{I_0}\right),$$

where $I_0 = 10^{-12}$ W/m^2

For amplified rock music,

$\beta = 110$, that is, $\log(I/I_0) = 11$

Thus, $\quad I = I_0 \times 10^{11}\ \text{W/m}^2 = 0.1\ \text{W/m}^2$

For a jet plane at take off, $\beta = 150$, or $\log\left(\dfrac{I}{I_0}\right) = 15$

Thus, $I = I_0 \times 10^{15}\ \mathrm{W/m^2} = 10^3\ \mathrm{W/m^2}$

For a rocket engine, $\beta = 180$, or $\log(I/I_0) = 18$

Thus, $I = I_0 \times 10^{18}\ \mathrm{W/m^2} = 10^6\ \mathrm{W/m^2}$

Thus we see that the noise levels due to a jet plane and rocket engine are well above the threshold of pain (120 dB = 1 $\mathrm{W/m^2}$). Such high noise levels may cause permanent hearing loss. Longer exposure to relatively lower noise levels such as those created by rock music may also damage hearing. For this reason legislation regarding the permissible noise exposure limits exists for persons working in noisy environments. For example, standards in US stipulate that the exposure to 90 dB sound should not exceed 8 hours per day, whereas the maximum permissible limit for 110 dB sound is half hour per day.

In Section 16.5 we studied the classical ray theory that serves as the basis of *architectural acoustics*—a branch pertaining to the study of acoustic waves in enclosures—with a view to optimise production and reception of speech and music. Sabine made pioneering contribution to modern architectural acoustics and the systematic incorporation of acoustical considerations in the design of concerts halls and auditoria started only after 1900. Today an acoustician is considered an important member of the design team of any theatre, concert hall or auditorium.

In Section 16.6 we studied ultrasonics. Ultrasonic waves have wider applications in industry compared to sonic waves. This became clear after the successful development of the piezoelectric transducer in 1920 for generation of ultrasonic waves. An ultrasonic wave is useful in two ways. Either it can be used as a probe to characterise the medium by studying its propagation behaviour in the medium or it can be used to bring about desired changes in the medium. In the first type, called *low amplitude applications*, low intensity ultrasound is employed, whereas high intensity waves are used for the second type of application. Table 16.3 illustrates this point.

Table 16.3 Applications of ultrasonic waves

Wave used	Before	After
Low intensity waves	(wave) + (medium)	= (medium) + (wave)′
High intensity waves	(wave) + (medium)	= (medium)′ + (wave)

Analytical applications of the first kind involve the study of *how fast the waves travel* (velocity of waves) and *how fast they decay* (attenuation of the waves) in the medium as a function of whatever parameters are of interest. The velocity of ultrasound depends on the elastic constants of a medium and the experimental determination of these constants provides useful information regarding the mechanical properties of the medium. The study of ultrasonic attenuation in solids is of interest to an engineer for assessing the suitability of materials for several practical problems such as the design of optimum sound transmission systems for ultrasonic delay lines (devices useful for storing information for certain periods of time).

Among the other low amplitude applications are the applications in devices such as mechanical filters (devices used for separating telephone communications sent simultaneously over one transmission line), inspectroscopes (devices used to transmit sound waves into metal casting or other solid pieces and determine the presence of flaws by reflection or by interruption), thickness gauges (devices for measuring thickness of pieces when one side is not accessible), and SAW (surface acoustic waves) devices.

High amplitude applications of ultrasonics include effects due to cavitation (Cavities can be created in liquids by high intensity waves. When a cavity collapses, extremely high pressure is produced leading to a number of mechanical, chemical and biological effects.), fatigue testing and metallurgical processing.

Acoustics had gained a lot from the already developed science of optics. The fields such as *acoustic microscopy, acoustic interferometry* and *acoustic holography* thus developed by drawing parallels with their existing optical counterparts. The existing knowledge that sound waves exhibit a change in frequency due to the relative motion between the source and the listener was extended by Doppler (1803–1853) to light waves. Optical Doppler effect was first observed by J Stark in 1905 and has important applications in astronomy, high energy physics and remote sensing. Acoustic Doppler effect finds application in medical diagnostics, for example, in echo cardiogram and in measurement of blood flow.

The symbiosis of acoustics and optics also resulted in the emergence of new fields of research such as *acoustooptics* and *photoacoustic spectroscopy*. Acoustooptical devices are based on the ability of sound to control the amplitude, direction and the frequency of light beams. With the advent of lasers acoustooptical devices have been increasingly used in image scanners, recorders, printers and projectors, and in a large number of signal processing devices and laser controllers. Photoacoustic spectroscopy is a technique for measuring small absorption coefficients in gaseous and condensed media involving the sensing of optical absorption by detection of sound. It is used in pollution monitoring. This symbiosis also gave a boost to the technology of sound recording and reproduction after 1970, when the optical disc, which reproduced sound with extremely high fidelity, was developed.

Apart from the branches of acoustics discussed above, *musical acoustics* (physics of musical instruments) and *underwater acoustics* (study of the sea using acoustic waves) have also developed into independent branches of acoustics. SONAR (sound navigation and ranging), SOFAR (sound fixing and ranging) and SODAR (sound detection and ranging) are techniques employed for the study and exploration of the sea (see Ex. 16.10). Other branches like *physiological acoustics, bioacoustics, animal acoustics and medical acoustics* relate to acoustics of living systems and are useful in fields such as life sciences, medical diagnostics and biomedical engineering.

Apart from the discussed well-defined branches of acoustics, the knowledge of sound waves also finds application in aerospace engineering (sonic boom, supersonic aircraft and wind tunnel; they relate to shock waves which arise when sound sources travel faster than the speed of sound), space technology (rocket combustion oscillations), nuclear engineering (nuclear blasts, controlled fusion in plasma) and earthquake technology (measurement of speed of seismic waves like surface acoustic waves, dilational waves and shear waves provide seismologists with valuable information about the structure of the earth, which in turn is useful for prediction of earthquakes, and prospecting of mines).

EXAMPLE 16.10

Determination of sea depth using ultrasonic waves

Ultrasonic *fathometer* (fathom = 6 ft) is an instrument that measures the depth of the sea by sending ultrasonic pulses of known velocity and recording the time of its travel from the transmitting transducer to the bottom of the sea and back. Find the depth of the sea if the recorded time is 0.8 s and the velocity of ultrasound in sea water is 1500 m/s.

Solution

Velocity of ultrasound, $v = \frac{D}{t}$

where D is the depth of the sea and $2t = 0.8$ s

Thus, $D = 1500 \text{ m/s} \times 0.4 \text{ s} = 600 \text{ m}$

17

Introduction to Nanotechnology

17.1 INTRODUCTION

A nanometre (nm) is a billionth part of a metre. How small is that? Figure 17.1, adopted from a UNESCO document, puts nanosizes in perspective. Matter of size 1–100 nm has many surprises to offer. At these sizes material properties are very different to what they are at larger and more familiar scales. In the last few decades the study of such novel behaviour has attracted a tremendous amount of attention from scientists and technologists of various persuations.

In literature nanotechnology is defined variously. A few examples should suffice to bring this out: Nanotechnology is the science and technology 'of creating nanoparticles', 'of materials and properties at 1 to 100 nm sizes', 'of building circuits and devices from single atoms and molecules', 'of manufacturing machines at 1 to 100 nm sizes', 'of molecular manufacturing', 'of self-organizing molecular assemblies', and so on. It is generally accepted that the field was presaged by Richard Feynman in a lecture given at Caltech in 1959 when he said that *there was plenty of room at the bottom.* What Feynman meant was that devices based on working of matter at atomic scales have a potential for a variety of hitherto unforeseen uses and applications. Much of this room is created because behaviour of matter is governed by the laws of quantum mechanics at these scales.

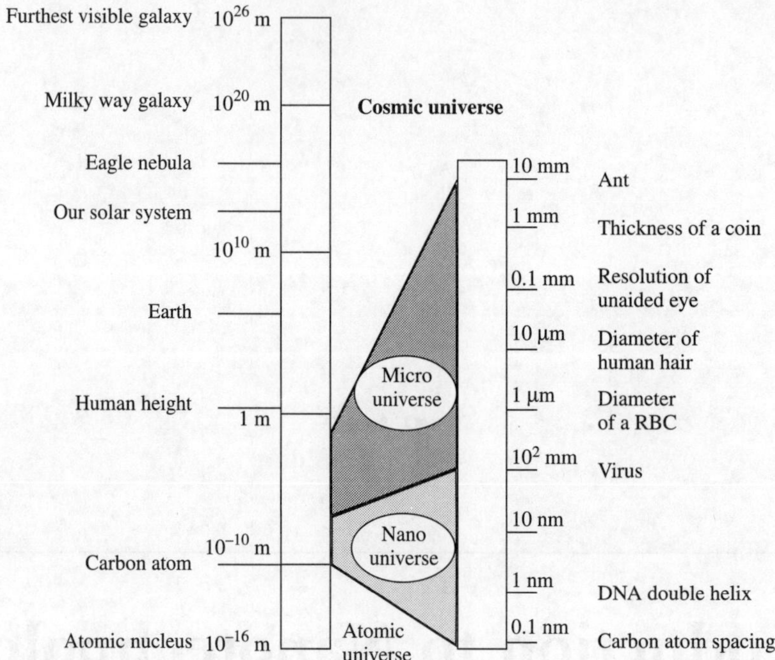

Figure 17.1 Relative size of the nanoworld

Nanotechnology is a field encompassing multiple disciplines from traditional and modern science and technology. These include physics, chemistry, materials science, biology, medicine, science of cognition, electronics, mechanical engineering, computer science and many others. No wonder that it has attracted so many working scientists and engineers over a relatively short span of time. Progress in nanotechnology seems to be driven by development of technologies enabling control of matter at atomic and molecular scales. Large financial resources allocated by governments for research in the field, and attention of big commercial establishments sensing a huge promise in the results of this research, have catapulted the field onto a steep upward trajectory.

This chapter, meant to introduce the student to nanoscience and nanotechnology, is different in a few respects from the other chapters in the book. All the discussions are at the level of an elementary introduction to subareas of arguably the most active scientific and technological domain at present. Also, there are no solved examples. Moreover, there are many references to resources on the internet scatterred all over the chapter. The student is urged to look at all these and then attempt a deeper delving into web-resources available. Section 17.2 is a brief outline of some methods of synthesis of nanomaterials. Section 17.3 introduces techniques used to make various types of measurements on nanomaterials and characterise them. Section 17.4 discusses graphene and carbon nanotubes. Section 17.5 is devoted to bringing out strange properties that nanomaterials exhibit. It also provides a glimpse at extant and potential applications of nanotechnology.

17.2 PREPARATION OF NANOMATERIALS

Nanomaterials are generally classified into two broad classes: Fullerenes and nanoparticles.

Fullerenes are allotropes of carbon. Carbon forms a hexagonal planar lattice structure with sp^2 hybrid molecular bonds between any two carbon atoms. This structure is a single-atom thick sheet and is known as *graphene*. Various nanostructures can be formed from graphene by rolling it into tubes. Spherical molecules such as C_{60}, known as *buckyball*, can also be formed. All these structures go by the common class name of *fullerenes*. Many of these materials can be formed by establishing a large current between two nearby graphite electrodes in an inert atmosphere. The resulting carbon plasma arc between the electrodes cools into a sooty residue from which many fullerenes can be isolated. Fullerenes are further discussed below in Section 17.4.

Nanoparticles are nanocrystals of metals, semiconductors and other chemical species. They show a striking novelty and variety of physical, chemical, electrical and other properties.

Nanomaterials may be synthesised in various forms such as thin films, nanoparticles, nanotubes, nanorods, nanobelts or nanoribbons, nanocomposites and so on. Many of the methods and processes designed and perfected to prepare these materials are specific to one or more of these forms. For example, a very common method to prepare gold nanoparticles in the form of a metallic colloidal dispersion is by reducing a solution of chlorauric acid by sodium citrate at 100°C. This process has been known for half a century. The general process for preparing metallic colloidal dispersions uses a precursor (such as chlorauric acid), a reducing agent (such as sodium citrate) and a polymeric stabiliser, which hinders growth of particle size by adhering to growth surfaces. Similarly, one-dimensional nanostructures can be prepared by processes specific to them. For example, silver nanowires are grown in solution using platinum nanoparticles as growth seeds. Silver is generated by reduction of silver nitrate. One-dimensional growth happens as a result of blocking of growth by selective adsorption of polyvinyl pyrrolidone (PVP) surfactant in the solution on some growth surfaces. For silver the growth directions are $[2\bar{1}\bar{1}]$ and $[01\bar{1}]$.

Thus, preparation methods can be classified separately for each of the forms of nanomaterials. For example, methods for producing one-dimensional nanostructures can be classified into:

- spontaneous growth techniques,
- template based synthesis,
- methods based on electrospinning, and
- lithography.

It is quite beyond our introductory purpose here to discuss details of such classification, or even to attempt to list them all. The short description in the preceding paragraph is meant only to hint at the detailed, complex and material-specific nature of methods of preparation of nanomaterials. Here attention is limited to a very broad classification of preparation methods.

Many of the processes used for preparation of nanomaterials have been known to materials scientists for a long time. These have been improvised and adopted to suit the goal of nanometre-sizes. Traditionally, methods for materials preparation are classified into two broad approaches:

(i) the top-down approach
(ii) the bottom-up approach.

This classification naturally extends to nanomaterial preparation too. A *top-down approach* is one in which macrocrystalline forms, or, macro-structures are progressively broken down to a nanocrystalline form, or nanostructures, that is, nanosizes are achieved by reducing macro-sizes by some means. On the other hand, a *bottom-up approach* is one in which nano-sized forms are built up atom by atom, molecule by molecule, cluster by cluster, or, monomer by monomer depending on the context. The above classification generally refers to the final step in the preparation of the material. It is quite common that, in a process classified as a bottom-up process, some initial stage of the entire process may involve a top-down step. Equally, a top-down process may involve an earlier bottom-up step. For example, in lithography, generally regarded as a top-down process, a bottom-up step of film deposition may precede the final lithographic reduction of the film to a nanostructure. Also, it may be noted that film deposition itself involves first a *top-down break-up* by dislodging of atoms from a large structure before their *bottom-up deposition* over a substrate! For this reason, lithography is sometimes regarded as a hybrid of both the approaches.

Box 17.1

Ball milling

A mechanical attrition method

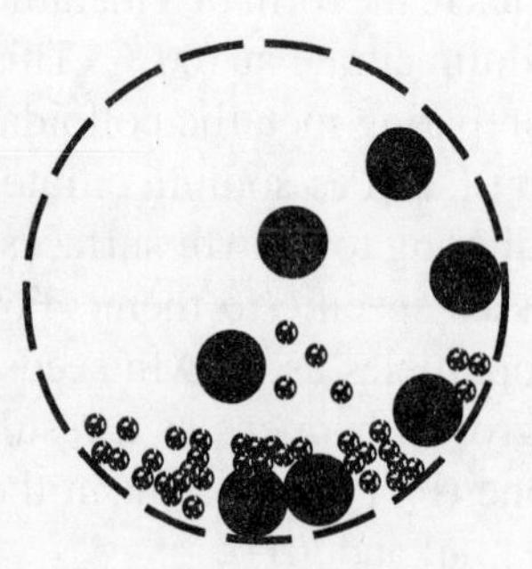

Figure 17.2 Cross section of cylindrical ball-mill

Ball milling is the best example of the top-down approach. Large sized particles are subjected to attrition, or reduction in size, by mechanical means. A ball mill is a cylindrical device used in grinding materials like ores, chemicals, ceramic raw material and paints. Ball mills rotate around a horizontal axis, partially filled with the material to be ground and a grinding tool, such as ceramic balls, flint pebbles and stainless steel balls. Ball mills are also used in pyrotechnics and manufacture of black powder, but cannot be used in the preparation of some pyrotechnic mixtures such as flash powder, because of their sensitivity to impact. High-quality ball mills are quite expensive and can grind particles to as small as 100 nm, thereby enormously increasing surface area and in consequence reaction rates.

Figure 17.2 shows the principle of a ball mill. The cylinder rotates subjecting the particles of the material being ground to repeated crushing impacts by the steel balls. Powders with nano-sized particles can be prepared by this method. For example, particles of size 30–80 nm of a yttrium-barium-copper-oxygen-silver composite are obtained by high-energy attrition of a mixture of metallic yttrium, metallic copper and barium titanate containing 5% by weight of silver. Twenty hours of milling is needed to reach those particle sizes.

Box 17.2

Electron beam lithography

A method borrowed from VLSI technology

Lithography is a printing method that uses chemical processes to create an image. This method is extensively used in the fabrication of semiconductor devices such as a junction diode, a transistor or an integrated circuit. There are two types of lithographic processes used in the silicon industry: photolithography and *electron-beam (e-beam) lithography*. Photolithography uses light to transfer a geometrical pattern from a photomask to a light sensitive chemical (photoresist) on the substrate. Depending on the pattern on the photo-mask, a structure can be generated on the substrate, which is further used to develop a silicon-based photonic and electronic device. Advanced photolithography tools use deep ultraviolet (DUV) light of wavelengths 248 and 193 nm, which allows minimum feature size of about 50 nm. Light is replaced by an electron beam in e-beam lithography. Figure 17.4 shows the successive steps in this process. Here an electron beam is used to *write* patterns on the resist-coated surface as it is scanned by the beam. Common resists used in e-beam lithography are PMMA (polymethyl methacrylate) and ZEP520. Both are *positive* resists, for which the portions exposed to e-beam become soluble in photoresist developer and are removed. PMMA allows for nanoscale resolution in the pattern created. ZEP520 allows higher resolution and is three times faster. However, adhesive strength of ZEP520 to the substrate is poor. The main advantage gained by use of electron beam as a pattern-writing tool is that, compared to light in photolithography, electrons have much smaller de Broglie wavelength. With smaller wavelengths it is possible to build much smaller sized patterns on the resist coat on the Si substrate. Originally developed for manufacturing integrated circuits, it is also used for creating nanotechnology architectures. The key limitation of e-beam lithography is its low throughput. It takes a very long time to expose an entire silicon wafer or glass substrate. Long exposure time makes the process vulnerable to beam drift.

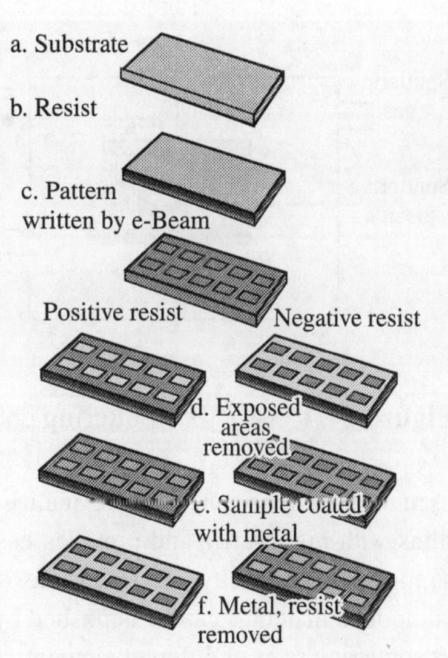

Figure 17.3 E-beam lithography

In Boxes 17.1 and 17.2, two examples of the top-down approach, ball-milling and lithography, are outlined. Lithography is the process responsible for very large-scale integration of electronic components, and in the form of electron beam lithography, it is used to fabricate nanosized electronic devices.

Top-down processes such as ball-milling produce materials with crystallographic defects, more so at the surface of the structure. In nanostructures surface effects are extremely important since at the nanoscales a much larger proportion of atoms in the entire structure is on the surface. Therefore, bottom-up approaches are generally the recourse for nanomaterial synthesis. The spirit of the bottom-up approach is akin to the spirit behind understanding biological processes such as protein synthesis (see protein synthesis related videos on youtube). In fact an influential definition of nanotechnology attributed to Eric Drexler is that it is the science and technology of self-organising molecular assemblies, which is much like protein synthesis inside cells. Boxes 17.3 and 17.4 provide an outline of two examples of the bottom-up approach: physical vapour deposition, or PVD and the sol-gel technique.

Box 17.3

Physical vapour deposition by sputtering

A physical bottom-up method

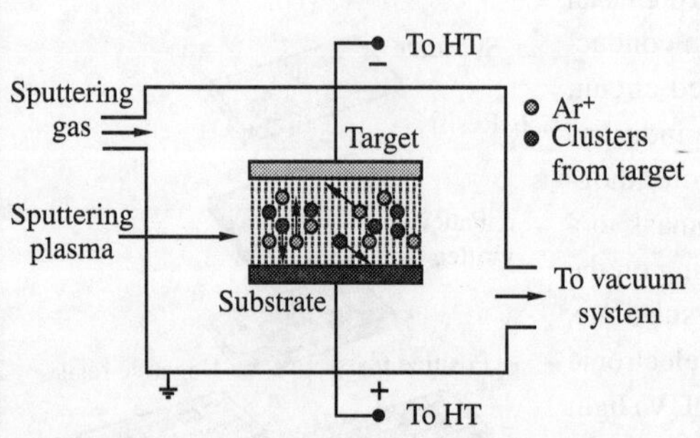

Figure 17.4 A PVD sputtering chamber

Sputtering is a physical vapour deposition process in which atoms in a solid target are dislodged by bombardment of the target by energetic ions of an inert gas like argon. The target is made of the material which is to be deposited on the substrate as a thin film. This technique is commonly used for thin film deposition. Standard physical sputtering is driven by energy exchange between the ions and atoms in the target material due to collisions. In sputtering a glow plasma discharge of non-reactive ions, like Ar^+, is created by applying a potential difference inside a vacuum chamber. The ions bombard the target and dislodge atoms. The sputtered atoms, shown by open circles in figure 17.4, move to deposit on the surface of the substrate. Sputtering is the most versatile of techniques used for the deposition of device quality thin films. As compared to other deposition techniques sputtering produces films with high purity and provides easier control over composition. It also provides films with good adhesive strength, homogeneity and uniformity. Initially, sputtering was used to deposit thin metal films, but now even compound materials can be deposited by this method. In these cases, target stoichiometry is the critical parameter as sputtering rates of different elements are in general different. Therefore, to get a proper film composition reactive sputtering techniques are generally used.

Box 17.4

Sol-gel technique

A chemical bottom-up method

Sol-gel is a wet chemical-based self-assembly process for nanomaterial synthesis. A sol is a colloidal suspension of solid particles with diameter of a few hundred nanometres. A gel is a macromolecule immersed in a solvent. The sol-gel process is a chemical transformation of a sol into a gel, that is, gelation of a sol. Gelation involves the evolution of networks of molecules. The precursors used for synthesising the colloids generally consist of metallic ions and ligands, which are elements surrounded by various reactive species. Metal alkoxides, such as aluminates, titanates, borates and cuprates, and alkoxysilanes, such as tetramethoxysilane (TMOS) and tetraethoxysilane (TEOS) are the most common precursors. The reason is they react readily with water. In case the alkoxide is immiscible with water some homogenising agents like alcohols are used as mutual solvents to enhance miscibility of these materials. In general sol-gel formation occurs in the following four steps: (i) hydrolysis and alcoholysis, (ii) condensation and polymerisation, (iii) particle growth and (iv) particle agglomeration. The figure depicts sol-gel technologies.

In general, all the four steps depend on several factors such as pH of the sol, temperature of the reaction, reagent concentrations, time of reaction, nature and concentration of catalyst used and the time of the gel formation. By controlling these parameters, it is possible to achieve wide variation in structural, electrical and optical properties of the material prepared.

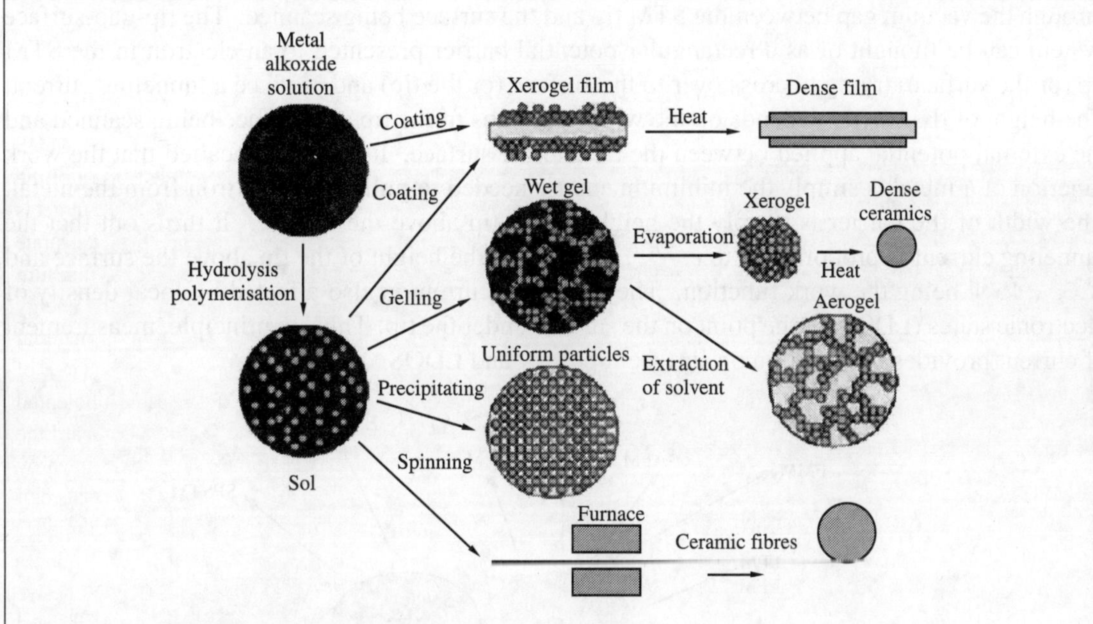

Figure 17.5 Sol-gel technologies

There are several ways to convert the gel (inorganic network) to the desired solid form. Depending on the deposition and drying processes and conditions, the gel can be converted into forms such as aerogel, xerogel, gelled spheres, nanopowders, thin film coatings, fibres and nanostructured layers.

17.3 CHARACTERISATION AND MEASUREMENT

When any material is prepared for use, it is necessary to determine whether it has been prepared as required before it is actually put to that use. Material preparation is always followed by material *characterisation*. Characterisation is a procedure where measurements are made to identify the material based on its known properties and/or determine its unknown properties. Due to the nanosize of the particles, special techniques have to be employed. The requirement is essentially that the tools used should be able to *see* at atomic scales. It is easy to understand that these tools are some sort of 'microscopes'. The methods of characterisation can be classified into two families: (i) Scanning probe microscopy (SPM), and (ii) Electron microscopy (EM).

17.3.1 Scanning Probe Microscopy

Scanning probe microscopy began with the development of the *scanning tunneling microscope* (STM) in 1981. The STM uses the phenomenon of quantum tunneling of electrons across a tiny vacuum gap between a sharp conducting probe (the tip) and the surface being scanned. Electron tunneling results in a measurable current between the tip and the sample, The magnitude of tunneling current is related to the probability of an electron in the tip or the surface of tunneling through the vacuum gap between the STM tip and the surface being scanned. The tip-gap-surface system can be thought of as a rectangular potential barrier presented to an electron in the STM tip (or the surface) trying to cross over to the surface (or the tip) and produce a tunneling current. The height of the barrier depends on the work functions of the tip and surface being scanned and the external potential applied between the tip and the surface. It might be recalled that the work function of a metal is simply the minimum energy needed to pull out an electron from the metal. The width of the barrier is simply the height of the tip above the surface. It turns out that the tunneling current is proportional to e^{-2Kd}, where, d is the height of the tip above the surface and $K \propto \sqrt{\Phi}$, Φ being the work function. The tunneling current is also affected by local density of electronic states (LDOS) at the point on the surface under the tip. Thus, in principle, measurement of current provides information on surface topology and LDOS.

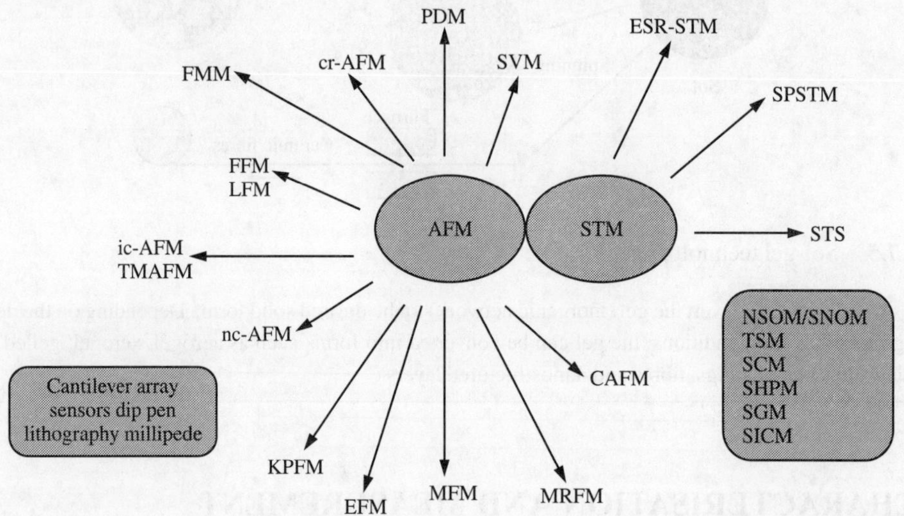

Figure 17.6 Scanning probe microscopies

The most important of the SPM family is *atomic force microscopy* (AFM), developed five years after the STM. This technique is based on measurement of the van der Waals force between the tip and the surface. AFM techniques have been developed to monitor various other type of forces and interactions, such as optical, electrical, magnetic as well as contact interactions like friction between the tip and the surface. Chart 17.1 describes some details of the STM and the AFM.

The SPM family is large and there are many types of scanning probe microscopes (see Figure 17.6). The acronyms indicate the type: for example, TMAFM is tapping mode AFM, nc-AFM is non-contact mode AFM and KPFM is Kelvin probe force microscopy. For others see the article *Scanning Probe Microscopy and Related Methods* by Ernst Meyer (available at http://dx.doi.org/10.3762/bjnano.1.18). As the name suggests, SPM generally uses a probe called a *tip* to scan the surface of a sample. The tip is very sharp, ideally ending in a single atom. Interaction between the tip and the surface is monitored as the tip moves over the surface in a predetermined manner, such as, for example, a raster scan. This movement can be extremely tiny.

To monitor the tiny movements of the tip, special arrangements such as an optical cantilever (see Fig. 17.7), are necessary. A narrow laser beam reflects from the cantilever. Tiny movements and twists of the lever are magnified in the form of large deflections of the beam recorded on a photodiode array. Figure 17.8 shows the detection system with more clarity for a 2×2 photodiode array. As the tip moves, or the cantilever twists, the reflected intensities measured by the four diodes change. The differential changes in response of diodes in the array can be used to compute up-down and lateral tip movements.

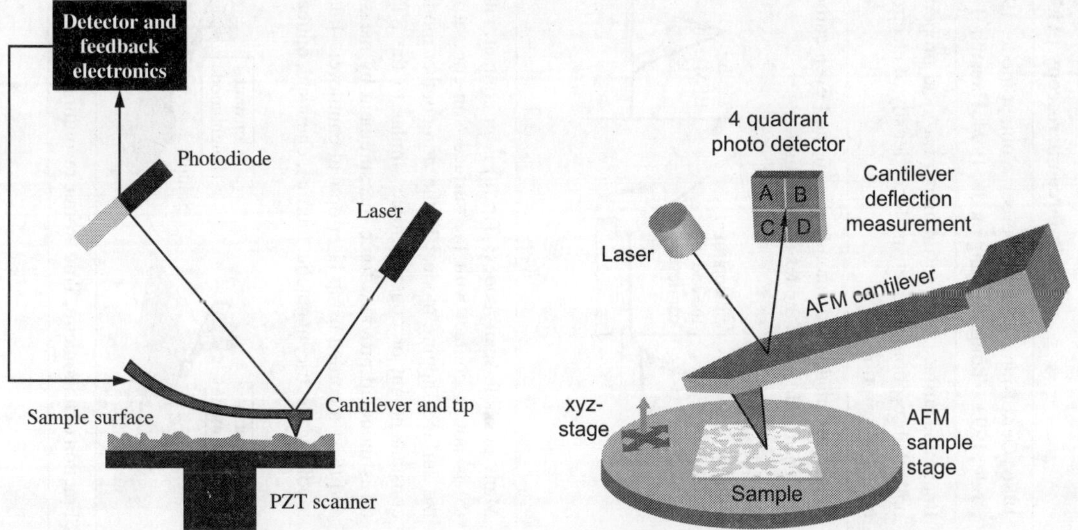

Figure 17.7 The optical cantilever in AFM

Figure 17.8 The photoarray in an optical cantilever

Extremely sensitive displacement sensors such as piezoelectric sensors like PZT (lead zirconate titanate, $Pb[Zr_xTi_{1-x}]O_3,\ 0 \le x \le 1$) are also used to record tip movement. The charge sensitivity of PZT is two orders of magnitude larger than quartz, the commonly used piezoelectric sensor material–350 picocoulomb per Newton of force for PZT compared to 2.2 pC/N for quartz.

17.3.2 Electron Microscopy

In electron microscopy the probe is a high-energy electron beam. All electron microscopes use systems of electromagnetic lenses to appropriately manipulate and focus the electron beam generated by an electron gun. This is done inside an evacuated metal chamber with very low pressures like 10^{-4} torr. When this beam encounters the sample, various types of interactions take place between the two. These include x-ray emission, cathodoluminescence, secondary electron emission, Auger electron emission, backscattering and elastic and inelastic electron scattering. Electron microscopes of different types use different groups of interactions from among these. For example, a transmission electron microscope (TEM) uses elastically as well as inelastically scattered electrons transmitted through the sample. An image of the sample is constructed on the basis of spectroscopic analysis of electrons or x-rays produced by these interactions.

Chart 17.1 Scanning probe microscopy (SPM)

	Scanning tunneling microscopy (STM)	Atomic force microscopy (AFM)
Discovery	1981; Gerd Binnig, Heinreich Rohrer, 1986 physics Nobel prize	1986; Gerd Binnig, Quate, Gerber, Contact mode; 1987, Martin, Non-contact mode; 1993, Zhong, Tapping mode; 1994, Hansma, Tapping mode in liquids.
Principle	Image the surface by measurement of electron current due to quantum tunneling between tip and surface	Image the surface by measurement of a local property like height of surface, optical absorption, magnetism, local density of electronic states, friction force, optical absorption.
Probe	Sharp conducting tip—ideally terminating in a single atom—made of tungsten, platinum-iridium. A tungsten tip for STM	Sharp tips made of diamond, silicon, and even carbon nanotubes. Different designs for different modes of operation. 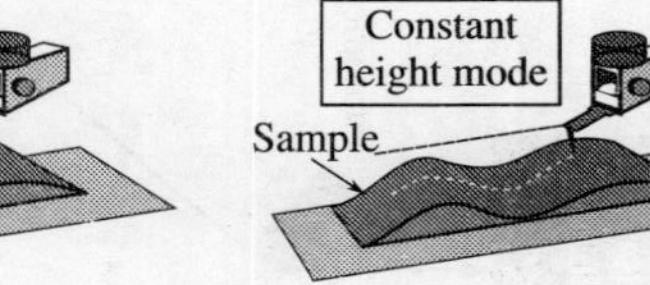
Modes of operation	Tip is moved to scan the surface. Two scan modes: (a) Constant height and (b) Constant current. Tunneling current, tip-sample voltage, tip displacements parallel and perpendicular to surface are monitored. Images give information on topology, LDOS.	Many possible scan-modes: Tip (a) is in contact with surface-Contact mode, (b) is not in contact with the surface-Non-contact mode, (c) keeps tapping the surface-Tapping mode. In the non-contact mode the scan mode can be constant height or constant force similar to the STM. Cantilever on which tip is mounted provides force measurement by measurement of its deflection and twist recorded using laser optical cantilever. Images give information on topology, electric, magnetic, thermal, chemical, optical, dielectric properties
Resolution	Lateral 0.1 nm, Depth 0.01 nm	Fraction of nanometre; True atomic resolutions in ultrahigh-vacuum conditions.

Electron microscopes give high-resolution images which are also substantially magnified. Besides scanning electron microscope (SEM) and TEM, there are other versions of the electron microscope like the reflection electron microscope (REM), the high-resolution TEM (HRTEM), the scanning transmission EM (STEM), the environmental SEM (ESEM) and low voltage EM (LVEM). Chart 17.2 describes some details of the SEM and the TEM.

17.3.3 Characterisation Tools as Fabrication Tools

The microscopes described above and in Charts 17.1 and 17.2 probe nanostructures by interacting with them with a nanosized probe. At nanoscales this also means that it is possible to use these instruments to *manipulate* matter! An electron microscope with its electron beam optics systems has all the hardware and instrumentation necessary for e-beam lithography. The electron beam can be used to write nanostructure patterns instead of probing the sample surface. For research purposes it is common to convert electron microscopes into e-beam lithography systems. Dedicated e-beam lithographs are comparatively far costlier. STM tips can manipulate and drag atoms on the sample surface to arrange them into patterns. Figure 17.9 depicts this schematically. Figure 17.10 is a realistic example of this. It shows progressive stages of a process in which atoms are arranged as in the final pattern in lower right image in the figure. Such a pattern is called a quantum coral. The first real demonstration of this type of atomic manipulation was by IBM researchers who wrote the letters 'IBM' in atoms using an STM tip (see http://www.almaden.ibm.com).

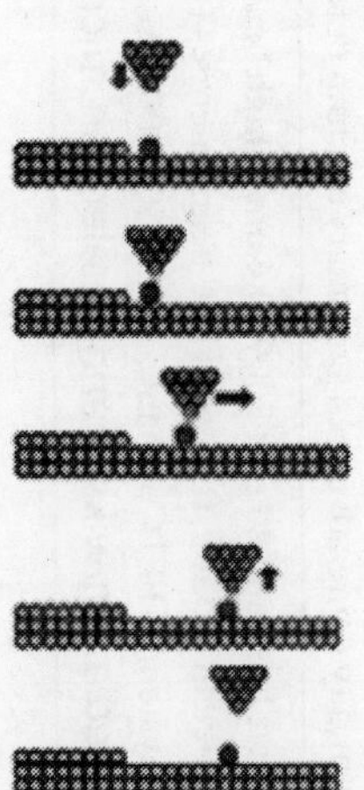

Figure 17.9 Schematic of atom manipulation by an STM tip

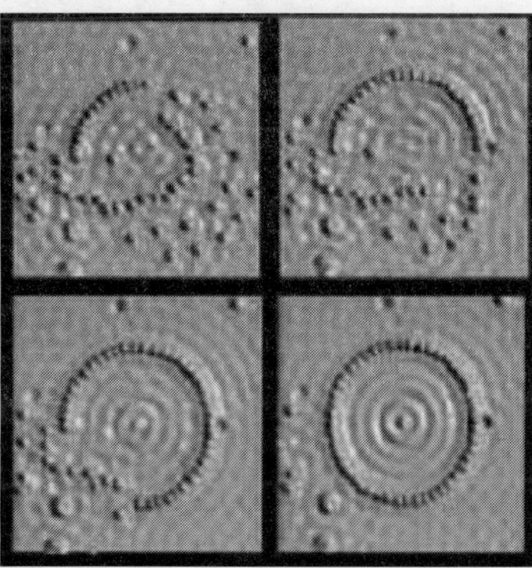

Figure 17.10 A quantum coral

The STM tip and AFM cantilever-tip system can be used to dent and etch surfaces also. In this the tip has to make contact with the surface as against the manipulation of atoms by STM tip where the tip only exerts pulling forces to drag an atom to the desired position. AFM tips can also achieve removal of a portion of atomic or molecular layer, called *nanoshaving*, and replacement of a portion of an atomic layer by different atoms, called *nanografting*.

Chart 17.2 Electron microscopy

	Scanning electron microscope	**Transmission electron microscope**
Discovery	1937, Manfred von Ardenne	1931, Ernst Ruska and Max Knoll; 1939, Siemens, first commercial TEM
Probe	Electron beam (energy 0.2 keV to 40 keV, de Broglie wavelength 0.19 to 2.75 nm) focussed at a point and scanning sample surface	Electron beam (energy 100 keV to 300 keV, de Broglie wavelength 0.071 to 0.123 nm) illuminating the entire sample
	In both SEM and TEM the electron source may be thermionic type with tungsten or lanthanum hexaboride element, or field emission type with tungsten single crystal emitters, or with zirconium oxide emitters.	
Schematic and working principle	The chambers in schematics (adapted from: http://dehistology.blogspot.in/2011/06/electron-microscopy.html) of both SEM and TEM given below are evacuated to 100 microTorr to avoid electron beam interactions with ambient gases. 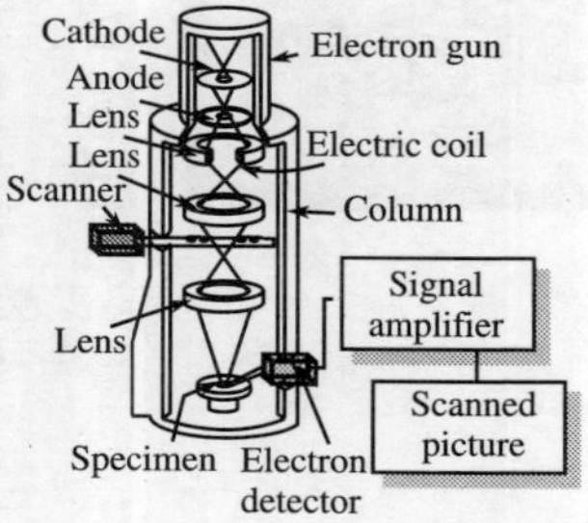 Electron beam is focussed by condenser lenses to a spot (radius 0.4 nm to 5 nm) on the sample. Spot raster scans rectangular surface area as controlled by scanning coils. Beam-sample interaction results in reflection of high-energy electrons by elastic scattering and emission of secondary electrons by inelastic scattering. Backscattered and secondary electrons are separately processed to build the SEM image.	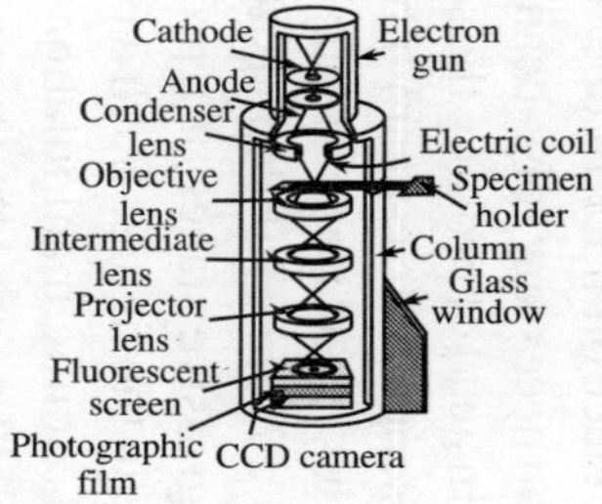 There are three lensing stages: condenser, objective and projector. Condenser lenses form the beam falling on the sample, objective lenses focus the beam transmitted through the sample and projector lenses expand the beam to fall on the imaging screen. Detailed lens designs differ substantially with manufacturers.
Interactions used	Uses backscattered and secondary electrons. REM uses reflected elastically scattered electrons. Environmental SEM (ESEM) is capable of working in 1–50 Torr pressure gaseous environment and in high relative humidity using advanced secondary electron detectors	Uses electrons elastically and inelastically scattered and transmitted through the sample. STEM uses transmitted beam and scanning of sample surface
Resolution, Magnification	Generally between 1 nm and 20 nm; 2009: 0.4 nm at high beam energies (30 kV) Hitachi S-5500, 0.9 nm at low-beam energies (1 kV) Magellan XHR system of FEI Company; 10X to 500,000X (see also http://en.wikipedia.org/wiki/Scanning_electron_microscope)	With advanced aberration corrector design resolution 50 pm at magnifications 50,000,000X (see http://en.wikipedia.org/wiki/Transmission_electron_microscopy)

An AFM tip can transfer a molecular layer onto a substrate in a process called dip-pen lithography. These operations achieve what is called *functionalising of surfaces*.

17.4 FULLERENES, GRAPHENE AND CARBON NANOTUBES

As we have seen fullerenes are allotropes of carbon. The first to be discovered in 1985 at Rice University, USA, was a spherical fullerene C_{60}, called *buckminsterfullerene*, or, *buckyball*. It is named in honour of Buckminster Fuller, who created the famous geodesic dome structures, the architecture of which the molecule resembles (see Figure 17.11). Buckyball has a molecular weight of 720.6 g/mol. The football-like molecule is prepared in helium by passing current of 150 A through a carbon rod and extracting the soot with benzene. The resulting magenta solution contains C_{60} and C_{70}. Spherical fullerenes with 76, 84, 90 and 96 carbon atoms, and much larger fullerenes with 180, 190, 240 and 540 carbon atoms have also been found. C_{540} has an icosahedral shape (see also http://www.encyclopedia.com/doc/1E1-buckmins.html).

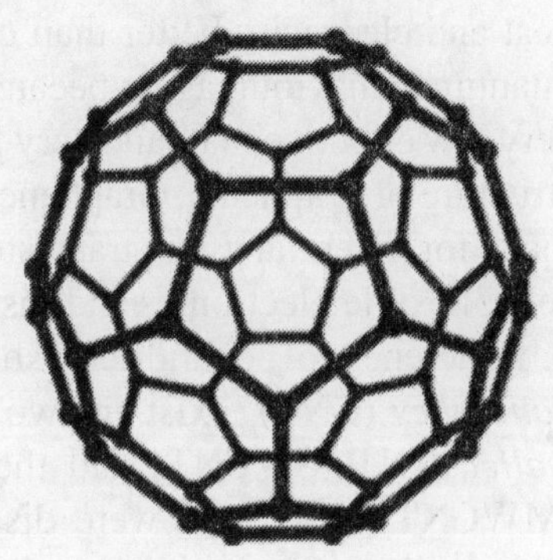

Figure 17.11 A ball and stick model of C_{60}

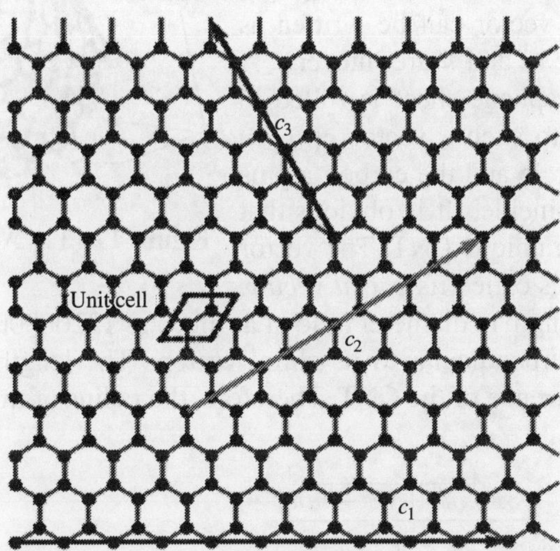

Figure 17.12 A graphene lattice showing three chiral vectors

Graphene is the thinnest and the strongest material known to us. It was first obtained in 2004 by Andre Geim and Konstantin Novoselov by taking graphite and using sticky tape to peel off carbon layers from it. They repeatedly folded the sticky part of the tape onto the initial carbon peel until a single layer was obtained. Geim and Novoselov received the Nobel Prize in 2010 for this work. Graphene sheets stack to form graphite with an inter-planar spacing of 0.335 nm. Figure 17.12 shows the planar structure of a graphene sheet. Also shown is a rhombic unit cell of the lattice with two lattice points in each cell. The lattice unit vectors are $\mathbf{a} = x\sqrt{3}\,\hat{\mathbf{i}}$ and $\mathbf{b} = x[\sqrt{3}/2\hat{\mathbf{i}} + 3/2\hat{\mathbf{j}}]$, where $x = 0.142$ Å is the graphene bond length and $\hat{\mathbf{i}}$ and $\hat{\mathbf{j}}$ are unit vectors pointing respectively to the right and upward in the figure.

Due to its two-dimensional lattice graphene has the largest surface area relative to its mass and volume. This is important for chemical reactions occurring on the surface of the material. Although it is only one carbon atom thick, it is impermeable to gases or liquids. It conducts heat and electricity better than copper. It has striking electronic properties. Electrons show full quantum behaviour, again because the material is just one atom thick. Electrons in graphene have very low effective massand very large mean free path. This happens because of the faultless lattice structure of graphene. Graphene has the highest known electron mobility. Such novel properties make for potential use in transistors which are faster than those made from silicon, improved solar cells, flexible electronics and faster computers.

Fullerenes of cylindrical shape, called *carbon nanotubes* (CNT), exist in two forms: the *single-walled* CNT (SWCNT) and the *multi-walled* CNT (MWCNT). SWCNTs were discovered in 1991 by Sumio Iijima. These cylindrical forms can be thought of as being formed from a graphene sheet by rolling it in various directions.

Figure 17.12 shows three vectors connecting two carbon atoms. Any such vector can be written as $\mathbf{c}(m, n) = m\mathbf{a} + n\mathbf{b}$, where m and n are integers. A CNT is obtained if the graphene sheet is rolled in a direction perpendicular to such a vector $\mathbf{c}(m, n)$, which itself rolls into a circle and the carbon atoms at the end of the vector coincide. It is obvious that any such vector specifies a unique CNT. The vector $\mathbf{c}(m, n)$ specifying a CNT is called its *chiral vector*. Carbon nanotubes with a length to diameter ratio of as much as 132000000 : 1 have been reported. The axis of the CNT is perpendicular to its chiral vector. The length $\sqrt{3}x\sqrt{m^2 + n^2 + mn}$ of $\mathbf{c}(m, n)$ equals the circumference of the CNT. Therefore, the radius of the CNT with chiral vector $\mathbf{c}(m, n)$ is given by,

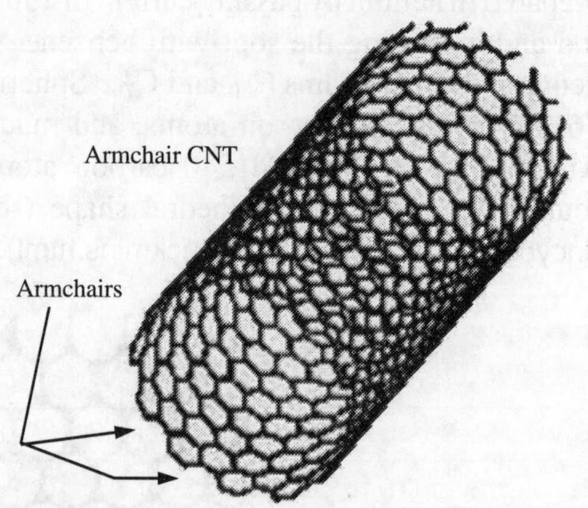

Figure 17.13 An armchair CNT

$$r_{\mathbf{c}(m,n)} \equiv r(m, n) = \frac{\sqrt{3}x\sqrt{m^2 + n^2 + mn}}{2\pi} \qquad (17.1)$$

If the vector is like $\mathbf{c}_2$ in figure 17.12, the CNT is called an *armchair* CNT. If it is like $\mathbf{c}_1$ or, $\mathbf{c}_3$ in figure 17.12, we obtain the *zigzag* form of CNT. Figure 17.13 shows an armchair CNT with chiral vector $\mathbf{c} = -12\mathbf{a} + 24\mathbf{b}$. As can be seen from the figure, the open end of this CNT has carbon atoms connected to each other to look like the profile of an armchair seat. Figure 17.14 shows a zigzag CNT with $\mathbf{c} = 18\mathbf{a}$. The carbon atoms at the open end form a zigzag as indicated in the figure. Other more general chiral vectors give CNTs with the hexagons on the surface of the cylinder twisting round the axis.

The strength of the sp^2 carbon–carbon bonds is responsible for mechanical properties of CNTs. Young's modulus of a CNT can be as high as 1000 GPa, about five times that of steel. The tensile strength can go up to 63 GPa, about fifty times that of steel. In spite of their great mechanical strength, the specific gravity of CNTs is just one-sixth of steel. This makes CNTs extremely suitable for applications in aviation and space technologies. Graphene and CNTs have novel elecrical properties. Graphene is a semi-metal with a zero band gap between its valence and conduction bands arising from π–bonds. CNTs can be metallic or semiconducting depending on their structure. Armchair CNTs and zigzag CNTs with $\mathbf{c} = 3p\,\mathbf{a},\ p = 1, 2, \cdots$ are always metallic. Chiral CNTs can be either metallic or semiconducting. Some nanotubes have conductivities higher than that of copper, and can sustain up to thousand times larger current densities (10^9 A/cm^2) than copper without melting! Other CNTs behave more like silicon and their electronic properties can be controlled by doping.

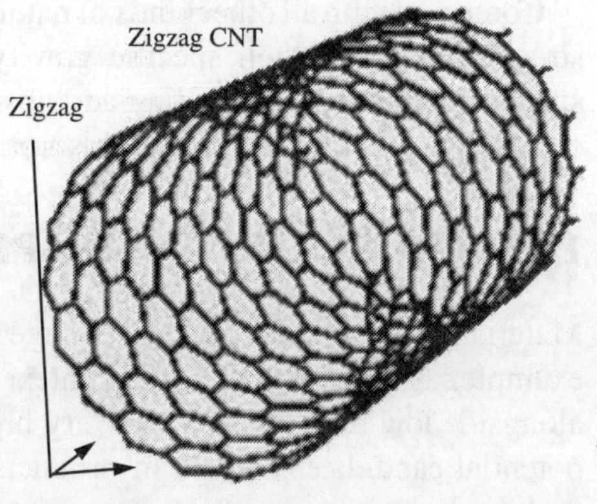

Figure 17.14 A zigzag CNT

Such properties make it possible to fabricate nanoscale electronic devices using CNTs. For example, a CNT can replace the conducting channel of a nano-MOSFET in devices called *single electron transistors* (SET). CNTs are being used as probes in SPMs, in flat-panel displays and in sensing devices.

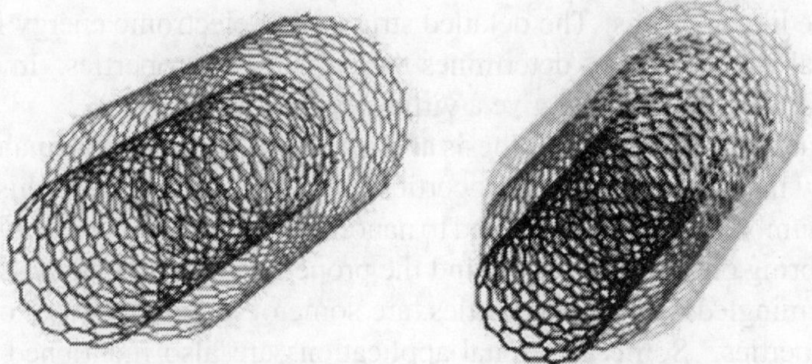

Figure 17.15 Multi-walled CNTs

Figure 17.15 shows two multi-walled CNTs. A multi-walled nanotube has multiple coaxial nanotubes precisely nested within one another. The inner nanotube core can slide almost without friction within its outer nanotube shell creating an atomically perfect linear or rotational bearing. The world's smallest rotational motor based on this principle is reported in literature. MWCNTs may find future applications as components in gigahertz mechanical oscillators. Carbon and silicon multi-walled nanotubes can be used to produce supercapacitors and Li-ion batteries with large electrical energy. The nanotubes provide very large surface area for Li ions to adhere to, thus increasing the total stored charge. Batteries constructed on this principle are expected to have longer life before they require to be re-charged.

Compared with all other kinds of natural materials, carbon nanotubes have the highest theoretical strength, although their specific gravity is only one-sixth of steel. Carbon nanotubes provide special advantages in shielding and absorbing electromagnetic radiation, field emission, thermal conductivity, hydrogen storage, adsorption and catalysis.

17.5 PROPERTIES AND APPLICATIONS

Materials at nanoscales possess strange and unexpected combinations of physical properties. For example, as noted above in the context of CNTs, it is possible to find very high elastic strength alongside low mass density, or, very high current densities with no heat loss. This makes them potential candidates for use in products ranging from paints and textiles to medical diagnostic tools and components of a future quantum computer. In general, properties of materials can be different at the nanoscale for two main reasons:

(i) Nanomaterials can have a relatively much larger fraction of atoms on the surface
(ii) Quantum effects can begin to dominate the behaviour of matter at the nanoscale.

Surface effects. A particle of size 30 nm has 5% of its atoms on its surface, at 10 nm, 20% of its atoms, and at 3 nm, 50% of its atoms. Thus, nanoparticles have a much larger number of atoms on their surface per unit mass. As a consequence, phenomena which take place at the surface dominate material behaviour. This can drastically alter their chemical reactivity, as well as elastic, optical, electrical and magnetic properties.

Quantum effects. In nanomaterials the de Broglie wave nature of particles like electrons can have far reaching effects. For example, electronic energies no longer form a quasi-continuum but show discrete nature like in atoms. The detailed structure of electronic energy levels in zero, one and two dimensional nanostructures determines their electrical properties. In certain situations, electrons can couple with photons and give a virtual particle-pair.

Intense research activity in design, synthesis and characterisation of nanomaterials is leading to the discovery of new materials and novel properties each day. Some idea of this can be gained by browsing the large number of websites devoted to nanomaterials and nanotechnology. The material here is intended to bring out the physics behind the properties of nanomaterials. Also mentioned alongside, and intermingled with the properties, are some of the applications of nanotechnology based on those properties. Some additional applications are also mentioned at the end of the section. The material is expected to serve as a guide and to encourage the student to indulge in some nano-browsing.

17.5.1 Confinement

Nanomaterials are generally regarded as those with at least one dimension less than 100 nm. Quantum wells are examples of nanosytems with one nano-dimension. A thin GaAs sheet sandwiched between two AlGaAs blocks is an example of a quantum well. Quantum wells can be used to fabricate semiconductor lasers. The channel between the source and the drain in a nano-MOSFET is also a quantum well. Examples of one-dimensional nanostructures are metallic nanowires and nanotubes of carbon, or silicon. An electron in a spherical nanoparticle, or a quantum dot, experiences confinement in all the three dimensions. Confinement *in* one, two and three dimensions (or, confinement *to* two, one and zero dimensions) gives different structures of energy levels available for an electron. This leads to different types of density of electron states (DOS). Table 17.1 summarises the situation. The wave functions and energy levels in the table are a straightforward generalisation of the particle in a rigid-box functions. It may be noted that in comparison with entries in the fourth column, the density of states for the bulk material is proportional to $\sqrt{E}$.

Table 17.1 Electron wavefunctions in systems confined to (in) zero (three), one (two) and two (one) dimension(s)

Confined in / to	**Energies,** E	**Wave functions,** Ψ	**DOS,** $\rho(E)$	**Examples**
Three/ zero dimension	$E_{n_x,n_y,n_z} = \frac{h^2}{8m}\left(\frac{n_x^2}{\ell_x^2}+\frac{n_y^2}{\ell_y^2}+\frac{n_z^2}{\ell_z^2}\right)$, $n_x, n_y, n_z = 1, 2, \cdots$	$\Psi_{n_x,n_y,n_z} = \sqrt{\frac{8}{\ell_x\ell_y\ell_z}} \sin\left(\frac{n_x\pi}{\ell_x}x\right)\sin\left(\frac{n_y\pi}{\ell_y}y\right) \sin\left(\frac{n_z\pi}{\ell_z}z\right)$	$\propto \sum\delta(E - E_{n_x,n_y,n_z})$	Quantum dot, nanoparticles
Two / one dimension	$E_{k_z,n_x,n_y} = \frac{\hbar^2k_z^2}{2m} + \frac{h^2}{8m}\left(\frac{n_x^2}{\ell_x^2}+\frac{n_y^2}{\ell_y^2}\right)$, any k_z, and $n_x, n_y = 1, 2, \cdots$	$\Psi_{k_z,n_x,n_y} = \sqrt{\frac{8}{\pi\ell_x\ell_y}} \sin\left(\frac{n_x\pi}{\ell_x}x\right)\sin\left(\frac{n_y\pi}{\ell_y}y\right)e^{ik_zz}$	$\propto E^{-1/2}$	Quantum wire, nanorods, nanotubes
One / two dimension	$E_{k_y,k_z,n_x} = \frac{\hbar^2k_y^2}{2m}+\frac{\hbar^2k_z^2}{2m}+\frac{h^2n_x^2}{8m\ell_x^2}$, any k_y, k_z and $n_x = 1, 2, \cdots$	$\Psi_{k_y,k_z,n_x} = \sqrt{\frac{8}{\pi^2\ell_x}}\sin\left(\frac{n_x\pi}{\ell_x}x\right)e^{ik_yy}e^{ik_zz}$	$\propto E^0$	Graphene, quantum wells

Quantum confinement effect is observed if the particle size is comparable to the width of the electron wave function. In such a case, electronic and optical properties deviate substantially from those of bulk materials. A particle is said to be free when the confining dimension is very large compared to the wavelength associated with the particle. Under this situation free electron gas-like behaviour leads to normal energy bands and band gaps. However, as the confining dimension decreases to nanoscales, the energy spectrum turns to discrete. Hence the inter-band and inter-level energy gaps are size dependent.

As noted in Table 17.1, nanostructures have different density of states compared to the bulk state. This feature determines the confinement criterion for a nanomaterial. Consider an atom in a semiconductor lattice losing an electron. This gives a conduction band ('free') electron and an unfilled state in the valence band, which can be considered to be a virtual positively charged ('free') particle called a hole. The electron and the hole can bind together due to Coulombic forces into a hydrogen electron-like bound state, called an *exciton*. This state is different from the donor, or the acceptor states given by the fifth electron, or, by the lack of a fourth electron in the dopant atom. In case of donor states the positive charge sits with the donor atom and the fifth electron is loosely bound to it. In the acceptor state, the positive charge sits on the *hole* loosely bound to the negatively charged acceptor impurity atom. In the case of an exciton none of the constituents of the pair is a lattice atom.

17.5.2 Electrical Properties

Electrical conductivity decreases with reduced size due to increased surface scattering. However, electrical conductivity of nanomaterials such as polymeric fibrils, could also be enhanced appreciably due to the better ordering in microstructure. For example, as was seen above (Section 17.3), electrons in graphene move very easily without scattering, a phenomenon known as *ballistic transport*. It can be used to fabricate ballistic transistors. Graphene is used in transparent electrodes for solar cells, LCD, robust non-volatile atomic switches, chemical and biological sensors and in spintronic devices.

Electron transport in nanostructures is *quantum transport* as the electron must be described by a quantum wave function. For example, in a nano-MOSFET the thin channel between the source and the drain leads is a quantum well. To understand electrical conduction single electron transfers from source to channel to drain have to be studied. This means that if conduction is to take place then there must be available empty energy levels in the channel (drain) region for the electron to go from source (channel) region to channel (drain) region. Which levels are available is determined by the bias voltages. As the bias voltage is changed new set of levels may become available opening what is called a new *transport channel*. The current is determined by how many channels are available. The important concepts in this context are *Coulomb blockade* and *quantum of conductance*. Quantum transport makes it possible to fabricate single electron transistors (SET). Such devices involve motion of a very small number of electrons and characteristics of single electron transfers define the device behaviour.

Semiconducting nanowires are known to be highly versatile optoelectronic components for a wide variety of applications such as polarisation-sensitive photodetectors and arrays with sub-wavelength resolution, polarisation-sensitive nano-APD (with gains up to 105); optical modulators and nano-waveguides, nano-LEDs and nano-lasers, solar cells and biomedical sensors.

17.5.3 Optical Properties

Just as for the bulk material, optical properties (that is, interaction with electromagnetic waves) of nanomaterials are determined by the internal electron energy level structure. As is clear from Table 17.1, this structure depends both on the *number* (1, 2 or 3) of nano-dimensions as well as on the actual *size* (values of $\ell_x,\ \ell_y,\ \ell_z$) in those dimensions. The number determines the form of the density of states. The size determines the values of inter-level energy gaps. In quantum wells the energy levels separate into bands with the gap between the bands determined by the size of the single nano-dimension. In quantum dots the energy levels are all discrete and determined by the quantum numbers n_x, n_y, n_z. Absorption spectra are determined by energy gaps between two energy levels. As the size of nanoparticles decreases, energy gaps increase and the absorption spectrum of the material changes. In the bulk state the absorption spectrum is unique, while as the size of the nanoparticle decreases, the absorption spectrum undergoes a *blueshift*.

The energy gaps are also determined by exciton effects referred to earlier. The frequency of the photon which induces an electron in the highest valence band energy level in a *semiconductor quantum dot* of radius R to jump to the lowest conduction band level is given by the Brus equation,

$$h\nu = E_g + \frac{h^2}{8R^2}\left(\frac{1}{m_e^\star} + \frac{1}{m_h^\star}\right) - \frac{1.8e^2}{4\pi\epsilon_r\epsilon_o R} \tag{17.2}$$

where $m_e^\star$ and $m_h^\star$ are effective electron and hole masses, E_g is the energy gap in bulk material and ϵ_r is the dielectric constant of the dot material. The last term is related to excitonic effects.

As noted above, effects of confinement become important when the electron de Broglie wavelength becomes comparable to the quantum dot size. The important electron energy is Fermi energy, E_F. The de Broglie wavelength for this energy is $\lambda_F = h/\sqrt{2m_e^\star E_F}$. If $\lambda_F \ll R$ confinement effects are not seen. If $\lambda_F \geq R$ confinement effects become important. For semiconducting QD, E_F is low and $\lambda_F \approx 10^1 - 10^2$ nm. The length important for excitonic effects to show up is the exciton radius

$$a_x = \frac{4\pi\epsilon_r\epsilon_o h^2}{e^2}\left(\frac{1}{m_e^\star} + \frac{1}{m_h^\star}\right) \tag{17.3}$$

If a_x is much less than the nano-dimension R of the structure, confinement is weak. Otherwise, confinement energies are large. This affects electrical properties of nanomaterials. Excitonic behaviour leads to applications such as fast switches. Strong absorption peaks are observed near the absorption edge in the absorption spectra of the materials. In highly pure bulk materials such behaviour is observed at very low temperatures. For nanoparticles, excitonic behaviour is observed even at the room temperature.

For a metallic QD since E_F is high and λ_F is low, size effects show up only for QD of sizes below 1–2 nm. At larger sizes optical properties are governed by *plasmon resonances*. Plasmons are oscillations of electron gas in the metal. The plasmon resonance effect is relatively insensitive to the size of nanoparticles so long as it is larger than λ_F. As particle size continues to increase toward the bulk limit, surface plasmon resonance wavelengths gradually move into the IR portion of the spectrum and most visible wavelengths are reflected, giving the nanoparticles a clear or translucent colour. The surface plasmon resonance can be tuned by varying the size or shape of the nanoparticles, leading to particles with tailored optical properties for different applications.

The optical properties of quantum dots allow them to be used as biological markers. The dot is coated with a material allowing it to bind selectively with certain biological structures like cancer cells. Such coated dots often have cadmium sulfide, cadmium selenide and cadmium telluride as core materials. Zinc sulfide is used as coat material. Such coated particles are called functionalised quantum dots. The cancer cells can be identified using fluorescence absorption and subsequent emission of photons. Functionalised metallic quantum dots are also used for marking biological structures. For example, functionalised gold nanoparticles can be obtained by coating silver nanoparticles with gold and dissolving silver out. Such gold nanoshells can attach to cancer cells and can also receive energy selectively at their resonant frequencies in order to deliver sufficient heat to kill the cell.

17.5.4 Magnetic Properties

Magnetic materials contain magnetic domains separated by boundaries called Bloch walls. Inside any one domain all the atomic magnetic dipoles are strongly coupled and aligned parallel to each other. Each domain, by itself, has a large magnetic moment. In bulk materials, the domains themselves are randomly oriented so that the net magnetic moment per unit volume, called magnetisation, is zero. If the bulk material is subjected to an alternating magnetic field it shows the phenomenon known as magnetic hysteresis (see http://en.wikipedia.org/wiki/Hysteresis#Magnetic_hysteresis). The magnetic properties of importance for materials showing hysteresis are *remanence* and *coercitivity*. Remanence measures magnetisation remaining after the magnetising field is reduced to zero. Coercivity measures the magnitude of opposite magnetic field needed to coerce the remanent magnetisation to zero. For bulk material both remanence and coercivity are independent of size. As the size is reduced to nanosizes single magnetic domains come to cover individual particles. The result is large increase in coercitivity and remanence. A further reduction in size results in decrease of both of these to zero because thermal fluctuations do not allow stable magnetisation in very small particles. This is called *superparamagnetism*. Thus, there is a range of particle sizes for which remanence and coercivity are large. Particles with sizes in this range are used for magnetic data storage.

17.5.5 Elastic Properties

The elastic properties of nanomaterials are thought to be determined by two factors accompanying size reduction: relative domination of surface effects and reduction in density of defects, imperfections and impurities. It is found that elastic constants vary significantly when the sample size decreases down to nanoscale. These variations of the modulii of elasticity can be radically different in different nanomaterials. The elastic constants can *decrease or increase with decreasing size*.

Reduction in number of defects and imperfections generally tends to strengthen the material. At nanosizes the defects move to the surface on thermal annealing, leading to an enhancement in chemical stability and the mechanical properties of the material. For example, superior mechanical properties of carbon nanotubes are achieved by reducing imperfections in the material. Ceramic and metallic composites made from particles smaller than 11 nm can be much stronger. Metals with 10-nm grains are as much as seven times harder and tougher than those with 100-nm grains.

On the other hand domination of surface energy tends to reduce the elastic constants as surface elasticity is weaker, there being lesser number of bonds per surface atom. Properties like bulk modulus have part contribution from infinite-size value of bulk modulus and part from surface elasticity. For cadmium and zinc sulphide it has been observed that the speeds of both longitudinal and transverse elastic waves decrease with decrease in size. The same is true of elastic modulii.

Some materials have different physical properties such as density and boiling point at the nanoscale. Nanoparticles have lower melting and boiling points because greater percentage of atoms at the surface implies that less energy is required to overcome intermolecular attractions. With extreme reduction in size, the concepts of melting and boiling lose sense.

17.5.6 More Applications

Many applications of nanomaterials and nanotechnology have already been mentioned in the last two sections. In the following a few more are noted, some of them already existing and some potential ones for the future.

Cosmetics and textile. Zinc oxide and titanium oxide nanoparticles provide the same sun protection as their normal sized counterparts. In addition, they do not scatter visible light and appear clear on the skin. Nanowhiskers on clothes create a cushion of air around the fabric so that liquids cannot stain them. Nanoparticles of silver embedded into fibers have anti-microbial action. Silver nanoparticles are also incorporated in apparel, footwear, paints, wound dressings, appliances, cosmetics and plastics for their antibacterial properties.

Health and medicine. Nanotechnology has direct applications in areas of drug manufacturing, drug delivery and diagnostics. Some of these applications have already been mentioned above. Nanodevices are being developed to keep track of daily changes in glucose and cholesterol levels in blood. Some surfactants spontaneously conglomerate to form capsules that can attach to, or encapsulate pharmaceuticals and serve as carriers of drugs to target areas in the body. Such capsules can be coated with smart materials capable of recognising intended targets in a relatively shorter time. Nanotubes made of magnetic materials and carrying drugs inside them can be used to deliver them to desired areas in the body. Mechanised nano-based drug synthesis and delivery using micro electromechanical systems (MEMS) have been demonstrated. Such nanofactories can collect raw material, synthesise the drug and deliver it according to a time and dosage schedule. Silver nanoparticles are used in biosensors as biological tags for quantitative detection. A certain graphene-based composite is used for rust-proofing of steel. It is a nontoxic alternative to coatings containing hexavalent chromium, which has carcinogenic properties (see also http://www.buffalo.edu/news/13401).

Energy and environment. New kind of solar cells that are cheaper to produce and use nanoparticles of titanium oxide coated with dye molecules to capture the energy of visible light and convert it into electricity have been developed. Special filters using nanomaterials can remove objects as tiny as viruses from water. In a recent research report (http://www.basqueresearch.com/) it is claimed that nanoenhanced paints can reduce emitted nitrogen oxides by 90%, hydrocarbons by 80%, and carbon monoxide by 75% in the atmosphere. One of the few alternative technologies that produce pure white light is white-light quantum dots. Cadmium selenide QD that can convert the blue light produced by an LED into a warm white light with a spectrum similar to that of incandescent light with a high efficiency of 45% have been reported recently (see also http://news.vanderbilt.edu/2012/05/quantum-dots/). A novel catalytic nanosheet form of a nickel-molybdenum-nitride, a thousand times cheaper than traditional platinum, is the new model for harvesting hydrogen from water for use as fuel (see also http://www.bnl.gov/bnlweb/pubaf/pr/).

It has been discovered that adding a little boron to carbon while creating nanotubes produces solid, spongy, reusable blocks that can absorb large quantities of oil spilled in water (see also http://news.rice.edu/2012/04/13/nanosponges-soak-up-oil-again-and-again/).

Electronics. New optical technologies using nanostructured layers of silver titanium oxide and quantum dots (see also http://www.purdue.edu/newsroom/research/2012/120524 NarimanovMeta.html) capable of ultra-efficient transmission of light, with potential applications in advanced solar cells and quantum computing are under development. Self-assembly of nanoparticles into device-ready materials using a technique based on blending nanoparticles with block co-polymer supramolecules has been achieved. A technique called nanolithography allowing microchips to be made much smaller has been developed. Complex three-dimensional structures using self-assembling polymer materials that form tiny wires and junctions have been developed. Such structures can produce features narrower than those achieved using light-based fabrication systems (see also http://phys.org/news/2012-06-d-self-assembling-microchips-devices-video.html). Gold thin films with potential applications in computer memory storage, energy harvesting, energy storage, remote-sensing, catalysis, light management and plasmonics have been developed (see also http://newscenter.lbl.gov/feature-stories/2012/4/27/gold-thin-films/). Nanoscale single crystals of a single material that can produce the red, green, or blue laser light needed in digital displays have been discovered (http://news.brown.edu/pressreleases/2012/04/nanolasers).

Construction industry. *Nanoenhanced construction ceramics* such as floor and wall tiles and sanitaryware have self-cleaning, anti-bacterial, hygienic and scratch resistant features. *Nanoenhanced cement* obtained by addition of nanoparticles like nano silica (silica fume), nanostructured metals, carbon nanotubes (CNTs) and carbon nanofibres (CNFs) gives stronger, more durable, self-healing, air purifying, fire resistant, easy to clean and quick compacting concrete. Nanotechnology based *smart windows* have energy-saving, easy cleaning, UV controlling and photovoltaic properties. Addition of fumed silica nanoparticles makes glasses fire resistant. Nanotechnology can facilitate different forms of *insulation solutions* like coatings, vacuum insulation glazing, and nanofoams. Addition of nanoparticles makes *paint* scratchproof, easy cleaning, air purifying, UV resistant, water repellent, flame resistant and anti-bacterial. Nanoenhanced paints also have a better penetration of the coated surface. Adding nanoclays to asphalt makes for safer, longer-lasting roadways (see also http://www.mtu.edu/news/stories/2012/may/story67607.html).

EXERCISES

1. What are the different definitions of the term *nanotechnology*? Try to trace who coined each of these definitions?
2. What are the reasons for rapid developments in the field of nanoscience and nanotechnology?
3. What is the roadmap for nanotechnology? Visit the website http://www.foresight.org/
4. What are the different methods of making nanomaterials? Try to make as long a list as possible from information on the internet. Attempt to classify the methods in your list.
5. What is scanning probe microscopy? How does scanning tunneling microscopy differ from atomic force microscopy?
6. What are the different types of electron microscopes? Make a list and differentiate each microscope in the list from others by listing its specific features.
7. What makes the instruments used to characterise nanomaterials suitable for their manipulation also? Discuss examples.
8. What are the main reasons for the novel properties of nanomaterials?
9. How are nanomaterials and nanotechnology of use in the medical field?
10. What is nanoelectronics? How do nanoelectronic devices differ from conventional electronic devices?

PROBLEMS

1. Justify the wave functions and energy levels listed in Table 17.1. Also, justify the conditions listed for n_x, n_y, n_z, k_x and k_y in the second column. Make graphical sketches showing energy spectra in the three cases (listed in Table 17.1).
2. Find the radii of the CNTs shown in Figs 17.14 and 17.15.
3. Consider a quantum well of 20 nm width. The other two dimensions are very large compared to this. If the integer quantum number n_x (See Table 17.1) defines the different *energy bands*, find the gap between the lowest two energy bands.
4. Take Fermi energy to mean the energy of the lowest unoccupied energy level. Fifteen non-interacting electrons are trapped in a three-dimensional box of size 10 nm $\times$ 10 nm $\times$ 5 nm. Two electrons can occupy any one level. Find the Fermi energy of the system.
5. The following data give respectively the effective mass of the electron, that of the hole and the dielectric constant in the materials indicated. The masses are in units of the electron mass $m_e = 9.1 \times 10^{-31}$ kg. (i) Si (heavy band): 0.26, 0.50, 11.7; (ii) Si (light band): 0.26, 0.24, 11.7; (iii) GaAs (heavy band): 0.067, 0.50, 13.13; (iv) GaAs (light band): 0.067, 0.08, 13.13; (v) InSb (light band): 0.015, 0.20, 17.88; (vi) CdSe: 0.13, 0.45, 9.4. In each case find the ground state exciton radius.

Index